Lecture Notes in Electrical Engineering

Volume 529

Lecture Notes in Electrical Engineering (LNEE) is a book series which reports the latest research and developments in Electrical Engineering, namely:

- Communication, Networks, and Information Theory
- Computer Engineering
- Signal, Image, Speech and Information Processing
- Circuits and Systems
- Bioengineering
- Engineering

The audience for the books in LNEE consists of advanced level students, researchers, and industry professionals working at the forefront of their fields. Much like Springer's other Lecture Notes series, LNEE will be distributed through Springer's print and electronic publishing channels.

More information about this series at http://www.springer.com/series/7818

Yingmin Jia · Junping Du
Weicun Zhang
Editors

Proceedings of 2018 Chinese Intelligent Systems Conference

Volume II

Springer

Editors
Yingmin Jia
Beihang University
Beijing, China

Junping Du
Beijing University of Posts and Telecommunications
Beijing, China

Weicun Zhang
University of Science and Technology Beijing
Beijing, China

ISSN 1876-1100 ISSN 1876-1119 (electronic)
Lecture Notes in Electrical Engineering
ISBN 978-981-13-4759-7 ISBN 978-981-13-2291-4 (eBook)
https://doi.org/10.1007/978-981-13-2291-4

This Springer imprint is published by the registered company Springer Nature Singapore Pte Ltd.
The registered company address is: 152 Beach Road, #21-01/04 Gateway East, Singapore 189721, Singapore

Contents

Multi-model Modeling of Heating Furnace System Based on FCM and GA Optimization ElasticNet-SVR

Zhengguang Xu, Weijian Kong and Mushu Wang

Abstract Aiming at the characteristics of non-linear, time-varying and wide-ranging working conditions of heating furnace, with the improvement of control requirements for actual system prediction, single model modeling has the problems of large amount of calculation and poor accuracy. A multi-model modeling method is proposed in this paper. This method first divides thc actual data of the heating furnace system into training set, validation set and test set, and uses FCM clustering to divide the training set into different working conditions; The Elastic Network (ElasticNet) and support Vector Machine regression (SVR) models are established in each local condition, and the optimal model of each local condition is selected from the two models by the validation set; use genetic algorithm (GA) to obtain the optimal weight of each local model, finally construct a model suitable for the global. This modeling method has a good global adaptability to the identification process. The veracity of the model is verified on the test set, and good results are obtained.

Keywords Multi-model modeling · FCM clustering · ElasticNet-SVR model
Genetic algorithm

1 Introduction

The actual industrial production process is often very complicated. The heating furnace is one of the key main energy-consuming equipment in the production line, it is a typical non-linear, large-lag, large-inertia, time-varying complex Control object

Z. Xu · W. Kong (✉) · M. Wang
School of Automation & Electrical Engineering,
University of Science and Technology Beijing, Beijing, China
e-mail: kongweijian12345@163.com

Z. Xu
e-mail: xzg_1@263.net

M. Wang
e-mail: 460809297@qq.com

Y. Jia et al. (eds.), *Proceedings of 2018 Chinese Intelligent Systems Conference*, Lecture Notes in Electrical Engineering 529,
https://doi.org/10.1007/978-981-13-2291-4_1

[1]. It is generally considered that the control object is linear in the whole control process or at a certain stage, and the identification and control are realized by the relatively mature linear system design method [2, 3]. With the improvement of the control precision of reheating furnace, it is difficult to achieve the precise control goal if the method is used to analyze and control the reheating furnace [4]. In Refs. [5, 6], the model of reheating furnace was established by neural network. In Refs. [7, 8], the nonlinear system is identified by using least square support vector machine (LS-SVM). However, in the actual production process, with the influence of complex working conditions, noise, disturbance, pressure fluctuation and so on, based on a single model, it is difficult to meet the requirements of industrial production for the performance of the model. The multi-model modeling strategy based on decomposition and composition rule can overcome the above problems to some extent [9]. In Ref. [10], the multi-model fusion method based on weighted performance index, the global fitting ability and local features of the model are not considered. In Refs. [11, 12], many models for nonlinear systems have been established, but the single model is mostly used locally. How to select appropriate weighted combination strategy and better local model is particularly critical.

In this paper, multi-model modeling of reheating furnace is studied. Section 2 introduce the principles of Elastic Net and SVR. Section 3 introduce the Multi-model modeling of heating furnace. Section 4 introduce the modeling and case Analysis of heating Furnace. Section 5 is the conclusion.

2 Principles of ElasticNet and SVR Models

ElasticNet and SVR models need to be established in each local working condition, and the better model is selected as the local model through the validation set.

2.1 The Principle of ElasticNet Model

ElasticNet regression is based on least squares and contains 1-norm and 2-norm regularization. The following objective function [13, 14]:

$$\min_{w} J(w) = \frac{1}{2}\sum_{i}^{N} ||(w)^T X_i - y_i||_2^2 + \lambda\left(\beta_1||w||_1 + \beta_2||w||_2^2\right) \tag{1}$$

2.2 The Principle of SVR Model

In Refs. [15, 16], good results were obtained by using SVR modeling for nonlinear systems. It is expressed as follows:

$$y = w^T \phi(x) + b \tag{2}$$

According to the KTT condition, the function of SVR can be expressed as:

$$y = \sum_{i=1}^{N} (\alpha_i^* - \alpha_i) K(x_i, x) + b \tag{3}$$

3 Multi-model Modeling of Heating Furnace

The heating section of each model is described as follows:

$$\begin{aligned} y(k) = f((y(k-1), \ldots, y(k-n_a), u(k-d), \\ \ldots, u(k-d-n_b+1)) + \zeta_k \end{aligned} \tag{4}$$

where, n_a, n_b is Model order, d system delay, ζ_k is noise. From the principal component analysis, it can be known that the furnace temperature is mainly affected by fuel flow, y is furnace temperature, u is fuel flow.

The multi-model modeling of reheating furnace is mainly divided into three parts: The overall structure is shown in Fig. 1.

a. *Data processing and working condition partitioning*: data preprocessing and partition into training set, validation set and test set.
b. *The establishment of Local Model*: the training set is divided into different working conditions by FCM clustering, ElasticNet and SVR models need to be established in each local working condition, and the better model is selected as the local model through the validation.

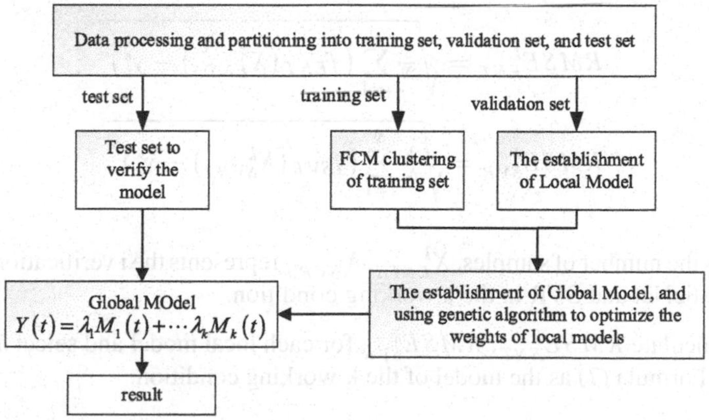

Fig. 1 Multi-model modeling structure for reheating furnace

c. *The establishment of Global Model based on genetic algorithm*: the local model is weighted into a global model as the heating furnace system model. The genetic GA is used to optimize the weight of the local model through the validation set RMSE to determine the optimal weight value.

3.1 Data Processing and Working Condition Partitioning

The data is de-noised, normalized and filtered. At the same time, the training set, the validation set and the test set are divided. The training set data is clustered into different working conditions.

Step 1: The number of clusters is divided into different groups. The objective function can be found in the Formula (5):

$$J(U, V) = \sum_{k=1}^{C} \sum_{i=1}^{n} u_{ki}^{m} d_{ki}(x_k, v_i)^2, \text{m} > 1 \tag{5}$$

Step 2: The clustering number of FCM is determined and the clustering number of FCM is determined by inflection point.

3.2 The Establishment of Local Model

On the training set, ElasticNet regression model and SVR model are established for FCM1, FCM2, FCM3 and FCM4 respectively.

Step 1: The accuracy of the model is evaluated with mean square error (RMSE):

$$\begin{aligned} RMSE_{ENT}^{k} &= \sqrt{\frac{1}{N}\sum_{i=1}^{N}\left(f_{ENT}\left(X_{ENT,i}^{k}\right) - y_i^k\right)^2} \\ RMSE_{SVR}^{k} &= \sqrt{\frac{1}{N}\sum_{i=1}^{N}\left(f_{SVR}\left(X_{SVR,i}^{k}\right) - y_i^k\right)^2} \end{aligned} \tag{6}$$

where N is the number of samples, $X_{ENT,i}^{k}$, $X_{SVR,i}^{k}$ represents the i verification sample of the ElasticNet and SVR in the k working condition.

Step 2: Calculate $RMSE_{ENT}^{k}$, $RMSE_{SVR}^{k}$ for each local model and select a smaller model by Formula (7) as the model of the k working condition.

$$M_k = \min_{k}\left(RMSE_{SVR}^{k}, RMSE_{ElasticNet}^{k}\right)|_{k=1,2\ldots4} \tag{7}$$

3.3 Establishment of Global Model Based on GA

After Sect. 3.2 steps, the best model for each local condition is selected as the current model. Then the genetic algorithm is used to find a set of optimal weight coefficients and to form a global model weighted. As follows:

$$Y(t) = (\lambda_1 M_1(t) + \cdots \lambda_k M_k(t))/p = W^T M(t) \tag{8}$$

where, $q = \lambda_1 + \cdots \lambda_k,\ \lambda_i \geq 0,\ i = 1 \ldots k,\ W^T = (\lambda_1, \ldots, \lambda_k)$ is weight vector, $\lambda_i,\ i = 1, \ldots k$, represents the weight of each model, $M_1(t) \ldots M_k(t)$ represents each local model.

The optimal W^T minimizes the RMSE of the model $Y(t)$ in the validation set:

$$\min_{W^T} J = \sqrt{\frac{1}{N} \sum_{i=1}^{N} (Y_i(t) - y_i)^2} \tag{9}$$

The key steps in genetic algorithms are as follows:

(1) using binary coding method, weight coefficients of each local model account for 8 bits, C working conditions a total of 8 * C position.
(2) The fitness function of RMSE reciprocal to population P(t) is calculated.

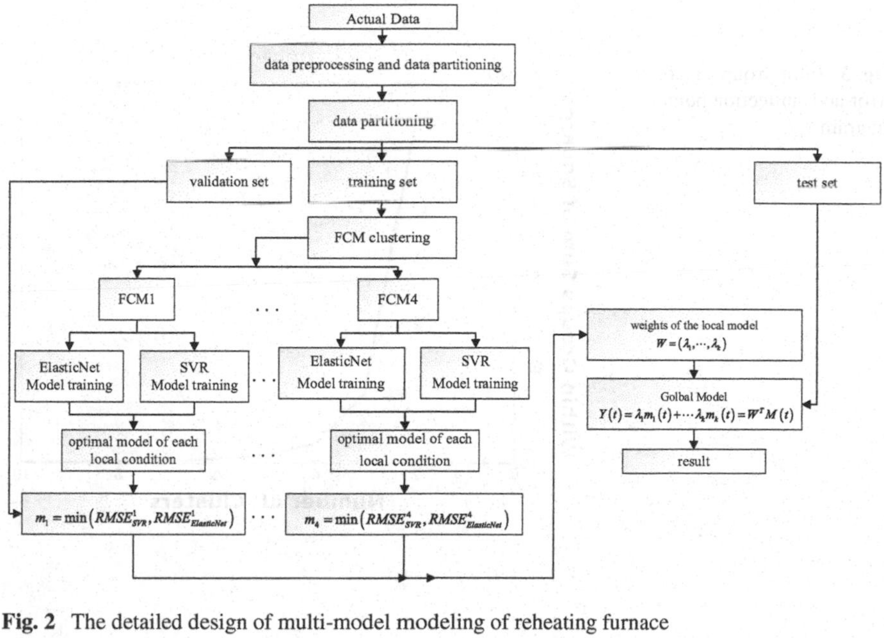

Fig. 2 The detailed design of multi-model modeling of reheating furnace

To sum up, the detailed design of multi-model modeling of reheating furnace is shown in the Fig. 2.

4 Modeling and Case Analysis of Heating Furnace

The heating furnace is divided into a preheating section, a heating section and a soaking section, the actual data on the left of the heating section is adopted, the sampling time T is 5 s, the delay of the correlation analysis system is 15 s, and the order of the system is 2 steps.

The sampled data total 15,000 time series points, the first 8998 is the training set, the 9001–12,000 is the validation set, and the 12,000–15,000 is the test set.

4.1 *FCM Clustering*

Initialization of FCM clustering parameters, according to the square error within the group and from Fig. 3, the optimal number of clustering C is 4.

Among them, the fuzzy weight value $\beta = 4.5$, the threshold value $\varepsilon = 1e{-}7$, the maximum iteration number T = 2000, the initialization of the initial cluster center is divided into $V^{(1)} = (800, 1000, 1100, 1200)$. The clustering results are as follows in Table 1.

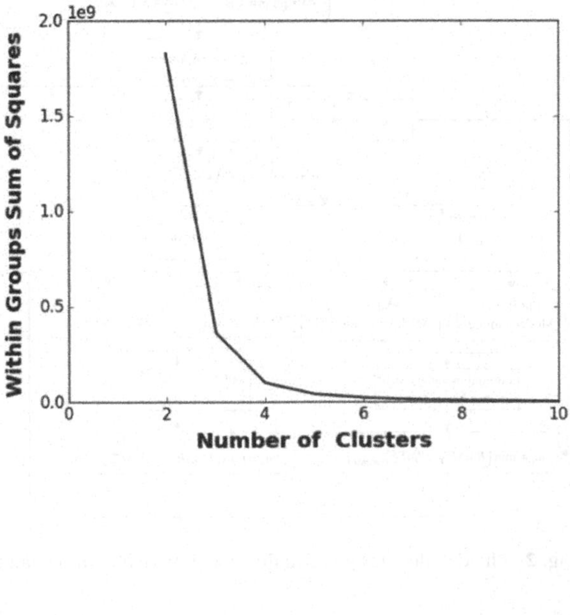

Fig. 3 Intra-group square error and -inflection point diagram

Table 1 Training data FCM clustering results

	FCM1	FCM2	FCM3	FCM4
$y(k-1)$	690.0	1136.8	1178.0	1181.1
$y(k-2)$	690.0	1136.7	1177.9	1181.2
$u(k-d)$	679.1	4963.8	4079.1	2012.0
$u(k-d-1)$	679.1	4964.0	4079.2	2011.7

4.2 Multi-model Modeling

Step 1: The local models of FCM1, FCM2, FCM3, FCM4 are established respectively through Formula 7.
Step 2: Multi-model modeling. Using genetic algorithm to optimize the weight of each condition, finally, the global model of multiple models is formed by Formula 8. Comparing with NN and LS and four local models. (In the neural network model, each neural network has three hidden layers and each hidden layer has 20 neurons), as shown in Table 2. Where RMSE is shown in Formula 6, and MAPE as follows (10).

$$MAPE = \sum_{i=1}^{N} \frac{(|Y_i(t) - y_i|{*}100)}{y_i} / \mathrm{N} \tag{10}$$

The iteration of genetic algorithm optimization on the validation set is shown in Fig. 4, the horizontal axis is the iteration number, the longitudinal axis is the RMSE trend of the verification set, and the initial value is equal. In the final selection, the chromosome number was 55, the maximum iteration number was 100, the crossover rate was 0.6, and the variation rate was 0.05. Because each weight is 8-bit binary, the ultimate optimal weight coefficient matrix is:

$$\mathrm{W} = [39.0, 63.0, 135.0, 160.0]$$

And applying the best weights to test the accuracy of the model in the test set, multi-model has the best effect, as shown in Table 3.

Table 2 Comparison of validation set simulation results

Result	Multi-model	Local model				LS	NN
		M1	M2	M3	M4		
RMSE	0.0581	0.152	0.364	0.118	0.129	0.092	0.20
MAPE	3.8e−05	10.5e−5	27.8e−5	8.3e−05	8.6e−05	7.8e−05	8.2e−05

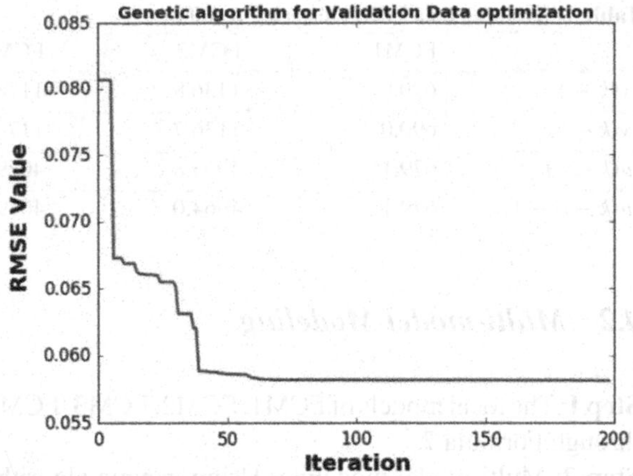

Fig. 4 Genetic algorithm optimized local model weights iteration diagram

Table 3 Comparison of test set simulation results

Result	Multi-model	Local model				LS	NN
		M1	M2	M3	M4		
RMSE	0.0452	0.131	0.338	0.105	0.148	0.085	0.11
MAPE	3.09e−05	8.7e−05	25.8e−5	6.9e−05	9.6e−05	5.1e−05	6.3e−5

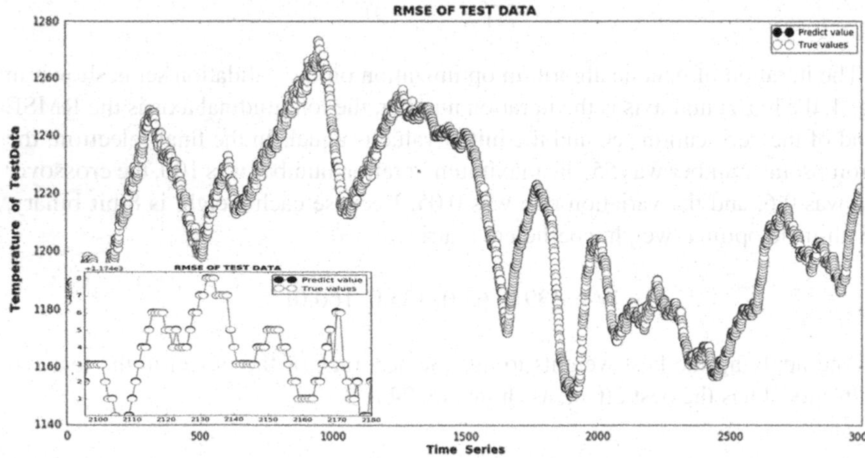

Fig. 5 The fitting effect of multi-model on test set

In order to more clearly observe the fitting effect of the global model constructed by multi-model, the curves of True value and Predict value are drawn. The bottom left corner is a local magnification diagram, as shown in Fig. 5.

In summary, the multi-model has a good degree of fitting to the system model, and the global model approximates the nonlinear property of the system well. The simulation accuracy of nonlinear process is greatly improved.

5 Conclusion

In view of the reheating furnace, a multi-model modeling scheme for ElasticNet-SVR heating furnace system based on FCM and genetic algorithm optimization is proposed in this paper. The methods are as follows: (1) the data is divided into training set, validation set and test set. The training set is divided into different working conditions by using FCM clustering method. (2) The ElasticNet and SVR models are established locally in each working condition, and the local optimal model is selected according to the validation set, which overcomes the singularity of the single model used locally. The generalization ability is improved. (3) Genetic Algorithm is used to obtain the optimal weight value of multi-model fusion. Finally, a global model is constructed.

Finally, the effectiveness of the proposed multi-model modeling method is verified by real data simulation, and good results are obtained.

References

1. Q. Cai, *Furnace* (Metallurgical Industry Press, Beijing, 2007)
2. L. Wang, Z. Zhao, System Identification: new patterns, challenges and opportunities. Acta Automatica Sin. **397**, 933–942 (2013)
3. G. Tang, Z. Zhuang, Analysis on energy saving scheme of refinery heating furnace. Chem. Mach. **27**, 352–355 (2000)
4. S. Liu, Y. Qi, L. Wang et al., Hybrid modeling and optimization of coal consumption and NOx emission of utility boiler. Petrochemical Auto-chemical Equip. **1**, 30–34 (2016)
5. H. Li, Q. Li, Y. Tang, A neural network model for heating furnace. Min. Metall. **12**(3), 67–69 (2003)
6. Z. Wang, T. Chai, C. Shao, Prediction model of reheating furnace steel temperature based on RBF neural network. Rep. Syst. Simul. 181–185 (1999)
7. C. Mu, R. Zhang, C. Sun, A particle swarm optimization based least squares support vector machine predictive control method for nonlinear systems. Control Theory Appl. (2010)
8. Y. Wang, D. Huang et al., Nonlinear predictive control technique based on LS-SVM. Control Decis. Making **4**, 383–387 (2004)
9. T.A. Johansen, B.A. Foss, A NARMAX model representation for adaptive control based on local models. Model. Ident. Control **13**(1), 25–39 (1992)
10. R. Murray-Smith, T.A. Johansen, Local learning in local model networks, in *International Conference on Artificial Neural Networks* (IET, 1995), pp. 40–46
11. C. Huang, Y. Gao, L. Peng, Study on modeling and identification of industrial process with multiple models. Comput. Eng. Appl. **52**(20), 251–262 (2016)
12. J. Lin, J. Shen, Y. Li, Multi-model modeling method for thermal processes based on hierarchical G-K clustering. J. Electr. Eng. China (CAE) **26**(11), 23–28 (2006)

13. H. Zou, T. Hastie, Regularization and variable selection via the elastic net. J. R. Stat. Soc. Ser. B (Stat. Methodol.) **67**(2), 301–320 (2005)
14. C. De Mol, E. De Vito, L. Rosasco, Elastic-net regularization in learning theory. J. Complex. **25**(2), 201–230 (2009)
15. Y. Cai, C. Hu, Study on multi-output ε-SVR model for identification of nonlinear MIMO systems. Control Decis. **237**, 813–816 (2008)
16. H. Zhang, Z. Han, C. Li, Nonlinear system identification based on support vector machine. J. Syst. Simul. **15**(1), 119–121 (2003)

User Interaction Based Bursty Topic Model for Emergency Detection

Zhijian Li, Junping Du, Wanqiu Cui and Pinpin Zhu

Abstract When an emergency suddenly occurs, people usually share information and feelings in the social network. Therefore, it is of great significance to detect emergencies by analyzing and mining messages posted by users. Considering social network contains a mass of user interaction behavior, in this paper, we proposed a novel bursty topic model for emergency detection, named User Interaction based Bursty Topic Model (UIBTM). To overcome the problem of short text sparsity and ambiguity, UIBTM first uses comment texts and the amount of users liking the microblog to enrich the semantic of microblog, then generates the bursty topic model for bursty topic discovery and emergency detection. Comprehensive experiments on the dataset of Sina Microblog show that UIBTM can effectively overcome the sparsity of short text and detect emergencies efficiently.

Keywords Short text · Semantic enrichment · Burst topic model
Emergency detection

1 Introduction

In recent years, with the rapid development and popularity of mobile terminals, microblog has become one of the most popular social network platforms for people to get and share information. In a short period after an emergency occurred, a large amount of data related to the topic will be generated in the social network. Thus, discovering bursty topics by analyzing and mining microblog contents posted by users can help us collect data more accurately and track the emergency more effectively.

Z. Li · J. Du (✉) · W. Cui
Beijing Key Laboratory of Intelligent Telecommunication Software and Multimedia,
School of Computer Science, Beijing University of Posts and Telecommunications, Beijing 100876, China
e-mail: junpingdu@126.com

P. Zhu
Xiaoi Research, Shanghai Xiaoi Robot Technology Co., Ltd, Shanghai 201803, China

Y. Jia et al. (eds.), *Proceedings of 2018 Chinese Intelligent Systems Conference*, Lecture Notes in Electrical Engineering 529,
https://doi.org/10.1007/978-981-13-2291-4_2

However, discovering bursty topic from social network confronts many challenges. Texts posted by users are usually short and not strictly syntactical, which make most traditional topic model not work well for microblog data. In order to solve the problem of short text sparsity and ambiguity, previous methods mainly use external resources to enrich the semantic of short texts, like WordNet [1], Wikipedia [2] and Probase [3]. In addition, users can post messages of various topics in the social network, a large amount of data of general topics may overwhelm emergency-related data. For this problem, researchers propose various methods to evaluate the acceleration of word tuples [4] or the bursty probability of biterms [5]. These methods have solved these problems to a certain extent, but there are still some limitations. We observed that besides of texts posted by users, there is also much user interaction information in the social network, such as commenting, forwarding and liking behaviors. User interaction information is beneficial for enriching the semantic of short text and evaluating the burstiness of biterms. Therefore, in this paper, we propose a novel bursty topic model based on user interactions. The main contributions of this paper are as follows:

- We propose a semantic enriching method for social network short texts based on user commenting and forwarding behaviors, which can effectively solve the semantic sparsity problem and improve the effectiveness of emergency detection.
- We improve the measurement method of the burstiness of biterms by introducing the amount of users liking the microblog and propose a novel User Interaction based Bursty Topic Model (UIBTM), which can detect emergencies more accurately and more sensitively.
- We conduct extensive experiments on the tasks of bursty topic discovery, evaluate the accuracy, coherence and novelty of the bursty topics discovered to verify the superiority of the proposed UIBTM.

The rest section of this paper is organized as follow. The related works about short text semantic enrichment and emergency detection are described in Sect. 2. We present the detail of our topic model UIBTM in Sect. 3. The experiments settings and results analysis are shown in Sect. 4. Section 5 presents the conclusion of this paper.

2 Related Works

2.1 Semantic Enrichment

In order to facilitate short text understanding, various semantic enrichment methods have been proposed in previous studies. The typical approach to semantic enrichment is through the introduction of external resources. Gao et al. [1] propose an edge-counting and information content theory based semantic enrichment and similarity measurement method. Yu et al. [5] enrich the short texts by obtaining the concept of each term from a probabilistic knowledge base. Hua et al. [6] use the lexical semantic knowledge provided by semantic network for short text understanding to enrich the semantic. These methods can achieve the task of semantic enrichment to some extent, but also have limits. On the one hand, the effect of semantic enrichment depends on the coverage of the external resources, which may cause that the newly generated pop words cannot be recognized. On the other hand, knowledge used by these methods doesn't take the feature of social networks into consideration.

2.2 *Emergency Detection*

For the task of emergency detection, previous studies can be mainly categorized into two groups: clustering based methods and topic modeling based methods. Clustering based methods first detect bursty features like words or biterms and then cluster these features to get bursty topics in a specific time slice [8], while topic modelling based methods use post-processing techniques [9] or temporal and spatial information [10, 11] to assist conventional topic models in bursty topic discovering tasks. However, due to the bursty features may be noisy and ambiguous, it is difficult for clustering based methods to get accurate topics and representations. And the topic modeling based methods may confront the problem that most of the discovered topics are not real emergencies.

In this paper, we focus on the problem of short text semantic enrichment and bursty topic discovery. Considering social network platforms contains a mass of user interaction behavior, the key idea of our approach is to extend microblog contents with texts posted by users when they comment and forward a microblog, and improve the burstiness measuring method by taking into account of the amount of users liking the microblog.

3 User Interaction Based Bursty Topic Model (UIBTM)

In this section, we first introduce how to use user's comment and forwarding behavior to implement short text semantic enrichment. Then we illustrate the improved measurement method of the burstiness of biterms and the generating process of UIBTM.

3.1 *User Interaction Based Semantic Enrichment*

Since commenting and forwarding are critically different, we propose semantic enrichment methods for these two types of user behaviors.

The commenting behavior in social networks generates a large number of comment texts for the corresponding microblog content, which mainly contains users' personal opinions and feelings about the incident. However, due to the uneven quality of comment texts, we use some strategies to find out comment texts which could effectively achieve semantic extension. First, we preprocess the microblog content and comment texts by splitting sentences and removing stop words, then the comment texts which contain less than k Chinese words are pruned. After completing the above two steps, we compute the Jaccard distance between microblog content T_i and the corresponding comment text $C_{i,j}$ by Eq. (1).

$$J(T_i, C_{i,j}) = \frac{|T_i \cap C_{i,j}|}{|T_i \cup C_{i,j}|} \tag{1}$$

where T_i denotes the word collection of microblog text, $C_{i,j}$ denotes the word collection of comment and $|\bullet|$ denotes the number of elements in the word collection. Finally, we filter the comment texts that not match the condition $\alpha < J(T_i, C_{i,j}) < \beta$. The remaining comment texts are valid texts which can be used for semantic enrichment.

As most of the comment texts in social networks contain very few words and even only some emoticons, it is helpful to pruning the comment texts which length is less than k, because these comment texts are not useful for semantic enrichment. In subsequent experiments, we set the value of k to 7. After this pruning step, we consider that the remaining comment texts contain complete semantics, then the Jaccard distance between microblog content and the comment texts can effectively measure their semantic similarity. For comment text $C_{i,j}$, if $J(T_i, C_{i,j}) < \alpha$, we suppose that the semantics of the comment text is irrelevant to the microblog content, so it will distort the meaning of the microblog content. If $J(T_i, C_{i,j}) > \beta$, we view that the semantics of the comment text is very similar to the microblog content, so it is helpless for semantic enrichment. Based on the above considerations, we filter out the comment texts which not match the condition $\alpha < J(T_i, C_{i,j}) < \beta$. In addition, considering that some microblog contents contain too many comment texts that satisfy the above conditions, we just use the top 5 comment texts with the most amount of users liking it. Since comments in microblog platforms are usually ordered by the amount of users liking it, we can get these comment texts easily.

After above steps, we append the comment texts to microblog contents, and extend the microblog content T_i to a tuple $(T_i, C_{i,1}, C_{i,2}, \ldots, C_{i,n})$, in which comment texts $C_{i,j}$ are ordered by publication time. For $x < y$, it satisfies $t_x <= t_y$, where t_x denotes the publication time of the comment text $C_{i,x}$.

Since users may post some comment when forwarding a microblog, we also use the same preprocessing method described above to capture the forwarding text $F_{i,j}$. For the forwarding texts which match the condition $\alpha < J(T_i, F_{i,j}) < \beta$, we generate a new microblog content tuple $(T_i, F_{i,j})$ to extend the dataset. However, as the Sina Microblog platform allows users to forward a microblog recursively, which may cause our method fall into the over complexity problem, we limit T_i as original microblog content which is posted by users instead of forwarded.

3.2 Generative Process of UIBTM

When an emergency occurs, relevant biterms usually appear more frequently than usual. Our bursty topic model uses bursty probability to quantify the burstiness of biterms, which is defined by Eq. (2):

$$\varphi_{b_i}^{(t)} = \frac{(c_{b_i}^{(t)} - \bar{c}_{b_i})_+}{c_{b_i}^{(t)}} \quad (2)$$

where $\varphi_{b_i}^{(t)}$ is the burstiness of biterm b_i, $c_{b_i}^{(t)}$ denotes the number of occurrences of biterm b_i during time period t. $\bar{c}_{b_i}$ indicates the average number of occurrences of biterm b_i during all time periods, and $(m)_+$ is $\max(m, \varepsilon)$ where ε is a small number used to avoid the probability to be zero. As $\bar{c}_{b_i}$ cannot be calculated directly, it is approximately calculated by the last k time periods, i.e., $\bar{c}_{b_i} = \frac{1}{K}\sum_{k=1}^{K} c_{b_i}^{(t-k)}$.

Since many users neither post nor comment microblog contents, the amount of users liking the microblog content can reflect how many people are following this topic and agree with this microblog. We improve the definition of $c_{b_i}^{(t)}$ as Eq. (3):

$$c_{b_i}^{(t)} = n \cdot \log\left(\sum_{p=1}^{P} l_p + 2\right) \quad (3)$$

where n denotes the number of occurrences of biterm b_i, l_p is the amount of users liking microblog content p which contains biterm b_i. To avoid $c_{b_i}^{(t)}$ less than zero, we add a constant 2 in the logarithmic function.

Considering that a biterm can be used either in normal topic or in bursty topic, we define a binary variable χ_i which denotes the topic type of biterm b_i. $\chi_i = 0$ indicates biterm b_i is used in a normal topic, while $\chi_i = 1$ indicates biterm b_i is used in a bursty topic. Followed by [4], we propose our generative process of UIBTM as following:

(1) For the collection, choose a bursty topic distribution $\theta \sim Dirichlet(\alpha)$ and a background word distribution $\phi_0 \sim Dirichlet(\beta)$.
(2) For each bursty topic k, choose a word distribution $\phi_k \sim Dirichlet(\beta)$.
(3) For each biterm $b_i \in$ B, choose a distribution $\chi_i \sim Bernoulli(\varphi_{b_i})$,

- if $\chi_i = 0$, draw two words $w_{i,1}, w_{i,2} \sim Multi(\phi_0)$.
- if $\chi_i = 1$, first draw a bursty topic $u \sim Multi(\theta)$, and then draw two words $w_{i,1}, w_{i,2} \sim Multi(\phi_u)$.

Due to the parameters in θ are intractable to determine exactly, we use the collapsed Gibbs sampling algorithm for approximate estimation. There are two latent variables in UIBTM, named χ_i and u_i, so we draw them jointly according to the following conditional distribution:

$$P\left(\chi_i = 0 | \chi^{-i}, u^{-i}, B, \varphi\right) \propto (1 - \varphi_{b_i}) \cdot \frac{\left(c_{0,w_{i,1}}^{-i} + \beta\right)\left(c_{0,w_{i,2}}^{-i} + \beta\right)}{\left(\sum_{w=1}^{W} c_{0,w}^{-i} + W\beta\right)\left(\sum_{w=1}^{W} c_{0,w}^{-i} + 1 + W\beta\right)} \quad (4)$$

$$P\left(\chi_i = 1, u_i = k | \chi^{-i}, u^{-i}, B, \varphi\right) \propto \varphi_{b_i} \cdot \frac{\left(c_k^{-i} + \alpha\right)}{\left(\sum_{w=1}^{W} c_{k,w}^{-i} + K\alpha\right)}$$

$$\cdot \frac{\left(c_{k,w_{i,1}}^{-i} + \beta\right)\left(c_{k,w_{i,2}}^{-i} + \beta\right)}{\left(\sum_{w=1}^{W} c_{k,w}^{-i} + W\beta\right)\left(\sum_{w=1}^{W} c_{k,w}^{-i} + 1 + W\beta\right)} \quad (5)$$

4 Experiment and Evaluation

In this section, we evaluate the performance of the proposed topic model UIBTM from three aspects. First, we evaluate the accuracy of bursty topics detected by *Precision@K* metric. Then, we evaluate the quality of bursty topics detected by the coherence of topics. Finally, we evaluate the sensitivity of bursty topics detected by the novelty of topics.

4.1 Description of Dataset and Baseline Methods

We focused on several national security emergencies happened in recent years and crawled relevant data from Sina Weibo. These events are "rainstorm in Hubei", "terrorist attack in Kunming railway station" and "Tianjin warehouse explosion". In order to simulate the real social network environment, we also randomly crawled about 200 thousand microblogs as a background dataset, which are used when evaluating the novelty of bursty topics. The crawled data mainly contains the original microblog texts posted by users, at most 100 hot comment texts of each microblog text, the number of users liking the microblog and the time when the microblog posted. The details of the dataset are shown in Table 1.

We compare our topic model against the following four baseline methods:

Table 1 Statistics of the social network datasets

Dataset name	Number of microblogs	Average number of comments	Average number of likes	Number of vocabularies
Rainstorm	31,524	21	64	9970
Terrorist	23,648	24	56	11,307
Explosion	25,798	13	85	15,227
Background	202,458	7	29	28,512

Twevent [8]: it first detects event segments by detecting bursty tweet segments, then clusters the event segments into events considering both frequency and content similarity between different event segments. In order to make a fair comparison, we treat individual Chinese words as microblog segments.
OLDA [9]: The method first uses online LDA to learn topics in each time slice, then measures the Jensen-Shannon divergence before and after the update to detect bursty topics.
UTM [11]: it supposes that bursty topics follow a time-dependent topic distribution, while non-bursty topics follow a user-dependent topic distribution. To ensure the topics discovered to be bursty, UTM heuristically boosts the probability of bursty words in temporal topics.
BBTM [5]: it uses the burstiness of biterms as prior knowledge to extend Biterm Topic Model so that it can automatically discover bursty topics in microblogs.

4.2 Hyper-parameters Selection

First of all, in order to find the value of α and β, we merge the text data of three emergency datasets and try different values on the classification tasks. For each value of pair of (α, β), we extract the feature vector of texts enriched by user commenting texts and forwarding texts by word embedding [13]. Then we build a softmax classifier to classify the three types of data and evaluate the classification results by F-measure.

Figure 1 shows the F-measure on different pair of (α, β), it can be seen intuitively that the softmax classifier achieves the highest F-measure when $\alpha = 0.35$ and $\beta = 0.7$. Based on this set of pre-experiments, we just use the commenting and forwarding texts match the condition $0.35 < J(T_i, C_{i,j}) < 0.7$.

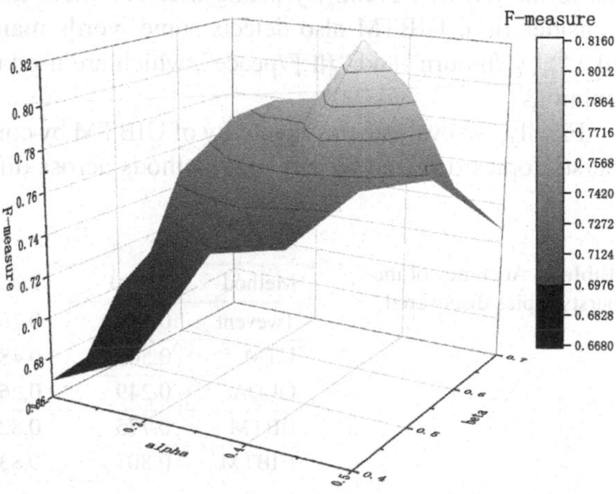

Fig. 1 F-measure on different hyper-parameter alpha and beta

4.3 Evaluation of Bursty Topics Discovered

In this section, we evaluate the quality, coherence and novelty of the bursty topics discovered.

We first evaluate the accuracy of bursty topics discovered by different methods based on average precision at K ($P@K$), the proportion of correctly detected topics in all K topics detected. We set the number of bursty topics to 10, 20, 30 and 50 respectively, then experiment on three different datasets and list the average results in Table 2.

As shown in Table 2, we find that UIBTM achieved the highest $P@K$ score, which is always greater than 0.8. Both UIBTM and BBTM achieve much higher scores than other methods because UIBTM and BBTM take into account the burstiness of biterms by introducing bursty probability computing. As UIBTM improved the bursty probability measuring formula by the amount of like, it performs better than BBTM.

Next, we evaluate the semantic modeling ability of UIBTM based on the coherence of topics. Following [12], we use PMI-Score (Pointwise Mutual Information) as the evaluation metric, larger PMI-Score indicates higher coherence.

By observing and analyzing the comparison results in Fig. 2, we find that the PMI-Score of UIBTM and BBTM are comparable and much higher than other baseline methods, which indicates that bursty topics discovered by UIBTM and BBTM are of great coherence. The PMI-Score of Twevent is almost always the lowest, which shows that the topics detected by simply clustering bursty segments might be pretty noisy and incoherent.

For more intuitive observation and analysis of the busty topics detected by different methods, we take the terrorist dataset as an example and select the topic closest to the empirical word distribution of the real topic in each method. The top 10 words of each topic detected are listed in Table 3.

From Table 3 we can see that bursty words discovered by UIBTM are the closest to the terrorist event. By taking user comment texts and forwarding texts into consideration, UIBTM also detects some words mainly appear in the comments like "默哀/mourn" and "和平/peace", which are used to describe user's moods and opinions.

Finally, we evaluate the sensitivity of UIBTM by comparing the novelty [3] of the bursty topics detected by different methods across different time slice. In order to

Table 2 Accuracy of the bursty topics discovered

Method	P@10	P@20	P@30	P@50
Twevent	0.612	0.701	0.673	0.658
UTM	0.565	0.488	0.459	0.453
OLDA	0.249	0.268	0.272	0.229
BBTM	0.775	0.823	0.809	0.816
UIBTM	0.807	0.831	0.824	0.813

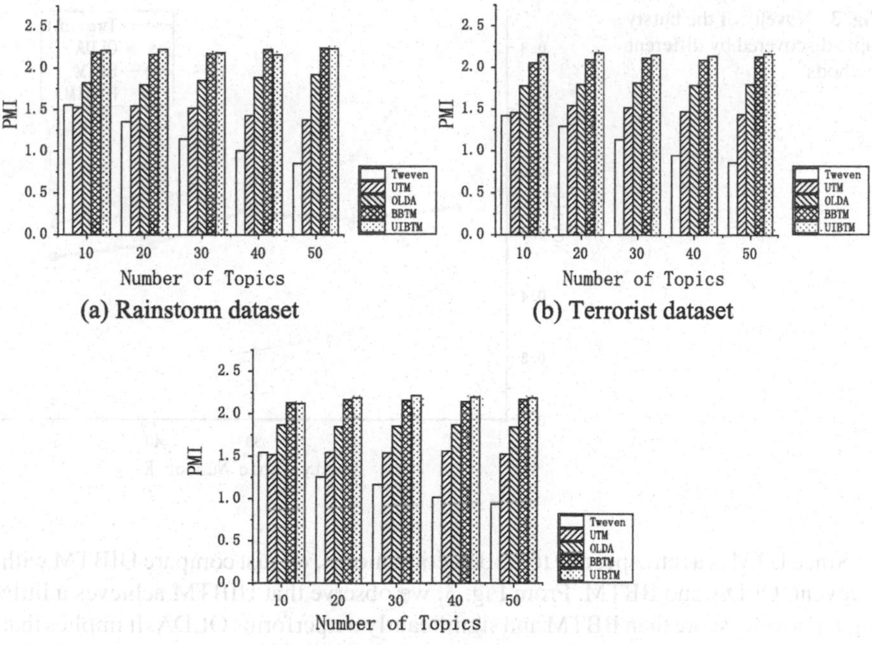

(a) Rainstorm dataset

(b) Terrorist dataset

(c) Explosion dataset

Fig. 2 Coherence of the bursty topics discovered on three datasets

Table 3 Top 10 words of the bursty topic discovered most relates to the terrorist event

Twevent	UTM	OLDA	BBTM	UIBTM
暴徒/mob, 事件/event, 火车站/railway station, 昆明/Kunming, 受伤/injured, 目标/target, 逝者/the dead, 记者/journalist, 人/people, 生命/life	暴徒/mob, 火车站/railway station, 砍/cut, 发生/happen, 逝者/the dead, 新闻/news, 无辜/innocent, 昆明/Kunming, 恐怖/terror, 人员/people	暴徒/mob, 火车站/railway station, 砍/cut, 微博/microblog, 生命/life, 逝者/the dead, 新闻/news, 昆明/Kunming, 无辜/ innocent, 路/road	暴徒/mob, 火车站/railway station, 暴徒/rioters, 昆明/Kunming, 受伤/injured, 事件/event, 男子/man, 报道/report, 逝者/the dead, 无辜/innocent	暴徒/mob, 火车站/railway station, 昆明/Kunming, 砍/cut, 受伤/injured, 默哀/mourn, 事件/event, 逝者/the dead, 无辜/innocent, 生命/life

simulate the real microblog environment, we merge each emergency dataset with the background dataset respectively and generate three new emergency datasets. When merging the emergency dataset with the background dataset, we keep the post time distribution of emergency data in new dataset consistent with the original distribution.

Fig. 3 Novelty of the bursty topic discovered by different methods

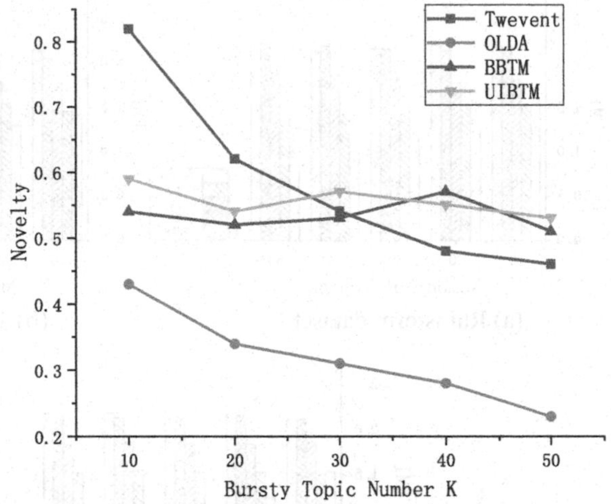

Since UTM is a retrospective topic detection model, we just compare UIBTM with Twevent, OLDA and BBTM. From Fig. 3, we observe that UIBTM achieves a little higher novelty score than BBTM and significantly outperforms OLDA. It implies that by introducing the amount of like into the measuring method of burstiness, UIBTM is more sensitive to bursty topics in the social network than BBTM and OLDA. Since Twevent summarizes bursty topics only with burst words, the novelty of Twevent is very high when K is small but decreases quickly with K increases.

5 Conclusions

In this paper, we propose a novel topic model for emergency detection based on user interactions in the social network, named UIBTM. First, it can solve the problem of semantic sparsity by enriching short texts using users' commenting and forwarding texts. Second, when measuring the burstiness of biterms, it takes the amount of users liking microblogs into account, which improves the accuracy and quality of topics discovered. The experiment results show that compared with the state-of-the-art emergency detection methods, UIBTM can effectively improve the accuracy and quality of the bursty topic discovered.

Acknowledgements This work is supported by the National Natural Science Foundation of China (No. 61532006, No. 61320106006, No. 61772083).

References

1. J.B. Gao, B.W. Zhang, X.H. Chen, A WordNet-based semantic similarity measurement combining edge-counting and information content theory. Eng. Appl. Artif. Intell. **39**, 80–88 (2015)
2. Y. Jiang, W. Bai, X. Zhang et al., Wikipedia-based information content and semantic similarity computation. Inf. Process. Manage. **53**(1), 248–265 (2016)
3. W. Wu, H. Li, H. Wang et al., Probase: a probabilistic taxonomy for text understanding (ACM, 2012), pp. 481–492
4. W. Xie, F. Zhu, J. Jiang et al., TopicSketch: real-time bursty topic detection from twitter, in *IEEE International Conference on Data Mining* (IEEE, 2013), pp. 837–846
5. X. Yan, J. Guo, Y. Lan et al., A probabilistic model for bursty topic discovery in microblogs, in *Twenty-Ninth AAAI Conference on Artificial Intelligence* (AAAI Press, 2015), pp. 353–359
6. Z. Yu, H. Wang, X. Lin et al., Understanding short texts through semantic enrichment and hashing. IEEE Trans. Knowl. Data Eng. **28**(2), 566–579 (2016)
7. C. Li, A. Sun, A. Datta, Twevent: segment-based event detection from tweets, in *ACM International Conference on Information and Knowledge Management* (ACM, 2012), pp. 155–164
8. J. Lau, N. Collier, T. Baldwin, On-line trend analysis with topic models: #twitter trends detection topic model online (2012)
9. Q. Diao, J. Jiang, F. Zhu et al., Finding bursty topics from microblogs, in *Meeting of the Association for Computational Linguistics: Long Papers* (Association for Computational Linguistics, 2012), pp. 536–544
10. H. Yin, B. Cui, H. Lu et al., A unified model for stable and temporal topic detection from social media data, in *IEEE International Conference on Data Engineering* (IEEE Computer Society, 2013), pp. 661–672
11. T. Mikolov, K. Chen, G. Corrado et al., Efficient estimation of word representations in vector space. Comput. Sci. (2013)
12. Y. Zuo, J. Wu, H. Zhang et al., Topic modeling of short texts: a pseudo-document view, in *ACM SIGKDD International Conference on Knowledge Discovery and Data Mining* (ACM, 2016), pp. 2105–2114
13. W. Hua, Z. Wang, H. Wang et al., Short text understanding through lexical-semantic analysis, in *IEEE, International Conference on Data Engineering* (2015), pp. 495–506

Microblog Search Based on Deep Reinforcement Learning

Xu Yao, Junping Du, Nan Zhou and Chengcai Chen

Abstract The big data and real-time publish make microblog search a challenge. We imitate mechanism of the behavior of playing Atari to realize microblog search based on reinforcement learning (RL). In this paper, we propose the Reinforcement Search Deep Q-Network (RSDQN) for microblog search, which is combined by Deep Q Network (DQN) and Long Short Term Memory Networks (LSTM). RSDQN sequentially takes action according to the current state of the environment. With state translating, the model parameters are trained by instant reward and total reward. We evaluate the instant reward as Q-Value and the total search results with normalize discounted cumulative gain (NDCG). Experimental results on Sina Weibo dataset showed better performance on the whole.

Keywords Microblog search · Reinforcement learning · Social network

1 Introduction

As one of famous social network platforms, Sina Weibo is increasingly popular in China. All types of information, including texts, images, and comments, generate uninterruptedly. The writing habits of microblog content lead to the semantic sparsity of language features, which makes it hard to search useful information. Effective processing of semantic noise has become a key research point for microblog content search.

Microblog search has become a hot topic in the field of information search. A variety of intelligent algorithms of machine learning emerge in recent years, one of

---

X. Yao · J. Du (✉) · N. Zhou
Beijing Key Laboratory of Intelligent Telecommunication Software and Multimedia, School of Computer Science, Beijing University of Posts and Telecommunications, Beijing 100876, China
e-mail: junpingdu@126.com

C. Chen
Xiaoi Research, Shanghai Xiaoi Robot Technology Co., Ltd, Shanghai 201803, China

Y. Jia et al. (eds.), *Proceedings of 2018 Chinese Intelligent Systems Conference*, Lecture Notes in Electrical Engineering 529,
https://doi.org/10.1007/978-981-13-2291-4_3

which is Reinforcement Learning (RL). RL algorithms are models which are designed to take actions in environments to maximize cumulative reward. The reward is defined as effective feedback from the environment, and is used to filter noise and is the vital element of the learning process.

The search results of each state are regarded as a frame in Atari. In the training phase, we design a DQN frame to estimate current Q-Value and target Q-Value. Then, loss function is redefined based on the mean square. Next, we define two types of reward as the instant reward and NDCG reward. Furthermore, we adopt next behavior depending on both rewards. In the microblog search environment, we come to process with epsilon.

The main contributions of our paper are as follows:

(1) A Reinforcement Search Deep Q Network model is proposed to optimize microblog search performance.
(2) We attach importance to both the single microblog document and the entirety of search results.
(3) Experimental evaluations on the dataset show that our method has better performance than the other comparison algorithms.

Our work is organized as follows, Sect. 2 introduces related works. Section 3 states the proposed microblog search algorithm based on reinforcement learning (RSDQN). Section 4 shows experimental results and analysis. And Sect. 5 gives the conclusions.

2 Related Work

The noisy information predominates microblogs. It can be a considerable challenge to extract valid information from microblog [1]. Peng [2] has proposed a method for preprocessing the input documents by time series aiming to get results of higher quality. Atefeh [3] indicates that the content of social network is shown as the information stream. As a result, the content's publish time must be emphasized. What's more, the increment of microblog deserves more attention, the microblog search based on time series is significant [4].

Reinforcement learning (RL) was brought out in 1992 [5], the initial aim of RL is the learner's being able to determine its behavior by the input-output pair. The best-known success of RL is TD-gammon, TD-gammon is trained by specifically TD-lambda [6].

Beyond the method of intelligent control, RL returned to people's vision in 2015 [7]. As a part of machine learning, Mnih develops a novel agent to combine RL with the deep neural network. More recently, there has been a revival of interest in combining deep learning with RL. RL is split into action-based and value-based by way of choosing an action [8]. The value-based RL is represented by Q-Value [9], Q-Value is an indicator of the model free algorithm to make actions explicit, and its importance lies how to build a Q-Value table. To deal with massive data, Deep Q-Network (DQN) is imported to generate Q-Value [10]. Furthermore, Hasselt [11] use Double Deep Q-Network to demonstrate the overestimation of DQN. Another action is to adopt policy optimization [12].

Recurrent neural networks with long short-term memory (LSTM) show great power on dealing with several problems which concern to sequential data, including time series microblog documents [13]. As a promotion of RNN, LSTM's additive and forget branches can control what and how much to store, LSTM is more efficient [14]. Reservoir Sampling Algorithm (RSA) [15] is helpful to simulate the input form of microblog as steam better, which is fair for all microblog documents to be chosen.

We use the entirety of all results as reward, it backwards to adjust the predefined network. And we designed an LSTM cell to measure the long-term effects of each search.

3 Microblog Search Based on Reinforcement Learning

We treat microblog search as the scene of playing Atari and split the model training into three stages. The first stage is waiting, which is to add enough candidate documents into search set. The second stage is observation, which is to explore the environment with a randomly chosen action. The third stage is training. In this work, we regard the search method as which document to choose or skip. By means of RSA, all documents are chosen with equal probability.

3.1 Microblog Search Problem Definition

The scene of microblog search is abstracted to the RL model, which consists of states, actions, transition, and reward. The detailed definitions are stated as follows.

States: S is defined as the set of states which describe the search environment. For each step of the search process, an appropriate candidate document is selected for the current ranking position. The input documents list is defined as $D = \{d_1, d_2 \ldots, d_e, \ldots\}$ and output results list is $O = \{O_1, O_2, \ldots O_e \ldots, O_k\}$. The search target is to get top K search results at episode e for the current search position. Therefore, D and O consist a State $S_e = \{D_e, O_e\}$.

Actions: A is defined as the set of discrete actions which can be chosen. For the state S_t, we choose an action from A. At the epsilon e, $A(S_e)$ selects a document D_e for the ranking position e + 1. An action set $A = \{A_c, A_s\}$ for each incoming state, A_c means choosing the document while A_s means skipping it. To ensure the possibility of action chosen for all documents, we adopt the Reservoir Sampling Algorithm (RSA) to calculate the possibility of A_c and A_s.

Transition: T is defined as a transition T: $S_e \times A_e \rightarrow S_{e+1}$ in the form of T(S, A). After adopting the selected action A_e with the state S_t, the state transits into the new state S_{e+1}.

Reward: We defined two types of reward. The first is the immediate reward r(S, A) known as Q-Value. The reward can be considered as an evaluation of the quality of the selected document. It is natural to define the reward function by the IR evaluation measures. The other reward is R(S, A) defined by the NDCG value of all selected documents.

In this paper, we define the reward received in response to choosing an action at as the promotion of the NDCG:

$$G_e = R_{e+1} + \lambda R_{e+2} + \cdots = \sum_{k=0}^{\infty} \lambda^k R_{e+k+1} \tag{1}$$

where G_e is NDCG in each episode, R_e is the instant reward in S_e, λ is the discount factor.

To obtain the total accumulative reward, we conduct the method to approach less loss function value. According to analyzing current state and next state, the loss function can be yielded:

$$L(\theta) = \mathrm{E}(r_1 + \gamma r_2 + \gamma^2 r_3 + \cdots | \pi(, \theta)) \tag{2}$$

where $L(\theta)$ is the objective value under the state of θ, r is instant reward in its state, γ is the discount factor, π is the policy under the state of θ.

3.2 Deep Q Network for Microblog Search

We use Reservoir Sampling Algorithm (RSA) to deal with the microblog text. Faced with massive social network data, the simple Q-Learning algorithm [16] will bring a dimensional explosion. We conduct the neural network [10] to calculate the incoming Q-Value instead of q-table of native Q-Learning. For single item reward, we define the process as follows.

As shown in Eq. (3), we define two type of objective reward, where r_e is the instant reward for each epsilon and R_e is the NDCG reward for current chosen all microblog documents

$$r_{e_0} = \sum_{e=e_0}^{\infty} \gamma^{e-e_0} r'_e \tag{3}$$

As shown in Eq. (4), we input preprocessed labeled microblog data into the deep Q neural network and mark current state.

$$Q^\pi(s, a) = r + \gamma Q^\pi(s', a') \tag{4}$$

Algorithm 1 Reinforcement Search with Deep Q-Network

1.Initialize defined neural network in the action-value function Q
2.for e = 1, 2, · · · E do
3. for t = 1, 2, · · · T do
4. Initialize D_t, O_t
5. $D_t = [D_{t+1}, D_{t+2}, \cdots]$
6. $O_t = [O_1, O_2, \cdots O_t]$
7. $S_t = [\{D_1, O_1\}, \{D_2, O_2\}, \cdots \{D_t, O_t\}]$
8. Compute: $Q(S_t, A_c, \theta)$ and $Q(S_t, A_s, \theta)$
9. Action: Randomly select an action for A_t with ε(probability) and RSA;
Otherwise $A_t = \max(Q(S_t, A_s; \theta))$. Execute A_t
10. Observe: instant reward r_t and total reward R_t
Set state $S_{t+1} = [S_t, A_t; \theta]$.
11 Update Q-Network by performing episode on the loss function
12. end for
13.end for
14.end

As shown in Eq. (5), we get current Q-Value error.

$$\delta = Q(s, a) - (r + \gamma \max_{a} Q(s', a')) \tag{5}$$

As what used in the DeepMind system [3], a deep convolutional neural network is used to represent Q with layers of convolutional. The algorithm is stated as Algorithm 1.

4 Experimental Results and Analysis

4.1 Dataset

The dataset is crawled from Sina Weibo by the official application program interface. We collect random microblog users and microblog documents between January 1, 2010 and December 31, 2015. After getting rid of useless data, we obtained about 129,000 original microblog documents.

4.2 Parameters Settings of Microblog Search Experiment

Each microblog content is segmented into words. After removing stop-words, each word is mapped into a word vector with the dimension of 100 with the help of word2vec. For a single microblog, we extract 60 words. Therefore, the input of RSDQN is (100, 60) to get top K results. We define the size of the output of RSDQN as (100, 60, K).

4.3 Evaluation Indexes

NDCG (normalized discount cumulative gain), Precision (precision for results), MAP (mean average precision), and ERR (Expected Reciprocal Ranking), etc. are used to evaluate search results for corresponding tasks. In this paper, we take the NDCG, MAP, and Precision to evaluate the search results.

4.4 Comparison and Analysis

The notations of the proposed model and the corresponding explanations are described in Table 1.

We set an LSTM network layer as a comparison task to estimate the time sequence of RSDQN. The reward decides the action, and we set different reward as well. So the training phase is organized into four groups, which is divided by reward and LSTM. The situation is split into four groups which are choosing positive documents, choosing negative documents, skipping positive documents and skipping negative documents. The details are shown in Table 2.

Table 1 Parameters of RSDQN

Symbol	Description
ε_0	1.0; The original epsilon
ε_E	0.1; The final epsilon
T_1	100; The number of episode of wait
T_2	1000; The number of episode of exploration
T_3	12,000; The number of episode of training
γ	0.95; The discount factor
B	10; The number of training batch
K	5, 10, 15, 20; The amount of search results

Table 2 Model group set

Model set	LSTM	Reward			
A	Yes	Choose	Positive: 1 Negative: −1	Skip	Positive: 1 Negative: 0
B	Yes	Choose	Positive: 1 Negative: −1	Skip	Positive: −0.5 Negative: 1
C	No	Choose	Positive: 1 Negative: −1	Skip	Positive: 1 Negative: 0
D	No	Choose	Positive: 1 Negative: −1	Skip	Positive: −0.5 Negative: 1

We use relevant queries for different datasets to evaluate the results. Different methods are deployed to experiment on the same dataset to compare with the proposed method shown as follows.

Table 3 reports the performances of our approach and all of the baseline methods regarding the NDCG of searching K results. RSDQN performs better on NDCG than LDA, DSSM, BM25, and RSDQN with LSTM. For RSDQN is designed to get better search results on the whole task. LSTM is aimed to avoid long-term dependency, which denotes previous input is filtered, positive documents included.

Table 4 reports the performance of our approach and all of the baseline methods in terms of the MAP of searching K results. RSDQN performs best in the five algorithms on MAP. Owing to the amount of candidate positive documents is more than K. And all algorithms perform better on MAP with the value of K increasing.

Table 5 reports the performances of our approach and all of the baseline methods in terms of the Precision of searching K results. We can see RSDQN and RSDQN (With LSTM) are slightly lower than LDA, DSSM and BM25 on P@K. This phenomenon is because RSDQN is devoted to getting better the whole instead of the single issue.

As it can be seen from Tables 2, 3 and 4, with K increasing, all algorithms are gaining better performance. At K = 20, the results of our model are 0.7371, 6.18% better than LDA, 9.28% better than BM25 and 4.83% better than DSSM. Comparing on MAP, the value of MAP@20 in our model is 0.7048, 0.02 higher than BM25 algorithm, 0.02 higher than LDA, and 0.03 higher than DSSM. However, our method is slightly lower than DSSM. It can be referred that RSDQN performs better on the whole.

Table 3 Search NDCG on Tianjin dataset

Method	NDCG@5	NDCG@10	NDCG@15	NDCG@20
LDA	0.5966	0.6185	0.6421	0.6942
DSSM	0.5974	0.6355	0.6531	0.7032
BM25	0.5741	0.6214	0.6288	0.6745
RSDQN	0.6233	0.6566	0.6874	0.7371
RSDQN (with LSTM)	0.6107	0.6465	0.6751	0.7011

Table 4 Search MAP on Tianjin dataset

Method	MAP@5	MAP@10	MAP@15	MAP@20
LDA	0.6384	0.6281	0.6471	0.6901
DSSM	0.6524	0.6205	0.6364	0.6879
BM25	0.6101	0.6118	0.6544	0.6841
RSDQN	0.6717	0.6926	0.6820	0.7048
RSDQN (with LSTM)	0.6450	0.6703	0.6151	0.6753

Table 5 Search precision on Tianjin dataset

Method	P@5	P@10	P@15	P@20
LDA	0.6805	0.6501	0.6384	0.6831
DSSM	0.6871	0.6399	0.6521	0.6941
BM25	0.6765	0.6752	0.6074	0.6412
RSDQN	0.6734	0.6699	0.6747	0.6641
RSDQN (with LSTM)	0.6650	0.6603	0.6712	0.6739

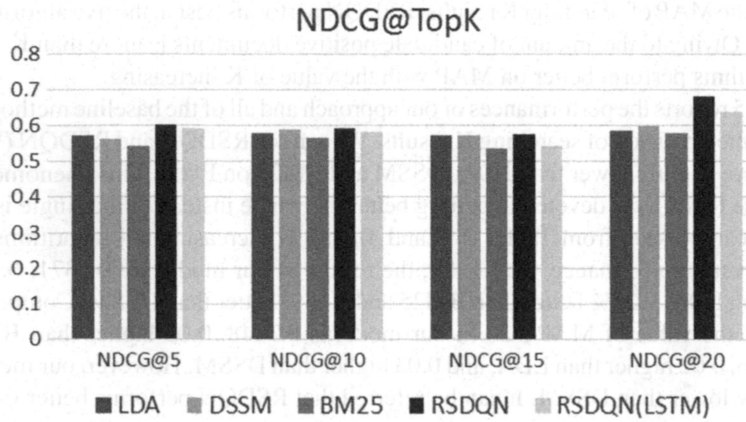

Fig. 1 NDCG value of top K

For dataset Wuhan Storm, the result shows below. The dataset consists of 16,975 items, and only 15 items are positive.

In Fig. 1, RSDQNs are slightly higher than LDA, DSSM, and BM25 when K = 5, 10 and 15. However, with K aiming to 20, RSDQN is remarkably higher than LDA, DSSM, and BM25. Due to the number of positive items is 15, RSDQNs are intended to find more related items to consist of search results.

In Fig. 2, the MAP values of RSDQN is higher than LDA and BM25, which indicates that RSDQN has a better performance. Compared with DSSM, when K is 20, the MAP of two algorithms decline significantly.

In Fig. 3, the tendency is the same as the shown in Fig. 2. And when K is 5, 10 and 15, RSDQN is quietly closed to LDA, DSSM, and BM25. It means that RSDQN can perform as well as the comparison algorithms, when the candidate positive documents are coming close to K. On the other hand, when the search amount is beyond K, the precision of RSDQN decreases more sharply.

As shown in Figs. 1, 2 and 3, RSDQN has better performance than LDA, DSSM and BM25 on NDCG and MAP. For all NDCG@K, RSDQN shows better NDCG value. When it refers to NDCG@20, its value reaches 0.68. As for MAP, RSDQNs perform better. And comparing the search result on Precision, RSDQN is not as accurate as LDA, DSSM and BM25.

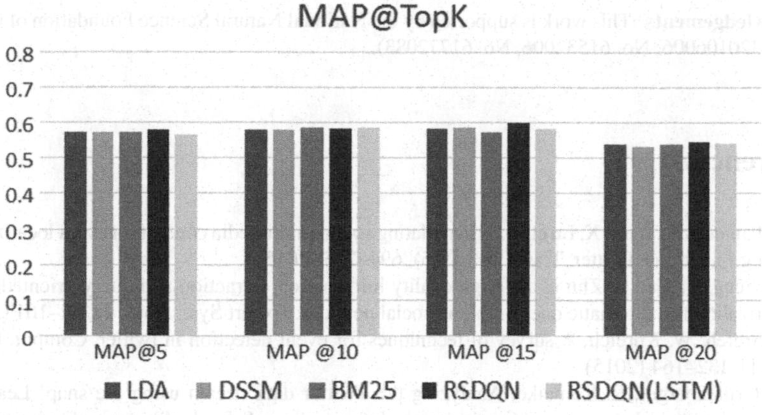

Fig. 2 MAP of top K

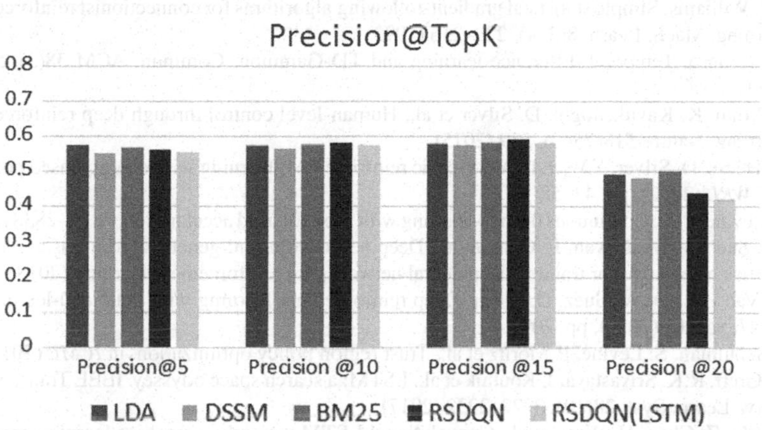

Fig. 3 Precision of top K

5 Conclusions

In this paper, we propose a microblog search algorithm based on reinforcement learning. First, we establish the environment of reinforcement learning in microblog search. With constructing the deep Q-network, we use labeled single microblog text as input and output its Q-Value. According to the delta of two action value, we will choose which action to take in next episode. We can obtain the NDCG of current search results. The experiment results indicates that our proposed algorithm in this paper improves NDCG and MAP for microblog search. RSDQN offers several advantages: utilizing both instant rewards and total rewards, and high NDCG in microblog search.

Acknowledgements This work is supported by the National Natural Science Foundation of China (No. 61320106006, No. 61532006, No. 61772083).

References

1. G. Panteras, S. Wise, X. Lu et al., Triangulating social multimedia content for event localization using Flickr and Twitter. Trans. GIS **19**(5), 694–715 (2015)
2. M. Peng, B. Gao, J. Zhu et al., High quality information extraction and query-oriented summarization for automatic query-reply in social network. Expert Syst. Appl. **44**, 92–101 (2016)
3. F. Atefeh, W. Khreich, A survey of techniques for event detection in twitter. Comput. Intell. **31**(1), 132–164 (2015)
4. A. Grillenberger, R. Romeike, Analyzing the Twitter data stream using the snap! Learning environment, in *International Conference on Informatics in Schools: Situation, Evolution, and Perspectives* (Springer International Publishing, 2015), pp. 155–164
5. R.J. Williams, Simple statistical gradient-following algorithms for connectionist reinforcement learning. Mach. Learn. **8**(3–4), 229–256 (1992)
6. G. Tesauro, Temporal difference learning and TD-Gammon. Commun. ACM **38**(3), 58–68 (1995)
7. V. Mnih, K. Kavukcuoglu, D. Silver et al., Human-level control through deep reinforcement learning. Nature **518**(7540), 529 (2015)
8. N. Heess, D. Silver, Y.W. Teh, Actor-critic reinforcement learning with energy-based policies, in *EWRL* (2012), pp. 43–58
9. S. Levine et al., Continuous deep Q-learning with model-based acceleration, 2829–2838 (2016)
10. F.P. Such, V. Madhavan, E. Conti et al., Deep neuroevolution: genetic algorithms are a competitive alternative for training deep neural networks for reinforcement learning (2018)
11. H. Van Hasselt, A. Guez, D. Silver, Deep reinforcement learning with double Q-learning, in *AAAI*, vol. 16 (2016), pp. 2094–2100
12. J. Schulman, S. Levine, P. Moritz et al., Trust region policy optimization, in *ICML* (2015)
13. K. Greff, R.K. Srivastava, J. Koutník et al., LSTM: a search space odyssey. IEEE Trans. Neural Netw. Learn. Syst. **28**(10), 2222–2232 (2017)
14. X. Shi, Z. Chen, H. Wang et al., Convolutional LSTM network: a machine learning approach for precipitation nowcasting **9199**, 802–810 (2015)
15. M. Al-Kateb, B.S. Lee, Adaptive stratified reservoir sampling over heterogeneous data streams (Elsevier Science Ltd., 2014)
16. C.J.C.H. Watkins, P. Dayan, Q-learning. Mach. Learn. **8**(3–4), 279–292 (1992)

A Cross-Modal Short Text Semantic Expansion Method for Microblog Search

Yansong Shi, Junping Du, Feifei Kou and Chengcai Chen

Abstract Image is an important part of microblog, and its visual information can offer additional semantics besides the textual information. To overcome short text's semantic sparsity problem and fully utilize the semantics of text and image, we propose a cross-modal short text expansion method for microblog search in this paper. First, we expand short texts using the distributed representations of words, and then based on deep neural network, we extract related information of images and append them to the original short text. The expanded pseudo-documents contain richer semantics, and by turning pseudo-documents into vectors, we can achieve accurate microblog search. Experiments on real-world datasets show that the proposed cross-modal short text expansion method can effectively extract the semantics of microblogs and improve search performance.

Keywords Short text · Cross modal · Image caption · Microblog search

1 Introduction

In recent years, social networks are becoming popular media. People usually share their feelings on the social network and publish their opinions. Sina Weibo is one of the major social media in China. It has a large number of active users and has accumulated huge amounts of user data, which has extremely high research value.

Y. Shi · J. Du (✉) · F. Kou
Beijing Key Laboratory of Intelligent Telecommunication Software and Multimedia, School of Computer Science, Beijing University of Posts and Telecommunications, Beijing 100876, China
e-mail: junpingdu@126.com

C. Chen
Xiaoi Research, Shanghai Xiaoi Robot Technology Co., Ltd, Shanghai 201803, China

Y. Jia et al. (eds.), *Proceedings of 2018 Chinese Intelligent Systems Conference*, Lecture Notes in Electrical Engineering 529,
https://doi.org/10.1007/978-981-13-2291-4_4

How to satisfy the users' microblog search needs from such massive data becomes an issue to be solved. The limit of 140 words makes microblog short text have inherently semantic sparsity, which poses a challenge to the traditional topic model based semantic analysis methods.

Image microblog refers to a microblog published with images, and its multimedia form have attracted more researchers' attention. Zhao et al. [1] find that on Sina Weibo microblogs that contain multimedia information such as images and videos are more likely to be forwarded than microblogs with only text information. Chen et al. [2, 3] proposed that there are three kinds of relationships between image and text: visual relevance, emotional relevance, and irrelevance.

In order to solve the short text sparsity and better understand the semantics of short texts, short texts are often expanded to generate longer pseudo-documents. Typical expansion methods include using Wordnet, aggregating words with co-occurrence relations or semantic similarities, or finding similar expressions using neural word embedding. In addition, microblogs also have image information, and this part hasn't been used to expand short text yet. If we can understand the semantic information of images in microblog and add it to the text, we can simultaneously solve the problem of the lack of effective use of multimedia information in search and the issues of short texts.

So as we have discussed, in order to solve the sparseness of short texts and achieve accurate search, we propose a cross-modal short text semantic expansion method for microblog search. We use word2vec to find similar expressions of the original text, and use a deep neural image caption model to transform image semantics to literal semantics, then expand the short text to pseudo-document with the meaningful and related part of the image-generated text. Then we transform pseudo-document into a vector to accomplish accurate search. We did experiments on our microblog security event dataset to validate the quality of our expansion method and search performance.

2 Related Works

Typical short text expansion methods need to use external knowledge bases. Chen et al. [4] use Wordnet to expand short texts. As the external knowledge introduced is often too general, and the microblog texts are usually very colloquial, Zuo et al. [5] aggregated words with co-occurrence relations. Aletras et al. [6] add words semantic similarities to expand texts. Bicalho et al. [7] propose a general method to expand short texts using word vectors including word2vec [8] and Glove.

Image caption is a method to "translate" image semantics to literal semantics. Early approaches [9] are able to describe images "in the wild", but require heavily hand-designed features and become rigid when generating text. With the development of deep learning, researchers begin to use deep neural networks to understand images and generate text. Mao et al. [10] first combine CNN with RNN to solve the image caption problem. Vinyals et al. [11] proposed a neural image caption model which uses LSTM instead of RNN.

3 Cross-Modal Semantic Expansion Algorithm (CSE)

In this section, we briefly introduce the basic flow of cross-modal semantic expansion algorithm and the implementation of each part.

3.1 *Algorithm Framework*

As shown in Fig. 1, we divide the original microblog data into two parts: text and image, and operate on the two parts separately. For the text, we use word2vec (CBOW) and find possible expansions of the text based on word vectors. For the image part, we use a deep neural network based image caption method. The image caption model contains two parts: encoder and decoder. The encoder is implemented by a deep convolutional neural network. This part extracts the feature vector of the image and serves as the initial input of the decoder. The decoder is implemented using a recurrent neural network, which predicts the T_{i+1}th word based on the image feature and the T_ith word. Thus the text generated is both related to the semantics of the image and maintains a readable sentence. In order to make the generated text semantically related to the original text of the microblog. We select the words belonging to $S_I \cap E_T$ to generate expanded pseudo-document.

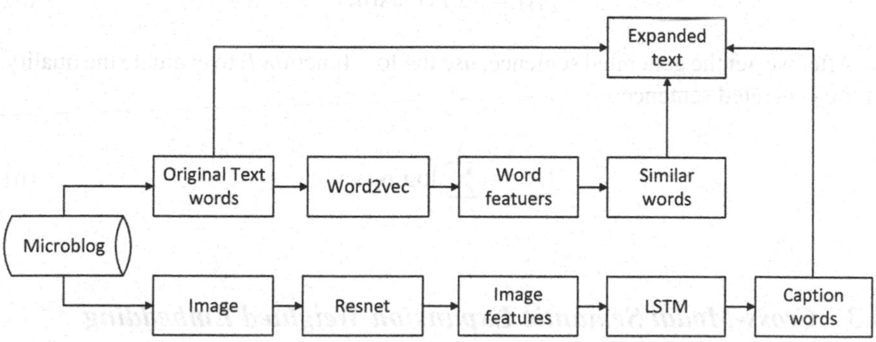

Fig. 1 Framework of the CSE algorithm

3.2 From Image to Text Semantics

In order to extract the semantic information in the image, we use a model called image caption. This model is used in the following situation: given a pair of image and text. According to the image, a natural language statement that can describe the content of the image is generated. In our microblog cross-modal expansion algorithm, we take the text and regard it as the caption or description of the image.

We used Resnet [12] in the encoder section. The basic idea of this model is to create a shortcut of the gradient so that the gradient can be passed directly from the last layer to the first few layers of the network.

We remove the last softmax layer and use the feature vector before softmax as the initial input of the recurrent neural network.

$$x_{-1} = \text{Resnet}(I) \tag{1}$$

For each word in a sentence, we look up the embedding in the vocabulary for the word

$$x_t = v_i w_i, \quad t \in \{0 \ldots N-1\} \tag{2}$$

The recurrent neural network used to generate sentences is implemented with long short-term memory network. This structure has been proved to be effective in learning the long-term dependencies of sentences. Three threshold control gates are used: input gate i_t, forget gate f_t and output gate o_t. Then we generate the next predicted word based on the currently entered word in the following way:

$$c_t = i_t \odot h(W_{ch} h_{t-1} + W_{cx} x_t) + f_t \odot c_{t-1} \tag{3}$$

$$h_t = o_t \odot c_t \tag{4}$$

$$p_{t+1} = Softmax(h_t) \tag{5}$$

After we get the generated sentence, use the loss function L to evaluate the quality of the generated sentence:

$$L = -\sum_{t=1}^{N} \log p_t(s_t) \tag{6}$$

3.3 Cross-Modal Semantic Expansion Weighted Embedding (CSEWE)

Before we expand the text, we need to use word2vec to generate a vector for each word in the corpus. After we expand the text, we want to get the vector of the whole microblog to use it for search. The embedding method is proposed by Arora et al.

[13] We describe the whole cross-modal semantic expansion weighted embedding algorithm as follows:

Algorithm 1 Cross-modal semantic expansion weighted embedding algorithm

Input: original microblog text and image $T + I$
Output: expanded microblog embedding v_E

1. For each $i \in [1, \ldots, N_m - 1]$ Do
2. compute the 2-gram vector r_i of word T_i and T_{i+1}: $r_i = \lambda v_i + (1 - \lambda)v_{i+1}$
3. find m most similar words of r_i in V according to (10)
4. End For
5. Sort the similarity in descending order, Append first k words to T as E_T
6. Generate sentence S from I
7. Append $S \cap E_T$ to T as pseudo document after expansion E
8. Calculate sentence vector v_E for E as (11)
9. Update v_E as (12)
10. Return v_E

The similarity between two vectors is defined as follows:

$$\mathrm{sim}(v_i, v_j) = 1 - \frac{(v_i \cdot v_j)}{\pi \|v_i\| \|v_j\|} \tag{7}$$

Instead of simply averaging all word vectors, we calculate the sentence embedding as follows:

$$v_E := \frac{1}{|E|} \sum_{w \in E} \frac{a}{a + p(w)} v_w \tag{8}$$

a is a smooth parameter, $p(w)$ is estimated probability of the words.

Then compute the first principle component u of all sentence and for each sentence do:

$$v_E := v_E - uu^{\mathrm{T}} v_E \tag{9}$$

We use the final v_E to represent a microblog content and use it to search.

4 Experiment and Analysis

We have verified the superiority of the proposed method from two aspects. In this section, we will give detailed information on experimental data sets, evaluation metrics, results, and analysis.

Table 1 Dataset description

Dataset number	Related event	Number of microblogs
Dataset 1	Explosion accident	10,137
Dataset 2	Heavy rainstorm	12,993
Dataset 3	Malaysia airlines	11,431
Dataset 4	General topic	53,827

4.1 Dataset

We crawled three security topics related data from Sina Weibo, which is one of the most popular social network platforms in China. The three topics are "Heavy Rainstorm", "Explosion Accident", and "Malaysia Airlines". Besides, In order to evaluate the performance of the proposed method under different data characteristics, we also crawled some general microblogs without a certain topic. We preprocessed the data as follows to obtain cleaner datasets: (1) keep the original microblog message and delete the forwarded data; (2) delete the microblog message containing less than 10 characters; (3) filter advertisements; (4) use ICTCLAS to segment words and then remove stop words; (5) manually remove the microblogs whose image are mainly composed of text. In Table 1, we list the description of the dataset that is used in this paper. We mix the four datasets together to conduct following experiments.

4.2 Experiment Settings and Evaluation Metric

First, as our designed cross-modal semantic expansion method (CSE) is the core component of the proposed CSEWE, we conducted experiments to evaluate the effectiveness of CSE. We use the expanded data that generated by CSE as the input of topic models, and then verify qualities of the generated topics. Two typical topic models: Latent Dirichlet Allocation (LDA) and Biterm topic model (BTM) are chosen in this experiment. As the evaluation metric: pointwise mutual information (PMI) can reflect the consistency and understandability of topics, we use it as the evaluation metric in this experiment. Given a topic and N most possible words, use $p(w_i)$ to represent the possibility of word w_i and $p(w_i, w_j)$ to represent the co-occurrence possibility of two words, the PMI value of a topic can be calculated as (10):

$$\mathrm{PMI} = \frac{2}{(N-1)N} \sum_{1 \le i < j \le N} \log \frac{p(w_i, w_j)}{p(w_i)p(w_j)} \tag{10}$$

Second, we want to evaluate the search performance of our microblog embedding. Therefore, we conducted the microblog search experiments. To evaluate the results of the experiment objectively and impartially, we invited several volunteers to assess the search results. The MAP (mean average precision), and NDCG (normalized discount cumulative gain) are used as the evaluation metrics to evaluate the performance of microblog search.

Ignoring the words that appear only once, and the parameters used in the experiments are as follow: image feature dimension is 512, word vector dimension is 256, LSTM hidden layer dimension is 512, the training epoch is 10, the learning rate is 0.001 and decays every epoch, the smooth parameter is 10^{-4}.

4.3 Result and Analysis

We use the topic models to generate the topic representation for the expanded microblog text and use the PMI to evaluate the quality of the generated topic. We expand each microblog with cross-modal semantic expansion method and use LDA to generate sample topics, then compare the PMI value with topics generated by LDA and BTM algorithm using the original text. The results are shown in Table 2 and Fig. 2.

In Table 2 and Fig. 2, Top10, Top15, and Top20 refer to the number of Top words in a topic. From Table 2 and Fig. 2, we can see that BTM performs better than LDA because it improves topic quality by introducing biterm representation of the text, which is specifically designed for short texts. Our CSE method outperforms both of the two methods. This is because our method uses both the semantics of microblog image and text. The comprehensiveness of cross-modal information helps to implement microblog semantics.

When topic number increases, the semantic sparsity of LDA will be more obvious, and the quality of the generated topic will be worse. After we expand the corpus, the quality of the topics is significantly improved, especially when the number of topics is large. CSE+BTM works best because it extends the corpus while introducing word pairs. It also shows that our expansion method can improve the quality of the topics on both two topic models.

We performed a microblog content search experiment on the expanded microblog text using the CSEWE method described in Sect. 3.3, and compare the results with BTM and BM25 algorithm. The search performance is evaluated with MAP and NDCG metrics. In actual use case of microblog search, people usually only consider

Table 2 PMI values with different topic numbers

Topic	10		30		50	
Algorithm	Top10	Top20	Top10	Top20	Top10	Top20
LDA	1.57	1.51	1.79	1.67	1.67	1.58
BTM	1.94	1.87	2.00	1.97	2.02	1.94
CSE+LDA	2.16	2.11	2.18	2.12	2.15	2.02
CSE+BTM	2.38	2.21	2.41	2.34	2.42	2.33

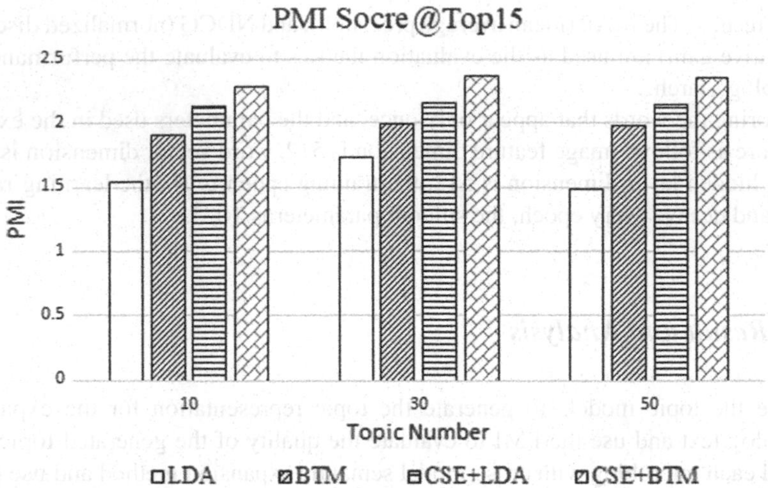

Fig. 2 PMI value @ Top15 under different topic numbers

the first n results they get, so recall rate is not what we are going to care about. The results are shown in Table 3.

Since under most circumstances, there are not so many unexpected events, Weibo data just have general topics. In order to verify the effectiveness of our search algorithm in such cases, we conducted separate search experiments on the dataset 4, which has no specific topics. The search results are shown in Fig. 3.

As Table 3 and Fig. 3 shows, when we only consider the top 10 results, the proposed CSEWE reaches a decent performance. This means our microblog search method can satisfy users' requirements. When we consider more search results, CSEWE's advantage becomes more obvious. We contribute this to that CSEWE can supplement the semantics of the microblogs that have similar images but with different literal expressions. The top @N values of MAP gradually decreases, while the top @N values of the NDCG keep increasing, which is consistent with the definition of these two metrics.

Table 3 Top @N MAP and NDCG search result of CSEWE

Metric	MAP				NDCG			
Algorithm	@10	@20	@30	@50	@10	@20	@30	@50
BTM	0.59	0.51	0.50	0.48	0.59	0.61	0.63	0.68
BM25	0.64	0.59	0.54	0.50	0.65	0.70	0.70	0.71
CSEWE	0.67	0.60	0.58	0.57	0.70	0.72	0.73	0.74

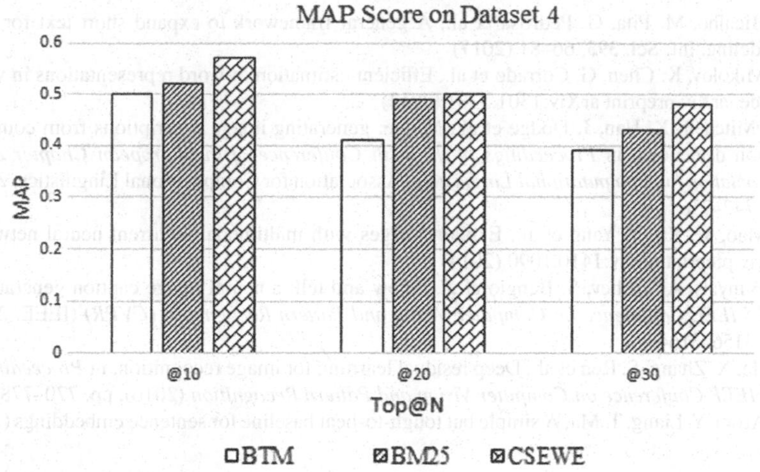

Fig. 3 MAP value on general topic dataset

5 Conclusions

In this paper, we present a method for semantic expansion of short text search across modal microblogs. We extract text semantic meanings from images to generate pseudo-documents. Experiments prove that the extracted image semantics can compensate for the short-text semantic sparseness. Pseudo-documents can effectively improve the quality of the topic and improve search performance.

Acknowledgements This work is supported by the National Natural Science Foundation of China (No. 61320106006, No. 61532006, No. 61772083).

References

1. X. Zhao, F. Zhu, W. Qian et al., *Impact of multimedia in Sina Weibo: Popularity and life span*, Semantic Web and Web Science (Springer, New York, NY, 2013), pp. 55–65
2. T. Chen, D. Lu, M.Y. Kan et al., Understanding and classifying image tweets, in *Proceedings of the 21st ACM International Conference on Multimedia* (ACM, 2013), pp. 781–784
3. T. Chen, H.M. SalahEldeen, X. He et al., VELDA: relating an image Tweet's text and images, in *AAAI* (2015), pp. 30–36
4. Z. Chen, A. Mukherjee, B. Liu et al., Discovering coherent topics using general knowledge, in *Proceedings of the 22nd ACM International Conference on Information & Knowledge Management* (ACM, 2013), pp. 209–218
5. Y. Zuo, J. Zhao, K. Xu, Word network topic model: a simple but general solution for short and imbalanced texts. Knowl. Inf. Syst. **48**(2), 379–398 (2016)
6. N. Aletras, M. Stevenson, Evaluating topic coherence using distributional semantics, in *Proceedings of the 10th International Conference on Computational Semantics (IWCS 2013)*–Long Papers (2013), pp. 13–22

7. P. Bicalho, M. Pita, G. Pedrosa et al., A general framework to expand short text for topic modeling. Inf. Sci. **393**, 66–81 (2017)
8. T. Mikolov, K. Chen, G. Corrado et al., Efficient estimation of word representations in vector space. arXiv preprint arXiv:1301.3781 (2013)
9. M. Mitchell, X. Han, J. Dodge et al., Midge: generating image descriptions from computer vision detections, in *Proceedings of the 13th Conference of the European Chapter of the Association for Computational Linguistics* (Association for Computational Linguistics, 2012), pp. 747–756
10. J. Mao, W. Xu, Y. Yang et al., Explain images with multimodal recurrent neural networks. arXiv preprint arXiv:1410.1090 (2014)
11. O. Vinyals, A. Toshev, S. Bengio et al., Show and tell: a neural image caption generator, in *2015 IEEE Conference on Computer Vision and Pattern Recognition (CVPR)* (IEEE, 2015), pp. 3156–3164
12. K He, X. Zhang, S. Ren et al., Deep residual learning for image recognition, in *Proceedings of the IEEE Conference on Computer Vision and Pattern Recognition* (2016), pp. 770–778
13. S. Arora, Y. Liang, T. Ma, A simple but tough-to-beat baseline for sentence embeddings (2016)

New Personnel Positioning Algorithm in Mine Based on PSO-GSA

Haibo Liu, Pengxin Liu and Fuzhong Wang

Abstract Aiming at the problem that the localization algorithm based on received signal strength indication is difficult to dynamically track the parameter change in the complex environment of coal mines. In order to improve the accuracy of the mine personnel positioning, this paper proposes a method to use the improved gravitational search algorithm to position the underground personnel in the weighted centroid positioning. Utilizing the long distance path loss model got the distance between the beacon nodes and unknown nodes, and then through the weighted centroid localization algorithm performed the unknown node positioning. Finally, the improved GSA-PSO optimized the preliminary location results and parameters. Experimental results show the proposed method can improve both the positioning accuracy effectively and the adaptive ability of changeful environment.

Keywords Personnel positioning · RSSI · Gravitational search algorithm
Centroid localization · Particle swarm optimization

1 Introduction

With the rapid development of coal mines, underground safety production has received more and more attention. The changing nature of the underground environment and the limitations of monitoring systems can pose a major threat to mining personnel. Therefore, it is of great significance to establish effective real-time personnel positioning systems. At present, the domestic and foreign positioning systems are mainly divided into two categories based on non-ranging and ranging-based positioning. The former uses information such as network connectivity to achieve

H. Liu (✉) · F. Wang
School of Electrical Engineering and Automation, Henan Polytechnic University, Jiaozuo 454000, China
e-mail: liuhaibo09@hpu.edu.cn

P. Liu
Basic Department, Air Force Communication NCO Academy, Dalian 116100, China

Y. Jia et al. (eds.), *Proceedings of 2018 Chinese Intelligent Systems Conference*, Lecture Notes in Electrical Engineering 529,
https://doi.org/10.1007/978-981-13-2291-4_5

positioning, but the accuracy is low; the latter realizes positioning by actually measuring the distance or angle between nodes, and the positioning accuracy is relatively high. The ranging-based algorithms are commonly used for TOA, TDOA, AOA, and RSSI [1]. Among them, based on the RSSI positioning method, the distance is calculated using the empirical transmission loss model, which is easy to implement without additional hardware devices, but it is greatly influenced by the environment. Now many researchers have optimized it to improve the accuracy. The literature [2] uses the genetic algorithm to improve the centroid algorithm to locate, but there are some defects such as large amount of computation and time-consuming. Literature [3] uses particle swarm optimization to optimize the centroid localization. Particle swarm optimization is easier to implement than genetic algorithm, the computational complexity is smaller, and convergence is faster, but the accuracy is not greatly improved.

Based on the above problems, this paper introduces the improved gravitational search algorithm (GSA) into the triangular weighted centroid positioning method based on RSSI. Firstly, the distance from the node to the unknown node is obtained by using the logarithmic distance path loss model, and the estimated position is obtained from the weighted centroid. Then, the obtained coordinates and parameters are sent to the GSA as the initial value of each particle. In order to avoid the system falling into a local optimum, the particle swarm optimization algorithm is used to improve the GSA. The simulation results show that the method is feasible and accurate.

2 Weighted Centroid Algorithm

(1) RSSI Ranging

The beacon node set in the lane sends its own node information periodically and is received by the unknown node, and the measured signal strength value is calculated as the node distance through the attenuation empirical formula. The RSSI ranging principle is based on the relationship between the transmitted power and the received power of a wireless signal, as shown in the following formula.

$$P_R = P_S/d^n \tag{1}$$

Among them: P_R is the received power of wireless signal; P_S is the transmitting power of wireless signal; d is the distance between transmitting and receiving elements; n is the attenuation factor, which means the path loss increases with the distance. Substituting the known node transmit power into Eq. (1) and taking logarithms on both sides, the transformation can be given by the following formula [4].

$$10 \lg P_R = A - 10n \lg d \tag{2}$$

where: A is the constant obtained by substituting $10\lg P_S$ into the transmit power, and $10\lg P_R$ is the expression that the received signal power is converted to dBm, so Eq. (2) can be expressed as $P_R = A - 10n\lg d$.

Experimental studies have shown that in the underground wireless propagation environment, the signal will be attenuated due to non-line-of-sight propagation reflection, diffraction, scattering, multipath propagation, and the like. In addition, real-time movement of peripheral electronic devices and objects to be measured will cause certain interference, and will also cause varying degrees of loss in the propagation of radio waves. The long-distance attenuation of the channel obeys the lognormal distribution, which is often represented by the logarithmic distance path loss model. The signal attenuation model is:

$$P_L(d) = P_L(d_0) + 10n\ \lg(d/d_0) + X_\sigma \tag{3}$$

where: $P_L(d)$ is the power of the received signal at a distance d from the transmitting node; $P_L(d_0)$ is the received signal power when the reference distance is d_0; The attenuation factor n is usually in the range of 2–5. X_σ is the random noise value of Gauss distribution whose mean value is zero and obeys the standard deviation of σ.

Let the signal transmission power be P_t, and then the signal strength value at the receiving point d is:

$$RSSI(d) = P_t - P_L(d) \tag{4}$$

Therefore, the received signal strength A_0 from the transmitting node d_0, $A_0 = P_t - P_L(d_0)$.

$$P_L(d_0) = P_t - A_0 \tag{5}$$

Substituting Eqs. (4) and (5) into Eq. (3) yields the following equation.

$$RSSI(d) = A_0 - 10n\ \lg(d/d_0) - X_\sigma \tag{6}$$

According to Eq. (6), the distance between the beacon node and the unknown node can be calculated.

(2) Triangle weighted centroid positioning algorithm

The traditional centroid positioning uses the polygons formed by the beacon nodes adjacent to the unknown node to achieve the positioning, and is entirely dependent on the network connectivity and the error are large. Based on RSSI, this paper adopts traditional three-side positioning method to select neighboring nodes to form polygons. Taking the three beaconing nodes in the three-sided positioning as the center of the circle, the distance from the beaconing node to the unknown node obtained by the signal path loss model is radiuses, and then an equation is established to solve the coordinates of the intersections of the three circles [5].

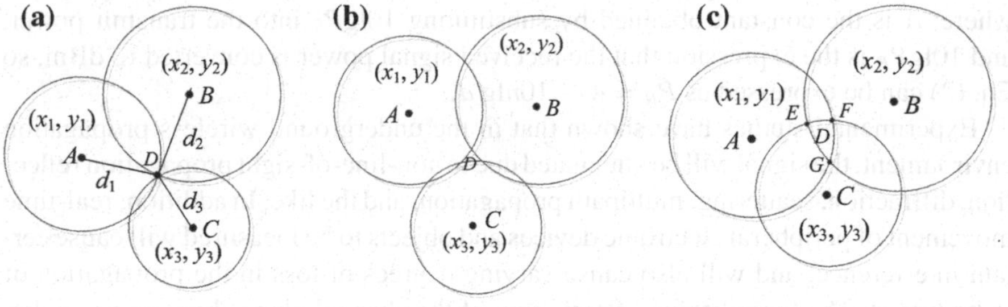

Fig. 1 Three conditions for three-sided positioning

Usually, the three conditions shown in Fig. 1 will occur at the position $D(x, y)$ to be measured due to the actual error. Figure 1a is the only intersection in the ideal case, (b) and (c) are two cases where the three circles meet each other.

The radius between the beacon node and unknown node is the radius of the circle.

$$(x - x_i)^2 + (y - y_i)^2 = d_i^2 \tag{7}$$

where x_i and y_i are the coordinates of the center of circle of i circle.

As shown in Fig. 1, the coordinates of all the intersections where the circles A, B, and C intersect, can be solved by solving equations for any two circles. According to the three-circle intersection region, the distance from any point to the center of the circle is equal to or less than the radius of any circle, and three points $E(a_1, b_1,)$, $F(a_2, b_2,)$, $G(a_3, b_3,)$ are enclosed into an approximation. The location area of the triangle takes the centroid of the triangle as the estimated position of the node to be positioned.

$$x, y = \left(\frac{a_1 + a_2 + a_3}{3}, \frac{b_1 + b_2 + b_3}{3}\right) \tag{8}$$

However, the intensity values received by the known nodes are different, in order to reflect the contribution of the nodes and the information of different weights, the greater the influence of the RSSI nodes to the unknown nodes, and the inverse relation to the distance, the reciprocal of the distance between the nodes is taken as their shadow factor, so the principle of the weighted algorithm is obtained.

$$\begin{cases} x' = \left[a_1\left(\frac{1}{d_2} + \frac{1}{d_3}\right) + a_2\left(\frac{1}{d_1} + \frac{1}{d_3}\right) + a_3\left(\frac{1}{d_1} + \frac{1}{d_2}\right)\right] \Big/ 2\left(\frac{1}{d_1} + \frac{1}{d_2} + \frac{1}{d_3}\right) \\ y' = \left[b_1\left(\frac{1}{d_2} + \frac{1}{d_3}\right) + b_2\left(\frac{1}{d_1} + \frac{1}{d_3}\right) + b_3\left(\frac{1}{d_1} + \frac{1}{d_2}\right)\right] \Big/ 2\left(\frac{1}{d_1} + \frac{1}{d_2} + \frac{1}{d_3}\right) \end{cases} \tag{9}$$

The position (x', y') is the estimated unknown node coordinates.

3 Underground Personnel Positioning Algorithm Based on PSO-GSA

Under the same environment, the above method can be used to locate better results. However, in the actual underground environment, the roadway is complex and variable, and it is difficult to describe the channel by using a constant value of n when the RSSI attenuation model is used to calculate the distance. Therefore, the unknown node coordinates must be corrected. In this paper, PSO-GSA algorithm is used to optimize the positioning results and parameters.

(1) GSA

The GSA algorithm considers a set of particles that operate in space as a solution to the problem to be optimized. Suppose there are N particles in a gravitational system and define the position of particle i is $X_i = (x_i^1, x_i^2, \ldots, x_i^d, \ldots, x_i^n)$. According to Newton's law, the gravity of particle j on particle i at time t can be expressed as

$$F_{ij}^d(t) = G(t)\frac{M_i(t)M_j(t)}{R_{ij}(t) + \varepsilon}(x_j^d(t) - x_i^d(t)) \tag{10}$$

where: $M_i(t)$ and $M_j(t)$ are the inertial masses of particles i and j, respectively; $G(t)$ is the gravitational constant at time t, which decreases with the age of the universe [6].

$$G(t) = G_0 e^{-\alpha \frac{t}{T}} \tag{11}$$

Among them: G_0 is the initial gravitational constant of the universe, which takes 1 or 100; α takes 20 or 23; T is the maximum number of iterations.

Let the gravitational mass be equal to the inertial mass and define the particle mass as the fitness function is:

$$m_i(t) = \frac{f_i(t) - f_{worst}(t)}{f_{best}(t) - f_{worst}(t)} \tag{12}$$

where, $f_i(t)$ is the fitness function of particle i at time t; $f_{worst}(t)$ is the value of the worst fitness function in the population; $f_{best}(t)$ is the value of the optimal fitness function. The gravitational mass generally uses the unit value:

$$M_i(t) = m_i(t) / \sum_{j=1}^{N} m_j(t) \tag{13}$$

Each particle's position and update speed during iteration is:

$$x_i^d(t+1) = x_i^d(t) + v_i^d(t+1) \tag{14}$$

$$v_i^d(t+1) = rand_j \times v_i^d(t) + a_i^d(t) \tag{15}$$

$$a_i^d(t) = F_i^d(t)/M_i(t) \tag{16}$$

Among them: $x_i^d(t)$, $v_i^d(t)$, $a_i^d(t)$ are the position, velocity and acceleration of particle i in d-dimensional space at time t; rand_j is a random number 0–1; d = 1, 2, …, D, D is the search space dimension; $F_i^d(t)$ and $M_i(t)$ represent the force that particle i receives in dimension d at t and its own inertial mass.

(2) Particle swarm optimization

PSO is researched by the bird predation. In the PSO, each particle has a memory, each particle separately retains the optimal solution information it finds in the search process, while the population retains the current group optimal solution information. PSO starts from a random solution and has excellent global search performance and fast convergence ability [7].

The particle i-mobility equation for the particle swarm algorithm is

$$v_i^d(t+1) = \omega(t)v_i^d(t) + c_1 r_{i1}(p_i^d - x_i^d(t)) + c_2 r_{i2}(g_{best}^d - x_i^d(t)) \tag{17}$$

$$x_i^d(t+1) = x_i^d(t) + v_i^d(t+1) \tag{18}$$

where: r_{i1} and r_{i2} are random numbers within [0, 1]; c_1, c_2 are non-negative constants, choosing the appropriate value makes the system less susceptible to local optima. $\omega(t)$ is the inertia weight, which is in the range of 0.1–0.9; p_i is the best location experienced by particle i; g_{best} is the best location that all particles have ever experienced [8].

(3) Optimization estimation coordinates based on PSO-GSA

In this paper, PSO and GSA are used to optimize the estimated position, and dynamic tuning parameters and position coordinates are achieved to improve the accuracy of the algorithm. For ease of description, take d_0 in Eq. (6) as 1 m, set $a = 10n$, $b = A_0$, get the following equation:

$$RSSI(d) = b - a \lg d - X_\sigma \tag{19}$$

where: $a \in (20, 50)$, $b \in (30, 50)$.

Let the final coordinates of the unknown node be (x, y), the estimated position (x', y') is obtained by Eq. (9), select an integer ε as the adjustment factor to adjust the optimization range, $x \in (x' - \varepsilon, x' + \varepsilon)$, $y \in (y' - \varepsilon, y' + \varepsilon)$, $\varepsilon = 20$.

Randomly generate N particles in the range of x, y, a, b, $X_i = (x_i^1, x_i^2, x_i^3, x_i^4)$, where $x_i^1, x_i^2, x_i^3, x_i^4$ correspond to the requested x, y, a, b, respectively, using GSA to iteratively calculate these four dynamic quantities. Formula (20) as its fitness function,

$$fitness(x,y,a,b) = \sum_{i=1}^{m} (RSSI_i + a\lg d_i - b)^2 \\ = \sum_{i=1}^{m} \left(RSSI_i + 0.5a\lg\left[(x-x_i)^2 + (y-y_i)^2\right] - b\right)^2 \quad (20)$$

In order to prevent the system from falling into a local optimum, based on the GSA, combined with the characteristics of the particle swarm algorithm, the particles are added with memory and social information exchange capabilities. This results in a particle swarm optimization algorithm (PSO-GSA). The particle velocity equation is as follows.

$$v_i^d(t+1) = r_{i1}v_i^d(t) + a_i^d(t) + c_1 r_{i2}(p_i^d - x_i^d(t)) + c_2 r_{i3}(g_{best}^d - x_i^d(t)) \quad (21)$$

where: r_{i1}, r_{i2} and r_{i3} are random numbers in the interval [0, 1]; c_1, c_2 are weight constants used to adjust the degree of influence of gravitational rules, memory, and social information exchanges during particle motion. It can be seen from Eq. (21) that when $c_1 = c_2 = 0$, speed equation is reduced to GSA algorithm [9].

4 Experimental Simulation and Result Analysis

Unknown nodes and beacon nodes are randomly distributed in a 200 m * 200 m MATLAB simulation area. PSO, GSA, and PSO-GSA algorithm are used to optimize weighted centroid localization. Simulation experiments were conducted in the same simulation environment and the results were compared and analyzed. The average positioning error was used as the evaluation criteria of the positioning algorithm.

$$e_{\mathrm{Err}} = \frac{\sum_{i=1}^{M} \left|x_i' - x_{\mathrm{ireal}}\right|}{M \times R} \times 100\%$$

where: M is the number of unknown nodes; x_i' is the estimated position of the unknown node; x_{ireal} is the actual position of the unknown node; R is the communication radius of the node.

In addition, the parameters are set to: $G_0 = 100$, $a = 20$, and the number of particle groups is 50; to adjust the maximum step size toward the individual optimal particle and the global optimal particle direction, $c_1 = c_2 = 0.85$; In general, stable convergence results can be obtained after about 100 iterations, and the search accuracy will increase slowly with the number of iterations. Therefore, the maximum number of iterations T is 400. The communication radius is set to 15 m and the wireless signal carrier frequency is 2.4 GHz. Figure 2 compares the error of the above three optimization algorithms under the same communication radius.

As shown in Fig. 2, the average positioning errors of the three optimization algorithms tend to decrease as the proportion of beacon nodes increases. The positioning performance of the PSO-GSA algorithm shows obvious advantages. The average error of the PSO-GSA algorithm is the smallest in the process of increasing the proportion of the entire beacon node. When the proportion of beacon nodes accounts for 40%, the error compared the PSO is reduced by 10.7%. It can be seen that the PSO-GSA algorithm can effectively reduce the positioning error and improve the positioning accuracy.

Figure 3 shows the comparison of average positioning error curves of various optimization algorithms by setting different communication radii when the proportion of beacon nodes accounts for 20%. The simulation results show that when R is small, the average positioning error decreases with increasing radius; when the communication radius is greater than 30 m, the error increases due to the increase of network connectivity, but the PSO-GSA algorithm is still the least error, such as $R = 50$ m, the average positioning error is 7.6% lower than the GSA algorithm, which is 13.5% lower than PSO.

In order to further verify the feasibility of PSO-GSA, CC2430 chip based on 8015 microprocessor core produced by TI Company was used to conduct experiments under the coal mine. An experiment was performed using six locations at different

Fig. 2 Comparison of three optimization method errors

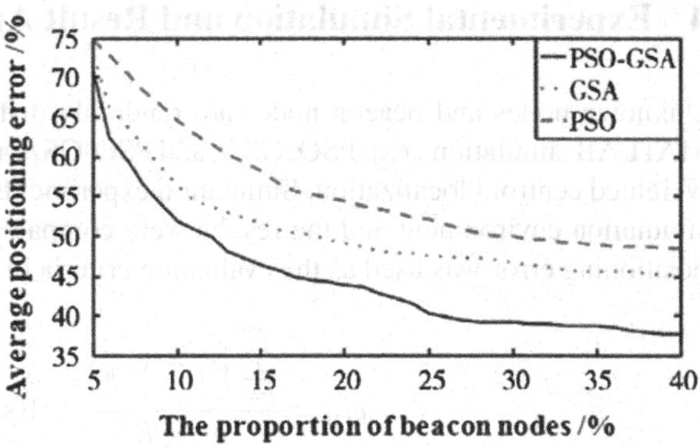

Fig. 3 Comparison of errors at different communication radius

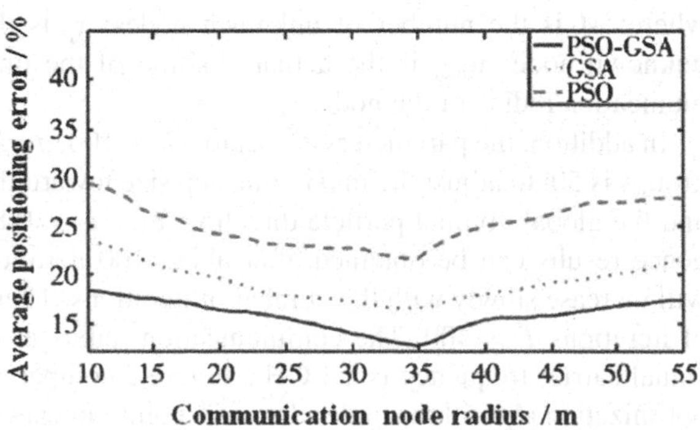

Table 1 Comparison of three algorithm measurement results

Actual coordinates (*x*, *y*)/m	Measured coordinates (x′, y′)/m		
	PSO	GSA	PSO-GSA
(1.45, 2.56)	(0.92, 1.76)	(1.83, 3.02)	(1.55, 2.36)
(9.23, 4.14)	(8.35, 3.26)	(9.95, 4.86)	(9.15, 4.06)
(21.47, 6.56)	(19.45, 5.26)	(20.25, 5.16)	(21.05, 6.91)
(31.40, 8.11)	(28.45, 6.76)	(29.45, 7.53)	(32.02, 8.83)
(41.49, 9.13)	(44.85, 11.1)	(39.45, 7.56)	(40.59, 8.24)
(48.55, 9.82)	(45.0, 6.56)	(46.45, 8.01)	(46.95, 8.83)

distances from AP. After the weighted centroid positioning was used to obtain the estimated position, the parameters were input into PSO, GSA and PSO-GSA three optimization models to obtain the final measured coordinates. The experimental data is shown in Table 1.

As can be seen from Table 1, the positioning error is smaller when distance is closer to the access point. Under the same experimental environment, the position obtained through PSO-GSA optimization is closer to the actual value. Therefore, the optimization algorithm proposed in this paper has smaller positioning error and great practical significance.

5 Conclusion

A new optimization model is proposed to improve the accuracy of coal mine underground positioning based on the triangular weighted centroid localization algorithm. The method obtains the distance from the node to the unknown node through the logarithmic distance path loss model, and uses the triangular weighted centroid localization algorithm to locate the unknown node coordinates. Taking into account the randomness of underground personnel moving in real time, and the received signal strength values are dealing with the influence of underground environment. The PSO-GSA optimization model was introduced to optimize the position estimation coordinates and parameters, reduce the dependence of the mathematical model on the environment, and improve the positioning accuracy.

The experimental results show that the average positioning error is controlled within 2 m to meet the real-time and accurate positioning of underground coalmine personnel. The system is also adapted to the positioning of underground cranes, rubber tires, and other engineering equipment. The system is simple in design, easy to operate, and has practical application value.

Acknowledgements This work is supported by Youth Foundation for Henan Polytechnic University under Grant Q2016-2A, Key Laboratory of Control Engineering of Henan Province under Grant KG2016-17 and Key Project of Science and Technology of Education Department of Henan Province (19B120002).

References

1. F. Wang, C. Sang, Three-dimensional positioning algorithm based on TDOA and AOA in coal mine underground. Ind. Mine Autom. **41**(5), 78–82 (2015)
2. D.S. Han, W. Yang, A weighted centroid localization algorithm based on received signal strength indicator for underground coal mine. J. China Coal Soc. **38**(7), 523–528 (2013)
3. X. Wang, B. Zhang, Y. Feng, Improved weighted centroid localization algorithm based on particle swarm optimization. Comput. Eng. **38**(1), 90–95 (2012)
4. L. Ma, L. Liu, Analysis and improvement of gravitational search algorithm. Microelectron. Comput. **32**(9), 76–80 (2015)
5. N. Li, G. Wang, New personnel localization algorithm in mine based on improved genetic algorithm. J. Cent. South Univ. **30**(10), 258–262 (2016)
6. H. Liu, Y. Dong, Y. Ai, The optimal positioning algorithm based on RSSI of WiFi. Int. J. Smart Home **10**(12), 203–212 (2016)
7. R. Yu, Z. Sun, Si, Research and implement of staff localization system for underground coal mine. Coal Mine Mach. **37**(10), 69–71 (2016)
8. Y. Jia, Alternative proofs for improved LMI representations for the analysis and the design of continuous-time systems with polytypic type uncertainty: a predictive approach. IEEE Trans. Autom. Control **48**(8), 1413–1416 (2003)
9. Y. Jia, Robust control with decoupling performance for steering and traction of 4WS vehicles under velocity-varying motion. IEEE Trans. Control Syst. Technol. **8**(3), 554–569 (2000)

Research on the Values Tendency Analysis of the Micro-blogging User Based on Social Networks

Zhongbao Liu, Changfeng Fu and Chia-Cheng Hu

Abstract With the development of Internet, the public opinion on the Internet is paid more and more attentions. Micro-blogging, as a popular social platform, is prone to cause some group unexpected events. In view of this, Value Tendency Analysis based on Social Network (VTASN) is proposed in this paper. VTASN is based on Schwartz values and it includes three parts: value vector space generation, value vector computation and individual value priority evaluation. The simulated experiments verify the effectiveness of VTASN.

Keywords Social networks · Micro-blogging user · Values · Tendency analysis

1 Introduction

Nowadays, with the development of Internet technology, micro-blogging has gradually become one of the important platforms for people to communicate and learn. On the one hand, because the micro-blogging had real-time and interactive characteristics, people could easily find the information of interest from the micro-blogging; On the other hand, some micro-blogging users would be able to express their views when they had a certain understanding of social events, which was easy to cause group unexpected events. In order to solve the above problems, a feasible way was to analysis the user's behavior and got the value tendency of micro-blogging users, which would help to monitor the key users effectively and avoided the group unexpected events. In recent years, with the continuous development of social network technology, new technology and product emerge in endlessly. In view of this, this paper put forward a method of Value Tendency Analysis based on Social Network

Z. Liu (✉) · C. Fu · C.-C. Hu
School of Software, Quanzhou University of Information Engineering, Quanzhou 362000, China
e-mail: liu_zhongbao@hotmail.com

Z. Liu
School of Software, North University of China, Taiyuan 030051, China

Y. Jia et al. (eds.), *Proceedings of 2018 Chinese Intelligent Systems Conference*, Lecture Notes in Electrical Engineering 529,
https://doi.org/10.1007/978-981-13-2291-4_6

(VTASN), which can analysis the tendency of values according to the user's speech or forwarding on the social network.

Investigation by the authors, in the last few years, the researchers mainly focused on the construction of library and reader's values in the field of library and information. Meanwhile, there was no research results related to the values tendency analysis of the micro-blogging users. However, the existed research results on values and micro-blog user behavior could provide some references for further research in this study. For this purpose, this paper reviewed the above two aspects, the concrete content is as follows.

1.1 The Study of Values

Sang Liangzhi discussed the core values and the basic theories of library science, as well as researched on the relationship between the core values of library and the professional knowledge and behavior [1]; Gao Fang reviewed the current situation of the library core values in China, made an analysis about the challenges and problems in the construction process at the same time [2]; Hu Ming and Chen Jianmin took the analysis of library value in the world as a point of entry, exposed the connotation of library value, and they put forward a point of view: in this day and age, it was the mainstream value of library to acquire knowledge in a free, equal and convenient way [3]; Based on the core value of library, Cheng Peng discussed the basic concept of the library value and how to established the approaches and methods of library value [4]; Wang Jianfang made some reflections on the core value of library viewed from the ideological dimension of reader's dignity, point out: reader's dignity, which is expressed as a kind of fundamental value for human beings, should become the core value for the library [5]; Lai Xiaolin took the value of library as an angle, and then, in-depth analysis the relationship between university library and values of campus cultural, put forward the development goal of the values of university library [6]; Ji Qinghua discussed the connotation of the library values and core value of library values, and put forward the thinking and positioning principle of library value construction [7]; Liang Canxin summed up the relationship between the research results belonging to the core values of library, in-depth discussion of the nature and structure of library core values [8]; Zhang Li discussed in depth some issues of the values in the Chinese library, such as the connotation and extension of library values, the values of library people, the core values of libraries, etc. [9]; Based on the library values of Bacon and Shera, Yin Hongbo proposed values of library in the new era [10]; Liu Guojun et al. reviewed the research on the core values of library in all countries and discussed on the value difference in Core Value research of the libraries in different Countries [11]; Wang Xiaoming et al. researched on the several common collection development values, compared their advantages and disadvantages, considered the questions which form some scholars and experts [12]; LvChaobin deeply analyzed the relationship between readers' reading values and library construction, pointed out that the library should pay more attention to understand the readers' reading values enough so as to for majority of reader with better services [13]; By analyzed the relation between values and invisible knowledge, Wang Nan. Put forward for constructing the values in medical staff in view of their value construction process and its influencing factors [14]; From the perspective of cognition, Han Jizhang analyzed the two values in library science: scientific values and value-Oriented values [15]; Yuan Jinli analyzed the forming process and internal causes of the reading values of university students, expounded the external factors influencing the formation and development of the reading values of university students, and put forward some measures for guiding university students to overcome the reading bias and established healthy reading values [16].

1.2 Analysis on User Behavior in Micro-blogging Communities

At present, the researchers in the world had obtained some results by the way of research on the user's behavior in micro-blogging communities. By the research of the Twitter, Kshay Java et al. draw the conclusion that the main purpose of the micro-blogging user is to talk about daily activities and obtain information that they are interested in [17]; Brian G. Smith, through in-depth analysis, pointed out that Twitter played an important role in the communication of information between users in disaster relief [18]; Kaye D. Sweetser in-depth analyzed of the relationship between motivation, leadership and social media from the perspective of the user [19]; Schwab et al. obtained the interest of the user by analyzing keywords in blog posts [20]; Maloof et al. discussed the issue of user interest drift in micro-blogging based on the forgetting theory [21]; Xia Yuhe took the Sinamicro-blogging as the research object, in-depth analyzed of blog posts topics, comments and forwarding and other information, discussed the user's behavior [22]; Based on the network topology relationship between micro-blogging users, Ping Liang and Zon Liyong analyzed the relationship of micro-blogging network [23]; Based on the analysis of the basic structure and the model of information transmission of micro-blogging, Wang Xiaoguang analyzed the relationship between the number of follower, the number of following, and the number of blog posts [24].

It can be seen that the research of the users in the micro-blogging is constantly being carried out, and the results obtained can help improve the service level of micro-blogging. However, the popularity of micro-blogging must be cause some group unexpected events. Therefore, it is necessary to conduct in-depth analysis of the value tendency of microblog users in order to effectively monitor the online behavior of key users, and to avoid the occurrence of social group events.

2 Research Method

Values are people's understanding of the importance of different things in life, and it plays a major role in personal or organizational behavior and attitudes. The value model established by Schwarz had been proved to have high general applicability in large-scale verification tests in 82 countries. And after more than a hundred tests, it can be confirmed that there is a strong relationship between values and behavior choices, so values provide an important evidence of the choice of individual or organizational behavior.

The basic idea of Value Tendency Analysis based on Social Network (VTASN) is to automatically analyze Schwartz values based on the public comments of users in Social Network Service. Schwartz values define ten types of values from the basic needs of human survival in society, namely Hedonism, Benevolence, Universalism, Power, Achievement, Tradition, Conformity, Security, Self-Direction, and Stimulation. Each value class contains a number of specific value. The existing research had shown that there are significant differences in the topic and wording of personal comments due to different value priorities. In view of this, Value Tendency Analysis based on Social Network (VTASN) is proposed in this paper, and it includes three parts: value vector space generation, value vector computation and individual value priority evaluation.

2.1 Value Vector Space Generation

Firstly, using the keyword extraction technology based on classification documents to obtain specific words related to the specific value in the current social context; Next, using Baidu News Search Engine to build a Dynamic Corpus to search keywords as category markers, ensure corpus automatic classification, and keep in sync with the current time and social background. In order to ensure a high correlation between corpus and value concept, a three-level tree structure of search keywords is constructed: The first layer of tree structure is the 10 value concept words, the second is the concrete value word, the third layer is the synonym extended from the second layer. According to the tree structure, each search used three search key words, respectively: a value concept word, a concrete value word and a synonym.

Used Chi-Square Statistics as a basis for feature extraction, it shown the relationship between word and value classes. The high-power squared value represents a strong correlation between word and value class, and it could be a feature word for this category. Extracted feature word was divided into three steps: first, use the NLPIR/ICTCLAS word segmentation system of the Chinese Academy of Sciences for new word discovery and word segmentation; and then calculate the high-power squared value of each word in each category; finally, select the top word in each category, remove duplicate words and generate feature index of vector space.

2.2 *Value Vector Computation*

Calculate the weight of each feature word in the feature index in the 10 categories to generate a value vector. The weight of each feature word is calculated by multiplying the word's Document Frequency, the word's inverse Document Frequency logarithm, and the word's Kullback-Leibler divergence. Based on this, the vector representations of 10 value categories could be obtained. These vectors are the specific mapping of value concept to social network language in the current social context, followed by the evaluation of individual values.

2.3 *Individual Value Priority Evaluation*

The individual value priority evaluation is composed of the following steps.
Step 1: Get personal speech. Get the latest personal speech from Sina micro-blogging, Tencent micro-blogging and forums;
Step 2: Calculate personal value vectors. Map the personal speech into value vector space, the weight of each feature word is calculated by multiplying the word's Document Frequency, the word's inverse Document Frequency logarithm, and the word's Kullback-Leibler divergence;
Step 3: Calculate the similarity measurement between individual values vectors and 10 values vectors of Schwartz values. Calculate 10 similarities for each person by calculating the cosine distance between the individual values vector and the 10 types of values vectors, and then sort the 10 similarities from the highest to the lowest, that is, get the individual values evaluation results.

3 Experiment Analysis

In order to prove the feasibility of the VTASN method, 92 Sina micro-blogging users (Randomly selected 30 verify users and 62 ordinary users) were analyzed using the VTASN method. Calculated the values evaluation results of ordinary users and verify users, and then processed with equalization. The results were shown in Table 1.

From Table 1, it can be seen that the verify user and the ordinary user had consistency in some values, and there are essential differences in other values. Specifically, in the values of "Benevolence" and "Achievement", verify users and ordinary users had similar tendencies; Neither of the two types of users had "Tradition", "Stimulation", "Security", "Self-Direction", and "Conformity" values; In the sense of "Hedonism", "Universalism" and "Power" there were essential differences between

Table 1 Comparison table between ordinary users and verify users

Values	Ordinary user	Verify user
Hedonism	0.5223	−0.3412
Benevolence	1.4645	1.2311
Tradition	−0.2037	−0.3936
Stimulation	−0.2465	−0.6003
Security	−0.5028	−0.3148
Achievement	0.1978	0.2125
Universalism	−0.2310	0.4317
Self-Direction	−1.1039	−0.7433
Conformity	−0.1978	−0.1215
Power	−0.1245	0.5738

the two types of users: most of the ordinary users had the value of "Hedonism", but not the values of "Universalism" and "Power". Ordinary users tend to prefer "Hedonism" values and are less interested in "Universalism" and "Power" values. Instead, verify users focus more on "Universalism" and "Power" values than on "Hedonism" values.

4 Conclusion

The judgment of the values tendency of the users in micro-blogging is beneficial to the monitoring of internet public sentiment and to prevent the group unexpected events. In view of this, this paper puts forward the Value Tendency Analysis based on Social Network (VTASN) method. VTASN is based on Schwartz values and it includes three parts: value vector space generation, value vector computation, individual value priority evaluation. The simulation experiments on Sina micro-blogging show that the VTASN method is effective in the tendency analysis of the micro-blogging.

Acknowledgements This work is supported by the National Science Foundation of Shanxi (201601D011042), the Program for Outstanding Innovative Team of High Learning Institutions of Shanxi (2016) and the Outstanding Youth Funds of North University of China.

References

1. Z. Sang, Core values and basic theory of library science. New Century Libr. **5**, 5–7 (2007). (in Chinese)
2. F. Gao, The challenges and problems faced by the research on the core values of domestic libraries. Inf. Documentation Serv. **6**, 21–23 (2007). (in Chinese)
3. M. Hu, J. Chen, The evolution of view on library value. J. Intell. **11**, 159–161 (2007). (in Chinese)

4. P. Cheng, My opinion on the core values of the library. J. Mod. Inf. **5**, 25–27 (2008). (in Chinese)
5. J. Wang, The ideological dimension of reader's dignity-thinking about the core values of the library. J. Acad. Libr. Inf. Sci. **26**(5), 3–5 (2008). (in Chinese)
6. X. Lai, On the value of university library and campus culture. Libr. Theor. Pract. **4**, 67–68 (2009). (in Chinese)
7. Q. Ji, Talking about the construction of library's values. Sci.-Tech. Inf. Dev. Econ. **19**(9), 39–41 (2009). (in Chinese)
8. C. Liang, Analysis of the nature and the structure of the core values of library. Libr. Inf. **4**, 1–5 (2009). (in Chinese)
9. L. Zhang, A study of the library's core values. Libr. Theor. Pract. **5**, 6–8 (2009). (in Chinese)
10. H. Yin, On library values and core values. Sci.-Tech. Inf. Dev. Econ. **20**(35), 9–14 (2010). (in Chinese)
11. G. Liu, J. Yu, Discussion on the value differences in the study of library's core values in various countries. Sci.-Tech. Inf. Dev. Econ. **22**(18), 27–28 (2012). (in Chinese)
12. X. Wang et al., Research on different values of collection development and its question. Libr. Dev. **3**, 30–35 (2012). (in Chinese)
13. LvChaoBin, Reader's Reading Values and Library Construction. Libr. J. Henan **34**(7), 14–16 (2014). (in Chinese)
14. N. Wang, Z. Liu, Medical staff values construction based on tacit knowledge. Chin. J. Med. Libr. Inf. Sci. **23**(2), 32–34 (2014). (in Chinese)
15. J. Han, Discussion on value. Libr. Work Coll. Univ. **35**(1), 87–89 (2015). (in Chinese)
16. Yuan Jinli, Research on the factors influencing the reading values of university. Sci.-Tech. Inf. Dev. Econ. **20**(25), 39–4190 (2016). (in Chinese)
17. A. Java, X. Song, T. Finin et al., Why we twitter: understanding micro-blogging usage and communities, in *Proceedings of the 9th Web KDD and 1st SNA-KDD 2007 Workshop on Web Mining and Social Network Analysis* (2007), pp. 56–65
18. B.G. Smith, Socially distributing public relations: Twitter, Haiti, and interactivity in social media. Pub. Relat. Rev. **36**(4), 329–335 (2010)
19. K.D. Sweetser, T. Kelleher, A survey of social media use, motivation and leadership among public relations practitioners. Pub. Relat. Rev. **37**(4), 425–428 (2011)
20. I. Schwab, W. Pohl, Learning user profiles from positive examples, in *Proceedings of the Acai'99 Workshop on Machine Learning in User Modeling* (1970)
21. M. Maloof, S. Michalski, Selecting examples for partial memory learning. Mach. Learn. **41**, 27–52 (2000)
22. Y. Xia, The structure and mechanism of microblog interaction-based on the empirical study of Sina microblog. J. Commun. **4**, 60–69 (2010). (in Chinese)
23. L. Ping, L. Zon, Research on microblog information dissemination based on SNA centrality analysis–a case study with Sina microblog. Doc. Inf. Knowl. **6**, 92–97 (2010). (in Chinese)
24. X. Wang, An empirical analysis of microblog users' behavioral characteristics and relationship characteristics-taking Sina microblog as an example. Libr. Inf. Serv. **54**(14), 66–70 (2010). (in Chinese)

A Method of Semantic Image Inpainting with Generative Adversarial Networks

Zhe Wang and Hongpeng Yin

Abstract Semantic image inpainting focuses on the completing task of high-level missing regions at the basis of the uncorrupted image. The classical methods of image inpainting can only deal with low-level or mid-level missing regions due to the lack of representation of the image. In the essay, we conclude a new method of semantic image inpainting. It's based on the generative model with learning the representation of image database. We propose an architecture of completion model using perceptual loss and contextual loss based on generative adversarial networks after having trained generative model using DCGAN. We qualitatively and quantitatively explore the effect of missing regions of different types and sizes on image inpainting. Our method successfully completes inpainting tasks in large missing regions and results looks realistic with extensive experiments. We conclude that the performance of our model mostly is good when completing image corrupted with the mask with an area of less than 50% as well as with center or random masks.

Keywords Generative model · Image inpainting · Deep neural networks

1 Introduction

Semantic image inpainting is a common image editing operation that designers and photographers use to fill in masked or missing regions of images with plausibly synthesized content based on image semantics [1]. Not only the generated contents is similar with the context, but the generated image looks visually realistic [2]. Because of high-level missing regions or complex scenes, this task is significantly more

Z. Wang · H. Yin (✉)
School of Automation, Chongqing University, Chongqing, China
e-mail: yinhongpeng@cqu.edu.cn

Z. Wang
e-mail: wzhtomg@gmail.com

Y. Jia et al. (eds.), *Proceedings of 2018 Chinese Intelligent Systems Conference*, Lecture Notes in Electrical Engineering 529,
https://doi.org/10.1007/978-981-13-2291-4_7

difficult than classical completion which concentrates on completing small missing regions in image. The key to solving this problem is to learn to represent images as samples from a high-dimensional probability distribution. Instead of getting the true probability distribution of image since it's difficult, we can only collect samples and learn to generate new samples. When we know some values in the image and want to complete the missing value. Regard it as a maximization problem where we search over all the possible missing values through the generative method [3]. The completion will be the most probable image.

In the essay, we conclude a new way to semantic image inpainting. We divide semantic image inpainting into three steps. First, Interpret images as being samples from a probability distribution. Second, Generate fake images through learning from the interpretation. Last, Select the best fake image for completion.

The paper makes three main contributions summarized as follows. First, we conclude the steps of solving the problem of high-dimensional missing regions or complex scenes completion. Second, we propose an architecture of completion model using Perceptual loss and Contextual loss based on generative adversarial networks after having trained generative model using DCGAN [4]. Third, we demonstrate our model's validity of image inpainting performs well, which could successfully predict the content of large missing regions and generate realistic images.

2 Related Work

Classical image inpainting methods always use local information to complete the image. For example, diffusion equation is used to iteratively propagate low-level features along the boundaries of the missing regions [5]. Texture synthesis can fill the missing region through searching similar texture from the source image [6]. PatchMatch (PM) [7] searches similar patches from known image region. However, all single local image inpainting methods require sufficient information including similar pixels, structures, or patches to be contained in the source image.

As for large image inpainting task of missing region, no-local methods are applied for predicting the missing regions using external data, for example, Hays selects the patch from database [8]. Whyte finds appreciate patch to replace a target region through Internet-based retrieval [9]. The completion is hard to achieve when the source image is different from any image in database or Internet.

We concentrate on the more difficult task of semantic image inpainting, which could fill large missing regions in the image. This mission cannot be handled by classical inpainting or texture synthesis approach because the missing region is too large to fix. Context Encoder [10] is close to our work, which is a CNN that aims to predict the information for arbitrary missing regions according to its surroundings. In contrast, our algorithm is able to perform semantic image inpainting without such strict constraints.

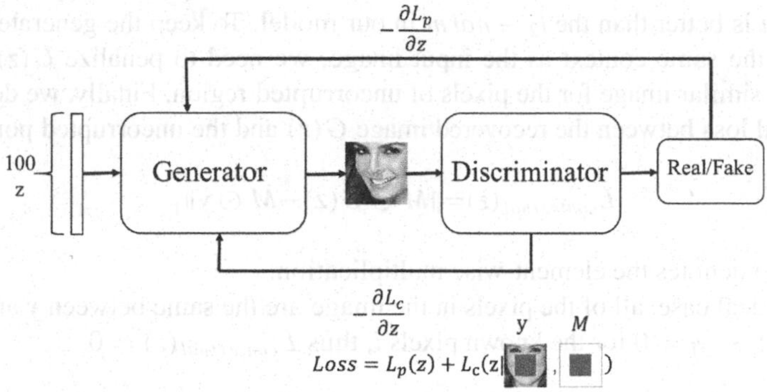

Fig. 1 The architecture of our semantic image inpainting model

3 Semantic Image Inpainting

To solve the image inpainting task of filling large missing regions, we use generator G and discriminator D. G and D are trained with uncorrupted images. The architecture of generator is a deep neural network which takes vector on input and returns an image. Meanwhile, the architecture of discriminator is also a deep convolution neural networks which takes image generated from generator G on input and returns True or False. After training, the generator G can select a point z which is sampled from $p(z)$ then G generate a fake image which is as close as possible samples from p_{data}. Therefore, our next goal is to find the encoding $\hat{z}$ "closest" to the corrupted image and is constrained in the mainfold. The architecture of our semantic image inpainting model is illustrated in Fig. 1. After we get the $\hat{z}$, G can generate the missing regions.

We transform the process into an optimization problem. Regard y as the corrupted image. M representative the binary matrix that has the same size of image. M indicates the missing regions.

we define loss function in completion with a combination of the contextual and perceptual losses and we will find the "closest" encoding $\hat{z}$ when the loss function is minimized.

$$\hat{z} = \underset{z}{arg\ min}\{L_c(z|y, M) + L_p(z)\} \tag{1}$$

where L_c represents the Contextual loss, L_p represents the Perceptual loss.

3.1 Contextual Loss

Contextual loss captures such information from remaining available regions. A common choice for the contextual loss is simply the $l_2{-}norm$ norm between the generated image $G(z)$ and the uncorrupted regions of the input corrupted image. We find the

$l_1 - norm$ is better than the $l_2 - norm$ in our model. To keep the generated image matching the same context as the input image, we need to penalize $G(z)$ for not creating a similar image for the pixels of uncorrupted region. Finally, we define the Contextual loss between the recovered image $G(z)$ and the uncorrupted portion y.

$$L_{contextual}(z) = \|M \odot G(z) - M \odot y\|_1 \tag{2}$$

Here, $\odot$ denotes the element-wise multiplication.

In the ideal case, all of the pixels in the image are the same between y and $G(z)$. Then $G(z)_i - y_i = 0$ for the known pixels i, thus $L_{contextual}(z) = 0$.

3.2 Perceptual Loss

To recover an image that looks real, we use the Perceptual loss to make the discriminator properly convince that the recovered image is similar to the source image and looks realistic.

The goal of Perceptual loss is to penalize unrealistic images. The discriminator D is trained to differentiate generated images from real images as proposed in GANs. We'll define Perceptual loss with the same criterion used in training the DCGAN for training discriminator D.

$$L_{perceptual}(z) = \lambda \log(1 - D(G(z))) \tag{3}$$

In the equation, λ is to balance the two losses. In the mapping from y to z, the existence of L_p make sure z is updated to maximize the probability of D and make generated image more realistic.

3.3 Inpainting

With the perceptual loss and contextual loss at hand, After generating $G(\hat{z})$, The reconstructed image fills in the missing region values of y with $G(\hat{z})$.

$$x_{reconstructed} = M \odot y + (1 - M) \odot G(\hat{z}) \tag{4}$$

Then we use Poisson blending [11] to rebuild our image. Poisson blending is a way to keep the gradients of $G(\hat{z})$ to preserve image details. Our final results $\hat{x}$ can be obtained by:

$$\hat{x} = \underset{x}{arg\ min}\ \|\nabla x - \nabla G(\hat{z})\|_2^2$$
$$s.t.\ \ x_i = y_i \ \text{ for } \ M_i = 1 \tag{5}$$

where ∇ is the gradient operator. The minimization problem contains a quadratic term [11].

4 Experiments

We evaluate our method qualitatively and quantitatively. We trained DCGANs on three datasets, We train on three datasets: CelebA [12], LFW [13] and MNIST [14].

4.1 *Qualitative Analysis*

We first train the DCGANs in 25 epochs based on MNIST and LFW databases, the generative images is shown in Fig. 2. The results show that our model performs well in learning representation of image in the databases and is able to generate convinced image that looks really realistic.

Based on the ability of generating fake but realistic images, we train the DCGANs in the celebA database in the same way in order to learn facial features and prepare for the next step. After training, we add different types of masks to corrupt the source image and send the corrupted image into our model to complete inpainting. The completion process contains 1000 iterations and we selected 64 uniformly nodes in an iterative process for visualization. We choose 64 source images aligned with 8×8 and corrupt them with center mask to recover, the visual inpainting samples drawn from different epoch are shown in Fig. 3. In this figure, we can see the missing regions recover the facial feature.

We apply five different masks to the source image, such as, central block masks, grid type masks, left block masks, random noisy pattern masks, low resolution masks and add a comparison without masks. Meanwhile, We choose center mask as an example and change the size among 64, 36, 16, 4% of source image area. The completion results can be seen in Fig. 4. As we can see, center and random type mask can be fixed well and the results looks like real image completely, which can

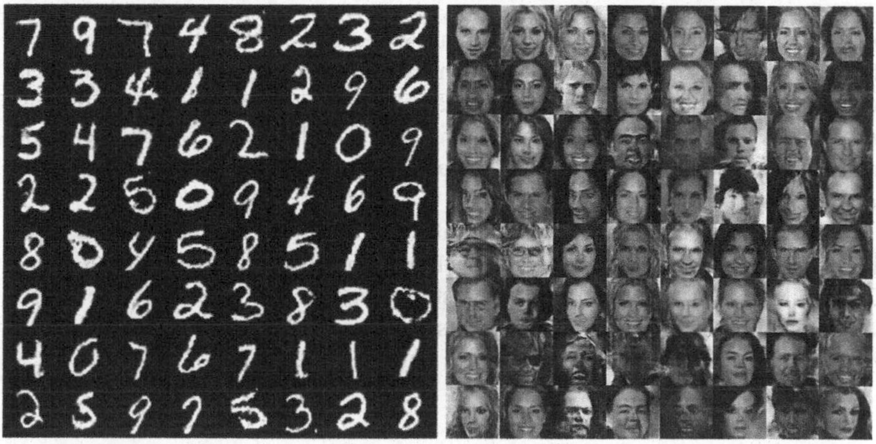

Fig. 2 Generations from the trained DCGANs after 25 epochs

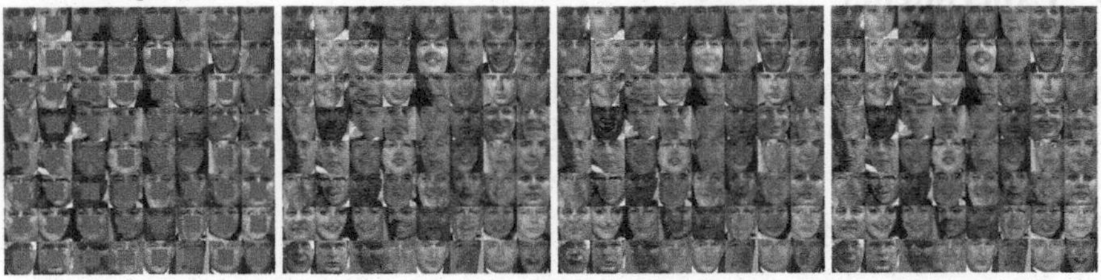

Fig. 3 The visualization of the process of image inpainting in 1st, 5th, 11th, 20th epoch. As the epoch increases, the missing regions is filled with more realistic facial feature

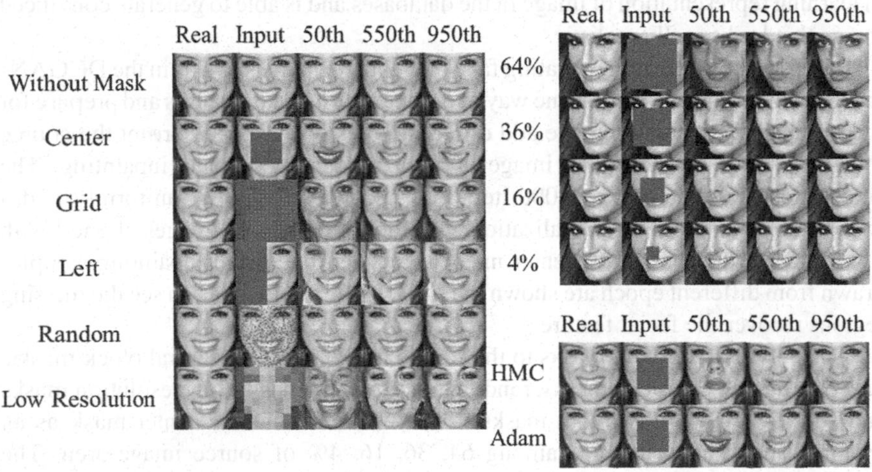

Fig. 4 Comparisons with different types of masks and different sizes of mask on image inpainting and the results using HMC and Adam algorithm in the optimization

be demonstrated in the section of quantitative analysis. As for other type of mask, the performance is still promising and generate realistic images that indicate facial feature sharply. It's clear that the performance is getting worse as the size of mask increasing. When the mask of 64% area is added to the source image, the output looks like another woman's face, while our model generate image looking like the same face in other three examples.

We also use the different optimization algorithm to train our model. In this paper, the author recommends the Adam algorithm. Compared with Hamiltonian Monte Carlo (HMC) [15], we found Adam is faster when both algorithms are used in the same condition. The Adam spends 10.8 s training while the HMC spends 216 s. What's more, the performance of Adam is not inferior to HMC.

Table 1 The PSNR and SSIM values between output images with different types of masks and the source image

Type of masks	PSNR	SSIM
Without masks	54.0938	0.9999
Center	26.2236	0.9689
Grid	20.4394	0.8950
Half	14.3176	0.8470
Random	27.7281	0.9780
Low resolution	16.4770	0.7776

Table 2 The PSNR and SSIM values between output images with different size of masks and the source image

Size of masks (%)	PSNR	SSIM
64	16.6389	0.7972
36	21.6757	0.9202
16	26.7048	0.9687
4	32.2650	0.9906

4.2 Quantitative Analysis

In addition to qualitative analysis, we also make quantitative analysis using PSNR and SSIM. PSNR directly measures the difference in pixel values. SSIM [16] estimates the holistic similarity between two images. These two metrics are computed between the inpainting image and the source face images. Tables 1 and 2 show our results. The higher value of PSNR means more similar in pixel values between two images. The higher value of SSIM means the two images have more similar structure.

There are higher PSNR values and SSIM values in center and random masks compared to other types of masks, which is corresponding to qualitative analysis. There is a large increase between the 64% size of masks and 36, 16, 4%, which also demonstrate that our model perform well when the size of masks is under 50%.

4.3 Discussion

We conclude that the performance of our model is excellent when complete image corrupted with mask with an area of less than 50% as well as with center or random masks. While the results are good at most time, our method still has obvious limitation. Our model's performance deeply rely on the performance of generative model as well as training procedure. The failure samples is shown in Fig. 5. Images with structural feature like faces can be easily generate using GAN, but it is not enough to learn the representation of complex images in this world.

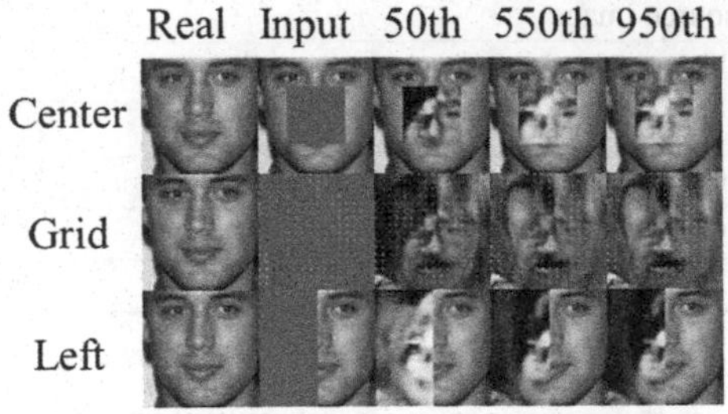

Fig. 5 Several failure examples of model

5 Conclusion

In the essay, we conclude a new way to semantic inpainting. Different from classical methods, the new method can learn the representation of image database, and can predict information for missing regions in corrupted images. Our method has major advantages on the capacity of completing arbitrary missing regions in the image compared with CE. Our results look more realistic with sharper edges. In experiments, the results also demonstrated its better performance on high-level missing image inpainting, especially when completing image corrupted with mask with an area of less than 50% as well as with center or random masks.

Acknowledgements This work is supported by center of Artificial Intelligence Laboratory of automation school in Chongqing university. We would like to acknowledge Han Zhou for helpful guidance in training model. Finally, we would like to thank Wenhui Li for meticulous support constantly during the research.

References

1. R.A. Yeh, C. Chen, T.Y. Lim et al., Semantic image inpainting with deep generative models, in *Proceedings of the IEEE Conference on Computer Vision and Pattern Recognition* (2017), pp. 5485–5493
2. Y. Li, S. Liu, J. Yang et al., Generative face completion, in *Proceedings of the IEEE Conference on Computer Vision and Pattern Recognition*, vol. 1, issue 3 (2017), p. 6
3. I. Goodfellow, J. Pouget-Abadie, M. Mirza et al., Generative adversarial nets, in *Advances in Neural Information Processing Systems* (2014), pp. 2672–2680
4. A. Radford, L. Metz, S. Chintala, Unsupervised representation learning with deep convolutional generative adversarial networks. arXiv preprint arXiv:1511.06434 (2015)
5. M. Bertalmio, G. Sapiro, V. Caselles et al., Image inpainting, in *Proceedings of the 27th annual conference on Computer graphics and interactive techniques* (ACM Press/Addison-Wesley Publishing Co., 2000), pp. 417–424
6. A.A. Efros, T.K. Leung, Texture synthesis by non-parametric sampling, in *The Proceedings of the Seventh IEEE International Conference on Computer Vision, 1999*, vol. 2 (IEEE, 1999), pp. 1033–1038
7. C. Barnes, E. Shechtman, A. Finkelstein et al., PatchMatch: a randomized correspondence algorithm for structural image editing. ACM Trans. Graph. TOG **28**(3), 24 (2009)

8. J. Hays, A.A. Efros, Scene completion using millions of photographs, in *ACM Transactions on Graphics (TOG)*, vol. 26, issue 3 (ACM, 2007), p. 4
9. O. Whyte, J. Sivic, A. Zisserman, Get Out of my Picture! Internet-based inpainting, in *BMVC*, vol. 2, issue 4 (2009), p. 5
10. D. Pathak, P. Krahenbuhl, J. Donahue et al., Context encoders: feature learning by inpainting, in *Proceedings of the IEEE Conference on Computer Vision and Pattern Recognition* (2016), pp. 2536–2544
11. P. Pérez, M. Gangnet, A. Blake, Poisson image editing. ACM Trans. Graph. (TOG) **22**(3), 313–318 (2003)
12. Z. Liu, P. Luo, X. Wang et al., Deep learning face attributes in the wild, in *Proceedings of the IEEE International Conference on Computer Vision* (2015), pp. 3730–3738
13. E. Learned-Miller, G.B. Huang, A. RoyChowdhury et al., Labeled faces in the wild: a survey, in *Advances in Face Detection and Facial Image Analysis* (Springer, Cham, 2016), pp. 189–248
14. Y. LeCun, L. Bottou, Y. Bengio et al., Gradient-based learning applied to document recognition. Proc. IEEE **86**(11), 2278–2324 (1998)
15. H. Niederreiter, A. Winterhof, *Quasi-Monte Carlo Methods. Encyclopedia of Quantitative Finance* (Wiley, New York, 2010), pp. 185–306
16. Z. Wang, A.C. Bovik, H.R. Sheikh et al., Image quality assessment: from error visibility to structural similarity. IEEE Trans. Image Process. **13**(4), 600–612 (2004)

Adaptive Neural Network Control for Uncertain Robotic Manipulators with Output Constraint Using Integral-Barrier Lyapunov Functions

Tengfei Zhang and Yingmin Jia

Abstract In this paper, an adaptive neural network (NN) output tracking control approach is presented for uncertain robotic manipulators with the output constraint. Integral-barrier Lyapunov functions (iBLF) are adopted to prevent the output from violating the given constraint. And adaptive neural networks, which are capable of approximating the arbitrary continuous function at any precision, are employed in handling uncertainties and disturbances. By appropriately choosing design parameters, the proposed method can guarantee the semi-global uniformly ultimate boundedness of the output error, and all signals of the closed-loop system remain bounded. The effectiveness and performance of the proposed control method are illustrated through a numerical simulation example.

Keywords Adaptive control · Neural networks (NN) · Uncertain robotic manipulator systems · Output constraint · Integral-barrier Lyapunov functions (iBLF)

1 Introduction

In recent years, the control problem of the constrained robotic manipulator has attracted considerable attention [1–3]. Generally, most physical systems are subject to various constraints such as physical stoppages, dead-zone, saturation and so on, which may result in performance degradation or hazards [4, 5]. Therefore, many fulfilling control methods pertinent to nonlinear constrained systems have been presented, including the model predictive control (MPC) [6], reference governors [7],

---

T. Zhang · Y. Jia (✉)
The Seventh Research Division and the Center for Information and Control,
School of Automation Science and Electrical Engineering,
Beihang University (BUAA), Beijing 100191, China
e-mail: ymjia@buaa.edu.cn

T. Zhang
e-mail: tfzhang@buaa.edu.cn

Y. Jia et al. (eds.), *Proceedings of 2018 Chinese Intelligent Systems Conference*, Lecture Notes in Electrical Engineering 529,
https://doi.org/10.1007/978-981-13-2291-4_8

the extremum seeking control [8], set invariance notions [9, 10], etc. These methods have obtained reputable control performance. However, the problem of state constraints was omitted in aforementioned research results.

Recently, barrier Lyapunov Functions (BLF) have been employed in handling with state constraints, which leads to several attractive outcomes [11–14]. Tee et al. [11] resolved the partial-state-constraint problem by BLF, in which the problem of full-state constraints and output constraint can be viewed as special situations of the partial-state-constraint. When the output constraint is time-varying, Tee et al. [12] dealt with this problem through a change of coordinates for the tracking error. In [13], Liu et al. extended the BLF-based control problem of the constrained uncertain strict-feedback system to pure-feedback system. He et al. [14] applied the adaptive neural network control to the uncertain full-state constrained robotic manipulator using BLF and fulfilled the desired control objective.

A slight drawback of BLF leads to the conservativeness of BLF-based methods, because BLFs are always constructed by tracking errors rather than original states. To eliminate the conservativeness, an adaptive control method based on integral-barrier Lyapunov functions (iBLF), which consist of original states and state errors, was proposed for strict-feedback systems with state constrains in [15]. Subsequently, many excellent results have sprung up. Kim et al. [16] dealt with the control problem of nonlinear pure-feedback systems with the output constraint using iBLF. In [17], He et al. proposed a top tension control algorithm for a flexible marine riser with the boundary output constraint.

Inspired by above fantastic work, an adaptive neural network (NN) output tracking control method will be presented for robotic manipulators considering uncertainties and constraints simultaneously in this paper. The rest of the paper is organized as follows. The problem formulation, some useful assumptions and lemmas are given in Sect. 2. In Sect. 3, the design procedure of the adaptive NN output tracking controller is provided, and a theorem with its proof is demonstrated. Section 4 illustrated a numerical simulation example to verify the effectiveness and performance of the proposed method. Section 5 offers the conclusion of this paper.

2 Problem Formulation and Preliminaries

2.1 Problem Formulation

The dynamics of an n-link rigid robotic manipulator can be expressed in the following Lagrangian form:

$$M(q)\ddot{q} + C(q, \dot{q})\dot{q} + G(q) + F(\dot{q}) + D(t) = u \tag{1}$$

where $q \in R^n$ represents the vector of joint positions, $u \in R^n$ is the control vector of applied joint torques, $M(q) \in R^{n\times n}$ is the inertia matrix, $C(q, \dot{q}) \in R^{n\times n}$ denotes the matrix of the Coriolis and centrifugal torques, and $G(q) \in R^n$, $F(\dot{q}) \in R^n$ and $D(t) \in R^n$ represent the vectors of gravity, friction and external disturbance, respectively. In practice, most of these terms such as $M(q), C(q,\dot{q}), G(q), F(\dot{q}), D(t)$ are unknown, which will cause difficulties for the controller design. The robot dynamics (1) have the following important properties:

Property 1 *The inertia matrix $M(q)$ is symmetric and positive definite.*

Property 2 *The matrix $\dot{M}(q) - 2C(q,\dot{q})$ is skew symmetric, i.e.,*

$$z^T\left[\dot{M}(q) - 2C(q,\dot{q})\right]z = 0$$

The control objective is to design suitable control torques u for the uncertain robotic manipulator system (1) so that the output q tracks the desired trajectory. Moreover, the output q is bounded.

2.2 Useful Assumptions and Lemmas

Assumption 1 The desired trajectory is a known bounded smooth function.

Assumption 2 The external disturbance $D(t)$ is uniformly bounded, i.e., there exists a constant $\bar{D} \in R^+$, satisfying

$$\|D(t)\| \le \bar{D}, \forall t \in [0, \infty)$$

Lemma 1 [2] *For any continuous function $h(X): R^n \to R$ on a compact set $U \subset R^n$ and an arbitrary small positive constant ε, there exists $h^*(X) = W^T S(X)$ satisfying* $\sup_{X\to U} |h^*(X) - h(X)| \le \varepsilon$ *with $X \in R^n$ being the input vector. The continuous function $h(X)$ can be approximated by*

$$h(X) = W^T S(X) + \varepsilon(X)$$

where $W \in R^m$, m and $\varepsilon(X)$ being the ideal weight vector, the NN node number and the approximation error, respectively. In principle, $\varepsilon(X)$ are bounded by constant ε^*.

Lemma 2 [14] *For bounded initial conditions, a C^1 continuous positive definite Lyapunov function $V(x)$ satisfies*

$$\gamma_1(\|x\|) \le V(x) \le \gamma_2(\|x\|)$$
$$\dot{V}(x) = \frac{\partial V}{\partial x}\dot{x} \le -\rho V(x) + c$$

where $\gamma_1, \gamma_2 : R^n \to R$ are class κ functions and ρ, c are positive constants. Then, the solution $x(t)$ is uniformly ultimate bounded.

Lemma 3 [15] *For* $|x| < k$*, the iBLF* $V(z,\alpha) = \int_0^z \frac{\tau k^2}{k^2-(\tau+\alpha)^2} d\tau$ *satisfies*

$$\frac{1}{2}z^2 \le V(z,\alpha) \le \frac{k^2 z^2}{k^2 - x^2}$$

where $z = x - \alpha$*, k is a positive constant.*

3 Control Design

Let $x_1 = [q_1, q_2, \ldots, q_n]^T$ and $x_2 = [\dot{q}_1, \dot{q}_2, \ldots, \dot{q}_n]^T$, the robot dynamics (1) can be transformed into

$$\begin{cases} \dot{x}_1 = x_2 \\ \dot{x}_2 = M^{-1}\left[u - C(x_1, x_2)x_2 - G(x_1) - F(x_2) - D(t)\right] \\ y = x_1 \end{cases} \tag{2}$$

Define the following error variables:

$$\begin{cases} z_1 = x_1 - y_d \\ z_2 = x_2 - \alpha \end{cases} \tag{3}$$

where $y_d = [y_{d1}, y_{d2}, \ldots, y_{dn}]^T$ represents the desired trajectory, and $\alpha = [\alpha_1, \alpha_2, \ldots, \alpha_n]^T$ is the virtual controller which will be designed in *step 1*. The output constraint requires

$$|x_{1i}| < k_i, i = 1, \ldots n \tag{4}$$

Denote the compact set $\Omega := \{x_1 \in R^n : |x_{1i}| < k_i, i = 1, \ldots, n\} \subset R^n$
Step 1. Consider the following iBLF candidate:

$$\begin{aligned} V_1 &= \sum_{i=1}^{n} V_{1i} \\ V_{1i} &= \int_0^{z_{1i}} \frac{\tau k_i^2}{k_i^2 - (\tau + y_{di})^2} d\tau \end{aligned} \tag{5}$$

The time derivative can be written as

$$\begin{aligned} \dot{V}_1 &= \sum_{i=1}^{n} \left[\frac{k_i^2 z_{1i} \dot{z}_{1i}}{k_i^2 - x_{1i}^2} + \frac{\partial V_{1i}}{\partial y_{d_i}} \dot{y}_{d_i} \right] \\ &= \sum_{i=1}^{n} \left[\frac{k_i^2 z_{1i}}{k_i^2 - x_{1i}^2} \left(z_{2i} + \alpha_i - \dot{y}_{d_i}\right) + \frac{\partial V_{1i}}{\partial y_{d_i}} \dot{y}_{d_i} \right] \end{aligned} \tag{6}$$

Utilize the substitution $\tau = \omega z_{1i}$ and integration by parts

$$\begin{aligned}\frac{\partial V_{1i}}{\partial y_{d_i}}\dot{y}_{d_i} &= \dot{y}_{d_i}\int_0^{z_{1i}}\frac{2\tau k_i^2\left(\tau+y_{d_i}\right)}{\left[k_i^2-\left(\tau+y_{d_i}\right)^2\right]^2}d\tau \\ &= \dot{y}_{d_i}\int_0^{z_{1i}}\tau d\left(\frac{k_i^2}{k_i^2-\left(\tau+y_{d_i}\right)^2}\right) \\ &= \dot{y}_{d_i}\left(\frac{k_i^2 z_{1i}}{k_i^2-x_{1i}^2}-\int_0^1\frac{z_{1i}k_i^2}{k_i^2-\left(\omega z_{1i}+y_{d_i}\right)^2}d\omega\right) \\ &= \left(\frac{k_i^2}{k_i^2-x_{1i}^2}-\beta_i\left(z_{1i},y_{d_i}\right)\right)z_{1i}\dot{y}_{d_i}\end{aligned} \tag{7}$$

Design the virtual controller α as

$$\alpha=-\begin{bmatrix}a_1 z_{11}-\frac{k_1^2-x_{11}^2}{k_1^2}\dot{y}_{d_1}\\ \vdots \\ a_n z_{1n}-\frac{k_n^2-x_{1n}^2}{k_n^2}\dot{y}_{d_n}\end{bmatrix} \tag{8}$$

where a_i, $i=1,\ldots,n$ are design positive constants.

Substitute (7) and (8) into (6), we have

$$\dot{V}_1=-\sum_{i=1}^{n}\frac{a_i k_i^2 z_{1i}^2}{k_i^2-x_{1i}^2}+\sum_{i=1}^{n}\frac{k_i^2 z_{1i}z_{2i}}{k_i^2-x_{1i}^2} \tag{9}$$

Step 2. Choose the second iBLF candidate as

$$V_2=V_1+\frac{1}{2}z_2^T M\left(x_1\right)z_2 \tag{10}$$

The time derivative of V_2 can be expressed as

$$\dot{V}_2=\dot{V}_1+z_2^T M\left(x_1\right)\dot{z}_2+\frac{1}{2}z_2^T\dot{M}\left(x_1\right)z_2 \tag{11}$$

Substitute (2) and (3) into (11), we obtain that

$$\begin{aligned}\dot{V}_2=\dot{V}_1+z_2^T&\left[u-C\left(x_1,x_2\right)\alpha-C\left(x_1,x_2\right)z_2-G\left(x_1\right)\right.\\ &\left.-F\left(x_2\right)-D\left(t\right)-M\left(x_1\right)\dot{\alpha}+\tfrac{1}{2}\dot{M}\left(x_1\right)z_2\right]\end{aligned} \tag{12}$$

where $\dot{\alpha}$ is calculated as

$$\dot{\alpha}=\frac{\partial\alpha}{\partial x_1}\dot{x}_1+\frac{\partial\alpha}{\partial y_d}\dot{y}_d+\frac{\partial\alpha}{\partial\dot{y}_d}\ddot{y}_d \tag{13}$$

Utilize the Property 2,

$$\begin{aligned}\dot{V}_2 = &-\sum_{i=1}^{n}\frac{a_i k_i^2 z_{1i}^2}{k_i^2-x_{1i}^2}+\sum_{i=1}^{n}\frac{k_i^2 z_{1i} z_{2i}}{k_i^2-x_{1i}^2}+z_2^T\left[u-C\left(x_1,x_2\right)\alpha-G\left(x_1\right)\right.\\ &\left.-F\left(x_2\right)-D\left(t\right)-M\left(x_1\right)\dot{\alpha}\right]+\tilde{\theta}\dot{\tilde{\theta}}\end{aligned} \tag{14}$$

Step 3. In fact, as mentioned in *Problem Formulation*, most terms of the robotic manipulator system are unknown. Therefore, according to Lemma 1, a radial basis function NN is employed to approximate the unknown function as

$$C\left(x_1,x_2\right)\alpha+G\left(x_1\right)+F\left(x_2\right)+M\left(x_1\right)\dot{\alpha}=WS\left(X\right)+\varepsilon\left(X\right) \tag{15}$$

where $X=\left[x_1, y_d, \dot{y}_d, \ddot{y}_d\right]^T$, W, $S(X)$ and $\varepsilon(X)$ are the ideal weight, the Gaussian basis function and the approximation error, respectively. Besides, there exists positive constant satisfying $|\varepsilon(X)| \le \varepsilon^*$.

Then, substitute (15) into (14) yields

$$\begin{aligned}\dot{V}_2 = &-\sum_{i=1}^{n}\frac{a_i k_i^2 z_{1i}^2}{k_i^2-x_{1i}^2}+\sum_{i=1}^{n}\frac{k_i^2 z_{1i} z_{2i}}{k_i^2-x_{1i}^2}+z_2^T u\\ &-z_2^T\left(WS\left(X\right)+\varepsilon\left(X\right)\right)-z_2^T D\left(t\right)\end{aligned} \tag{16}$$

Consider the final iBLF candidate as

$$V_3 = V_2+\frac{1}{2}\tilde{\theta}^2 \tag{17}$$

where $\tilde{\theta}=\hat{\theta}-\theta$, $\theta=\|W\|^2$. And $\hat{\theta}$, the estimation of θ, is the adaptive parameter which helps to reduce the online calculation of the NN weight.

The time derivative of V_3 is

$$\dot{V}_3=\dot{V}_2+\tilde{\theta}\dot{\tilde{\theta}}=\dot{V}_2+\tilde{\theta}\dot{\hat{\theta}} \tag{18}$$

According to Young's inequality and Assumption 2, the following formulas are established.

$$-z_2^T WS\left(X\right)\le\frac{1}{2b_1^2}z_2^T z_2\theta\|S\left(X\right)\|^2+\frac{b_1^2}{2} \tag{19}$$

$$-z_2^T\varepsilon\left(X\right)\le\frac{1}{2b_2^2}z_2^T z_2+\frac{b_2^2}{2}\varepsilon^{*2} \tag{20}$$

$$-z_2^T D\left(t\right)\le\frac{1}{2b_3^2}z_2^T z_2+\frac{b_3^2}{2}\bar{D}^2 \tag{21}$$

where b_1, b_2 and b_3 are design positive constants.

we choose the adaptive NN controller u as

$$u = -\begin{bmatrix} \frac{k_1^2 z_{11}}{k_1^2 - x_{11}^2} \\ \vdots \\ \frac{k_n^2 z_{1n}}{k_n^2 - x_{1n}^2} \end{bmatrix} - K z_2 - \frac{1}{2b_1^2} z_2 \hat{\theta} \|S(X)\|^2 - \frac{1}{2b_2^2} z_2 - \frac{1}{2b_3^2} z_2 \tag{22}$$

Substitute (19)–(22) into (18) yields

$$\begin{aligned} \dot{V}_3 = &- \sum_{i=1}^{n} \frac{a_i k_i^2 z_{1i}^2}{k_i^2 - x_{1i}^2} - z_2^T K z_2 + \frac{b_1^2}{2} + \frac{b_2^2}{2} \varepsilon^{*^2} + \frac{b_3^2}{2} \bar{D}^2 \\ &- \tilde{\theta} \left(\frac{1}{2b_1^2} z_2^T z_2 \|S(X)\|^2 - \dot{\hat{\theta}} \right) \end{aligned} \tag{23}$$

Design the adaptive law as

$$\dot{\hat{\theta}} = -\lambda \hat{\theta} + \frac{1}{2b_1^2} z_2^T z_2 \|S(X)\|^2 \tag{24}$$

where λ is a design constant.

Similarly utilizc the Young's inequality

$$\begin{aligned} -\lambda \tilde{\theta} \hat{\theta} &= -\lambda \tilde{\theta}^2 - \lambda \tilde{\theta} \theta \\ &\le -\frac{\lambda}{2} \tilde{\theta}^2 + \frac{\lambda}{2} \theta^2 \end{aligned} \tag{25}$$

Then, thc time derivative of V_3 can be written as

$$\dot{V}_3 = -\sum_{i=1}^{n} \frac{a_i k_i^2 z_{1i}^2}{k_i^2 - x_{1i}^2} - z_2^T K z_2 - \frac{\lambda}{2} \tilde{\theta}^2 + \frac{b_1^2}{2} + \frac{b_2^2}{2} \varepsilon^{*^2} + \frac{b_3^2}{2} \bar{D}^2 + \frac{\lambda}{2} \theta^2 \tag{26}$$

Utilize the Lemma 3, $\dot{V}_3$ is expressed as

$$\begin{aligned} \dot{V}_3 \le &- \sum_{i=1}^{n} a_i \int_0^{z_{1i}} \frac{\tau k_i^2}{k_i^2 - (\tau + y_d)^2} d\tau - z_2^T K z_2 - \frac{\lambda}{2} \tilde{\theta}^2 \\ &+ \frac{b_1^2}{2} + \frac{b_2^2}{2} \varepsilon^{*^2} + \frac{b_3^2}{2} \bar{D}^2 + \frac{\lambda}{2} \theta^2 \\ \le &-\rho V_3 + C \end{aligned} \tag{27}$$

where

$$\rho = \min\left\{a_i, \frac{2\lambda_{\min}(K)}{\lambda_{\max}(M)}, \lambda\right\}$$
$$C = \frac{b_1^2}{2} + \frac{b_2^2}{2}\varepsilon^{*2} + \frac{b_3^2}{2}\bar{D}^2 + \frac{\lambda}{2}\theta^2 \tag{28}$$

According to the above design procedures, the following theorem is established by Lyapunov analysis method.

Theorem 1 *Consider the robotic manipulator system (1) under Assumptions 1 and 2, with constructing the adaptive law (24), choosing appropriate parameters a_i, b_i, λ, and designing the virtual controller (8) and the adaptive NN controller (22). If the initial condition satisfies $x_1(0) \in \Omega$, i.e., the initial condition $x_1(0)$ is bounded, the following properties will be easily obtained.*

1. *The error signals $z_{1i}(t), i = 1, \ldots, n$, $z_2(t)$ and $\tilde{\theta}$ are semi-global uniformly ultimate bounded, and $\forall t \geq 0$, these signals will remain within the compact sets $\Omega_{z_{1i}}, i = 1, \ldots, n$, Ω_{z_2}, $\Omega_{\tilde{\theta}}$, which are defined by*

$$\Omega_{z_{1i}} := \left\{z_{1i} \in R : |z_{1i}| \leq \sqrt{2V_3(0) + \frac{2C}{\rho}}\right\}, i = 1, \ldots, n$$
$$\Omega_{z_2} := \left\{z_2 \in R^n : \|z_2\| \leq \sqrt{\left(2V_3(0) + \frac{2C}{\rho}\right)\Big/\lambda_{\min}(M)}\right\} \tag{29}$$
$$\Omega_{\tilde{\theta}} := \left\{\tilde{\theta} \in R : \left|\tilde{\theta}\right| \leq \sqrt{2V_3(0) + \frac{2C}{\rho}}\right\}$$

2. *For all $t > 0$, the output $y = x_1(t)$ remains in the constrained compact set Ω.*
3. *All signals of closed-loop system, such as the adaptive NN controller $u(t)$, the virtual controller $\alpha(t)$ and the adaptive law $\hat{\theta}(t)$, are bounded for all $t > 0$.*

Proof 1. Multiplying $e^{\rho t}$ on both sides of (27) yields

$$e^{\rho t}\frac{dV_3}{dt} \leq -\rho V_3 e^{\rho t} + Ce^{\rho t} \Leftrightarrow \frac{d}{dt}\left(V_3 e^{\rho t}\right) \leq Ce^{\rho t} \tag{30}$$

And integrating it over $[0, t]$,

$$V_3 e^{\rho t} - V_3(0) \leq \frac{C\left(e^{\rho t} - 1\right)}{\rho} \Leftrightarrow V_3(t) \leq \left(V_3(0) - \frac{C}{\rho}\right)e^{-\rho t} + \frac{C}{\rho} \tag{31}$$

According to the Lemma 3, $\frac{1}{2}z_{1i}^2 \leq V_3$. Therefore,

$$
\begin{aligned}
|z_{1i}| &\leq \sqrt{2\left(V_3(0) - \frac{C}{\rho}\right)e^{-\rho t} + \frac{2C}{\rho}} \leq \sqrt{2V_3(0) + \frac{2C}{\rho}} \\
\|z_2\| &\leq \sqrt{\left(2V_3(0) + \frac{2C}{\rho}\right)\Big/\lambda_{\min}(M)} \\
\left|\tilde{\theta}\right| &\leq \sqrt{2V_3(0) + \frac{2C}{\rho}}
\end{aligned} \tag{32}
$$

Since design parameters can be arbitrarily chosen and the boundary of output constrains can be arbitrarily large, we can draw the conclusion that signals $z_{1i}(t), i = 1, \ldots, n, z_2(t)$ and $\tilde{\theta}$ are semi-global uniformly ultimate bounded from Lemma 2.

2. By reduction, we assume that there exist some $t = T$ and some $i \in \{1, \ldots, \text{n}\}$ satisfying $|x_{1i}(T)| = k_i$, $x_1(0) \in \Omega$. Then, according to Lemma 3 and (31), we can obtain

$$
\begin{aligned}
F(T) &= \int_0^{z_{1i}} \frac{\tau k_i^2}{k_i^2 - (\tau + y_{d\,i})^2} d\tau \,|_{t=T} \\
&= \int_0^{z_{1i}} \frac{(\tau + y_{d\,i})k_i^2}{k_i^2 - (\tau + y_{d\,i})^2} d\tau \,|_{t=T} - \int_0^{z_{1i}} \frac{y_{d\,1}k_i^2}{k_i^2 - (\tau + y_{d\,i})^2} d\tau \,|_{t=T} \\
&= \frac{k_i^2}{2}\ln\frac{k_i^2 - {y_{d\,i}}^2}{k_i^2 - x_{1i}^2} \,|_{t=T} + \frac{k_i y_{d\,i}}{2}\ln\frac{(k_i + y_{d\,i})(k_i - x_{1i})}{(k_i - y_{d\,i})(k_i + x_{1i})} \,|_{t=T} \\
&\leq V(0) + \frac{C}{\rho}
\end{aligned} \tag{33}
$$

Substituting that $|x_{1i}(T)| = k_i$, we can obtain that $F(T)$ is unbounded, which contradicts the result $F(T) \leq V(t) \leq V(0) + \frac{C}{\rho}$.
Hence, $|x_{1i}(T)| \neq k_i$, $\forall t > 0$, $i = 1, \ldots, n$. Then, we conclude that $x_1(t) \in \Omega$, $\forall t > 0$ if $x_1(0) \in \Omega$.

3. Because $\tilde{\theta}(t)$, and $\theta(t)$ are bounded, the boundedness of the adaptive law $\hat{\theta}(t)$ is established. Then, according to above results, $z_1(t), z_2(t), x_1(t), \theta(t)$ are bounded, we can obtain that the adaptive NN controller $u(t)$ and the virtual controller $\alpha(t)$ are bounded from (8) and (22), i.e., the property *iii* is established. ■

4 Numerical Simulation Example

In this section, a numerical simulation example is provided to verify the effectiveness of our proposed control method. Considering the 2-DOF robotic manipulator system, the mass and length of link 1 are 1 kg and 0.8 m, and these properties of link 2 are 1 kg and 0.7 m. The bounded external disturbance is chosen as $D(t) = \left[1 - e^{-t}\cos(1.5t),\ 1.2\left(1 - e^{-0.5t}\cos(2t)\right)\right]^T$. The friction $F(\dot{q})$ is chosen by the Coulomb friction model $F(\dot{q}) = \left[0.8 + 0.4\tanh(0.6\dot{q}),\ 1 + 0.5\tanh(0.5\dot{q})\right]^T$. The initial positions and adaptive laws of the manipulator are $q_0 = \left[0.4,\ 0.3\right]^T$, $\dot{q}_0 = \left[0.2,\ 0.3\right]^T$ and $\hat{\theta}_0 = 0.2$.

The desired trajectory is given as $q_d = \left[0.8\sin(2.5t),\ 0.5\sin(2t)\right]^T$. And output constraint parameters are set as $k_1 = 1, k_2 = 0.7$. For NN, the node number is chosen by $m_1 = 10$, the width is $d_1 = 2$. Then, we choose the design parameters as

$$K = \begin{bmatrix} 22 & 0 \\ 0 & 20 \end{bmatrix}; a_1 = 15, a_2 = 20; b_1 = 0.6, b_2 = 1, b_3 = 0.8; \lambda = 10$$

From the simulation result, the tracking performance of output $x_1(t)$ are shown in Fig. 1, and it is obvious that positions $x_{11}(t), x_{12}(t)$ of link 1 and 2 track the desired trajectories $y_{d_1}(t), y_{d_2}(t)$ successfully without violating the output constraint. Figures 3 and 4 indicate the error signals $z_1(t), z_2(t)$ are bounded. Moreover, the

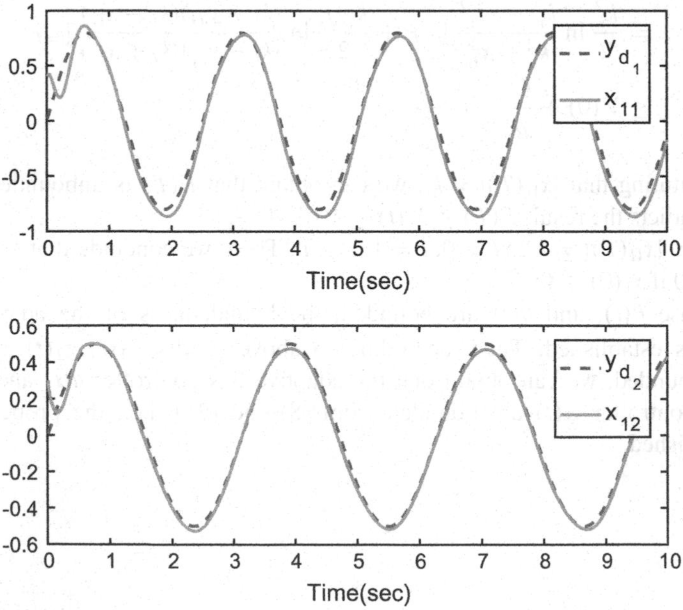

Fig. 1 Trajectories of the x_{11}, y_{d_1} and x_{12}, y_{d_2}

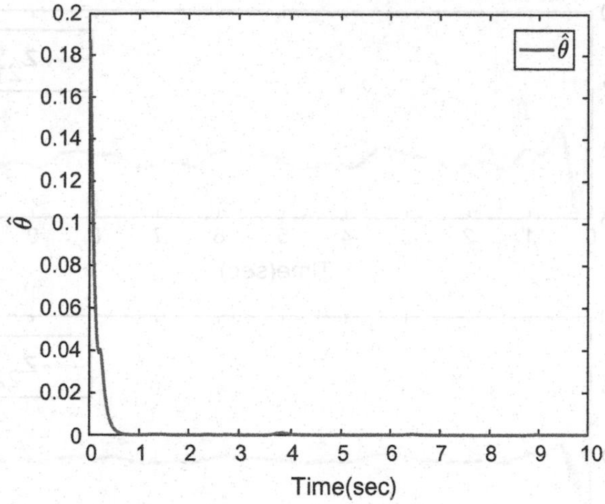

Fig. 2 Trajectory of $\hat{\theta}(t)$

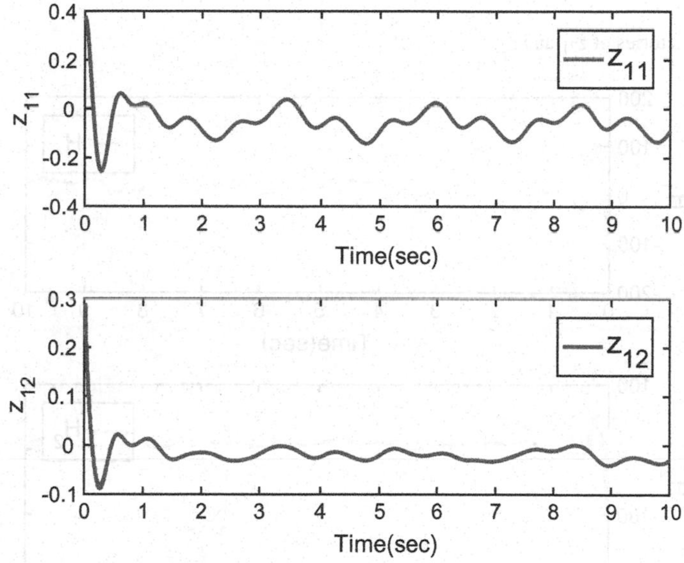

Fig. 3 Trajectories of z_{11} and z_{12}

boundedness of the input $u(t)$ and virtual controller $\alpha(t)$ are manifested by Figs. 5 and 6, which means our method could guarantee all closed-loop signals are bounded. At last, we can draw the conclusion that the chosen NN commendably compensates for the uncertainty of the manipulator system, because the adaptive law $\hat{\theta}(t)$ is bounded from Fig. 2.

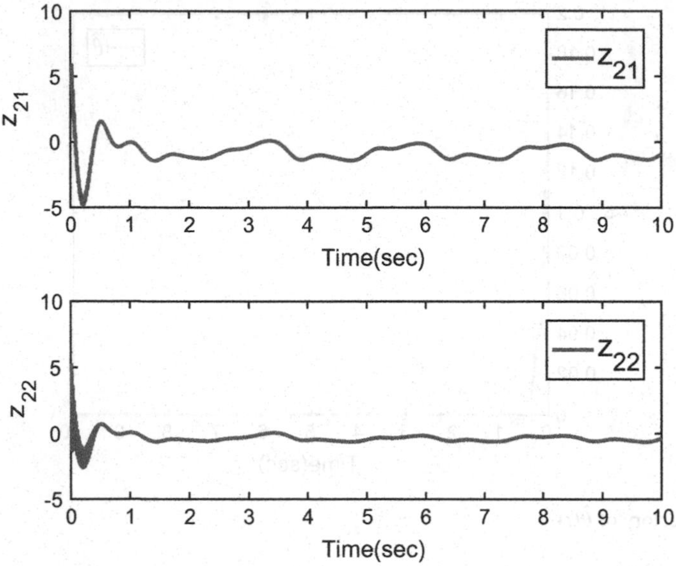

Fig. 4 Trajectories of z_{21} and z_{22}

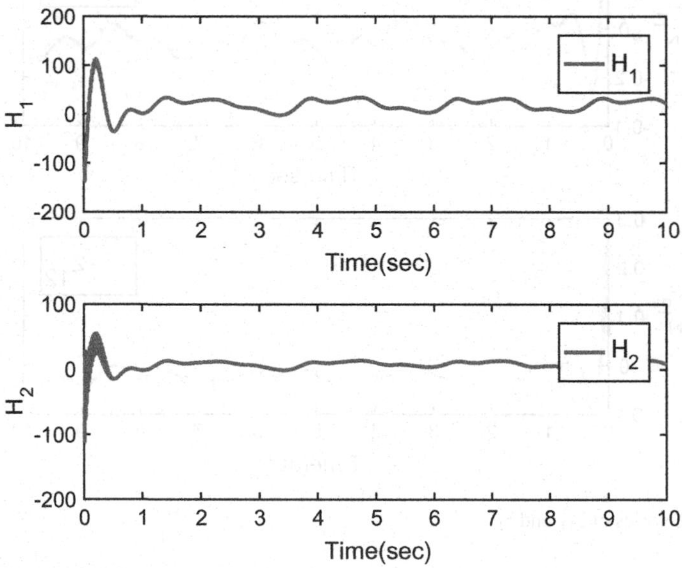

Fig. 5 Trajectories of u_1 and u_2

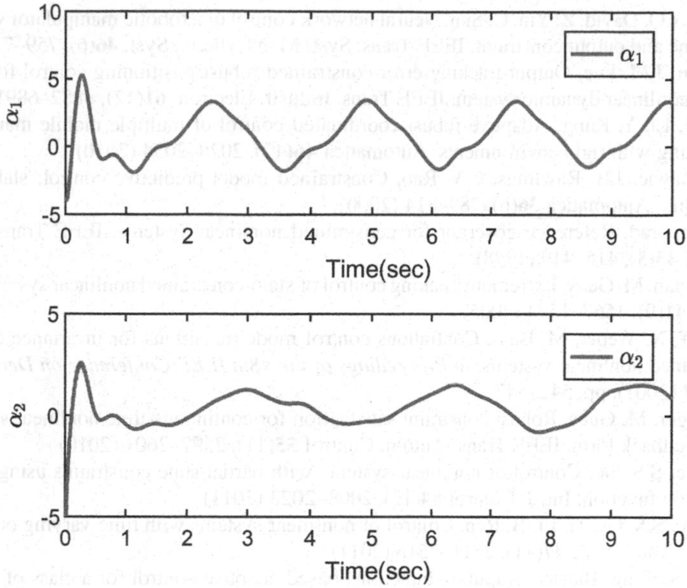

Fig. 6 Trajectories of α_1 and α_2

5 Conclusion

The adaptive NN output tracking control method is capable of accomplishing the control objective superiorly. Using the original output mixed with the error term, the iBLF-based control method alleviates the conservativeness of BLF without violating the output constraint. Utilizing the neural network, unknown functions of the robotic manipulator system are approximated well with less adaptive parameters in the controller design procedure. Moreover, our method could guarantee the semi-global uniformly ultimate boundedness of the output error and the boundedness of all closed-loop system signals.

Acknowledgements This work was supported by the NSFC (61327807, 61521091, 61520106010, 61134005) and the National Basic Research Program of China (973 Program: 2012CB821200, 2012CB821201).

References

1. H. Yang, X. Fan, P. Shi, C. Hua, Nonlinear control for tracking and obstacle avoidance of a wheeled mobile robot with nonholonomic constraint. IEEE Trans. Control Syst. Technol. **24**(2), 741–746 (2016)
2. Y.-J. Liu, S. Lu, S. Tong, Neural network controller design for an uncertain robot with time-varying output constraint. IEEE Trans. Syst. Man Cybern.: Syst. **47**(8), 2060–2068 (2017)

3. W. He, A.O. David, Z. Yin, C. Sun, Neural network control of a robotic manipulator with input deadzone and output constraint. IEEE Trans. Syst. Man Cybern.: Syst. **46**(6), 759–770 (2016)
4. S.I. Han, J.M. Lee, Output-tracking-error-constrained robust positioning control for a nonsmooth nonlinear dynamic system. IEEE Trans. Industr. Electron. **61**(12), 6882–6891 (2014)
5. Z. Li, J. Li, Y. Kang, Adaptive robust coordinated control of multiple mobile manipulators interacting with rigid environments. Automatica **46**(12), 2028–2034 (2010)
6. D.Q. Mayne, J.B. Rawlings, C.V. Rao, Constrained model predictive control: stability and optimality. Automatica **36**(6), 789–814 (2000)
7. A. Bemporad, Reference governor for constrained nonlinear systems. IEEE Trans. Autom. Control **43**(3), 415–419 (1998)
8. D. DeHaan, M. Guay, Extremum-seeking control of state-constrained nonlinear systems. Automatica **41**(9), 1567–1574 (2005)
9. J. Wolff, X. Weber, M. Buss, Continuous control mode transitions for invariance control of constrained nonlinear systems, in *Proceedings of the 48th IEEE Conference on Decision and Control* (2007), pp. 542–547
10. M. Burger, M. Guay, Robust constraint satisfaction for continuous-time nonlinear systems in strict feedback form. IEEE Trans. Autom. Control **55**(11), 2597–2601 (2010)
11. K.P. Tee, S.S. Ge, Control of nonlinear systems with partial state constraints using a barrier Lyapunov function. Int. J. Control **84**(12), 2008–2023 (2011)
12. K.P. Tee, S.S. Ge, H. Li, B. Ren, Control of nonlinear systems with time-varying output constraints. Automatica **47**(11), 2511–2516 (2011)
13. Y. Liu, S. Tong, Barrier Lyapunov Functions-based adaptive control for a class of nonlinear pure-feedback systems with full state constraints. Automatica **64**(C), 70–75 (2016)
14. W. He, Y. Chen, Z. Yin, Adaptive neural network control of an uncertain robot with full-state constrains. IEEE Trans. Cybern. **46**(3), 620–629 (2016)
15. K.P. Tee, S.S. Ge, Control of state-constrained nonlinear systems using Integral Barrier Lyapunov Functionals, in *IEEE Conference on Decision and Control* (2012), pp. 3239–3244
16. B.S. Kim, S.J. Yoo, Adaptive control of nonlinear pure-feedback systems with output constraints: integral barrier Lyapunov functional approach. Int. J. Control Autom. Syst. **13**(1), 249–256 (2015)
17. W. He, C. Sun, S.S. Ge, Top tension control of a flexible marine riser by using integral-barrier Lyapunov function. IEEE Trans. Mechatron. **20**(2), 497–505 (2014)

Consensus of a Class of Heterogeneous Networked System with Sampled-Data

Huanyu Zhao, Liangping Shi, Xi Zhang and Zhongyi Tang

Abstract This paper studies the consensus problem for a class of heterogeneous networked system. We first discretize the continuous-time networked system by sampled-data method. Then we convert the networked system into the reduced-order error system by a system transformation. Based on algebraic graph theory and matrix theory, a sufficient condition for the networked system to achieve consensus is obtained by analyzing the stable problem of the reduced-order system. Simulation example will be given to show the usefulness of the results.

Keywords Heterogeneous system · Networked system · Consensus · Sampled-data system

1 Introduction

Recently, the heterogeneous multi-agent systems have attracted the attentions of a lot of researchers. The heterogeneous multi-agent systems are used to describe the systems with different model structure or different parameter, such as the time-delay. In [1], the authors studied the coordination problem of the continuous-time heterogeneous multi-agent system with a leader, where the interaction topologies were switching jointly connected. In [2], the authors considered the output consensus problem of heterogeneous multi-agent systems by proposed a new event-triggered protocol. The output regulation problem of heterogeneous multi-agent systems consisted of some followers and one leader was studied in [3], where the differential graphical game was used. Based on output regulation, the containment control problem of a class

H. Zhao (✉) · L. Shi · X. Zhang · Z. Tang
Faculty of Automation, Huaiyin Institute of Technology,
Huai'an 223001, Jiangsu, People's Republic of China
e-mail: hyzhao@163.com

H. Zhao
School of Automation, Southeast University,
Nanjing 210096, People's Republic of China

Y. Jia et al. (eds.), *Proceedings of 2018 Chinese Intelligent Systems Conference*, Lecture Notes in Electrical Engineering 529,
https://doi.org/10.1007/978-981-13-2291-4_9

of heterogeneous multi-agent system was considered in [4]. On the other hand, the group consensus of heterogeneous multi-agent systems also can be found [5, 6]. In [5], the authors studied the group consensus of heterogeneous multi-agent systems under both fixed topology and switching topology. In [6], a new consensus algorithm for group consensus of heterogeneous multi-agent system was proposed based on the feature of agents. In [7], the finite-time consensus problem of the high-order heterogeneous multi-agent system was studied. The authors proposed two kinds of consensus algorithms based on state feedback and output feedback, respectively. In [8], the authors used the sampled-data method to study the consensus problem of heterogeneous multi-agent systems, where a sampling delay was considered.

Motivated by the aforementioned literature, we will consider the consensus problem for a class of sampled-data heterogeneous networked system. We suppose that the system considered consist of first-order integrator and second-order integrator. We transform the continuous-time system to the discrete-time system by using the sampled-data method. Moreover, we convert the multi-agent system into the reduced-order error system. Then we obtain the conditions of consensus by analyzing the stable problem of the reduced-order system.

Notation: Let $\mathbb{R}$ denote the real number set. Let $\rho(M)$ denote the spectral radius of the matrix M. Let S_n denote a index set $\{1, \ldots, n\}$. Let $|A|$ denote the determinant of matrix A. Let I_n and $\mathbf{0}$ denote, respectively, the $n \times n$ identity matrix and zero matrix with appropriate dimension. $\mathrm{Re}(\cdot)$ and $\mathrm{Im}(\cdot)$ represent, respectively, the real and imaginary parts of a number.

2 Problem Formulations

We first give some necessary graph theory notion before the problem is proposed. Let $\mathcal{G} = (\mathcal{V}, \mathcal{E}, \mathcal{A})$ be a directed graph of order n, where $\mathcal{V} = \{v_1, \ldots, v_n\}$ and $\mathcal{E}$ represent the node set and the edge set, respectively. $\mathcal{A} = [a_{ij}] \in \mathbb{R}^{n\times n}$ is the adjacency matrix associated with $\mathcal{G}$, where $a_{ij} > 0$ if $(v_i, v_j) \in \mathcal{E}$, otherwise, $a_{ij} = 0$. An edge $(v_i, v_j) \in \mathcal{E}$ if agent j can obtain the information from agent i. We say agent i is a neighbor of agent j. Let $N_i = \{v_j \in \mathcal{V} : (v_i, v_j) \in \mathcal{E}\}$ denote the neighbor set of agent i. The (nonsymmetrical) Laplacian matrix $\mathcal{L}$ associated with $\mathcal{A}$ and hence $\mathcal{G}$ is defined as $\mathcal{L} = [l_{ij}] \in \mathbb{R}^{n\times n}$, where $l_{ii} = \sum_{j=1, j\neq i}^{n} a_{ij}$ and $l_{ij} = -a_{ij}$, $\forall i \neq j$. A directed path is a sequence of edges in a directed graph in the form of (v_{i_1}, v_{i_2}), (v_{i_2}, v_{i_3}), …, where $v_{i_k} \in \mathcal{V}$. A directed spanning tree of $\mathcal{G}$ is a directed tree that contains all nodes of $\mathcal{G}$. A directed graph has or contains a directed spanning tree if there exists a directed spanning tree as a subset of the directed graph, that is, there exists at least one node having a directed path to all of the other nodes.

Suppose that the heterogeneous networked system consists of m second-order integrators and $n - m$ first-order integrators ($m < n$). The second-order integrator is given as follows:

$$\dot{x}_i(t) = v_i(t), \quad \dot{v}_i(t) = u_i(t), \quad i \in S_m, \tag{1}$$

The first-order integrator is given as follows:

$$\dot{x}_i(t) = u_i(t), \quad i \in S_{n-m}. \tag{2}$$

According to the direct discretization in [9], the discretized dynamics of (2) and (1) are respectively,

$$x_i[k+1] = x_i[k] + Tu_i[k], \quad i \in S_{n-m}, \tag{3}$$

$$\begin{cases} x_i[k+1] = x_i[k] + Tv_i[k] + \frac{T^2}{2}u_i[k] \\ v_i[k+1] = v_i[k] + Tu_i[k] \end{cases} i \in S_m. \tag{4}$$

In this paper, we choose the following consensus algorithm for the first-order agent,

$$u_i[k] = \alpha \sum_{j \in N_i} a_{ij}(x_j[k] - x_i[k]), \qquad i \in S_{n-m}, \tag{5}$$

where $\alpha > 0$ is the control gain. And the consensus algorithm for the second-order agent is as follows,

$$u_i[k] = \alpha \sum_{j \in N_i} a_{ij}(x_j[k] - x_i[k]) - \beta v_i[k], i \in S_m \tag{6}$$

where $\alpha > 0$ and $\beta > 0$ are the control gains.

Denote $X_1[k] = [x_1[k], \dots, x_m[k]]^T$, $X_2[k] = [x_{m+1}[k], \dots, x_n[k]]^T$, $V_1[k] = [v_1[k], \dots, v_m[k]]^T$, then the systems (3) and (4) with algorithm (5) and (6) respectively can be rewritten in matrix form:

$$\begin{bmatrix} X_1[k+1] \\ X_2[k+1] \\ V_1[k+1] \end{bmatrix} = \Xi_1 \begin{bmatrix} X_1[k] \\ X_2[k] \\ V_1[k] \end{bmatrix} \tag{7}$$

where

$$\Xi_1 = \begin{bmatrix} I_m - \frac{\alpha T^2}{2}\mathcal{L}_{11} & -\frac{\alpha T^2}{2}\mathcal{L}_{12} & (T - \frac{\beta T^2}{2})I_m \\ -\alpha T\mathcal{L}_{21} & I_{(n-m)} - \alpha T\mathcal{L}_{22} & \mathbf{0} \\ -\alpha T\mathcal{L}_{11} & -\alpha T\mathcal{L}_{12} & (1-\beta T)I_m \end{bmatrix}$$

and the definitions of $\mathcal{L}_{11}, \mathcal{L}_{12}, \mathcal{L}_{21}, \mathcal{L}_{22}$ are as follows:

$$\mathcal{L} = \begin{bmatrix} \mathcal{L}_{11} & \mathcal{L}_{12} \\ \mathcal{L}_{21} & \mathcal{L}_{22} \end{bmatrix}$$

with

$$\mathcal{L}_{11} = \begin{bmatrix} l_{11} & \dots & l_{1m} \\ \vdots & \dots & \vdots \\ l_{m1} & \cdots & l_{mm} \end{bmatrix}$$

$$\mathcal{L}_{12} = \begin{bmatrix} l_{1(m+1)} & \dots & l_{1n} \\ \vdots & \dots & \vdots \\ l_{m(m+1)} & \cdots & l_{mn} \end{bmatrix}$$

$$\mathcal{L}_{21} = \begin{bmatrix} l_{(m+1)1} & \dots & l_{(m+1)m} \\ \vdots & \dots & \vdots \\ l_{n1} & \cdots & l_{nm} \end{bmatrix}$$

$$\mathcal{L}_{22} = \begin{bmatrix} l_{(m+1)(m+1)} & \dots & l_{(m+1)n} \\ \vdots & \dots & \vdots \\ l_{n(m+1)} & \cdots & l_{nn} \end{bmatrix}.$$

Definition 1 The heterogeneous multi-agent systems (7) is said to reach consensus if for any initial conditions, we have

$$\lim_{k\to\infty} \|x_i[k] - x_j[k]\| = 0, \forall i, j \in S_n,$$
$$\lim_{k\to\infty} \|v_i[k] - v_j[k]\| = 0, \forall i, j \in S_m.$$

3 Consensus Analysis

In this section, we will analyze the consensus problem of the multi-agent system. First we will convert the multi-agent system into a reduced-order system by a system transformation. Let $y_i = x_i - x_1$, $i = 2, \dots, m$, $y_j = x_j - x_{m+1}$, $j = m+2, \dots, n$, $z_i = v_i - v_1$, $i = 2, \dots, m$. Denote

$$Y = [y_2, \dots, y_m, y_{m+2}, \dots, y_n, z_2, \dots, z_m]^T.$$

One has

$$Y[k+1] = \Xi_2 Y[k] \tag{8}$$

where

$$\Xi_2 = \begin{bmatrix} I_{m-1} - \frac{\alpha T^2}{2}\overline{\mathcal{L}}_{11} & -\frac{\alpha T^2}{2}\overline{\mathcal{L}}_{12} & (T - \frac{\beta T^2}{2})I_{m-1} \\ -\alpha T\overline{\mathcal{L}}_{21} & I_{(n-m-1)} - \alpha T\overline{\mathcal{L}}_{22} & \mathbf{0} \\ -\alpha T\overline{\mathcal{L}}_{11} & -\alpha T\overline{\mathcal{L}}_{12} & (1-\beta T)I_{m-1} \end{bmatrix}$$

and the definitions of $\overline{\mathcal{L}}_{11}, \overline{\mathcal{L}}_{12}, \overline{\mathcal{L}}_{21}, \overline{\mathcal{L}}_{22}$ are as follows:

$$\overline{\mathcal{L}} = \begin{bmatrix} \overline{\mathcal{L}}_{11} & \overline{\mathcal{L}}_{12} \\ \overline{\mathcal{L}}_{21} & \overline{\mathcal{L}}_{22} \end{bmatrix}$$

with

$$\overline{\mathcal{L}}_{11} = \begin{bmatrix} l_{22} - l_{12} & \dots & l_{2m} - l_{1m} \\ \vdots & \dots & \vdots \\ l_{m2} - l_{12} & \cdots & l_{mm} - l_{1m} \end{bmatrix}$$

$$\overline{\mathcal{L}}_{12} = \begin{bmatrix} l_{2(m+2)} - l_{1(m+2)} & \cdots & l_{2n} - l_{1n} \\ \vdots & \dots & \vdots \\ l_{m(m+2)} - l_{1(m+2)} & \cdots & l_{mn} - l_{1n} \end{bmatrix}$$

$$\overline{\mathcal{L}}_{21} = \begin{bmatrix} l_{(m+2)2} - l_{(m+1)2} & \dots & l_{(m+2)m} - l_{(m+1)m} \\ \vdots & \dots & \vdots \\ l_{n2} - l_{(m+1)2} & \cdots & l_{nm} - l_{(m+1)m} \end{bmatrix}$$

$$\overline{\mathcal{L}}_{22} = \begin{bmatrix} l_{(m+2)(m+2)} - l_{(m+1)(m+2)} & \cdots & l_{(m+2)n} - l_{(m+1)n} \\ \vdots & \dots & \vdots \\ l_{n(m+2)} - l_{(m+1)(m+2)} & \cdots & l_{nn} - l_{(m+1)n} \end{bmatrix}.$$

Before giving the main results, the following assumptions and lemma are necessary.

Assumption 1 The interaction topology $\mathcal{G}$ has a directed spanning tree.

Assumption 2 If the $(m+1)$th agent can obtain the information of the kth agent, then the jth agent can obtain the information of the kth agent $k \in \{2, \dots, m\}$, $j \in \{m+2, \dots, n\}$, i.e., $\overline{\mathcal{L}}_{21} = 0$.

Lemma 1 [10] *If the graph $\mathcal{G}$ has a directed spanning tree, then all the eigenvalues of $\overline{\mathcal{L}}_{11}$ and $\overline{\mathcal{L}}_{22}$ associated with $\mathcal{G}$ have positive real parts.*

Theorem 1 *Under the Assumptions 1 and 2, the heterogeneous multi-agent system (7) can achieve consensus if α, β and T satisfy the conditions:*

$$(i) \quad \alpha < \min_{i=1,\dots,n-m-1} \left\{ \frac{2Re(\eta_i)}{T|\eta_i|^2} \right\}$$

$$(ii) \quad f_j(\alpha, \beta, \mu_j, T) < 256, \; j = 1, \dots, m-1 \tag{9}$$

where

$$f_j(\alpha, \beta, \mu_j, T) = (\alpha T^2 Re\mu_j + 2\beta T - 4)^2 \pm \alpha T^2 Im\mu_j (2\sqrt{m_{2j}^2 + 4m_{1j}} + 2m_{2j})^{\frac{1}{2}}$$
$$\mp (\alpha T^2 Re\mu_j + 2\beta T - 4)(2\sqrt{m_{2j}^2 + 4m_{1j}} - 2m_{2j})^{\frac{1}{2}}$$

$$+\sqrt{m_{2j}^2+4m_{1j}+\alpha^2T^4Im^2\mu_j}$$
$$m_{1j}=\alpha^2T^4(2\beta T+\alpha T^2Re\mu_j-8)^2Im^2\mu_j$$
$$m_{2j}=4\beta^2T^2+4\alpha T^2(\beta T-4)Re\mu_j+\alpha^2T^4Im^2\mu_j,$$

and μ_j is the jth eigenvalue of $\overline{\mathcal{L}}_{11}$, η_i is the ith eigenvalue of $\overline{\mathcal{L}}_{22}$.

Proof It followers from the aforementioned discussion that the heterogeneous multi-agent system (7) can achieve consensus if and only if the reduced system (8) is asymptotically stable. Hence, we only need to prove that $\rho(\Xi_2)<1$, Ξ_2 is as defined in (8). It is not difficult to see that the matrix Ξ_2 can be decomposed into

$$\Xi_2=I_{n+m-3}+\Xi_3 \tag{10}$$

with

$$\Xi_3=\begin{bmatrix}-\frac{\alpha T^2}{2}\overline{\mathcal{L}}_{11} & -\frac{\alpha T^2}{2}\overline{\mathcal{L}}_{12} & (T-\frac{\beta T^2}{2})I_{m-1}\\ -\alpha T\overline{\mathcal{L}}_{21} & -\alpha T\overline{\mathcal{L}}_{22} & \mathbf{0}\\ -\alpha T\overline{\mathcal{L}}_{11} & -\alpha T\overline{\mathcal{L}}_{12} & -\beta TI_{m-1}\end{bmatrix}.$$

Then we have $\lambda(\Xi_2)=1+\lambda(\Xi_3)$. Now we will calculate the eigenvalue of matrix Ξ_3. Let $|\lambda I_{n+m-3}-\Xi_3|=0$. That is

$$\begin{vmatrix}\lambda I_{m-1}+\frac{\alpha T^2}{2}\overline{\mathcal{L}}_{11} & \frac{\alpha T^2}{2}\overline{\mathcal{L}}_{12} & -(T-\frac{\beta T^2}{2})I_{m-1}\\ \alpha T\overline{\mathcal{L}}_{21} & \lambda I_{n-m-1}+\alpha T\overline{\mathcal{L}}_{22} & \mathbf{0}\\ \alpha T\overline{\mathcal{L}}_{11} & \alpha T\overline{\mathcal{L}}_{12} & (\lambda+\beta T)I_{m-1}\end{vmatrix}=\mathbf{0} \tag{11}$$

Let

$$P_1=\begin{bmatrix}I_{m-1} & \mathbf{0} & -\frac{T}{2}I_{m-1}\\ \mathbf{0} & I_{n-m-1} & \mathbf{0}\\ \mathbf{0} & \mathbf{0} & I_{m-1}\end{bmatrix},\ Q_1=\begin{bmatrix}I_{m-1} & \mathbf{0} & \frac{T}{2}I_{m-1}\\ \mathbf{0} & I_{n-m-1} & \mathbf{0}\\ \mathbf{0} & \mathbf{0} & I_{m-1}\end{bmatrix},$$
$$Q_2=\begin{bmatrix}I_{m-1} & \mathbf{0} & \mathbf{0}\\ \mathbf{0} & I_{n-m-1} & \mathbf{0}\\ \frac{\lambda}{T}I_{m-1} & \mathbf{0} & I_{m-1}\end{bmatrix},\ Q_3=\begin{bmatrix}\mathbf{0} & \mathbf{0} & I_{m-1}\\ \mathbf{0} & I_{n-m-1} & \mathbf{0}\\ I_{m-1} & \mathbf{0} & \mathbf{0}\end{bmatrix}.$$

It follows that $|P_1(\lambda I_{n+m-3}-\Xi_3)Q_1Q_2Q_3|=0$. Then, by some computations, one has

$$\begin{vmatrix}-TI_{m-1} & \mathbf{0} & \mathbf{0}\\ \mathbf{0} & \lambda I_{n-m-1}+\alpha T\overline{\mathcal{L}}_{22} & \mathbf{0}\\ (\lambda+\beta T)I_{m-1}+\frac{\alpha T^2}{2}\overline{\mathcal{L}}_{11} & \alpha T\overline{\mathcal{L}}_{12} & W_1\end{vmatrix}=\mathbf{0} \tag{12}$$

where

$$W_1=\alpha T\overline{\mathcal{L}}_{11}+\frac{\lambda}{T}[\frac{\alpha T^2}{2}\overline{\mathcal{L}}_{11}+(\lambda+\beta T)I_{m-1}].$$

Then we have

$$\left|(\lambda I_{n-m-1} + \alpha T\overline{\mathcal{L}}_{22})\right| \left|(\frac{\lambda^2}{T} + \lambda\beta)I_{m-1} + (\frac{\alpha T}{2}\lambda + \alpha T)\overline{\mathcal{L}}_{11}\right| = 0. \quad (13)$$

Denote the jth eigenvalue of $\overline{\mathcal{L}}_{11}$ as μ_j $(j = 1, \ldots, m-1)$ and the ith eigenvalue of $\overline{\mathcal{L}}_{22}$ as η_i $(i = 1, \ldots, n-m-1)$, respectively. According to the Lemma 1, one has $Re\mu_j > 0$, and $Re\eta_i > 0$. Then we have

$$\prod_{i=1}^{n-m-1} (\lambda + \alpha T\eta_i) \prod_{j=1}^{m-1} [\frac{1}{T}\lambda^2 + \beta\lambda + (\frac{\alpha T}{2}\lambda + \alpha T)\mu_j] = 0. \quad (14)$$

It follows that

$$\lambda + \alpha T\eta_i = 0, i = 1, \ldots, n-m-1 \quad (15)$$
$$2\lambda^2 + (2\beta T\lambda + \alpha T^2\mu_j)\lambda + 2\alpha T^2\mu_j = 0, j = 1, \ldots, m-1. \quad (16)$$

In order to guarantee $|\lambda(\Xi_2)| < 1$, it follows from (15) that

$$|1 - \alpha T\eta_i| < 1, i = 1, \ldots, n-m-1.$$

That is

$$(1 - \alpha TRe\eta_i)^2 + \alpha^2T^2Im^2\eta_i < 1. \quad (17)$$

Note $Re\eta_i > 0$, $\alpha, \beta > 0$, one has $\alpha < \min_{j=1,\ldots,n-m-1}\left\{\frac{2Re(\eta_i)}{T|\eta_i|^2}\right\}$, which is the condition (i) in Theorem 1.

It follows from (16) that

$$|4 - 2\beta T - \alpha T^2\mu_j \pm \sqrt{(2\beta T + \alpha T^2\mu_j)^2 - 16\alpha T^2\mu_j}| < 16, j = 1, \ldots, m-1. \quad (18)$$

Let $\sqrt{(2\beta T + \alpha T^2\mu_j)^2 - 16\alpha T^2\mu_j} = c_j + id_j$, here i is the imaginary unit. By some computations, one has

$$\begin{cases} c_j^2 - d_j^2 = 4\beta^2T^2 + 4\alpha T^2(\beta T - 4)Re\mu_j + \alpha^2T^4Im^2\mu_j \\ c_jd_j = \alpha T^2(2\beta T + \alpha T^2Re\mu_j - 8)Im\mu_j \end{cases}. \quad (19)$$

Denote

$$m_{1j} = \alpha^2T^4(2\beta T + \alpha T^2Re\mu_j - 8)^2Im^2\mu_j$$
$$m_{2j} = 4\beta^2T^2 + 4\alpha T^2(\beta T - 4)Re\mu_j + \alpha^2T^4Im^2\mu_j.$$

It is not difficult to obtain that

$$\begin{cases} c_j^2 = \frac{\sqrt{m_{2j}^2+4m_{1j}}+m_{2j}}{2} \\ d_j^2 = \frac{\sqrt{m_{2j}^2+4m_{1j}}-m_{2j}}{2} \end{cases}. \tag{20}$$

It follows from (18) that

$$(\alpha T^2 Re\mu_j + 2\beta T - 4)^2 \mp 2c(\alpha T^2 Re\mu_j + 2\beta T - 4)$$
$$+\alpha^2 T^4 Im^2\mu_j \pm 2d\alpha T^2 Im\mu_j + c^2 + d^2 < 256 \tag{21}$$

which combine with (20) imply the conditions (ii) in Theorem 1. This completes the proof.

Remark 1 Theorem 1 provides a sufficient condition for multi-agent system (7) to achieve consensus. The consensus algorithms employed in this paper can be found in some existing literature. However, we consider the heterogeneous networked system with sampled-data, which is different from the results in existing literature. The results are based on algebraic graph theory and linear system theory. It gives how the control gains and the interaction topology affect the consensus of the system.

4 Simulations

In this section, we will give an example to show the effectiveness of the presented results. For simplicity, suppose that $a_{ij} = 1$ if $(i, j) \in \mathcal{E}$, otherwise $a_{ij} = 0$.

Suppose that the multi-agent systems consist of 8 agents. According to the Assumptions 1 and 2, we choose the interaction topology for system (8) as Fig. 1. Let $T = 0.1s$. According to the conditions in Theorem 1, we obtain a feasible $\alpha = 0.32$, $\beta = 0.67$. Then we get the trajectories of the agents as shown in Figs. 2 and 3. Which are consistent with the theoretical results.

Fig. 1 Topology $\mathcal{G}$

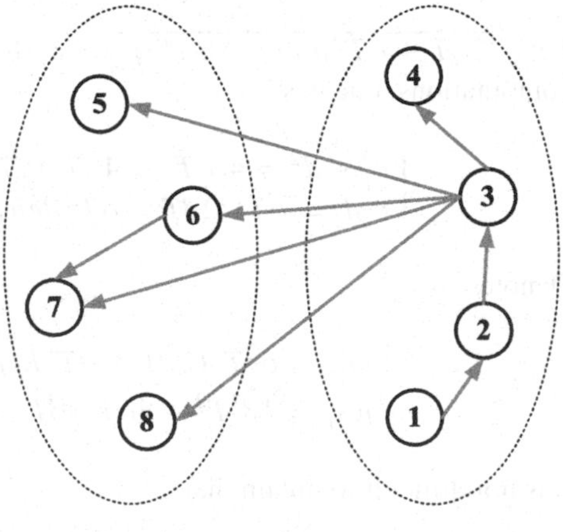

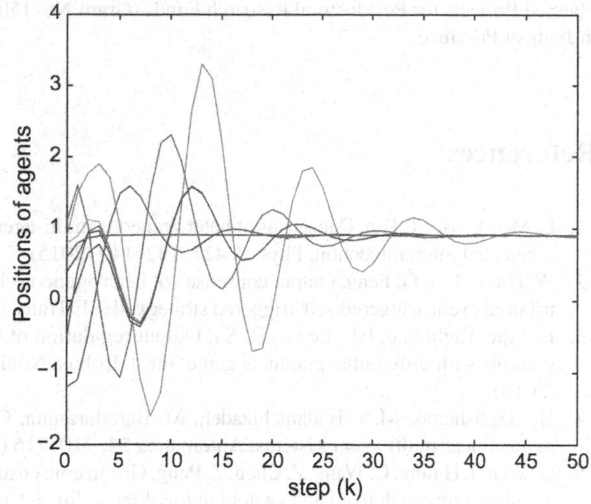

Fig. 2 Position trajectories of case 1

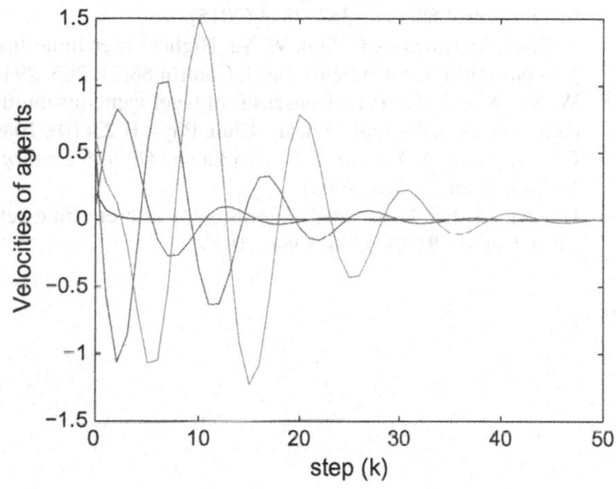

Fig. 3 Velocity trajectories of case 1

5 Conclusions

In this paper, the consensus problem for a class of heterogeneous multi-agent system has been considered. The continuous-time system has been converted into the discrete-time system by sampling. Then the multi-agent system has been transformed to a reduced-order error system. A sufficient condition for the multi-agent system to achieve consensus has been obtained. A simulation example has been given to show the usefulness of the obtained results.

Acknowledgements This work was supported by the Natural Science Foundation of Jiangsu Province of China (Grant No. BK20151290), China Postdoctoral Science Foundation (Grant No. 2015M581699), Six Talent Peaks Project in Jiangsu Province (Grant No. 2015-DZXX-029), Jiangsu

Planned Projects for Postdoctoral Research Funds (Grant No. 1501040B) and "333 Talent Project" in Jiangsu Province.

References

1. L. Mo, Y. Niu, T. Pan, Consensus of heterogeneous multi-agent systems with switching jointly-connected interconnection. Phys. A **427**, 132–140 (2015)
2. W. Hu, L. Liu, G. Feng, Output consensus of heterogeneous linear multi-agent systems by distributed event-triggered/self-triggered strategy. IEEE Trans. Cybern. **47**(8), 1914–1924 (2017)
3. F. Adib Yaghmaie, F.L. Lewis, R. Su, Output regulation of heterogeneous linear multi-agent systems with differential graphical game. Int. J. Robust Nonlinear Control **26**(10), 2256–2278 (2016)
4. H. Haghshenas, M.A. Badamchizadeh, M. Baradarannia, Containment control of heterogeneous linear multi-agent systems. Automatica **54**, 210–216 (2015)
5. G. Wen, J. Huang, C. Wang, Z. Chen, Z. Peng, Group consensus control for heterogeneousmulti-agent systems with fixed and switching topologies. Int. J. Control **89**(2), 259–269 (2016)
6. Y. Zheng, L. Wang, A novel group consensus protocol for heterogeneous multi-agent systems. Int. J. Control **88**(11), 2347–2353 (2015)
7. Y. Zhou, Y. Xinghuo, C. Sun, W. Yu, Higher order finite-time consensus protocol for heterogeneous multi-agent systems. Int. J. Control **88**(2), 285–294 (2015)
8. W. Na, W. Zhi-Hai, P. Li, Consensus of heterogeneous multi-agent systems based on sampled data with a small sampling delay. Chin. Phys. B **23**(10), 108901-1-9 (2014)
9. G.F. Franklin, J.D. Powell, M. Workman, *Digital Control of Dynamic Systems* (Addison-Wesley, Reading, MA, 2006)
10. H. Zhao, S. Fei, Distributed consensus for discrete-time heterogeneous multi-agent systems. Int. J. Control **91**(6), 1376–1384 (2018)

Research on Location Strategy of Multi-mobile Robots Based on Gaussian Plume Model

Liangping Shi, Huanyu Zhao and Xi Zhang

Abstract This paper simulates the concentration distribution of leakage gas in the leakage area based on the Gaussian plume model which is more suitable for continuous point source diffusion. The research about multi-mobile robots track and locate leak source is completed by using formation control algorithm, concentration gradient method and headwind search method. The simulation results show that the Gaussian plume model can directly simulate the plume distribution of the leaking gas, the multi-mobile robots can be controlled by formation, and the leakage point can be tracked and located by using the concentration gradient method and the headwind search method.

Keywords Gaussian plume model · Multi-mobile robots · Tracking and locating Formation control · Headwind search method

1 Introduction

Recent years, there have been a lot of researches on the tracking and locating of odor source by the way where mobile robot detects odor plume. In the early stage, Jatmiko [1] and Marques [2] carried out the simulation study on the search and location of odor source based on the plume model of Farrell [3] and Nielsen [4] by using mobile robot, respectively. Cui et al. [5] used a continuous plume model approximate to Gauss distribution to search and locate the odor source by using mobile robot. Because of its ability to work together and share information with each other, multi-mobile robots can not only expand the range of information perception but also improve work efficiency. Hence, it has great research and application value in odor source search task. Hayes [6] used multi-mobile robots to study the search and location of odor source by combining spiral search method with headwind search method. Then Loutfi et al. [7] joined the visual information to make more detailed study about the

L. Shi · H. Zhao (✉) · X. Zhang
Faculty of Automation, Huaiyin Institute of Technology, Huai'an 223001, Jiangsu, China
e-mail: hyzhao@163.com

Y. Jia et al. (eds.), *Proceedings of 2018 Chinese Intelligent Systems Conference*, Lecture Notes in Electrical Engineering 529,
https://doi.org/10.1007/978-981-13-2291-4_10

applications in this field. Ishida et al. [8, 9] studied the location of odor source by switching concentration gradient method and along wind search method. Balch [10] combined the formation behavior method with the navigation behavior method to complete the research of locating the target point by using multi-mobile robots in a certain formation. In our country, Meng [11] combined ant colony algorithm with headwind search method, and used Farrel model to study the process of searching for odor source by multi-mobile robots in turbulent plume environment.

Based on the above research results, the Gaussian plume model which is more suitable for continuous point source diffusion is used to simulate the concentration distribution of leakage gas in the leakage area. Tracking and positioning strategy of multi-mobile robots is proposed which is more suitable for this model. In other words, the simulation study of the plume will be completed by combining formation control algorithm, concentration gradient method and headwind search method.

2 Establishment of Gaussian Plume Model

The Gaussian plume model simulates the concentration distribution of the gas diffusing along the downwind direction of the leakage source. The establishment of the model requires that the stability of the atmosphere and the magnitude and direction of the wind speed do not change with the change of time and location during the diffusion process, and the total reflection of the gas is required. The model satisfies the following formula:

$$C(x, y, z, H) = \frac{Q}{2\pi u \sigma_y \sigma_z} \exp\left(-\frac{y^2}{2\sigma_y^2}\right)\left[\exp\left(-\frac{(z-H)^2}{2\sigma_Z^2}\right) + \exp\left(-\frac{(z+H)^2}{2\sigma_z^2}\right)\right] \tag{1}$$

where, $C(x, y, z, H)$ represents gas diffusion concentration, the unit is mg/m^3; Q represents release rate (source strength), the unit is mg/s; H represents the height of the leak source from the ground, the unit is m; u represents average wind speed, the unit is m/s; σ_y represents horizontal diffusion coefficient; σ_z represents vertical diffusion coefficient. $\sigma_y = \gamma_1 x^{\alpha_1}, \sigma_z = \gamma_2 x^{\alpha_2}$, where, $\gamma_1, \gamma_2, \alpha_1, \alpha_2$ can be found from the table of relationship between atmospheric stability and diffusion parameters. If z is set as $z = 0$, the following formula will be obtained based on concentration distribution of the leakage gas which be detected on the ground:

$$C(x, y, 0, H) = \frac{Q}{\pi u \sigma_y \sigma_z} \exp\left(-\frac{y^2}{2\sigma_y^2} - \frac{H^2}{2\sigma_z^2}\right) \tag{2}$$

After the leakage happened, the multi-mobile robots located in the leakage area can easily measure the wind speed, atmospheric stability, concentration and coordinate information of the current environment. From the above formula, if we want to obtain a complete plume diffusion model, we will need to know the source strength of the leakage source and the height of the leakage point relative to the ground that can not be directly measured. Only two groups of measured quantities can be obtained to calculate the values of H and Q, then, the diffusion model of the plume can be obtained. In this paper, the concentration values and coordinates of two mobile robots located in different positions in the leakage region are measured to determine the values of H and Q.

Assuming that two mobile robots can detect the plume and the positions on the ground which can be described as (x_1, y_1) and (x_2, y_2), the measured concentrations can be expressed as follows:

$$C_1(x_1, y_1, 0, H) = \frac{Q}{\pi u \sigma_{y_1} \sigma_{z_1}} \exp\left(-\frac{y_1^2}{2\sigma_{y_1}^2} - \frac{H}{2\sigma_{z_1}^2}\right) \tag{3}$$

$$C_2(x_2, y_2, 0, H) = \frac{Q}{\pi u \sigma_{y_2} \sigma_{z_2}} \exp\left(-\frac{y_2^2}{2\sigma_{y_2}^2} - \frac{H^2}{2\sigma_{z_2}^2}\right) \tag{4}$$

According to (3) and (4), the computational formula of H can be concluded as follows:

$$H = \sqrt{\frac{2\sigma_{z_1}^2 \sigma_{z_2}^2}{\sigma_{z_1}^2 - \sigma_{z_2}^2}\left[\ln \frac{\sigma_{y_1}\sigma_{z_1} C_1(x_1, y_1, 0, H)}{\sigma_{y_2}\sigma_{z_2} C_2(x_2, y_2, 0, H)} - \frac{y_2^2\sigma_{y_1}^2 - y_1^2\sigma_{y_2}^2}{2\sigma_{y_1}^2\sigma_{y_2}^2}\right]} \tag{5}$$

We can calculate the value of H by substituting formula (5) with the measured known quantities. Then the value of Q can be obtained by substituting formula (3) or (4) with the value of H.

3 Gas Diffusion Concentration Distribution in Leakage Region

In this paper, the concentration distribution of leakage gas is simulated by selecting a 100 m × 100 m two-dimensional plane as the leakage region. Taking chlorine leakage as an example, the concentration distribution of chlorine diffusion was modeled by Gaussian plume model, and visualized simulation by MATLAB. The atmospheric stability is selected as category D. The values of the standard variance σ_y and σ_z of category D can be obtained from the table of the relationship between the degree of stability and the diffusion parameters. The table is accessible in the handbook of environmental evaluation criteria. σ_y and σ_z can be expressed as:

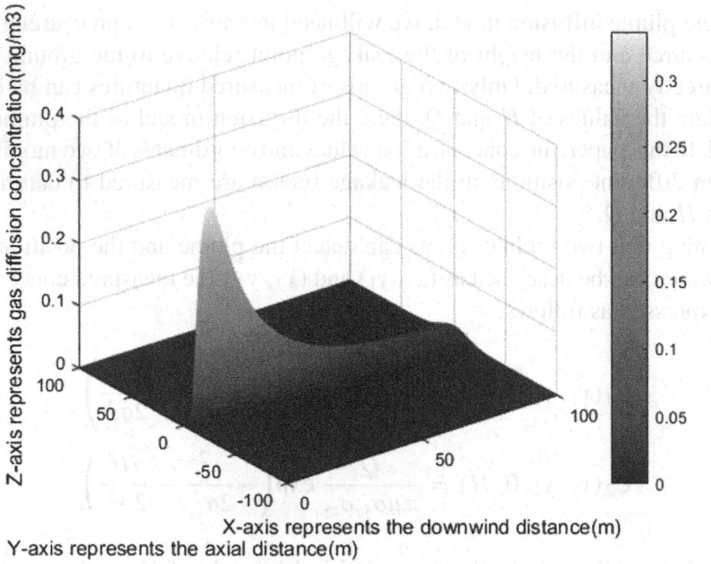

Fig. 1 Three-dimensional diagram of simulating chlorine leakage concentration distribution by gaussian plume model

$$\begin{cases} \sigma_y = 0.209x^{0.897} \\ \sigma_z = 0.22x^{0.8} \end{cases} \tag{6}$$

In order to simplify the difficulty of the simulation calculation, H and Q can be set as known quantities, where $H = 2\,\text{m}$ and $Q = 10\,\text{mg/m}^3$. Current wind speed u can be set as $u = 1.5\,\text{m/s}$. So, the distribution of chlorine concentration in the leaking area can be obtained. Figure 1 shows a three-dimensional pattern of chlorine diffusion concentration distribution simulated by Gaussian plume model. Figure 2 shows the diffusion of chlorine concentration based on the surface level. Setting $z = 0$ and $y = 0$, then the distribution of chlorine concentration on the basis of the downwind diffusion of x-axis shown in Fig. 3 is obtained. The parameter value of the maximum concentration point of the leaking gas in the whole diffusion process can be seen directly from Fig. 3. The coordinate of maximum concentration point which is a specific position in three dimensional space can be expressed as $T(9.9, 0, 0.33653)$.

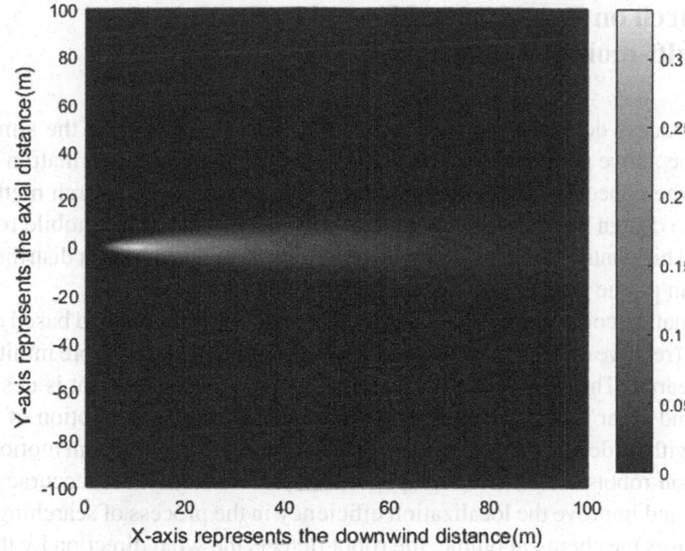

Fig. 2 Two-dimensional diagram of chlorine diffusion based on the ground

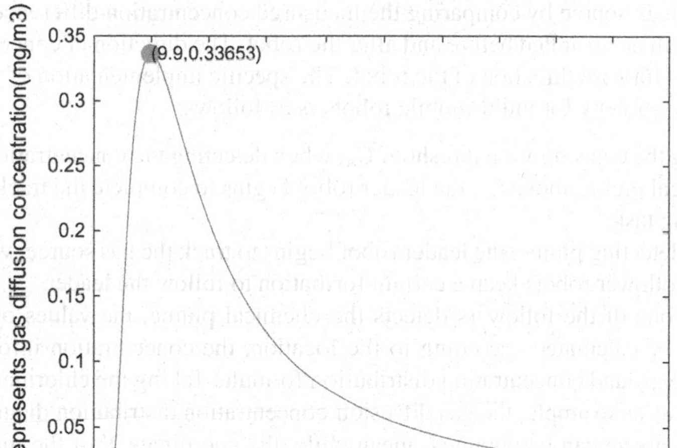

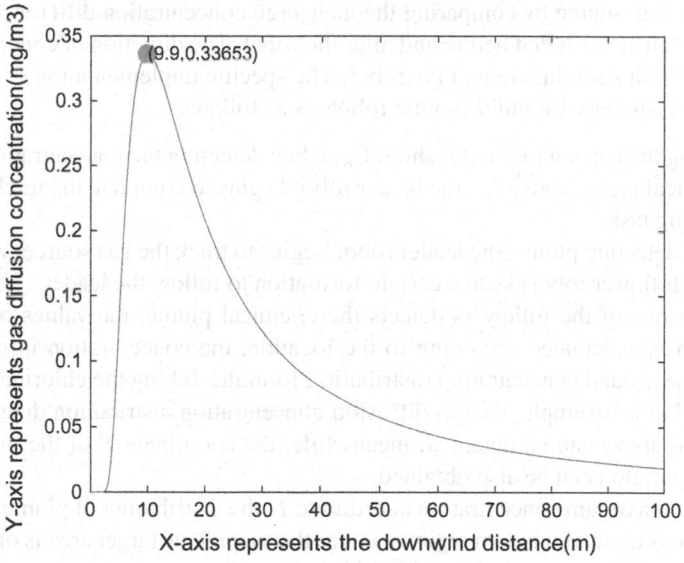

Fig. 3 Concentration distribution of chlorine along the downwind direction of x-axis

4 Research on Tracking and Location Strategy of Multi-mobile Robots

The total process does not consider obstacle avoidance. Based on the simulation results of the above plume model, the methods which include the formation control algorithm, the concentration gradient method and the headwind search method are combined to design the tracking and location strategy of the multi-mobile robots in this paper. This control strategy is more suitable for the concentration distribution of the Gaussian plume model.

The formation control algorithm adapts the leader-follower method based on $l-\varphi$ parameters (relative distance and relative azimuth angle), which is more intuitive and easy to research. The realization ideas are as follows: a certain robot is designated as leader, and other robots are used as its followers to keep track motion of desired formation with leader. Compared with the group robots with no uniform motion rules, the formation robots increase the cooperation, systematization and accuracy of the search task and improve the localization efficiency in the process of searching target. When it senses the chemical plume, the robot detects the wind direction by the wind direction sensor which makes the robot track the leakage source against the wind until the leak source is found. The law of concentration gradient is that the robot tracks and locates the leak source by comparing the measured concentration difference of two sensors which are installed before and after the robot. The direction of concentration difference is forward direction of the robot. The specific implementation of tracking and location strategy for multi-mobile robots is as follows:

(1) Setting the concentration threshold C_ε, when detecting the concentration of the chemical plume above C_ε, the leader robot begins to complete the tracking and locating task.
(2) After detecting plume, the leader robot begins to track the gas source, while the other follower robots keep a certain formation to follow the leader.
(3) When one of the followers detects the chemical plume, the values of H and Q can be calculated according to the location, the concentration information and the ground concentration distribution formula. Taking the chlorine leakage model as an example, the gas diffusion concentration distribution diagram discussed above can be obtained, meanwhile, the coordinate T of the maximum concentration can be also obtained.
(4) At the maximum concentration coordinate T, the distribution of plume concentration is divided into two regions where the suspicious target area is on the left and the target search area is on the right.
(5) In the tracking process, the leader continuously detects the concentration information of the leaking gas. When the detected concentration increases, the robot is judged to be located in the target search area. In this case, the robot uses the concentration gradient method to track the chemical plume until the concentration reaches the maximum. When the detected concentration decreases, the leader robot is judged to be in the suspicious target area and uses the headwind search method to further track the plume until it does not detect the plume in

the direction of the headwind. That is, when the detected concentration below C_{ε}, multi-mobile robots stop in a certain formation near the leak point which is located at the height H from the ground.

The reason why the Gaussian plume model is divided into two regions and different search methods are used to track the plume is that compared with other plume models which take the point near maximum gas concentration as emission point, it improves the accurate of tracking. When simulating the elevated leakage source, the model of gas projected to the ground has a process of growing out of nothing near the odor source. Sometimes the short process has little effect on the positioning accuracy, and sometimes the simulated model is farther away from the actual odor source. If the leakage source is determined by locating the emission point near the maximum concentration of leakage gas, the location will be inaccurate. Figure 2 shows that the maximum concentration point of the gas has a deviation of nearly 10 meters from the actual leakage source. But wind direction is more stable in this range because it is close to the leaking gas. At this time, the headwind search method is used to improve the efficiency and accuracy of the location. Because of the complexity of the environment, the concentration gradient method is more accurate and effective than the headwind search in the target search area.

Suppose that the sampling position of the robot be C and define $C_{\max}$ as the maximum concentration. n indicates the position of the current robot, and $n-1$ indicates the sampling position of the previous step of the current robot. C_n is the sampling concentration of the current robot, and C_{n-1} is the sampling concentration of the previous step of the current robot. The selection process of the location method can be summarized as follows:

repeat($C_{\varepsilon} < C_n$ and ($C_n > C_{n-1}$ or $C_n < C_{n-1}$))
{if ($C_n > C_{n-1}$)
then

using the headwind search method

else

using concentration gradient method

}

Assume that there are three mobile robots that require a triangular formation to track the leakage source. Based on the chlorine leakage simulation case, we can use the concentration gradient method to search the target as shown in Fig. 4. The schematic diagram of the headwind search for the leakage source is shown in Fig. 5. It can be seen that multi-mobile robots can track and locate the target points after the formation is completed and the desired formation will be maintained.

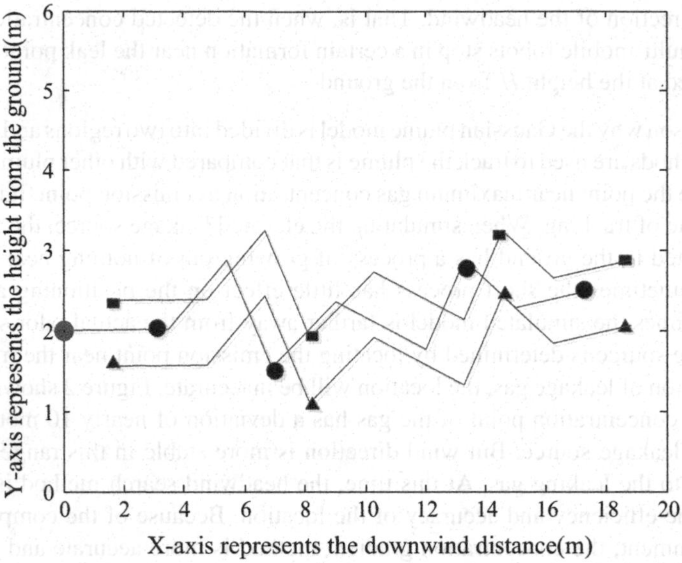

Fig. 4 Schematic diagram of concentration gradient search strategy based on formation behavior

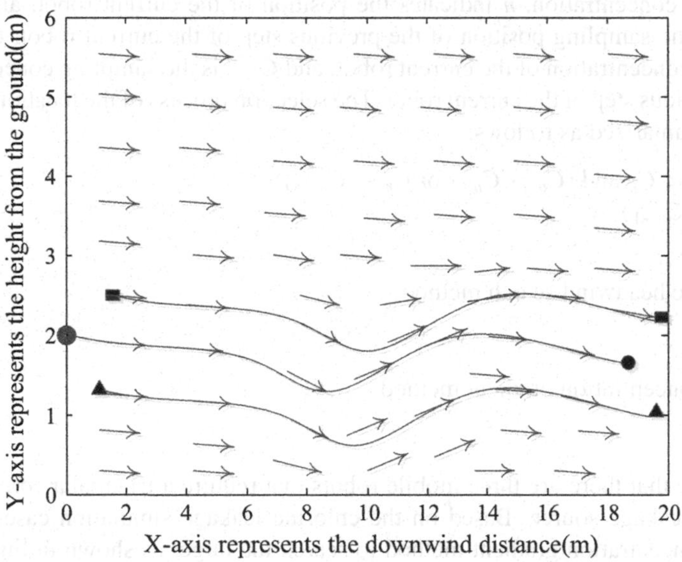

Fig. 5 Schematic diagram of headwind search strategy based on formation behavior

5 Conclusion

Based on the Gaussian plume model, the simulation research on the location of leakage gas plume is completed by combining the formation control algorithm, concentration gradient method and headwind search method. However, in practical applications, there are still many problems to be solved in the process of realizing multi-mobile robots tracking target points at a certain formation which need further study by scholars.

Acknowledgements This work was supported by the Natural Science Foundation of Jiangsu Province of China (Grant No. BK20151290), Six Talent Peaks Project in Jiangsu Province (Grant No. 2015-DZXX-029).

References

1. W. Jatmiko, K. Sekiyama, T. Fukuda, A mobile robot PSO-based for odor source localization in dynamic advection diffusion environment, in *Proceedings of the 2006 IEEE/RSJ International Conference on Intelligence Robots and Systems* (IEEE, Piscataway, USA, 2006), pp. 4527–4532
2. L. Marques, U. Nunes, A.T. Almeida, Particle swarm-based olfactory guided search. Auton. Robot **20**(3), 277–287 (2006)
3. J.A. Farrell, J. Murlis, X.Z. Long, W. Li, R.T. Carde, Filament-based atmospheric dispersion model to achieve short time-scale structure of odor plumes. Environ. Fluid Mech. **2**(1–2), 143–169 (2002)
4. M. Nielsen, P. Chatwin, H.E. Jøgensen, N. Mole, R.J. Munro, S. Ott, Concentration Fluctuations in Gas Releases by Industrial Accidents—Final Report, Technical Report R-1329(EN) (Risø National Laboratory, Denmark, 2002)
5. X. Cui, T. Hardin, R.K. Ragade, A.S. Elmaqhraby, A swarm-based fuzzy logic control mobile sensor network for hazardous contaminants localization, in *Proceedings of IEEE International Conference on Mobile Ad-hoc and Sensor Systems* (IEEE, Piscataway, USA, 2004), pp. 194–203
6. A.T. Hayes, A. Martinoli, R.M. Goodman, Distributed odor source localization. IEEE Sens. J. **2**(3), 260–271 (2002)
7. A. Loutfi, M. Broxvall, S. Coradeschi et al., Object recognition: a new application for smelling robots. Robot. Auton. Syst. **52**, 272–289 (2005)
8. H. Ishida, G. Nakayama, T. Nakamoto et al., Controlling a gas/odor plume-tracking robot based on transient responses of gas sensors, in *Proceedings of the IEEE International Conference on Sensors* (IEEE, Piscataway, NJ, USA, 2002), pp. 1665–1670
9. H. Ishida, G. Nakayama, T. Nakamoto et al., Controlling a gas/Odor Plume-tracking robot based on transient responses of gas sensors. IEEE Sens. J. **5**(3), 537–545 (2005)
10. T. Balch, R.C. Arkin, Behavior-based formation control for multi-robot teams. IEEE Trans. Robot. Autom. **14**(6), 926–939 (1998)
11. Q. Meng, F. Li, M. Zhang et al., Implementation of multi-robot active olfaction in turbulent plume environment. J. Autom. **34**(10), 1281–1290 (2008)

Research on Obstacle Avoidance Control Strategy of Networked Systems Based on Leader-Follower Formation Tracking

Liangping Shi, Huanyu Zhao and Xi Zhang

Abstract This paper studies the obstacle avoidance control of networked system in completing leader-follower formation tracking task. By using methods of the behavior along the wall, the artificial potential field and the hybrid autonomous obstacle avoidance control, the obstacle avoidance control strategy of possible obstacles in different environments is studied. In addition, the cooperative obstacle avoidance control method is used to discuss the obstacle avoidance in specific environment. Finally, the simulation results show that the obstacle avoidance control strategy studied in this paper is flexible and effective.

Keywords Networked systems · Leader-follower formation tracking
Obstacle avoidance

1 Introduction

Recent years, cooperative control technology of networked systems has been widely used, such as multi-agent obstacle avoidance control, dynamic target formation tracking and so on [1]. The scope of research covers all aspects of post-disaster search and rescue, military operations, underwater exploration and so on. The commonly used multi-agent obstacle avoidance methods are artificial potential field method, genetic algorithm, fuzzy logic algorithm and so on [2–4]. The artificial potential field method has the advantages of simple structure, smooth and reliable path, and is more suitable

L. Shi · H. Zhao (✉) · X. Zhang
Faculty of Automation, Huaiyin Institute of Technology, Huai'an 223001, Jiangsu, China
e-mail: hyzhao@163.com

Y. Jia et al. (eds.), *Proceedings of 2018 Chinese Intelligent Systems Conference*, Lecture Notes in Electrical Engineering 529,
https://doi.org/10.1007/978-981-13-2291-4_11

for obstacle avoidance in practical environment [5]. However, there are some limitations of local minimum problem. In reference [6], the artificial potential field method is improved to adapt it better. In this paper, a hybrid autonomous obstacle avoidance control strategy is designed by combining the artificial potential field method and the behavior along the wall method. When it is necessary to avoid obstacle by changing formation in special environment, the control strategy of cooperative obstacle avoidance is adapted.

2 Brief Introduction of Formation Tracking Method

In this paper, the formation model of multi-agent system using leader-follower method [7] is established to study the obstacle avoidance of multi-agent system in formation tracking. In a multi-agent system, one agent is designated as leader, and the rest as followers which keep a certain distance l and angle track φ leader agent [8] when $t \to +\infty$, l and φ converge and remain at a set value, even if $l \to l_d$, $\varphi \to \varphi_d$, the formation can be formed.

3 Obstacle Avoidance Control Strategy

This paper mainly discusses two kinds of obstacle avoidance control strategies. One is hybrid autonomous obstacle avoidance control, and the other is cooperative obstacle avoidance control. These two kinds of obstacle avoidance method can be switched at any time in different states of multi-agent motion and different external environments to achieve the best obstacle avoidance effect.

3.1 Hybrid Autonomous Obstacle Avoidance Control Strategy

Hybrid autonomous obstacle avoidance mainly combines the behavior along the wall with the artificial potential field method. When an agent is trapped in a potential field caused by a local minimum problem, it will stop moving. It is very likely that the communication between agents will be interrupted, and the formation and tracking of formation will be interfered, which makes the task of formation tracking and control impossible. At this time, obstacle avoidance control will switch from the artificial potential field method to the behavior along the wall. After obstacle avoidance is completed, new pose information will be obtained according to the leader-follower formation model, and multi-agent will continue to complete the formation tracking task. The obstacle avoidance control in this paper divides the environment around the obstacle into three regions: I, II and III. According to the distance information returned by the agent to detect the obstacle, the region where the agent is located is

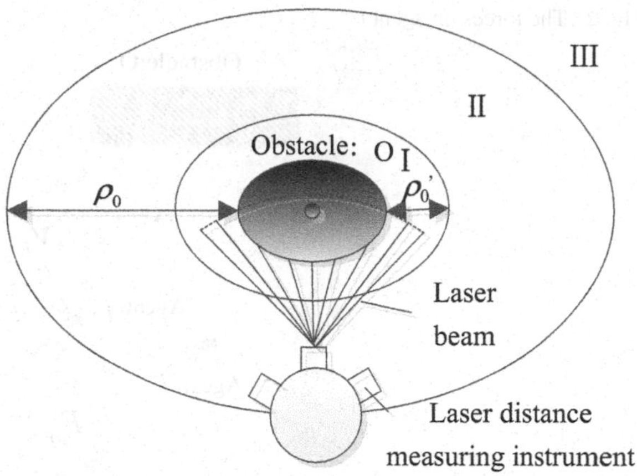

Fig. 1 Regional distribution of obstacle avoidance

judged and the optimal obstacle avoidance mode is chosen. Regional distribution of obstacle avoidance is shown in Fig. 1.

In region I, the agent is close to the obstacle. By setting a dangerous area value ρ_0', when the value returned by the sensor is within the value range of this region, the tracking control command is ignored and the control decision of the behavior along the wall is made directly.

In region II, by setting an area value ρ_0, the artificial potential field method works when the value returned by the sensor lies between ρ_0' and ρ_0. If the agent falls into a local minimum, it will use the behavior along the wall to avoid obstacles. If not, artificial potential field method will be used to avoid them.

In region III, obstacles do not pose a danger to agents, and agents do not need to avoid obstacles and continue to track the target points in formation.

3.1.1 Artificial Potential Field Control Strategy

Multi-agent obstacle avoidance based on artificial potential field is an obstacle avoidance method that is widely used. The gravitational poles and repulsive poles in the potential field represent obstacles and target points respectively, and the corresponding potentials are generated by the potential function around them. The negative gradient direction of the corresponding potential represents the direction of the abstract force acting on the agent. This kind of abstract force causes the agent to bypass obstacles and move towards the target point. Assume that there are only two agents. The forces on the agent i under the influence of artificial potential field are shown in Fig. 2.

When the artificial potential field method is used to avoid obstacles, the potential field produced by the obstacles can be repulsive to the agent. At this time, the following repulsive potential function can be selected as follows:

Fig. 2 The forces on agent i

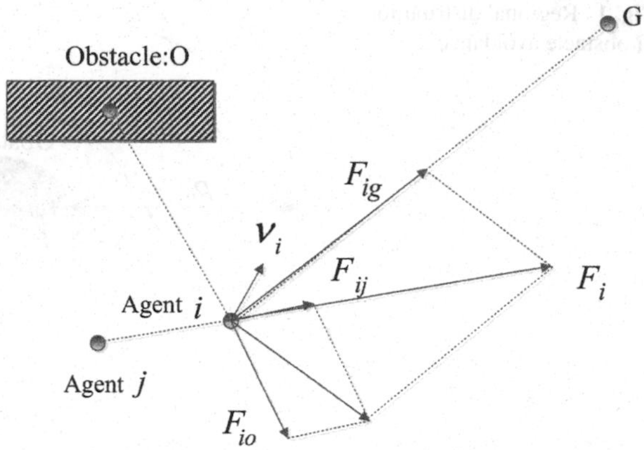

$$U_r(p_i) = \begin{cases} \frac{1}{2}k_r\left(\frac{1}{\|p_i - p_o\|} - \frac{1}{\rho_0}\right)\|p_i - p_g\|^2, & \rho'_0 \le \|p_i - p_o\| \le \rho_0 \\ 0, \|p_i - p_o\| > \rho_0 \end{cases} \quad (1)$$

Among that, p_i, p_o and p_g meaning $p_i(x_i, y_i)$, $p_o(x_o, y_o)$ and $p_g(x_g, y_g)$ represent the positions of the agent i, the obstacle o and the goal point G respectively. k_r represents the repulsive force coefficient, ρ_0 represents the distance of which the obstacle o affects the agent i. Then the repulsive force of the obstacle o acting on the agent i can be expressed as:

$$\begin{aligned} F_r &= -grad[U_r(p_i)] \\ &= \begin{cases} \frac{1}{2}k_r\|p_i - p_o\|^{-2}\|p_i - p_g\|^2 - k_o\left(\frac{1}{\|p_i - p_o\|} - \frac{1}{\rho_0}\right)\|p_i - p_g\|, & \rho'_0 \le \|p_i - p_o\| \le \rho_0 \\ 0, \|p_i - p_o\| > \rho_0 \end{cases} \end{aligned} \quad (2)$$

The potential field generated by the target point has a gravitational effect on the agent. The gravitational potential function is chosen as follows:

$$U_g(p_i) = \frac{1}{2}k_g\|p_i - p_g\|^2 \quad (3)$$

Among that, k_g represents gravitational force coefficient. Then the gravitational force of the target point acting on the agent i can be expressed as:

$$F_g = -grad[U_g(p_i)] = -k_g\|p_i - p_g\| \quad (4)$$

For the potential field force between multi-agent, sometimes the force of the agent i that is exerted by the agent j can be gravitational force, sometimes it is repulsive force. The relative gravitational potential function between agents can be expressed as:

$$U_{gij} = k_g \left(\frac{l_d^2}{\|\rho_{ij}\|^2} + \log\|\rho_{ij}\|^2 \right) \tag{5}$$

Then the gravitational force of agent j on agent i can be expressed as:

$$F_{gij} = -grad(U_{gij}) = -k_g \left(-\frac{2l_d^2}{\|\rho_{ij}\|^3} + \frac{2}{\|\rho_{ij}\|} \right) \tag{6}$$

l_d represents the expected distance between agents, and ρ_{ij} represents the actual distance between agent i and agent j.

The relative repulsion potential function between agents can be expressed as:

$$U_{rij} = k_r \left(\frac{1}{\rho_{ij}} - \frac{1}{l_d} \right) \|p_i - p_g\|^2 \tag{7}$$

Then the repulsive force of the agent j on the agent i can be expressed as:

$$F_{rij} = -grad(U_{rij}) = k_r \rho_{ij}^{-2} \|p_i - p_g\|^2 - 2k_r \left(\frac{1}{\rho_{ij}} - \frac{1}{l_d} \right) \|p_i - p_g\| \tag{8}$$

The potential function between agents acts as gravitational force or repulsive force, which mainly satisfies the following condition: $F_{ij} = \begin{cases} F_{gij}, \ \rho_{ij} \geq l_d \\ F_{rij}, \ \rho_{ij} < l_d \end{cases}$.

The resultant force on agent i in the potential field can be expressed as follows:

$$F_i = F_r + F_g + F_{ij} \tag{9}$$

When there are multiple follower agents in a multi-agent system, it is expressed as the resultant force of other follower agents on leader agent i. F_r and F_g represent the repulsive force and the gravitational force are measured under the action of multiple sensors, respectively.

3.1.2 Behavior Control Strategy Along Wall

The behavior along the wall means that the agent moves along the contour of the wall and keeps a certain distance from the wall.

(1) **The concept of virtual adjoint agent**

Based on the obstacle avoidance control of the behavior along the wall, Saber [9] introduced the concept of virtual adjoint agent. The virtual adjoint agent is the projection of the physical agent on the obstacle. Its actual position is the coordinate value of the projection point which is the shortest distance returned by the physical agent in detecting obstacles (approximate to the orthogonal projection). The angle

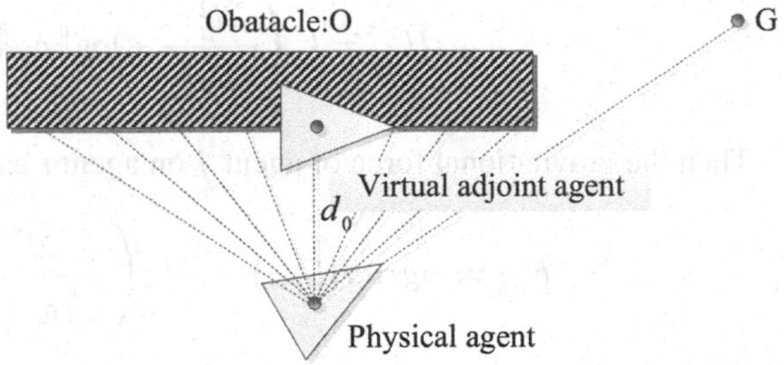

Fig. 3 The schematic diagram of virtual adjoint agent obstacle avoidance

of attack of virtual adjoint agent is the angle of the surface of the obstacle. When one of obstacle avoidance conditions of the behavior along the wall is satisfied, virtual adjoint agent is generated, and the physical agent continues moving toward the target point in the direction deviating from the obstacle. The schematic diagram of virtual adjoint agent obstacle avoidance is shown in Fig. 3.

(2) **Metope modeling algorithm**

When the virtual adjoint agent is generated, the obstacle avoidance method of the behavior along the wall is adapted, and the first step is to establish the metope model algorithm. Metope modeling is mainly divided into two steps. The first step is to identify whether the wall is a one-sided wall or multi-sided wall. The second step is to determine the direction of the wall.

(1) Identification wall

Based on the method of laser ranging, the obstacle is scanned. The coordinate sequence of laser points of obstacle that be obtained by scanning, that is (o_{x_i}, o_{y_i}), $i \in [1, N]$, $\rho_i = \sqrt{(o_{x_{i+1}} - o_{x_i})^2 + (o_{y_{i+1}} - o_{y_i})^2}$.

The identification of wall type can be distinguished by this formula:

$$\begin{cases} \text{one-sided wall}, & \forall \rho_i | \{\rho_i \in (0, 2(r_i + D))\} \\ \text{multi-sided wall}, & \exists \rho_i | \{\rho_i \in [2(r_i + D), +\infty]\} \end{cases}$$

Among that, ρ_i represents the distance between laser point i and laser point $i+1$, r_i represents the maximum radius of agent i, D represents the safe collision avoidance distance of agent i, N represents number of laser points represented by reflection, and $N \geq 2$. When $N = 1$, the direction of the agent's motion does not change.

(2) Determining the direction of the wall

For one-sided wall, the movement of agent i along the wall can be described as:

$$\phi_i' = \begin{cases} \phi_i, \ \|\phi_i - \theta_i\| < \frac{\pi}{2} \\ \phi_i + \pi, \ \|\phi_i + \pi - \theta_i\| < \frac{\pi}{2} \end{cases} i \in [1, N] \tag{10}$$

Among that, $\phi_i = \frac{\sum_{i=1}^{N-1}\delta_i}{N-1}$, $\delta_i = \arctan\left(\frac{o_{y_{i+1}} - o_{y_i}}{o_{x_{i+1}} - o_{x_i}}\right)$ represents the angle between laser point i and laser point $i + 1$.

For multi-sided wall, the movement of agent i along the wall can be described as:

$$\phi_i = \frac{\sum_{i=1}^{M-1}\delta_i}{M-1}, \quad M = \min(i \in \{\rho_i | \rho_i \in [2(r_i + D), +\infty]\}) \tag{11}$$

The hybrid autonomous obstacle avoidance process is shown as follows:

Step 1: Judging whether the agent needs to avoid obstacle.
Step 2: If the condition of the behavior along the wall is satisfied, the virtual adjoint agent is generated to move along the wall when the obstacle avoidance is required; otherwise, the artificial potential field method is applied.
Step 3: Judging whether obstacle avoidance is completed. After the obstacle avoidance is completed, the latest pose and tracking information are obtained and the agent continues to move towards the target point; otherwise, the conditions of the obstacle avoidance are judged until the obstacle avoidance task is completed. If the agent encounters obstacles in the process, repeat steps 1 through 3.

3.2 *Cooperative Obstacle Avoidance Control Strategy*

In the process of multi-agent formation tracking task, when the obstacle is too large or the narrow channel is encountered, adopting autonomous obstacle avoidance may lead to the dispersion of tracking formation or the loss of agents, result in the interruption of communication between agents, and affect the smooth completion of formation tracking task. If multi-agent can avoid obstacles in a cooperative way and avoid large obstacles successfully and pass through a narrow channel by changing the formation, the efficiency of collision avoidance can be improved to some extent.

3.2.1 Formation Transformation Control Conditions

In formation tracking motion, there are two kinds of cases which need to change formation to avoid obstacles. As shown (a) and (b) in Fig. 4.

(a) in Fig. 4 simulates the distribution of multi-agent when they encounter larger obstacle. In this case, the use of autonomous obstacle avoidance will cause agents R_3, R_4 to move toward the right of the obstacle, while the other agents will move to the left to avoid the obstacle to track the target point. At this time, the communication between agents will be interfered by the obstacle, which makes agents R_3, R_4 unable to communicate with the other agents. (b) in Fig. 4 simulates a narrow channel. In this case, autonomous obstacle avoidance will lead to all agents moving to the middle of the narrow channel, which makes it easy to collide and oscillate between agents. Therefore, in both cases, the formation should be changed to avoid obstacles.

3.2.2 Formation Transformation Control Strategy

When no obstacle is detected, the agents use leader-follower method to track the formation. By perceiving the pose information, the agents set the $l-\varphi$ parameters of the desired formation to form and maintain the formation, such as regular polygon. At this point, followers will follow leader to conduct tracking movement of the desired formation. Then when the agents detect obstacles ahead and belong to conditions as shown (a) and (b) in Fig. 4. In this case, it is necessary to use the leader-follower method to switch the formation into a straight line or a like straight line so that the agents can bypass the obstacle and smoothly pass through the narrow channel. After the formation of the new formation, the following relation of multi-agent in the formation is the same as before the transformation formation, and the formation mode is the same, while the difference is the setting of the $l-\varphi$ parameters.

The obstacle avoidance process in the formation tracking task is shown as follows:

Step 1: During the formation tracking task, determining whether multi-agent need to avoid obstacles constantly.
Step 2: If obstacles are detected, it is necessary to change the formation to avoid obstacles. The methods of judgement are as follows:

(1) The agent i is connected with the target point G as a straight line, and the angle δ_{ig} of the agent i relative to the target point is calculated;

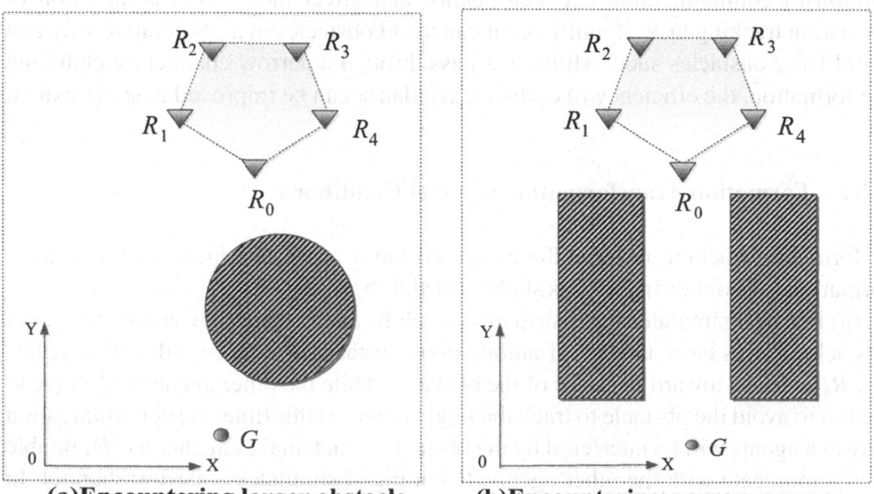

Fig. 4 Two kinds of cases which need to change formation to avoid obstacles

(2) Assuming the direction of the resultant force F_i of the agent i is δ_{if}, and $\Delta = \delta_{ig} - \delta_{if}$;
(3) If the Δ of all agents, have different positive and negative signs and maintain a certain number of steps, the hybrid autonomous obstacle avoidance method will be adopted. If the signs are the same and keep a certain number of steps, then the formation transformation is carried out, and the control function of the transformation formation is determined by leader-follower method.

Step 3: Judging whether obstacle avoidance is completed. If it is completed, the target point will continue to be tracked according to the original formation; otherwise, the judgment of the conditions of formation transformation will continue to avoid obstacles until the obstacle avoidance is completed. If the agents encounter obstacles in the process, repeat steps 1 through 3.

4 Simulation and Analysis

To verify the effectiveness of the control algorithm, the obstacle avoidance control algorithms in different environments are simulated. The simulation results show that the different control strategies are effective and feasible.

(a) and (b) in Fig. 5 show the simulation results of two different cases of obstacle avoidance control using the behavior along the wall method. It can be seen from the results that for the artificial potential field with minimal problem, the obstacle avoidance mode that automatically switches to the behavior along the wall method is more flexible and practical.

(a) and (b) in Fig. 6 show the simulation results in two cases where the artificial potential field method is used to accomplish the obstacle avoidance control task. The results show that the artificial potential field method can effectively avoid obstacles for a single agent or a multi-agent system.

(a) and (b) in Fig. 7 show the simulation results of two specific environments for obstacle avoidance control by transforming multi-agent formation. The results

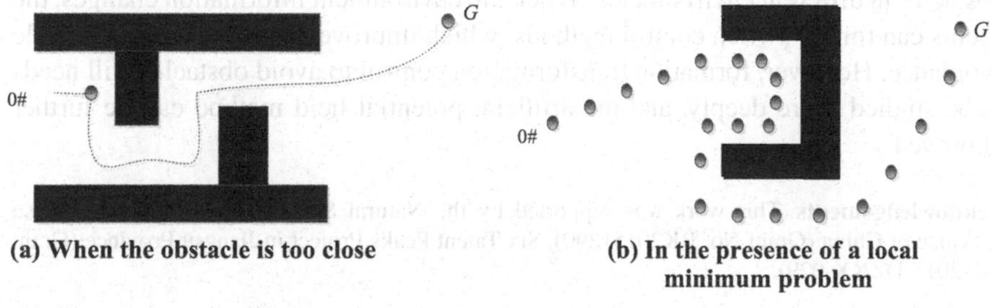

(a) When the obstacle is too close **(b) In the presence of a local minimum problem**

Fig. 5 Two cases of obstacle avoidance control using the behavior along the wall

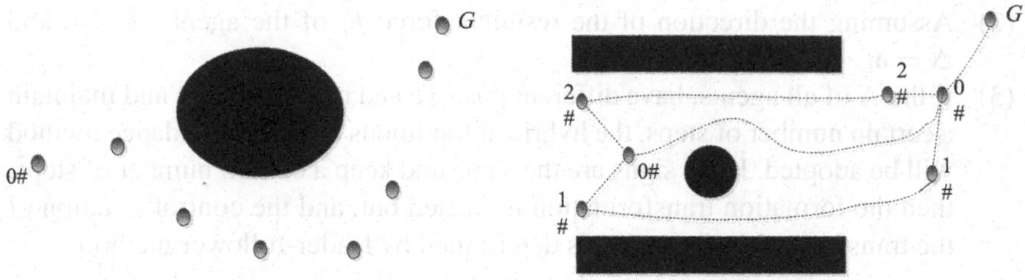

(a) Autonomous obstacle avoidance without formation **(b) Autonomous obstacle avoidance after formation completion**

Fig. 6 Two cases of obstacle avoidance control using artificial potential field method

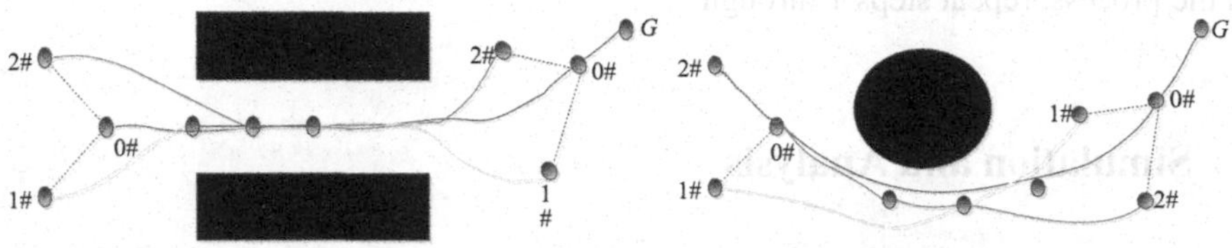

(a) The multi-agent encounters narrow channel **(b) The obstacle distribution wider than multi-agent**

Fig. 7 Two cases of obstacle avoidance control by transforming multi-agent formation

show that multi-agent can bypass the obstacle and smoothly pass through the narrow channel by transforming the formation into a straight line or a like straight line.

5 Conclusion

Multi-agent obstacle avoidance control strategy based on leader-follower formation tracking has been studied in this paper. The methods of the behavior along the wall, artificial potential field and formation transformation are combined to deal with obstacles in different environments. When the environment information changes, the agents can timely switch control methods, which improve the efficiency of obstacle avoidance. However, formation transformation control to avoid obstacles still needs to be studied more deeply, and the artificial potential field method can be further improved.

Acknowledgements This work was supported by the Natural Science Foundation of Jiangsu Province of China (Grant No. BK20151290), Six Talent Peaks Project in Jiangsu Province (Grant No. 2015-DZXX-029).

References

1. B. Cui, J. Song, Obstacle avoidance and dynamic target tracking of robot in unknown environment. J. Shenyang Univ. Technol. (2018)
2. O. Khatib, Real-time obstacle avoidance for manipulators and mobile robots. Int. J. Robot. Res. **5**(1), 90–98 (1986)
3. C.J. Kim, D. Chwa, Obstacle avoidance method for wheeled mobile robots using interval type-2 fuzzy neural network. IEEE Trans. Fuzzy Syst. **23**(3), 677–687 (2015)
4. Z. Su, J. Lu, Research on path planning of mobile robot by fuzzy logic method. J. Beijing Univ. Technol. **23**(3), 290–293 (2003)
5. N. Xiao, *Intelligent Robot* (South China University of Technology Press, Guangzhou, 2008), pp. 76–87
6. F. Xu, Research on obstacle avoidance and path planning of robot based on improved artificial potential field method. Comput. Sci. **43**(12), 293–296 (2016)
7. J. Shao, G. Xie, L. Wang, Leader-following formation control of multiple mobile vehicles. IET Control Theory Appl. **1**(2), 545–552 (2007)
8. D. Xu, W. Gui, Multi-agent tracking control based on network consistency. Control Eng. **17**(03), 304–308+355 (2011)
9. R.O. Saber, R.M. Murray, Flocking with obstacle avoidance: cooperation with limited communication in mobile networks, in *Proceedings of IEEE Conference on Decision and Control*, Maui, HI, USA (IEEE, Piscataway, 2003), pp. 2022–2028

Research of Driving Cycle Construction for Electric Drive Mining Truck Based on Travelling Analysis Method

X. Zhang, Sufang Wang and Weicun Zhang

Abstract This paper is concerned with the driving cycle construction for the electric drive mining truck based on travelling analysis method. Firstly, the complete operation data of the dump truck is collected, which is based on the continuous transportation test on an open pit road. Secondly, according to the travelling analysis method, a single duty cycle data separated from the multi-group dump truck duty cycle data, is divided into short travelling kinematic sequences. Next, based on the driving characteristics, ten characteristic parameters are selected. Through the principal component analysis and density peak clustering analysis, the data dimension is reduced to three principal component, and then the classification of the kinematics sequences is obtained. Further, a typical driving cycle for mining truck can be built by extracting kinematic sequences from each category. Finally, compared with the actual data, the typical driving cycle can strongly reflect the operating characteristic of electric drive mining truck on the open pit road.

Keywords Electric drive mining truck · Driving cycle
Principal component analysis · Density peak clustering analysis

1 Introduction

The bigger the mining truck is, the better it has become the consensus of its development [1]. Today, most of the mines with more than 100-ton mines are driven by

X. Zhang
BBMG Beishui Environmental Protection Technology Co. Ltd, Beijing 102200, China
e-mail: zhangxiao_miracle@126.com

X. Zhang
Tangshan Jidong Equipment and Engineering Co. Ltd, Tangshan 063000, China

S. Wang · W. Zhang (✉)
School of Automation and Electrical Engineering,
University of Science and Technology Beijing, Beijing 100083, China
e-mail: weicunzhang@ustb.edu.cn

Y. Jia et al. (eds.), *Proceedings of 2018 Chinese Intelligent Systems Conference*, Lecture Notes in Electrical Engineering 529,
https://doi.org/10.1007/978-981-13-2291-4_12

electric drive [2, 3]. Compared to mechanical and hydraulic transmission, the electric drive system has the advantages of low speed, large torque, stepless speed regulating, high transmission efficiency, and excellent driving performance [4].

Mining trucks are heavy-duty dump trucks used in open pit mines to accomplish rock earthwork stripping and ore transportation tasks, and are important transportation equipment for mine production. Most of its engine power is used in the driving system. The main performance indicators are fuel consumption rate and production efficiency [5]. Electric drive mining truck is currently playing an important role in the world's major open pit mine production equipment [6]. Therefore, reducing the fuel consumption rate of dump trucks is of great significance to improving the economic benefits of mines [7].

In the process of studying the fuel consumption rate of vehicles, the research on the driving cycle of vehicles is usually involved [8]. The driving cycle of the vehicle is the variation rule of the vehicle speed v with time t for a certain model in a certain road network [9]. At present, people at home and abroad put forward such as the federal testing program (FTP75), European ECE15 and Japan 10 operating conditions mainly aiming at passenger vehicles, buses and other road vehicles [10]. However, these conditions cannot be applied to the research of off-road vehicles such as electric dump trucks. The working characteristics of mining dump trucks in the production system of open pit mines are: a short cycle in a single cycle, large load changes in the cycle, complex transportation environment, and long continuous working hours. Taking the 220t mining dump truck as an example, the fuel cost almost occupies half of the total operating costs [11]. Even if it is not obvious that the fuel economy performance is optimized, the reduced production cost through accumulation is also enormous. Therefore, it is necessary to construct the driving cycle of the electric dump truck to provide working conditions for subsequent optimization.

In this paper, the mine test of a 220t electric drive dump truck is planned, continuous operating data are obtained, the data of a single work cycle is divided into reasonable short-stroke segments, and characteristic parameters are extracted. Then principal component analysis and density peak clustering are used to classify the kinematic sequences, select the appropriate sequences, and make a reasonable combination. Finally, the representative working conditions of the electric dump trucks for electric transmission were fitted.

2 Preparation of Driving Cycle

In this paper, a travelling analysis method is used to construct the dump truck driving cycle [12]. The specific flow chart shown in Fig. 1.

Through an on-site inspection of the road surface conditions of the open pit mine where the research object was located, one of the open pit mine transport sections from the loading point to the unloading point was selected. The road surface adhesion coefficient is the average level of open pit mines. The length of this section is approximately 3.5 km, including a long ramp and two short ramps. The road surface is compacted gravel road. The rolling friction coefficient of this type of road surface

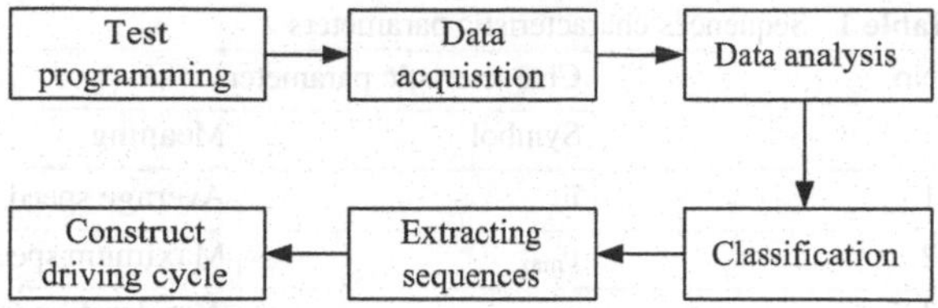

Fig. 1 Flow chart of driving cycle construction

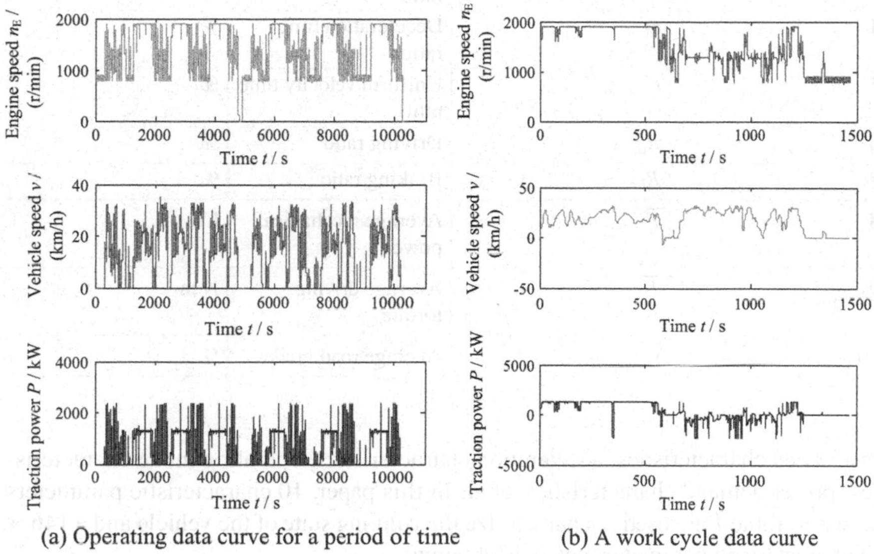

Fig. 2 Field data curves

is about 3%. The operating data for a period of time is shown in Fig. 2a. Through preliminary observation, it can be seen that the data includes six dump truck working cycles and one dump truck idle transfer process. A work cycle data is intercepted and detailed data processing is performed as the basic data for the construction of the cycle conditions in this paper, as shown in Fig. 2b.

Using travelling analysis method to construct the dump truck cycle conditions, the whole cycle process needs to be divided into several short stroke kinematic sequences. In this paper, 10 s as the short-stroke period is selected and the single-cycle data is divided to obtain 146 kinematic sequences.

Due to the many parameters that can reflect the characteristics of the cycle conditions in the operation process of the mine electric wheel dump truck, such as the kinematics parameters reflecting the movement rules of the dump truck, and the electrical parameters reflecting the working state of the electric transmission system. Therefore, it is necessary to select reasonable characteristic parameters. Too few feature parameters cannot characterize the effective information of kinematic sequences. Too many feature parameters increase the difficulty of data processing. In general, characteristic parameters for studying vehicle driving cycles can be divided

Table 1 Sequences characteristic parameters

No.	Characteristic parameters		
	Symbol	Meaning	Unit
1	$\overline{v}$	Average speed	km/h
2	$v_{\max}$	Maximum speed	km/h
3	r_a	Acceleration time ratio	%
4	r_d	Deceleration time ratio	%
5	r_c	Uniform velocity time ratio	%
6	R_{dr}	Driving ratio	%
7	R_{br}	Braking ratio	%
8	$\overline{P}$	Average mechanical power	kW
9	$\overline{T}$	Average driving torque	N m
10	$\overline{i}$	Average road grade	%

into: speed characteristics, acceleration characteristics, operating mode characteristics, power demand characteristics, et al. In this paper, 10 characteristic parameters shown in Table 1 are used to characterize the running state of the vehicle and a 146 × 10 characteristic parameter matrix is obtained.

3 Driving Cycle Construction

The usual way to build driving cycle is principal component analysis (PCA) and cluster analysis [13].

3.1 Principal Component Analysis

In order to effectively extract working condition information on the basis of ensuring accuracy, the process of dimensionality reduction of data is indispensable. In the process of constructing driving cycle, the PCA method is used to reduce the dimension of the feature parameters of kinematic sequences.

PCA is an effective multivariate analysis method in data mining technology. By constructing a series of linear combinations of the original variables, the principal component analysis reflects the information of the original variables as much as possible on the premise that the correlation is as small as possible [14].

Taking the data of driving cycle construction in this paper as an example. There are 146 kinematic sequences and 10 feature parameters, which form a 146 × 10 sample matrix: $\boldsymbol{X}_{146\times 10}$. All variables are standardized. The purpose of the processing is to eliminate the dimension effects, so that the average value of variables is 0 and the standard deviation is 1.

$$y_{i,j} = \frac{x_{i,j} - u_j}{\sqrt{\sigma_j}}(i = 1 - 146,\ j = 1 - 10) \tag{1}$$

where $u_j = E(x_j)$, $\sigma_j = D(x_j)$ are the average value and variance of each characteristic parameter in the sample matrix $Y_{146\times 10}$, respectively. Then the correlation coefficient matrix between the characteristic parameters is obtained.

$$r_{i,j} = \frac{\sum_{p=1}^{146}(y_{p,i} - \overline{y_i})(y_{p,j} - \overline{y_j})}{\sqrt{\sum_{p=1}^{146}(y_{p,i} - \overline{y_i})^2 \sum_{p=1}^{146}(y_{p,j} - \overline{y_j})^2}} \tag{2}$$

where $r_{i,j}$ is the correlation coefficient between y_i and y_j in the sample matrix, so as to obtain the correlation coefficient matrix.

Let $|\lambda \boldsymbol{E}\text{-}\boldsymbol{R}| = 0$, find the eigenvalue λ of the matrix $\boldsymbol{R}_{146\times 10}$ and arrange λ in descending order to get $\Lambda = [\lambda_1, \lambda_2, \ldots, \lambda_{10}]$. For each $\lambda_k(k = 1\text{–}10)$, according to $\boldsymbol{R}\boldsymbol{b} = \lambda_k \boldsymbol{b}$, the corresponding eigenvector $\boldsymbol{b}_k$ is obtained, then the j-th component $\boldsymbol{b}_{k,j}$ of the eigenvector $\boldsymbol{b}_k$ represents the load coefficient of the original j-th feature parameter on the k-th principal component. The larger the load coefficient, the higher the correlation coefficient between the feature parameter and the principal component. Calculate principal component contribution rate and cumulative contribution rate.

$$\varphi_k = \frac{\lambda_k}{\sum_{k=1}^{10}\lambda_k} \tag{3}$$

Calculate the variance contribution rate of the kth component, which reflects the amount of information that this principal component can reflect the original 10 feature parameters.

$$\Psi_m = \frac{\sum_{k=1}^{m}\lambda_k}{\sum_{k=1}^{10}\lambda_k}(m < 10) \tag{4}$$

The cumulative contribution rate of the top m principal components is obtained, that is, the proportion of the former m principal components in all principal components. The contribution rate reflects the total amount of information of the original characteristic parameters of the first m principal components. Generally, when the cumulative contribution rate reaches 80%–85 or more, it means that these components can collectively represent most of the information contained in the original feature parameter matrix and can be used to form a principal component matrix.

Table 2 Total variance of principal component analysis

No.	Eigenvalue	Variance contribution rate (%)	Cumulative contribution rate (%)
1	4.131	41.310	41.310
2	2.985	29.848	71.158
3	1.591	15.908	87.066
4	0.487	4.874	91.940
5	0.353	3.525	95.465
6	0.240	2.403	97.868
7	0.096	0.957	98.825
8	0.069	0.690	99.515
9	0.039	0.387	99.902
10	0.010	0.098	100.000

Solve the principal component matrix, and get a matrix consisting of m ($m<10$) principal components: $\mathbf{Z}_{146\times m.}$ The linear equations are

$$z_1 = \sum_{j=1}^{10} l_{1,j}\cdot \boldsymbol{x}_j \quad z_2 = \sum_{j=1}^{10} l_{2,j}\cdot \boldsymbol{x}_j \quad \cdots \quad z_m = \sum_{j=1}^{10} l_{m,j}\cdot \boldsymbol{x}_j \tag{5}$$

where $l_{n,j} = \sqrt{\lambda_n} b_{n,j}$ ($n = 1, 2, \ldots, m;\ j = 1, 2, \ldots, 10$) represents the principal component coefficient.

PCA is used to process the original characteristic parameter matrix, and the principal component eigenvalue, contribution rate and cumulative contribution rate are obtained as shown in Table 2. The principal component eigenvalue represents the information content of the original feature parameter contained in the principal component. The larger the eigenvalue, the more information the principal component contains. The eigenvalue greater than 1 should be retained. The eigenvalues of the first 3 principal components are all greater than 1, and the cumulative contribution rate is 87.066%. The new matrix consisting of these 3 principal components is only 3 dimension, and the matrix $\mathbf{Z}_{146\times 3}$ is obtained.

3.2 *Density Peak Clustering Analysis*

For the driving cycle construction, kinematic sequences with similar information are grouped into one category by clustering algorithm. Based on the different categories, we can construct driving cycle for the corresponding time length.

Density peak clustering is a new clustering algorithm proposed in 2014 [15]. The algorithm selects cluster centers only based on the distance of every data point and the local density of data points' distribution. This algorithm assumes that in the space formed by the sample data, the density of the clustering center points are all higher than the points around them. And these points are closer to the center of the cluster than others [16].

All row vectors in matrix $\mathbf{Z}_{146\times3}$ of the principal component can be regard as data points in a three-dimensional space. The distance from the ith data point to the jth data point is $d_{i,j}$. Then, we can obtain the local density of each data point.

$$\rho_i = \sum_{j=1}^{146} \chi\left(d_{i,j} - d_c\right) \tag{6}$$

where d_c represents the cutoff distance and $\chi(x) = \begin{cases} 1 & x < 0 \\ 0 & x > 0 \end{cases}$.

Using the density decision function, the local density of the ith data point, ρ_i, $i = 1, 2, \ldots, 146$, is the number of data points whose distance $d_{i,j}$ to z_i are less than the truncation distance in all data points. For any data point z_i, if z_i doesn't have the maximum local density, the covering range δ_i is

$$\delta_i = \min_{j:\rho_j>\rho_i} (d_{i,j}) \tag{7}$$

If z_i has the largest local density, the covering range is

$$\delta_i = \max(d_{i,j}) \tag{8}$$

For a data point that is a cluster center, its local density and coverage range are larger than others around it. By constructing the ρ–δ coordinate system, a clustering chart is obtained, which is shown in Fig. 3a. The data points with large ρ and δ values on the clustering chart are cluster centers. As we know from Fig. 3a, there are three isolated data points in the upper right area, whose density and coverage range are larger than others. So, they are selected as clustering centers. Further, the Fig. 3b shows the cluster center and clustering results' 3D distribution.

According to the peak density clustering result, 146 short driving cycle are divided into 3 categories, and the clustering centers are respectively No. 22, No. 69 and No. 75 sequences. The short travelling clustering results based on the original data are shown in Fig. 4. From the clustering results, we know that for the category 2 of driving cycle obtained by the clustering analysis, termed as C2 driving cycle, the dump truck is in an idle state, and the electric drive system doesn't provide power. For category 1, termed as C1 driving cycle, the traction system is in a drive state for most of the time, the mechanical power is high, and the driving torque is high. Therefore, it has a certain demand for vehicle speed and a great demand for power and driving torque. Similarly, for C3 driving cycle, the traction system is braked, whose speed is high and the driving power and driving torque are low.

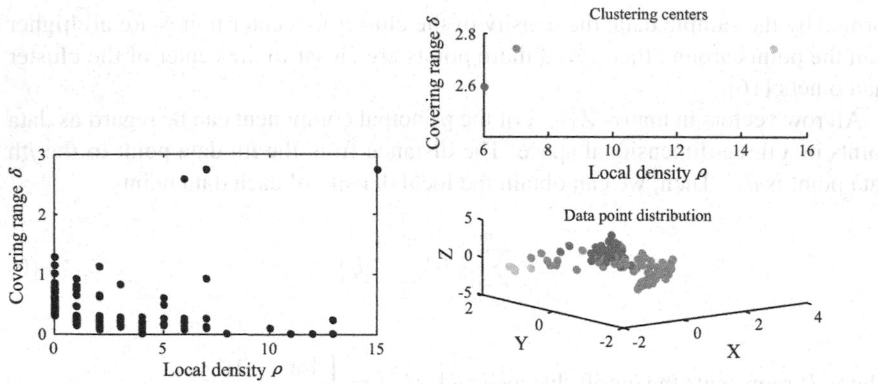

(a) Clustering chart (b) Clustering centers and data point distribution

Fig. 3 Clustering results

Fig. 4 Short travelling clustering results

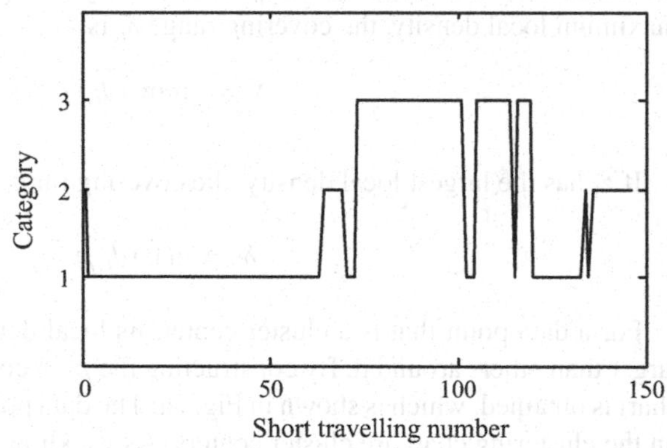

3.3 *Driving Cycle Construction*

The total operation cycle of the mining truck is composed of the fixed working time and the driving operation time, which mainly includes full-load forward time and no-load return time. For driving cycle construction, the full-load forward time and no-load return time should be appropriately increased or decreased in proportion, whose journeys are approximately equal.

In this paper, we construct driving cycle with a total time length of 1500 s. According to the data obtained from the clustering analysis, the time of C2 driving cycle is fixed, that is approximately 170 s. According to the corresponding proportion, the kinematic sequences, which belong to the C1 driving cycle and the C3 driving cycle, are extracted and then combined to form the driving cycle with a time length of 1500 s. Some characteristic parameters are shown in Fig. 5.

In the driving cycle, the time for C1 driving cycle is 930 s, the time for C2 driving cycle is fixed as 170 s, and the time for C3 driving cycle is 400 s. The mileage of

Fig. 5 Some characteristic parameters for driving cycle

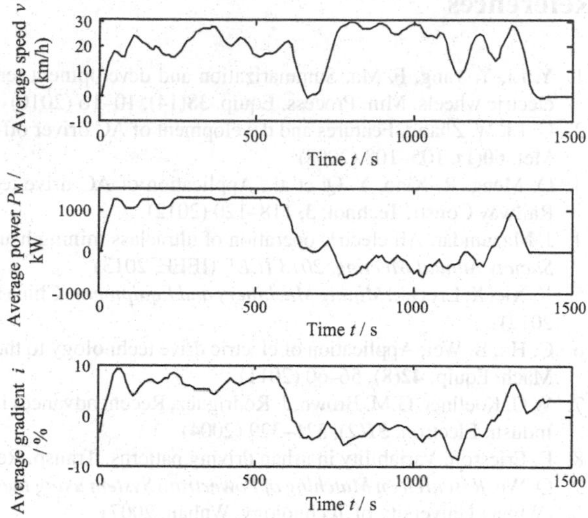

the mining truck is estimated to be 3.4 km for full-load and 3.25 km for no-load, which are approximately the same. Therefore, the driving cycle can accurately and reasonably reflect the normal driving conditions.

4 Conclusion

In this paper, based on travelling analysis method, the driving cycle is constructed for the electric drive mining truck. By designing the running test in the open pit mining environment, the operation data has the representation of the actual working environment. Through the principal component analysis and density peak clustering analysis, the classification of the kinematics sequences is obtained, whose data dimension of characteristic parameters is reduced effectively. Then, a typical driving cycle for mining truck is built, which can accurately and reasonably reflect the normal driving conditions on the open pit road. In a practical sense, the driving cycle construction for the electric drive mining truck lays the foundation for improving the performance indexes by studying and optimizing the driving control strategies.

Acknowledgements This work was supported by National Natural Science Foundation of China (No. 61520106010; 61741302).

References

1. Y. Li, Y. Yang, F. Ma, Summarization and development tendency of mine dump truck with electric wheels. Min. Process. Equip. **38**(14), 10–16 (2010)
2. G. Li, W. Zhang, Features and development of AC driver off-high way haul truck. Nonferrous Met. **60**(1), 105–108 (2008)
3. Q. Meng, R. Xiao, Y. Qi et al., Application of AC drive technology in mining dump truck. Railway Constr. Technol. **3**, 118–120 (2012)
4. J. Mazumdar, All electric operation of ultraclass mining haul trucks, in *Industry Applications Society Annual Meeting, 2013 IEEE* (IEEE, 2013)
5. X. Xie, X. Li et al., *Mining Machinery and Equipment* (China Mining University Press, Xuzhou, 2012)
6. C. Hu, B. Wei, Application of electric drive technology to the construction equipment. Constr. Mach. Equip. **42**(8), 56–60 (2011)
7. W.G. Koellner, G.M. Brown, J. Rodríguez, Recent advances in mining haul trucks. IEEE Trans. Industr. Electron. **51**(2), 321–329 (2004)
8. E. Ericsson, Variability in urban driving patterns. Transp. Res. Part D **5**(5), 337–354 (2000)
9. Q. Wu, *Research on Matching of Powertrain System using the Urban Bus Driving Cycle Method* (Wuhan University of Technology, Wuhan, 2007)
10. J. Zhang, M. Li, G. Ai et al., A study on the features of existing typical vehicle driving cycles. Autom. Eng. **27**(2), 220–224, 245 (2005)
11. E. Penman, Ultra-class truck tire operation and management, in *3rd Annual Driving Down Operating Costs Conference, Australian Journal of Mining* (2002)
12. H. Yao, *Traction Dynamics Overview and Research Review of Heavy Vehicle* (China Communication Press, Beijing, 2004)
13. X. Zhu, M. Li, Z. Ma et al., Methodology for driving cycles development. J. Jiangsu Univ. (Nat. Sci. Ed.) **26**(2), 110–113 (2005)
14. Z. Ma, M. Li, F. Zhang et al., A study on the application of PCA to the development of vehicle real driving cycle. J. WUT (Inf. Manage. Eng.) **26**(4), 32–35 (2004)
15. A. Rodriguez, A. Laio, Clustering by fast search and find of density peaks. Science **344**(6191), 1492–1496 (2014)
16. J. Xie, H. Gao, W. Xie et al., Robust clustering by detecting density peaks and assigning points based on fuzzy weighted K-nearest neighbors. Inf. Sci. **354**, 19–40 (2016)

Unit Commitment and Load Distribution Within a Power Plant

Jinhui Zhu, Ming Guo, Sufang Wang and Weicun Zhang

Abstract With the rapid development of electricity market, a power plant or power supply producer involved in the market need to face the economic dispatch problem, which includes unit commitment and load distribution. These problems were formerly resolved by the dispatching department of the state power grid manager. Taking into account the new situation in the electricity market, this paper presents a new solution to this problem, i.e., dynamic programming method combined with equal incremental cost criteria.

Keywords Electricity market · Economic dispatch · Unit commitment
Load distribution

1 Introduction

Electricity power market brings new challenging problems as well as new chances, for power producer (plant). Unit commitment and load distribution is one of these challenging problems, which, before the implementation of electricity power market, was undertaken by dispatching department of power grid. Unit commitment and load distribution conducted by dispatching department of power grid should consider transmission line loss, as a result, only recursive optimization methods can be used to solve the problem. Correspondingly, the weak points include heavy burden of

J. Zhu
State Grid Hebei Electric Power Supply Co., LTD, Hebei 050000, China
e-mail: zjh@he.sgcc.com.cn

M. Guo
State Grid Hebei Electric Power Company & Telecommunication Branch, Hebei, China
e-mail: sjzgm@163.com

S. Wang · W. Zhang (✉)
School of Automation and Electrical Engineering, University of Science and Technology Beijing, Beijing 100083, China
e-mail: weicunzhang@263.net

Y. Jia et al. (eds.), *Proceedings of 2018 Chinese Intelligent Systems Conference*, Lecture Notes in Electrical Engineering 529,
https://doi.org/10.1007/978-981-13-2291-4_13

calculation and possible low accuracy. Unit commitment and load distribution conducted within power plant should not consider transmission line loss, thus accurate optimization solution is possible.

2 Problem Description

Unit commitment and load distribution within a power plant can be described as following [1].

The objective of load distribution is to minimize the total cost of power generation, i.e.,

$$F = \sum_{m=1}^{M} F_m(P_m) \rightarrow \min \tag{1}$$

Subject to

$$\sum_{m=1}^{M} P_m - P_D = 0 \tag{2}$$

where F is the total cost of power generation; F_m is the cost of each generation unit, which is obviously a function of P_m, the active power of unit m; m = 1, 2, … M, M is the total number of units; P_D is the power generation quantity obtained through bidding.

The objective of unit commitment is

$$F = F_o + F_s = \sum_{t=1}^{T}\sum_{m=1}^{M} [f_{m,t}(P_{m,t}, S_{m,t}) + f_{s,m,t}(S_{m,t}, S'_{m,t})] \rightarrow \min \tag{3}$$

where F is the total cost during the generation period; F_o is the total generation cost during the generation period; F_s is the total start-up cost during the generation period; $P_{m,t}$ is the active generation power of unit m in period t. $S_{m,t}$ is the start or stop state of unit m in period t: $S_{m,t} = 1$ means unit m is working, $S_{m,t} = 0$ means unit m is down. $S'_{m,t}$ is the start or stop state of unit m before period t: $S'_{m,t} = 0$ means unit m is working before period t, $S'_{m,t} \neq 0$ represents how many time periods unit m has been down. $f_{m,t}$ is the generation cost of unit m in time period t: $f_{m,t} \neq 0$ (when $S_{m,t} = 1$), $f_{m,t} = 0$ (when $S_{m,t} = 0$).

$f_{s,m,t}$ is the starting cost of unit m in time period $t = 0$, (when $S_{m,t} = 1, S_{m,t-1} = 1$ or $S_{m,t} = 0$), the start or stop state of unit m in period t remains unchanged;

$f_{s,m,t} \neq 0$ (when $S_{m,t} = 1, S_{m,t-1} = 0$) means unit m starts at period t, is related to $S'_{s,m,t}$.

We have the following constraint conditions

(1) Load balance

$$\sum_{m=1}^{M} P_{m,t} - P_{D,t} = 0 \quad (t = 1, 2, \ldots, T) \tag{4}$$

where $P_{m,t}$ is the active generation power of unit m in period t; $P_{D,t}$ is the load in period t.

(2) Reservation requirements

$$\sum_{m\in 0} P_{\min,m,t} \leq P_{D,t} + P_{L,t} \leq \sum_{m\in 0} P_{\min,m,t} - P_{r,t} \quad (t = 1, 2, \ldots, T) \tag{5}$$

where $P_{r,t}$ is the reservation requirement in period t;

(3) Unit power generation limits

$$P_{\min,m,t} \leq P_{m,t} \leq P_{\max,m,t} \quad (m = 1, 2, \ldots, M; t = 1, 2, \ldots, T) \tag{6}$$

Besides, there are two more other constraint conditions, i.e., shortest unit downtime limit and the number limit of start units at the same time period. In this paper these two constraint conditions will not be considered.

3 Unit Commitment

There are many methods to unit commitment problem [2–8], such as dynamic programming (DP). To be adapted to DP, unit commitment problem can be considered as a multi-step decision problem, i.e., one day can be divided into 48 or 96 time periods, to decide the unit commitment (unit start/stop schedule) and load distribution of each time period. The contribution of this paper is as follows: (1) to consider all possible unit combination states without simplification; (2) the solution of load distribution is accurate without iteration process. The principle of DP can be found in Ref. [1]. We only give the calculation procedure. For convenience, some notes of matrices and vectors will be introduced.

(1) State matrix A_i: Each row represents a unit combination state and meet the need of generation schedule of the current time. There are three kinds of state matrix, i.e., initial state, current state and next state. Maximum possible number of row, i.e., the maximum possible state numbers (combinations) is $2^n - 1$ (the situation that all the unit are down, is not in consideration), n is unit numbers, the number of ranks is n, there is only one possible initial state, i.e., only one row. We need to save A_i and A_{i+1}, i is the current time period number.

(2) Load distribution matrix P_i with same dimensions as state matrix A_i, represents the load distribution of n units in time period i.
(3) Optimal state transition matrix B_{ij}, i is the current time period number, j is state number.
(4) Start-up cost (for each unit) vector C with dimension $n \times 1$.
(5) Stop cost (for each unit) vector C_1 with dimension $n \times 1$.
(6) Operation cost (for each state) vector D with dimension $2^n - 1$.
(7) Cumulative minimum cost (for each state) vector E with dimension $2^n - 1$.
(8) On-grid energy (scheduled generation quota, obtained through bidding) vector by time period with dimension t, $t = 24, 48, 96$.
(9) Minimum output power vector (for each unit) G with dimension n.
(10) Maximum output power vector (for each unit) H with dimension n.
(11) Reserve capacity vector (for each unit) R with dimension n.
(12) Cost function matrix M, represents cost function parameters of all units, in this paper we consider quadratic cost function, then for each unit, there are three parameters, i.e., a, b, c. Specifically we have

$$M = \begin{bmatrix} a_1, b_1, c_1 \\ a_2, b_2, c_2 \\ \cdots\cdots\cdots \\ a_n, b_n, c_n \end{bmatrix} \text{ with dimension } n \times 3$$

Following are the solution procedures:

Each day we need to calculate the unit commitment and load distribution schedule for each time period of next day. The initial state of calculation is the state of the last time period of current day.

(1) Set the time period $i = 1$;
(2) Call sub procedure to calculate all possible units combination states of current time period and save the results to state matrix A_i, in which element "1" represents "ON", "0" represents "OFF" of corresponding unit;
(3) Calculate the admissible unit combination states according to F, G, H, R and the condition:

$$\sum_{j=1}^{n} A_{ij} G_j \leq P_i \leq \sum_{j=1}^{n} A_{ij}(H_j - R_j)$$

Save the results to A_i and A_{i+1};
(4) For each state in A_i do load distribution (details will be described in the next section);
(5) Calculate the state transition cost (from initial state to each state in A_i), from "OFF" to "ON" there is a start-up cost, from "ON" to "OFF" there is a stop cost;

(6) Calculate the operation cost D for each state in A_i, according to cost function and P_i, $i = 1, \ldots n$.
(7) Calculate the minimum cumulative cost for each state E;
(8) Fill in the optimal state transition matrix B_{ij};
(9) $i = i + 1$, return to (2) until $i = t$, $t = 24, 48, 96$;
(10) Choose the minimum E_l from E, note down the subscript l, then correspondingly the content of B_{tl} is the optimal transition path, the content of P_i is the optimal load distribution results.

The calculation of A_i:

(1) Set $i = 2$, calculate state matrix $A(2^i, i)$;
(2) Add an all "0" rank to $A(2^i, i)$ then we get A_1; Add an all "1" rank to $A(2^i, i)$ then we get A_2;
(3) Obtain the state matrix for i + 1 units

$$A(2^{i+1}, i+1) = \begin{bmatrix} A_1 \\ A_2 \end{bmatrix}$$

(4) If $n = i + 1$, then finish, otherwise set $i = i + 1$ and go back to (2).

3.1 Load Distribution

Consider the cost function for each unit

$$F_m(P_m) = a_m P_m^2 + b_m P_m + c_m \tag{7}$$

Then the total generation cost

$$F = \sum_{m=1}^{M} F_m(P_m) = \sum_{m=1}^{M} (a_m P_m^2 + b_m P_m + c_m) \tag{8}$$

We have by Lagrange multiplier method

$$\begin{aligned} \bar{F} &= F + \lambda\left(\sum_{m=1}^{M} P_m - P_D\right) \\ &= \sum_{m=1}^{M} (a_m P_m^2 + b_m P_m + c_m) + \lambda\left(\sum_{m=1}^{M} P_m - P_D\right) \end{aligned} \tag{9}$$

The optimization condition is

$$\begin{cases} 2a_1 p_1 + b_1 = \lambda \\ 2a_2 p_2 + b_2 = \lambda \\ \dots\dots\dots\dots\dots \\ 2a_{n-1} p_{n-1} + b_{n-1} = \lambda \\ 2a_n p_n + b_n = \lambda \\ p_1 + p_2 + \cdots + p_n = P_D \end{cases} \tag{10}$$

Substitute $2a_n p_n + b_n = \lambda$ into other n $-$ 1 equations, we obtain

$$\begin{cases} 2a_1 p_1 - 2a_n p_n = b_n - b_1 \\ 2a_2 p_2 - 2a_n p_n = b_n - b_2 \\ \dots\dots\dots\dots\dots \\ 2a_{n-1} p_{n-1} - 2a_n p_n = b_n - b_{n-1} \\ p_1 + p_2 + \cdots + p_n = P_D \end{cases} \tag{11}$$

Solve this equation set then we get the load distribution result P_m.

Next we regulate the load distribution result according to reserve condition and power output limit condition.

From the optimization condition we know

$$2a_m(P_m + \Delta P_m) + b_m = \lambda + \Delta\lambda \tag{12}$$

Thus we have

$$2a_m \Delta P_m = \Delta\lambda \tag{13}$$

If the total regulation needed is ΔP.

Then

$$\sum \Delta P_m = \sum \frac{\Delta\lambda}{2a_m} = \Delta P \tag{14}$$

Consequently

$$\Delta\lambda = \frac{\Delta P}{\sum \frac{1}{2a_m}}, \quad \Delta P_m = \frac{\Delta\lambda}{2a_m} \tag{15}$$

In detail, if the output limit condition is $[P_{m\max}, P_{m\min}]$, then regulation procedures are as follows

(1) Set $\Delta P = 0$

(2) If $P_m \geq P_{m\max}$, then set $P_m = P_{m\max}$, $\Delta P + = P_m - P_{m\max}$
Otherwise if $P_m \leq P_{m\min}$, then set $P_m = P_{m\min}$, $\Delta P + = P_m - P_{m\min}$

(3) Calculate ΔP_m for each unit according to (14) and (15).

4 Conclusions

In this paper, a practical solution to the unit commitment and load distribution problem within a power plant is presented. It is rather different from the same problem undertaken by the dispatching department of the grid, which should consider the power loss on the transmission line.

Acknowledgements This work was supported by National Natural Science Foundation of China (Nos. 61520106010; 61741302).

References

1. Y. Erkeng, *Energy Management System* (Science Press, Beijing, 1998)
2. Q. Jinlong, Y. Erkeng et al., *Economic Dispatchment of Power System* (Publishing Press of Harbin Institute of Technology, Haerbin, 1993)
3. C. Haoyong, W. Xifan, A survey of the optimization-based methods for unit commitment. Autom. Electric Power Syst. **23**(5), 51–56 (1999)
4. W. Feng, Z. Yiying, B. Xiaomin, Study of GA-based unit commitment. Autom. Electric Power Syst. **27**(6), 36–41 (2003)
5. M. Wang, B. Zhang, Q. Xia, A novel unit commitment method considering various operation constraints. Autom. Electric Power Syst. **24**(12), 29–35 (2000)
6. H. Jin, S. Libao, Z. Jiaqi, An ant colony optimization algorithm with random perturbation behavior for unit commitment problem. Autom. Electric Power Syst. **26**(23), 23–28 (2002)
7. H. Jiasheng, G. Chuangxin, C. Yijia, A hybrid particle swarm optimization method for unit commitment problem. Proc. CSEE. **24**(4), 24–28 (2004)
8. W. Zhe, Y. Yixin, Z. Hongpeng, Social evolutionary programming based unit commitment. Proc. CSEE. **24**(4), 12–17 (2004)

Urban Street Image Matching Method Based on Improved SURF Algorithm

Shuoshi Wang and Lin Li

Abstract The Speeded Up Robust Features (SURF) algorithm was too cumbersome to extract the feature points, so it could not meet the needs of more rapid image matching. This paper takes the urban street image as the research object and proposes a more efficient and rapid image matching strategy which is based on SURF. Firstly, use Haar wavelet for image preprocessing to remove high frequency information and extract the low-frequency part of the image. Secondly, use Brisk algorithm to extract feature points from the processed image instead of the step of extracting the feature points by SURF. Finally, use surf's feature point description method and matching strategy to match. The simulation results showed that the improved algorithm increased the matching rate and reduces matching time, the effect is more prominent when the image affect by the influence of light.

Keywords Feature matching · SURF algorithm · Brisk algorithm · City street

1 Introduction

Civil drones have been commonly used in aerial photography field nowadays. Players take photos for streets by manipulating drones. Photographs of the same things will be deformation due to different lighting, angles, distances or other factors which cause some phenomena such as low matching rate or spend a lot of time even mismatch.

Image matching technology is divide into two categories: based on gray value and feature information. The classical algorithm based on gray information was normalized cross correlation. The method was slid the template image on image, the area with the largest cross-correlation value during the sliding process was the matching

S. Wang (✉) · L. Li
Department of Control Science and Engineering, University of Shanghai for Science and Technology, Shanghai 200093, China
e-mail: wangss_smile@163.com

L. Li
e-mail: lilin0211@163.com

Y. Jia et al. (eds.), *Proceedings of 2018 Chinese Intelligent Systems Conference*, Lecture Notes in Electrical Engineering 529,
https://doi.org/10.1007/978-981-13-2291-4_14

area of the two images. In addition, there also have Mean Absolute Deviation algorithm which need to traverse each pixel, that's caused the speed of match slow, so it is less applicable in practice [1].

Another type of image matching is based on image feature information. Moravec proposed the concept of feature points which main idea was to slide a $(2n+1) \times (2n+1)$ window with one of the pixels as the center in the image. Calculate the sum of squares of the corresponding pixel differences in the four directions, then select the minimum value as the response function CRF of the pixel (x, y). The pixel is the feature point when the CRF value is greater than a certain threshold and is a local maximum [2]. Harris proposed the Harris algorithm based on the Moravec algorithm. Harris introduced the Gaussian template in the process of extracting feature points which has improved the feature point location. However, they were sensitive to noise and did not have rotation invariance. The SURF algorithm is still cumbersome and other issues on feature points extraction. For the above, this paper uses the urban street image as an example and improves the surf algorithm to increase the matching rate and reduces the matching time.

2 SURF Algorithm

Bay et al. proposed SURF algorithm that alleviates image mismatch and let feature points added the information scale [3]. The steps of the surf algorithm can be summarized as follows: construct hessian matrix, extract key points, determine the main direction, generate feature descriptors and match.

3 Extract Feature Points by SURF

The first step of the SURF algorithm is to construct the Hessian matrix. And Gaussian filtering needs to be add before constructing the Hessian matrix. The Hessian matrix transform after adding Gaussian filtering is:

$$H(x,\sigma) = \begin{vmatrix} L_{xx}(x,\sigma) & L_{xy}(x,\sigma) \\ L_{xy}(x,\sigma) & L_{yy}(x,\sigma) \end{vmatrix} \tag{1}$$

$L_{xx}(x, \sigma)$ is the convolution of the Gaussian second-order differential at the x-point with the image, the remaining three $L_{xy}(x, \sigma)$, $L_{yy}(x, \sigma)$ and $L_{xx}(x, \sigma)$ are similar. The Gaussian filter needs to be discretized in calculation, so Bay proposed to replace the Gaussian second-order differential filter template with a block filter. The L_{xx}, L_{xy} and L_{yy} become D_{xx}, D_{xy} and D_{yy}. Hessian's value also becomes $\det(H) = D_{xx} \cdot D_{yy} - (0.9D_{xy})^2$. 0.9 is an approximate coefficient prescribed by Bay.

The SURF algorithm constructs many block filters by changing the value of the scaling function σ, to building scale spaces. The filters can be divided into multiple groups. Each group of filters has multiple layers. The size of the filters in the group was different. The specific formula is:

$$d = 3 \times (2^{o} \times s + 1) \quad \sigma = 1.2 \times \frac{d}{9} \tag{2}$$

where O is the group index, s is the index of the inner layer of the group, each index counts from 1 and the σ of first layer on the first layer is 1.2. Each pixel in the scale space is compared with a total of 26 adjacent pixels. The point is the feature point when the hessian value is the maximum or minimum.

4 Improved SURF Algorithm

From the above, the process of SURF extracting feature points requires a lot of time because, it was not possible to make image matching more quickly.

Due to the picture contains a lot of noise, therefore the first step of the improve algorithm is to use wavelet to preprocess the image to reduces noise impact, secondly use Brisk algorithm to extracts feature points. Finally, use surf's feature point description method and matching strategy to match. The algorithm flow chart is shown by Fig. 1.

5 Image Preprocess by Haar Wavelet

The noise mainly exists in the high frequency area, but more useful information of the image are stored in the low frequency area. In order to filter out the noise and the low-frequency part could be obtained, this paper select haar wavelet transform. The wavelet function contains two features. The first is that any wavelet function $\varphi_{ab}(t)$ is computed by the mother wavelet $\varphi(t)$ [4]. Where a is the scale factor and b is the translation factor. The second feature is that it has a scale feature that makes it

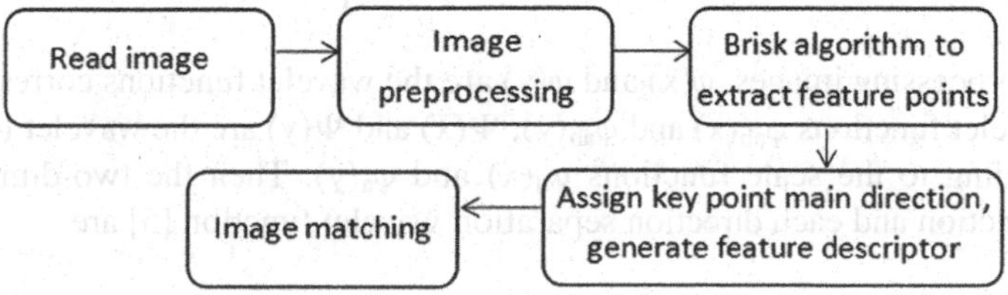

Fig. 1 Improved algorithm flow chart

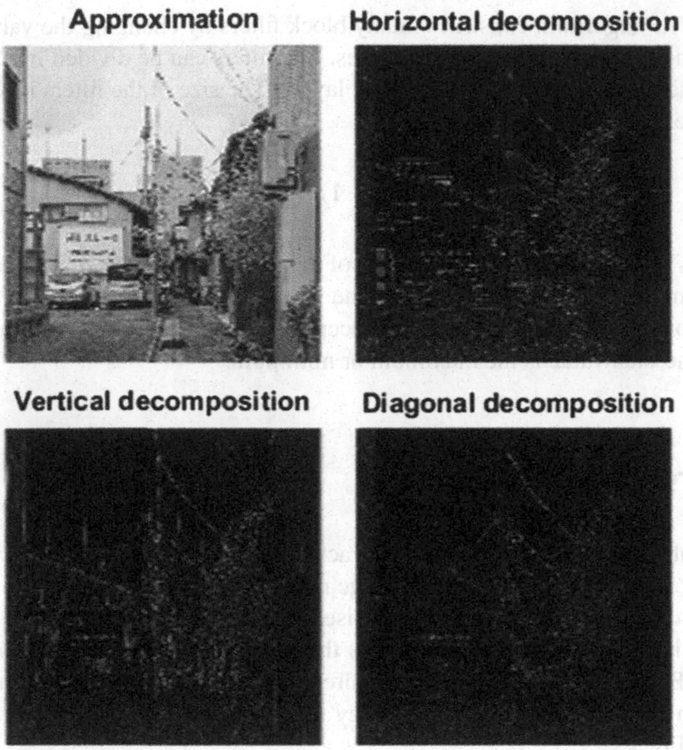

Fig. 2 Image obtained by wavelet one-level decomposition

easier to describe the local features of the signal. The haar wavelet function and it's mother wavelet and scale function are:

$$\varphi(\mathrm{t}) = \begin{cases} 1 & 0 \leq \mathrm{t} < \frac{1}{2} \\ -1 & \frac{1}{2} \leq t < 1 \\ 0 & \text{others} \end{cases} \tag{3}$$

$$\varphi_{a,b}(t) = 2^{\frac{a}{2}} \psi\left(2^a t - b\right) \quad \Psi_{\mathrm{a}}(t) = \begin{cases} 1 \; 0 \leq t < 1 \\ 0 \text{ others} \end{cases} \tag{4}$$

When processing images, $\varphi(x)$ and $\varphi(y)$ are the wavelet functions corresponding to the wavelet functions $\varphi_{ab}(x)$ and $\varphi_{ab}(y)$, $\Psi(x)$ and $\Psi(y)$ are the wavelet functions corresponding to the scale functions $\varphi_a(x)$ and $\varphi_a(y)$. Then the two-dimensional scaling function and each direction separation wavelet function [5] are:

$$\begin{aligned}\varphi(x, y) &= \varphi(x)\varphi(y)\\ \psi^H(x, y) &= \varphi(x)\psi(y)\\ \psi^V(x, y) &= \psi(x)\varphi(y)\\ \psi^D(x, y) &= \psi(x)\psi(y)\end{aligned} \quad (5)$$

φ(x, y) is The approximate value of the image, ψ^*(x, y) respectively are horizontal, vertical, and diagonal details function. Wavelet transform includes dual low frequency channels, dual high frequency channels, mix channels (the order of high and low frequencies are different). Therefore, get up to 4 images (in Fig. 2) after through each wavelet transform. Respectively are approximations, horizontal details, vertical and diagonal details. Only the approximate image is the result of the image be processed after dual low frequency channels. The approximate image filters high frequency part and contains more useful information to extract the feature points. The wavelet transform can be decomposed by an arbitrary number of times, but in the experiment we find that the approximate image obtained by multiple decomposition that cannot find or find very few key points, so the wavelet transform is only performed once in the improved algorithm.

6 Extract Feature Points

The BRISK algorithm [6] is an extension of the FAST algorithm [7]. The FAST algorithm is introduced before the discussion of the Brisk [8] algorithm. The FAST algorithm uses a pixel point as it's center in the process of extracting feature points, take 3 pixels (excluding the center point) for the radius to make a circle, as shown in Fig. 3.

There are 16 pixels on the outline of the circle (if take 1 pixel as radius, there are 8 pixels on the outline) [9]. If there are consecutive 12 pixels' gray value are greater than the sum of the center pixel and threshold or less than the difference between the center pixel and threshold, then the center pixel is the feature point. The judgment method at this time is 12–16 masks. The Brisk algorithm selects 9–16 masks and 5–8 masks. There is also a more rapid method, from vertical and horizontal to select 4 points 1, 5, 9 and 13, if any 3 points are larger or less than the gray value of the center pixel, the center pixel is a key point.

$$S_{p\to x} = \begin{cases} d, \; I_{p\to x} \le I_p - t & \text{(darker)} \\ s, \; I_p - t < I_{p\to x} < I_p + t & \text{(similar)} \\ b, \; I_p + t \le I_{p\to x} & \text{(brighter)} \end{cases} \quad (6)$$

where t is a threshold set to 10, I_p is center pixel gray value, $I_{p\to x}$ is pixel value on the rounded edge, the points on the edge can be divided into three categories: d, s, b.

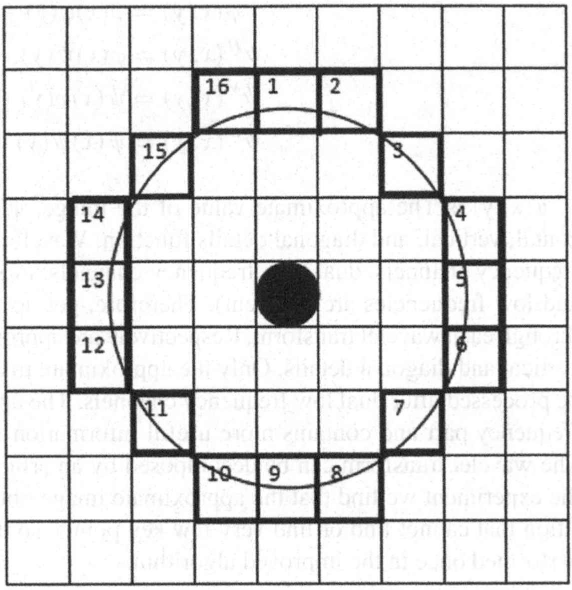

Fig. 3 Center pixel and pixels on the edge of the neighborhood

7 Extract Feature Points by BRISK

In the Brisk algorithm, the original image is decomposed into four C-layers and four D-layers, where C_0 is the original image and C_1 is the 2 down-sampling of the C_0 layer. Each layer is 2 times lower than the previous layer. The D_0 layer is 1.5 times lower sampling of the C_0 layer, each layer is the previous layer of 2 down-sampled. The Fast 9–16 form of feature point detection is adopted for the every layer image. The Fast 5–8 form of feature point detection needs to be performed once again on the C_0 layer. The final result is 9 images including feature points.

The non-maximum suppression of feature points need to consider the adjacent 8 points of the layer where the key points are located and 18 points of the upper and lower layers. The feature points need to be larger than the fast values of the 26 points. If the fast value is small, the feature points are canceled. The Fast value refers to the sum of the absolute value of the difference between each pixel and the pixels on the circular edge. The two-dimensional quadratic function interpolation is performed on the scale value of the extremum point and the score value corresponding to the upper and lower layers to obtain the exact position of the extremum point. Then the one-dimensional interpolation of the scale direction is performed to obtain the final scale value [10, 11]. At this point, the Brisk algorithm extracts feature points finish.

8 The Feature Descriptors and Match

Make a radius of 6σ circular neighborhood with the center of the feature point. Through a sector Z of 60° traversed the entire circle in steps of 0.2 radians. Calculate the haar response of the horizontal and vertical directions d_x and d_y of each pixel in the neighborhood. Next, accumulate the response values of all points in the sector area. The main direction of the feature point is the corresponding direction of the largest accumulated value [12] (Fig. 4).

$$m = \sum_z d_x + \sum_z d_y \quad \theta = \tan^{-1}\left(\sum_z d_x \Big/ \sum_z d_y\right) \tag{7}$$

Select a rectangle with an edge length of 20σ centered on the feature point, divide the rectangle into 16 sub-regions, count the pixels in each sub-area, and apply Gaussian weights, then calculation the haar response value d_x, $|d_x|$, d_y and $|d_y|$ of the points and the corresponding sum $\sum d_x$, $\sum |d_x|$, $\sum d_y$ and $\sum |d_y|$. Four response values generate a four-dimensional vector. Last, the normalization process be used to reduce the influence of light on the image. In the end, each feature point has a 64-dimensional descriptor.

Select a feature point x0 in the image, then find the first shortest point x1 and the second shortest point x2 with the Euclidean distance of the feature point on the image to be matched, these distances respectively are d1 and d2. If the ratio of d1 and d2 is less than the threshold 0.6, then x0 and x1 matching successfully, otherwise it does not hold.

9 Experimental Results and Analysis

In order to verify the improved algorithm and compare the Surf, Brisk, and Harris algorithms for matching rate and matching time, this experiment uses a memory 4G, CPU 2.50 Hz PC to run, the tool used was MATLAB 2017a. Select 5 group city street pictures, in which the 4th group of images was taken at night and under the

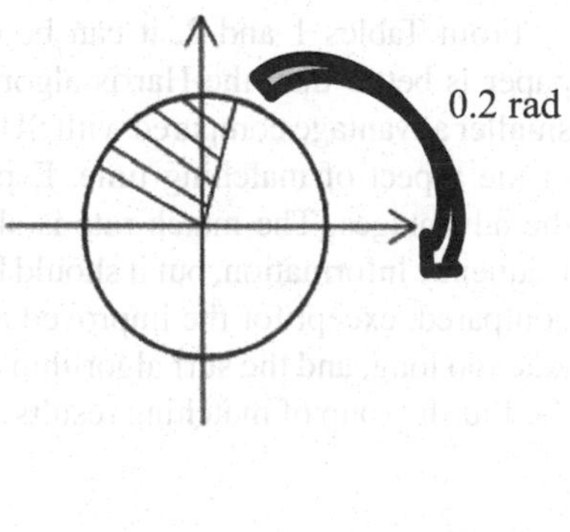

Fig. 4 Traversing a circular neighborhood

Table 1 Match rate comparison

	Group 1 (%)	Group 2 (%)	Group 3 (%)	Group 4 (%)	Group 5 (%)
Improve algorithm	94.56	94.49	95.68	94.20	94.82
Brisk	93.32	92.74	95.49	91.25	94.50
SURF	94.12	94.47	95.88	94.83	94.67
Harris	93.00	92.83	95.00	0	91.36

Table 2 Time-consuming comparison

	Group 1	Group 2	Group 3	Group 4	Group 5
Improve algorithm	2.6250	3.1094	2.6875	5.3130	2.4375
Brisk	6.3750	5.4844	10.0938	69.2969	8.8594
SURF	6.7969	5.0469	9.8750	79.4844	9.2344
Harris	5.9063	4.6094	9.8125	74.4844	10.2969

Unit second

Fig. 5 Group 4 and Group 5 images

influence of strong light. The 5th group (from Baidu) of the to-be-matched images was processed by a rotation of 50°, and the remaining 3 groups were not rotated and images that are not affected by strong light sources. The matching rate is accurate to percentile. The experimental results are as follows.

From Tables 1 and 2, it can be concluded that the improved algorithm in the paper is better than the Harris algorithm in the aspect of matching rate, and has a smaller advantage compared with SURF and Brisk, but has a significant improvement in the aspect of matching time. Especially in the third group can better highlight the advantages. The match rate is slightly lower due to we filtered too much high frequency information, but it should be noted that when the fourth group of images is compared, except for the improved algorithm in the text, the other three algorithms take too long, and the surf algorithm has obvious mis-matching feature point pairs in the fourth group of matching results. The Brisk algorithm has obvious mis-matching

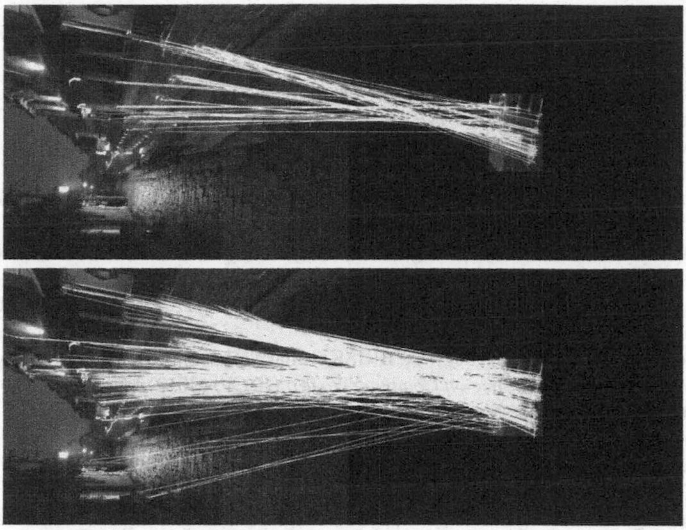

Fig. 6 The Improved algorithm (up) and SURF (down) matched the result of the 4th group

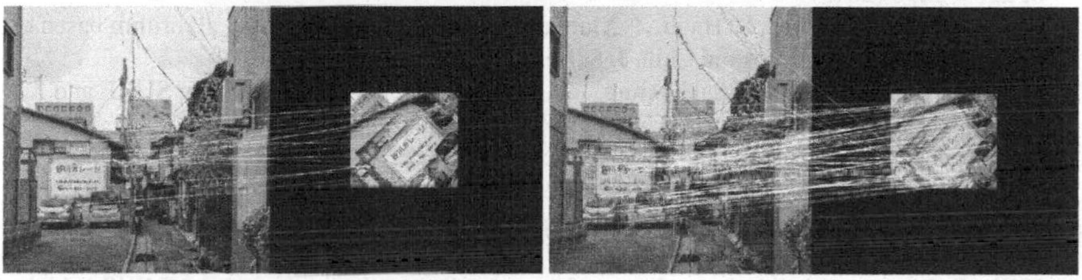

Fig. 7 The Improved algorithm (left) and Brisk (right) matched the results of the 5th group

feature points in the fifth group of images, and the Harris algorithm does not find the matching feature point pairs (Figs. 5, 6 and 7).

10 Conclusion

The algorithm uses Haar wavelet transform to extract the low-frequency part of the image before image matching, and uses the brisk algorithm to extract the feature points of the processed image, instead of the step of extracting the feature point by the SURF, and then describes according to the feature point of the SURF. Make a match. The simulation results show that the improved algorithm improves the matching rate in a small range while significantly reducing the time for matching. More rapid and accurate image matching will be the direction of my next effort.

References

1. Q. Wang, Y. Shen, F. Liu, Layered image matching based on adaptive genetic algorithm. J. Zhejiang Univ. Technol. **6**, 599–603 (2003)
2. Z. Weichuan, C. Dong, Z. Lei, Multi-scale corner detection based on anisotropic Gaussian kernel. J. Electron. Meas. Instrum. **1**, 37–42 (2012)
3. H. Bay, T. Tuytelaars, L. Van Gool, SURF: speeded up robust features. Comput. Vis. Image Underst. **110**(3), 346–359 (2008)
4. M. Sonka, V. Hlavac, R. Boyle, *Image Processing, Analysis and Machine Vision* (Tsinghua University Press, Beijing, 2003), pp. 47–49
5. Y. Dan, Z. Haibin, L. Zhe, *Detailed MATLAB Image Processing Examples* (Tsinghua University Press, Beijing, 2013), pp. 372–373
6. S. Leutenegger, M. Chli, R.Y. Siegwart, BRISK: binary robust invariant scalable keypoints, in *IEEE International Conference on Computer Vision (ICCV)* (2012), pp. 1589–1596
7. E. Rosten, T. Drummond, Machine learning for high-speed corner detection, in *European Conference on Computer Vision (ECCV)* (2006), pp. 430–443
8. D. Qiang, L. Jinghong, W. Chao, Z. Qianfei, Image mosaic algorithm based on improved BRISK. J. Electron. Inf. Technol. **2**, 444–450 (2017)
9. P. Xingcheng, T. Shaofeng, Z. Yi, Mobile robot navigation based on improved FAST algorithm. CAAI Trans. Intell. Syst. **8**, 419–424 (2014)
10. H. Jingshuang, An improved BRISK algorithm for fusion depth information. J. Comput. Appl. **8**, 2285–2290 (2015)
11. Z. Qiguang, Z. Pengzhen, L. Haoli, Z. Xianjiao, C. Ying, Image matching algorithm based on global and local feature fusion. Chin. J. Sci. Instrum. **1**, 170–176 (2016)
12. Z. Lulu, G. Guohua, L. Kang, H. Ajing, Image matching algorithm based on SURF and fast approximate nearest neighbor search. Appl. Res. Comput. **3**, 921–923 (2013)
13. H.P. Moravec, Toward automatic visual obstacle avoidance, in *Proceedings of 5th International Joint Conference on Artificial Intelligence* (1977), p. 584
14. G. David, Object recognition from local scale-invariant features. Int. Conf. Comput. Vis. **9**(2), 1150–1157 (1999)
15. Y. Jia, Robust control with decoupling performance for steering and traction of 4WS vehicles under velocity-varying motion. IEEE Trans. Control Syst. Technol. **8**(3), 554–569 (2000)
16. Y. Jia, Alternative proofs for improved LMI representations for the analysis and the design of continuous-time systems with polytopic type uncertainty: a predictive approach. IEEE Trans. Autom. Control **48**(8), 1413–1416 (2003)

Towards End-to-End Gesture Recognition with Recurrent Neural Networks

Tong Du and Xuemei Ren

Abstract With the development of smart devices, gesture recognition is used in more and more fields. The current gesture recognition devices on the market are inconvenient and expensive. Human motion analysis and recognition based on attitude sensor is a new field. The algorithm based on the recurrent neural network takes into account the timing information of the actions and can better resolve the uncertainty of the human motion in time, but as the training sample increases, the efficiency becomes lower. This paper proposes an action recognition method based on Connectionist temporal classification for sequence learning. This method realizes end-to-end recognition of gestures.

Keywords Gesture recognition · Connectionist temporal classification · End-to-end

1 Introduction

There are many kinds of expressions of human movements, the common one is the expression of gestures [1–4]. Human body movement is a macroscopic reflection of many comprehensive motor functions such as the human musculoskeletal system and the neural control system. Through the identification of human body motion information, the positions and trajectories of various parts of the body during the exercise process are recorded, and can be obtained through appropriate processing. Kinematics and dynamics information, it has a wide range of applications in medical electronics, elderly care and smart homes.

There are two modes of human motion information recognition: vision-based human motion pattern recognition [5] and sensor-based human motion pattern recognition. At present, a lot of research work has been done on vision-based human motion pattern recognition. However, because of the high dimensional characteristics of the

---

T. Du · X. Ren (✉)
Beijing Institute of Technology, Beijing 100081, China
e-mail: xmren@bit.edu.cn

Y. Jia et al. (eds.), *Proceedings of 2018 Chinese Intelligent Systems Conference*, Lecture Notes in Electrical Engineering 529,
https://doi.org/10.1007/978-981-13-2291-4_15

human body and the randomness and high complexity of human motion [6], detecting and estimating human motion has always been a challenge. The disadvantages of vision-based human motion pattern recognition are that it is too dependent on the external environment [7] and it is difficult to target. Motion pattern recognition based on gesture sensors greatly enriches the content of human-computer interaction [8]. Perceiving human motion and understanding the meaning of human limb movement is the basic function of human motion based on the gesture sensor.

In the past, the motion was often identified by the physical characteristics of the angle and acceleration signals, or the motion was identified by the geometric characteristics of the acceleration signals (period, peak, trough) [9]. The recognition effect was not satisfactory. So we proposed a method of deep learning to determine the behavioral classification [10]. Deep learning has a good feature extraction and classification functions, as a simulation of biological neural network algorithm framework, has great research value [11].

2 Network Architecture

The purpose of the RNN is to use to process sequence data. RNN is called recurrent neural network. The recurrent neural network has a certain memory function, but it does not handle the long-term dependency problem well. Long-term dependency is such a problem that it is difficult to learn relevant information when the predic-

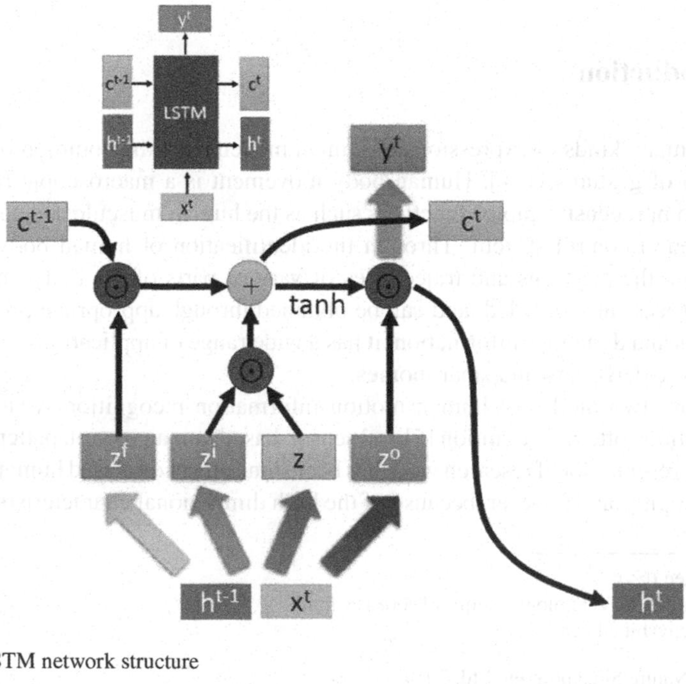

Fig. 1 LSTM network structure

tion point is far from the relevant information of the dependence. Long Short Term Mermory network (LSTM) can solve long-term dependencies [12]. The network structure of LSTM is shown in Fig. 1. The formula for this network is derived as follows:

$$z = tanh(w \cdot [x^t, h^{t-1}] + b^c) \tag{1}$$

$$z^i = \sigma(w^i \cdot [x^t, h^{t-1}] + b^i) \tag{2}$$

$$z^f = \sigma(w^f \cdot [x^t, h^{t-1}] + b^f) \tag{3}$$

$$z^o = \sigma(w^o \cdot [x^t, h^{t-1}] + b^o) \tag{4}$$

$$c^t = z^f \odot c^{t-1} + z^i \odot z \tag{5}$$

$$h^t = z^o \odot tanh(c^t) \tag{6}$$

$$y^t = \sigma(w^{'} h^t) \tag{7}$$

x^t represents the input of t = 1, 2, 3 …,
h^t, c^t is the hidden layer step t, it is the network's memory unit.
y^t is the output of step t.

3 Connectionist Temporal Classification

3.1 From Outputs to Labellings

Neural networks are typically trained as sequence classifiers in gestures recognition. We need to consider the alignment between the input and output sequences. Connectionist Temporal Classification (CTC) [13] is an objective function. It does not require any pre-alignment between the input sequence and the target sequence [14].

The output layer outputs an identification tag for each cell, plus an extra cell called "blank" that corresponds to null emissions. Assuming that the input is a sequence x of length T, the output y_t is processed with the softmax function.

$$P(k, t|x) = \frac{exp(y_t^k)}{\sum_{k'} exp(y_t^{k'})} \tag{8}$$

where y_t^k is element k of y_t, $P(\pi|x)$ of a is the probabilities at every time-step:

$$P(\pi|x) = \prod_{t=1}^{T} P(\pi_t, t|x) \tag{9}$$

For a sequence, there are as many alignments as possible. Defining a many-to-one map B that removes first the repeated labels, then the blanks from alignments (e.g. $B(a-a-b-) = B(-aa--abb) = aab$, $-$ indicates 'blank'). The total output probability is equal to the sum of the probabilities:

$$P(l|x) = \sum_{\pi \in B^{-1}(l)} P(\pi|x) \tag{10}$$

The probability that the output label sequence is the probability sum of multiple paths.

3.2 *Forward-Backward Algorithm*

Considering the probability that $P(l|x)$ needs to calculate many paths, as the input length increases exponentially, a forward-backward algorithm similar to HMM can be introduced to calculate the probability value. To import the blank node, insert the blank node at the beginning and end of the label. If the original length of the label sequence is U, then U' = 2U + 1.

For label l, the forward variable $a(t, c)$ is the sum probability of all length t paths pre-defined by the length c/2 of B mapped to l. Let the set $V(t, c) = \{\pi \in A'^t : B(\pi) = l_{1:c/2}, \pi_t = l'_c\}$, then we define $a(t, c)$, $p(l|x)$ as :

$$a(t, c) = \sum_{\pi \in V(t,c)} \prod_{i=1}^{t} y^i_{\pi_i} \tag{11}$$

$$p(l|x) = a(T, U') + a(T, U' - 1) \tag{12}$$

All correct paths must begin with the first symbol in either blank(b) or l(l_1), the initial conditions are as follows:

$$a(1, 1) = y^1_b \tag{13}$$

$$a(1, 2) = y^1_{l_1} \tag{14}$$

$$a(1, c) = 0, \forall c > 2 \tag{15}$$

After that, variables can be recursively calculated:

$$a(t, c) = y^t_{l'_c} \sum_{i=f(c)}^{u} c(t-1, i) \tag{16}$$

where

$$f(c) = \begin{cases} c-1 \ if \ \ l'_c = \text{blank} \ \ or \ \ l'_{c-2} = l'_c \\ c-2 \ \text{otherwise} \end{cases} \tag{17}$$

$$a(t, c) = 0 \ \ \forall c < U' - 2(T - t) - 1 \tag{18}$$

$$a(t, 0) = 0 \ \ \forall t \tag{19}$$

Let $W(t, c) = \pi \in A'^{T-t} : B(\hat{\pi} + \pi) = 1 \forall \hat{\pi} \in V(t, c)$. Then

$$b(t, c) = \sum_{\pi \in W(t,c)} \prod_{i=1}^{T-t} y_{\pi_i}^{t+i} \tag{20}$$

Initialization and recursion rules for backward variables

$$b(T, U') = b(T, U' - 1) = 1 \tag{21}$$

$$b(T, c) = 0, \forall c < U' - 1 \tag{22}$$

$$b(t, c) = \sum_{i=c}^{g(c)} b(t+1, i) y_{l'_i}^{t+1} \tag{23}$$

where

$$f(c) = \begin{cases} c-1 \ \ if \ \ l'_c = \text{blank} \ \ or \ \ l'_{c-2} = l'_c \\ c-2 \ \ \text{otherwise} \end{cases} \tag{24}$$

$$b(t, c) = 0 \ \ \forall c > 2t \tag{25}$$

$$b(t, U' + 1) = 0 \ \ \forall t \tag{26}$$

3.3 Loss Function

The CTC loss function L(S) is defined as follows:

$$L(S) = -ln \prod_{(g,z) \in S} p(g|x) = - \sum_{(x,g) \in S} lnp(g|x) \tag{27}$$

$$L(x, g) = -lnp(g|x) \tag{28}$$

According to the forward and backward variables, we can obtain:

$$p(g|x) = \sum_{c=1}^{|g'|} a(t, c) b(t, c) \tag{29}$$

$|g'|$ denotes the U of the label length corresponding to g, and $a(t, c)b(t, c)$ denotes the probability sum of all paths passing through the node c at time t.

$$L(x, g) = -ln \sum_{g=1}^{|g'|} a(t, c)b(t, c) \tag{30}$$

4 Experimental Results

This paper tested 4 gestures. The definition of 4 gestures is shown in Fig. 2. In order to verify the adaptability and accuracy of the algorithm to individual differences, 10 experimenters (5 males and 5 females) were selected to meet the requirements of the gesture definition, and each sequence was operated 25 times in the manner and intensity of their respective habits. Each sequence action contains about 10 defined actions. We randomly selected 185 sets of data as training samples and 55 sets of data as test samples. We use a two-layer neural network structure with 128 nodes per layer. For each neuron output, the last tag is calculated by the CTC algorithm. Thus, the actions in the sequence are identified. This paper uses the LSTM model and the CTC algorithm to train the model parameters on the training data set, and then validates the test data set. After iteration, we get training Loss, training accuracy and test accuracy, they are shown in Figs. 3, 4 and 5. The accuracy of using end-to-end gesture recognition reached 81.3%. Due to the small number of our samples, overfitting occurred. The result is that the accuracy of the test is lower than the correct rate of training. When the amount of data increases, I believe our results will be better. As shown in Fig. 6, the result of applying this algorithm. The following figure output

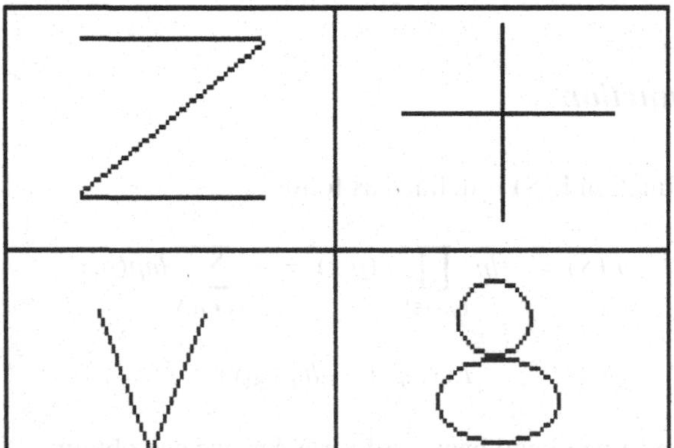

Fig. 2 Gesture type

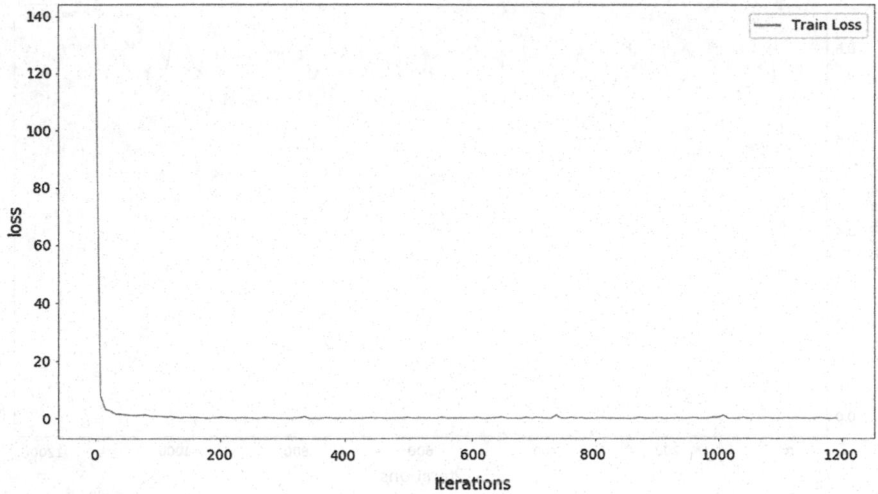

Fig. 3 Training loss

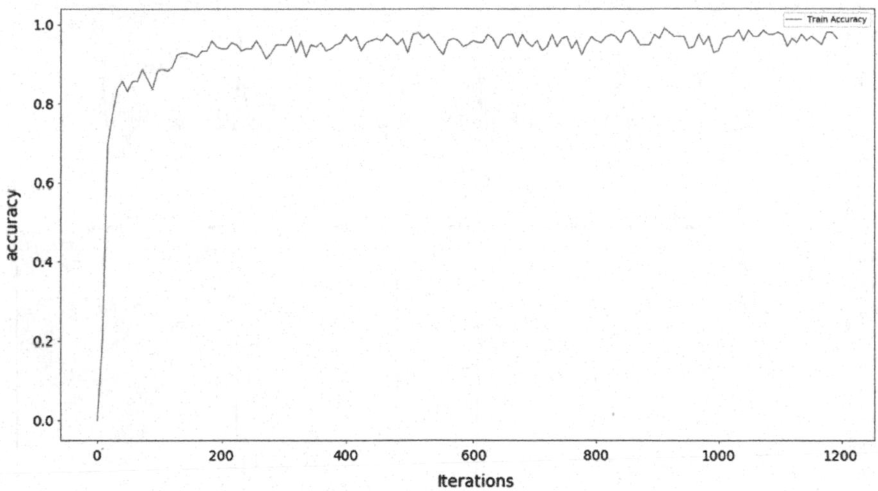

Fig. 4 Training accuracy

signal types. We randomly made five actions, three of which were valid, two were invalid, and all were correctly identified.

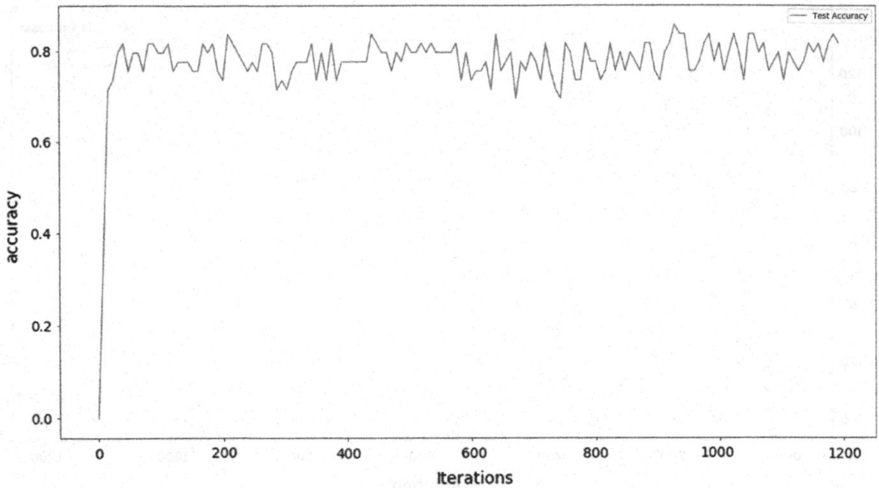

Fig. 5 Testing accuracy

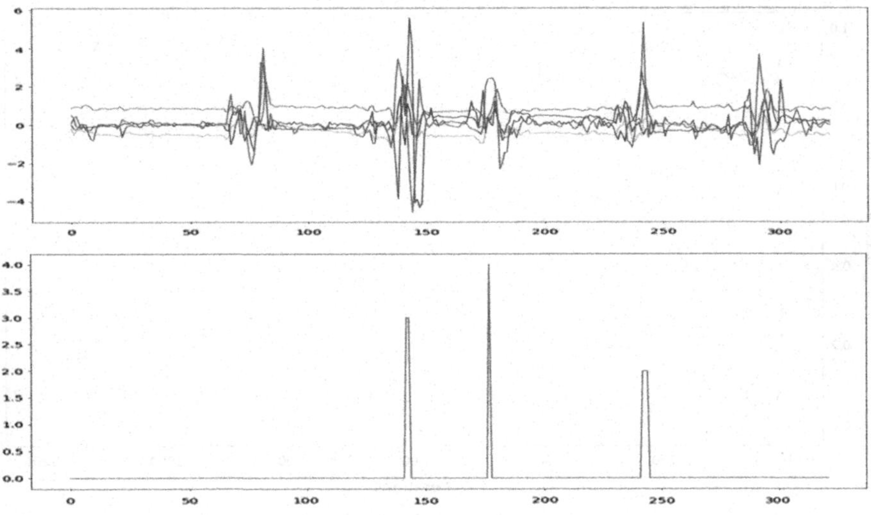

Fig. 6 Online test effect chart

5 Conclusions

This paper designs and implements an end-to-end gesture recognition algorithm based on gesture sensors. In this article we propose a sequential learning action recognition method based on connectionist temporal classification. Using the LSTM model and CTC algorithm to achieve gesture recognition. The recognition accuracy reached 81.3%, and most of the defined actions can be correctly identified. However, for very similar actions, the recognition effect is poor, which is an important direction for future work.

Acknowledgements The work was supported by National Natural Science Foundation of China (No. 61433003, No. 61573174, and No. 61273150).

References

1. C. Maggioni, A novel gestural input device for virtual reality, in *IEEE Virtual Reality Annual* (Seattle, Wash, 1993), pp. 118–124
2. K. Vaananen, K. Boehm, Gesture driven interaction as a human factor in virtual environments, an approach with neural networks, in *Virtual Reality Systems*, Chap. 7 ed. by R. Earnshaw, M. Gigante, H. Jones (Academic Press, Cambridge, MA, 1993), pp. 93–106
3. J. Davis, M. Shah, Visual gesture recognition, in *Visualization and Image Signal Processing* (Apr 1994), pp. 101–106
4. C. Cedras, M. Shah, Motion based recognition: a survey, in *Image and Vision Computing* (Mar 1995), pp. 129–155
5. M. Andriluka, L. Pishchulin, P. Gehler, B. Schiele, 2d human pose estimation: new benchmark and state of the art analysis, in *2014 IEEE Conference on Computer Vision and Pattern Recognition (CVPR)* (IEEE, New York, 2014), pp. 3686–3693
6. B. Sapp, B. Taskar, Modec: multimodal decomposable models for human pose estimation, in *2013 IEEE Conference on Computer Vision and Pattern Recognition (CVPR)* (IEEE, New York, 2013), pp. 3674–3681
7. J. Dai, K. He, J. Sun, Convolutional feature masking for joint object and stuff segmentation, in *Proceedings of the IEEE Conference on Computer Vision and Pattern Recognition* (2015), pp. 3992–4000
8. P. Zappi, T. Stiefmeier, Activity recognition from on-body sensors by classier fusion: sensor scalability and robustness, in *IEEE Intelligent Sensors, Sensor Networks and Information* (Dec 2007), pp. 281–286
9. J. Choi, K. Song, S. Lee, Enabling a gesture based numeric input on mobile phones, in *IEEE International Conference on Consumer Electronics*, Xi-an, China (2011), pp. 151–152
10. Y. Jia, Robust control with decoupling performance for steering and traction of 4WS vehicles under velocity-varying motion. IEEE Trans. Control Syst. Technol. **8**(3), 554–569 (2000)
11. Y. Jia, Alternative proofs for improved LMI representations for the analysis and the design of continuous-time systems with polytopic type uncertainty: a predictive approach. IEEE Trans. Autom. Control **48**(8), 1413–1416 (2003)
12. N. Nishida, H. Nakayama, Multimodal gesture recognition using multi-stream recurrent neural network, in *Pacific-Rim Symposium on Image and Video Technology* (2015)
13. A. Graves, S. Fernandez, F. Gomez, J. Schmidhuber, Connectionist temporal classification: labelling unsegmented sequence data with recurrent neural networks, in *ICML*, Pittsburgh, USA (2006)
14. H. Sak, A. Senior, K. Rao, O. Irsoy, A. Graves, F. Beaufays, J. Schalkwyk, Learning acoustic frame labeling for speech recognition with recurrent neural networks, in *2015 IEEE International Conference on Acoustics, Speech and Signal Processing (ICASSP)* (IEEE, New York, 2015), pp. 4280–4284

Automatic Liver and Tumor Segmentation of CT Based on Cascaded U-Net

Xiaorui Feng, Chaoli Wang, Shuqun Cheng and Lei Guo

Abstract Automatic segmentation of liver and tumor plays a crucial role in medical-aided diagnosis. At present, neural networks have been widely used in medical image processing. There are many FCN-based methods used for the automatic segmentation of the liver and the tumor, but results are not precise enough to the details in the images. In this paper, we use cascaded U-Net to segment livers and tumors automatically. The first U-Net is used to segment livers, and the livers are the input of the second U-Net. We perform experiments on the published 3DIRCAD dataset and the dataset provided by medical institutions. Medical institutions provide CT of patients with advanced liver cancer. Compared with FCN, U-Net is more accurate. When the false positive rate is the same, U-Net's true positive is higher. The accuracy of segmentation of the liver is 91.3 and 89.8%, respectively, and the accuracy of segmentation of the tumor reaches 82.4 and 86.6%.

Keywords Medical image · Segmentation · U-Net · Deep learning · FCN

X. Feng · C. Wang (✉)
School of Optical-Electrical and Computer Engineering, University of Shanghai for Science and Technology, Shanghai 200093, China
e-mail: clwang@usst.edu.cn

X. Feng
e-mail: fengxiaorui@yahoo.com

S. Cheng · L. Guo
Department of Hepatic Surgery, Eastern Hepatobiliary Surgery Hospital, Second Military Medical University, Shanghai 200433, China
e-mail: chengshuqun@aliyun.com

L. Guo
e-mail: goto.st@outlook.com

Y. Jia et al. (eds.), *Proceedings of 2018 Chinese Intelligent Systems Conference*, Lecture Notes in Electrical Engineering 529,
https://doi.org/10.1007/978-981-13-2291-4_16

1 Introduction

Liver cancer is a common malignancy. Early detection and diagnosis of liver cancer is the key to rational treatment. However, the boundaries of tumors are not obvious and the number is uncertain, so radiologists will be difficult and time-consuming to judge. Computer-aided diagnosis has important clinical significance.

Convolutional neural networks have achieved excellent results in computer vision tasks. Due to the large number of training samples, the accuracy rate is very high. In recent years, convolutional neural networks have been widely used in medical image processing. Hamidian et al. [1] converted the 3D CNN to 3D FCN to automatically detect pulmonary nodules in chest CT. Akkus et al. [2] used a patch-based convolutional neural network for brain tumor segmentation. Havaei et al. [3] used a deep convolutional network to achieve brain tumor segmentation. They used a cascaded architecture in which the output of the basic CNN was considered as an additional information for subsequent CNNs. Ravishankar et al. [4] used FCN to segment the kidney. The author adds shape information to the FCN framework. It is necessary to achieve a certain degree of accuracy with few medical images. Therefore, further improvement needs to be studied.

In this paper, the convolutional neural network is applied to detect liver tumors. Due to the differences in the contrast agents and devices, the CT we obtained may have different gray levels. The liver CT have heterogeneities problems and the boundaries are not obvious, so the algorithm used in medical image-assisted diagnosis must be robust.

In recent years, Li et al. [5] used a deep convolutional network to segment the liver and tumor in CT automatically. The patch was the input to predict the image. Since the network must run for each patch, the patch has overlap and a large amount of the redundancy, so the speed was very slow. The patch-based method can only obtain local information, and cannot apply the feature relationships of the tumor and its surrounding areas. It limited performance of classification and reduced accuracy rate. Ben-Cohen et al. [6] used a full convolutional network to segment the liver and tumor. In this paper, FCN was cascaded to perform liver segmentation and tumor segmentation. FCN lost a lot of detail in the up-sampling process. Compared to U-Net network, FCN is not sensitive to details and the result is not precise. The accuracy is low. Christ et al. [7] used cascade U-Net to segment the liver and tumor automatically. This paper mentioned that the liver occupies 7% of the entire CT, and the tumor accounts for 0.25% of the entire CT. Therefore, they used the first U-Net network for liver segmentation and excluded other organs, and then trained a U-Net network to segment the tumor. In addition, there is an algorithm in the algorithm field, which can increase the robustness of the system and can be used to analyze the system's uncertainty interference [8, 9]. The algorithm can solve the uncertainty interference on the picture to some extent and increase the system robustness.

In this paper, according to the methods proposed by Christ et al. [7], we use the first U-Net to segment the liver from the CT as ROI to segment tumors. The proportion of liver and tumor in the entire CT is small, and the dataset is relatively

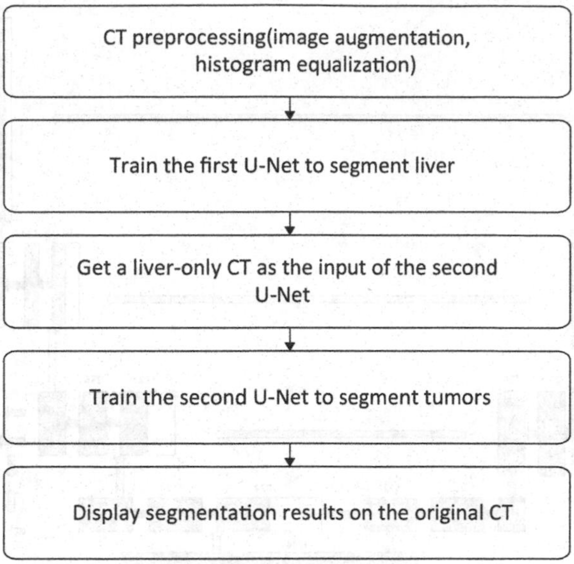

Fig. 1 The flow chart is a brief description of training and testing. Preprocessing the image is to enhance the diversity of the image and improve the robustness of the network. The first U-Net is used to segment the liver from the CT which prevents other organs from causing interference, and then segment the tumor

small. We augment the current images in terms of cropping, rotating, enhancing brightness, elastic distorting, and finally perform histogram equalization to enhance the contrast. We use U-Net to perform automatic liver segmentation and automatic tumor segmentation on the published 3DIRCAD dataset and the dataset provided by medical institutions. The accuracy of segmentation of the liver is 91.3 and 89.8%, respectively, and the accuracy of segmentation of the tumor reaches 82.4 and 86.6%, respectively.

2 Methods

According to the method proposed by us, we show the workflow in Fig. 1. First, we prepare data including data augmentation and histogram equalization. Then the first U-Net (U-Net as shown in Fig. 2) is trained to divide the liver and put only the liver. The liver-only CT is used to train the second U-Net. Finally, the results are shown in the original CT.

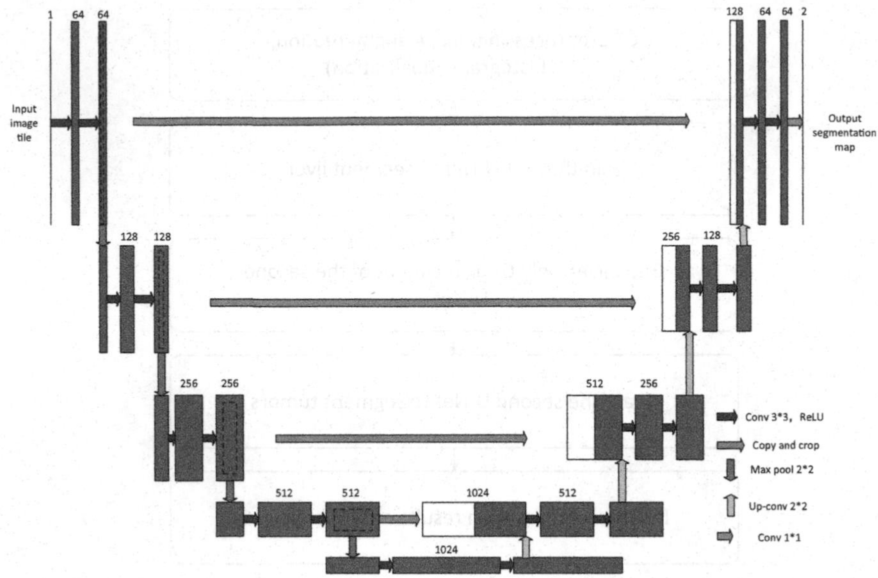

Fig. 2 The U-Net framework is shown in the Fig. 2. The blue box corresponds to the multi-channel feature map. The blue arrow indicates the convolution and activation function. The red arrow indicates pooling. The light green scissors indicate up-sampling progress. The conv 3 × 3 indicates that the convolution kernel is 3 × 3. The grey arrows indicate that the down-sampled features are copied and cropped to the corresponding convolutional layer

2.1 Data Preparation

Since datasets we have are small dataset. If we use the datasets directly to train the neural network, overfitting will occur. The CT obtained by different contrast media and different devices may have differences in gray values, and the liver shape of different people is different too, so the model we get must be robust enough to get high accuracy.

At present, there are many ways to augment the dataset. In [10], the author proposes three ways to artificially increase the number of datasets, namely to crop the original picture, flip horizontally, and change the intensity of RGB channels. The pictures obtained by these methods are far from enough. Ronneberger et al. used excessive data augmentation by applying elastic deformations to the available training images [11]. Dsovitskiy et al. has shown the feasibility of image augmentation, and they use translation, scaling, rotation and enhanced contrast to augment the picture [12].

We find that the gray value of many organs in the CT of the liver is similar to the gray value of the liver. If horizontal flip is used, other organs have a great influence on the detection of the liver. Therefore, we will not use the horizontal flip method for image augmentation. In liver segmentation and tumor segmentation, we use the same method for image augmentation. Each transformation is a composition of elementary transformations from the following list:

Crop: Here a region of a size specified by us is cropped at random from the original image. We could combine this with a resize operation, so that the images returned are the same size as the images of the original, pre-augmented dataset.

Rotation: The value, in this case between $-3°$ and $3°$, is chosen at random.

Brightness enhancement: We change the brightness of the image within a certain range.

Elastic distortions: We make distortions to an image while maintaining the image's aspect ratio.

When we collate datasets, we find that many CT do not have tumors, so in the tumor segmentation experiment, we remove some CT without tumors.

2.2 Cascaded U-Net

Our segmentation network is the U-Net architecture [11]. This network solves the problem of small datasets. It is widely used in medical image processing and is one of the most successful and popular networks.

The FCN classifies the image at the pixel level. While the convolutional network only uses the features of partial convolution layers in the up-sampling process, the FCN-8S performs much better than the FCN-32S, but the result of the up-sampling is still relatively blurry and insensitive.

The U-Net network is an improvement of the FCN. From Fig. 2 we can see that U-Net also provides a good design pattern for jumping connections between different stages of a neural network. The U-Net network combines spatial and context information to achieve pixel-by-pixel classification. At the same time, we found that the results obtained using U-Net retained more details. The input of the U-Net network we use are images with 512×512 pixels.

2.3 Class Balancing

According to the dataset, if the tumor is small, the proportion of the tumor in the CT is very small. If there is no class balance, the trained network is impossible to divide small tumors. We added a weighting factor to the FCN's cross-entropy loss function:

$$L = -\frac{1}{n}\sum_{i=1}^{N} \omega^{class}[\hat{P}_i \log P_i + (1 - \hat{P})\log(1 - P_i)] \quad (1)$$

P_{i} is the probability that voxel i is the foreground. $\omega_i^{class} = \frac{\sum_i 1-\hat{P}_i}{\sum_i \hat{P}_i}$, if $\hat{P}_i = 1$, otherwise, $\omega^{\mathrm{class}} = 1$.

3 Experiments and Results

We applied our approach on the open dataset 3DIRCAD [13] and the dataset provided by medical institutions respectively. The 3DIRCAD dataset provides CT with more diversity and complexity of the liver and its tumor. The other dataset consists of CT of 18 patients from medical institutions. We selected CT of seventeen patients as the training set and one patient's CT as the validation set.

3.1 Parameter Settings

The neural network was implemented through the caffe framework. We use a stochastic gradient descent optimizer. The learning rate is 0.0001 and the momentum is 0.8.

3.2 Quality Measures

There are three criteria for measuring the experimental results in this paper. They are the FCN common metrics Overall accuracy and Mean accuracy [14]. Another is VOE.

Let n_{ij} be the number of pixels in class i whose pixels are predicted to be class j, n_{cl} represent the number of classes, and $t_i = \sum_j n_{ij}$ denote the total number of pixels in class i.

Overall accuracy:

$$\sum_{\mathrm{i}} n_{ii} / \sum_{\mathrm{i}} t_i \tag{2}$$

Mean accuracy:

$$(1/n_{cl}) \sum_{i} n_{ii}/t_i \tag{3}$$

VOE:

$$VOE(A, B) = 1 - \frac{|A \cap B|}{|A \cup B|} \tag{4}$$

Overall accuracy indicates the ratio of the pixels classified correctly to the total pixel. Mean accuracy represents the proportion of pixels classified correctly in each class, and then calculate the average of all classes. VOE is supplement of Jaccard coefficient.

4 Results

Based on the two datasets, we first performed image augmentation on CT, as described in METHODS. Then use the cascaded U-Net network to train the model and finally display the segmentation results on the original image.

First, we performed liver segmentation to extract the liver and rule out the influence of other organs on tumor detection.

From Table 1, we can see that using the U-Net can achieve better segmentation results than FCN. The U-Net network we use can get a better segmentation result for tumor segmentation.

The ROC curve shows the overall performance of the two networks. The lower the false positive, the higher the true positive, indicating that the network works better, as shown in Fig. 3. We can see that U-Net's overall results are better. Despite the fact that FCN's true positivity is higher at the beginning of our own dataset, the lower true positivity is not practical.

Table 1 The accuracy of automatic segmentation of liver

Dataset	Approach	Overall accuracy (%)	Mean accuracy (%)	VOE (%)
3DIRCAD	U-Net	91.3	90.1	11.5
The dataset provided by medical institutions	U-Net	89.8	88.7	13.1
3DIRCAD	FCN	85.2	83.5	18.1
The dataset provided by medical institutions	FCN	82.9	81.7	20.7

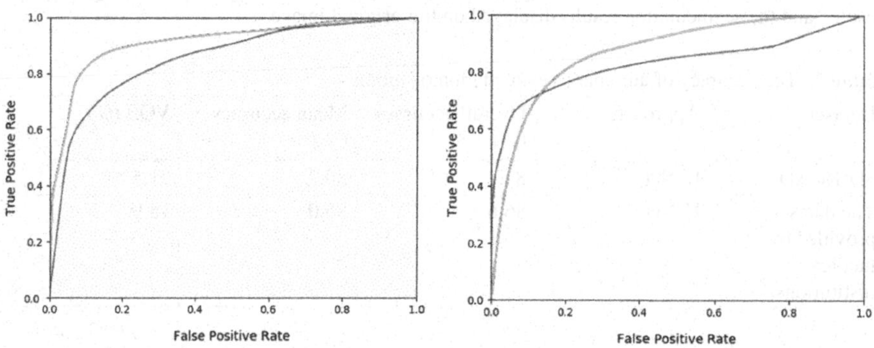

Fig. 3 The first ROC plot shows the detection results of the FCN and U-Net networks on the 3DIRCAD dataset and the dataset provided by medical institutions. The green line represents U-Net, and the blue line represents FCN

The experimental results are as follows (Fig. 4).

Second, we used the liver-only CT to train the second U-Net to segment the tumor. According to the experiment of segmenting the liver, we can obtain that U-Net's segmentation effect is better than that of FCN. Therefore, when performing tumor segmentation experiments, we only use U-Net.

From Table 2, We know that the accuracy of the published dataset is low because the tumors in the dataset are small.

The experimental results are as follows (Fig. 5).

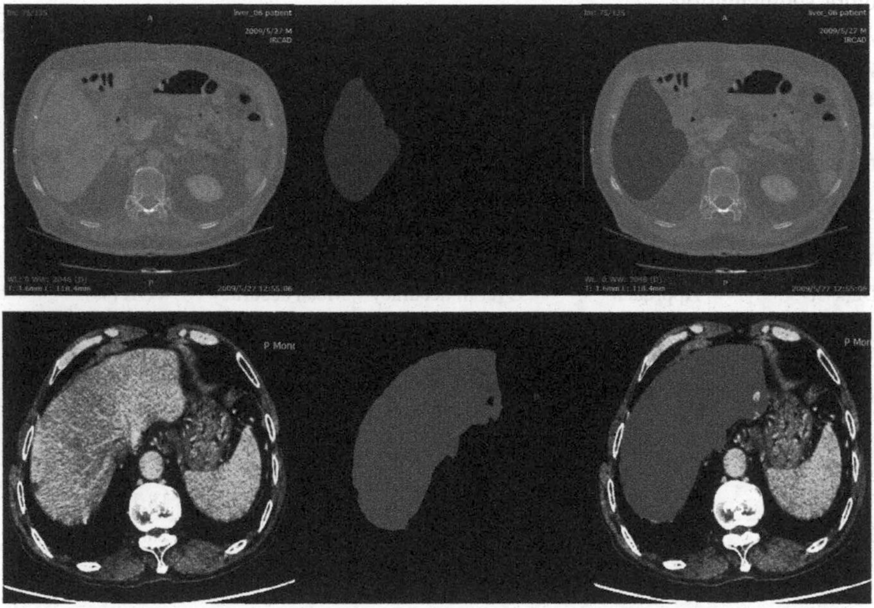

Fig. 4 From left to right are the CT images of 3DIRCAD and medical institutions, the segmentation results, and the segmentation results displayed on the original image

Table 2 The accuracy of automatic segmentation of tumor

Dataset	Approach	Overall accuracy (%)	Mean accuracy (%)	VOE (%)
3DIRCAD	U-Net	82.4	80.2	21.5
The dataset provided by medical institutions	U-Net	86.6	86.0	15.9

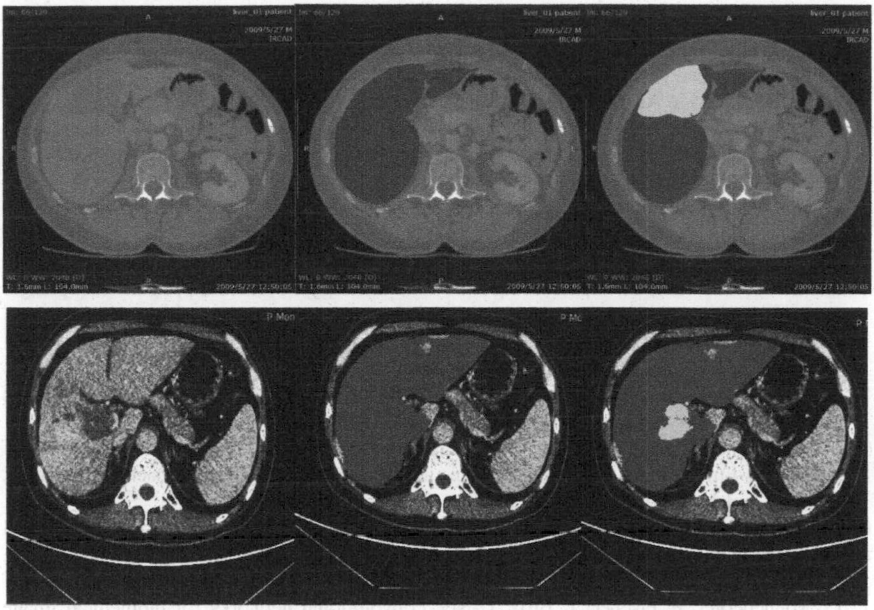

Fig. 5 From left to right are the CT images of 3DIRCAD and medical institutions the liver segmentation results on the original image, and the segmentation results of the tumor on the original image

5 Conclusion

We used a cascaded U-Net network to perform liver and tumor segmentation on 3DIRCAD dataset and the dataset provided by medical institutions. Compared to methods using FCN for liver and tumor segmentation, the segmentation result based on cascade U-Net is more accurate than FCN. Our first U-Net segment the liver from CT automatically, excluding the influence of other organs on the tumor segmentation, and then using the second U-Net segment tumor. This paper shows that using cascaded U-Net improves accuracy and reduces false positives. In future work, we will consider 3D information, such as using 3D CNN and FCN architectures [15, 16], and adding 3D conditional randomization [17] to improve the accuracy of the network.

References

1. S. Hamidian, B. Sahiner, N. Petrick et al., 3D convolutional neural network for automatic detection of lung nodules in chest CT, in *Proceedings of SPIE—The International Society for Optical Engineering*, vol. 10134 (2017), pp. 1013409

2. Z. Akkus, A. Galimzianova, A. Hoogi et al., Deep learning for brain MRI segmentation: state of the art and future directions. J. Digit. Imaging **30**(4), 1–11 (2017)
3. M. Havaei, A. Davy, D. Warde-Farley et al., Brain tumor segmentation with deep neural networks. Med. Image Anal. **35**, 18–31 (2017)
4. H. Ravishankar, R. Venkataramani, S. Thiruvenkadam et al., Learning and incorporating shape models for semantic segmentation, in medical image computing and computer assisted intervention—MICCAI (2017), p. 2017
5. W. Li, F. Jia, Q. Hu, Automatic segmentation of liver tumor in CT images with deep convolutional neural networks. J. Comput. Commun. **03**(11), 146–151 (2015)
6. A. Ben-Cohen, I. Diamant, E. Klang et al., Fully convolutional network for liver segmentation and lesions detection, in *Deep Learning and Data Labeling for Medical Applications* (Springer International Publishing, Berlin, 2016), pp. 77–85
7. P.F. Christ, F. Ettlinger, F. Grün et al., Automatic liver and tumor segmentation of CT and MRI volumes using cascaded fully convolutional neural networks (2017)
8. Y. Jia, Robust control with decoupling performance for steering and traction of 4WS vehicles under velocity-varying motion. IEEE Trans. Control Syst. Technol. **8**(3), 554–569 (2000)
9. Y. Jia, Alternative proofs for improved LMI representations for the analysis and the design of continuous-time systems with polytopic type uncertainty: a predictive approach. IEEE Trans. Autom. Control **48**(8), 1413–1416 (2003)
10. A. Krizhevsky, I. Sutskever, G. Hinton, Imagenet classification with deep convolutional neural networks, in *Annual Conference on Neural Information Processing Systems (NIPS)* (2012), pp. 1106–1114
11. O. Ronneberger, P. Fischer, T. Brox, U-Net: convolutional networks for biomedical image segmentation. **9351**, 234–241 (2015)
12. A. Dosovitskiy, J.T. Springenberg, M. Riedmiller, T. Brox, Discriminative un-supervised feature learning with convolutional neural networks, in *NIPS* (2014)
13. L. Soler, A. Hostettler, V. Agnus et al., 3D image reconstruction for comparison of algorithm database: a patient-specific anatomical and medical image database (2012). http://www-sop.inria.fr/geometrica/events/wam/abstract-ircad.pdf
14. J. Long, E. Shelhamer, T. Darrell, Fully convolutional networks for semantic segmentation, in *Computer Vision and Pattern Recognition* (IEEE, New York, 2015), pp. 3431–3440
15. F. Milletari, N. Navab, S.A. Ahmadi, V-Net: fully convolutional neural networks for volumetric medical image segmentation, in *Fourth International Conference on 3d Vision* (IEEE, New York, 2016), pp. 565–571
16. Ö. Çiçek, A. Abdulkadir, S.S. Lienkamp et al., 3D U-Net: learning dense volumetric segmentation from sparse annotation (2016), pp. 424–432
17. K. Kamnitsas, C. Ledig, V.F. Newcombe et al., Efficient multi-scale 3D CNN with fully connected CRF for accurate brain lesion segmentation. Med. Image Anal. **36**, 61 (2016)

CT Recognition for Liver and Lesions

Lin Li, Chaoli Wang, Shuqun Cheng and Lei Guo

Abstract This article uses a two-step method for liver and lesions recognition, and our main purpose is to identify liver lesions. Therefore, the first step does not need much resources. After the segmentation of the liver is performed, and the recognition of the lesions in the second step requires fine identification. Therefore, it is necessary to identify with U-net, which has a high recognition accuracy. This paper proposes a method that can effectively reduce the complexity of the algorithm and ensure the accuracy. The first step is to use the traditional segmentation algorithm level set to segment the liver, and the second step to increase the segmentation effect by increasing the depth of the U-net and reducing the size of the convolution kernels. 89.3% of the lesions segmentation accuracy can be obtained on commercial medical institutions' CT, and the lesions segmentation accuracy on the open data set can reach more than 90%, which may meet the needs of the assistant doctors.

Keywords Liver segmentation · Lesion identification · Level set · U-Net

L. Li · C. Wang (✉)
School of Optical-Electrical and Computer Engineering, University of Shanghai for Science and Technology, Shanghai 2000093, China
e-mail: clwang@usst.edu.cn

L. Li
e-mail: bup@126.com

S. Cheng · L. Guo
Department of Hepatic Surgery, Eastern Hepatobiliary Surgery Hospital, Second Military Medical University, Shanghai 200433, China
e-mail: chengshuqun@aliyun.com

L. Guo
e-mail: goto.st@outlook.com

Y. Jia et al. (eds.), *Proceedings of 2018 Chinese Intelligent Systems Conference*, Lecture Notes in Electrical Engineering 529,
https://doi.org/10.1007/978-981-13-2291-4_17

1 Introduction

Yann Lecun published an article in *nature* [1], deep learning draws extensive attention, and a large number of researchers have poured into the field of deep learning. Most of the field have introduced deep learning algorithms and achieved good results. The cascading method has greatly improved the accuracy of a single algorithm. For example, Felzenszwalb introduced cascading to object detection, which is higher than the original method [2]. In the field of face detection, some used cascading method to increase the accuracy of the original [3–5]. Later, cascade technology was introduced into the field of liver lesion segmentation. An improved dual FCN network was used to identify the hepatic lesions, which was greatly improved from single network [6]. In summary, the addition of cascading can effectively improve the accuracy of a single network. This paper used the traditional level set algorithm and the U-net method in deep learning, experiments results show that can effectively eliminate interference and improve segmentation accuracy. The first step is to segment the liver from the CT, which we used level set. The second step is to segment the lesion from the hepatic region segmented in the first step. In this step we used an improved FCN network.

2 Related Work

Since Kass proposed the Snakes model in 1987, various image segmentation understanding and recognition methods based on active contour lines have arose. The basic idea of Snakes model is very simple. It uses some control points that make up a certain shape as a template (outline). Through the elastic deformation of the template itself, it is matched with the local features of the image to achieve the harmony, and a certain energy function is minimized. The purpose of constructing Snakes model is to reconcile the contradiction between upper layer knowledge and underlying image features. Since Long J published in 2014, the segmentation algorithm based on FCN structure has developed rapidly [7]. Later, since a single network needs to perform too many tasks and it is not good to directly deepen the network, they put two networks together. In series, the front network performs some simple tasks, and the back network performs more detailed network segmentation. The author designed three networks in series [3]. Three networks detect different sizes of faces to achieve multi-scale detection of faces, and the third network finally outputs detected face frames. The author used three networks to detect faces [4, 5], and implemented detection of different sizes of faces by using different input sizes and network depths. Based on the experience in the field of face recognition, the cascade method was later introduced into the field of liver segmentation. Many scholars have connected the FCN in series to achieve multi-level segmentation. The authors connected two FCNs in series and implemented segmentation of liver lesions based on the addition of 3D-CRF [8]. In this article, the lesion segmentation was performed step by step. The first step was to segment the liver from the CT. Out of the second step, the segmented liver is sent to the second network and the lesion is detected in the segmented liver area. The results of the experiment can be known, compared to the single network to segment lesions, the effect of non-hepatic regions on segmentation can be effectively reduced. Based on the same idea, the author published Survival Net [9], in which the author added a network based on the literature [8] to segment the segmented liver lesions, and plus some artificially extracted features into a new network to predict the survival time of cancer patients. The author also adopted the idea of cascading and achieved better results than a single network [6]. The original method can improve the original performance through improvement [10, 11].

Based on the inspiration of the above literatures, this paper proposes a new idea of segmentation of liver lesions, and divides the liver lesion segmentation step by step. Firstly, the liver was segmented by human intervention, and then the segmented liver image was input into the second-step network for lesion identification. Different from the above methods, this article uses different methods to perform liver and lesion segmentation on CT.

In comparison to prior works, we develop a method to segment lesions. Our contribution in this work is two-fold. First, we proposed a new method by which we can segment the liver from the CT. Secondly, through experiments, this method can be used to process liver CT images in actual commercial medical institutions, which has practical significance and can help doctors to assist in diagnosis.

3 Dataset

Our experimental data consists of two parts. The first part is the liver CT provided by commercial medical institutions, those patients are in middle or late stage. The segmentation is complex and difficult than that in the early stage. The other part is derived from 3Dircadb. This dataset is a public dataset [12] on the Internet. The other is derived from the CT map of the liver and gallbladder in the east. We invited professional doctors to mark them and invited other professional doctor to make a judgment. A total of 18 individual patients' CT were used for this experiment. The CT thickness was 1.5 mm. Energy levels range from 40 to 140 keV for a total of 11 energy levels, and each with three energy levels. The single-tap CT resolution is 512×512 pixels and no data enhancement.

4 Experiment

In comparison to prior works, we develop a method to segmentation lesions. Our contribution in this work is two-fold. Firstly, we proposed a new method by which we can segment the liver from the CT, and the lesion can be recognized. Secondly, this method can be used to process liver CT images in actual commercial medi-

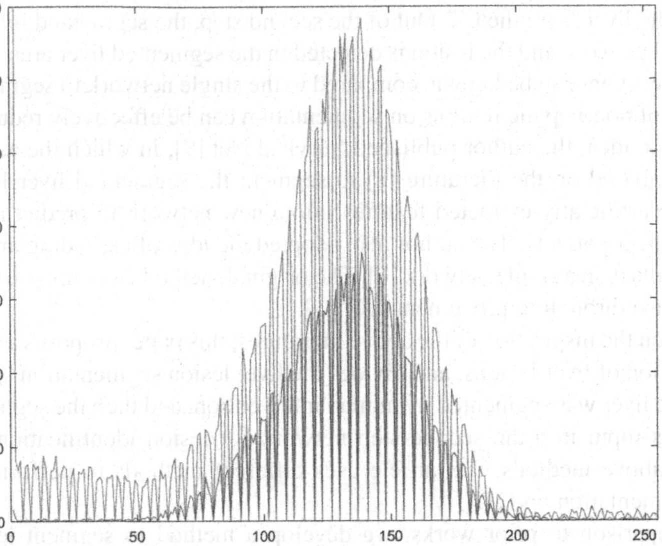

Fig. 1 Original CT liver and non-hepatic gray histogram. It can be seen from this figure that the gray values of the liver (blue) and non-liver (orange) areas are very large and cannot be segmented by setting a grayscale threshold

cal institutions, which has practical significance and can help doctors to assist in diagnosis.

There are many methods for image segmentation. Threshold segmentation can be used for segmentation with high contrast images. A new FCN segmentation method based on convolutional neural network is proposed [7]. Another method widely used in medical image segmentation is U-NET [6, 8, 9]. In this section we mainly explore threshold segmentation based on gray information and active contour segmentation based on level set algorithm.

Before segmentation, we preprocess the pictures. The global contrast is increased by histogram equalization. Especially for liver CT images, the gray values of the images are similar, the brightness is low, and the boundary is not conspicuous. By histogram equalization, we can make the brightness better in the histogram. In the upper distribution, the local contrast of the processed liver region is enhanced without affecting the overall contrast.

The gray histograms of the original liver and non-liver regions are shown in Fig. 1. As we can see from Fig. 1, there are many disturbances in the original image. We use mean filter to process the original image, we re-draw the gray distribution map of the liver and the gray distribution map of the entire image (Fig. 2).

As can be seen from the above experiment, we cannot perform liver segmentation by setting a threshold. We need to extract other more advanced features to segment the liver.

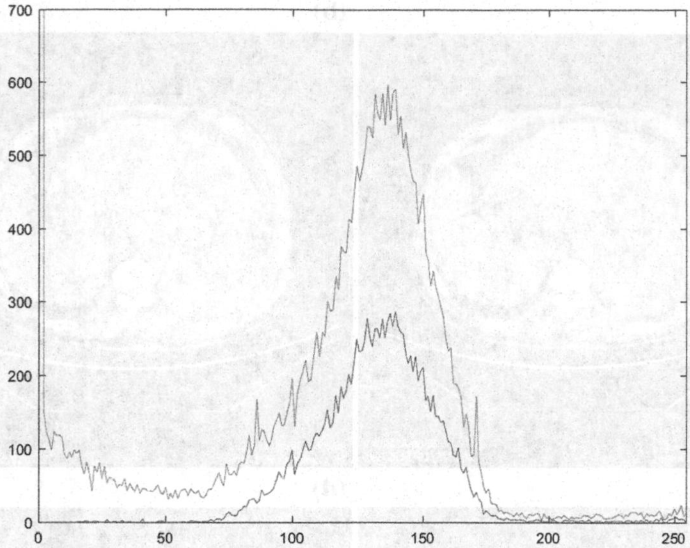

Fig. 2 It can be seen that there is not a certain degree of differentiation between the liver (blue) and non-hepatic regions (orange)

In this paper, the level set segmentation method is used to segment the liver. Firstly we draw an initial closed outline in the liver. Then, the algorithm iterates and the contour line expands to the boundary. When the contour line reaches the boundary, it stops the iteration. The position of the contour line is the boundary of the liver that we need to divide. The first step of the liver segmentation is over. The following pictures are our original picture and the results of our liver segmentation (Fig. 3).

This section mainly describes the use of level set algorithm for segmentation of the liver. The correction segmentation of the liver for the subsequent recognition of the lesion is significant, and the correct segmentation of the liver, you can avoid other organs on the lesion segmentation. From the above experiment, we can see that our first step of liver segmentation can be segmented well with prior knowledge. Compared to single threshold segmentation and FCN segmentation, using level set can be effective. And it can remove the interference points and segment the liver.

In order to segment the lesion, we use a neural network based on U-NET. The network inputs the first part of the segmented liver image into a well-designed full convolutional neural network. Finally, outputs the position of the lesion and mark the outline in the original image.

This article uses the FCN-based improved network U-net [13]. The architecture are as follows (Fig. 4).

The network structure is as shown in the figure above. It contains four convolutional layers, four pooled layers, and two fully connected layers. There is also a corresponding up sampling layer. To detect tiny lesions in the network, we set the size of the convolution kernels in the network to 3 × 3 pixels and the pooled layers

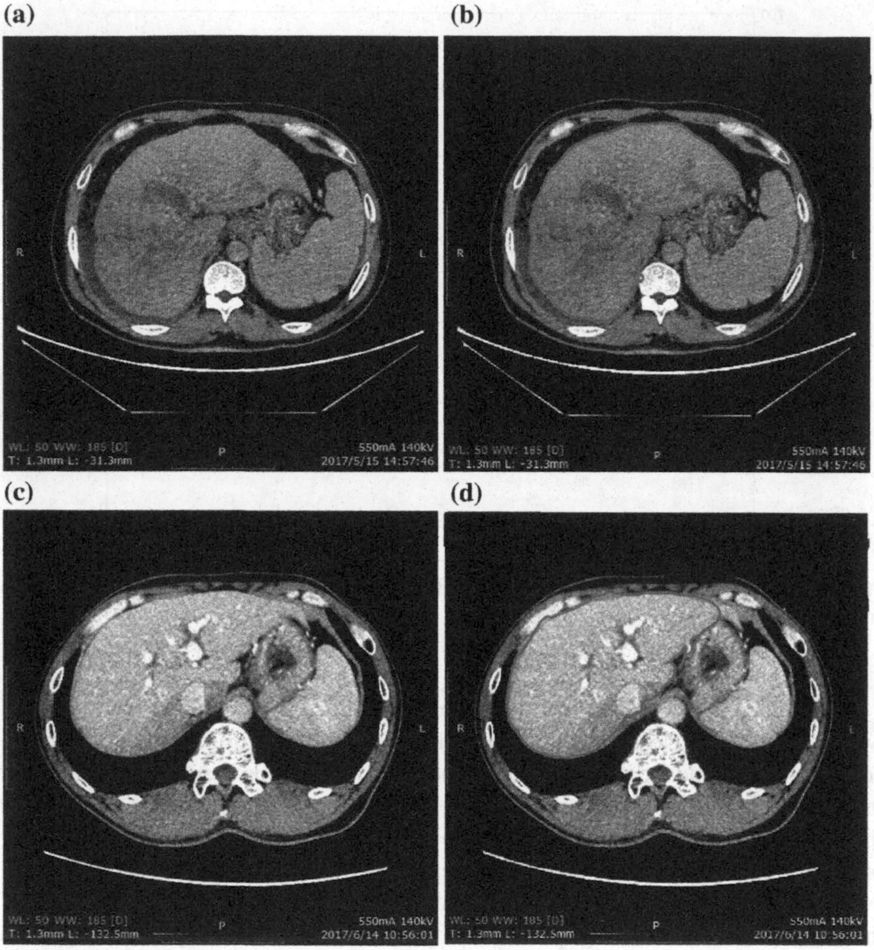

Fig. 3 The image **a** and **c** are the original CT, and the **b** and **d** are the liver segmentation results. From the above figure, we can see that the segmentation effect is satisfactory, and only some of the boundary parts have some interference5. Analysis of experimental results

to 2×2 pixels. Other network parameters are shown above. To ensure that the input and output sizes are the same, we mirrored the input image boundary, and the output image was finally resized to a size of 512×512 pixels.

The input at this stage is the liver segmented image from the first step, except for the portion outside the liver. Image size is 512×512 pixels. The images are as follows.

Figure 5 are the result of experiments performed on data obtain from the commercial medical institutions. From A we can see that the input of the second step is only liver regions. Reduce the interference to the segmentation. After the final lesion was segmented, we marked it on the second step of the liver and finally displayed

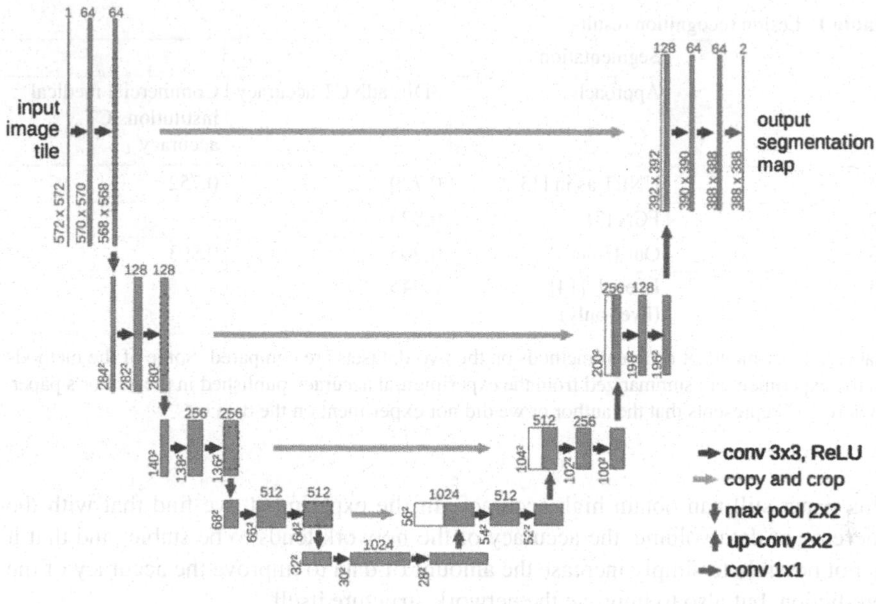

Fig. 4 U-net architecture

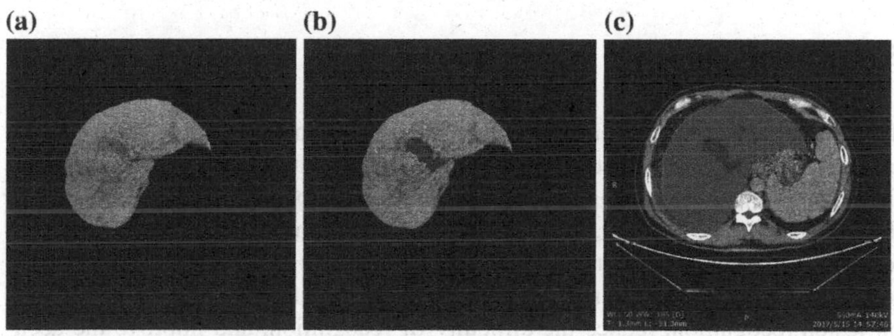

Fig. 5 **a** Shows the image of the liver cut from the first step, **b** shows the result of the lesion segmentation. **c** is the recognition result of lesion and liver marked in the original image. From the figure we can see that compared with the standard template, the partitioning of our algorithm is very good except some boundary processing is not very good. No interference points appear outside the lesion. Can be used as a reference for doctors to judge liver lesions

it on the original CT. It can be seen that the effect of its segmentation is somewhat different from the boundary and the given label, and the other general coincidence rate is still very high. We also conducted experiments on published datasets [12]. The experimental results show that the coincidence rate of lesions segmented with our method by more than 90% is almost comparable to that of professional physicians.

We can see from Table 1 that on the same data set, our network is superior to other methods in the accuracy of segmentation of lesions, and the experimental results of

Table 1 Lesion recognition results

	Segmentation		
	Approach	3Dircadb CT accuracy	Commercial medical institutions CT accuracy
1	UNET as in [13]	0.729	0.752
2	FCN [7]	0.825	–
3	Our U-net	0.905	0.893
4	Li et al. [14] (liver-only)	0.945	–

Table 1 experiments of different methods on the two datasets are compared. Some of the methods in the experiment are summarized from the experimental accuracy published in the author's paper. Where "–" represents that the author or we did not experiment on the data set

this paper still can obtain high accuracy. In the experiment, we find that with the increase of data volume, the accuracy of the network tends to be stable, and that it is not possible to simply increase the amount of data to improve the accuracy of the prediction, but also to improve the network structure itself.

5 Conclusion

This article proposes a new idea to solve the liver lesion recognition, combines the traditional segmentation method with the latest deep learning methods, and identifies liver lesions in two steps. In the first step, we use prior knowledge to segment the liver. In the second step, only the segmented region of the first step is used to identify the liver lesion. Compared with other methods for performing a one-time lesion identification, you can effectively exclude the influence of non-liver organs on lesion recognition, reduce the difficulty of recognition, and improve the accuracy of recognition. Because of the number of CT images used in this paper, there are some limitations. The next step is to increase the robustness and generalization performance of the model with increasing data volume.

References

1. Y. Lecun, Y. Bengio, G. Hinton, Deep learning. Nature **521**(7553), 436 (2015)
2. P.F. Felzenszwalb, R.B. Girshick, D. Mcallester, Cascade object detection with deformable part models, in *Computer Vision and Pattern Recognition* (IEEE, New York, 2010), pp. 2241–2248
3. H. Li, Z. Lin, X. Shen et al., A convolutional neural network cascade for face detection, in *Computer Vision and Pattern Recognition* (IEEE, New York, 2015), pp. 5325–5334
4. H. Qin, J. Yan, X. Li et al., Joint training of cascaded CNN for face detection, in *Computer Vision and Pattern Recognition* (IEEE, New York, 2016), pp. 3456–3465

5. K. Zhang, Z. Zhang, Z. Li et al., Joint face detection and alignment using multitask cascaded convolutional networks. IEEE Signal Process. Lett. **23**(10), 1499–1503 (2016)
6. M. Bellver, K.K. Maninis, J. Ponttuset et al., Detection-aided liver lesion segmentation using deep learning, in *NIPS* (2017)
7. J. Long, E. Shelhamer, T. Darrell, Fully convolutional networks for semantic segmentation. IEEE Trans. Pattern Anal. Mach. Intell. **39**(4), 640–651 (2017)
8. P.F. Christ, M.E.A. Elshaer, F. Ettlinger et al., Automatic liver and lesion segmentation in CT using cascaded fully convolutional neural networks and 3D conditional random fields, in *CVPR* (2016), pp. 415–423
9. P.F. Christ, F. Ettlinger, G. Kaissis et al., SurvivalNet: predicting patient survival from diffusion weighted magnetic resonance images using cascaded fully convolutional and 3D convolutional neural networks (2017), pp. 839–843
10. Y. Jia, Robust control with decoupling performance for steering and traction of 4WS vehicles under velocity-varying motion. IEEE Trans. Control Syst. Technol. **8**(3), 554–569 (2000)
11. Y. Jia, Alternative proofs for improved LMI representations for the analysis and the design of continuous-time systems with polytopic type uncertainty: a predictive approach. IEEE Trans. Autom. Control **48**(8), 1413–1416 (2003)
12. L. Soler, A. Hostettler, V. Agnus, A. Charnoz, J. Fasquel, J. Moreau, A. Osswald, M. Bouhadjar, J. Marescaux, 3D image reconstruction for comparison of algorithm database: a patient-specific anatomical and medical image database (2012). http://www-sop.inria.fr/geometrica/events/wam/abstract-ircad.pdf
13. O. Ronneberger, P. Fischer, T. Brox, U-Net: convolutional networks for biomedical image segmentation, in *Medical Image Computing and Computer-Assisted Intervention—MICCAI 2015* (Springer International Publishing, New York, 2015), pp. 234–241
14. C. Li, X. Wang, S. Eberl et al., A likelihood and local constraint level set model for liver tumor segmentation from CT volumes. IEEE Trans. Bio-med. Eng. **60**(10), 2967–2977 (2013)

Adaptive Neural Control of Stochastic Nonlinear System with Dynamic Uncertainties

Meizhen Xia, Tianping Zhang and Ziwen Wu

Abstract In this paper, an adaptive neural dynamic surface control (DSC) scheme is developed for a class of stochastic nonlinear systems in the presence of input and state unmodeled dynamics. A dynamic signal is employed to handle the state unmodeled dynamics. A normalization signal and a novel adjustable parameter are used to deal with the input unmodeled dynamics. By theoretical analysis, it is shown that all the signals in the closed-loop system are bounded in probability.

Keywords Stochastic nonlinear systems · Unmodeled dynamics · Dynamic surface control

1 Introduction

Since the backstepping design in [1] was first proposed, it has been widely used in the controller design of the nonlinear system. In recent years, it has also been extended to the controller design for the stochastic nonlinear system [2, 3]. By using the quartic Lyapunov function instead of the classical quadratic one, the stabilization problem of stochastic nonlinear systems were solved in [4–7]. In [8], the fuzzy adaptive control was proposed for stochastic nonlinear systems with unknown virtual control gain function. With the backstepping scheme and the minimum learning parameters algorithm, an adaptive neural control method was proposed to devise the controller for the stochastic nonlinear strict-feedback systems with multiple time-varying delays in [9]. But, there exists a serious drawback of backstepping scheme, which is called explosion of complexity or computational burden. In order to make up for the disadvantage of the backstepping, the dynamic surface control (DSC) [10] was proposed, which can also avert the circular argument in the course of stability analysis. In [11, 12], adaptive DSC technique was used in controller designing for the nonlinear systems with unknown dead zone or uncertainties. In [13], the dynamic surface

M. Xia · T. Zhang (✉) · Z. Wu
College of Information Engineering, Yangzhou University, Yangzhou 225127, China
e-mail: tpzhang@yzu.edu.cn

Y. Jia et al. (eds.), *Proceedings of 2018 Chinese Intelligent Systems Conference*, Lecture Notes in Electrical Engineering 529,
https://doi.org/10.1007/978-981-13-2291-4_18

control approach was first introduced to solve the stabilization problem of stochastic nonlinear systems with the output feedback form. As we all know, many uncertain factors exist in the control systems, such as unmodeled dynamics, external or internal disturbances and so on. Much attention has been focused on the unmodeled dynamics. In [14], the unmodeled dynamics was handled through introducing an available dynamic signal, adaptive backstepping design procedure was used to devise the controller. Furthermore, it can also be solved by the description of Lyapunov function in [15]. In [16, 17], the input modeled dynamics was analysed and come to the conclusion that the input modeled dynamics can bring to systems destabilizing effects even though it is in a stable linear form. In [18], an adaptive neural output feedback scheme was investigated for the nonlinear systems with uncertainties including state and input unmodeled dynamics. In [19], combining DSC or backstepping method with stochastic small-gain theorem, adaptive control schemes were presented for a class of stochastic nonlinear systems with unmodeled dynamics or unknown dead zone.

In this paper, an adaptive dynamic surface control scheme is proposed for a class of stochastic strict-feedback nonlinear systems with input unmodeled dynamics and dynamic uncertainties. The design makes the dynamic surface control method be extended to the stochastic nonlinear systems. The dynamic signal and a normalization signal are introduced to handle the unmodeled dynamics and the input unmodeled dynamics, respectively. By the theoretical analysis, it is shown that all signals in the closed-loop system are bounded in probability.

2 Problem Statement and Assumptions

Consider a class of stochastic strict-feedback nonlinear systems with unmodeled dynamics described by

$$\begin{cases} \dot{\varsigma} = \chi(t, \varsigma, x_1) \\ dx_i = (f_i(\bar{x}_i) + g_i(\bar{x}_i)x_{i+1} + d_i(x, \varsigma))dt + h_i^T(\bar{x}_i)dw \\ i = 1, \ldots, n-1 \\ dx_n = (f_n(\bar{x}_n) + g(\bar{x}_n)v + d_n(x, z))dt + h_n^T(\bar{x}_n)dw \\ y = x_1 \end{cases} \tag{1}$$

where $x = [x_1, \ldots, x_n]^T \in R^n$, $\bar{x}_i = [x_1, \ldots, x_i]^T$ is the states, $\varsigma \in R^{n_1}$ is the unmodeled dynamics; $y \in R$ is the output of the system; w is an r-dimensional standard Brownian motion defined on the complete probability space (Ω, F, P) with Ω being a sample space, F being a σ field and P being a probability measure. $d_i(\cdot)$, $\chi(\cdot)$, $h_i(\cdot)$, $f_i(\cdot)$ are unknown continuous functions and $g(\cdot)$ is also an unknown continuous function, $d_i(\cdot)$ is dynamic disturbances. The input unmodeled dynamics subsystem is described as follows:

$$\begin{cases} \dot{\xi} = A_\Delta(\xi) + B_\Delta u \\ v = C_\Delta(\xi) + D_\Delta u \end{cases} \tag{2}$$

where $\xi \in R^{n_2}$ is the state of unmodeled dynamics driven by the input; $v \in R$ is the output of the ξ-subsystem. $A_\Delta(\xi)$, $C_\Delta(\xi)$ are unknown functions; B_Δ is an unknown vector and D_Δ is an unknown constant. The control objective is to design adaptive control signal u for system (1) such that the output y follows the specified desired trajectory y_d.

Assumption 1 The sign of $g_i(\cdot)$ is known, and there exist constants $g_{\min}$ and $g_{\max}$, such that $0 < g_{\min} \leq |g_i(\cdot)| \leq g_{\max}$, $i = 1, \ldots, n$. Without loss of generality, we shall assume that

$$0 < g_{\min} \leq g_i(\cdot) \leq g_{\max}, i = 1, \ldots, n.$$

Assumption 2 For each $1 \leq i \leq n$, there exist unknown continuous functions $\varphi_{i1}(\cdot) \geq 0$ and $\varphi_{i2}(\cdot) \geq 0$ such that

$$|d_i(x, \varsigma)| \leq \varphi_{i1}(\|\bar{x}_i\|) + \varphi_{i2}(\|\varsigma\|) \tag{3}$$

Assumption 3 The unmodeled dynamics is exponentially input-state-practically stable (exp-Isps), that is, for the ς-system, there exist class k_∞-functions $\underline{a}, \bar{a}$ and a Lyapunov function $W(\varsigma)$ such that

$$\underline{a}(\|\varsigma\|) \leq W(\varsigma) \leq \bar{a}(\|\varsigma\|) \tag{4}$$

$$\frac{\partial W(\varsigma)}{\partial \varsigma}\chi(t, \varsigma, x_1) \leq -cW(\varsigma) + \gamma(|x_1|) + d \tag{5}$$

where $\gamma(\cdot) \in k_\infty$ is a known function and $c > 0$, $d > 0$ are known constants.

Assumption 4 The desired trajectory vectors are continuous and available, $x_d = [y_d, \dot{y}_d, \ddot{y}_d]^T \in \Omega_d$ with known compact set $\Omega_d = \left\{x_d : y_d^2 + \dot{y}_d^2 + \ddot{y}_d^2 \leq B_0\right\} \subset R^3$, whose size B_0 is a known positive constant.

Assumption 5 The input unmodeled dynamics subsystem has relative degree zero, that is $D_\Delta \neq 0$, without loss of generality, assume $D_\Delta > 0$, and there exists an unknown nonnegative constant $\bar{c}$ such that

$$|C_\Delta(\xi(t))| \leq \bar{c}\,\|\xi(t)\|\,, \forall \xi \in R^{n_2} \tag{6}$$

Assumption 6 The input unmodeled dynamics subsystem is globally exponential with the convergence rate δ_0 when $u = 0$, and the input unmodeled dynamics subsystem is also exponential stable when $A_\Delta(\cdot)$ is globally Lipschiz, that is, there exists a Lyapunov function $U(\xi)$ satisfying

$$b_1\,\|\xi\|^2 \leq U(\xi) \leq b_2\,\|\xi\|^2 \tag{7}$$

$$\frac{\partial U}{\partial \xi}A_\Delta(\xi) \leq -2\delta_0 U(\xi) \tag{8}$$

$$\left|\frac{\partial U}{\partial \xi}\right| \leq b_3\,\|\xi\| \tag{9}$$

where b_1, b_2, b_3 are positive constants, and δ_0 is a known positive constant.

Begin with the nonlinear system described by the following stochastic differential equation

$$dx = f(t,x)dt + h(t,x)dw \tag{10}$$

where $x \in R^n$ is the system state, w is r-dimensional independent standard Winner process. $f : R^+ \times R^n \to R^n$ and $h : R^+ \times R^n \to R^{n\times r}$ are locally bounded and locally Lipschitz continuous in $x \in R^n$ for all $t \geq 0$. For any given $V(t,x) \in C^{1,2}$, associated with the system (10), we have

$$dV(t,x) = LV(t,x) + \frac{\partial V(t,x)}{\partial x^T} h dw \tag{11}$$

where the infinitesimal generator L is defined as follows

$$LV(t,x) = \frac{\partial V(t,x)}{\partial t} + \frac{\partial V(t,x)}{\partial x^T} f + \frac{1}{2} tr[h^T \frac{\partial^2 V(t,x)}{\partial x^T \partial x} h] \tag{12}$$

Lemma 1 *For any stochastic process $\{x(t)\}$, if there exists integer p and a positive constant C_0 such that $E|x(t)|^p \leq C_0, \forall t \geq 0$, then $\{x(t)\}$ is bounded in probability.*

Lemma 2 *Consider system (10) and suppose there exists a C^2 function $V(t,x(t))$: $R^n \times R \to R^+$, two constants $d_1 > 0$, $d_2 \geq 0$, class κ_∞ functions μ_1, μ_2 such that*

$$\mu_1(\|x\|) \leq V(t,x) \leq \mu_2(\|x\|) \tag{13}$$

$$LV \leq -d_1 V + d_2 \tag{14}$$

for all $x \in R^n$ and $t > t_0$. Then, (i) for any initial state $x_0 \in R^n$, there exists a unique strong solution $x(t)$ for system (10); (ii) the solution $x(t)$ of system (10) is bounded in probability; (iii)

$$E[V(t_0,x)] \leq EV(x_0,t_0)e^{-c_1 t} + c_1^{-1}c_2, \quad \forall t \geq t_0 \tag{15}$$

Lemma 3 *If $W(\varsigma)$ is an exp-Isps Lyapunov function for the subsystem $\dot{\varsigma} = q(t,\varsigma,x_1)$, i.e. inequality (4), (5) hold, then for any constant $\bar{c} \in (0,c)$ and initial condition $\varsigma_0 = \varsigma(t_0)$, $t_0 > 0$, for any function $\bar{\gamma}(|x_1|) \geq \gamma(|x_1|)$ and $r_0 > 0$, there exist a finite $T_0 = \max\{0, \ln[\frac{W(\varsigma_0)}{r_0}]/(c-\bar{c})\} \geq 0$, a non-negative function $D(t,t_0)$, defined for all $t \geq t_0$ and a signal described by*

$$\dot{r} = -\bar{c}r + \bar{\gamma}(|x_1|) + d, r(t_0) = r_0 \tag{16}$$

such that $D_0(t,t_0) = 0$ for all $t \geq t_0 + T_0$ and $W(\varsigma) \leq r(t) + D(t,t_0)$ for all $t \geq t_0$ with $D(t,t_0) = \max\{0, e^{-c(t-t_0)}W(\varsigma_0) - e^{-\bar{c}(t-t_0)}r_0\}$.

Lemma 4 *Consider system (2) and construct a first-order system* $\dot{\bar{m}} = -\delta_0 \bar{m} + |u|$. *If Assumption 6 holds and supposes* $u(t) \in L_\infty[0, T]$, *then there exist constants* $\alpha_1, \alpha_2 > 0$ *satisfying,*

$$\|\xi(t)\| \le \alpha_1(\|\xi(0)\| + |\bar{m}(0)|)e^{-\delta_0 t} + \alpha_2 |\bar{m}(t)| . \tag{17}$$

where δ_0 *is given in (8),* ξ *is determined by (2).*

In this paper, unknown smooth nonlinear function $\psi_i(Z_i) : R^{n_i} \to R$ *will be approximated on a compact set* Ω_{Z_i} *by the following radial basis function neural networks*

$$\psi_i(Z_i) = W_i^{*T} \phi_i(Z_i) + \varepsilon_i(Z_i), \forall Z_i \in \Omega_{Z_i} \tag{18}$$

where $\varepsilon_i(Z_i)$ *is the approximation error,* $\phi_i(Z_i) = [\phi_{i1}(Z_i), \ldots, \phi_{il_i}(Z_i)]^T : \Omega_{Z_i} \to R^{l_i}$ *is a known smooth vector function with the neural network node number* $l_i > 1$, *the basis functions* $\phi_{ij}(Z_i), 1 \le j \le l_i$ *are chosen as the commonly used Gaussian functions with the form* $\phi_{ij}(Z_i) = \exp(\frac{-\|Z_i-\mu_{ij}\|^2}{d_{ij}^2}), j = 1, \ldots, l_i$ *where* μ_{ij}, d_{ij} *are the center and the width of the basis function* $\phi_{ij}(Z_i)$, *respectively.* Z_i *and* $\psi_i(Z_i)$ *will be given later. The optimal weight vector* $W_i^* = [w_{i1}^*, \ldots, w_{il_i}^*]^T$ *is defined as*

$$W_i^* = \arg \min_{W_i \in R^{l_i}} \{ \sup_{Z_i \in \Omega_{Z_i}} \left|\psi_i(Z_i) - W_i^T \phi_i(Z_i)\right|\} \tag{19}$$

Let $\lambda_i = g_{min}^{-1} \|W_i^*\|^2$, $i = 1, \ldots, n$, $\hat{\lambda}_i$ *is the estimation of the unknown constant* λ_i *and the estimation error satisfies* $\tilde{\lambda}_i = \hat{\lambda}_i - \lambda_i \bar{\hat{\lambda}}_i = [\hat{\lambda}_1, \ldots, \hat{\lambda}_i]^T$.

3 Adaptive Controller Design

In this section, the adaptive controller is designed.

Step 1 ($i = 1$). Let $\omega_1 = y_d$, $s_1 = x_1 - \omega_1$. Combined with (1) and (11), it follows that

$$ds_1 = (f_1(\bar{x}_1) + g_1(\bar{x}_1)x_2 + d_1(x, \varsigma) - \dot{\omega}_1)\, dt + h_1^T(\bar{x}_1)dw \tag{20}$$

Define some Lyapunov functions as follows:

$$V_{s_1} = \frac{1}{4}s_1^4 \tag{21}$$

$$V_{s_r} = V_{s_1} + \frac{r}{\lambda_0} \tag{22}$$

$$V_1 = V_{s_r} + \frac{g_{\min}}{2\beta_1}\tilde{\lambda}_1^2 \tag{23}$$

According to (12) and (21), we obtain

$$LV_{s_1} = s_1^3(f_1(\bar{x}_1) + g_1(\bar{x}_1)x_2 + d_1(x, \varsigma) - \dot{\omega}_1) + \frac{3}{2}s_1^2 h_1^T(\bar{x}_1)h_1(\bar{x}_1) \quad (24)$$

Based on Assumption 2, we have

$$\left|s_1^3 d_1(x, \varsigma)\right| \leq \left|s_1^3\right| \varphi_{11}(\|\bar{x}_1\|) + \left|s_1^3\right| \varphi_{12}(\|\varsigma\|) \quad (25)$$

According to Assumption 3 and Lemma 3, we obtain

$$\|\varsigma\| \leq \underline{a}^{-1}(r(t) + D(t, t_0)) \quad (26)$$

In view of the monotone-increasing function $\varphi_{12}(\cdot)$, we have

$$\varphi_{12}(\|\varsigma\|) \leq \varphi_{12}(\underline{a}^{-1}(r(t) + D(t, t_0))) \quad (27)$$

Utilizing Young's inequality, the following inequalities hold

$$\left|s_1^3\right| \varphi_{11}(\|\bar{x}_1\|) \leq \frac{3}{4}s_1^4 \varphi_{11}^{\frac{4}{3}}(\|\bar{x}_1\|) + \frac{1}{4} \quad (28)$$

$$\left|s_1^3\right| \varphi_{12}(\|\varsigma\|) \leq \frac{3}{4}s_1^4 \varphi_{12}^{\frac{4}{3}}(\underline{a}^{-1}(r(t) + D(t, t_0))) + \frac{1}{4} \quad (29)$$

$$\frac{3}{2}s_1^2 h_1^T(\bar{x}_1)h_1(\bar{x}_1) \leq \frac{9}{16}s_1^4 \|h_1(\bar{x}_1)\|^4 + 1 \quad (30)$$

Let

$$\begin{aligned}\psi_1(Z_1) = {} & f_1(\bar{x}_1) + \frac{3}{4}s_1\varphi_{11}^{\frac{4}{3}}(\|\bar{x}_1\|) + \frac{3}{4}s_1\varphi_{12}^{\frac{4}{3}}(\underline{a}^{-1}(r(t) + D(t, t_0))) \\ & + \frac{9}{16}s_1 \|h_1(\bar{x}_1)\|^4 + \frac{\bar{\gamma}(|x_1|)}{\lambda_0 \cdot s_1^3} - \dot{\omega}_1\end{aligned} \quad (31)$$

with $Z_1 = [s_1, r, \omega_1, \dot{\omega}_1]^T \in R^4$. So

$$LV_{sr} \leq s_1^3 g_1(\bar{x}_1)x_2 + s_1^3\psi_1(Z_1) - \frac{\bar{c}}{\lambda_0}r + \frac{d}{\lambda_0} + \frac{3}{2} \quad (32)$$

Design ω_2 as follows

$$\tau_2\dot{\omega}_2 + \dot{\omega}_2 = \alpha_1, \quad \omega_2(0) = \alpha_1(0) \quad (33)$$

where τ_2 is a design constant, which will be given later.

Let $y_2 = \omega_2 - \alpha_1$, based on (33), we have $\dot{w}_2 = -\frac{y_2}{\tau_2}$, $x_2 = s_2 + y_2 + \alpha_1$.

Applying Young's inequality leads to

$$s_1^3 g_1(\bar{x}_1)x_2 = s_1^3 g_1(\bar{x}_1)(s_2 + y_2 + \alpha_1)$$
$$\leq [\frac{3}{4}s_1^4 + \frac{1}{4}g_{\max}^4 s_2^4] + [\frac{3}{4}s_1^4 + \frac{1}{4}g_{\max}^4 y_2^4] + s_1^3 g_1(\bar{x}_1)\alpha_1 \tag{34}$$

Using RBFNNs to approximate the unknown function $\psi_1(Z_1)$ with $Z_1 \in \Omega_{Z_1}$, so the following result holds

$$s_1^3\psi_1(Z_1) = s_1^3(W_1^{*T}\phi_1(Z_1) + \varepsilon_1(Z_1))$$
$$\leq \frac{g_{\min}}{2a_1}\lambda_1 s_1^6 \|\varphi_1(Z_1)\|^2 + \frac{a_1}{2} + \frac{3}{4}s_1^4 + \frac{1}{4}\varepsilon_1^4(Z_1) \tag{35}$$

Choose the virtual control α_1 and the adaptive law of $\hat{\lambda}_1$ as follows:

$$\alpha_1 = -k_1 s_1 - \frac{1}{2a_1}s_1^3\hat{\lambda}_1 \|\phi_1(Z_1)\|^2 \tag{36}$$

$$\dot{\hat{\lambda}}_1 = \beta_1(\frac{1}{2a_1}s_1^6 \|\phi_1(Z_1)\|^2 - \sigma_1\hat{\lambda}_1) \tag{37}$$

where $k_1, \beta_1, \sigma_1, a_1$ are positive design constants.

Based on (23), (32), (34)–(37), we obtain

$$LV_1 \leq (-k_1 g_{\min} + \frac{9}{4})s_1^4 + \frac{1}{4}g_{\max}^4(s_2^4 + y_2^4)$$
$$-\frac{1}{2}g_{\min}\sigma_1\tilde{\lambda}_1^2 - \frac{c}{\lambda_0}r + \frac{1}{4}\varepsilon_1^4(Z_1) + C_1 \tag{38}$$

where $C_1 = \frac{1}{2}g_{\min}\sigma_1\lambda_1^2 + \frac{d}{\lambda_0} + \frac{a_1}{2} + \frac{3}{2}$.

The infinitesimal generator of α_1 satisfies

$$L\alpha_1 = \sum_{j=1}^{2}\frac{\partial\alpha_1}{\partial x_1}(f_1(\bar{x}_1) + g_1(\bar{x}_1)x_2 + d_1(x,\varsigma)) + \frac{\partial\alpha_1}{\partial y_d}\dot{y}_d + \frac{\partial\alpha_1}{\partial\dot{y}_d}\ddot{y}_d$$
$$+\frac{\partial\alpha_1}{\hat{\lambda}_1}\dot{\hat{\lambda}}_1 + \frac{\partial\alpha_1}{r}\dot{r} + \frac{1}{2}\sum_{p,q=1}^{2}\frac{\partial\alpha_1}{\partial x_{p,q=1}}h_p^T(\bar{x}_p)h_q^T(\bar{x}_q) \tag{39}$$

$$d\alpha_1 = L\alpha_1 dt + \sum_{j=1}^{2}\frac{\partial\alpha_1}{\partial x_j}h_j^T(\bar{x}_j)dw \tag{40}$$

$$Ly_2 = \dot{\omega}_2 - L\alpha_1 = -\frac{y_2}{\tau_2} - L\alpha_1 \tag{41}$$

$$dy_2 = Ly_2 dt - \sum_{j=1}^{2}\frac{\partial\alpha_1}{\partial x_j}h_j^T(\bar{x}_j)dw \tag{42}$$

From (12), we obtain

$$\frac{1}{4}Ly_2^4 = -\frac{y_2^4}{\tau_2} - y_2^3 L\alpha_1 + \frac{3}{2}y_2^2[\sum_{j=1}^{2}\frac{\partial\alpha_1}{\partial x_j}h_j^T(\bar{x}_j)][\sum_{j=1}^{2}\frac{\partial\alpha_1}{\partial x_j}h_j^T(\bar{x}_j)] \tag{43}$$

There exists two nonnegative smooth functions $\eta_2(\cdot)$ and $\kappa_2(\cdot)$ such that the following results hold

$$\left|Ly_2 + \frac{y_2}{\tau_2}\right| = |-L\alpha_1| \le \eta_2(\bar{s}_2, y_2, r, \hat{\lambda}_1, y_d, \dot{y}_d, \ddot{y}_d) \tag{44}$$

$$\frac{3}{2}y_2^2[\sum_{j=1}^{2}\frac{\partial\alpha_1}{\partial x_j}h_j^T(\bar{x}_j)][\sum_{j=1}^{2}\frac{\partial\alpha_1}{\partial x_j}h_j^T(\bar{x}_j)] \le \kappa_2(\bar{s}_2, y_2, \hat{\lambda}_1, r, y_d, \dot{y}_d, \ddot{y}_d) \tag{45}$$

With the help of (43)–(45), we get

$$\frac{1}{4}Ly_2^4 \le -\frac{y_2^4}{\tau_2} + \frac{3}{4}y_2^4 + \frac{1}{4}\eta_2^4 + \kappa_2 \tag{46}$$

Step $i(2 \le i \le n-1)$. Let $s_i = x_i - \omega_i$, then we have

$$ds_i = (f_i(\bar{x}_i) + g_i(\bar{x}_i)x_{i+1} + d_i(x, \varsigma) - \dot{\omega}_i)dt + h_i^T(\bar{x}_i)dw \tag{47}$$

Define the Lyapunov functions of the following form

$$V_{s_i} = \frac{1}{4}s_i^4 \tag{48}$$

$$V_i = V_{s_i} + \frac{g_{\min}}{2\beta_i}\tilde{\lambda}_i^2 \tag{49}$$

Repeat the same approach used in step 1, we have

$$LV_{s_i} \le s_i^3 g_i(\bar{x}_i)x_{i+1} + s_i^3\psi_i(Z_i) + \frac{3}{2} \tag{50}$$

where

$$\psi_i(Z_i) = f_i(\bar{x}_i) + \frac{3}{4}s_i\varphi_{i1}^{\frac{4}{3}}(\|\bar{x}_i\|) + \frac{3}{4}s_i\varphi_{i2}^{\frac{4}{3}}((\underline{a}^{-1}(r(t) + D(t, t_0))) + \frac{9}{16}s_i\|h_i(\bar{x}_i)\|^4 - \dot{\omega}_i \tag{51}$$

with $Z_i = [\bar{x}_{i+1}, r, \omega_i, \dot{\omega}_i]^T$.

Define ω_{i+1} in such a way that

$$\tau_{i+1}\dot{\omega}_{i+1} + \omega_{i+1} = \alpha_i, \quad w_{i+1}(0) = \alpha_i(0) \tag{52}$$

where τ_{i+1} is a design constant that we will choose later.

Let $y_{i+1} = \omega_{i+1} - \alpha_i$, based on (52), we have

$$\dot{\omega}_{i+1} = -\frac{y_{i+1}}{\tau_{i+1}}, x_{i+1} = s_{i+1} + y_{i+1} + \alpha_i$$

Choose the virtual control α_i and the adaptive law of $\hat{\lambda}_i$ as follows

$$\alpha_i = -k_i s_i - \frac{1}{2a_i} s_i^3 \hat{\lambda}_i \left\|\phi_i(Z_i)\right\|^2 \tag{53}$$

$$\dot{\hat{\lambda}}_i = \beta_i\left(\frac{1}{2a_i} s_i^6 \left\|\phi_i(Z_i)\right\|^2 - \sigma_i \hat{\lambda}_i\right) \tag{54}$$

where $k_i, \beta_i, \sigma_i, a_i$ are positive design constants that we will give later. Through (12), we obtain

$$LV_i \le \left(-k_i g_{\min} + \frac{9}{4}\right)s_i^4 + \frac{1}{4} g_{\max}^4 (s_{i+1}^4 + y_{i+1}^4) - \frac{1}{2} g_{\min}\sigma_i \tilde{\lambda}_i^2 + \frac{1}{4}\varepsilon_i^4(Z_i) + C_i \tag{55}$$

where $C_i = \frac{1}{2} g_{\min}\sigma_i \lambda_i^2 + \frac{a_i}{2} + \frac{3}{2}$.

$$\begin{aligned} L\alpha_i = & \sum_{j=1}^{i+1} \frac{\partial \alpha_i}{\partial x_j}(f_j(\bar{x}_j) + g_j(\bar{x}_j)x_{j+1} + d_j(x, \varsigma)) + \frac{\partial \alpha_j}{\partial y_d}\dot{y}_d + \frac{\partial \alpha_j}{\partial \dot{y}_d}\ddot{y}_d + \frac{\partial \alpha_1}{r}\dot{r} \\ & + \sum_{j=1}^{i} \frac{\partial \alpha_i}{\partial \hat{\lambda}_i}\dot{\hat{\lambda}}_1 + \sum_{j=2}^{i} \frac{\partial \alpha_i}{\partial \omega_i}\dot{\omega}_i + \frac{1}{2}\sum_{p,q=1}^{i+1} \frac{\partial \alpha_1}{\partial x_p \partial x_q} h_p^T(\bar{x}_p) h_q^T(\bar{x}_q) \end{aligned} \tag{56}$$

$$d\alpha_i = L\alpha_i dt + \sum_{j=1}^{i+1} \frac{\partial \alpha_i}{\partial x_j} h_j^T(\bar{x}_j) dw \tag{57}$$

$$Ly_{i+1} = \dot{\omega}_{i+1} - L\alpha_i = -\frac{y_{i+1}}{\tau_{i+1}} - L\alpha_i \tag{58}$$

$$dy_{i+1} = Ly_{i+1}dt - \sum_{j=1}^{i+1} \frac{\partial \alpha_i}{\partial x_j} h_j^T(\bar{x}_j) dw \tag{59}$$

From (12), we obtain

$$\frac{1}{4}Ly_{i+1}^4 = -\frac{y_{i+1}^4}{\tau_{i+1}} - y_{i+1}^3 L\alpha_i + \frac{3}{2}y_{i+1}^2[\sum_{j=1}^{i+1}\frac{\partial\alpha_i}{\partial x_j}h_j^T(\bar{x}_j)][\sum_{j=1}^{i+1}\frac{\partial\alpha_i}{\partial x_j}h_j^T(\bar{x}_j)] \quad (60)$$

There exist two smooth functions $\eta_{i+1}(\cdot)$ and $\kappa_{i+1}(\cdot)$ such that the following results hold

$$\left|Ly_{i+1} + \frac{y_{i+1}}{\tau_{i+1}}\right| = |-L\alpha_i| \le \eta_{i+1}(\bar{s}_{i+1}, y_{i+1}, r, \hat{\lambda}_i, y_d, \dot{y}_d, \ddot{y}_d) \quad (61)$$

$$\frac{3}{2}y_{i+1}^2[\sum_{j=1}^{i+1}\frac{\partial\alpha_i}{\partial x_j}h_j^T(\bar{x}_j)][\sum_{j=1}^{i+1}\frac{\partial\alpha_i}{\partial x_j}h_j^T(\bar{x}_j)] \le \kappa_{i+1}(\bar{s}_{i+1}, y_{i+1}, \hat{\lambda}_i, r, y_d, \dot{y}_d, \ddot{y}_d) \quad (62)$$

Step $n(i = n)$. Let $s_n = x_n - \omega_n$, we have

$$ds_n = (f_n(\bar{x}_n) + g_n(\bar{x}_n)v + d_n(x, \varsigma) - \dot{\omega}_n)dt + h_n^T(\bar{x}_n)dw \quad (63)$$

Define the Lyapunov functions of the following form

$$V_{s_n} = \frac{1}{4\tilde{D}_\Delta}s_n^4 \quad (64)$$

$$V_n = V_{s_n} + \frac{1}{2\beta_n}\tilde{\lambda}_n^2 + \frac{1}{2r_1}\tilde{H}^2 \quad (65)$$

where $\tilde{D}_\Delta = g_{\min}D_\Delta$.

$$\begin{aligned} LV_{s_n} = &\frac{1}{\tilde{D}_\Delta}s_n^3[f_n(\bar{x}_n) + d_n(x, \varsigma) - \dot{\omega}_n] + \frac{1}{\tilde{D}_\Delta}s_n^3 g_n(\bar{x}_n)[C_\Delta(\xi) + D_\Delta u] \\ &+ \frac{3}{2\tilde{D}_\Delta}s_n^2 h_n^T(\bar{x}_n)h_n(\bar{x}_n) \end{aligned} \quad (66)$$

With Assumption 5 and Lemma 4, the following inequalities hold

$$|C_\Delta(\xi)| \le \bar{c}\,\|\xi(t)\| \le \bar{c}\alpha_1(\|\xi(0)\| + |\bar{m}(0)|)e^{-\delta_0(t)} + \bar{c}\alpha_2\,|\bar{m}(t)| \quad (67)$$

$$\frac{|C_\Delta(\xi)|}{1 + |\bar{m}(t)|} \le \frac{\bar{c}\alpha_1(\|\xi(0)\| + |\bar{m}(0)|)e^{-\delta_0(t)} + \bar{c}\alpha_2\,|\bar{m}(t)|}{1 + |\bar{m}(t)|} \le H_{\bar{m}} \quad (68)$$

where $H_{\bar{m}} = \max\{\bar{c}\alpha_1(\|\xi(0)\| + |\bar{m}(0)|), \bar{c}\alpha_2\}$.
Thus, we have

$$\frac{1}{\tilde{D}_\Delta}\left|s_n^3 g_n(\bar{x}_n)C_\Delta(\xi)\right| \le \frac{3}{4}s_n^4 + \frac{1}{4\tilde{D}_\Delta^4}g_{\max}^4 H_{\bar{m}}^4(1 + |\bar{m}(t)|)^4 \quad (69)$$

Let $P_{\bar{m}} = (1 + |\bar{m}(t)|)^4$, $H_c = \frac{1}{4\tilde{D}_\Delta^4} g_{\max}^4 H_{\bar{m}}^4$, so

$$\frac{1}{\tilde{D}_\Delta} \left| s_n^3 g_n(\bar{x}_n) c_\Delta(\xi) \right| \leq \frac{3}{4} s_n^4 + P_{\bar{m}} H_c \leq \frac{3}{4} s_n^4 + s_n^4 P_{\bar{m}} H + (1 - \frac{s_n^4}{\varepsilon^*}) P_{\bar{m}} H_c \tag{70}$$

where $H = \frac{H_c}{\varepsilon^*} \varepsilon^*$ is a positive design constant. From (67)–(70), we have

$$LV_{s_n} \leq s_n^3 \psi_n(Z_n) + \frac{3}{4} s_n^4 + s_n^4 H P_{\bar{m}} + \frac{g_n(\bar{x}_n)}{g_{\min}} s_n^3 u + \frac{3}{2} \tag{71}$$

where

$$\psi_n(Z_n) = \frac{1}{\tilde{D}_\Delta} (f_n(\bar{x}_n) + \frac{9}{16\tilde{D}_\Delta} s_n \|h_n(\bar{x}_n)\|^4 + \frac{3}{4} s_n \tilde{D}_\Delta^{7/3} \varphi_{n1}^{\frac{4}{3}}(\|\bar{x}_n\|)$$
$$+ \frac{3}{4} s_n \tilde{D}_\Delta^{7/3} \varphi_{n2}^{\frac{4}{3}}(\underline{a}^{-1}(r(t) + D(t, t_0))) - \dot{\omega}_n) + (\frac{1}{s_n^3} - \frac{s_n}{\varepsilon^*}) P_{\bar{m}} H_c \tag{72}$$

with $Z_n = [\bar{x}_n^T, \omega_n, \dot{\omega}_n]^T$.

Choose the control law u and the adaptive laws of $\hat{\lambda}_n, \hat{H}$ as follows

$$u = -k_n s_n - \frac{1}{2a_n} s_n^3 \hat{\lambda}_n \|\phi_n(Z_n)\|^2 - s_n P_{\bar{m}} \hat{H} \tag{73}$$

$$\dot{\hat{\lambda}}_n = \beta_n (\frac{1}{2a_n} s_n^6 \|\phi_n(Z_n)\|^2 - \sigma_n \hat{\lambda}_n) \tag{74}$$

$$\dot{\hat{H}} = r_1 (s_n^4 P_{\bar{m}} - \delta_1 \hat{H}) \tag{75}$$

where $k_n, \beta_n, \sigma_n, a_n, \delta_1, r_1$ are positive design constants which we will be given later, $\hat{\lambda}_n, \hat{H}$ are the estimates of λ_n, H, $\tilde{\lambda}_n = \hat{\lambda}_n - \lambda_n$, $\tilde{H} = \hat{H} - H$.

$$LV_{s_n} \leq (-k_n + \frac{3}{2}) s_n^4 - \frac{1}{2a_n} \tilde{\lambda}_n s_n^6 \|\phi_n(Z_n)\|^2 - s_n^4 P_{\bar{m}} \tilde{H}$$
$$+ \frac{1}{4} \varepsilon_n^4 + \frac{a_n}{2} + \frac{3}{2} \tag{76}$$

$$LV_n \leq (-k_n + \frac{3}{2}) s_n^4 - \frac{1}{2} \sigma_n \tilde{\lambda}_n^2 - \frac{1}{2} \delta_1 \tilde{H}^2 + \frac{1}{4} \varepsilon_n^4(Z_n) + C_n \tag{77}$$

where $C_n = \frac{1}{2}\sigma_n \lambda_n^2 + \frac{1}{2} r_1 H^2 + \frac{a_n}{2} + \frac{3}{2}$.

Theorem 1 *For system (1), the controller (73), and adaptation laws (54), (74) and (75), and bounded initial conditions, if Assumptions 1–6 are true, then there exist* *constants* $k_i > 0$, $\tau_{i+1} > 0$, $\beta_i > 0$, $\sigma_i > 0$ *satisfying* $V(0) \leq c$, *such that all of the signals in the closed-loop system are bounded in probability, and* k_i *and* τ_{i+1} *satisfy*

$$\begin{cases} k_1 \geq \frac{1}{g_{\min}}(\frac{9}{4} + \frac{\alpha_0}{4}), k_i \geq \frac{1}{g_{\min}}(\frac{9}{4} + \frac{1}{4}g_{\max}^4 + \frac{\alpha_0}{4}), i = 2, \ldots, n-1 \\ k_n \geq \frac{3}{2} + \frac{\alpha_0}{4}, \frac{1}{\tau_{i+1}} \geq \frac{1}{4}g_{\max}^4 + \frac{3}{4} + \frac{\alpha_0}{4}, i = 1, \ldots, n-1 \\ \alpha_0 \leq \min(\beta_1\sigma_1, \ldots, \beta_n\sigma_n, \delta_1 r_1) \end{cases} \quad (78)$$

where $c > 0$ *is a positive constant,* V *will be given later.*

Proof To save space, the proof and simulation results are omitted.

4 Conclusion

Using the radial basis function networks, an adaptive dynamic surface control scheme is proposed for a class of stochastic nonlinear systems with unmodeled dynamics. Unmodeled dynamics is dealt with through the dynamic signal. The designed adaptive controller can guarantee that all signals in the closed-loop system are bounded in probability, theoretical analysis and simulation results are given to illustrate the effectiveness of the proposed control scheme.

Acknowledgements This work was partially supported by the National Natural Science Foundation of China (61573307 and 61473250) and Yangzhou University Top-level Talents Support Program (2016).

References

1. I. Kanellakopoulos, P.V. Kokotovic, S. Morse, Systematic design of adaptive controllers for feedback linearizable systems. IEEE Trans. Autom. Control **36**(11), 1241–1253 (1991)
2. Z.G. Pan, T. Basar, Backstepping controller design for nonlinear stochastic systems under a risk-sensitive cost criterion, in *Proceedings of the American Control Conference*, Albuquerque, New Mexico (1997), pp. 1278-1282
3. Z.G. Pan, T. Basar, Adaptive controller design for tracking and disturbance attenuation in parametric strict-feedback nonlinear systems. IEEE Trans. Autom. Control **43**(8), 1066–1083 (1998)
4. H. Deng, M. Krstic, Stochastic nonlinear stabilization-I : a backstepping design. Syst. Control Lett. **32**(3), 143–150 (1997)
5. H. Deng, M. Krstic, Stochastic nonlinear stabilization- II: inverse optimality. Syst. Control Lett. **32**(3), 151–159 (1997)
6. H. Deng, M. Krstic, output-feedback stochastic nonlinear stabilization. IEEE Trans. Autom. Control **44**(2), 328–333 (1999)
7. H. Deng, M. Krstic, Stabilization of stochastic nonlinear systems with unknown covariance. IEEE Trans. Autom. Control **46**(8), 1237–1253 (2001)

8. Y.C. Wang, H.G. Zhang, Y.Z. Wang, Fuzzy adaptive control of stochastic nonlinear systems with unknown virtual control gain function. Acta Automatica Sinica **32**(2), 170–178 (2006)
9. G.Z. Cui, T.C. Jiao, Y.L. Wei, G.F. Song, Y.M. Chu, Adaptive neural control of stochastic nonlinear systems with multiple time-varying delays and input saturation. Neural Comput. Appl. **25**(3–4), 779–791 (2014)
10. D. Swaroop, J.K. Hedrick, P.P. Yip, Dynamic surface control for a class of nonlinear systems. IEEE Trans. Autom. Control **45**(10), 1893–1899 (2000)
11. T.P. Zhang, S.S. Ge, Adaptive dynamic surface control of nonlinear systems with unknown dead zone in pure feedback form. Automatica **44**(7), 1895–1903 (2008)
12. D. Wang, J. Huang, Neural network-based adaptive dynamic surface control for a class of uncertain nonlinear systems in strict-feedback form. IEEE Trans. Neural Network **16**(1), 195–202 (2005)
13. W.S. Chen, L.C. Jiao, Z.M. Du, Output-feedback adaptive dynamic surface control of stochastic nonlinear systems using neural network. IET Control Theory Appl. **4**(12), 3012–3021 (2010)
14. X.N. Xia, T.P. Zhang, J.M. Zhu, Y. Yang, Adaptive output feedback dynamic surface control of stochastic nonlinear systems with state and input unmodeled dynamics. Int. J. Adapt. Control Signal Process. **1**, 1–13 (2015)
15. Z.F. Li, T.S. Li, B.B. Miao, Adaptive NN control for a class of stochastic nonlinear systems with unmodeled dynamics using DSC technique. Neurocomputing **149**, 142–150 (2015)
16. Z.P. Jiang, L. Praly, Design of robust adaptive controllers for nonlinear systems with dynamic uncertainties. Automatica **34**(7), 825–840 (1998)
17. Z.P. Jiang, D.J. Hill, A robust adaptive backstepping scheme for nonlinear systems with unmodeled dynamics. IEEE Trans. Autom. Control **44**(9), 1705–1711 (1999)
18. M. Krstic, J. Sun, P.V. Kokovic, Robust control of strict and output feedback system with unmodeled dynamics, in *Proceedings of the 34th IEEE Conference on Decision and Control*, New Orleans (1995), pp. 2257–2262
19. S.C. Tong, Z.P. Jiang, J.F. Zhang, Global output feedback stabilization for a class of stochastic non-minimum-phase nonlinear systems. Automatica **44**(8), 1944–1957 (2008)

Tumor Recognition in Liver CT Images Based on Improved CURE Clustering Algorithm

Xinyi Zhu, Chaoli Wang, Shuqun Cheng and Lei Guo

Abstract Spectral Computed Tomography (CT) images can help doctors diagnose the lesions of the organs and the types of organ lesions. According to the gray level information and spatial information of the spectral CT image of the liver, the characteristics of the image are selected. Using the improved Clustering Using Representatives (CURE) unsupervised clustering algorithm to cluster the image features to automatically identify liver tumors, not only does it not need to manually mark a large number of training samples, but also does not require long training on the classification model. This paper has two improvements to the CURE algorithm: (1) Liver is divided into multiple categories, and then combining the multiple categories into two categories according to certain rules instead of being divided into two categories directly by CURE. (2) When the liver in the spectral CT image is healthy, in order to meet the practical application, analyze the image before classification to avoid separating the normal liver into two categories. The experimental results show that the location of liver tumors is well marked based on the improved CURE clustering algorithm. It has a good clinical guidance value after being evaluated by clinicians and imaging doctors.

Keywords CT images · Feature extraction and selection
Improved CURE clustering algorithm · Tumor recognition

X. Zhu · C. Wang (✉)
School of Optical-Electrical and Computer Engineering,
University of Shanghai for Science and Technology, Shanghai 2000093, China
e-mail: clwang@usst.edu.cn

X. Zhu
e-mail: usstzxy@163.com

S. Cheng · L. Guo
Department of Hepatic Surgery, Eastern Hepatobiliary Surgery Hospital,
Second Military Medical University, Shanghai 200433, China
e-mail: chengshuqun@aliyun.com

L. Guo
e-mail: goto.st@outlook.com

Y. Jia et al. (eds.), *Proceedings of 2018 Chinese Intelligent Systems Conference*, Lecture Notes in Electrical Engineering 529,
https://doi.org/10.1007/978-981-13-2291-4_19

1 Introduction

China is a country with a high incidence of liver cancer. On February 3, 2014, the World Cancer Organization released the "Global Cancer Report 2014" showing that the number of new cases and deaths of liver cancer in China ranks first in the world, and liver cancer is one of the diseases with high mortality rate and high death rate in cancer [1, 2].

1.1 Related Work

At present, in the field of hepatic tumor-assisted diagnosis, Brightness mode images, spectrum CT images, and nuclear magnetic resonance images are mainly used as diagnostic data. Zhang et al. used support vector machine to segment the liver tumors in CT images and to segment the liver tumors accurately and effectively [3, 4]. Chen et al. selected the shape, grayscale, texture, gradient as image features and used support vector machine to detect the lung nodule in chest radiographs. As all we know, a large number of training samples need to be manually generated before the tumor is segmented, so that it can accurately identify the tumor, which takes a long time and is greatly influenced by subjective factors [5]. Li et al. used the K-means to automatically extract CT lung tumors. The experimental data showed that the method was almost consistent with the manually extracted regions. Ahmed Afifi et al. used unsupervised detection to identify liver lesions, demonstrating the effectiveness and feasibility of unsupervised algorithms [6, 7]. Patrick Ferdinand Christ et al. used a multi-cascade FCN full convolutional neural network to segment tumors in the liver. To accurately identify hepatic tumors, a professional team was required to label a large number of training samples, and the classification model was trained for a long time. The rate reached more than 90% [8, 9]. Although the accuracy is high, it does not apply to our data set. We used the author's method for our own dataset and the experimental results are as follows:

As shown in Fig. 1, the left picture is the original picture, the middle picture shows the liver site (the red part), and the right picture shows the position of the tumor (the yellow part). It can be seen that the experimental results are not ideal.

In recent years, neural network identification and cluster analysis and recognition technologies have been rapidly developed and widely used [10–13]. The advantage of the latter is that there is no need to label training samples before identifying tumors. Although the latest methods have many advantages, the accuracy of the training samples is high. Considering the uncertainty and robustness in the experiment, this paper chooses the traditional method and improves it [14, 15].

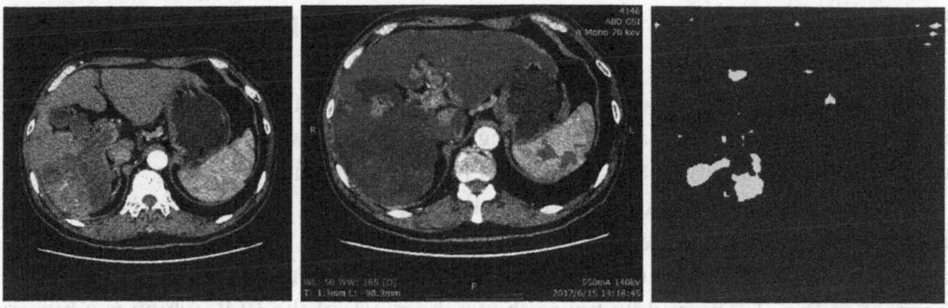

Fig. 1 Experimental results

1.2 The Work of This Paper

In this paper, the subject of the study was the CT image of the liver. The research question is to use the improved CURE clustering algorithm to perform cluster analysis on the liver parts in the spectral CT images, distinguish between the normal region and the tumor region. The classification results were compared with the professional doctors' diagnostic results to determine the accuracy of the classification. Finally, the algorithm network was determined to provide an assistant for doctors to diagnose liver tumors. There are mainly three steps in image classification and recognition. Firstly, determine the ROI (region of interest) of the image; Secondly, extract and select the image features; Thirdly, the images were classified and the locations of liver tumors were marked.

1.3 Contribution

The contribution of this paper is to improve the CURE clustering algorithm for our dataset. There are two improvements to the CURE clustering algorithm in the experiment: (1) The improved CURE clustering algorithm classifies all the small modules into multiple groups, and then merge the multiple categories into two categories according to certain rules instead of directly dividing all modules into two categories. (2) If the liver in the spectral CT image is healthy, it is not correct to divide it into the normal region and the tumor region. Therefore, it is necessary to analyze the spectral CT image before deciding whether to separate it into two categories.

2 Region of Interest and Experimental Samples

In order to improve the recognition rate of tumors, classification experiments were performed within the liver, and a suidu region of interest was selected by MATLAB to perform tumor identification. As shown in the left picture of Fig. 2.

The spectrum CT image size used in the experiment is 512 pi × 512 pi. After the selected region of interest on the spectrum CT image, the ROI is divided into a large number of small modules of equal size as experimental samples, as shown in the right of Fig. 2. The middle figure shows the mask of the ROI, which is a binary image. The experiment proved that the 10 pi × 10 pi module is more appropriate.

3 Feature Extraction and Selection

Various features of medical images and practical experience of clinicians should be fully combined in the extraction and selection of image features. After feature extraction experiments, some texture features were selected as experimental features.

The mean of the average gray value of each small module:

$$Mean = \frac{\sum_{i=0}^{N_g-1} i}{N} \tag{1}$$

The variance describes the fluctuation of the gray value of the image on the mean:

$$Variance = \frac{\sum_{i=0}^{N_g-1} (i - Mean)^2}{N} \tag{2}$$

The Angular Second Moment (ASM) describes the uniformity of the grayscale distribution of the image and the thickness of the texture:

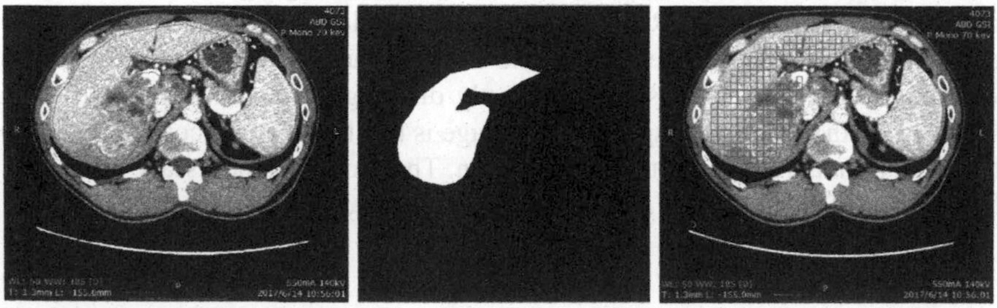

Fig. 2 Region of interest and experimental samples (modules)

$$ASM = \sum_i \sum_j \{p(i, j)\}^2 \quad (3)$$

The contrast describes the depth and the visual clarity of the image:

$$Contrast = \sum_{n=0}^{N_g-1} (i - j)^2 \left\{ \sum_{i=1}^{N_g} \sum_{j=1}^{N_g} p(i, j) \right\} \quad (4)$$

The correlation coefficient reflects the similarity of the image in the horizontal or vertical direction:

$$Correlation = \frac{\sum_i \sum_j (ij)p(i, j) - \mu_x \mu_y}{\sigma_x \sigma_y} \quad (5)$$

In the above formulas, *Ng* shows the number of different gray levels in the image, *N* is the total number of pixels in a single small module, *i* and *j* both represent pixel values in the module. The gray-level co-occurrence matrix is a probability matrix [16]. Each element *p(i, j)* in the matrix represents the probability that one point with gray level *i* in the image reaches another point with a fixed position *d* and the gray level of this point is *j*. *d* can be understood as a vector representing the distance and direction between two pixels, where $d = 1$, $\theta = 0°, 45°, 90°$, and $135°$.

The p_x, and p_y are the row and column sums of each element *p(i, j)* in the matrix:

$$p_x(i) = \sum_{j=0}^{N_g-1} p(i, j) \text{ and } p_y(j) = \sum_{i=0}^{N_g-1} p(i, j) \quad (6)$$

The μ_x and σ_x are the mean and variance of p_x respectively:

$$\mu_x = \sum_{i=0}^{N_g} i p_x(i) \text{ and } \sigma_x = \sqrt{\sum_{i=0}^{N_g-1} \{p_x(i) - \mu_x(i)\}^2} \quad (7)$$

The μ_y and σ_y are the mean and variance of p_y, respectively:

$$\mu_y = \sum_{j=0}^{N_g} j p_y(j) \text{ and } \sigma_y = \sqrt{\sum_{j=0}^{N_g-1} \{p_y(j) - \mu_y(j)\}^2}. \quad (8)$$

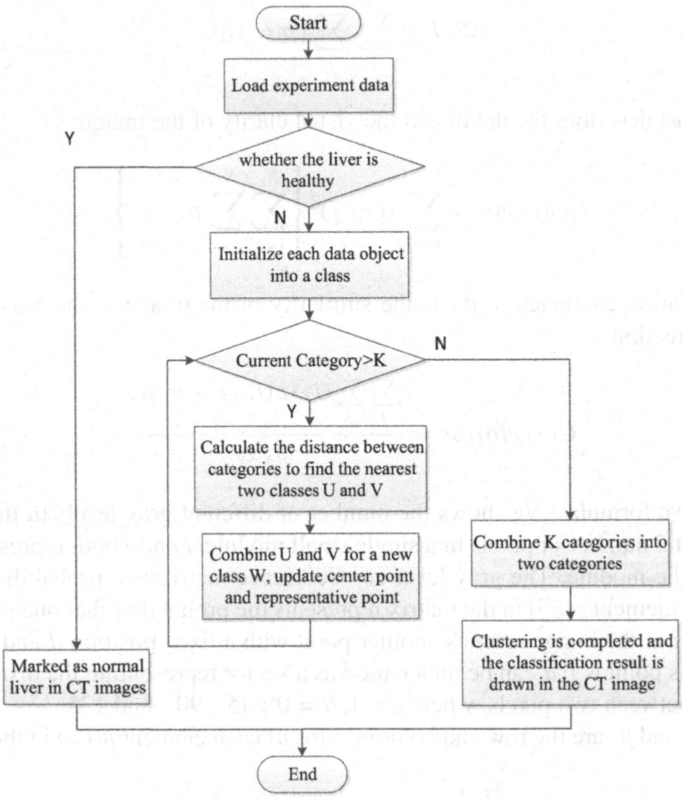

Fig. 3 The improved CURE clustering algorithm flow chart

4 Implementation of Improved CURE Clustering Algorithm

The CURE clustering algorithm can be used to merge similar modules into a class to distinguish the normal region and the tumor region from the liver. Through experimental analysis, the classic CURE clustering algorithm cannot directly distinguish between the tumor region and the normal region, so the CURE clustering algorithm needs to be improved. At present, many authors have improved the cure algorithm. Basically, they improved the clustering process to improve the clustering efficiency, it is not the same as this paper [17].

CURE is a hierarchical clustering algorithm that uses multiple data points to represent a class [18]. The improved CURE clustering algorithm flow chart is shown Fig. 3.

There are two improvements to the CURE clustering algorithm: (1) All small modules are divided into multiple categories at first, and then multiple categories are merged into two categories according to certain rules instead of classifying all modules into two categories directly. (2) If the liver in the spectral CT image is healthy, it is not correct to divide it into the normal region and the tumor region. Therefore, it is necessary to analyze whether it needs to be classified.

The distance between the data object $x=(x_1,x_2,\ldots,x_p)$ and $y=(y_1,y_2,\ldots,y_p)$ is:

$$dist(x, y) = \sqrt{(x_1 - y_1)^2 + (x_2 - y_2)^2 + \cdots + (x_p - y_p)^2} \quad (9)$$

The distance between the two clusters u, v is:

$$dist(u, v) = \min_{p \in u.rep, q \in v.rep} dist(p, q) \quad (10)$$

Combining the clusters u and v into a new cluster w and the center point of w:

$$w.mean = \frac{u.mean \times count(u) + v.mean \times count(v)}{count(u) + count(v)} \quad (11)$$

Representative point of w ($\alpha =$ is the contraction factor mentioned above):

$$w.rep = p + \alpha \times (w.mean - p) \quad (12)$$

5 Experimental Results

The experimental images of this paper is from the spectrum CT sequence image of liver tumor patients from third-party medical institutions. The experimental platform is personal notebook computer and the software MATLAB and PyCharm are used in the experiment.

From the Fig. 4, the left and middle figures show the classification results of the normal liver region and the tumor by using original CURE clustering algorithm and the improved CURE clustering algorithm. The right figure shows the boundary between the normal region and the tumor region drawn by the clinician in the ROI, the green line in the picture. The improved CURE clustering algorithm can distinguish between the tumor region and the normal region well.

When only the normal liver is selected in the range of ROI, Fig. 5 shows the right result of classification as below:

Randomly select 10 pictures from one person for testing at first. To judge the feasibility of this method, it is also necessary to consider its generalization. In addition, 5 patients were selected in this study to perform tumor identification on liver CT images. A total of 20 CT images were selected randomly for testing, and the accuracy of recognition was about 89%, indicating a good clinical guidance value. The results are shown as Table 1.

Table 1 Detection results for different pictures

10 pictures of the same person										
No.	1	2	3	4	5	6	7	8	9	10
Number of samples	196	186	200	187	169	181	215	220	183	197
Accuracy (%)	91	90	85	86	87	84	88	82	84	87
Total number	1934			Total accuracy (%)			86.4			
20 pictures from different patients										
No.	1	2	3	4	5	6	7	8	9	10
Number of samples	165	239	257	220	146	208	197	192	194	182
Accuracy (%)	85	92	94	90	96	90	87	84	91	85
No.	11	12	13	14	15	16	17	18	19	20
Number of samples	164	139	120	158	142	255	234	160	165	203
Accuracy (%)	84	87	95	90	86	93	85	91	88	91
Total number	3740			Total accuracy (%)			89.2			

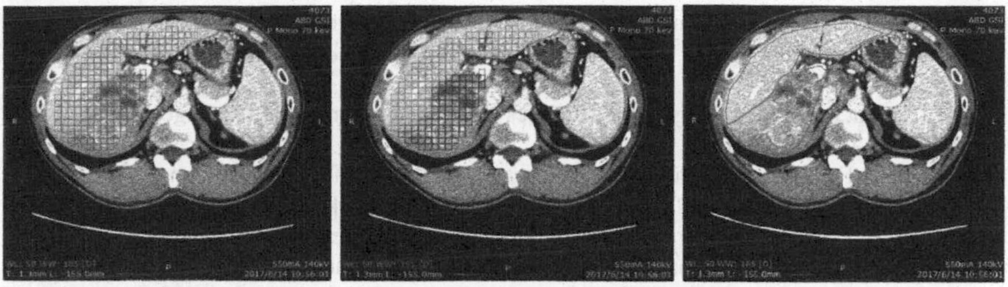

Fig. 4 Comparison of before and after the improvement of the algorithm

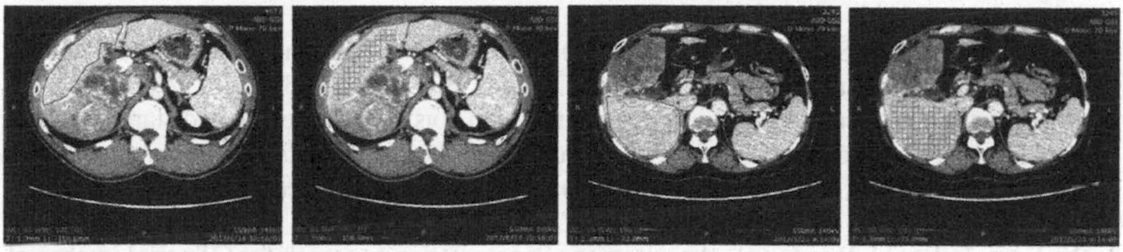

Fig. 5 The ROI and classification result of normal liver

6 Conclusion

The improved CURE clustering algorithm proposed in this paper makes full use of the gray information and spatial information of liver CT images to select the features of the image to distinguish between the normal liver region and the tumor region without artificially marking the training samples, and when there is no tumor in the liver, it is not divided into two categories. The experimental results were identified by clinicians. This method is feasible and has good clinical guidance value.

References

1. J. Ferlay, H.R. Shin, F. Bray et al., Cancer incidence and mortality worldwide: IARC CancerBase No. Int. J. Cancer J. Int. Du Cancer **136**(5), 359–386 (2010)
2. S. Mcguire, World Cancer Report 2014. Geneva, Switzerland: World Health Organization, International Agency for Research on Cancer, WHO Press, 2015. Adv. Nutr. **7**(2), 418–419 (2016)
3. X. Zhang, J. Tian, D. Xiang et al., Interactive liver tumor segmentation from ct scans using support vector classification with watershed, in *International Conference of the IEEE Engineering in Medicine and Biology Society*, vol. 2011, no. 4 (IEEE, 2011), pp. 6005–6008
4. J. Weston, C. Watkins, Multiclass support vector machines, in *Support Vector Machines for Pattern Classification*, vol. 102, no. 479 (Springer London, 2005), pp. 83–128
5. S. Chen, K. Suzuki, H. Macmahon, Development and evaluation of a computer-aided diagnostic scheme for lung nodule detection in chest radiographs by means of two-stage nodule enhancement with support vector classification. Med. Phys. **38**(4), 1844–1858 (2011)

6. A. Afifi, T. Nakaguchi, Unsupervised detection of liver lesions in CT images, in *2015 International Conference of the IEEE Engineering in Medicine and Biology Society* (2015), pp. 2411–2414
7. L. Li, H. Yu, Retrieval of lung tumor ct images based on k-means clustering algorithm. J. Gansu Sci. **27**(1), 058–061 (2015)
8. P.F. Christ, F. Ettlinger, F. Grün et al., Automatic liver and tumor segmentation of CT and MRI Volumes using Cascaded Fully Convolutional Neural Networks (2017), pp. 1530–1534
9. P.F. Christ, F. Ettlinger, F. Grün et al., Automatic Liver and Lesion Segmentation in CT Using Cascaded Fully Convolutional Neural Networks and 3D Conditional Random Fields (2016), pp. 415–423
10. B.N. Li, C.K. Chui, S.H. Ong et al., Integrating FCM and level sets for liver tumor segmentation, in *13th International Conference on Biomedical Engineering* (Springer Berlin Heidelberg, 2009), pp. 202–205
11. B.N. Li, C.K. Chui, S. Chang et al., Integrating spatial fuzzy clustering with level set methods for automated medical image segmentation. Comput. Biol. Med. **41**(1), 1–10 (2011)
12. P. Yugander, G.R. Reddy, Liver tumor segmentation in noisy CT images using distance regularized level set evolution based on fuzzy C-means clustering, in *IEEE International Conference on Recent Trends in Electronics, Information & Communication Technology* (IEEE, 2017), pp. 1530–1534
13. X. Jiang, L. Luo, J. Wang et al., An automatic segmentation approach for CT serial images of lung tumors. J. Image Graph. **8**(9), 1028–1033 (2003)
14. Yingmin Jia, Robust control with decoupling performance for steering and traction of 4WS vehicles under velocity-varying motion. IEEE Trans. Control Syst. Technol. **8**(3), 554–569 (2000)
15. Yingmin Jia, Alternative proofs for improved LMI representations for the analysis and the design of continuous-time systems with polytopic type uncertainty: a predictive approach. IEEE Trans. Autom. Control **48**(8), 1413–1416 (2003)
16. R.M. Haralick, K. Shanmugam, I. Dinstein, Textural features for image classification. IEEE Trans. Syst. Man Cybern. **smc-3**(6), 610–621 (1973)
17. C.Y. Gao , H.J. Wang, J. Wang, An improved CURE algorithm based on the uncertainty of mobile user data clustering[J]. Computer Engineering & Science (2016)
18. S. Guha, R. Rastogi, K. Shim et al., CURE : an efficient clustering algorithm for large databases. Inf. Syst. **26**(1), 35–58 (1998)

Inspection of Rail Surface Defects Image Based on Histogram Processing by the Judgment Threshold

Yidi Wu and Lin Li

Abstract It is a challenge to accurately detect the defects under the influence of light, reflection, shadow and rust. This paper optimizes the background difference method as preprocessing by reducing the template so that the influence of shadow can be reduced, and presents a histogram processing algorithm based on the judgment threshold to detect defects of the rail images. The judgment threshold is obtained by Otsu method. As we all know, the applicable condition of the Otsu method is the proportion of the target and background close to each other, and in order to achieve this kind of condition, the histogram processing is to remove partial histogram which is obtained by the judgment threshold. The histogram processing is performed cyclically and a new threshold for judging is obtained by Otsu method on the remaining histogram after the every loop. The constraint formula is introduced to make the threshold converge to the fixed value. Finally, the image is segmented by this threshold. The experimental results show the benefits of the proposed algorithm comparing to the existing algorithms.

Keywords Rail surface defect · Background differencing · Otsu method
Histogram processing

1 Introduction

With the advent of the high-speed railway, the safety of railway operation must be guaranteed. The detection of rail surface defects is a very important part. It relates to personal safety and the smoothness of rails [1].

Y. Wu (✉) · L. Li
Department of Control Science and Engineering, University of Shanghai for Science and Technology, Shanghai 200093, China
e-mail: yidiwu@yeah.net

L. Li
e-mail: lilin0211@163.com

Y. Jia et al. (eds.), *Proceedings of 2018 Chinese Intelligent Systems Conference*, Lecture Notes in Electrical Engineering 529,
https://doi.org/10.1007/978-981-13-2291-4_20

The surface of the rail is affected by light, reflection, shadow and rust, so the difficulty of detecting defects is self-evident. In 1992, Zhangjiang Wu treated the transforming of indistinct images into clear ones and derived a wide-line inspecting operator algorithm, which has found a new direction for rail defect detection [2]. Ze Liu proposed dynamic threshold segmentation and defect extraction algorithm, which can accurately extract defects and demarcate their positions [3]. Research on Robust Fast Algorithm of Rail Surface Defect Detection was proposed by Shengwei Ren which used the maximum entropy to separate the defects. It effectively eliminates the interference of light and noise [4]. However, the above algorithms cannot eliminate shadow interference and they are very easy to classify shadows as defects.

In view of the above situation, this paper presents a histogram processing algorithm based on the judgment threshold. Firstly, the background difference method is optimized by reducing the template so that the influence of shadow is reduced. Most of the defects have lower gray levels than the image of rail surface, and the defect accounts for a small proportion of the total area of the rails. However, for the Otsu algorithm, it is more suitable for the same area of the target and background in the image [5]. In order to achieve this condition, this paper uses the threshold obtained by Otsu method as the judgment threshold, and then finds the proportion of the defect based on this threshold. The removing partial histogram based on the proportion makes the target and the background area close to each other. After that the new histogram is used to calculate the new threshold by Otsu algorithm, and the part of new histogram will be removed further based on the new threshold. In order to make the threshold converge, the threshold value will tend to a fixed value after a few cycles by constructing constraint formula. Finally, the image is segmented by this threshold.

2 Preliminaries

2.1 Background Difference

According to the characteristics that the gray value of rail image changes little in the direction of the train, Zhendong He proposed a rail defect segmentation algorithm based on background difference [6]. The background model is established by the average gray value of the each row of rail image. The background model of row y is defined as

$$I_B(y) = mean(I_x(y)) \tag{1}$$

where $I_x(y)$ represents a vector that contains all gray values in row y, and $mean(I_x(y))$ produces a new vector which is obtained by averaging all values of $I_x(y)$. And the difference image is defined as

$$I_f = I(x, y) - I_B(x, y) \tag{2}$$

where $I(x, y)$ denotes the original image and $I_B(x, y)$ represents the image of background model. The algorithm can better highlight defects, and reduce the influence of illumination, rail gray-scale difference and reflection irregularity, but cannot eliminate the interference of shadow.

2.2 Otsu

The Otsu algorithm divides the image into target and background based on the grayscale feature of the image [5]. The larger the between-class variance of backgrounds and targets means the greater the difference between the two parts of the image. The between-class variance of backgrounds and targets is given by

$$g(t) = \omega_0(t)\omega_1(t)(\mu_0(t) - \mu_1(t))^2 \tag{3}$$

where ω_0 and μ_0 denote the proportion and the average gray value of targets, ω_1 and μ_1 represent the proportion and the average gray value of backgrounds, t denotes the threshold for dividing targets and backgrounds. So the optimal threshold is defined as

$$T = \mathrm{argmax}(g(t)) \tag{4}$$

where argmax() represents the variable t that makes g(t) get the maximum value.

3 Histogram Processing Based on Judgment Threshold

In this section, firstly, in order to highlight the rail defects and reduce interference of image shadow, the original image is preprocessed by using the optimized background difference. Secondly, the Otsu algorithm is applied to get the judgment threshold. And then, based on this judgment threshold, the histogram processing is performed cyclically. Finally, morphology operations are used on the image, which include open and close operations, so that noise can be eliminated.

3.1 Background Differential Preprocessing

In order to better display the difference image, and maintain the low gray level of the defect, we rewrite the difference image as

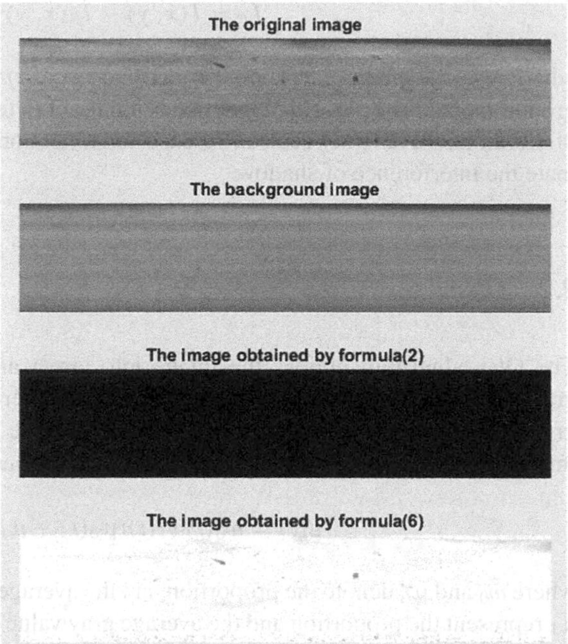

Fig. 1 Comparison of difference images obtained by Formulae (2) and (6)

$$I_d = I_B(x, y) - I(x, y) \tag{5}$$

$$I_f = \mathrm{E} - I_d \tag{6}$$

where I_d denotes Intermediate variables, and E denotes an unit matrix. The gray values of images are normalized in this paper. As is shown in Fig. 1, from the image obtained by Formula (2), we can see that the characteristics of the rail surface defects are changed from lower gray values to higher gray values. And the grey-scale feature of the defects is maintained in the image obtained by Formula (6).

In general, the greatest disturbance in the defect detection is caused by shadow. Therefore, we rewrite the background model as

$$I_B(y) = mean\left(I_{x/3}(y)\right) \tag{7}$$

where $I_{x/3}(y)$ denotes a vector of 1/3 length of original vector which is in Formula (1). In Fig. 2, the difference image obtained by Formula (7) has a clear de-shadowing result compared to the one obtained by Formula 1. So this optimization can greatly remove the interference caused by the shadow under the premise of ensuring the defect features.

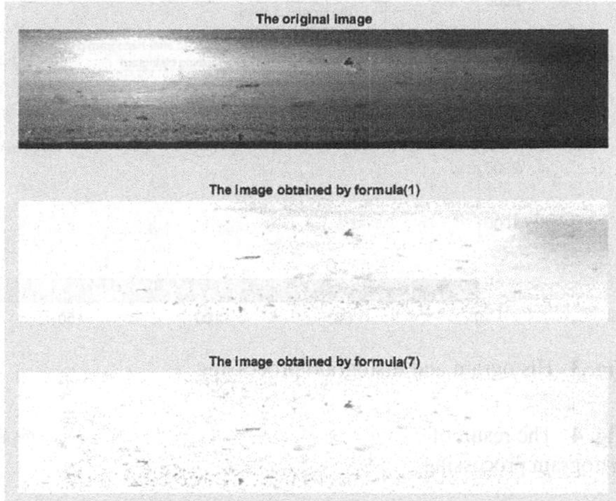

Fig. 2 Comparison of difference images obtained by Formulae (1) and (7)

3.2 *Histogram Processing*

Considering that the gray level of the defect is generally lower than the gray level of the rail surface, and from Formula (9), we can see that for the Otsu algorithm, it is more suitable for the same proportion of the target and background in the image, therefore, this paper proposes a histogram processing algorithm based on judgment threshold to make the proportion of the target and background close to each other to obtain an accurate threshold. The threshold is obtained by Otsu method from histogram and this histogram is cut to leave a low gray part by judgment threshold. The removed histogram does not have any effect on the detection of defects and can further eliminate the impact of strong reflection. After processing, the proportion of the background and the target is close, which can be more suitable for Otsu.

Firstly, the judgment threshold T is obtained by Formula (4). With this threshold, the proportion of target part in the image P_j can be calculated. The sum of proportion of the retention target and the background is given by

$$P_r = 2P_j \tag{8}$$

which means to remove $1 - P_r$ of the background. After that, the target and background have same proportion P_j in remaining histogram. The resulting histogram is given by

$$Hist_r = Hist_{im}(st : nd) \tag{9}$$

where $Hist_{im}$ denotes the histogram of the original image, st represents the smallest gray level in the image, nd denotes the gray level based on the proportion P_r, that is,

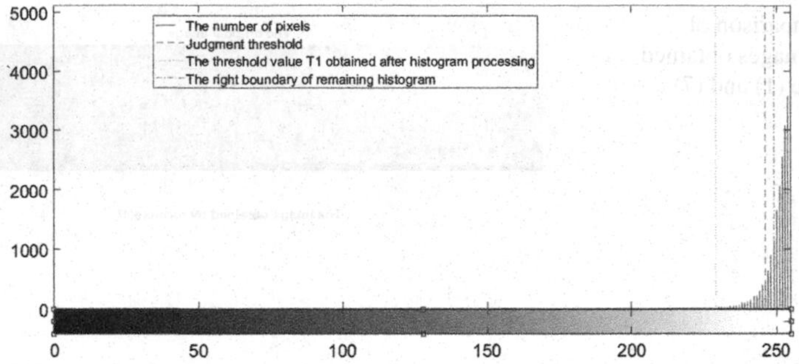

Fig. 3 Histogram and histogram processing

Fig. 4 The result of histogram processing

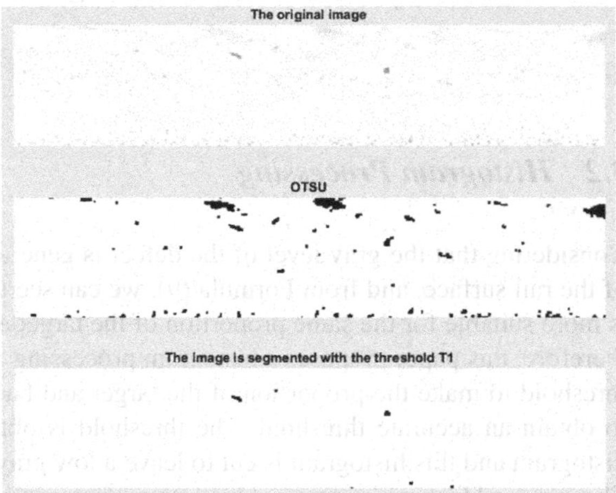

when $sum(P(0:nd))$ equals P_r, the gray value nd will be obtained. Then, the Otsu is applied to the resulting histogram to obtain the threshold T_1.

As shown in Fig. 3, the threshold value T_1 obtained after histogram processing is much lower than that obtained by Otsu algorithm. It can also be seen from Fig. 4 that this method extracts defects very well and also excludes most of the subtle rust.

3.3 Loop Histogram Processing

Since the histogram processing in the previous step is based on the judgment threshold, it is clear that there is still a big gap between the target proportion obtained by the judgment threshold and the true proportion. In order to achieve a threshold that matches the true ratio, further histogram processing is required. For loop histogram processing, the threshold is always smaller than the previous one because part of the background will always be discarded. According to the histogram processing described above, each remaining histogram is always smaller than the histogram of

Fig. 5 Cyclic processing of histogram result

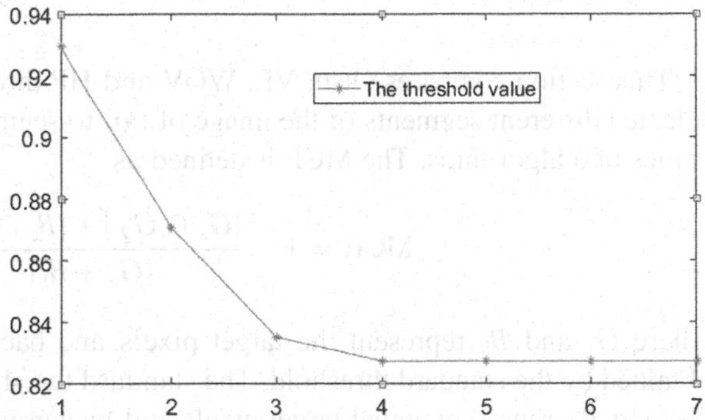

Fig. 6 The threshold value of each cyclic processing of histogram

the previous cycle. So the threshold of the result of the last loop is definitely zero. In order to get the ideal value through the loop histogram processing, this paper achieves the goal by adjusting the proportion retained each time. When the proportion obtained each time is smaller than the proportion obtained for the first loop, the constraint formula is introduced to get more histograms. And the formula is given by

$$P_r = nP_{lr}\left(P_j - P_{lr} + 1\right) \tag{10}$$

where P_r denotes the retention proportion after adjusting, P_{lr} denotes the proportion calculated based on the threshold obtained by this cycle histogram processing, P_j denotes the proportion based on the judgment threshold, and n is a coefficient. The bigger n is, the greater the binding force is, and the faster T_1 converges. Through experiments, the convergence value reaches the ideal threshold when n is 4.

From Fig. 5, we can see that the segmented image, obtained by the threshold value after loop histogram processing, shows the interference of rust is eliminated. Defects are accurately segmented. In Fig. 6, the threshold of experiment is completely convergent when the cycle is repeated five times. This algorithm guarantees speed of segmentation and accurately segments defects.

4 Experimental Results

In order to test the segmentation effect, the Mis-classification Error (MCE) evaluation standard was introduced [7]. It is the evaluation of the segmentation effect of a single image. At the same time, the two criteria, detection rate and false alarm rate

[8] are introduced to comprehensively evaluate the detection effect. We compared Histogram Processing (HP) with Otsu, Valley-emphasis Otsu (VE) [9] and Weighted Object Variance (WOV) [10].

4.1 MCE

This section compares Otsu, VE, WOV and HP through the MCE standard. We selected different segments of the image of rail to segment and compared the MCE values of 5 algorithms. The MCE is defined as

$$\mathrm{MCE} = 1 - \frac{|G_s \cap G_T| + |B_s \cap B_T|}{|G_s + B_s|} \tag{11}$$

where G_s and B_s represent the target pixels and background pixels respectively obtained by the standard threshold. The standard threshold is determined manually. G_T and B_T represent actual target pixels and background pixels, respectively. The MCE value is between [0, 1], and the smaller the MCE value is, the lower the misclassification rate is, the better the segmentation effect is.

The background differential preprocessing of this paper was used in other algorithms. After segmentation, all the segmentation images obtained by the algorithms were denoised.

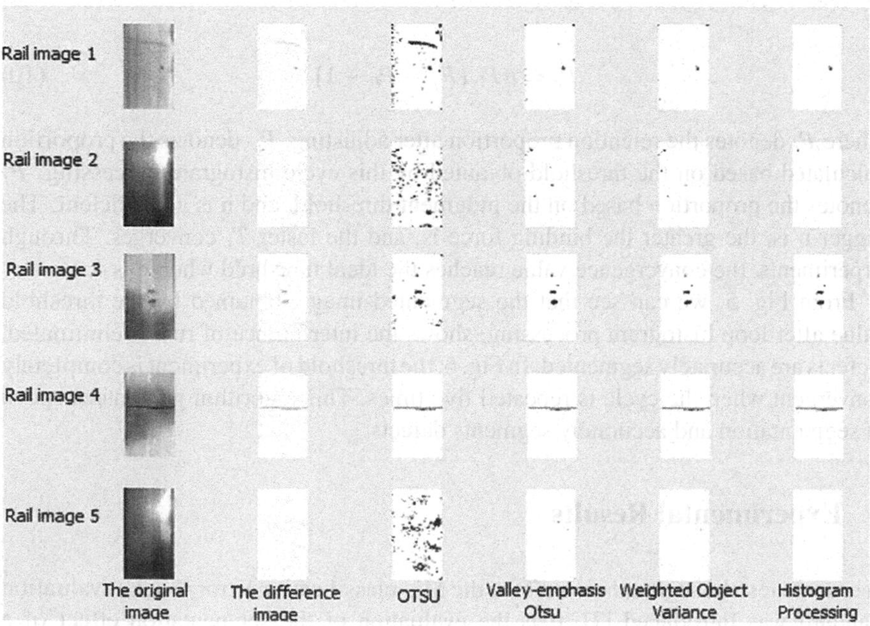

Fig. 7 Segmentation of results using different methods with preprocessing

In Fig. 7, after the preprocessing, the segmentation results of various algorithms are very good. Table 1 shows that the MCE values of various algorithms are close to zero. The classic OTSU algorithm divides most of the rust into defects, which is not conducive to detect. Its segmentation effect is the worst among the four algorithms, and the MCE value is also the largest. WOV is generally slightly better than VE. It basically eliminates the effects of rust, but there are still minor misclassifications compared to HP. It can be seen from the rail image 2 in Fig. 7 that the noise-like black point is the wrong division. As shown in Fig. 7, HP performs excellently in a defect-free image, and no misclassification can be seen from the graph. The MCE value of HP is also lower than other algorithms. Based on the figures, HP performs well, and the rust and other disturbances can be basically eliminated.

4.2 Detection Rate and False Alarm Rate

This section shows the comparisons of detection rate and false alarm rate of OTSU, VE, WOV and HP based on taking 1000 images of rails [8]. The detection rate and false alarm rate are defined as

$$\text{Detection Rate} = \frac{Number\ of\ Samples\ Defective\ Correctly\ Detected}{Total\ Number\ of\ Defective\ Samples} \tag{12}$$

$$\text{False Alarm Rate} = \frac{Number\ of\ Defect-free\ Samples\ Detected\ as\ Defective}{Total\ Number\ of\ Defect-free\ Samples} \tag{13}$$

After the defect was segmented, the defect was marked with the connected region. If the number of pixels in the connected region was greater than 10, the defect was detected. Since there was no specific number and size of defects to determine each image, separating the background into defects was also considered as detecting the defect. Otsu is likely to separate the background into defects, so the detection rate reaches 100%, and the false alarm rate reaches 100%. Its segmentation is not good. As can be seen from Table 2, HP is superior to the other two algorithms in both detection rate and false alarm rate, which conforms to the requirements of the application of rail defect detection.

Table 1 MCE values of five images by four methods

Number of rail	MCE			
	OTSU	VE	WOV	HP
1	0.0703	0.003	0.001	0.0022
2	0.106	0.0067	0.0036	0.0005
3	0.0269	0.0022	0.0003	0
4	0.0003	0	0.0002	0
5	0.1191	0.004	0.0013	0.0007

Table 2 Detection rates and false alarm rates of defect by four methods

	Otsu (%)	VE (%)	WOV (%)	HP (%)
Detection rate	100	79.2	76.6	92.2
False alarm rate	100	9.8	11.4	6.8

5 Conclusion

In this paper, a method based on the judgment threshold histogram processing has been proposed by analyzing the characteristics of rails and combining the characteristics of Otsu. The optimized background difference reduces the influence of shadows on the segmentation while minimizing the interference caused by illumination and uneven reflection. In the experiment, multiple evaluation criteria were introduced: MCE, detection rate and false alarm rate. The segmentation effects of Otsu, DB, VE, WOV and HP are compared comprehensively through these evaluation criteria. The experimental results show that the proposed method has the best segmentation effect. Due to the effect of removing shadows by reducing the size of the template in this paper, for defects that are close to or larger than the length of the template in the direction of the train, this article will mistake it as a shadow and eliminate it. We will continue to study solutions in the future.

References

1. R. Clark, The importance of rail flaw detection to heavy haul operations, in *International Heavy Haul Association Conference*, Australia: Brisbane (2001)
2. Z. Wu, X. Li, The application of computer image process to railway field. China Acad. Railway Sci. **14**(1), 36–41 (1992)
3. Z. Liu, W. Wang, P. Wang, Design of machine vision system for inspection of rail surface defects. J. Electr. Meas. Instrum. **11**, 1012–1017 (2010)
4. S. Ren, Q. Li, G. Xu et al., Research on robust fast algorithm of rail surface defect detection. China Railway Sci. **32**(1), 25–29 (2011)
5. N. Otsu, A threshold selection method from gray-level histograms. IEEE Trans. Syst. Man Cybern. **9**(1), 62–66 (1979)
6. Z. He, Y. Wang, J. Liu et al., Background differencing-based high-speed rail surface defect image segmentation. Chin. J. Sci. Instrum. **37**(3), 640–649 (2016)
7. W.A. Yasnoff, J.K. Mui, J.W. Bacus, Error measures y for scene segmentation. Pattern Recogn. **9**(4), 217–231 (1977)
8. H.Y.T. Ngan, G.K.H. Pang, N.H.C. Yung, Automated fabric defect detection—a review. Image Vis. Comput. **29**(7), 442–458 (2011)
9. J.L. Fan, B. Lei, A modified valley-emphasis method for automatic thresholding. Pattern Recogn. Lett. **33**(6), 703–708 (2012)
10. X. Yuan, L. Wu, H. Chen, Rail image segmentation based on Otsu threshold method. Opt. Precis. Eng. **24**(7), 1772–1781 (2016)

Finite-Time Stochastic Stabilization for Markovian Jump One-Sided Lipschitz Systems

Jun Huang, Lei Yu and Xiang Ma

Abstract This paper considers the finite-time stochastic stabilization problem for Markovian jump one-sided Lipschitz systems. The transition rates matrix of the system is partly unknown. The feature of the studied systems lies that the nonlinearities satisfy the property of one-sided Lipschitz. Firstly, the definitions of one-sided Lipschitz function and finite-time stochastic stability are introduced. Then, the feedback laws are designed to make the closed-loop systems finite-time stochastically stable. Finally, a numerical example is used to demonstrate the effectiveness of the designed controllers.

Keywords Finite-time stochastic stabilization · One-sided Lipschitz systems · Markovian jump · linear matrix inequalities

1 Introduction

As a kind of important hybrid systems, Markovian jump (MJ) systems have draw considerable attention in the past decades [1, 2]. MJ systems can be used to describe many practical systems, such as economic systems, networked control systems, neural networks and so on. As a dominant factor, the transition rates (TRs) determine the final result of the change from one mode to the next mode in the MJ systems. When the TRs are completely known, the works of the (MJ) systems can be found in [3–5]. Due to complexity in control process, all the information of the TR matrix (TRM) can not be accessible, then the investigation of MJ systems with partly unknown TRs (PUTRs) has become a hot topic [6–9]. Ma et al. [6] investigated the Markovian jump neural networks with PUTRs and gave the design method of synchronization. Qi and Gao [7] concerned with finite-time stabilization for stochastic time-delayed MJ systems with PUTRs. Li and Cao [8] studied the finite-time stochastically sta-

J. Huang (✉) · L. Yu · X. Ma
School of Mechanical and Electrical Engineering,
Soochow University, Suzhou 215021, China
e-mail: cauchyhot@163.com

Y. Jia et al. (eds.), *Proceedings of 2018 Chinese Intelligent Systems Conference*, Lecture Notes in Electrical Engineering 529,
https://doi.org/10.1007/978-981-13-2291-4_21

bilization for MJ memristive neural networks with PUTRs. Ma et al. [9] addressed the finite-time saturated control problem for singular MJ systems with PUTRs. It is worth pointing out that the nonlinear functions in the mentioned Refs. [6–9] are all assumed to be Lipschitz.

On the other hand, one-sided Lipschitz nonlinear systems have been studied extensively [10–12]. The one-sided Lipschitz systems can stand for a large family of real plants because the region of one-sided Lipschitz constant is much larger than that of Lipschitz constant. In order to give the systematic design method for one-sided Lipschitz systems, [13] defined the standard one-sided Lipschitz condition as well as the quadratic inner bounded condition. Zhang et al. [14, 15] employed the S-procedure to deal with the cross terms and unified the asymptotical observer framework for continuous and discrete one-sided Lipschitz systems. Under the P-one-sided Lipschitz condition, [16] designed exponential observers by passive method. For the stabilization problem, [17] presented the analysis method for the stabilization problem for one-sided Lispschitz systems. While, [18] designed the H^{∞} controllers for one-sided Lispschitz systems, and closed-loop systems are finite-time bounded. Based on the observer, [19] gave the controller design method for one-sided Lispschitz systems and derived less conservative sufficient conditions. However, to the best of authors' knowledge, the finite-time stabilization for MJ one-sided Lipschitz systems with PUTRs has not been investigated.

Motivated by the above discussion, this paper investigates MJ one-sided Lipschitz systems with PUTRs and gives the finite-time stabilization controllers. The rest of the paper is organized as follows: Sect. 2 formulates problem statement and preliminaries. Section 3 derives the sufficient conditions for the existence of the stochastic observer. Section 4 provides numerical example to verify the effectiveness of the proposed method.

2 Problem Formulation and Preliminaries

Let us consider the following system

$$\dot{x}(t) = A_{r(t)}x(t) + B_{r(t)}u(t) + f_{r(t)}(x(t)), \tag{1}$$

where $x(t) \in R^n$ is the state, $u(t) \in R^m$ is the control input. $r(t), t \geq 0$ is a right-continuous Markov chain on the probability space taking values in a finite state space $S = \{1, 2, \ldots, N\}$ with generator $\Pi = (\pi_{ij})$ $(i, j \in S)$ defined by

$$P\{r(t+\Delta) = j | r(t) = i\} = \begin{cases} \pi_{ij}\Delta + o(\Delta), & j \neq i, \\ 1 + \pi_{ii}\Delta + o(\Delta), & j = i, \end{cases}$$

where $\Delta > 0$, and $\lim_{\Delta \to 0} \dfrac{o(\Delta)}{\Delta} = 0$, $\pi_{ij} \geq 0$ is the TR from mode i at time t to mode j at time $t + \Delta$ if $j \neq i$ and $\pi_{ii} = -\sum_{j \neq i} \pi_{ij}$. For $r(t) = i \in S$, A_i, B_i are given matrices, and $f_i(x(t)) \in R^n$ is one-sided Lipschitz function with $f_i(0) = 0$.

In this paper, we assume that the TRs of $r(t)$ are partly unknown. For instance, a system with three operation or four operation modes may have the following TRM:

$$\Pi_3 = \begin{bmatrix} ? & \pi_{12} & ? \\ ? & \pi_{22} & ? \\ \pi_{31} & ? & ? \end{bmatrix}, \Pi_4 = \begin{bmatrix} \pi_{11} & ? & ? & \pi_{14} \\ \pi_{21} & \pi_{22} & ? & ? \\ \pi_{31} & ? & ? & \pi_{34} \\ \pi_{41} & ? & ? & \pi_{44} \end{bmatrix},$$

where ? is the inaccessible elements. Denote that $S = S_k^i \cup S_{uk}^i$, where $S_k^i = \{j : \pi_{ij} \text{ is known}\}$ and $S_{uk}^i = \{j : \pi_{ij} \text{ is unknown}\}$.

In the sequel, the variable t is omitted. Then (1) can be written as:

$$\dot{x} = A_r x + B_r u + f_r(x). \tag{2}$$

We now introduce some basics about one-sided Lipschitz function, more details can be referred to [13].

Definition 1 The nonlinear function $f(x)$ is said to be one-sided Lipschitz if there exists $\alpha \in R$ such that $\forall\, x_1, x_2 \in D$

$$\langle f(x_1) - f(x_2), x_1 - x_2 \rangle \le \alpha \|x_1 - x_2\|^2. \tag{3}$$

Definition 2 The nonlinear function $f(x)$ is called quadratic inner-boundedness in the region $\tilde{D}$ if there exist $\beta, \gamma \in R$ such that $\forall\, x_1, x_2 \in \tilde{D}$

$$(f(x_1) - f(x_2))^T (f(x_1) - f(x_2)) \le \beta \|x_1 - x_2\|^2 + \gamma \langle x_1 - x_2, f(x_1) - f(x_2) \rangle. \tag{4}$$

The nonlinear function $f_i(x)$ is supposed to be one-sided Lipschitz in a region D_i and quadratically inner-bounded in a region $\tilde{D}_i$ for $i \in S$. Thus, if $\forall\, x, \hat{x} \in D_i \cap \tilde{D}_i$, there exist $\alpha_i, \beta_i, \gamma_i \in R$ such that

$$\langle f_i(x_1) - f_i(x_2), x_1 - x_2 \rangle \le \alpha_i \|x_1 - x_2\|^2, \tag{5}$$

$$(f_i(x_1) - f_i(x_2)^T (f_i(x_1) - f_i(x_2) \le \beta_i \|x_1 - x_2\|^2 + \gamma_i \langle x_1 - x_2, (f_i(x_1) - f_i(x_2)) \rangle. \tag{6}$$

Remark 1 If we choose $x_1 = x$ and $x_2 = 0$, then (5) and (6) imply that

$$\langle f_i(x), x \rangle \le \alpha_i \|x\|^2, \tag{7}$$

$$f_i^T(x) f_i(x) \le \beta_i \|x\|^2 + \gamma_i \langle x, f_i(x) \rangle. \tag{8}$$

Let us consider the open-loop system (2), i.e, $u = 0$,

$$\dot{x} = A_r x + f_r(x). \tag{9}$$

Definition 3 Denote that $x(0) = x_0$ is the initial condition of the system (9). The system (9) is said to be finite-time stochastic stable (FTSS) with respect to (c_1, c_2, R, T_f) if

$$x^T(0)Rx(0) < c_1 \Rightarrow E[x^T(t)Rx(t)] < c_2, \quad t \in [0, T_f],$$

where the constants $c_1 > 0$, $c_2 > 0$, $T_f > 0$, and matrix $R > 0$.

3 Main Results

In this section, we will design the feedback law, under which the system (2) is FTSS.

Theorem 1 *If there exist matrices $\tilde{P}_i \in R^{n\times n} > 0$, $Q_i \in R^{n\times n} > 0$, $K_i \in R^{m\times n}$, positive constants ε_i, μ_i, η, such that for any $i \in S$*

$$\begin{bmatrix} \Omega_i & \tilde{P}_i + \frac{\mu_i\gamma_i - \varepsilon_i}{2} I \\ * & -\mu_i I \end{bmatrix} < 0, \tag{10}$$

where $\Omega_i = \tilde{P}_i(A_i + B_iK_i) + (A_i + B_iK_i)^T\tilde{P}_i - \eta\tilde{P}_i + (\varepsilon_i\alpha_i + \mu_i\beta_i)I + \sum_{j\in S_k^i} \pi_{ij} (\tilde{P}_j - Q_i)$,

$$\tilde{P}_j - Q_i \geq 0, \ \forall j \in S_{uk}^i, \ j = i, \tag{11}$$

$$\tilde{P}_j - Q_i \leq 0, \ \forall j \in S_{uk}^i, \ j \neq i, \tag{12}$$

$$\tilde{P}_i = R^{\frac{1}{2}} P_i R^{\frac{1}{2}}, \tag{13}$$

then the system (2) *under the feedback law $u = K_r x$ is FTSS.*

Proof The closed-loop system is

$$\dot{x} = (A_r + B_rK_r)x + f_r(x). \tag{14}$$

Choose the Lyapunov function as $V(x, r) = x^T\tilde{P}(r)x$ and consider the mode $r = i$. Then, the operator $\mathcal{L}V(x, i)$ of the closed-loop system (14) can be calculated as

$$\begin{aligned} \mathcal{L}V(x, i) = {} & 2x^T\tilde{P}_i(A_i + B_iK_i)x + 2x^T\tilde{P}_i f_i(x) \\ & + \sum_{j\in S_k^i} \pi_{ij}x^T\tilde{P}_jx + \sum_{j\in S_{uk}^i} \pi_{ij}x^T\tilde{P}_jx. \end{aligned} \tag{15}$$

Since $\sum_{j\in S} \pi_{ij} = 0$, we can obtain that

$$-\sum_{j\in S}\pi_{ij}x^TQ_ix = 0. \tag{16}$$

It follows from (15) and (16) that

$$\begin{aligned}\mathcal{L}V(x,i) &= 2x^T\tilde{P}_i(A_i + B_iK_i)x + 2x^T\tilde{P}_if_i(x)\\ &\quad+\sum_{j\in S_k^i}\pi_{ij}x^T(\tilde{P}_j - Q_i)x + \sum_{j\in S_{uk}^i}\pi_{ij}x^T(\tilde{P}_j - Q_i)x.\end{aligned} \tag{17}$$

If $j \neq i$, then $\pi_{ij} \geq 0$, it can be deduced from (12) that $\sum_{j\in S_{uk}^i}\pi_{ij}x^T(\tilde{P}_j - Q_i)x \leq 0$. If $j = i$, then $\pi_{ij} < 0$, hence (11) means that $\sum_{j\in S_{uk}^i}\pi_{ij}x^T(\tilde{P}_j - Q_i)x \leq 0$. Thus, we can conclude that

$$\sum_{j\in S_{uk}^i}\pi_{ij}x^T(\tilde{P}_j - Q_i)x \leq 0. \tag{18}$$

Then (17) becomes

$$\begin{aligned}\mathcal{L}V(x,i) &\leq 2x^T\tilde{P}_i(A_i + B_iK_i)x + 2x^T\tilde{P}_if_i(x)\\ &\quad+\sum_{j\in S_k^i}\pi_{ij}x^T(\tilde{P}_j - Q_i)x.\end{aligned} \tag{19}$$

which is also written as

$$\mathcal{L}V(x,i) \leq \begin{bmatrix} x \\ f_i(x)\end{bmatrix}^T \begin{bmatrix} \Omega_i & \tilde{P}_i \\ * & 0\end{bmatrix}\begin{bmatrix} x \\ f_i(x)\end{bmatrix}, \tag{20}$$

where $\Omega_i = \tilde{P}_i(A_i + B_iK_i) + (A_i + B_iK_i)^T\tilde{P}_i + \sum_{j\in S_k^i}\pi_{ij}(\tilde{P}_j - Q_i)$. Using (7), one can get that

$$\alpha_ix^Tx - x^Tf_i(x) \geq 0,$$

which implies that for any $\varepsilon_i > 0$

$$\varepsilon_i\begin{bmatrix} x \\ f_i(x)\end{bmatrix}^T\begin{bmatrix}\alpha_iI & -\frac{I}{2} \\ * & 0\end{bmatrix}\begin{bmatrix} x \\ f_i(x)\end{bmatrix} \geq 0. \tag{21}$$

By (8), it holds for any $\mu_i > 0$

$$\mu_i\begin{bmatrix} x \\ f_i(x)\end{bmatrix}^T\begin{bmatrix}\beta_iI & \frac{\gamma_i}{2}I \\ * & -I\end{bmatrix}\begin{bmatrix} x \\ f_i(x)\end{bmatrix} \geq 0. \tag{22}$$

Substituting (21) and (22) into (20), together with (10), we obtain that

$$\mathcal{L}V(x,i) \leq \begin{bmatrix} x \\ f_i(x) \end{bmatrix}^T \begin{bmatrix} \eta\tilde{P}_i & 0 \\ 0 & 0 \end{bmatrix} \begin{bmatrix} x \\ f_i(x) \end{bmatrix} = \eta x^T \tilde{P}_i x. \tag{23}$$

It is deduced from (23) that

$$\mathcal{L}V(x,r) < \eta V(x,r). \tag{24}$$

By Gronwall-Bellman inequality, we have

$$E[V] < \exp\{\eta T_f\} V(0), \quad t \in [0, T_f], \tag{25}$$

which implies that

$$E[x^T(t)Rx(t)] < \frac{\lambda_2}{\lambda_1} \exp\{\eta T_f\} x^T(0)Rx(0), \quad t \in [0, T_f]. \tag{26}$$

where $\lambda_1 = \min_{i\in S}\{\lambda_{\min}(P_i)\}$, $\lambda_2 = \max_{i\in S}\{\lambda_{\max}(P_i)\}$, and $\lambda_{\min}(A)$ ($\lambda_{\max}(A)$) is the minimal (maximal) eigenvalue of the matrix A. Since $x^T(0)Rx(0) < c_1$, let $c_2 = \frac{\lambda_2}{\lambda_1} c_1 \exp\{\eta T_f\}$, then $E[x^T(t)Rx(t)] < c_2$. In view of Definition 3, the closed-loop system (14) is FTSS. □

Remark 2 It should be noted that the inequalities (10)–(12) can not be solved by LMI toolbox in Matlab, because the bilinear term $\tilde{P}_i B_i K_i + (B_i K_i)^T \tilde{P}_i$ is contained. Thus, it is essential to transform the inequalities (10)–(12) into the LMI forms.

In the sequel, without loss of generality, it is supposed that the set

$$S_k^i = \{i_1, i_2, \ldots, i_{r-1}, i_r, i_{r+1}, \ldots, i_k\},$$

where $k < n$. If $i \in S_k^i$, then $i_r = i$.

Theorem 2 *If there exist matrices $W_i \in R^{n\times n} > 0$, $Z_i \in R^{n\times n} > 0$, $Y_i \in R^{m\times n}$, positive constants ε_i, μ_i, η, such that*

$$\begin{bmatrix} \Lambda_{11} & \Lambda_{12} & \Lambda_{13} & \Lambda_{14} \\ * & \Lambda_{22} & 0 & 0 \\ * & * & \Lambda_{33} & 0 \\ * & * & * & \Lambda_{44} \end{bmatrix} < 0, \quad i \in S_k^i, \tag{27}$$

$$\begin{bmatrix} \bar{\Lambda}_{11} & \Lambda_{12} & \Lambda_{13} & \Lambda_{14} \\ * & \Lambda_{22} & 0 & 0 \\ * & * & \Lambda_{33} & 0 \\ * & * & * & \bar{\Lambda}_{44} \end{bmatrix} < 0, \quad i \in S_{uk}^i, \tag{28}$$

where

$$\Lambda_{11} = A_iW_i + W_iA_i^T + B_iY_i + Y_i^TB_i^T - \eta W_i + \pi_{ii}W_i - \sum_{j\in S_k^i}\pi_{ij}Z_i,$$

$$\bar{\Lambda}_{11} = A_iW_i + W_iA_i^T + B_iY_i + Y_i^TB_i^T - \eta W_i - \sum_{j\in S_k^i}\pi_{ij}Z_i,$$

$$\Lambda_{12} = I + \frac{(\mu_i\gamma_i - \varepsilon_i)}{2}W_i, \quad \Lambda_{13} = \sqrt{|\varepsilon_i\alpha_i + \mu_i\beta_i|}W_i,$$

$$\Lambda_{14} = [\sqrt{\pi_{ii_1}}W_i \;\; \sqrt{\pi_{ii_2}}W_i \;\; \ldots \;\; \sqrt{\pi_{ii_{r-1}}}W_i \;\; \sqrt{\pi_{ii_{r+1}}}W_i \;\; \ldots \sqrt{\pi_{ii_k}}W_i],$$

$$\Lambda_{22} = -\mu_i I, \quad \Lambda_{33} = -I,$$

$$\Lambda_{44} = diag\{W_{i_1}, W_{i_2}, \;\; \ldots \;\; W_{i_{r-1}} \;\; W_{i_{r+1}} \;\; \ldots W_{i_k}\},$$

$$\bar{\Lambda}_{44} = diag\{W_{i_1}, W_{i_2}, \;\; \ldots \;\; W_{i_k}\},$$

$$W_j - Z_j \geq 0, \; \forall\, j \in S_{uk}^i, \; j = i, \tag{29}$$

$$\begin{bmatrix} -Z_i & W_i \\ * & -W_j \end{bmatrix} \leq 0, \; \forall\, j \in S_{uk}^i, \; j \neq i, \tag{30}$$

then the system (2) *under the feedback law* $u = K_r x$ *is FTSS. Furthermore, the controller gain is determined by* $K_i = Y_iW_i^{-1}$.

Proof Let $W_i = \tilde{P}_i^{-1}, W_j = \tilde{P}_j^{-1}, Z_i = \tilde{P}_i^{-1}Q_i\tilde{P}_i^{-1}, Z_j = \tilde{P}_i^{-1}Q_j\tilde{P}_i^{-1}$ and $Y_i = K_iW_i$. Pre-and post-multiplying inequality (10) by the matrix $diag\{W_i, I\}$ yields

$$\begin{bmatrix} \Theta_i & I + \dfrac{(\mu_i\gamma_i - \varepsilon_i)}{2}W_i \\ * & -\mu_i I \end{bmatrix} < 0, \tag{31}$$

where $\Theta_i = A_iW_i + W_iA_i^T + B_iY_i + Y_i^TB_i^T - \eta W_i + (\varepsilon_i\alpha_i + \mu_i\beta_i)W_iW_i + \sum_{j\in S_k^i} \pi_{ij}W_i(W_j^{-1} - Q_i)W_i$. We argue in the following two cases.

(1) When $i \in S_k^i$, by using Shur complement, we can conclude that (27) means (31).

(2) When $i \in S_{uk}^i$, by using Shur complement, we can obtain that (28) means (31).

Then Pre-and post-multiplying inequalities (11) and (12) by the matrix W_i respectively, we have

$$W_j - Z_j \geq 0, \; \forall\, j \in S_{uk}^i, \; j = i, \tag{32}$$

and

$$W_iW_j^{-1}W_i - Z_i \leq 0, \; \forall\, j \in S_{uk}^i, \; j \neq i, \tag{33}$$

which imply that (29) and (30) are equivalent to (32) and (33). Thus, the proof is completed.

4 Numerical Example

Consider the system (2) with three modes:

$$A_1 = \begin{bmatrix} 2.22 & -16 \\ 1.9 & -14.3 \end{bmatrix},\ B_1 = \begin{bmatrix} 1.53 \\ 1.7 \end{bmatrix},\ A_2 = \begin{bmatrix} -16.9 & 1.84 \\ -19.5 & 2.46 \end{bmatrix},\ B_2 = \begin{bmatrix} 1.88 \\ 1.92 \end{bmatrix},$$

$$A_3 = \begin{bmatrix} -15.7 & 2.23 \\ -20.2 & 2.86 \end{bmatrix},\ B_3 = \begin{bmatrix} 2.22 \\ 2.35 \end{bmatrix},\ f_1 = f_2 = f_3 = -\begin{bmatrix} 0 \\ x_1(x_1^2 + x_2^2) \\ x_2(x_1^2 + x_2^2) \end{bmatrix}.$$

Thus, $\alpha_1 = \alpha_2 = \alpha_3 = 0$, $\beta_1 = \beta_2 = \beta_3 = -399$, $\gamma_1 = \gamma_2 = \gamma_3 = -40$. The TRM is chosen as:

$$\Pi = (\pi_{ij})_{3\times 3} = \begin{bmatrix} -0.1 & ? & ? \\ ? & ? & 0.2 \\ 0.3 & 0.1 & ? \end{bmatrix}.$$

By Theorem 2, we can obtain that

$$W_1 = \begin{bmatrix} 0.25 & 0.31 \\ 0.31 & 0.43 \end{bmatrix},\ W_2 = \begin{bmatrix} 0.44 & 0.52 \\ 0.52 & 0.57 \end{bmatrix},\ W_3 = \begin{bmatrix} 0.53 & 0.16 \\ 0.16 & 0.37 \end{bmatrix},$$

$$Z_1 = \begin{bmatrix} 0.1975 & 0.2615 \\ 0.2615 & 0.3463 \end{bmatrix},\ Z_2 = \begin{bmatrix} 0.4037 & 0.457 \\ 0.457 & 0.5174 \end{bmatrix},\ Z_3 = \begin{bmatrix} 0.121 & 0.0611 \\ 0.0611 & 0.0445 \end{bmatrix},$$

$$Y_1 = \begin{bmatrix} -8805 & -11724 \end{bmatrix},\ Y_2 = \begin{bmatrix} -8263.2 & -9359.8 \end{bmatrix},\ Y_3 = \begin{bmatrix} -87011 & -59933 \end{bmatrix}.$$

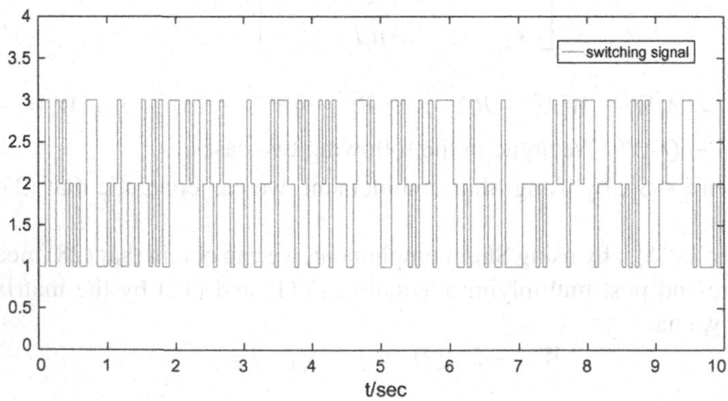

Fig. 1 Response of the switching signal

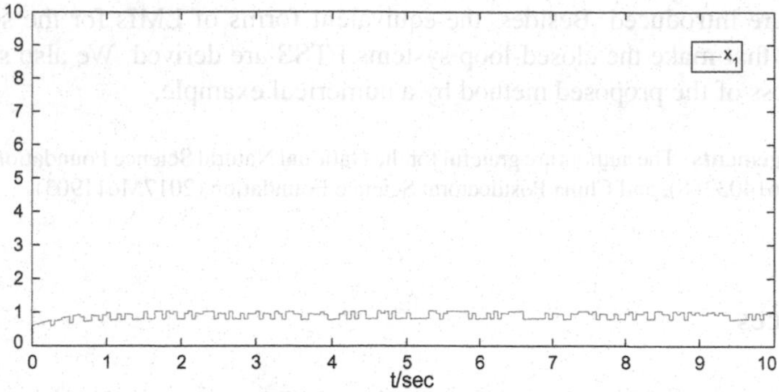

Fig. 2 Response of x_1 of the closed-loop system

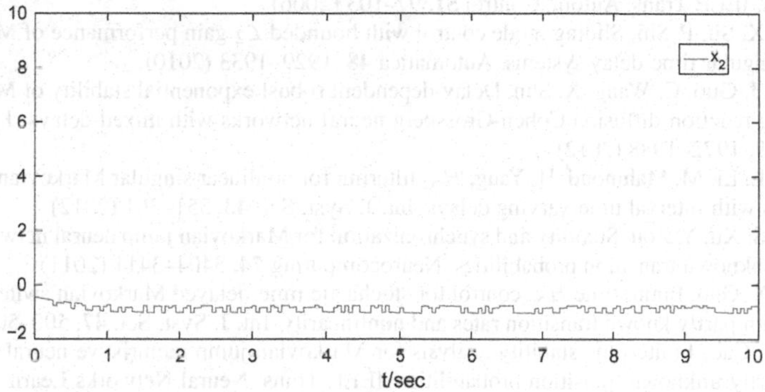

Fig. 3 Response of x_2 of the closed-loop system

Thus,

$$K_1 = \left[\,-13330\ -17654\,\right],\ K_2 = \left[\,-8015\ -9108.8\,\right],\ K_3 = \left[\,-13258\ -10465\,\right].$$

Denote that $x = [x_1\ \ x_2]^T$. Figure 1 shows the switching signal r. The initial states of the system (2) is chosen as [10 10]. The response of the closed-loop system trajectories is presented in Figs. 2 and 3. From the simulation result, we can conclude that the designed controller is effective.

5 Conclusion

In this paper, we deal with the FTSS problem for MJ one-sided Lipschitz systems with PUTRs. For the design of the controllers, we use the state feedback laws $u = K_r x$. In order to get more freedom of the sufficient conditions, the free-weight matrices

Q_i, $i \in S$ are introduced. Besides, the equivalent forms of LMIs for the sufficient conditions that make the closed-loop systems FTSS are derived. We also show the effectiveness of the proposed method by a numerical example.

Acknowledgements The authors are grateful for the National Natural Science Foundation of China (61403267, 61403268), and China Postdoctoral Science Foundation (2017M611903).

References

1. Z. Wang, H. Qiao, K. Burnham, On stabilization of bilinear uncertain time-delay stochastic systems with Markovian jump systems. IEEE Trans. Autom. Control **47**, 640–646 (2002)
2. P. Shi, Y. Xia, G. Liu, D. Rees, On designing of sliding-mode control for stochastic jump systems. IEEE Trans. Autom. Control **51**, 97–103 (2006)
3. L. Wu, X. Su, P. Shi, Sliding mode control with bounded $\mathcal{L}_2$ gain performance of Markovian jump singular time-delay systems. Automatica **48**, 1929–1933 (2010)
4. Y. Kao, J. Guo, C. Wang, X. Sun, Delay-dependent robust exponential stability of Markovian jumping reaction-diffusion Cohen-Grossberg neural networks with mixed delays. J. Franklin Inst. **349**, 1972–1988 (2012)
5. Y. Xia, L. Li, M. Mahmoud, H. Yang, $\mathcal{H}_\infty$ filtering for nonlinear singular Markovian jumping systems with interval time-varying delays. Int. J. Syst. Sci. **43**, 351–361 (2012)
6. Q. Ma, S. Xu, Y. Zou, Stability and synchronization for Markovian jump neural networks with partly unknown transition probabilities. Neurocomputing **74**, 3404–3411 (2011)
7. W. Qi, X. Gao, Finite-time $\mathcal{H}_\infty$ control for stochastic time-delayed Markovian switching systems with partly known transition rates and nonlinearity. Int. J. Syst. Sci. **47**, 500–508 (2016)
8. R. Li, J. Cao, Finite-time stability analysis for Markovian jump memristive neural networks with partly unknown transition probabilities. IEEE Trans. Neural Networks Learn. Syst. **28**, 2924–2935 (2017)
9. Y. Ma, X. Jia, D. Liu, Finite-time dissipative control for singular discrete-time Markovian jump systems with actuator saturation and partly unknown transition rates. Appl. Math. Model. **53**, 49–70 (2018)
10. G. Hu, Observers for one-sided Lipschitz non-linear systems. IMA J. Math. Control Inf. **23**, 395–401 (2006)
11. G. Hu, A note on observer for one-sided Lipschitz non-linear systems. IMA J. Math. Control Inf. **25**, 297–303 (2008)
12. Y. Zhao, J. Tao, N. Shi, A note on observer design for one-sided Lipschitz nonlinear systems. Syst. Control Lett. **59**, 66–71 (2010)
13. M. Abbaszadeh, H. Marquez, Nonlinear observer design for one-sided Lipschitz systems, in *Proceedings of the American Control Conference* (2010)
14. W. Zhang, H. Su, Y. Liang, Z. Han, Non-linear observer design for one-sided Lipschitz systems: an linear matrix inequality approach. IET Control Theory Appl. **6**, 1297–1303 (2012)
15. W. Zhang, H. Su, F. Zhu, D. Yue, A Note on observers for discrete-time Lipschitz nonlinear systems. IEEE Trans. Circuits Syst. II **59**, 123–127 (2012)
16. W. Zhang, H. Su, F. Zhu, S. Bhattacharyya, Improved exponential observer design for one-sided Lipschitz nonlinear systems. Int. J. Robust Nonlinear Control **26**(18), 3958–3973 (2016)
17. F. Fu, M. Hou, G. Duan, Stabilization of quasi-one-sided Lipschitz nonlinear systems. IMA J. Math. Control Inf. **30**, 169–184 (2013)
18. J. Song, S. He, Finite-time H_∞ control for quasi-one-sided Lipschitz nonlinear systems. Neurocomputing **149**(Part C), 1433–1439 (2015)
19. S. Ahmad, M. Rehan, On observer-based control of one-sided Lipschitz systems. J. Franklin Inst. **353**, 903–916 (2016)

A Method for Reutilization of Scrap Metal

Xiulin Hou and Zhongsuo Shi

Abstract With the development and progress of industrial technology, there will be a large amount of scrap metal produced. Discarding the scrap metal everywhere will cause environmental pollution and a great waste of resources. This paper presents a method for reutilization of scrap metal in order to recover and utilize the metal scrap. Firstly, identifying the reusable areas by using image processing algorithms. Then, processing available areas of the irregularly shaped scrap metal. Finally, producing parts of mechanical equipment or new types of handicrafts that meet human needs. This method not only makes use of metal scrap reasonably, but also protects environmental resources.

Keywords Scrap metal · Metal recycling · Image acquisition · Vision system
Image processing · Environmental protection

1 Introduction

Metal scrap is a kind of solid waste which is often produced in the mechanical industry. It includes chips, casting heads from metal processing, scrap castings, metal powders, surplus materials, waste tools, and obsolete machinery equipment. The reutilization of scrap metal is of great significance for ensuring sustainable resources, reducing environmental pollution, saving energy, and improving economic efficiency [1]. Every year in China, a large amount of scrap metals is discarded with rubbish instead of being recycled. Therefore, it is urgent to study how to recycle waste metal resources, which has aroused great concern from all walks of life.

With the renewal of science and technology, there are many methods for the disposal and utilization of scrap metal [2]. Commonly, there are classified collection, recycled smelting, and utilizing the out-of-date mental. It should be classified and

X. Hou (✉) · Z. Shi
School of Automation and Electrical Engineering, University of Science and Technology Beijing, Beijing 100083, China
e-mail: btbu1994@163.com

Y. Jia et al. (eds.), *Proceedings of 2018 Chinese Intelligent Systems Conference*, Lecture Notes in Electrical Engineering 529,
https://doi.org/10.1007/978-981-13-2291-4_22

collected according to different materials, and then processed and recycled according to different properties for the convenience of processing and utilization. It is an effective and reasonable way to make new industrial or civilian products from scrap metal materials. For example, metal leftover materials or defective products, can be directly used to manufacture mechanical equipment and all kinds of civil appliances. At the same time, all kinds of waste tools, such as chain knives, saws, milling knives, and broaches can be repaired or converted to other tools again. All parts of the old and obsolete machinery equipment should be dismantled as far as possible and classified according to their purposes, so that they can be reapplied to production [3].

This paper aims to machining a number of circular parts with a radius of 1 cm in the area where the aluminum scrap can be scrapped. Machine vision system that needs to programming with Visual Studio 2015 is introduced to deal with waste aluminum corners. Firstly, preprocessing the collected image to obtain a high-quality image which should be binarized and divided. Secondly, drawing a number of circles with a radius of 1 cm on the available area. Then, transmitting the obtained center information to the robot through computer. At last, making parts of mechanical equipment by processing, thereby realizing the reuse of waste materials. Due to the extremely irregular shape of the scrap metal leftover material, the introduction of machine vision system can accurately extract areas that can be processed and used to avoid waste. Compared with the traditional methods, on the one hand, it improves the utilization rate of scrap metal. On the other hand, it constructs an automatic and intelligent leftover system which has a strong practical value.

2 Image Acquisition

Image acquisition is the first step in image processing. In industrial environments, it is often necessary to select different cameras according to the requirements and resolution of the scene to collect images. High-quality images are easy to follow in order to achieve accurate and effective purposes. The image acquisition process uses an image sensor to convert the optical signal into an electrical signal firstly, which is quantized to obtain a digital signal. Then, the digital signal is passed to a storage device and finally processed by an image processing unit [4–6]. The image acquisition process is shown in Fig. 1.

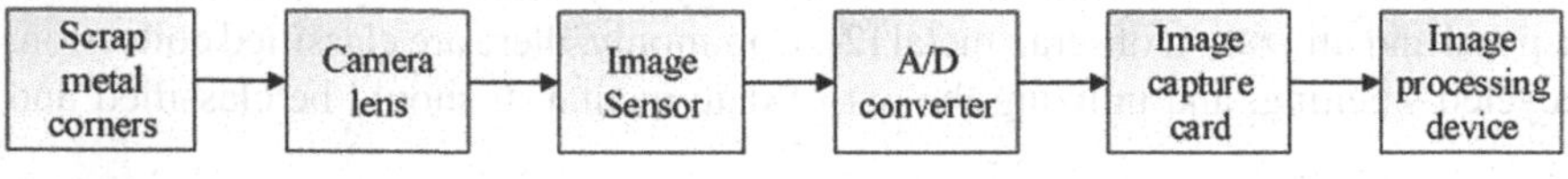

Fig. 1 Image acquisition process

3 Image Processing

For a piece of discarded metal scrap, due to the interference of external environment such as illumination and so on during the process of collection, transformation and transmission, the collected images will be mixed with noise and affect the quality of the image. In order to eliminate and suppress the useless information in images, enhance the detectability of the useful information, minimize the data volume, and improve the measurement accuracy and reliability of the vision system [7], it is necessary to perform a series of preprocessing on the original acquired image to obtain high quality images. The image processing process is shown in Fig. 2.

3.1 Image Graying

Placing the irregularly discarded aluminum corner materials on the image acquisition platform as shown in Fig. 3.

The camera captures a color image. Grayscale can improve the speed of image processing because of its simple texture and gray level. There are several ways to grayscale: selecting the maximum component of the color pixel value as the luminance value of the grayscale image; using the average value of the three components

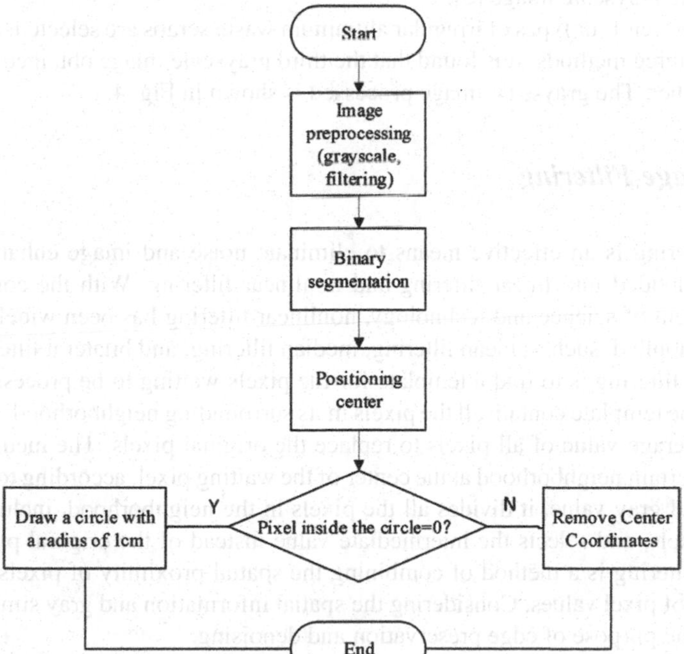

Fig. 2 Image processing flowchart

Fig. 3 Image acquisition platform

Fig. 4 Grayscale image of aluminum waste scraps

of the color pixel as the luminance value of the grayscale image; and adding the three components of the color pixel, the weighted average value is used as the luminance value of the grayscale image [8].

In this paper, four types of irregular aluminum waste scraps are selected, and tests the above three methods. It is found that the third grayscale image obtained is softer and smoother. The grayscale image processed is shown in Fig. 4.

3.2 Image Filtering

Image filtering is an effective means to eliminate noise and image enhancement, which is divided into linear filtering and nonlinear filtering. With the continuous advancement of science and technology, nonlinear filtering has been widely developed and applied, such as mean filtering, median filtering, and bilateral filtering [9]. The mean filtering is to find a template for the pixels waiting to be processed. The pixels in the template contain all the pixels in its surrounding neighborhood, and then use the average value of all pixels to replace the original pixels. The median filter selects a certain neighborhood at the center of the waiting pixel, according to the size of the pixel gray value, it divides all the pixels in the neighborhood, including the central pixels, and selects the intermediate value instead of the original pixel. The bilateral filtering is a method of combining the spatial proximity of pixels and the similarity of pixel values. Considering the spatial information and gray similarity, it achieves the purpose of edge preservation and denoising.

The mean filter causes the blurred edge of the image because of the new gray value. Although the median filter does not generate new gray values, it is easy to filter out images with more points and sharp details as noise points. The focus of this paper is to extract the available area from the aluminum waste scraps, so the pixels of the edge point are crucial and cannot be replaced at will. At the same time, due to the irregular shape, there will be more points and points. Therefore, the bilateral filtering is considered comprehensively, which not only has no effect on the edge features of the image, but also does not filter out the more details of the image points.

The pixel value g at the output position (i, j) in the bilateral filter depends on the weighted combination of the pixel value f in the neighborhood.

The formula is as follows: where k, l represent the neighborhood pixel position.

$$g(i, j) = \frac{\sum_{k,l} f(k, l)w(i, j, k, l)}{\sum_{k,l} w(i, j, k, l)} \tag{1}$$

The weight coefficient $w(i, j, k, l)$ depends on the product of the defined domain kernel d and the value domain kernel r:

$$d(i, j, k, l) = \exp\left(-\frac{(i-k)^2 + (j-l)^2}{2\sigma_d^2}\right) \tag{2}$$

$$r(i, j, k, l) = \exp\left(-\frac{\|f(i, j) - f(k, l)\|^2}{2\sigma_r^2}\right) \tag{3}$$

$$w(i, j, k, l) = \exp\left(-\frac{(i-k)^2 + (j-l)^2}{2\sigma_d^2} - \frac{\|f(i, j) - f(k, l)\|^2}{2\sigma_r^2}\right) \tag{4}$$

This method takes into account the difference between the spatial domain and the value domain and also plays a very good denoising effect.

3.3 *Image Binarization*

After separating useful information from high-quality images which are preprocessed, and then describe, judge, and analyze the characteristic information of these regions [10]. As it is shown in Fig. 4, the gray area of the aluminum scrap is obviously different in the gray value between the useful pixel area and the background, so it is convenient to extract the reused area by using threshold segmentation. By setting an appropriate threshold value, pixel points that are greater than or equal to the threshold value are divided into useful pixel areas, and those that are less than the threshold value are divided into background areas. Due to the influence of the external environment such as the intensity of the light source in the actual environment, only a single threshold method cannot obtain the binary image at different times. This will cause a poor adaptability. Therefore, this paper uses a local adaptive threshold segmentation to obtain binary images.

The local self-adaptive threshold determines the binarization threshold at the pixel location based on the pixel value distribution of the neighborhood block pixel. The binarization threshold of each pixel is not fixed after such processing, which is

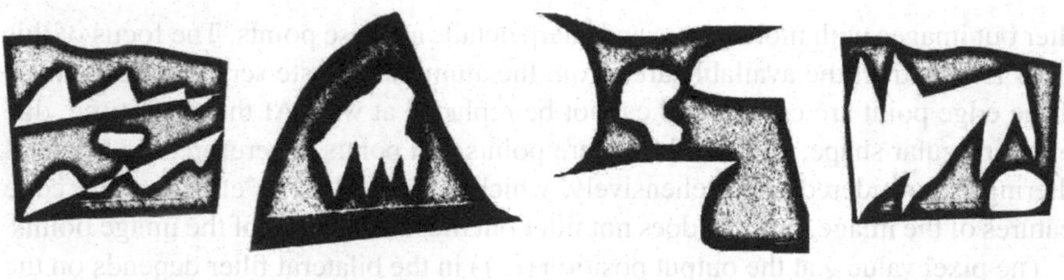

Fig. 5 Binary image of the smaller blockSize

Fig. 6 Binary image of the larger blockSize

determined by the distribution of its surrounding pixels. The threshold value of the image area with higher brightness is usually higher, and the threshold value of the area with lower brightness is smaller accordingly. The Opencv function library provides an adaptive threshold function for self-adaptive threshold.

The prototype of this function is as follows:

adaptive Threshold(Input Array src, OutputArray dst, double max Value, int adaptive Method, int threshold Type, int blockSize, double c)

From that, it can be seen that src and dst refer to the source image and the output image respectively. The value of maxValue is 255, adaptiveMethod refers to how to calculate the threshold in a neighborhood.

One is ADAPTIVE_THRESH_MEAN_C, it calculates the average of the neighborhood and minus the value of double c. Another way is ADAPTIVE_THRESH_GAUSSIAN_C, it calculates the gaussian average of the neighborhood and minus the value of double c.

The first method is used in this paper, blockSize refers to neighborhood blocks of pixels. In the process of experiment, it is found that the value of blockSize has a great influence on the result of image binarization. As shown in Fig. 5, the smaller the value is, the less obvious the binarization of the image is.

As shown in Fig. 6, the larger the value is, the more obvious the binarization of the image is. Since it is an adaptive threshold, double c is the amount of offset adjustment.

Fig. 7 The positioning center of aluminum scrap material

3.4 *Positioning Center*

It can be seen that the black area is the available area after the aluminum scrap is binarized. This article divides the whole image into 50 * 50 grids, followed by the traversal method to virtualize a circle with a radius of 1 cm with the center of the grid as the center, and then judge of all pixels within the circular value. If the values of all pixels are 0, it means that the circles are all black areas, then keep the center coordinates, and draw a circle with a radius of 1 cm. If all pixels are between 0 and 255 or all 255, then remove the center coordinates. The results after processing are shown in Fig. 7.

4 Experimental Results

It can been seen from Fig. 7 that due to the introduction of the vision system, firstly, processing the image of the aluminum scrap and drawing a circle with a radius of 1 cm as much as possible in the available area to realize the maximum utilization of scrap. Next, transmitting the information of the center coordinates to the truss robot shown in Fig. 8 through computer. Finally, transporting the aluminum scrap to the bottom of the processing mechanism to process circular parts to meet the needs of human production.

Fig. 8 Truss robot

5 Conclusions

The reutilization of scrap metal can not only solve the problem of limited and non-renewable mineral resources, but also can avoid the pollution caused by throwing away the scrap metal. It has a great economic and social benefits. This article describes in detail the application of image processing technology of aluminum scrap. It introduces the grayscale, filtering denoising and binary processing of irregular images and improves the utilization rate of leftover materials, which is worthy of further promotion and application.

References

1. B.K. Reck, T.E. Graedel, Challenges in metal recycling. Science **337**(6095), 690 (2012)
2. T.E. Graedel, A. Julian, B. Jean-Pierre et al., What do we know about metal recycling rates? J. Ind. Ecol. **15**(3), 355–366 (2011)
3. E.S. Kim, J.Y. Park, Apparatus for recycling metal scraps: US, US 20100330217 A1 (2010)
4. H.N. Yen, Y.J. Sie, Machine vision system for surface defect inspection of printed silicon solar cells, in *Consumer Electronics* (IEEE, 2012), pp. 422–424
5. J.K. Oh, C.H. Lee, Development of a stereo vision system for industrial robots, in *International Conference on Control, Automation and Systems* (IEEE, 2007), pp. 659–663
6. W. Wu, J. Xie, G. Chen et al., Visualization automatic programming system of bending machine based on machine vision, in *IEEE International Conference on Information and Automation* (IEEE, 2017), pp. 631–636
7. G. Lu, L. Pei, Q. Huang et al., Designation of automatic pointer meter calibration system based on machine vision, in *Fifth International Conference on Instrumentation and Measurement, Computer, Communication and Control* (IEEE, 2016), pp. 1310–1315
8. X. Zhang, X. Wang, Novel survey on the color-image graying algorithm, in *IEEE International Conference on Computer and Information Technology* (IEEE, 2017), pp. 750–753
9. Z. Li, J. Zheng, Z. Zhu et al., Weighted guided image filtering. IEEE Trans. Image Process. **24**(1), 120–129 (2014)
10. H. Kim, E. Ahn, S. Cho et al., Comparative analysis of image binarization methods for crack identification in concrete structures. Cem. Concr. Res. **99**, 53–61 (2017)

Synchronization Control for Multiple Nonholonomic Mobile Robots

Lu Dai, LiXia Liu, ZhongHua Miao and Jin Zhou

Abstract This brief investigates the problems of synchronization control for a group of multiple nonholonomic mobile robots. An integrated algorithm of a kinematic controller and a torque controller is proposed to solve synchronization tracking problem of multi-nonholonomic mobile robots based on backstepping technique, and its asymptotic stability is then guaranteed by the use of Lyapunov-like analysis. A distinctive feature of the proposed algorithm is to introduce the network topology with directed topology graph characterizing communication interaction among agents based on algebraic graph theory. An illustrate example and its simulation is finally provided to demonstrate the theoretical results.

Keywords Synchronization control · Nonholonomic mobile · Robots backstepping technique · Multi-agent systems

1 Introduction

Coordination control of multiple nonholonomic mobile robots (MNMRs) has currently been a widespread interest research topic in many areas of science and engineering with the great advancement of computer science, communication technique, and artificial intelligence, etc [1]. This is mainly due to the following elementary reasons. Firstly, multi-robot systems (MRSs) can effectively execute the cooperative sophisticated tasks that are impossible to be accomplished by a single robot. Secondly, the control of mobile robots which nonholonomic constraint is challenging but possesses the advantage of a low-cost and simple construction, which is espe-

L. Dai · L. Liu · J. Zhou (✉)
Shanghai Institute of Applied Mathematics and Mechanics,
Shanghai University, Shanghai 200072, China
e-mail: jzhou@shu.edu.cn

Z. Miao
School of Mechatronic Engineering and Automation,
Shanghai University, Shanghai 200072, China

Y. Jia et al. (eds.), *Proceedings of 2018 Chinese Intelligent Systems Conference*, Lecture Notes in Electrical Engineering 529,
https://doi.org/10.1007/978-981-13-2291-4_23

cially applicable to unstructured environments in practice [2]. More importantly, the cooperative control of multiple nonholonomic mobile robots can offers high flexibility,scalability, reliability and robustness, and so it has a wider range of engineering applications, including in spacecraft formation flight, multiple planet navigation, and unmanned autonomous vehicles, among others [3, 4].

As is generally known, coordination control of multiple nonholonomic mobile robots can be rephrased as the problems of the controlled synchronization or tracking consensus for multi-agent systems with non-holonomic constraint. Compared with multi-agent systems with single or double integrator dynamics [5], as well as fully-actuated Lagrange dynamics [6], the control of multiple nonholonomic agent systems is much more difficult because of taking into account not only the integration of both trajectory tracking and posture stabilization for each agent [7], but also the communication interaction among robots [8, 9]. As a consequence, the previous majority of research is mainly concentrated on tracking control problems of single nonholonomic mobile robot. For example, Fierro and Lewis proposed the backstepping control approach to perform the tracking control of a nonholonomic robot, and later used neural network control method to solve the tracking problem of nonholonomic robots [10]. However, to the best of our capacity, so far very little research has been devoted to cooperative control of multiple nonholonomic mobile robot. These observations further motivate the present research work to be reported.

With the aforementioned background, we are mainly interested in the problem of synchronization tracking control of multiple nonholonomic mobile robots from the view of networked multi-agent systems. The main objective here is to advance the tracking control algorithm of single nonholonomic mobile robot for studying synchronization tracking control of networked nonholonomic mobile robots under directed topology graph. An integrated control strategy of a kinematic controller and a torque controller will be proposed to solve synchronization tracking problem of multi-nonholonomic mobile robots based on backstepping technique. Subsequently, an illustrate example and its simulation is provided to demonstrate and visualize the theoretical results.

2 Preliminaries

2.1 Graph Theory

As usual, a weighted directed graph $\mathcal{G} = (\mathcal{V}, \mathcal{E}, \mathcal{A})$ is employed to describe the interaction among robots in a network of n robots, where the node set is $\mathcal{V} = \{1, 2, \ldots, n\}$, the edge set is $\mathcal{E} \in \mathcal{V} \times \mathcal{V}$, and a weighted adjacency matrix $A = [a_{ij}] \in \mathbb{R}^{n\times n}$ is defined as $a_{ij} = 0$ if $(j, i) \notin \mathcal{E}$, and $a_{ij} \neq 0$ otherwise. $(j, i) \in \mathcal{E}$ means that robot i can obtain the information from robot j, but not vice versa. Here, $a_{ii} = 0$ for $i \in \mathcal{V}$ means there is no edge between a node and itself. The Laplician matrix $\mathcal{L}_A = [l_{ij}] \in \mathbb{R}^{n\times n}$ associated with $A = [a_{ij}] \in \mathbb{R}^{n\times n}$ is defined as $l_{ii} = \sum_{j=1}^{n} a_{ij}$ and $l_{ij} = -a_{ij}, i \neq j$. A directed spanning tree in $\mathcal{G}$ is that there exists a root node which has a directed path to other nodes [5].

2.2 Multiple Nonholonomic Mobile Robots

Consider a network of n nonholonomic mobile robots, where the ith mobile robot subjected to m constraints is compactly expressed as [10]

$$M_i(q_i)\ddot{q}_i + V_i(q_i, \dot{q}_i)\dot{q}_i + F_i(\dot{q}_i) + G_i(q_i) = B_i(q_i)\tau_i - A^T(q_i)\lambda_i, i = 1, 2, \ldots, n \tag{1}$$

where $q_i \in R^p$ is the generalized coordinate vector, $M_i(q_i) \in R^{p\times p}$ is the symmetric positive definite inertia matrix, $V_i(q_i, \dot{q}_i) \in R^{p\times p}$ is the centripetal and coriolis matrix, $F_i(\dot{q}_i) \in R^p$ is the surface friction, $G_i(q_i) \in R^p$ is the gravitational vector, $\tau_i \in R^p$ is the input vector, $\lambda_i \in R^p$ is the vector of constraint forces, $A_i(q_i) \in R^{m\times p}$ is the constraint matrix.

Following [1], the network of n nonholonomic wheeled mobile robot systems (1) can be written as

$$\dot{q}_i = S(q_i)v_i. \tag{2}$$

$$\overline{M}_i(q_i)\dot{v}_i + \overline{V}_{mi}(q_i, \dot{q}_i)v_l + \overline{\tau_d} = \overline{B}_i(q_i)\tau_i. \tag{3}$$

where $S_i \in \mathbb{R}^{n\times(n-m)}$ is a full-rank matrix spanning the null space of $A_l(q_l)$, $\nu_i(t) \in \mathbb{R}^{n-m}$ is the velocity vector, $\overline{M}_i(q_i) = S_i^T M_i S_i$, $\overline{V}_{mi}(q_i, \dot{q}_i) = S_i^T(M_i\dot{S}_i + V_{mi}S_i)$, $\overline{\tau}_i = \overline{B}_i\tau_i$, and $\overline{B}_i = S_i^T B_i$ is a constant nonsingular matrix.

Accordingly, the motion models of multi-nonholonomic mobile robot systems can be classified into the kinematic steering system (2) and some additional dynamics (3). The following properties are presented to reflect the physical characteristics of the dynamics (3) of the transformed robot systems [9]:

Property 1 *$\overline{M}_i(q_i)$, the norm of the matrix $\overline{V}_{mi}(q_i, \dot{q}_i)$ and $\overline{\tau}_{di}$ are bounded.*

Property 2 *The matrix $\overline{M}_i - 2\overline{V}_{mi}$ is skew symmetric.*

It is usually assumed that each mobile robot in a network of multi-robot systems possesses two actuated wheels, and thus it has a typical property of nonholonomic constraints. Accordingly, the position and direction of robot i can be denoted by the coordinates $q_i = [X_{Ci}, Y_{Ci}, \theta_i]^T$, the robot only move along the direction of the front wheel, but there is no pure rolling and sliding phenomenon, as shown in Fig. 1.

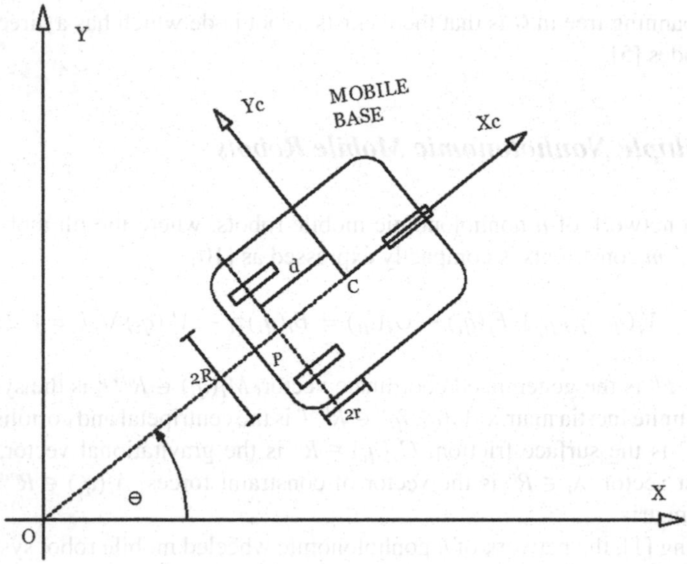

Fig. 1 A nonholonomic mobile platform

3 Synchronization Control

3.1 Control Objective

In general, an appropriate velocity is usually selected as a kinematic controller for nonholonomic mobile robot, but it is more realistic that the dynamical controller is actually torque. As a result, it is possible to convert such input velocity into a torque controller for actual robotic systems, and thereby yielding a desired torque controller to track a specific synchronization trajectory for multiple nonholonomic mobile robots.

To do so, let u_i be an auxiliary input for robot i, by using following dynamical controller

$$\tau_i = \overline{B}_i(q_i)^{-1}[\overline{M}_i(q_i)u_i + \overline{V}_{mi}(q_i, \dot{q}_i)v_i + \overline{F}_i(v_i) + \overline{\tau}_{di}], \tag{4}$$

then the model (1) of multi-nonholonomic mobile robots can be converted into the control problem of kinematic model below

$$\dot{q}_i = S(q_i)v_i, \tag{5}$$

$$\dot{v}_i = u_i. \tag{6}$$

Now we are ready to present the concept of synchronization tracking control for networked multinonholonomic mobile robots as follow [10].

Definition 1 The control protocols $\tau_i (i = 1, 2, \ldots, n)$ (or the torque inputs (4)) for each robot i are said to solve synchronization tracking problems for a network of n nonholonomic mobile robots (1) whose the kinematics (2) and the dynamics (3) with respect to a given synchronization reference cart being given by

$$\begin{cases} \dot{x}_r = v_r cos\theta_r \\ \dot{y}_r = v_r sin\theta_r \\ \dot{\theta}_r = w_r \end{cases}, \tag{7}$$

where $q_r = [x_r, y_r, \theta_r]^T$, with $v_r > 0$, $w_r > 0$. If each robot i can find a smooth velocity control input $v_{ic} = f_{ic}(e_i, v_r, K_i)$, such that $v_i \to v_{ic}$, then $\lim_{t\to\infty}(q_i - q_r) = 0$, $i = 1, 2, \ldots, N$, as $t \to 0$, where e_i is the tracking synchronization position errors, v_r is the reference velocity vector, and K_i is the control gain vector.

This paper focuses mainly on the synchronization tracking control problems for multi-nonholonomic mobile robots. The primary objective here is to design a torque controller for networked multi-nonholonomic mobile robots such that each of robots in the network can track a desired synchronization trajectory with respect to its perfect velocity matching.

3.2 Controller Design

The design of synchronization tracking algorithm for a network of multi-nonholonomic mobile robots is considered in this section.

In order to search for an ideal time-varying velocity input for steering system (2), we first introduce the tracking error vector of both position and direction with respect to the synchronization trajectory from mobile robot i as [10]:

$$e_i = \begin{bmatrix} e_{i1} \\ e_{i2} \\ e_{i3} \end{bmatrix} = \begin{bmatrix} \cos\theta_i & \sin\theta_i & 0 \\ -\sin\theta_i & \cos\theta_i & 0 \\ 0 & 0 & 1 \end{bmatrix} \begin{bmatrix} x_r - x_i \\ y_r - y_i \\ \theta_r - \theta_i \end{bmatrix}, \tag{8}$$

and its derivative then is

$$\dot{e}_i = \begin{bmatrix} \dot{e}_{i1} \\ \dot{e}_{i2} \\ \dot{e}_{i3} \end{bmatrix} = \begin{bmatrix} w_i e_{i2} - v_i + v_r \cos e_{i3} \\ -w_i e_{i1} + v_r \sin e_{i3} \\ w_r - w_i \end{bmatrix}. \tag{9}$$

Let an auxiliary velocity control input of steering system (2) for robot i, which is given by

$$v_{ic} = \begin{bmatrix} v_r \cos e_{i3} + k_1 e_{i1} \\ w_r + k_2 v_r e_{i2} + k_3 v_r \sin e_{i3} \end{bmatrix}, \tag{10}$$

where k_1, k_2 and k_3 are the positive parameters to be designed later.

Thus, by assuming that the linear and angular reference velocities are constants, and the corresponding derivative is also given by

$$\dot{v}_{ci} = \begin{bmatrix} k_1 & 0 & -v_r \sin e_{i3} \\ 0 & k_2 v_r & k_3 v_r \cos e_{i3} \end{bmatrix} \dot{e}_i. \tag{11}$$

Consider the Lyapunov functions for system (8) as

$$V_i = \frac{1}{2}(e_{i1}^2 + e_{i2}^2) + \frac{1 - cose_{i3}}{k_{i2}}, \tag{12}$$

and take the derivative of V_i as

$$\dot{V}_i = -k_{i1} e_{i2}^2 - \frac{k_{i3} sin^2 e_{i3}}{k_{i2}} \leq 0. \tag{13}$$

It can be concluded from Barbalate's lemma that e_{i1}, e_{i2} and e_{i3} asymptotically converge to zero. This implies that each robot can completely track a synchronization reference trajectory, i.e., $\lim_{t\to\infty}(q_r - q_i) = 0$ as $t \to \infty$. Therefore, the synchronization tracking problems for multi-nonholonomic robot systems are fully guaranteed by the smooth velocity control input $v_{ic} = f_{ic}(e_i, v_r, K_i)$ for robot i.

Next, we can choose the nonlinear feedback acceleration control input for robot i

$$u_i = \dot{v}_{ic} + \sum_{j=1}^{n} a_{ij}(v_j - v_i) + k_{i0}(v_i - v_{ic}), \tag{14}$$

where k_{i0} is the feedback control gain with respect to an ideal velocity input v_{ic}. Furthermore, we assume that there exists at least a $k_{i0} > 0$, and $k_{i0} = 0$ otherwise.

With the above preparations, the main results of synchronization tracking control for nonholonomic mobile robot systems (1) are introduced below.

Theorem 1 *Consider a directed graph network composed of n nonholonomic mobile robots (1), then the control protocols (or the torque inputs) (4) with the feedback acceleration control input (14) can always solve the synchronization protocol problems for networked multi-nonholonomic mobile robot systems in the sense of Definition 1 if the directed network graph has a spanning tree.*

Proof Define an auxiliary synchronization velocity error:

$$e_{ic} = v_i - v_{ic} = \begin{bmatrix} e_{i4} \\ e_{i5} \end{bmatrix} = \begin{bmatrix} v_i - v_r \cos e_{i3} - k_1 e_{i1} \\ w_i - w_r - k_2 v_r e_{i2} - k_3 v_r \sin e_{i3} \end{bmatrix}. \tag{15}$$

By using the nonlinear feedback acceleration control input, the synchronization velocity error system can be formulated in a vector form as

$$\dot{E}_c = -((L_A + K_0) \otimes I_2)E_c + W, \tag{16}$$

where $E_C = [e_{c1}, e_{c2}, \ldots, e_{cn}]^T$, $W = [w_1, w_2, \ldots, w_n]^T$ with $w_i = \sum_{j=1}^{n} a_{ij}(v_{ic} - v_{jc})$, $K_O = \mathrm{diag}\{k_{1O}, k_{2O}, \ldots, k_{nO}\}$, and $\otimes$ denotes Kronecker product.

It is easy to see that $L_A + K_O$ is Hurwitz stable if the directed network graph has a spanning tree. Therefore, the following auxiliary system

$$\dot{E}_c = -\big((L_A + K_O) \otimes I_2\big)E_c, \tag{17}$$

is global asymptotically exponentially stable with respect to zero solutions, and this implies that $v_i \to v_{ci}$ as $t \to +\infty$, which yields to $w_i \to 0$ as $t \to +\infty$.

By looking back the linear different dynamical systems (16), it can be also proved that $v_i \to v_{ci}$ as $t \to +\infty$, which leads to $q_i = \mathbf{S}_i \nu_i = \mathbf{S}_i \nu_{ic} \to q_r$ as $t \to +\infty$.

Consequently, we can conclude that all the robots will asymptotically converge to a specific synchronization trajectory with respect to perfect velocity matching. i. e., $\nu_i - \nu_{ic} \to 0$ as $t \to +\infty$. The proof of Theorem 1 is completed.

Remark 1 Theorem 1 is actually an important generalization of Theorem in the literature for the tracking control of a single mobile robot with nonholonomic constraints. In this brief, a dynamical torque controller in combination with a kinematic velocity controller is proposed to fully guarantee the synchronization tracking for multi-nonholonomic mobile robots [9]. A key feature of this work is to introduce the network topology characterizing communication interaction among robots into the proposed synchronization schemes, and the asymptotic stability of closed-loop control systems is ensured by the use of Lyapunov-like analysis. It is now widely suggested that communication interaction among agents can play a critical role in the cooperative control for network of multi-agent systems [10].

4 Simulation Results

To illustrate the effectiveness of the proposed control algorithms, a team of three nonholonomic mobile robots is taken as a simple simulation example in this section.

To do so, for a network of three nonholonomic mobile robots (1), we can choose

$$M(q_i) = \begin{bmatrix} m_i & 0 & m_i d_i \sin\theta_i \\ 0 & m_i & -m_i d_i \cos\theta_i \\ m_i d_i \sin\theta_i & -m_i d_i \cos\theta_i & I \end{bmatrix},$$

$$V(q_i, \dot{q}_i) = \begin{bmatrix} m_i d_i \dot{\theta}_i^2 \cos\theta_i \\ m_i d_i \dot{\theta}_i^2 \cos\theta_i \\ 0 \end{bmatrix}, \quad B(q_i) = \frac{1}{r}\begin{bmatrix} \cos\theta_i & \cos\theta_i \\ \sin\theta_i & \sin\theta_i \\ R & -R \end{bmatrix},$$

$$A^T(q_i) = \begin{bmatrix} -\sin\theta_i \\ \cos\theta_i \\ -d_i \end{bmatrix}, \quad \text{and} \quad S(q_i) = \begin{bmatrix} \cos\theta_i & -d_i \sin\theta_i \\ \sin\theta_i & d_i \cos\theta_i \\ 0 & 1 \end{bmatrix}.$$

Fig. 2 Desired (red) and actual trajectories

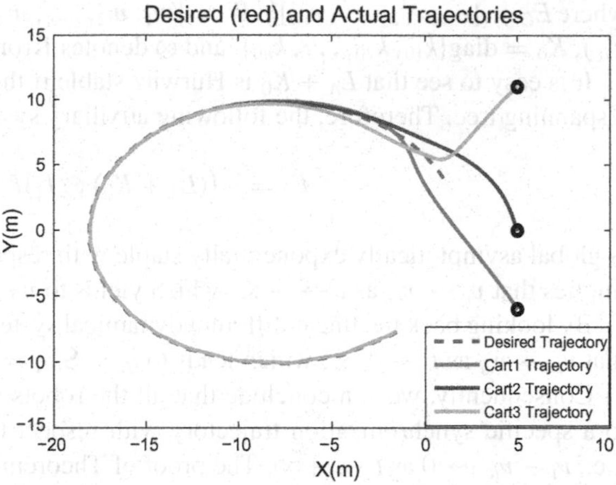

Fig. 3 Actual (1/2/3) linear velocities v (m/s)

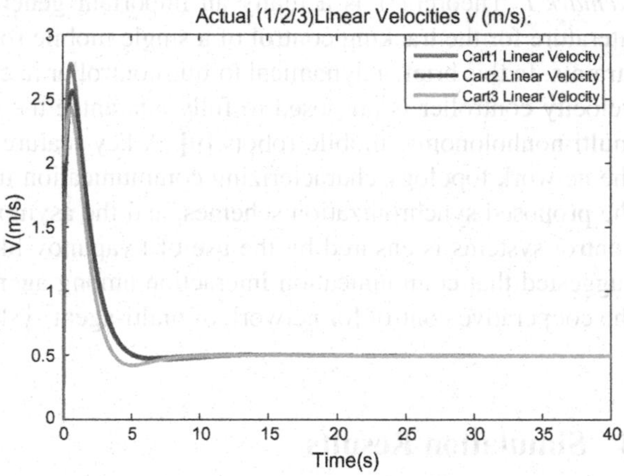

Accordingly, the Laplician matrix of the network is taken as $\mathcal{L}_A = \begin{bmatrix} -2 & 1 & 1 \\ 1 & -2 & 1 \\ 1 & 1 & -2 \end{bmatrix}$.

In addition, the other parameters are selected as $v_r = 0.5$; $w_r = 0.05$; $d_i = 0.5$; $k_1 = 10$; $k_2 = 5$; $k_3 = 1$; $k_0 = -1$; $c = 2$, respectively. In this simulation, we still choose the initial conditions of the three robots as $q_1(0) = (5, -6, 2)$, $q_2(0) = (5, 0, 1)$, $q_3(0) = (5, 11, 4)$. According to Theorem 1, the trajectories of the three cars exactly tracking the reference trajectory as shown in Fig. 2, which reveals that the proposed synchronization tracking scheme is very effective. Furthermore, the evolution of both linear velocities and angular velocities are presented in Figs. 3 and 4 respectively, which indicate that the linear and angular converge to the desired values, too.

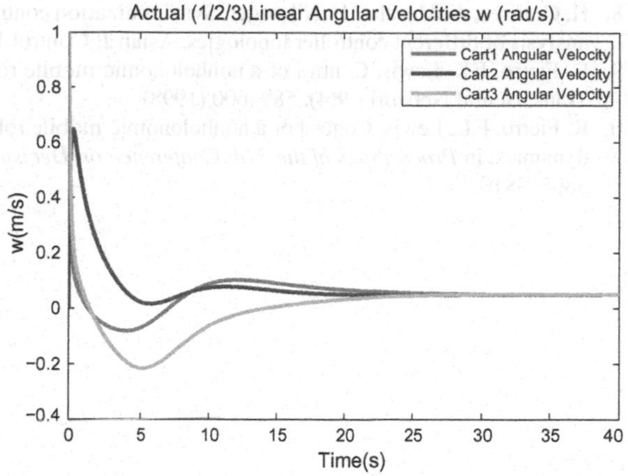

Fig. 4 Actual (1/2/3) linear angular velocities w (rad/s)

5 Conclusions

From the point of view of networked multi-agent systems, the problems of synchronization control for a group of multiple nonholonomic mobile robots have been studied in this brief. A synchronization control strategy with an integrated combination of a kinematic controller and a torque controller has been proposed for multi-nonholonomic robotic systems, and the convergence analysis of closed-loop control system has been implemented in combination with the Lyapunov-like analysis and the backstepping technique. To this end, an illustrate example and its simulations have been provided to demonstrate the theoretical results.

Acknowledgements The paper is supported by the National Natural Science Foundation of China (Grant Nos. 11672169 and 51875331).

References

1. Q. Wang, Z. Chen, Adaptive coordinated tracking control for multi-robot system with directed communication topology. Int. J. Adv. Rob. Syst. **14**(6), 1–10 (2017)
2. Q.K. Yang, H. Fang, M. Cao, Distributed trajectory tracking control for multiple nonholonomic mobile robots. Int. Fed. Autom. Control **49**(4), 31–36 (2016)
3. Y.C. Liu, N. Chopra, Controlled synchronization of heterogeneous robotic manipulators in the task space. IEEE Trans. Robot. **28**(1), 268–275 (2012)
4. D.K. Chwa, Sliding-mode tracking control of nonholonomic wheeled mobile robots in polar coordinates. IEEE Trans. Control Syst. Technol. **12**(4), 637–644 (2004)
5. A.R. Mehrabian, S. Tafazoli, K. Khorasani, Cooperative tracking control of Euler-Lagrange systems with switching communication network topologies, in IEEE vol. 2 (2010), pp. 756–761
6. S.J. Chung, J.J.E. Slotine, Cooperative robot control and concurrent synchronization of Lagrangian systems. IEEE Trans. Robot. **25**(3), 686–700 (2009)
7. E. Zergeroglu, D.M. Dawson, I. Walker, P. Setlur, Nonlinear tracking control of kinematically redundant robot manipulators. IEEE/ASME Trans. Mech. **9**(1), 129–132 (2004)

8. H. Gutirrez, A. Morales, H. Nijmeijer, Synchronization control for a swarm of unicycle robots: analysis of different controller topologies. Asian J. Control **19**(5), 1822–1833 (2017)
9. R. Fierro, F.L. Lewis, Control of a nonholonomic mobile robot using neural networks. IEEE Trans. Neural Networks **9**(4), 589–600 (1998)
10. R. Fierro, F.L. Lewis, Control of a nonholonomic mobile robot: backstepping kinematics into dynamics, in *Proceedings of the 34th Conference on Decision and Control*, vol. 2 (1995), pp. 3805–3810

Interval Observer Design for Nonlinear Switched Systems

Huangyuwei Lu, Jun Huang and Xiang Ma

Abstract This paper deals with the interval observer design problem for nonlinear switched systems. The nonlinearity is assumed to satisfy the property of Lispschitz. The interval observers are constructed and multiple linear copositive Lyapunov function is used to analyze exponential stability of the error systems. Different from the most of current works, the sufficient conditions for the existence of interval observers are derived by the forms of linear programming. Finally, a numerical example is simulated to show the efficiency of the proposed method.

Keywords Interval observers · Nonlinear switched systems · Linear programming

1 Introduction

Recently, the interval observer became a hot topic in the field of control. It is not an easy task to estimate the states exactly when the uncertainty exists in the real systems. The interval observer is constructed to give the estimation of upper and lower bound of the state [1]. The main methods for interval observer can be divided into two categories. One is linear programming (LP) method [2, 3], the other one is coordinate transformation method [4–6]. By using LP method, [2] considered the linear systems where the uncertainty is contained in the state equations, while [3] studied the systems whose state equations and output equations are both suffered from disturbances. Under the time-varying transformation, [4] designed the time-varying exponentially stable interval observers for the linear systems with disturbances. References [5] and [6] improved the former results. By the time invariant transformation, [5, 6] constructed the interval observers for time invariant and time-varying nonlinear systems, respectively. Besides, there also exist many works on the interval observer, we refer the readers to [7–9].

H. Lu · J. Huang (✉) · X. Ma
School of Mechanical and Electrical Engineering,
Soochow University, Suzhou 215131, China
e-mail: cauchyhot@163.com

Y. Jia et al. (eds.), *Proceedings of 2018 Chinese Intelligent Systems Conference*, Lecture Notes in Electrical Engineering 529,
https://doi.org/10.1007/978-981-13-2291-4_24

On the other hand, switched systems have been paid much attention since they are useful to describe many kinds of systems, such as traffic control systems, chemical processing systems, switching power converters, multi-agent systems and so on. The feature of switched systems lies that they consist of finite subsystems described by differential or difference equations, which are controlled by a switching law. For switched systems, the asymptotical observers have been established in the former works [10–12]. However, the interval observers for switched systems have not been studied fully except for [13–16]. He and Xie [13] designed the interval observers for switched systems by the feature of system matrices. By choosing proper transformation, [14, 15] investigated the interval observer design method for continuous and discrete switched systems. Actually, the conditions given in [13–15] are all linear matrix inequalities (LMIs). More recently, [16] studied linear switched systems and derived sufficient conditions in the forms of LP by multiple linear copositive Lyapunov function (MLCLF), which are more tractable than LMIs.

In this paper, just following the line of [16], we try to study the interval observer design approach for nonlinear switched systems. The rest of the paper is organized as follows. The problem formulation and some necessary preliminary are given in Sect. 2. Section 3 presents the main results, consisting of sufficient conditions that make the interval observer exponentially convergent. A numerical example is simulated to show the effectiveness of the designed interval observer in Sect. 4.

Throughout this paper, $||x||$ denotes the Euclidean norm of the vector x. $x > (\geq)0$ means that its components are positive (nonnegative), i.e., $x_i > (\geq)0$. $A > (\geq)0$ means that its components are positive (nonnegative), i.e., $A_{ij} > (\geq)0$. $\underline{\varepsilon}(x)$ and $\overline{\varepsilon}(x)$ are the minimum and maximum value of the elements of x, respectively.

2 Problem Formulation and Preliminaries

Let us consider the following system described by

$$\begin{cases} \dot{x}(t) = A_{\sigma(t)}x(t) + B_{\sigma(t)}u(t) + f_{\sigma(t)}(x(t)), \\ y(t) = C_{\sigma(t)}x(t), \\ x^-(0) \leq x(0) \leq x^+(0), \end{cases} \tag{1}$$

where $x(t) \in R^n$ is the state, $u(t) \in R^m$ is the control input, and $y(t) \in R^q$ is the output of the system. $\sigma(t)$ is a continuous mapping taking values in a finite set $S = \{1, 2, \ldots, N\}$. For any $\sigma(t) = i \in S$, $A_i \in R^{n\times n}$, $B_i \in R^{n\times m}$, $C_i \in R^{q\times n}$ are determined matrices, and $f_i(x(t)) \in R^n$ is the nonlinear function. The upper bound and lower bound of initial state $x(0)$ are $x^+(0)$ and $x^-(0)$, which are both known.

Definition 1 An interval observer for (1) is pair of upper and lower recovered states $\{\hat{x}^+(t), \hat{x}^-(t)\}$, which satisfy for any $t > 0$

$$\hat{x}^-(t) \leq x(t) \leq \hat{x}^+(t),$$

under the initial condition

$$\hat{x}^-(0) \le x(0) \le \hat{x}^+(0).$$

Definition 2 An interval observer for (1) is said to be exponentially convergent if there exist positive constants $\alpha_1, \alpha_2, \beta_1, \beta_2$ such that for any $t \ge 0$

$$||\hat{x}^+(t) - x(t)|| \le \alpha_1 \exp\{-\beta_1 t\}||\hat{x}^+(0) - x(0)||,$$

and

$$||x(t) - \hat{x}^-(t)|| \le \alpha_2 \exp\{-\beta_2 t\}||x(0) - \hat{x}^-(0)||.$$

For simplicity, the variable t is omitted in the expressions. In order to establish the main the result of the paper, we need the following assumptions.

Assumption 1 The nonlinear function $f_\sigma(x)$ is global Lipschitz function, and

$$f_i(x) = g_i(x) - h_i(x), \ \forall \sigma = i \in S,$$

where $g_i(x)$ and $h_i(x)$ are increasing Lipschitz functions.

Assumption 2 Nonlinear functions $g_i(x)$ and $h_i(x)$ satisfy

$$g_i(\hat{x}^+) - g_i(x) \le N_{i1}(\hat{x}^+ - x),\ h_i(\hat{x}^+) - h_i(x) \le N_{i2}(\hat{x}^+ - x),$$
$$g_i(x) - g_i(\hat{x}^-) \le N_{i3}(x - \hat{x}^-),\ h_i(x) - h_i(\hat{x}^-) \le N_{i4}(x - \hat{x}^-),$$

where $N_{i1}, N_{i2}, N_{i3}, N_{i4}$ are given matrices.

Then, the interval observer for the system (1) is designed as:

$$\begin{cases} \dot{\hat{x}}^+ = A_\sigma \hat{x}^+ + B_\sigma u + g_\sigma(\hat{x}^+) - h_\sigma(\hat{x}^-) + L_\sigma(y - C_\sigma \hat{x}^+), \\ \dot{\hat{x}}^- = A_\sigma \hat{x}^- + B_\sigma u + g_\sigma(\hat{x}^-) - h_\sigma(\hat{x}^+) + L_\sigma(y - C_\sigma \hat{x}^-), \\ \hat{x}^+(0) = x^+(0), \\ \hat{x}^-(0) = x^-(0), \end{cases} \tag{2}$$

where the observer gain $L_\sigma \in R^{n\times q}$ will be determined later. Comparing (2) with (1), we obtain the error system

$$\begin{cases} \dot{e}^+ = (A_\sigma - L_\sigma C_\sigma)e^+ + g_\sigma(\hat{x}^+) - g_\sigma(x) + h_\sigma(x) - h_\sigma(\hat{x}^-), \\ \dot{e}^- = (A_\sigma - L_\sigma C_\sigma)e^- + g_\sigma(x) - g_\sigma(\hat{x}^-) + h_\sigma(\hat{x}^+) - h_\sigma(x), \\ e^+(0) \ge 0,\ \ e^-(0) \ge 0, \end{cases} \tag{3}$$

where $e^+ = \hat{x}^+ - x$ and $e^- = x - \hat{x}^-$. In the sequel, we will review the definitions as well as properties of positive systems. Consider the following switched system:

$$\begin{cases} \dot{x} = M_\sigma x + \phi_\sigma(x), \\ x(0) = x_0 \ge 0, \end{cases} \tag{4}$$

where M_i is constant matrix and $\phi_i(x) > 0$ is nonlinear function for $\sigma = i \in S$.

Definition 3 The system (4) is said to be positive if the corresponding trajectory $x(t) \geq 0$ for any $t \geq 0$, $\sigma \in S$.

Definition 4 The matrix M_σ is said to be a Metzler matrix if all its off-diagonal entries are non-negative for any $\sigma \in S$.

Definition 5 ([17]) Consider the time interval $[t_1, t_2)$ where $t_1 \geq 0$. Let the switching number of σ on $[t_1, t_2)$ be $N_\sigma(t_1, t_2)$. If the following inequality holds

$$N_\sigma(t_1, t_2) \leq N_0 + (t_2 - t_1)/\tau^*,$$

where $N_0 \geq 0$ and $\tau^* > 0$, then τ^* is an ADT of the switching signal σ.

Lemma 1 [18] *The system* (4) *is positive if and only if M_σ is a Metzler matrix for any $\sigma \in S$.*

Lemma 2 [19] *M_σ is a Metzler matrix if and only if there exists a constant μ such that $M_\sigma + \mu I \geq 0$.*

3 Main Result

Theorem 1 *If there exist constants $\eta > 0$, $\rho > 1$, μ and vectors $v_i \in R^n > 0$, $v_j \in R^n > 0$, $z_i \in R^q$, $\forall\, i, j \in S$, $i \neq j$ such that*

$$(A_i^T + N_{i1}^T + N_{i4}^T + \eta I)v_i + C_i^T z_i \leq 0, \tag{5}$$

$$(A_i^T + N_{i2}^T + N_{i3}^T + \eta I)v_i + C_i^T z_i \leq 0, \tag{6}$$

$$\xi_i^T v_i(\xi_i^T v_i A_i + \xi_i z_i^T C_i + \mu I) \geq 0, \tag{7}$$

$$v_i \leq \rho v_j, \tag{8}$$

where $\xi_i \in R^n \neq 0$ is a prescribed vector. Then the observer gain is designed by

$$L_i = -\frac{\xi_i z_i^T}{\xi_i^T v_i}, \tag{9}$$

and ADT satisfying

$$\tau^* \geq \frac{\ln \rho}{\eta}, \tag{10}$$

the observer (2) *is an exponentially convergent interval observer for the system* (1).

Proof From Definitions 1 and 2, we will show the positivity as well exponentially stability of the error system (3). We first prove the positivity of the upper error system, i.e,

$$\begin{cases} \dot{e}^{+} = (A_\sigma - L_\sigma C_\sigma)e^{+} + g_\sigma(\hat{x}^{+}) - g_\sigma(x) + h_\sigma(x) - h_\sigma(\hat{x}^{-}), \\ e^{+}(0) \geq 0. \end{cases} \tag{11}$$

It is obvious that $g_i(\hat{x}^{+}) - g_i(x) \geq 0$ and $h_i(x) - h_i(\hat{x}^{-}) \geq 0$. By using (9), then

$$A_i - L_i C_i = A_i + \frac{\xi_i z_i^T}{\xi_i^T v_i} C_i, \;\; i \in S. \tag{12}$$

In view of (7), one get that

$$A_i + \frac{\xi_i z_i^T}{\xi_i^T v_i} C_i + \mu I \geq 0. \tag{13}$$

It follows from Lemma 2 that $A_i + \frac{\xi_i z_i^T}{\xi_i^T v_i} C_i$ is a Metzler matrix, i.e., $A_i - L_i C_i$ is a Metzler matrix. By Lemma 1, the system (11) is positive. Similarly, the lower error system is also positive. Then, we show the exponential stability of the error system (3). Without loss of generality, it is supposed that the switching sequence is $\{t_i, i = 1, 2, \ldots\}$, and $0 < t_1 < t_2 < \cdots$. Let $\sigma(t_k) = i \in S$, we choose the MLCLF as follows:

$$V_i = (e)^T v_i, \;\; i \in S, \tag{14}$$

where $e = [(e^{+})^T \; (e^{-})^T]^T, \bar{v}_i = [v_i^T \; v_i^T]^T$. Consider $t \in [t_k, t_{k+1})$, one can compute that

$$\begin{aligned} \dot{V}_i &= (\dot{e}^{+})^T v_i + (\dot{e}^{-})^T v_i \\ &= (e^{+})^T (A_i - L_i C_i)^T v_i + (g_i(\hat{x}^{+}) - g_i(x))^T v_i + (h_i(x) - h_i(\hat{x}^{-}))^T v_i \\ &\quad + (e^{-})^T (A_i - L_i C_i)^T v_i + (g_i(x) - g_i(\hat{x}^{-}))^T v_i + (h_i(\hat{x}^{+}) - h_i(x))^T v_i. \end{aligned} \tag{15}$$

By Assumptions 1 and 2, we obtain that

$$\begin{aligned} \dot{V}_i \leq\; & (e^{+})^T (A_i - L_i C_i + N_{i1} + N_{i4})^T v_i \\ & + (e^{-})^T (A_i - L_i C_i + N_{i2} + N_{i3})^T v_i. \end{aligned} \tag{16}$$

Substituting (9) into (16) yields

$$\begin{aligned} \dot{V}_i \leq\; & (e^{+})^T ((A_i^T + N_{i1}^T + N_{i4}^T) v_i + C_i^T z_i) \\ & + (e^{-})^T ((A_i^T + N_{i2}^T + N_{i3}^T) v_i + C_i^T z_i). \end{aligned} \tag{17}$$

It follows from (5) and (6) that

$$\dot{V}_i \leq -\eta(e^+)^T v_i - \eta(e^-)^T v_i = -\eta(e)^T \bar{v}_i = -\eta V_i. \quad (18)$$

Integrating both sides of (18) from t_k to t yields

$$V_i \leq \exp\{-\eta(t - t_k)\} V_i(t_k). \quad (19)$$

By using (8), we have

$$e^T(t_k)\bar{v}_i \leq \rho e^T(t_k)\bar{v}_j, \;\; \forall\, i, j \in S, \;\; i \neq j. \quad (20)$$

Let $\sigma(t_{k-1}) = j$, it can be deduced from (19) and (20) that

$$V_i \leq \rho \exp\{-\eta(t - t_k)\} V_j(t_k). \quad (21)$$

Repeating (20) and (21), one can get

$$V_i \leq \rho \exp\{-\eta(t - t_k)\} V_{\sigma(t_{k-1})}(t_k) \leq \cdots \leq \rho^k \exp\{-\eta t\} V_{\sigma(0)}(0). \quad (22)$$

In view of Definition 5, then $k = N_\sigma \leq N_0 + t/\tau^*$. Denote that $c_1 = \exp\{N_0 \ln \rho\}$, $c_2 = \eta - \dfrac{\ln \rho}{\tau^*}$, then (10) means that $c_2 > 0$. Since $\rho > 1$, it follows from (22) that

$$\begin{aligned} V_i &\leq \exp\{N_\sigma \ln \rho\} \exp\{-\eta t\} V_{\sigma(0)}(0) \\ &\leq \exp\{(N_0 + \frac{t}{\tau^*}) \ln \rho\} \exp\{-\eta t\} V_{\sigma(0)}(0) \\ &\leq c_1 \exp\{-c_2 t\} V_{\sigma(0)}(0), \end{aligned} \quad (23)$$

i.e.,

$$e^T \bar{v}_i \leq c_1 \exp\{-c_2 t\} V_{\sigma(0)}(0). \quad (24)$$

By the positivity of e, the following inequalities hold

$$e^T v_i = \sum_{m=1}^{n} e_m (v_i)_m \geq \underline{\varepsilon}(v_i) \sum_{m=1}^{n} e_m \geq \underline{\varepsilon}(v_i)||e||, \quad (25)$$

and

$$(e(0))^T v_{(\sigma(0))} = \sum_{m=1}^{n} e_m(0)(v_{\sigma(0)})_m \leq \bar{\varepsilon}(v_{\sigma(0)}) \sum_{m=1}^{n} e_m(0) \leq \sqrt{n}\bar{\varepsilon}(v_{\sigma(0)})||e(0)||. \quad (26)$$

Let $c_3 = \frac{\sqrt{n}\bar{\varepsilon}(v_{\sigma(0)})}{\underline{\varepsilon}(v_i)} c_1$, then substituting (25) and (26) into (24) yields

$$||e(t)|| \leq c_3 \exp\{-c_2 t\} ||e(0)||. \quad (27)$$

Thus, we can conclude that the error system (3) is positive and exponentially stable, i.e., (2) is an exponentially convergent interval observer for the system (1).

Theorem 2 *The constraints* (5)–(8) *are equivalent to the following conditions:*

$$(A_i^T + N_{i1}^T + N_{i4}^T + \eta I)v_i + C_i^T z_i \leq 0, \tag{28}$$

$$(A_i^T + N_{i2}^T + N_{i3}^T + \eta I)v_i + C_i^T z_i \leq 0, \tag{29}$$

$$\xi_i^T v_i \geq 0, \tag{30}$$

$$\xi_i^T v_i A_i + \xi_i z_i^T C_i + \mu I \geq 0, \tag{31}$$

$$v_i \preceq \rho v_j, \tag{32}$$

or

$$(A_i^T + N_{i1}^T + N_{i4}^T + \eta I)v_i + C_i^T z_i \leq 0, \tag{33}$$

$$(A_i^T + N_{i2}^T + N_{i3}^T + \eta I)v_i + C_i^T z_i \leq 0, \tag{34}$$

$$\xi_i^T v_i \leq 0, \tag{35}$$

$$\xi_i^T v_i A_i + \xi_i z_i^T C_i + \mu I \leq 0, \tag{36}$$

$$v_i \preceq \rho v_j, \tag{37}$$

Theorem 2 gives the standard LP forms of the constraints (5)–(8). Since the sufficient conditions in Theorem 2 are detailed forms of that in Theorem 1, the proof is just omitted here. In fact, we just need argue whether the scalar $\xi_i^T v_i$ is positive or negative.

4 Numerical Example

Consider the system (1) with

$$A_1 = \begin{bmatrix} -3.2 & 0.3 \\ 0.5 & -3.4 \end{bmatrix}, \quad B_1 = \begin{bmatrix} 0.1 & 0.2 \\ 0.3 & 0.4 \end{bmatrix}, \quad C_1 = \begin{bmatrix} 1 & 0 \\ 0 & 1 \end{bmatrix},$$

$$A_2 = \begin{bmatrix} -3.6 & 1 \\ 2 & -3.5 \end{bmatrix}, \quad B_2 = \begin{bmatrix} 0.2 & 0.3 \\ 0.4 & 0.5 \end{bmatrix}, \quad C_2 = \begin{bmatrix} 2 & 0 \\ 0 & 2 \end{bmatrix}, \quad f_1(x) = f_2(x) = sin(x).$$

It can be determined that

$$g_1(x) = g_2(x) = x, \quad h_1(x) = h_2(x) = x - sin(x)$$

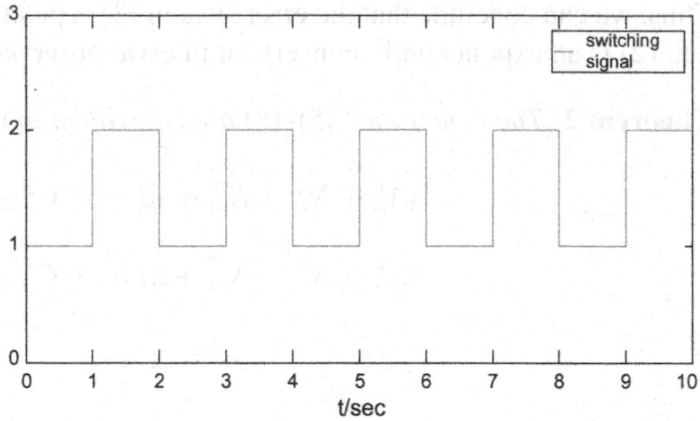

Fig. 1 Simulation of switching signal σ

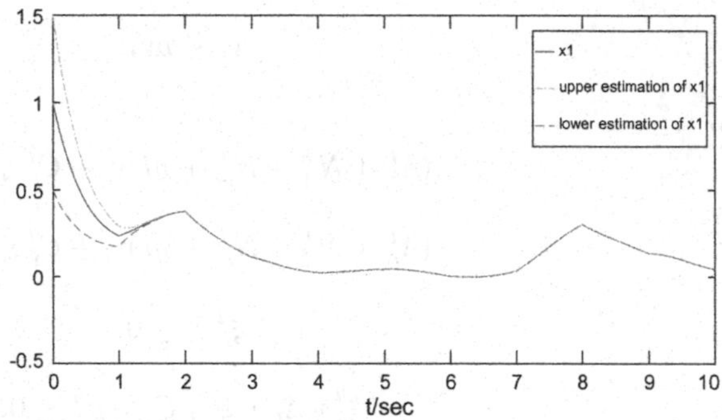

Fig. 2 Evolution of the real state $x_1(t)$ and the estimations $\hat{x}_1^+(t), \hat{x}_1^-(t)$

$$N_{11} = N_{13} = N_{21} = N_{23} = \begin{bmatrix} 1 & 0 \\ 0 & 1 \end{bmatrix}, \quad N_{12} = N_{14} = N_{22} = N_{24} = \begin{bmatrix} 2 & 0 \\ 0 & 2 \end{bmatrix}.$$

Given $\xi^{(1)} = [1; 2]$, $\xi^{(2)} = [2; 1]$, $\eta = 0.5$, $\rho = 1.2$, $\mu = 5$. By Theorem 2, we can obtain

$$v^{(1)} = \begin{bmatrix} 0 \\ 0.3901 \end{bmatrix}, \quad z^{(1)} = \begin{bmatrix} -0.1905 \\ -0.0931 \end{bmatrix}, \quad v^{(2)} = \begin{bmatrix} 0 \\ 0.4418 \end{bmatrix}, \quad z^{(2)} = \begin{bmatrix} -0.4418 \\ -0.0282 \end{bmatrix},$$

$$L_1 = \begin{bmatrix} 0.2499 & 0.1193 \\ 0.4999 & 0.2387 \end{bmatrix}, \quad L_2 = \begin{bmatrix} 2 & 0.1277 \\ 1 & 0.0638 \end{bmatrix}.$$

The inputs and initial conditions of system (1) and (2) are chosen as follows:

$$u = \begin{bmatrix} \sin t \\ \sin^2 t \end{bmatrix}, \quad x_0 = \begin{bmatrix} 1 \\ 2 \end{bmatrix}, \quad x_0^+ = \begin{bmatrix} 1.5 \\ 2.5 \end{bmatrix}, \quad x_0^- = \begin{bmatrix} 0.5 \\ 1.5 \end{bmatrix}.$$

The time response of switching signal σ is shown in Fig. 1. The results of simulation of the interval observer are presented in Figs. 2 and 3. It can be concluded that the designed interval observer is valid.

Fig. 3 Evolution of the real state $x_2(t)$ and the estimations $\hat{x}_2^+(t)$, $\hat{x}_2^-(t)$

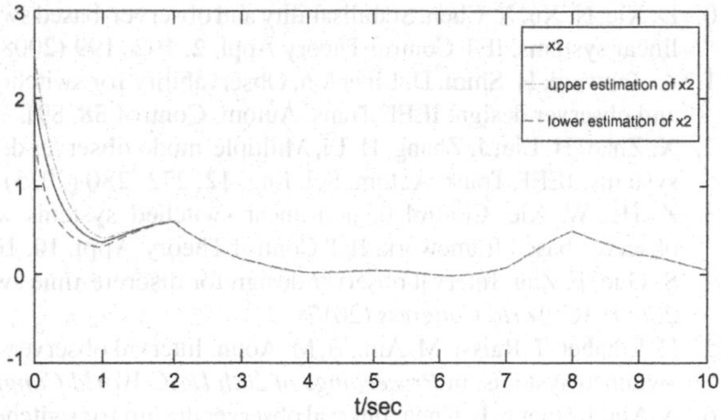

5 Conclusion

In this paper, we extend the interval observer design method to nonlinear switched systems without disturbances. The nonlinear function is supposed to satisfy the property of Lipschitz. The interval observers are designed and the sufficient conditions are given by LP forms. It is the first piece of work to design the interval observers for nonlinear switched systems by MLCLF. In the future, we will study nonlinear switched systems subjected to disturbances.

Acknowledgements The authors are grateful for the National Natural Science Foundation of China (61403267), China Postdoctoral Science Foundation (2017M611903), and Junzheng Fund Soochow.

References

1. J. Gouze, A. Rapaport, Z. Hadj-Sadok, Interval observers for uncertain biological systems. Ecol. Model **133**, 45–56 (2000)
2. M. Rami, C. Cheng, C. Prada, Tight robust interval observers: an LP approach, in *Proceedings of the 47th IEEE Conference on Decision and Control* (2008)
3. M. Bolajraf, M. Rami, A robust estimation approach for uncertain systems with perturbed measurements. Int. J. Robust Nonlinear Control **26**, 834–852 (2016)
4. F. Mazenc, O. Bernard, Interval observers for linear time-invariant systems with disturbances. Automatica **47**, 140–147 (2011)
5. T. Raissi, D. Efimov, A. Zolghadri, Interval state estimation for a class of nonlinear systems. IEEE Trans. Autom. Control **57**, 260–265 (2012)
6. D. Efimov, T. Raissi, S. Chebotarev, A. Zolghadri, Interval state observer for nonlinear time varying systems. Automatica **49**, 200–205 (2013)
7. G. Zheng, D. Efimov, F. Bejarano, W. Perruquetti, H. Wang, Interval observer for a class of uncertain nonlinear singular systems. Automatica **71**, 159–168 (2016)
8. K. Degue, D. Efimov, J. Ny, Interval observer approach to output stabilization of linear impulsive systems, in *Proceeedings of 20th IFAC* (2017)
9. H. Oubabas, S. Djennoune, M. Bettaye, Interval sliding mode observer design for linear and nonlinear systems. J. Process Control **61**, 12–22 (2018)

10. D. Xie, N. Xu, X. Chen, Stabilisability and observer-based switched control design for switched linear systems. IET Control Theory Appl. **2**, 192–199 (2008)
11. A. Tanwani, H. Shim, D. Liberzon, Observability for switched linear systems: characterization and observer design. IEEE Trans. Autom. Control **58**, 891–904 (2013)
12. X. Zhao, H. Liu, J. Zhang, H. Li, Multiple-mode observer design for a class of switched linear systems. IEEE Trans. Autom. Sci. Eng. **12**, 272–280 (2015)
13. Z. He, W. Xie, Control of non-linear switched systems with average dwell time: interval observer-based framework. IET Control Theory Appl. **10**, 10–16 (2016)
14. S. Guo, F. Zhu, Interval observer design for discrete-time switched system, in *Proceedings of 20th IFAC World Congress* (2017)
15. H. Ethabet, T. Raissi, M. Amairi, M. Aoun, Interval observers design for continuous-time linear switched systems, in *Proceedings of 20th IFAC World Congress* (2017)
16. X. Ma, J. Huang, L. Chen, Interval observer design for switched systems by linear programming method, *Proceedings of 30th CCDC* (2018)
17. X. Zhao, L. Zhang, P. Shi, M. Liu, Stability of a class of switched positive linear systems with average dwell time switching. Automatica **48**, 1132–1137 (2012)
18. L. Farina, S. Rinaldi, *Positive Linear Systems, Interscience Series* (Wiley, New York, 2000)
19. R. Horn, C. Johnson, *Topics in Matrix Analysis* (Cambridge Univ. Press, Cambridge, MA, 1991)

A Fuzzy Multi-objective Strategy of Polymer Flooding Based on Possibilistic Programming

Zhe Liu, Shurong Li and Lu Han

Abstract Against some practical problems for people to solve optimization of polymer flooding in oil exploitation, which include the uncertainty of crude oil price and differences of decision maker's satisfaction degree about the performance index, a fuzzy multi-objective optimal control model will be established and solved by an improved possibilistic programming algorithm based on Gaussian Probability Distribution in this paper. Gaussian Probability Distribution has comparative complex membership function, so the algorithm can process actual decision information better and reflect decision maker's subjectivity much better. And a full implicit finite-difference method will be used to solve complicated governing equations in the process of optimization. Additionally, we obtain a regular conclusion in which the decision maker can acquire different appropriate schemes by changing aspiration levels. The acquired optimal schemes in oil field exploitation verify the feasibility and effectiveness of the improved algorithm.

Keywords Dynamic programming · Gaussian probability distribution
Optimization of polymer flooding · Possibilistic programming algorithm

1 Introduction

The polymer flooding is an important enhanced oil recovery technology which is widely concerned in recent years, it can be used to increase net present value (NPV) and the yield which are usually regarded as expected objectives by people. A fuzzy multi-objective optimization problem while using polymer flooding technol-

Z. Liu (✉) · S. Li
Automation School, Beijing University of Posts and Telecommunications, Beijing 100876, China
e-mail: liuzheupc@163.com

L. Han
College of Information and Control Engineering, China University of Petroleum, Qingdao 266580, China

Y. Jia et al. (eds.), *Proceedings of 2018 Chinese Intelligent Systems Conference*, Lecture Notes in Electrical Engineering 529,
https://doi.org/10.1007/978-981-13-2291-4_25

ogy should be solved. In this paper, an improved possibilistic programming algorithm based on Gaussian Probability Distribution (GPD) will be proposed to solve it, and an analysis about optimal solution will be proposed at last.

Fuzzy mathematics has been intensively applied in our daily life and study, Zadeh proposed the fuzzy set theory, and the theory provided a highly efficient method of dealing with fuzzy data [1]. Zadeh also presented the possibilistic theory, which was connected with the fuzzy set theory [2]. He defined the concept of a possibility distribution as a fuzzy restriction. There have been many studies to deal with the imprecise coefficients in the objective functions and constraints in some literature as well. Yadav proposed a new mathematical model which was described by a nonlinear membership function [3]. Tanaka et al. used the weighted average of the upper and lower bounds as a substitute for fuzzy objective [4]. Luhandjula used the concept of α-level set to acquire a single-objective semi-infinite linear programming problem and in his paper a cutting plane method was proposed to tackle a kind of semi-infinite problem [5]. Rommelfanger et al. presented an algorithm to solve multi-objective linear programming by using several α-level sets and confirming membership functions of the upper and lower limit for every α-level set [6]. Pankaj Gupta and Mukesh Kumar Mehlawat proposed an improved possibilistic programming algorithm based on triangular fuzzy numbers which was used to solve fuzzy multi-objective assignment problem [7], Liu and Li [8] have worked out a fuzzy assignment problem in oil field exploitation by using the similar approach on the basis of Sun [9]. Compare with the previous work, the algorithm proposed in this paper has more accuracy and can tackle much more complicated fuzzy multi-objective optimization problems in oil exploitation.

In recent years, polymer flooding has played an important technology to enhance oil recovery and decrease water cut in oil exploitation. Wu [10] worked out an optimization problem of polymer flooding by SWIFT algorithm simply. Lei and Li [11] solved optimal control of polymer flooding based on maximum principle. After that, researches on polymer flooding progress fast and extensive, Lei and Li [12] studied dynamic optimization of polymer flooding based on iterative dynamic programming with variable stage lengths. In this paper, a more complicated and practical fuzzy multi-objective optimization problem concerned the uncertainty of crude oil price and differences of decision maker's satisfaction degree will be solved.

2 Model Description of Polymer Flooding

2.1 Objective Functions

We regard performance index NPV as the first objective function. Suppose that there are N_w injection wells and N_o production wells. The first objective function can be formulated as

$$\min \widetilde{J}_1 = \int_0^{t_f} \iint_\Omega \left[\xi_p q_{in} c_{in} - \widetilde{\xi}_o (1 - f_w) q_{out}\right] (1+\chi)^{-t} \, d\sigma dt, \tag{1}$$

where $\widetilde{\xi}_o$ represents the crude oil price, it is a fuzzy number due to the uncertainty and fluctuation of the international crude oil price, here we suppose that the pattern of Gaussian Probability Distribution (GPD) is adopted to represent the imprecise coefficient $\widetilde{\xi}_o \cdot f_w(x, y, t), (x, y) \in L_o$ represents the water cut of the production wells, so $(1 - f_w)$ is the oil cut of the production wells. ξ_p represents the price of polymer, $c_{in}(x, y, t)(x, y) \in L_w$ is the polymer concentration of the injection fluid, q_{in}, q_{out} are the flow velocity of injection and the production fluid. χ denotes the discount rate. We need maximize NPV in our actual exploitation process.

Decision makers hope to not only maximize NPV but also wish to maximize the yield of production. So the yield J_2 is what we normally think of as another performance index. In actual oil exploitation problems, in order to prevent the situation which the yield much fewer than MMV, fuzzy relation equations will be adopted here to help us establish an optimization mathematical model. And the second objective function will be formulated as

$$\max J_2 = \int_0^{t_f} \iint_\Omega (1 - f_w) q_{out} d\sigma dt \geq J_u, \tag{2}$$

where J_u is the yield value decision maker hope to obtain and Eq. (2) is the other objective function of fuzzy multi-objective optimization model. We need to minimize J_1 and maximize J_2 at the same time.

2.2 Governing Equations and Constraints

The optimization model has some governing equations because the objective functions (1) and (2) are not totally free but is constrained by the system process dynamics in actual exploitation. Let $c_p(x, y, t)$, $S_w(x, y, t)$ and $p(x, y, t)$ represent the polymer concentration, water saturation and the pressure of the oil field, respectively, the polymer concentration, water saturation and the pressure must satisfy the governing equations given by Lei and Li [12] and no further details are given here for the concise paragraph.

3 A Possibilistic Programming Algorithm Based on GPD

3.1 Modeling the Imprecise Data with GPD

Here we consider the uncertain of crude oil price $\tilde{\xi}_o$ and suppose it obey Gaussian Probability Distribution (GPD). The conversion formula of crude oil price between China and international can be shown as follows:

$$P_{oil} = \frac{P_b P_e}{0.159\rho_{oil}}, \tag{3}$$

where P_{oil} is the domestic crude oil price, P_b is the international crude oil price, P_e is the exchange rate, ρ_{oil} is the crude oil density and a bucket is equal to 0.159 cubic meters.

According to the fluctuation data of international crude oil price in the year of 2017 and Eq. (3), in order to facilitate the operation later, here we set the value $\sigma = 120$ in GPD to make the distribution curve closer to actual crude oil price. Then the formula of the mentioned GPD will be shown as

$$\mu_{\tilde{\xi}_o} = e^{-\frac{1}{2}\left(\frac{\xi_o - 2163.72}{120}\right)^2}. \tag{4}$$

We can also construct the probability distribution of crude oil price $\tilde{\xi}_o$ as Fig. 1.

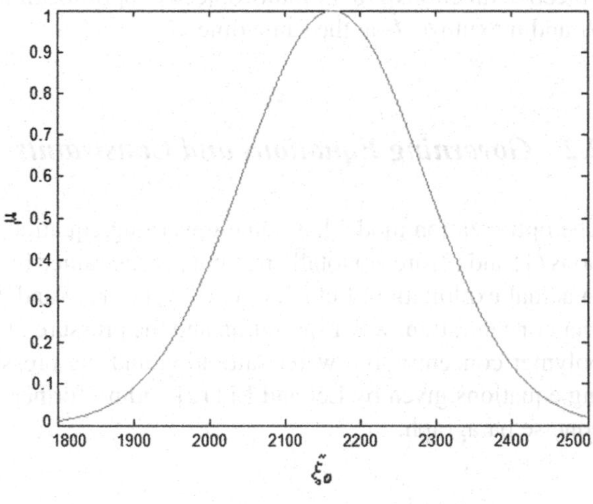

Fig. 1 Gaussian probability distribution of $\tilde{\xi}_o$

3.2 Development and Solving an Auxiliary Multi-objective 0–1 Optimization Model

In order to solve the fuzzy multi-objective optimization model, we use suitable strategies to convert it into equivalent forms. Six random data $\xi_o^1, \xi_o^2, \xi_o^3, \xi_o^4, \xi_o^5$ and ξ_o^6 will be chosen to help us carry out our calculations. Here, consider the objective function $\tilde{J}_1$ first, it can be written as:

$$\min \tilde{J}_1 = \int_0^{t_f} \iint_\Omega \left[\xi_p q_{in} c_{in} - \tilde{\xi}_o (1 - f_w) q_{out} \right] (1+\chi)^{-t} d\sigma dt$$
$$= \min\left(J_1^1, J_1^2, J_1^3, J_1^4, J_1^5, J_1^6\right), \qquad (5)$$

We consider the following auxiliary multi-objective functions corresponding to the fuzzy objective function (5):

$$\min J_1^1 = \int_0^{t_f} \iint_\Omega [\xi_p q_{in} c_{in} - (\xi_o)_\alpha^1 (1 - f_w) q_{out}] (1+\chi)^{-t} d\sigma dt$$
$$\min J_1^2 = \int_0^{t_f} \iint_\Omega [\xi_p q_{in} c_{in} - (\xi_o)_\alpha^2 (1 - f_w) q_{out}] (1+\chi)^{-t} d\sigma dt$$
$$\min J_1^3 = \int_0^{t_f} \iint_\Omega [\xi_p q_{in} c_{in} - (\xi_o)_\alpha^3 (1 - f_w) q_{out}] (1+\chi)^{-t} d\sigma dt$$
$$\min J_1^4 = \int_0^{t_f} \iint_\Omega [\xi_p q_{in} c_{in} - (\xi_o)_\alpha^4 (1 - f_w) q_{out}] (1+\chi)^{-t} d\sigma dt$$
$$\min J_1^5 = \int_0^{t_f} \iint_\Omega [\xi_p q_{in} c_{in} - (\xi_o)_\alpha^5 (1 - f_w) q_{out}] (1+\chi)^{-t} d\sigma dt$$
$$\min J_1^6 = \int_0^{t_f} \iint_\Omega [\xi_p q_{in} c_{in} - (\xi_o)_\alpha^6 (1 - f_w) q_{out}] (1+\chi)^{-t} d\sigma dt$$
$$s.t. \, governing \; equations \; and \; constraints. \qquad (6)$$

The auxiliary multi-objective functions in (6) represent different scenarios which have different likelihood of being to the set of available value. Further, the concept of α-levels and α-cuts will be introduced here, each fuzzy coefficient $\tilde{\xi}_o$ can be decomposed as

$$(\tilde{\xi}_o)_\alpha = ((\xi_o)_\alpha^1, (\xi_o)_\alpha^2, (\xi_o)_\alpha^3, (\xi_o)_\alpha^4, (\xi_o)_\alpha^5, (\xi_o)_\alpha^6), \tag{7}$$

where $(\xi_o)_\alpha^1$ and $(\xi_o)_\alpha^6$ are the boundary points of distribution curve which is cut by α-cuts and they represent the least possible value. $(\xi_o)_\alpha^4$ is the point $\xi_o = 2163.72$ and it denotes the most possible value. $(\xi_o)_\alpha^2$, $(\xi_o)_\alpha^3$ and $(\xi_o)_\alpha^5$ are random points chosen by decision maker.

After developing MOP (6), fuzzy programming approaches will be applied to solve MOP by working out the satisfaction degree of each objective function. First, we calculate the positive ideal solution (PIS) and negative ideal solution (NIS) at a given confidence level α. To exemplify, we use objective function J_1^1:

$$\begin{aligned}\left(J_1^1\right)^{PIS} &= \min J_1^1 \\ & s.t. \, governing\ equations \\ & and\ constraints.\end{aligned} \tag{8}$$

$$\begin{aligned}\left(J_1^1\right)^{NIS} &= \max J_1^1 \\ & s.t. \, governing\ equations \\ & and\ constraints.\end{aligned} \tag{9}$$

Next, a membership function of J_1^1 will be defined as follows because J_1^1 is also a fuzzy number obeys the same GPD with $\tilde{\xi}_o$:

$$\mu_1^1 = \begin{cases} e^{-\frac{1}{2}(\frac{J_1^1 - 2163.72}{120})^2}, & if \left(J_1^1\right)^{PIS} \le J_1^1 \le \left(J_1^1\right)^{NIS} \\ 0, & if\ J_1^1 < \left(J_1^1\right)^{PIS} or\ J_1^1 > \left(J_1^1\right)^{NIS}. \end{cases} \tag{10}$$

Similarly, we calculate the PIS and NIS of the objective functions J_1^2, J_1^3, J_1^4, J_1^5 and J_1^6 at the given confidence level α.Then we can get their membership functions μ_1^2, μ_1^3, μ_1^4, μ_1^5 and μ_1^6 by the above approach.

Next we consider the membership function of objective function J_2, it is a fuzzy function contains fuzzy relation equations from Eq. (2). In real-world exploitation, decision makers usually have a minimum margin value for J_2 and J_2 cannot be lower than the value. The membership function of J_2 can be constructed as

$$\mu_2 = \begin{cases} 0, & J_2 \le J_e, \\ \frac{J_2 - J_e}{J_u - J_e}, & J_e < J_2 < J_u\ , \\ 1, & J_2 \ge J_u \end{cases} \tag{11}$$

where J_u is the yield value which decision maker hope to obtain, J_e is the minimum margin value which decision maker can accept, $J_u - J_e$ is the tolerance at the worst. After getting the above membership functions, an equivalent single-objective 0–1 programming (SOP) as follows will be constructed to help us aggregate all the fuzzy sets and obtain the optimal solution:

$$
\begin{aligned}
\max\ & w_1\lambda_1 + w_2\lambda_2 \\
s.t.\ & \lambda_1 \le \mu_{1k}, \quad k = 1, 2, 3, 4, 5, 6 \\
& \lambda_2 \le \mu_{1k}, \quad k = 1, 2, 3, 4, 5, 6 \\
& \lambda_1 \le \mu_2, \quad \lambda_2 \le \mu_2, \\
& 0 \le \lambda_1 \le 1, \quad 0 \le \lambda_2 \le 1, \\
& 0 \le w_1 \le 1, \quad 0 \le w_2 \le 1, \\
& w_1 + w_2 = 1, \\
& \textit{governing equations and constraints},
\end{aligned}
\tag{12}
$$

where the auxiliary variable λ_1 and λ_2 represent the satisfaction degree for the overall determined objective values, w_1 and w_2 are the weight values of objective functions J_1 and J_2, respectively. By constructing SOP (12), a complicated fuzzy multi-objective optimization of polymer flooding has been converted to a relatively simple single objective optimization problem. The optimal solution will be acquired by solving (12) and we will obtain the optimal polymer injection scheme.

Notice that the governing equations in our fuzzy multi-objective optimization model are all nonlinear partial differential equations so the solving processes of model (12) is difficult. A full implicit finite-difference method will be adopted here to approximate treat the PDEs.

4 Numerical Example

4.1 Data Description

Now a numerical example of optimization for polymer flooding will be represented in this section. Consider a two-phase flow of oil and water in a two-dimensional oil field with $441(21 \times 21 \times 1)$ grid blocks. The exploitation model is a pattern with four injection wells and nine production wells. For the reservoir there are also some essential parameters and fluid data shown in the paper written by Lei [12].

4.2 Simulation Result

Next the possibilistic programming algorithm based on GPD and the full implicit finite-difference method will be used to solve the above actual numerical example. We select three different α values 0.1, 0.5 and 0.8 to indicate three different conditions of decision maker's judgments. The average water cut of the production wells will be shown as Fig. 2.

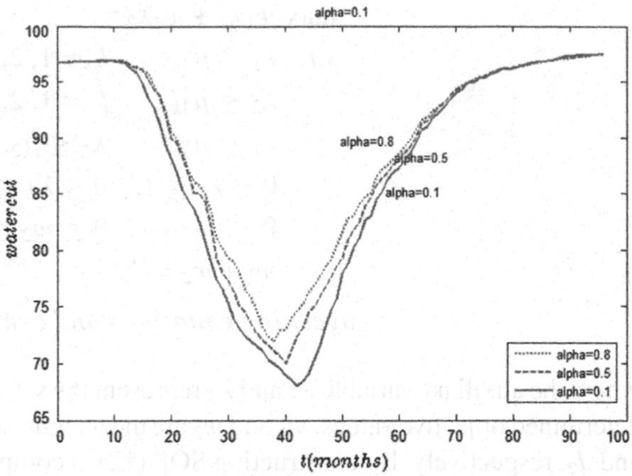

Fig. 2 The average water cut (%) of the production wells at different α

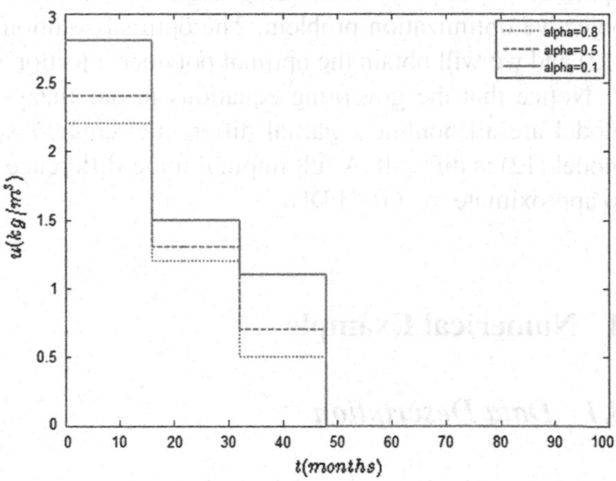

Fig. 3 The injection polymer concentration at different α

The values of optimal injection concentration $u(kg/m^3)$ acquired by calculation are $u_{\alpha=0.8} = [2.2, 1.2, 0.5]$, $u_{\alpha=0.5} = [2.4, 1.3, 0.7]$ and $u_{\alpha=0.1} = [2.8, 1.5, 1.1]$ while the values of α are 0.8,0.5 and 0.1,respectively. And the control variable u can be shown as Fig. 3.

In the process of computation, the polymer price is set as $\xi_p = 25{,}000(yuan/t)$ and the crude oil price is set as the mathematical expectation of the GPD in order to facilitate calculation. The values of objective functions with respect to different α will be shown in Table 1.

Table 1 The value of performance index

The value of α	0.1	0.5	0.8
Auxiliary variable λ_1	0.919	0.889	0.842
Auxiliary variable λ_2	0.901	0.863	0.826
NPV(ten thousand yuan)	1534.18	1547.47	1555.97
The yield of production $J_2(t)$	9576.2	9177.3	8986.4

4.3 Analysis of Result with Respect to Different

Table 1 clearly shows that we will obtain different optimization plans while choosing different α values. The value of α will be chosen depending on how much attention is attached to each objective function by decision makers. We can chose a smaller confidence level α if we feel like more yield of production. Similarly, we can chose a bigger value of α if we regard NPV as the most important index. Now we can find out another advantage of the improved solution algorithm is that if the decision makers are not satisfied with the obtained optimization scheme, several other optimization plans can be provided by changing the value of α, so we can solve complicated fuzzy optimization problems much more authentic and accurate.

5 Conclusions

The paper proposed the possibilistic programming algorithm by using Gaussian Probability Distribution (GPD) to increase accuracy of optimal solution when we solve fuzzy multi-objective optimization problem. Compare with triangular possibility distribution and linear membership function, GPD has more complex and authentic membership function, so we can process actual decision information better and reflect decision maker's subjectivity much better. Then we applied the improved possibilistic programming approach to work out a fuzzy multi-objective optimization problem for polymer flooding which contains imprecise objective functions and fuzzy relation equations. At last, the paper analyzed the differences among optimal solutions when decision makers choose diverse values of confidence level α. The acquired optimal exploitation schemes verify the feasibility and prospect of the improved algorithm in oi field exploitation.

References

1. J.A. Goguen, Zadeh L. A. Fuzzy sets. vol. 8 (1965), pp. 338–353; Zadeh L. A. Similarity relations and fuzzy orderings. vol. 3 (1971), pp. 177–200; Journal of Symbolic Logic, 1973, 38(4):656–657

2. L.A. Zadeh, Fuzzy sets as a basis for a theory of possibility. Fuzzy Sets Syst. **1**(1), 3–28 (1978)
3. P.K. De, B. Yadav, An algorithm to solve multi-objective assignment problem using interactive fuzzy goal programming approach. Int. J. Contemp. Math. Sci. (33),1651–1662 (2011)
4. H. Tanaka, H. Ichihashi, K. Asai, A formulation of fuzzy linear programming problem based on comparison of fuzzy numbers. Control Cybern. **13**(13), 41–52 (1984)
5. M.K. Luhandjula, Linear programming with a possibilistic objective function. Eur. J. Oper. Res. **31**(1), 110–117 (1987)
6. H. Rommelfanger, R. Hanuscheck, J. Wolf, Linear programming with fuzzy objectives. Fuzzy Sets Syst. **29**(1), 31–48 (1989)
7. P. Gupta, M.K. Mehlawat, A new possibilistic programming approach for solving fuzzy multiobjective assignment problem. IEEE Trans. Fuzzy Syst. **22**(1), 16–34 (2014)
8. Z. Liu, S. Li, Y. Ge, A possibilistic programming approach based on trapezoidal possibility distribution and its application in oil field exploitation, in *Control Conference* (IEEE, 2017), pp. 3014–3019
9. Q.C. Sun, SH. R. Li, *The Application of Fuzzy Programming in Oilfield Development Programming* (China University of Petroleum (East China), Shandong, 2005)
10. L.I. Shu-Rong, W.U. Yu-Xiao, X.D. Zhang, Solution of optimal injection strategies for polymer flooding based on SWIFT method. Syst. Eng. **26**(10), 107–111 (2008)
11. Y. Lei, S. Li, X. Zhang et al., Optimal control of polymer flooding based on maximum principle. J. Appl. Math. **2012**(1), 203–222 (2012)
12. Y. Lei, S. Li, X. Zhang et al., Optimal control of polymer flooding based on mixed-integer iterative dynamic programming. Int. J. Control **84**(11), 1903–1914 (2011)

Real-Time Classification of Steel Strip Surface Defects Based on Deep CNNs

Yan Liu, Jiahui Geng, Zhenfeng Su, Weicun Zhang and Jiangyun Li

Abstract Steel strip surface defects recognition is very important to steel strip production and quality control, in which correct classification of these surface defects is crucial. The surface defects of steel strips are classified according to various features, but it is hard for traditional methods to extract all these features and use them effectively. In this paper, we propose a method to deal with the problem of defect classification based on deep convolutional neural networks (CNNs). Wc adopt GoogLeNet, as our base model and add an identity mapping to it, which obtains improvement to some extent. At the same time, we establish a dataset of cold-rolled steel strip surface defects of six types and augment it in order to reduce over-fitting. Then we detect defects of six types with our network and reach an accuracy of 98.57%. Besides, our network achieves a speed of 125 FPS, which fully meets the real-time requirement of the actual steel strip production lines.

Keywords Surface defects · Steel strip · Convolutional neural networks
Defect classification

1 Introduction

There are many types of surface defects produced during the producing and processing of steel strips, such as scars, scratches, burrs, seams, and etc. These defects are caused by different factors which reflect different problems of the pro-

This work was supported by National Science Foundation of China (No. 61520106010, No. 61741302).

Y. Liu · J. Geng · Z. Su · W. Zhang · J. Li (✉)
School of Automation and Electrical Engineering,
University of Science and Technology Beijing, Beijing 100083, China
e-mail: leejy@ustb.edu.cn

Y. Liu · J. Geng · Z. Su · W. Zhang · J. Li
Key Laboratory of Knowledge Automation for Industrial Processes,
Ministry of Education, Beijing 100083, China

Y. Jia et al. (eds.), *Proceedings of 2018 Chinese Intelligent Systems Conference*, Lecture Notes in Electrical Engineering 529,
https://doi.org/10.1007/978-981-13-2291-4_26

duction line. We can analyze the problem of broken lines by classifying defects they produced. Therefore, it is very important to classify different surface defects in the quality assurance of steel strips. How to extract a set of better feature representations and design an appropriate classifier for surface defects has been a hot re-search topic for many years [1–4].

However, there are many factors that make accurate classification of steel strip surface defects particularly difficult, such as the high-speed production line, diversity and large scale changes of defects, random distribution and non-defective interferences. In addition, surface defects caused by different production lines tend to have different characteristics. Thus the classification algorithms for steel strip surface defects should not only have good generalization performance but satisfy real-time requirement.

The surface defects of steel strips are classified according to various features, such as texture, edge, spatial relationship and so on. It is hard for traditional methods to extract all these features and use them effectively. In this paper, we adopt GoogLeNet [5], a 22-layer convolutional neural network, as our base model to classify the defects images of steel strip, due to the powerful ability in image feature extraction of Deep CNNs. The network can extract large number of features automatically, so the classification accuracy tend to be higher than traditional methods. Furthermore, the CNN architecture does not require cumbersome steps and it provides an end-to-end solution to the defect classification of the steel strips. Besides, we add identity mapping inspired from [12] to our network to deal with the information loss when network goes deeper.

The remainder of this paper is organized as follows. In the second section, we re-view the traditional algorithms on defects classification and CNN. Then we established a dataset of steel strip surface defect of six types. In the fourth section the proposed methods are explained with more details. Finally, we evaluate our method on detect dataset which proposed, followed by the conclusion in Sect. 7.

2 Related Work

The traditional defects classification process are coarsely divided in three main stages: image preprocessing, feature extraction, and classification. And the classification results mainly depends on the extracted features and the classifiers. Hu et al. [6] extracted four kinds of defect features and transformed them to a 38-dimensional feature vector, then an optimized SVM (Support Vector Machine) classifier was trained to classify 5 types of 101 defect images. Ke et al. [7] pro-posed a Tetrolet-based surface defect identification methods. After extracting the sub-band characteristics of steel strip surface defects in different scales and directions, the SVM classifier was trained to classify the extracted features. Suvdaa et al. [8] classified the surface defect images of steel strip by extracting SIFT features, and a voting strategy was proposed for the final decision handling the problem of multiple outputs of a given image with a specific defect type.

However, these traditional algorithms need to design feature extractors artificially. The hand-crafted features is heavily dependent on the designer's experience and requires a lot of manpower, and this kind of features lack generality due to its inability to extract and organize discriminant information from the data. What's more, the 3-stage system is usually based on a set of pipelines with partial or no self-adjustable parameters which make fine-tuning process of this industrial systems cumbersome, requiring much more human invention than desired.

CNNs have impressive performance in feature extraction, and have achieved remarkable results in image classification. There have been some researches in the steel strip surface defect classification using CNNs. For example, Masci et al. [3] presented a Max-pooling convolutional neural network to perform supervised feature extraction directly. Yi et al. [4] proposed an end-to-end surface defects recognition system using a 7-layer convolutional neural network. But these methods with shallow networks did not fully utilize the powerful feature extraction ability of deep CNNs.

In this paper, we adopt CNNs to automatically extract various types of features, and the model can adaptively learn according to the changing data distribution. Besides, the CNN architectures provides an end-to-end solution to the defect classification task for the steel strip producing lines.

3 Defect Database

We established a dataset of cold-rolled steel strip surface defect of ten types, which are all from the cold-rolled steel strip production lines. But due to the limited number of defect images that can be collected on the producing lines, several types of defects are too few to extract features. In this paper, we mainly extract defect features of 6 types and classify them, including scar, scratches, inclusions, burrs, seams and iron scales.

The defect database contains 4260 surface defect images of 6 types. Each image was center cropped or resized into 256×256 before sent to network. The details of our dataset are shown in Table 1, and defect images of 6 types are shown in Fig. 1.

Table 1 Dataset of strip surface defect images

	Scar	Scratches	Inclusion	Burrs	Seams	Iron scales	Amount2
Train set	500	500	500	500	500	500	3000
Test set	210	210	210	210	210	210	1260
Amount1	710	710	710	710	710	710	4260

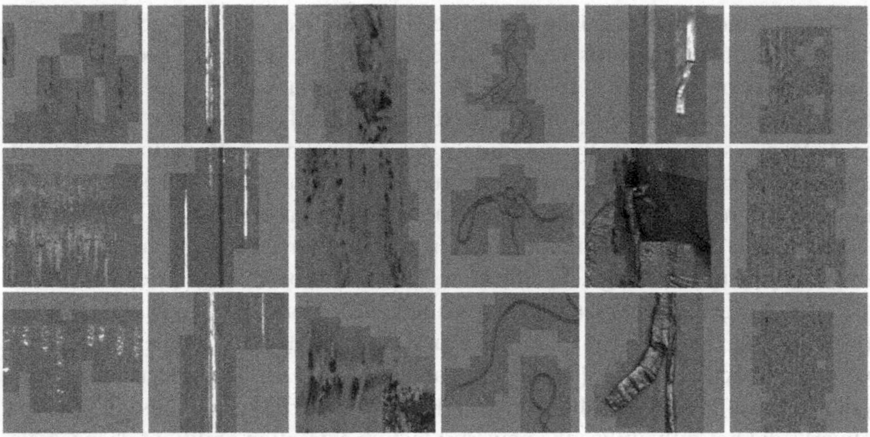

(a) Scar (b) Scratches (c) Inclusion (d) Burrs (e) Seams (f) Iron scales

Fig. 1 Steel strip surface defect of six types

4 Network Architecture

CNN was first proposed in [9] and achieved good performance in handwritten character recognition and face detection tasks in the early 1990s. The CNN architecture become popular after the work of [10] shown significant improvement on the ILSVRC2012. Except for AlexNet proposed in [10], deeper and broader networks are gradually proposed, such as VGG-16 [11], GoogLeNet [5], ResNet [12]. We adopted GoogLeNet as our base model, which is a tradeoff between model size and classification accuracy. And we modify the GoogLeNet slightly by combine it with the identity mapping in ResNet.

4.1 The Improved GoogLeNet Network

The most straightforward way to improve the performance of deep neural net-work is by increasing its depth and width. In this paper, we chose GoogLeNet, a 22-layer model which is wider than AlexNet [10], to extract the features of steel strip surface defects, and improved it to adapt to our task better with several ways.

GoogLeNet was the first network to stray away from the strategy of simply stacking convolutional and pooling layers. It adopt the Inception module to extract the image features. An Inception module consists of convolutional kernels of multiple sizes, as shown in Fig. 2. The 3×3 and 5×5 convolutions are used to extract different spatial features while 1×1 convolutions are applied to learn correlation between different channels. Thus the Inception module can extract spatial features of various scales. Inspired from ResNet, we add identity mapping to Inception. Identity map-

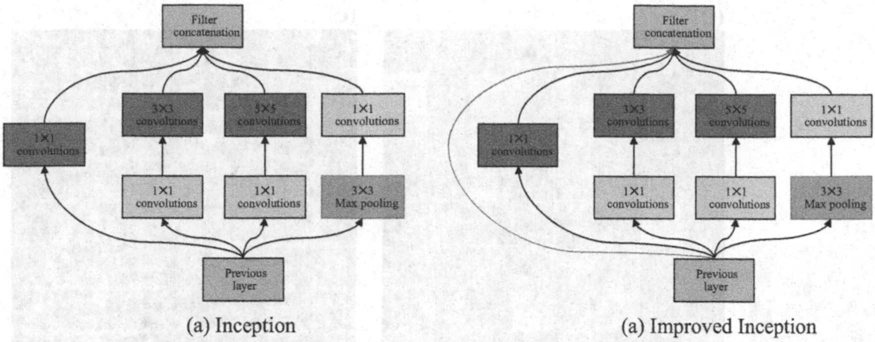

Fig. 2 Inception module with multiple sizes of convolutional kernels and the improved inception inspired from ResNet

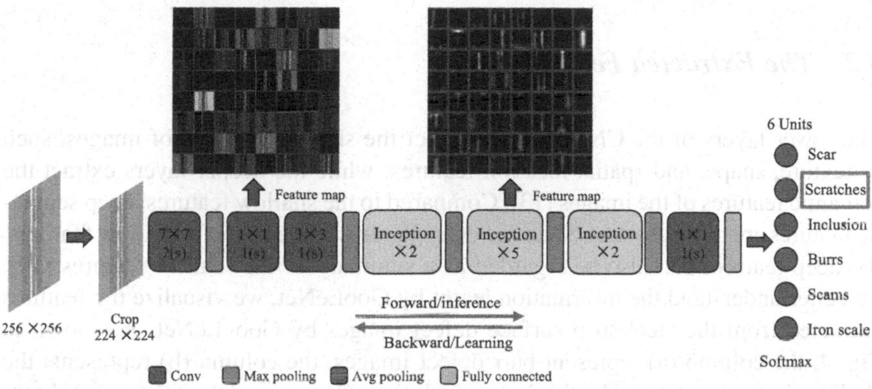

Fig. 3 The schematic diagram of the architecture of the model

ping is proposed in ResNet [12] to deal with a counter-intuitive problem that the accuracy falls down when the network architecture goes deeper. The identity mapping combines the output and the input of a module and inputs them into the next module. This combination introduces a straight path from the input to the output and reduces the information loss in the training procedure.

The GoogLeNet is a network consisting of Inception modules stacked upon each other, with occasional max-pooling layers with stride 2 to halve the resolution of the grid [5]. There are 9 Inception modules in the network, and as features of higher abstraction are captured by higher layers, the ratio of 3×3 and 5×5 convolutions in the Inception modules should increase as we move to higher layers. A fully connected layer with 1024 units and a linear layer with softmax classifier are followed by the last Inception module to classify the extracted features. We add two identity mappings to it between the Inception 4a and the Inception 4c. It provided 1% improvement in accuracy. The Network structure is shown in Fig. 3.

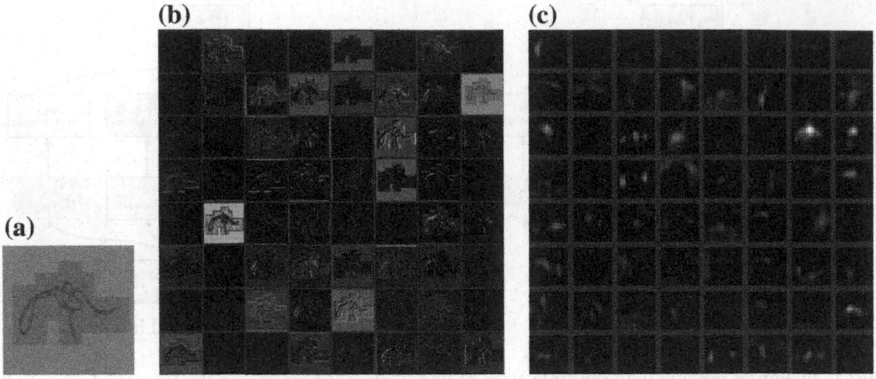

Fig. 4 Visualization at various (lower and deeper) layer of our network

4.2 *The Extracted Features*

The lower layers of the CNN tend to extract the shallow features of images, such as texture, shape, and spatial location features, while the deeper layers extract the semantic features of the images [13]. Compared to the shallow features, deep semantic features are more abstract. Since the deep features come from the shallow features, the deep features can also be regarded as a summary of the shallow features [14]. To better understand the information learnt by GooLeNet, we visualize the features extracted from the steel strip surface defect images by GoogLeNet. As shown in Fig. 4, the column (a) represent burr defect images, the column (b) represents the shallow features extracted by the 2nd convolutional layers, and the column (c) shows the deeper features extracted by the 3rd last Inception module.

5 Reduce Overfitting

Although the GoogLeNet has 12 × fewer parameters than AlexNet proposed in 2012, the size of the network is too big and may overfit on our dataset. We adopted several strategies to reduce overfitting when training the network.

Data Augmentation

Since CNN models take defect images as input in-stead of hand-craft features of defect, it is essential to have a bigger training dataset. Data augmentation allows networks to extract features better and prevent overfitting, which can greatly improve the performance of classification and generalization [10]. We used two affine transforms as follows on our dataset randomly: (1) Flipping: flip over our images in two directions, left and right, up and down; (2) Rotation: rotate our images at three degrees(90°, 180°, 270°). Through these methods, we greatly augmented our dataset and helping CNNs to extract features of different objects.

Dropout

The dropout, introduced in [15], consists of omitting the hidden units randomly from the network with a probability. The dropped neurons do not con-tribute to the forward pass and do not participate in back-propagation. This technique can reduces complex co-adaptations of neurons and thus reduce overfitting. We use dropout in the fully-connected layer with a ratio of 40% to reduce overfitting.

Batch Normalization

In the process of training the network, the network does not train all the images at the same time, but divides all the images into several batches. Our network utilizes Batch Normalization [16] to normalize the data for each batch, the equations are as followed. Here x_i denotes the activation input, and batch size is m.

$$\mu_\beta \leftarrow \frac{1}{\mathrm{m}} \sum_{i=1}^{m} x_i$$

$$\sigma_\beta^2 \leftarrow \frac{1}{\mathrm{m}} \sum_{i=1}^{m} \left(x_i - \mu_\beta\right)^2$$

$$\hat{x}_i \leftarrow \frac{x_i - \mu_\beta}{\sqrt{\sigma_\beta^2 + \varepsilon}}$$

$$y_i \leftarrow \gamma \hat{x}_i + \beta \equiv BN_{\gamma,\beta}(x_i)$$

Batch Normalization leads to a significant improvement in convergence while eliminating the need for other forms of regularization. It also helps regularize the model. By adding batch normalization on the convolutions layers we get more than 0.5% improvement in accuracy.

6 Experiment

We trained our network for 50,000 iterations on our dataset. Throughout training we use stochastic gradient descent with a batch size of 32, a momentum of 0.9 and a decay of 0.0002. The learning rate was initialized at 0.001. The training process took 6 h on two NVIDIA GTX 1080Ti GPUs. The train dataset consist of 8539 images in total after data augmentation. And 1260 images are used to evaluate the performance. The loss attenuation curve, as shown in Fig. 5.

Table 2 Confusion matrix of our surface defect test set using Deep CNNs

	Scar	Scratches	Inclusion	Burr	Seams	Iron scales
Scar	209	0	1	0	0	0
Scratches	0	209	0	0	1	0
Inclusion	0	0	208	0	2	0
Burr	1	1	0	208	0	0
Seams	0	0	1	0	209	0
Iron scales	0	0	0	0	2	208

Fig. 5 The loss decay curve

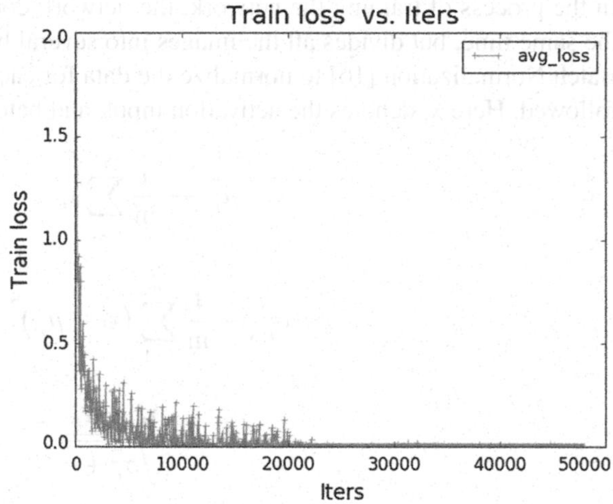

We test the testing data with our trained network and 210 images were clarified within 1.78 s, achieving 98% accuracy. The confusion matrix of experiment results is shown in Table 2.

Compared to other classification algorithm

We compare the experiment results of our network with other classification algorithm, including the 3-stage traditional methods and several other CNN architectures. The comparison results are shown in Table 3.

As shown in Table 3, in which the best results are bold, our network has a better performance than the hand-crafted features of [6, 17–20] and the shallow neural networks of [3] not only in the classification accuracy but also in the test time per image. There is no doubt that the richer features extracted from the CNN ensure the higher classification accuracy. Since the amount of defects images far exceeds that in these references, the defect features that our network extracted have better generalization.

Table 3 Compare with other algorithm

	Accuracy (%)	Amount of train	Inference time
M-pooling CNN [3]	93.03	2927	Unknown
HCGA [6]	95.04	351	0.158 s
HSVM-MC [17]	95.18	900	1.1044 s
Infrared Imaging [18]	95.42	1200	Unknown
Contourlet Transfotm [19]	96.46	868	0.103 s
Tetrolet Transform [20]	97.38	868	0.239 s
Ours	**98.57**	**4260**	**0.0085** s

Real-time analysis

The average inference time for our network to correctly classify an image is only 0.0085 s. It means that our network can detect 125 defect images per second. Supposing that the maximum speed of the production line is 30 m/s and the view field of a single CCD camera is 50–100 cm, this requires the defect classification have a speed of 30–60 FPS. The network we adopted achieves a speed of 125 FPS which fully meets the real-time requirement of the actual steel strip production lines.

7 Conclusion

We established a steel strip surface defect database which contains steel strip surface defects images of six types. We utilized the improved GoogLeNet to extract various features of the steel strip surface defect images and classified them with softmax classifier, achieving 98.57% classification accuracy. Besides, our network achieves a speed of 125 FPS which fully meets the real-time requirement of the actual steel strip production lines.

References

1. M. Sharifzadeh, S. Alirezaee, R Amirfattahi et al., Detection of Steel Defect Using the Image Processing Algorithms, in *IEEE International Multitopic Conference, INMIC 2008* (IEEE, 2008), pp. 125–127
2. K. Peng, X. Zhang, Classification Technology for Automatic Surface Defects Detection of Steel Strip Based on Improved BP Algorithm, in *Fifth International Conference on Natural Computation, 2009. ICNC'09*, vol. 1 (IEEE, 2009), pp. 110–114
3. J. Masci, U. Meier, D. Ciresan et al., Steel defect classification with max-pooling convolutional neural networks, in *The 2012 International Joint Conference on Neural Networks (IJCNN)* (IEEE, 2012), pp. 1–6
4. L. Yi, G. Li, M. Jiang, An end-to-end steel strip surface defects recognition system based on convolutional neural networks. Steel Res. Int. **88**(2), 176–187 (2017)

5. C. Szegedy, W. Liu, Y. Jia et al., Going Deeper with Convolutions, in *Cvpr* (2015)
6. H. Hu, Y. Liu, M. Liu et al., Surface defect classification in large-scale strip steel image collection via hybrid chromosome genetic algorithm. Neurocomputing **181**, 86–95 (2016)
7. K. Xu, Y. Xu, P. Zhou et al., Application of RNAMlet to surface defect identification of steels. Opt. Lasers Eng. **105**, 110–117 (2018)
8. B. Suvdaa, J. Ahn, J. Ko, Steel surface defects detection and classification using SIFT and vot-ing strategy. Int. J. Softw. Eng. Appl. **6**(2), 161–165 (2012)
9. Y. LeCun, B. Boser, J.S. Denker et al., Backpropagation applied to handwritten zip code recognition. Neural Comput. **1**(4), 541–551 (1989)
10. A. Krizhevsky, I. Sutskever, G.E. Hinton, Imagenet Classification with Deep Convolutional Neural Networks, in *Advances in Neural Information Processing Systems* (2012), pp. 1097–1105
11. K. Simonyan, A. Zisserman, Very deep convolutional networks for large-scale image recognition. arXiv preprint arXiv:1409.1556 (2014)
12. K. He, X. Zhang, S. Ren et al., Deep Residual Learning for Image Recognition, in *Proceedings of the IEEE Conference on Computer Vision and Pattern Recognition* (2016), pp. 770–778
13. M.D. Zeiler, R. Fergus, Visualizing and Understanding Convolutional Networks, in *European Conference on Computer Vision* (Springer, Cham, 2014), pp. 818–833
14. I. Hadji, R.P. Wildes, What do we understand about convolutional networks?. arXiv preprint arXiv:1803.08834 (2018)
15. N. Srivastava, G. Hinton, A. Krizhevsky et al., Dropout: a simple way to prevent neural networks from overfitting. J. Mach. Learn. Res. **15**(1), 1929–1958 (2014)
16. S. Ioffe, C. Szegedy, Batch normalization: accelerating deep network training by reducing internal covariate shift. arXiv preprint arXiv:1502.03167 (2015)
17. M. Chu, J. Zhao, R. Gong et al., Steel surface defects recognition based on multi-label classifier with hyper-sphere support vector machine, in *Control And Decision Conference (CCDC), 2017 29th Chinese* (IEEE, 2017), pp. 3276–3281
18. Xuwu Zhang et al., Vision inspection of metal surface defects based on infrared imaging. Acta Optica Sinica **31**(3), 0312004 (2011)
19. K. Xu, Y.H. Ai, P. Zhou et al., Recognition of surface defects in continuous casting slabs based on Contourlet transform. J. Univ. Sci. Technol. Beijing **35**(9), 1195–1200 (2013)
20. K. Xu, L. Wang, J. Wang, Surface defect recognition of hot-rolled steel plates based on Tetrolet transform. J. Mech. Eng. (2016)

A Planning Evaluation Method for Esophageal VMAT Based on Machine Learning

Muya Liu, Jiwei Liu, Jianfei Liu, Hui Yan and Ronghu Mao

Abstract To reduce the complexity associated with VMAT planning, we developed a model which can predict the dose volume histograms (DVHs) of organ-at-risk using the prior knowledge of the high quality esophageal VMAT plans and the distance to target histograms (DTHs). We extracted the anatomical information and dose information of patients from DICOM-RT files. With these information, the DTH and DVH curves were calculated. Principal component analysis was used to identify the main features of DTH and DVH curves. Then, least absolute shrinkage and selection operator regression was used to establish the functional relationship between the main features of DTH and DVH curves. In this study, the training dataset consists of 35 esophageal VMAT plans and the trained model was validated by 12 cases outside the training dataset. The experimental results demonstrated that the DTH/DVH curves can be effectively expressed by one or two principal components, the accuracy of the model in prediction is about 75%. These promising results suggest that this method can predict does distribution in the esophageal VMAT plans and assist the physicist to make plans by giving objective function and orientation, which can improve the efficiency and quality of plan making.

Keywords VMAT · Model training · Machine learning · PCA
LASSO regression

M. Liu · J. Liu (✉)
School of Automation and Electrical Engineering,
University of Science and Technology, Beijing, China
e-mail: liujiwei@ustb.edu.cn

J. Liu
Electrical Engineering and Automation, Anhui University, Hefei, China

H. Yan
Department of Radiation Oncology Cancer Hospital Chinese Academy of Medical Sciences, 2258, Beijing 100021, China

R. Mao
Department of Radiation Oncology, The Affiliated Cancer Hospital of Zhengzhou University, Henan Cancer Hospital, Zhengzhou, Henan, China

Y. Jia et al. (eds.), *Proceedings of 2018 Chinese Intelligent Systems Conference*, Lecture Notes in Electrical Engineering 529,
https://doi.org/10.1007/978-981-13-2291-4_27

1 Introduction

VMAT is an arc-based approach to IMRT delivery which is widely used in cancer treatment [1]. The tumor area is changeable and has complicated distribution, which lead to increased complexity of treatment planning. Even an experienced physicist cannot guarantee that the plan is optimal. In general, DVHs are applied as a tool to evaluate plan by comparing doses from different structures or plans [2]. The DVH has been adopted in dose evaluation and optimization in radiotherapy for a long time [3]. Using the fact that dose distribution in radiotherapy is closely constrained by the patient anatomical geometry [4, 5], a method for comparing treatment plans was previously reported using overlap volume histogram (OVH) to infer the likely dose volume levels [6]. Zhu et al. [7] used the distance-to-target histogram (DTH) to establish the correlation between the OAR-PTV anatomy and OAR DVHs. We note that DTH is equivalent to overlap volume histogram (OVH) defined by Wu et al. [6] when the Euclidean form of the distance function is used. Yuan et al. [8] identified several important factors that explain interpatient DVH variations in OARs based on patient anatomical features and their correlation with OAR dose sparing. But they can hardly get the estimation of the whole DVH curve and no one applied these methods to cure esophageal cancer. In recent years, more and more attention has been paid to the field of using the prior knowledge contained in radiotherapy plans to improve the efficiency of planning [9–12].

In this study, we developed a model for predicting DVH of OAR based on machine learning using the prior knowledge of the high-quality esophageal VMAT plans. The estimation model was trained with a supervised approach using the DVHs and DTHs from 35 high quality esophageal VMAT plans generated by experienced planners. PCA was applied to characterize main features of DVH and DTH and lasso regression was applied to model their correlation. The accuracy of the model was evaluated with 12 additional esophageal VMAT plans outside the training dataset.

2 Materials

A. Patient data

The dataset consists of 47 VMAT plans retrospectively selected from 47 esophageal cancer patients. The training dataset has 35 VMAT plans that are chosen from dataset randomly. The remaining 12 VMAT plans make up the testing dataset. Those plans were generated using the same procedures and shared the same beam setup. Using the Pinnacle treatment planning system (TPS), contours of planning target volume (PTV) and organ-at-risk (OAR) (left lung) were drawn by attending physicians.

B. Characteristics of VMAT plans: DVH versus DTH

DVH is a histogram linking radiation dose to organ volume in radiation therapy planning [5]. In the cumulative DVH, the value of vertical axis represents the volume of structure receiving greater than or equal to the corresponding dose in the horizontal axis.

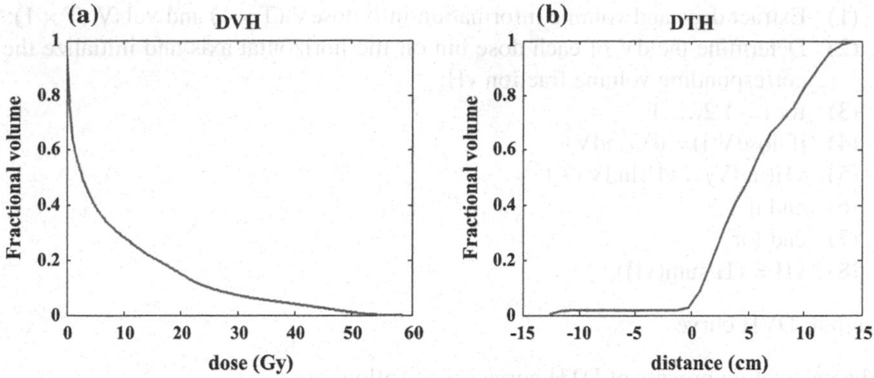

Fig. 1 **a** Presents the cumulative DVH of the left lung; **b** plots the cumulative DTH of the left lung

DTH is the fractional volume of the OAR within certain distance from the PTV surface [8]. It is included to represent the spatial relationship between the PTV and the OARs. In the Euclidean space, the value of DTH at a distance bin d is the fraction of OAR volume with its maximum distance to the PTV surface less than d [6]. The Euclidean form of the distance function r from an OAR voxel v^i_{OAR} to the PTV surface, $r(v^i_{OAR}, PTV)$ is [8, 13]

$$r(v^i_{OAR}, PTV) = \min_k\{||v^i_{OAR} - v^k_{PTV}||, v^i_{PTV} \in S_{PTV}\} \quad (1)$$

Negative signs are assigned to the distance values for OAR voxels inside PTV to indicate the intrusion of OAR into the PTV [6].

The cumulative DVH of the left lung is plotted in Fig. 1a. Figure 1b illustrates the DTH which provides a summarized characterization of organ shape relative to PTV.

3 Methods

A. Calculate DVH and DTH

DICOM-RT files have five parts: RT-Image, RT-Dose, RT-Structure set, RT-Plan and RT-Treatment Record. There are contours of organs in RT-Structure set and dose information in RT-Dose. The open source software CERR (Computational Environment for Radiotherapy Research) is built on the MATLAB platform and can be used to parse DICOM-RT files. We used the CERR to import VAMT plans to extract contours of PTV and OAR from RT-Structure set as anatomical information and dose information from RT-Dose and then carry out DVH and DTH calculations.

The calculation process of DVH curve are as follows:
Inputs: RT-Dose and RT-Structure set process:

(1) Extract dose and volume information into doseV (T × 1) and volsV (T × 1);
(2) Determine the dV of each dose bin on the horizontal axis and initialize the corresponding volume fraction vH;
(3) for i = 1,2,…,T
(4) if doseV(i) = dV(indV)
(5) vH(indV) = vH(indV) + 1
(6) end if
(7) end for
(8) vH = vH/sum(vH).

Output: DVH curve

The calculation process of DTH curve are as follows:
Input: RT-Structure set process:

(1) Acquire contour points and internal point sets of OAR into data1 and PTV contour point sets into data2
(2) Get the overlap points of data1 and data2, and use 'pos' to mark the position of overlap points in data1;
(3) for i = 1,2,…,size(data1)
(4) for j = 1,2,…,size(data2)
(5) dist = min(dist(data1(i),data2(j)));
(6) if (data1(i) = data1(pos))
(7) dist = 0 -dist;
(8) end if
(9) end for
(10) end for
(11) Calculate the volume fraction vH of each dist value.

Output: DTH curve

B. **PCA of DVH and DTH curves**

DVH curves have different distributions in dose as well as the distance of DTH. Therefore, those values (dose and distance) are needed to be normalized to make the original data in the same magnitude. In this paper, min-max normalization, also known as deviation normalization, was used to map the original data to the interval [0–1]. The conversion function is shown in Eq. (2):

$$x^* = \frac{x - \min}{\max - \min} \tag{2}$$

After the normalization process, 50 points in $x \in [0, 1]$ were sampled at the same interval. The y-coordinates of the 50 points were saved in y_i. All the DVH curves were processed in the same way to obtain the DVH sample set $D = \{y_1, y_2, \ldots, y_m\}$. Applying the same approach to DTH curves and obtained the DTH sample set $\hat{D} = \{x_1, x_2, \ldots, x_m\}$.

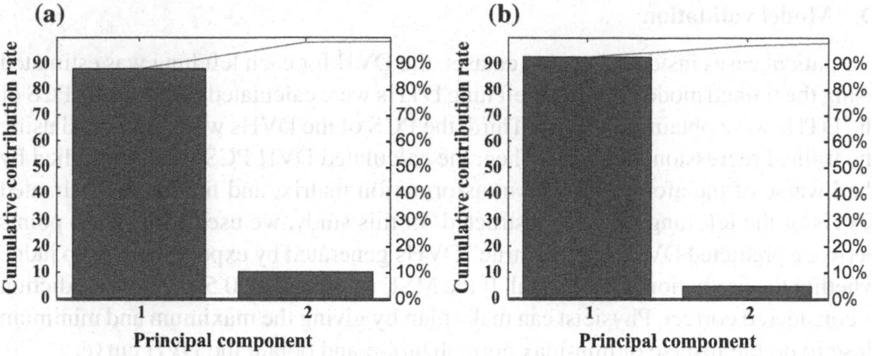

Fig. 2 Results of PCA applied to the training dataset

Principal component analysis (PCA) was used for representing DVH and DTH curves in greatly reduced dimensions. PCA is mostly used as a tool in extracting main features of data. It converts possibly correlated variables into smaller-sized linearly uncorrelated variables called principal components (PCS). In this study, each cumulative DVH or DTH was sampled by 50 points. We calculated a covariance matrix (50×50) for all feature points, and performed singular value decomposition on the covariance matrix to generate 50 eigenvectors and 50 corresponding eigenvalues. According to PCA theory, each principal component represents the dimensional variation represented by that component. By selecting the principal components corresponding to the larger eigenvalues, we can use smaller-scale DVH or DTH points to represent the DVH and DTH curves, thus achieving the goal of feature extraction.

The results of applying PCA to DVH and DTH curves in training dataset are shown in Fig. 2. The contribution ratio of each principal component is plotted in descending order using a bar graph. To cover >90% cumulative contribution rate, Fig. 2a shows that the first two principal components are needed for the left lung DTH. Figure 2b indicates the first principal component is needed for the left lung DVH.

C. **Lasso regression and model training**

Least absolute shrinkage and selection operator (LASSO) regression can perform both variable selection and regularization. It makes the prediction more accurate by transform the model fitting process to choose a subset of provided covariates for use in the final model [14]. Support vector regression (SVR) is a method commonly used in limited training samples, but it works badly when the independent variables have strong correlations while lasso can handle this issue by forcing certain fitting coefficients to be set to zero. So in this study, we chose the lasso regression to train the model. Choosing the regularization parameter (lambda) well is significant to perform lasso since it influences the strength of variable selection. In this study, 20-fold cross-validation was used to calculate the mean squared error (MSE) of the fitted model for each value of lambda. We selected the lambda corresponding to the minimum MSE and computed the fitting coefficients.

D. **Model validation**

For patient cases inside the testing dataset, the DVH for each left lung was estimated using the trained model. First, the left lung DTHs were calculated. Second, the PCS of the DTHs were obtained by PCA. Third, the PCS of the DVHs were calculated using the trained regression functions. Then, the calculated DVH PCS were multiplied by the inverse of the aforementioned transformation matrix, and finally, the estimated DVHs of the left lung were reconstructed. In this study, we used MSE of 50 points between predicted DVHs and the "true" DVHs generated by expert planners to judge whether the prediction is successful. If the MSE is lower than 0.5, then the prediction is considered correct. Physicist can make plan by giving the maximum and minimum dose to do the inverse of min-max normalization and obtain the DVH curve.

4 Results

We used MSE of 50 points between predicted DVHs and the "true" DVHs generated by expert planners to judge whether the prediction is successful. If MSE is lower than 0.5, then the prediction is considered correct.

Figure 3 shows the estimated DVHs for left lung of the 12 cases. The blue lines show the actual DVH curves generated by expert planners using Pinnacle treatment planning system. The predicted DVH curves using lasso regression are plotted with red lines. Green lines indicate the predicted DVH curves using SVR. We can easily see that SVR works worse than lasso regression since the correlations between the two PCS of DTH. Figure 4 shows the MSE of 50 points between predicted DVHs using lasso regression and the "true" DVHs generated by expert planners. Except for cases 8, 9, and 10, the estimations are successful for 9 out of 12 test plans. As Fig. 5 indicates, for case 8 and 10, the treatment regions are limited in the neck, while the treatment regions of training dataset cover the area from the neck to the whole chest, which enlarges predicted errors in these 2 cases.

5 Conclusion

In this paper, we developed a model for predicting DVH of OAR based on machine learning using the prior knowledge of the high-quality VMAT plans. The experimental results demonstrated that the DTH/DVH curves can be effectively expressed by one or two principal components, the accuracy of the model in prediction is about 75%. These promising results suggest that this method can predict does distribution in the esophageal VMAT plan and assist the physicist to make plans by giving objective function and orientation, which can improve the efficiency and quality of plan making.

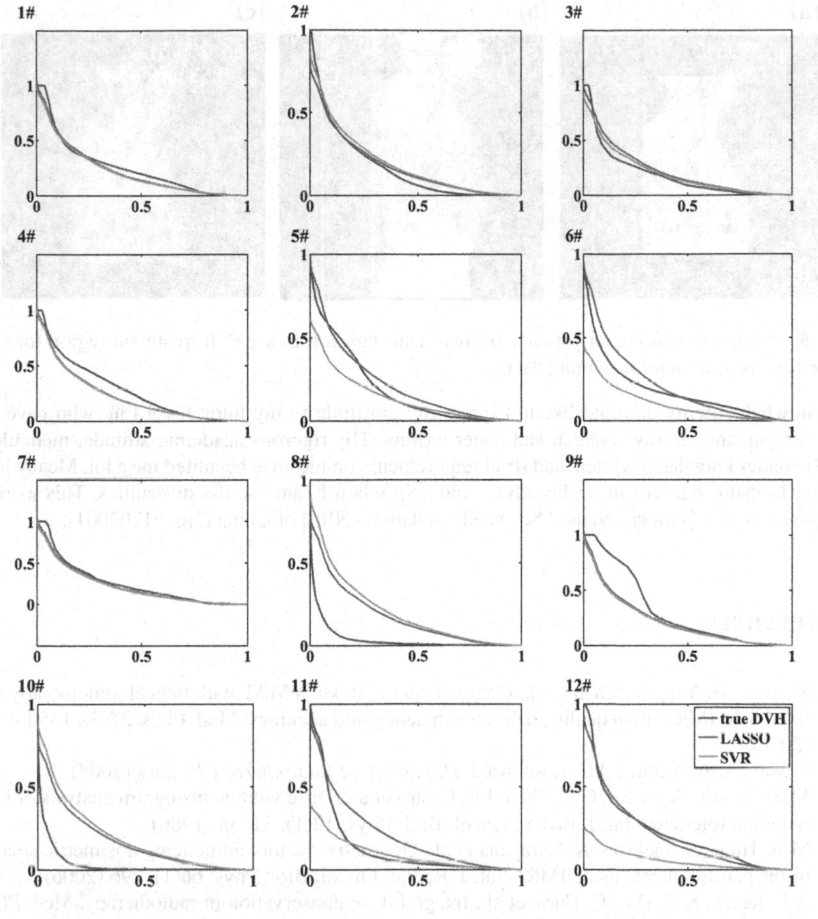

Fig. 3 Model evaluation on left lung using 12 cases outside the training pool

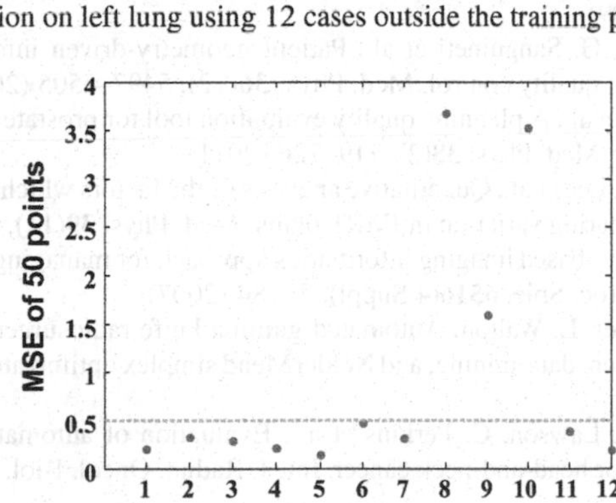

Fig. 4 MSE of 50 points between predicted values and true values for 12 test cases

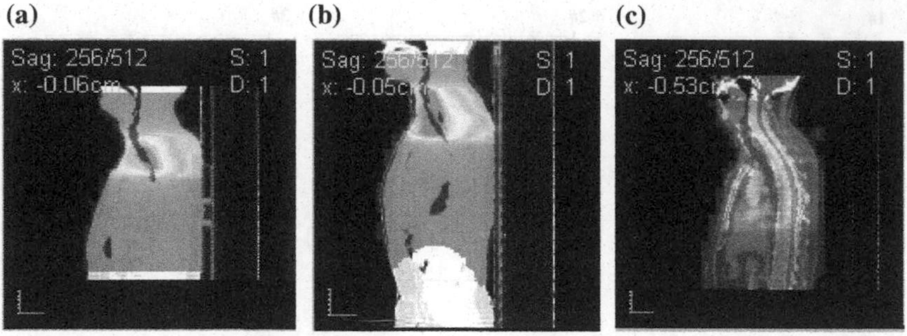

Fig. 5 Treatment regions for patients. **a** Treatment region for case 8; **b** treatment region for case 10; **c** treatment region for training data

Acknowledgements I would like to express my gratitude to my tutor Jiwei Liu, who gave me a lot of guidance in my research and paper writing. His rigorous academic attitude, meticulous mathematics knowledge system and strict requirements on me have benefited me a lot. Meanwhile, I want to thank Jianfei Liu for his advice and help when I came across difficulties. This work is supported by the National Natural Science Foundation (NSF) of China (No. 61702001).

References

1. M. Rao, W. Yang, F. Chen et al., Comparison of Elekta VMAT with helical tomotherapy and fixed field IMRT: plan quality, delivery efficiency and accuracy. Med. Phys. **37**(3), 1350–1359 (2010)
2. P. Mayles, A. Nahum, J.C. Rosenwald, *Handbook of Radiotherapy Physics* (2007)
3. M.M. Austin-Seymour, G.T. Chen, J.R. Castro et al., Dose volume histogram analysis of liver radiation tolerance. Int. J. Radiat. Oncol. Biol. Phys. **12**(1), 31–35 (1986)
4. M.A. Hunt, A. Jackson, A. Narayana et al., Geometric factors influencing dosimetric sparing of the parotid glands using IMRT. Int. J. Radiat. Oncol. Biol. Phys. **66**(1), 296 (2006)
5. A.S. Reese, S.K. Das, C. Curie et al., Integral dose conservation in radiotherapy. Med. Phys. **36**(3), 734–740 (2009)
6. B. Wu, F. Ricchetti, G. Sanguineti et al., Patient geometry-driven information retrieval for IMRT treatment plan quality control. Med. Phys. **36**(12), 5497–5505 (2009)
7. X. Zhu, Y. Ge, T. Li et al., A planning quality evaluation tool for prostate adaptive IMRT based on machine learning. Med. Phys. **38**(2), 719–726 (2011)
8. L. Yuan, Y. Ge, W.R. Lee et al., Quantitative analysis of the factors which affect the interpatient organ-at-risk dose sparing variation in IMRT plans. Med. Phys. **39**(11), 6868–6878 (2012)
9. B.J. Liu, A knowledge-based imaging informatics approach for managing proton beam therapy of cancer patients. Proc. Spie. **6516**(4 Suppl), 77–84 (2007)
10. K.J. Lee, D.C. Barber, L. Walton, Automated gamma knife radiosurgery treatment planning with image registration, data-mining, and NelderMead simplex optimization. Med. Phys. **33**(7), 2532 (2006)
11. L.J. Stapleford, J.D. Lawson, C. Perkins et al., Evaluation of automatic atlas-based lymph node segmentation for head-and-neck cancer. Int. J. Radiat. Oncol. Biol. Phys. **77**(3), 959–966 (2010)
12. G. Strassmann, S. Abdellaoui, D. Richter et al., Atlas-based semiautomatic target volume definition (CTV) for head-and-neck tumors. Int. J. Radiat. Oncol. Biol. Phys. **78**(4), 1270–1276 (2010)
13. R.E. Drzymala, R. Mohan, L. Brewster, Dose-volume histograms. Int. J. Radiat. Oncol. Biol. Phys. **21**(1), 71 (1991)
14. R. Tibshirani, Regression shrinkage and subset selection with the Lasso. J. R. Stat. Soc. (1996)

Edge Detection for Conveyor Belt Based on the Deep Convolutional Network

Yan Liu, Yaoping Wang, Chan Zeng, Weicun Zhang and Jiangyun Li

Abstract Conveyor belt deviation is the most common failure of the conveyor system. Timely and accurate detection of deviations from conveyor belt and rapid processing are the guarantee for the safe and stable operation of entire system. Based on the industrial scene of coal transport belt, this paper develop a new method of belt edge detection based on deep convolution network, addressing the problems that traditional mechanical anti-deviation treatment is not timely and the edge detection of the conveyor belt with machine vision is imprecise. We establish a new dataset of the conveyor belt edge, by contrast experiments on training FCN, Deeplab, HED, we selecte the HED model which is most suitable for this task, then compress the model and simplify its output. Finally, the processing speed of the single picture is 0.26 s and the error of conveyor belt edge detection is less than 16 mm. The method proposed in this paper is quick and simple, with high precision and strong anti-jamming capability. It can be used for all kinds of production scenarios.

Keywords Edge detection · Belt deviation · HED · Model compression

1 Introduction

Belt conveyors have been widely used in the industrial fields of mining, coal, and grain for their advantages of long distance, large capacity, and high efficiency. However, in some actual production processes, conveyor belt deviation often occurs when the

This work was supported by National Science Foundation of China (No. 61520106010, No. 61741302).

Y. Liu · Y. Wang · C. Zeng · W. Zhang · J. Li (✉)
School of Automation & Electrical Engineering,
University of Science and Technology Beijing, Beijing 100083, China
e-mail: leejy@ustb.edu.cn

Y. Liu · Y. Wang · C. Zeng · W. Zhang · J. Li
Key Laboratory of Knowledge Automation for Industrial Processes,
Ministry of Education, Beijing 100083, China

Y. Jia et al. (eds.), *Proceedings of 2018 Chinese Intelligent Systems Conference*, Lecture Notes in Electrical Engineering 529,
https://doi.org/10.1007/978-981-13-2291-4_28

conveyor is running, which causes the loss of material sprinkles, conveyer belt to wear and so on, even triggers the emergency switch in serious cases, leading to the operation system stopping which greatly affects the production efficiency [1].

In general, there are two ways to detect the deviation of the conveyor belt, which are contact and non-contact. Contact detection method includes installing correction roller or emergency switch for conveyor belt, these are mechanical measure means [2–4], they will work only when the conveyor belt touch the vertical roller or emergency switch. On the one hand, it causes the wear of the belt edge, on the other hand, the detection precision is poor, even results in the failure of the whole conveying system.

Non-contact detection is mainly based on the methods of edge detection in traditional machine vision. After collecting images of the conveyor belt from CCD camera, one method is detecting its belt edge by Canny, then judging whether it is deviant according to the geometric characteristics of the edge [5, 6]. Or the collected image is processed by Image Binarization, later the deviation angle and offset are extracted from the binary image [7]. Since both the Canny and the Image Binarization are based on the grayscale gradient of the image, these two methods are sensitive to the noise, and can detect all the edges in the whole picture.

In recent years, the method based on deep learning has made important breakthroughs in the fields of object proposal, segmentation, and 3D reconstruction. The method of deep learning can not only achieve end to end processing, but also have strong generalization and fast detection speed [8].

Based on the above problems and the advantages of deep learning, this paper chooses deep learning method to detect the edge of conveyor belt. First, make our own belt edge dataset (including 2500 train pictures with 2500 labels and 200 test pictures). The dataset can change the global edge detection model to a specific edge detection model. After that we use the dataset to train three networks, namely FCN [9], Deeplab [10], HED [11], and compare the results, then select the HED network model because it is most suitable for this task. Finally, in view of the processing speed and precision requirements, we compress the HED model and reduce its outputs. The detection speed of our network is improved from the original 0.42 s per picture to 0.26 s, and the detection error is less than 16 mm.

2 Approach

2.1 Dataset

Deep learning has an unparalleled ability for feature learning compared with traditional methods. For our task, deep convolution network needs to learn the edge features of the conveyer belt and ignore the other edge features, such as the cross beam and the coal on the conveyor belt, etc. Therefore, we establish a special dataset for the conveyor belt, including 2500 train pictures and 2500 labels, 200 test pictures.

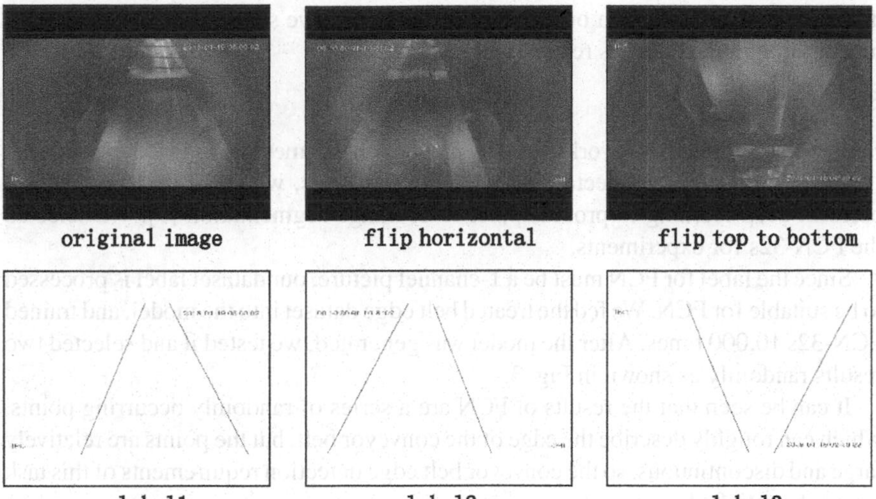

Fig. 1 Pictures and labels in dataset

From a coal mine conveyor video, we intercepted 700 images, 500 of them were selected as training set, and the remaining 200 pictures were taken as test set. The training set includes 300 pictures without coals and 200 pictures containing coals. For these 500 pictures, we line draw the edges of belt with one pixel wide and process the background into black to make each corresponding label. However, deep learning is a method based on large-scale data to fully study sample features to make better predictions. In order to prevent 500 training samples from over-fitting the model, by mirroring, reversing, flipping up and down, scale 1.5 and scale 0.5, these 500 original images have been expanded to 2500 original images and 2500 labels. The pictures and labels in the training set are shown in Fig. 1.

2.2 Model Select

With the explosion of deep learning in recent years, various network models are also growing blowout.

First of all, allowing for the implementation of end to end. Deep Contour [12] and Deep Edge [13] methods based on Deep Convolutional Network all preprocess data using traditional machine learning methods firstly, extract a class of specific features, such as the direction of the edge and so on. Then feed these features into the model as a classification problem. The corollary of such a processing is that detection is not very accurate even if not considering the end to end.

Because of the great similarity between the edge detection and image segmentation, we trained the classic segmentation models FCN, Deeplab and the global edge detection model HED with our dataset, after testing, we selected the most suitable model through the analysis results.

FCN

Fully Convolutional Networks, as the pioneer in segmentation in deep learning. It removed the fully connected layer for the first time, which made it possible to perform deep learning to process pixel-level image segmentation [9]. We selected the FCN-32s for experiments.

Since the label for FCN must be a 1-channel picture, our dataset label is processed to be suitable for FCN. We fed the treated belt edge dataset into the model, and trained FCN-32s 10,000 times. After the model was generated, we tested it and selected two results randomly as shown in Fig. 3.

It can be seen that the results of FCN are a series of randomly occurring points, which can roughly describe the edge of the conveyor belt, but the points are relatively large and discontinuous, so the conveyor belt edge detection requirements of this task cannot be satisfied.

Deeplab

At present, Depplab has become the most classic method for deep learning and processing image semantic segmentation [10].

This paper used Deeplab v2 for experiment. The required dataset is the same as FCN. Therefore, the processed belt edge dataset is fed to the Deeplab v2 for 10,000 times training. After the trained model is generated, we also tested it and chose two results accordingly as shown in Fig. 2. It is important to note that in the result graph after the Deeplab processing, the points of the belt edge are very small, with an average of one pixel value. In order to show their effects better, we used a 5-pixel point to depict them. So it looks like that FCN is similar to Deeplab, but in fact points in FCN are about 5 times bigger than Deeplab.

It can be seen that compared with FCN, Deeplab's result is more refined, but the points become less. In addition, like the FCN, the conveyor belt edge in Deeplab also becomes a series of discrete pixels, and the occurrence of these points are random and cannot be accurately measured.

For the random occurrence of the discontinuities in the FCN and Deeplab methods, the primary analysis is that the training set labels of semantic segmentation are all connected regions, but the labels in this paper are two disconnected lines, resulting that the category labels do not work well during the training process. While the Deeplab's pionts are finer but fewer are caused by Atrous Convolution and increased CRF.

HED

The Deep edge and Deep contour methods is not end-to-end processing and the FCN and Deeplab methods cannot detect the edge of the continuous conveyor belt, so they are all excluded. The HED (Holistically-Nested Edge) network published in ICCV has become the best choice. The next section will describe it in detail.

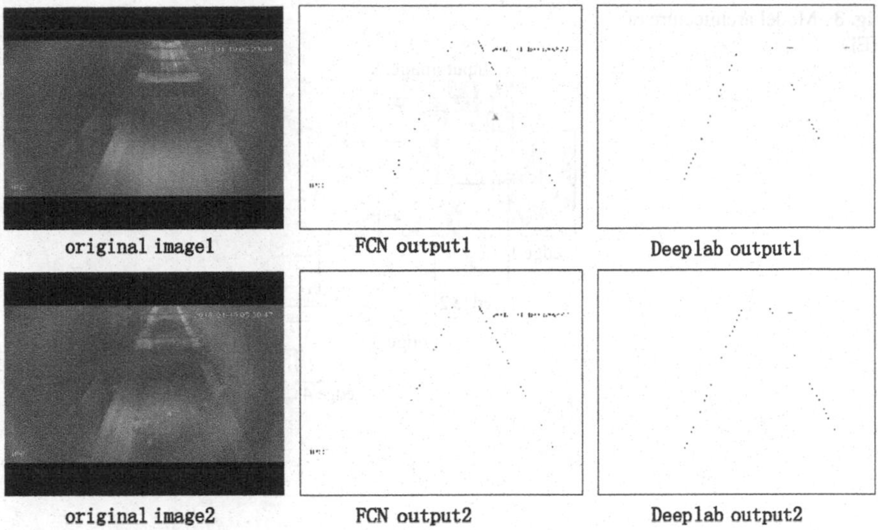

Fig. 2 Outputs of FCN and Deeplab

Table 1 Network architecture and stride of VGGNet

Layer	c1_2	p1	c2_2	p2	c3_3
Stride	1	2	2	4	4
Layer	p3	c4_3	p4	c5_3	p5
Stride	8	8	16	16	32

2.3 HED Network Architecture

HED was modified on the basis of the classic deep learning model VGGNet [14]. The network structure and stride of VGGNet are shown in Table 1.

HED made two modifications to VGGNet: (1) After each stage, the last convolutional layer is followed by a Side-output layer, which is behind conv1_2, conv2_2, conv3_3, conv4_3, conv5_3 layers. (2) Remove the fifth stage of the pooling layer and the subsequent full connected layer because the step size of the pooling5 layer is 32. Network architecture is shown in Fig. 3.

Training model

The training model has five stages, with 2, 2, 3, 3, 3 convolutional layers and 4 pooling layers. When image X is input, after the first stage, the detection results of the first stage is the edge 1. Edge 1 performs two major roles: (a) Continue to enter the second stage and generate the side-output edge 2 of the second stage. (b) Generate an image with the same size as the Ground Truth by bilinear interpolation and participates in the fusion with weighted fusion layers. Similarly, edge 2, edge 3, and edge 4 also have the same dual role as edge 1. They both participate in the

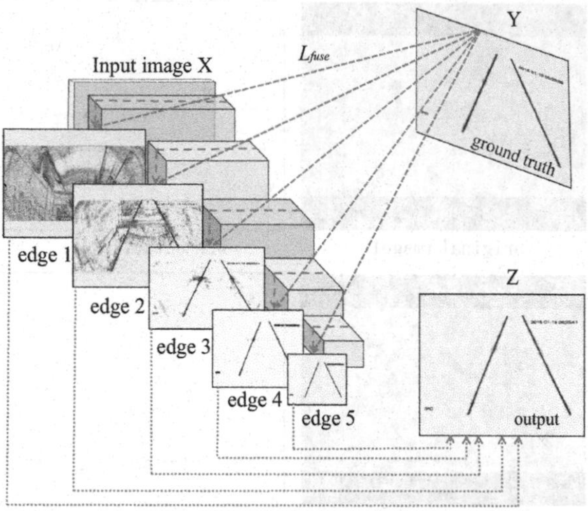

Fig. 3 Model architecture of HED

integration of weighted fusion layers and as the input of the next stage. Network will end after edge 5, so edge 5 is no longer used as the input for the next stage. This is also the origin of Holistically and Nested in the HED network. Finally, a fused image output will be produced by weighted fusion layer.

However, in the experiment, we found that the convolutional layers in stage 5 have much more parameters than other stages, resulting in the larger trained model and longer testing time. Therefore, we pruned this model. This will be introduced detailedly in Sect. 2.4.

Testing model

The test model mentioned above is a multi-output model with five side-outputs and one fusion layer output. For the conveyor belt edge detection task of this paper, only the best output is needed, so we modified the test model to show the only fusion output, eliminating the periodic output, which makes outputs no longer redundant, reduces the memory consumption of the model when running, and improves the detection speed of the conveyor belt edge.

2.4 Model Compression

In the test process, although the test Model was simplified, the processing speed of the original HED network model test image was 0.42 s per picture, which was little slow for our task. In order to meet the speed requirements of industrial real-time scenarios, we compressed the model according to the pruning method [15].

Network pruning greatly decreases the number of model parameters to reduce the model size, speed up the test, and to a certain extent prevent overfitting.

Table 2 Experimental results of model compression

Pruned (%)	Model size (M)	Average accuracy (pixel)	Speed (s)
0	56.15	9.5	0.42
60	39.95	10.1	0.29
70	37.25	10.3	0.26

The network structure of the HED is shown in Fig. 3. The number of convolution kernel in each stage multiplied, from the initial 64 to the 512, which led the parameters of the fifth stage to stagger. The amount of parameters in the last stage of a trained model even accounts for almost 50% of all parameters through our calculation, so the pruning operation for this model is selected in the fifth stage. After several pruning sensitivity experiments based on the layer in the fifth stage, we finally decided to prune the three 3×3 convolution filters.

Before model compression, the original HED network is trained as the reference accuracy and model size. First, the first 3×3 convolution is pruned, and the appropriate pruning rate is used to generate a new trained model. Comparing the accuracy after test with reference accuracy, after several experiments, we selected the pruning strategy that met the requirements, and then performed the same operation on the other second 3×3 convolution. The final model size was compressed by 30%, and the test achieved a speed of 0.26 s per picture, which met the industrial requirements. At the same time, the test accuracy was almost unchanged. The model size, edge accuracy, and test speed at the pruning rates of 60 and 70% are shown in Table 2.

3 Experimental Results

3.1 *Implementation*

The experiment was implemented on a server with an NVIDIA 1080Ti graphics card. The network model was built on Caffe.

The training phase used the compressed HED training model and VGG16 as a pre-training model to initialize the network parameters. Because this model is a pixel-level edge detection, parameter adjustment is more difficult, and there will be usually gradient explosions.

After a mass of experiments, some hyper-parameters include: the learning rate is 1e-7, training iterations are 10,000, learning rate decay is 0.1 after 5000 iterations, the initial weight of the fusion layer is 0.2, and the weight decay is 0.0002. For our dataset, the training time is 4 h.

100 pictures with coal and 100 pictures without coal of conveyor belts were selected for testing. After removing the output of the side output layer, the average test time for a picture is 0.42 s. Table 2 shows the results of the compressed model.

3.2 Results Comparison

The pictures chosen from the test results of compressed model are compared with the results of the Canny and they are shown in Fig. 4.

As we can see, Canny is a global edge detection algorithm. For a specific edge, such as the belt edge detection in our paper, although Canny can detect the edge of the conveyor belt more clearly, the other edges are completely interference to the task, and it cannot solve this problem.

Camera model: L is the maximum distance that the camera can detect, camera pixel is $v \times w$, its accuracy is per pixel $=$ L/v.

It can be seen that the measured L is closer to the camera, the higher the accuracy of the measured distance, so the edge of the conveyor belt at the bottom of the picture is used for accuracy calculation. In this paper, the camera pixel is 1000 × 800, the corresponding position L is 1530 mm, and the camera accuracy on this line is 1.53 mm. From the 200 test pictures, the average number of pixel on the edge of the conveyor belt for that line is 10.3. The edge accuracy is 15.76 mm. This error is completely within the permissible of the industrial scene.

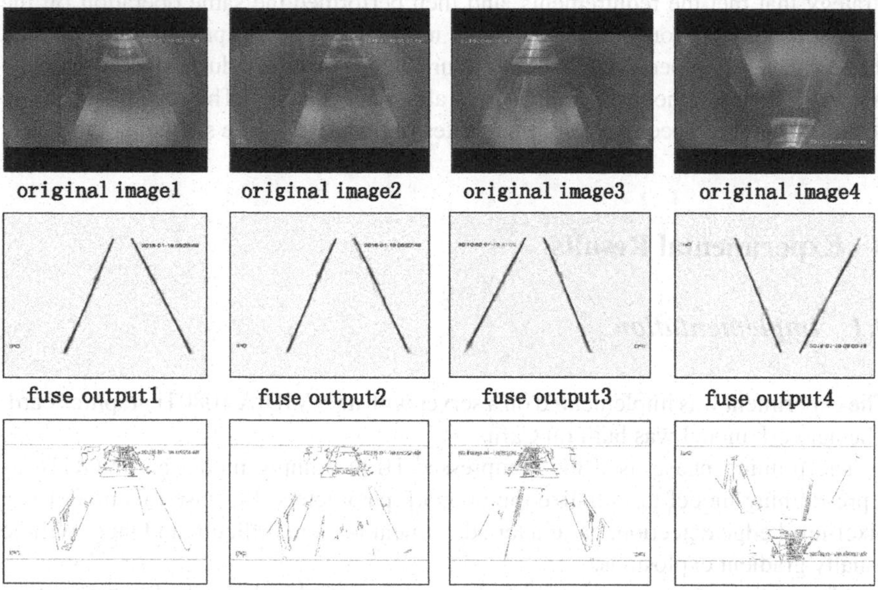

Fig. 4 Results comparison with compressed HED and Canny

4 Conclusion

This paper realizes the application of the deep convolutional network in the detection of the edge of the conveyor belt in the actual industrial scene, and achieves a detection error of less than 16 mm and a detection speed of 0.26 s per picture. We take the advantages of end-to-end, strong learning ability and good generalization of deep convolutional network, established the conveyor edge detection dataset, and select the appropriate model (HED) from the methods of FCN, Deeplab, and HED. What's more, we compress HED network to make it more suitable for industrial applications.

References

1. Y.W. Zhao, L.I. Yun-Hai, The present and future development of belt conveyer. Coal Mine Mach. (2004)
2. T.P. Han, Y.K. Chen, Y. Zheng et al., Analysis on reasons of lateral misalignment of belt conveyor and the technology avoiding misalignment. Agric. Equip. Veh. Eng. (2009)
3. G. Yong-bo, Analysis of the reasons for the deviation of the belt conveyor and its countermeasures. Shandong Coal Sci. Technol. **3**, 113–115 (2018)
4. F.U. Chou-Kui, D.U. Xin-Ji, Correct an error ways of conveyor's belt. Coal Technol. (2006)
5. A.C. Zhu, G. Hua, Y.X. Wang, The research on the detection method of belt deviation by video in coal mine, in *2011 International Conference on Mechatronic Science, Electric Engineering and Computer* (2011), p. 433
6. X.U. Huan, L.I. Zhenbi, Y. Jiang et al., Research of automatic detection algorithm of conveying belt deviation based on OpenCV. Ind. Mine Autom. (2014)
7. Y.L. Yang, C.Y. Miao, K. Kang et al., Machine vision inspection technique for conveyor belt deviation. J. North Univ. China **33**(6), 667–671 (2012)
8. Y. Lecun, Y. Bengio, G. Hinton, Deep learning. Nature **521**(7553), 436 (2015)
9. J. Long, E. Shelhamer, T. Darrell, Fully Convolutional Networks for Semantic Segmentation, in *Proceedings of the IEEE Conference on Computer Vision and Pattern Recognition* (2015), pp. 3431–3440
10. L.C. Chen, G. Papandreou, I. Kokkinos et al., Deeplab: Semantic image segmentation with deep convolutional nets, atrous convolution, and fully connected crfs. IEEE Trans. Pattern Anal. Mach. Intell. **40**(4), 834–848 (2018)
11. S. Xie, Z. Tu, Holistically-Nested Edge Detection, in *Proceedings of the IEEE International Conference on Computer Vision* (2015), pp. 1395–1403
12. W. Shen, X. Wang, Y. Wang et al., DeepContour: A Deep Convolutional Feature Learned by Positive-Sharing Loss for Contour Detection Draft Version, in *CVPR* (2015)
13. G. Bertasius, J. Shi, L. Torresani, in Deepedge: A Multi-Scale Bifurcated Deep Network for Top-Down Contour Detection, in *2015 IEEE Conference on Computer Vision and Pattern Recognition (CVPR)* (IEEE, 2015), pp. 4380–4389
14. K. Simonyan, A. Zisserman, Very deep convolutional networks for large-scale image recognition. arXiv preprint arXiv:1409.1556 (2014)
15. H. Li, A. Kadav, I. Durdanovic et al., Pruning Filters for Efficient Convnets. arXiv preprint arXiv:1608.08710 (2016)

Automatic Cell Segmentation and Signal Detection in Fluorescent in Situ Hybridization

Jing Wang, Jiwei Liu, Jianfei Liu, Hui Yan and Ronghu Mao

Abstract With the progress of science and technology, computer technology has also been developed, and there are a huge promotion and wider application in the field of medicine. This study is to determine the positive rate of breast cancer cells, mainly in cell segmentation and signal point recognition methods. Containing threshold iterative segmentation, edge detection, K-means clustering and watershed, four methods are used to segment the cells. But these methods do not allow good separation of adherent cells, therefore, we should improve the segmentation methods of adherent cells. The distance transformation is done first, then the watershed method is used. Finally, the watershed based on distance transformation is combined with the contour extraction to separate the adherent cells. And for the identification of signal points, using the Otsu achieved a more satisfactory result, and the accuracy rate has reached over 90%. According to the result of cell segmentation, the accuracy rate is basically more than 80%, in the subsequent study can also be improved to get more accurate results. Therefore, it is feasible to use computer technology for automatic segmentation to determine whether breast cancer cells are positive, and it can be popularized and applied.

Keywords Cell segmentation · Signal point recognition · Watershed Distance transformation · Otsu-algorithm

---

J. Wang · J. Liu (✉)
School of Automation and Electrical Engineering,
University of Science and Technology, Beijing, China
e-mail: liujiwei@ustb.edu.cn

J. Liu
Electrical Engineering and Automation, Anhui University, Hefei, China

H. Yan
Department of Radiation Oncology Cancer
Hospital Chinese Academy of Medical Sciences, 2258, Beijing 100021, China

R. Mao
Department of Radiation Oncology, The Affiliated Cancer
Hospital of Zhengzhou University, Henan Cancer Hospital, Zhengzhou, Henan, China

Y. Jia et al. (eds.), *Proceedings of 2018 Chinese Intelligent Systems Conference*, Lecture Notes in Electrical Engineering 529,
https://doi.org/10.1007/978-981-13-2291-4_29

1 Introduction

In recent years, women are increasingly at risk of breast cancer. Breast cancer has become a major threat to women's health. The clinicians mainly diagnose breast cancer from the point of cell position, morphology and positive rate [1], and the judgment of the positive rate is the main research direction of this study. Whether the cells are positive is mainly determined by the amplification of HER-2 gene. The cells are stained with fluorescent dyes, and the ratio of the red signal point to the green signal point is calculated to determine whether the cell is positive.

Because of the difference of images that each has its own characteristics, and the existing segmentation methods have some limitations. Therefore, although a variety of segmentation methods have been studied, no segmentation method is applicable to all images so far, and the appropriate segmentation method should be selected according to the different features of different images. Hongli Wei proposed a medical image threshold segmentation algorithm based on multi wavelets [2]. Mingwu Ren et al. put forward a new method to construct histograms that based on the edge mode [3]. An improved fuzzy enhancement edge extraction algorithm is proposed by Zhang [4]. Su et al. have proposed a local maximal clustering method [5]. Nath et al. proposed the coupled level set and the graph vertex coloring method [6]. The method of two-stage graph cut was proposed by Matula et al. [7]. Arslan et al. proposed a model-based algorithm for segmenting nuclei [8]. The method relied on the modeling of cell boundaries.

In this study, the initial segmentation of cells was carried out firstly. Because of the different cell morphology, individual cells can be easily segmented in segmentation, while adherent or overlapping cells are difficult to deal with. Therefore, improvements were made to segment the adherent cells after initial segmentation. Next, recognition and extraction of signal points were carried out, then counted the signal points located in the cell, and calculated the ratio of red and green signal points.

2 Basic Method

In this paper, we used the basic methods including threshold iteration [9], edge detection [10], k-means clustering and watershed to segment the cells. In the edge detection, Laplace and Canny edge detection were used. And then we used Otsu [11] to identify the signal points.

The some results of cell segmentation using threshold iteration, edge detection, K-means clustering and watershed segmentation are shown in Fig. 1.

Using the Otsu algorithm to extract and identify the signal highlights, the result is shown in Fig. 2.

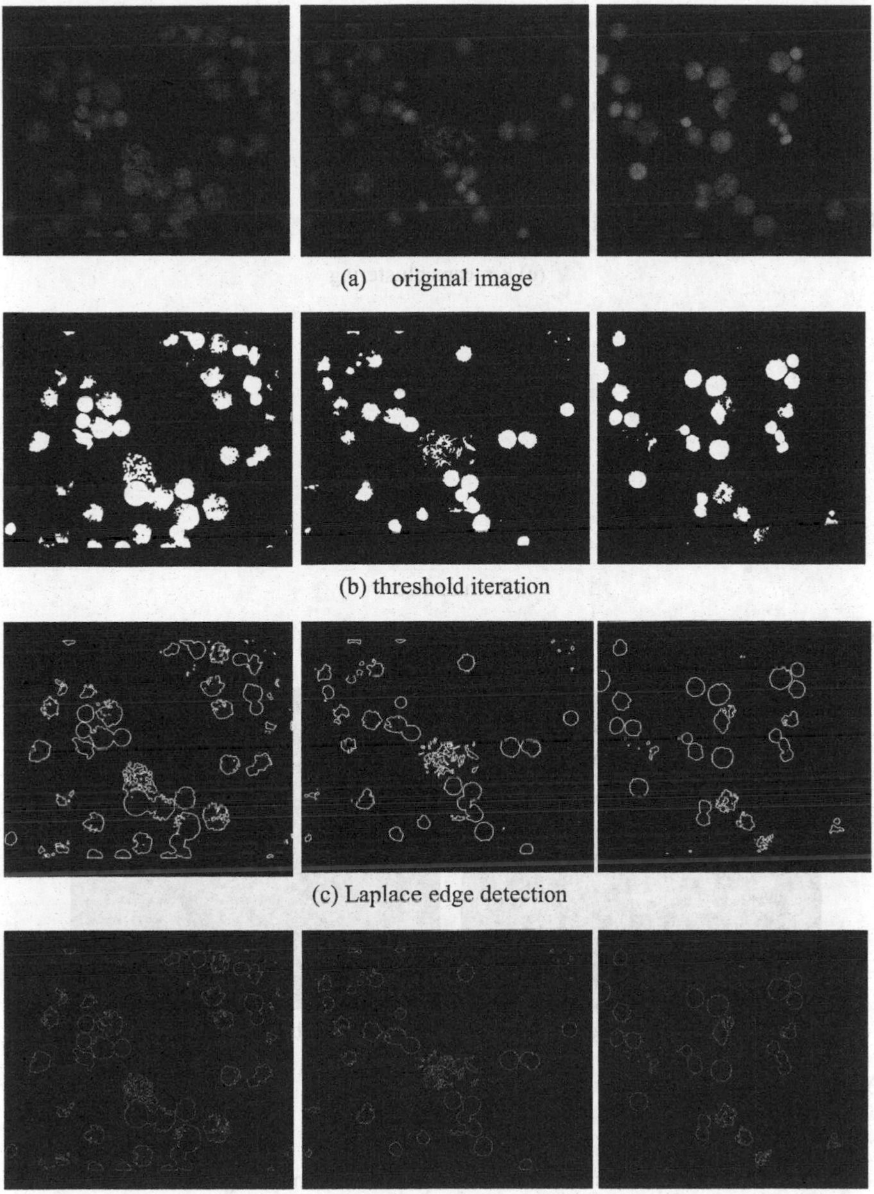

(a) original image

(b) threshold iteration

(c) Laplace edge detection

(d) Canny edge detection

Fig. 1 Results of cell segmentation

From the result, it can be seen that for the segmentation of signal points, the use of the Otsu algorithm can achieve ideal results, the highlights in the image have been basically detected and identified. For cell segmentation, the two methods of edge

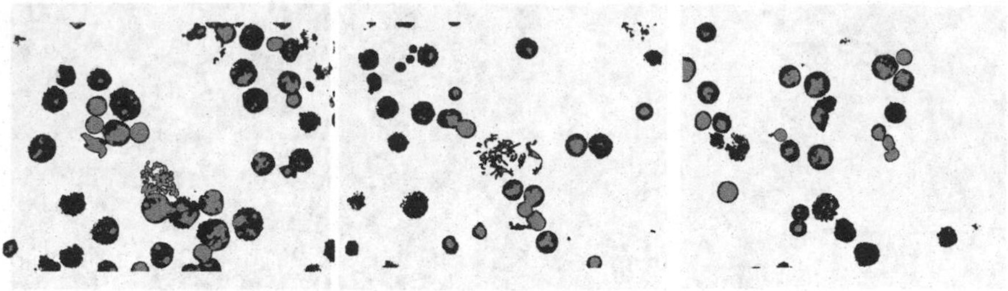

(e) k-means clustering

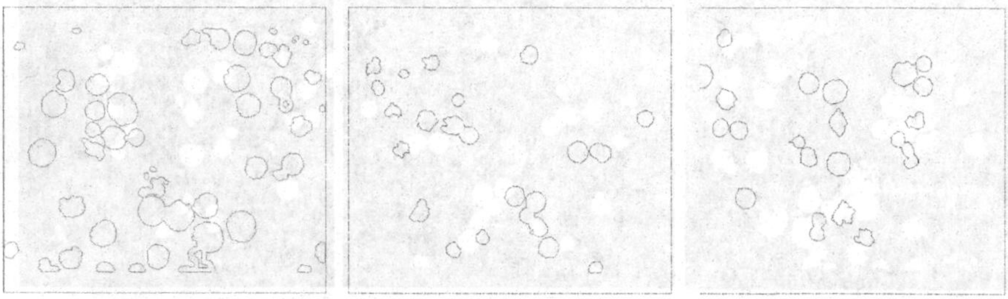

(f) watershed

Fig. 1 (continued)

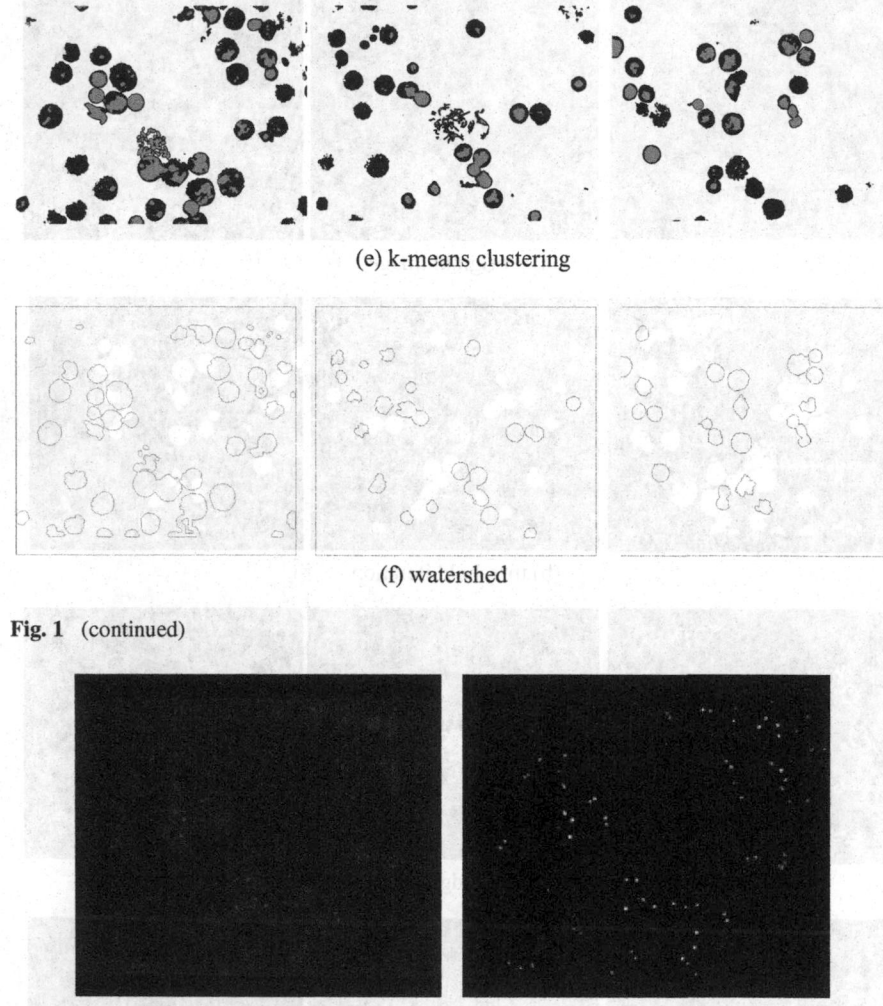

(a) original image (b) Otsu

Fig. 2 Results of signal point extraction

detection are basically consistent. But compared to Canny edge detection, Laplace edge detection produces two pixel wide edges, resulting in a more obvious edge.

Using these methods to deal with the image, the cell regions can be initially segmented preferably. But for some special cells, such as adherent cells, these methods can not well handle. It can be seen that in the processed image, the cell regions have been detected and identified, but the cell adhesions are not separated, which will affect the final judgment.

3 Improved Method

3.1 *Distance Transformation*

For a distance image, the gray value of each pixel is the distance between the pixel and the nearest background pixel, that is, the distance transform provides the distance from each pixel of the image to the nearest non-zero pixel.

3.2 *Watershed Based on Distance Transformation and Contour Extraction*

In the image after distance transformation, each cell produces a local maximum in its center position, and the center position is the local minimum after the distance image is reversed. In the subsequent watershed processing, the lowest point of the image is advanced water, and then the water gradually immerses into the whole valley basin. When the water level reaches the edge of the basin, it will overflow, and the dam will be built in the overflow area. The built dam will be the watershed to separate basin until the whole image is submerged in the water.

To segment cells, it is necessary to extract the contour of cells to identify the segmented cells. Contour extraction is used to draw the initial segmented cell contours. After the previous processing, the final segmentation image can be obtained by combining the watershed line image and the contour extracted image.

The result of using distance transformation is shown in Fig. 3. And the final segmentation results are shown in Fig. 4.

For images with obvious contrast and clear edge, the ideal segmentation result can be achieved by using the basic segmentation method, such as the threshold method, etc. But medical images are usually accompanied by fuzziness and without obvious edges. Therefore, the use of basic segmentation method may not achieve the desired results. The improved watershed segmentation method based on distance transformation is not only good for segmentation but also fast and easy to implement.

4 Signal Point Counts

In this study, we need to count the effective signal points. The signal points identified are detected. The signal points are counted if the signal points in the cell, while the signal point is an invalid signal point and does not count if the signal point is outside the cell. Therefore, draw the contour of the extracted signal and find out the center point of the contour. If the center point is within the contour of the cell, the point is judged to be the effective signal point and it is counted. Otherwise, it is not counted. The final display result is shown in Fig. 5.

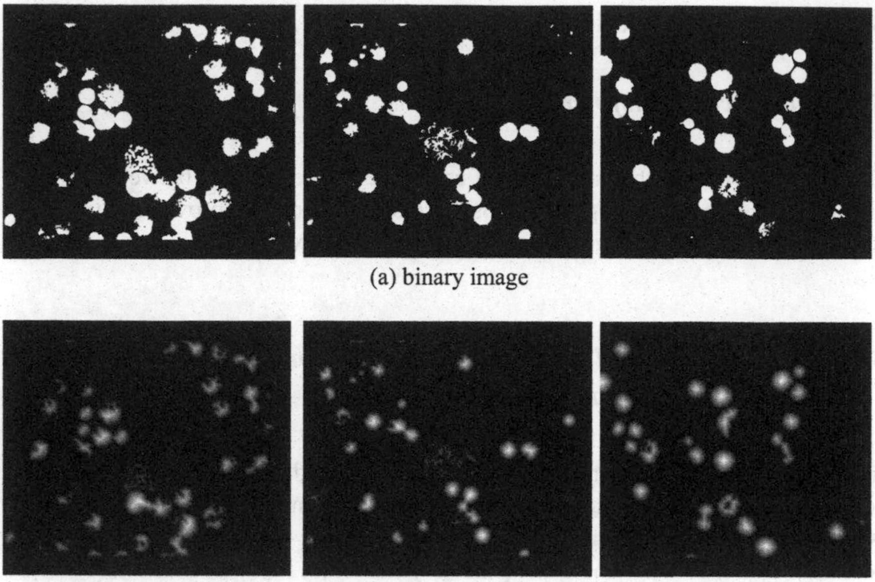

(a) binary image

(b) distance transformation

Fig. 3 Results of distance transformation

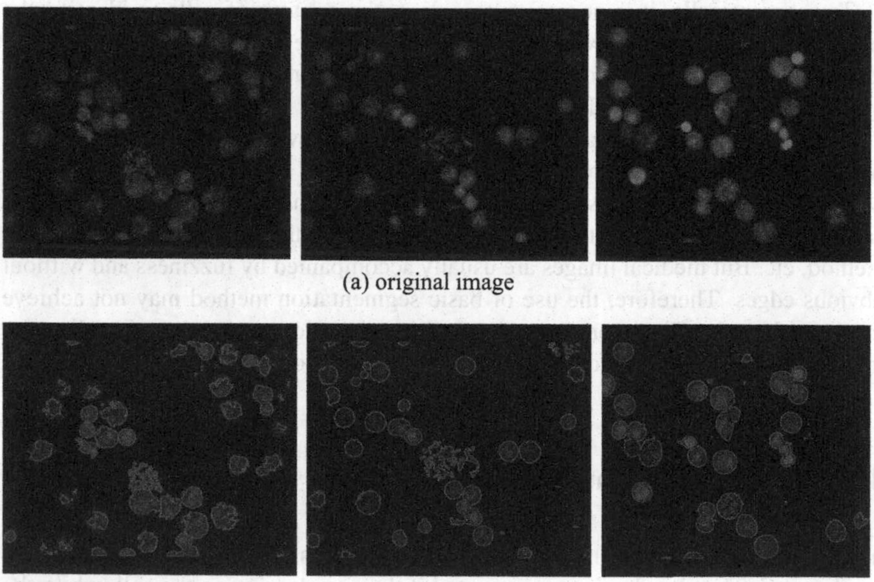

(a) original image

(b) segmentation result

Fig. 4 Results of the final segmentation

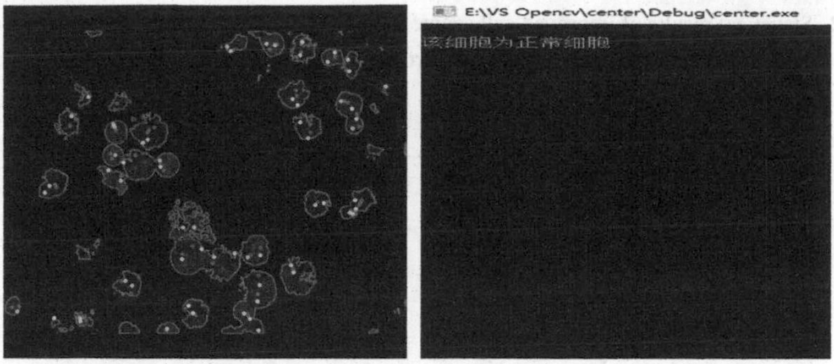

Fig. 5 Display of the final result

Table 1 Statistical results of cell segmentation

Image	Unrecognition rate	Error rate	Under-recognition rate (%)	Accuracy rate (%)
1	8%	2%	24	68
2	0	7%	22	78
3	0	6%	37	63
4	0	12%	24	76
5	0	3%	35	65
6	0	4%	29	71
7	0	11%	16	84
8	11%	0	18	71
9	11%	6%	28	61
10	11%	11%	7	82

5 Analysis of Results

In this study, we need to segment the cells and extract the signal highlights. In the process of implementation, we used blue channel images to try to segment cells in many ways. First, several traditional segmentation methods, such as threshold segmentation, edge detection, and watershed method, were used to deal with images. The result of cell segmentation is not ideal, and there is no appropriate segmentation in cell adhesions. The statistics of the segmentation results are shown in Table 1. It can be seen from the table that the accuracy of the segmentation is basically between 60–80%.

After the improvement, the segmentation results of the ten sets of data images were calculated, and the results are shown in Table 2. It can be found that the segmentation accuracy has been increased to over 80%, and the improved results have been greatly improved for each set of data.

Among them, unrecognition is the area of cells that are not identified. Identification error refers to the recognition of parts that do not need to be identified, that is, broken and invalid cell parts. Under-recognition means that the cell area is identified, but

Table 2 Statistical results of improved cell segmentation

Image	Unrecognition rate	Error rate	Under-recognition rate (%)	Accuracy rate (%)
1	8%	2%	11	81
2	0	7%	12	88
3	0	6%	19	81
4	0	12%	18	82
5	0	3%	15	85
6	0	4%	7	93
7	0	11%	3	97
8	11%	0	3	86
9	11%	6%	6	83
10	11%	11%	7	82

there are some inaccuracies, such as not identifying the complete cell area that only identifying a part of the cell or some non-divided adherent cells.

It can be seen from the statistical results that the accuracy rate after improvement has been significantly improved, but it is not ruled out that there are still a few parts of adherent cells that are not separated. This is a part that needs further improvement.

For the red channel and the green channel, the Otsu method was used to extract and identify the signal highlights. The results of the ten sets of data images are still statistically analyzed, and the final accuracy of each set of data can reach more than 95%, and most of them can even reach 97–99%. Therefore, using the Otsu method can be a good recognition of highlights.

6 Conclusion

This paper mainly introduces several methods of segmenting cell image. From the results of segmentation, four basic image segmentation methods including threshold iterative segmentation, edge detection, K-means clustering and watershed, can extract most of the cell regions and segment the cells initially. For the special adherent cells, the final method was to perform distance transformation on the image and then segmenting it to a watershed, then the watershed line was combined with the extracted cell contour to obtain the final segmentation results, which can better separate the adherent cells.

It is of great significance and feasibility to use the computer to perform automatic segmentation of cells instead of artificial and improve efficiency. However, it is necessary to improve the accuracy of the final result and ensure the accuracy, so that it can be promoted to the practical application. There is still a lot of work to be done in practice.

Acknowledgements This work was completed under the guidance of teacher Jiwei Liu and Jianfei Liu, and they provided me with many suggestions. I am very grateful to Jiwei Liu for his rigorous attitude towards academic and his strict requirements for us, which have benefited me greatly. At the same time, I would also like to express my heartfelt thanks to Jianfei Liu for his patience and guidance. This work is supported by the National Natural Science Foundation (NSF) of China (No. 61702001).

References

1. Research and Implementation of Key Technologies for Image Denoising and Contour Extraction of Breast Cancer Cells. Sichuan University (Sichuan, 2014)
2. H. Wei, X. Yu, W. Zhao et al., Threshold segmentation algorithm based on multiwavelets for medical image. Res. CT Theor. Appl. **18**(1), 8–15 (2009)
3. M. Ren, J. Yang, S. Han, A new method of histogram construction based on edge mode. Comput. Res. Dev. **38**(8), 972–976 (2001)
4. L. Zhang, F. Gu et al., Fuzzy Enhancement Edge Extraction Algorithm Based on transition Region, in *Image Graphics Technology and Application Progress: The Third Conference on Image Graphics Technology and Application Conference Proceedings*, vol. 11 (2008), pp. 165–168
5. W. Xiongwu, C. Yidong, B.R. Brooks, Y.A. Su, The local maximum clustering method and its application in microarray gene expression data analysis. EURASIP J. Appl. Sig. Process. **1**, 53–63 (2004)
6. S.K. Nath, K. Palaniappan, F. Bunyak, Cell segmentation using coupled level sets and graph-vertex coloring. Med. Image Comput. Assist. Interv. **9**(Pt 1), 101–108 (2006)
7. O. Daněk, P. Matula, C. OrtizdeSolórzano, A. MuñozBarrutia, M. Maška, Segmentation of Touching Cell Nuclei Using a Two-Stage Graph Cut Model, in *Scandinavian Conference on Image Analysis*, vol. 5575 (2009), pp. 410–419
8. S. Arslan, T. Ersahin, R. Cetin-Atalay, C. Gunduz-Demir, Attributed relational graphs for cell nucleus segmentation in fluorescence microscopy images. IEEE Trans. Med. Imag. **32**(6), 1121–1131 (2013)
9. M. Jinbao, *Research on Color Medical Image Segmentation* (Nanjing University of Science and Technology, Nanjing, 2012)
10. C. Xin, *Application of Wavelet Transform in Medical Image Processing* (Beijing University of Posts and Telecommunications, Beijing, 2008)
11. F. Zhongliang, Image threshold selection method—generalization of otsu method. Comput. Appl. **20**(5), 37–39 (2000)

Dynamic Scheduling Algorithm Considering Uncertain Service Time in Cloud Manufacturing Environment

Jian Wang, Lin Zhang, Longfei Zhou and Yuanjun Laili

Abstract Cloud manufacturing is an advanced production method in modern manufacturing. Cloud manufacturing can improve the production efficiency of a company by satisfying the diversified needs of customers by managing distributed resources in a centralized manner and rationally distributing and sharing resources according to production requirements. The production scheduling problem is the core issue of production in the cloud manufacturing environment. This paper first analyzes the characteristics of the cloud manufacturing mode of production and the new problems brought by these characteristics to the scheduling, and then analyzes the limitations of the traditional scheduling algorithm. Thirdly, based on the uncertainty of service time in cloud manufacturing environment, the paper proposes a new dynamic scheduling algorithm. Finally, the algorithm is verified by simulation experiments.

Keywords Cloud manufacturing · Uncertainty of service time
Dynamic scheduling algorithm

1 Introduction

Cloud Manufacturing is a new web-based, service-oriented and intelligent manufacturing model that integrates and develops emerging information technologies such as information-based manufacturing technologies, cloud computing, Internet

---

J. Wang · L. Zhang (✉) · L. Zhou · Y. Laili
School of Automation Science and Electrical Engineering,
Beihang University, Beijing 100191, China
e-mail: johnlin9999@163.com

J. Wang
e-mail: 1148322761@qq.com

L. Zhou
e-mail: zlfstudy@126.com

Y. Laili
e-mail: llyj0721@gmail.com

Y. Jia et al. (eds.), *Proceedings of 2018 Chinese Intelligent Systems Conference*, Lecture Notes in Electrical Engineering 529,
https://doi.org/10.1007/978-981-13-2291-4_30

of things, service computing, intelligent science, and high-performance computing [1]. Enterprises encapsulate manufacturing resources and manufacturing information into cloud platforms and become cloud services. Other companies make requests to the cloud platform according to their own needs. The cloud platform allocates manufacturing resources for manufacturing requests according to requests. Reasonable optimal scheduling of resources will directly affect the quick and accurate sharing of cloud resources, enabling enterprises to successfully complete tasks and improve the utilization of resources in different regions. Therefore, solving the problem of optimal configuration and scheduling of cloud manufacturing production resources is an important issue to be solved in the cloud manufacturing production model [2].

The scheduling of cloud manufacturing system has great differences in the scale of production, production methods and uncertainties.

First, the traditional mode of manufacturing production is the mode of large-scale production. Manufacturing systems, workshops or production lines can only produce specific products with fixed standards. However, the Manufacturing tasks in the cloud manufacturing system have the characteristics of diversification and individualization. There are a wide range of products and the demands are flexible and changeable [3, 4].

Secondly, in the cloud environment, the manufacturing resources distributed in different geographical locations are registered into the cloud manufacturing system through virtualization and servitization methods to form a large-scale production resource set. When designing scheduling algorithms for production scheduling, we must take into account the logistics time and logistics costs when manufacturing tasks are transferred between manufacturing resources in different geographic locations [5, 6].

Finally, the dynamics and uncertainty of the cloud manufacturing system are more prominent than the traditional manufacturing model, and interference events are more common [7]. Therefore, the process processing time is a common and uncertain factor in the production process.

2 Cloud Manufacturing Task Scheduling Model

2.1 Problem Description

The production scheduling problem in the cloud manufacturing environment can be summarized as follows:

The cloud manufacturing platform receives n manufacturing tasks $J = \{J_1, J_2, \ldots, J_n\}$ at a certain moment. The manufacturing task J_i has a specific production process sequence $S_i = \{S_{i,1}, S_{i,2}, \ldots, S_{i,k}\}$, and each $S_{i,k}$ in the production process sequence represents a certain type of production service. There are a large number of production resources in the production resource pool of the cloud manufacturing platform, which constitutes the total production resource set

$M = \{M_1, M_2, \ldots, M_m\}$. Each production resource can provide a series of different types of production services for manufacturing tasks. The set $C_i = \{S_1, S_2, \ldots, S_n\}$ is a set of production services that production resources M_i can provide. Production resources M_i can only provide one production service for one task at a time. The definition set $O(S_{i,k}) = \{M_{k,1}, M_{k,2}, \ldots, M_{k,n}\}$ is a set of production resources that can provide $S_{i,k}$ production services. Since a large number of production resources are integrated in the cloud manufacturing platform, there are many production resources that can provide certain production services in the cloud manufacturing environment, i.e., $|O(S_{i,k})| > 1$. When the task J_i wants to accept the production service $S_{i,k}$, it needs to concentrate on $O(S_{i,k})$ resources and perform resource selection according to the status of each production resource. The production scheduling in the cloud manufacturing environment requires the system to reasonably allocate production resources, production order, and production start time for manufacturing tasks according to the specific production process for each manufacturing task, thereby shortening the overall production time and increasing production efficiency [8, 9].

The specific features of cloud manufacturing production scheduling are as follows:

(1) Manufacturing Services

Cloud manufacturing services have strong flexibility and there is a relationship between services and services. In addition, dynamic changes in the scalc of cloud service pools need to be taken into account during cloud manufacturing task scheduling [10].

(2) Manufacturing resources

In the cloud manufacturing task scheduling process, not only the relevance between services must be taken into account, but also the logistics time and logistics cost required for tasks to be transferred between distributed manufacturing resources [11].

(3) Manufacturing tasks

In the cloud manufacturing service platform, the manufacturing task has the characteristics of individualization and diversification. This is also an important feature of cloud manufacturing that differs from the traditional manufacturing model. Therefore, cloud manufacturing task scheduling requires the use of smarter and more efficient methods [12].

(4) Fluctuation of service time

In the cloud manufacturing service platform, the time for each production resource to provide services for manufacturing tasks is not fixed, but it has certain fluctuations. Therefore, the scheduling method must consider the fluctuation of the service time, and cannot give the scheduling policy based on the fixed service time [1].

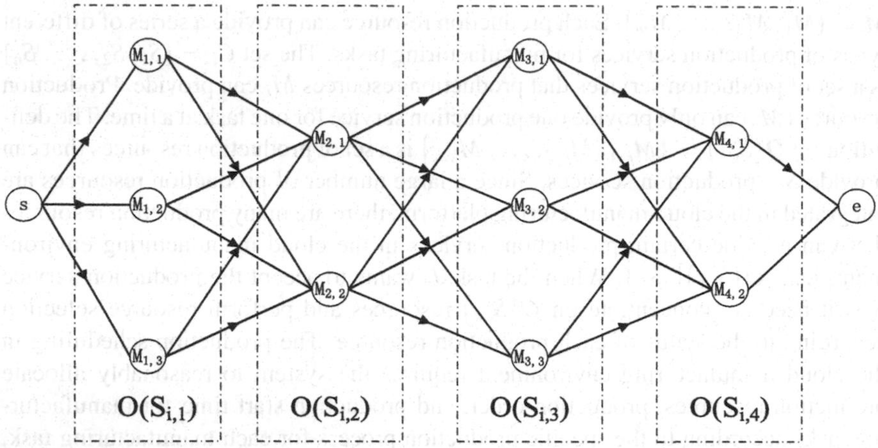

Fig. 1 Production scheduling model of manufacturing task J_i

2.2 Scheduling Model

This section will establish a scheduling mathematical model based on the characteristics of production and scheduling in the cloud manufacturing environment. As mentioned before, the manufacturing task J_i has the production process sequence $S_i = \{S_{i,1}, S_{i,2}, \ldots, S_{i,k}\}$, which needs to be strictly followed in order to accept the corresponding production service. For each production service $S_{i,k}$, there is a set of resources $O(S_{i,k}) = \{M_{k,1}, M_{k,2}, \ldots, M_{k,n}\}$ that can provide this service. Therefore, the production scheduling model under the cloud manufacturing environment is shown in Fig. 1.

The model shown in Fig. 1 is the production scheduling model of manufacturing task J_i. The model abstracts the production process of task J_i in the form of a network [13, 14]. The model will abstract each production resource into a node. Each node has specific coordinates and represents the specific location of the production resource. All nodes within each dotted box represent a set of production resources $O(S_{i,k})$ that can provide a certain production service. Set the node set of all nodes of this network model to V. According to the description of the production scheduling problem in the cloud manufacturing environment, the manufacturing task J_i has a strict sequence of production processes S_i. Therefore, in the scheduling model, the manufacturing tasks are transported between nodes and nodes of adjacent production resource sets. By connecting two nodes in the adjacent production resource sets, a series of directed edges is formed. Let the directed edge set consisting of all directed edges of the network model be E. These directed edges and nodes form multiple paths in the network model so that each path from the start node to the end node can be used as the production path for the manufacturing task J_i. Because the manufacturing tasks in the cloud manufacturing environment have multiple production paths, the cloud manufacturing platform needs to screen the production path and select the optimal production path. Each directed edge in the model has a weight, which represents the cost that the manufacturing task needs to pass through this directed edge. The scheduling algorithm will select the production path with the smallest weight for the manufacturing task J_i.

In this scheduling model, the weight of each directed edge is related to logistics time, queue time, and service time. Take the directional edge (a, b) as an example: The directed edge represents that the manufacturing task will be transferred from the production resource a to the production resource b and receive the production service of the production resource b. Since the production resources under the cloud manufacturing platform are distributed in different geographical locations, the logistics time of the directed edge (a, b) is the time consumed by the manufacturing task J_i from the production resource a to the production resource b. In order to simplify the model, suppose the logistics time is directly proportional to the distance between production resources. That is, the logistics time between production resources a and b is $t_l = k * \sqrt{(x_a - x_b)^2 + (y_a - y_b)^2}$.

The service time of the directed edge (a, b) is the time consumed by the production resource b to provide a specific production service for the manufacturing task. According to the specific characteristics of production under the cloud manufacturing environment, the time for production resources to provide services for manufacturing tasks is uncertain. At present, the mathematical description of service time uncertainty has three methods: random number, fuzzy number and interval number [15]. In actual production, it is difficult to collect a large amount of production sample data due to practical factors such as data collection methods, workload, safety, and randomness, and it is difficult to fit most service times into precise functions. Because the boundary or range of service time is more easily obtained, this article uses interval numbers to describe the fluctuation of service time. The modeling advantage of the interval number method is that it only needs to determine the lower boundary t_u and the upper boundary t_h of the service time, and does not need to consider the specific value or distribution law within the interval. This description method is closest to the actual production status [16, 17].

The uncertainty of service time is one of the reasons for the fluctuation of the weight of the directed edge. Since the service time is an interval number, this model adopts $t_s = \frac{t_u + t_h}{2}$ as the expectation of the service time of the manufacturing task receiving service in the production resource b. The actual service time will be known after the manufacturing task J_i has received the corresponding production service.

The queue time t_q of the directed edge (a, b) is the sum of the expected service times of all the manufacturing tasks waiting for service in the waiting queue of the production resource b. In the cloud manufacturing scheduling model, each production resource can only provide one production service for one manufacturing task at a time. When multiple manufacturing tasks are transported to the same production resource, the manufacturing tasks will be queued in the order of first in first out (FIFO), waiting to receive specific production services. The queue time t_q is the expected value of the waiting time of the manufacturing task J_i in the waiting queue of the production resource b. As with the actual service time, the actual waiting time needs to be known when the manufacturing task J_i completes waiting and is about to accept the production service.

This model defines the weight $V_{a,b} = t_l + t_s + t_q$ of the edge (a, b). This weight is the total time it takes for a manufacturing task to transfer from production resource a to production resource b to receive production services. The weight of the production path is the sum of the weights of all directed edges. When the system schedules production, it will face the choice of different production routes. Since the desire of the production path consumption time with the smallest weight is the smallest, the path is the optimal production path in the current state.

3 Dynamic Scheduling Algorithm Considering Time Uncertainty

3.1 Dynamic Scheduling

This topic proposes a new dynamic scheduling algorithm based on the scheduling rules method and the re-scheduling method. The algorithm can effectively deal with the changes brought about by the uncertainty of the service time. At the same time, the algorithm can reduce the time complexity of the scheduling calculation under the premise of ensuring the accuracy of the scheduling, so that the algorithm can handle large-scale production dynamic scheduling problems in the cloud manufacturing environment.

3.2 Dynamic Scheduling Algorithm

The scheduling algorithm is divided into two parts. The first part is that when the manufacturing task J_i is released, an optimal production path L_{se} from the source node s to the end node e is selected for the manufacturing task based on the current production environment, and the manufacturing task is sent to the target production resource to receive the production service $S_{i,1}$. The second part is to update the scheduling model and model parameters whenever the manufacturing task J_i completes the production service $S_{i,k}$. Based on the updated production scheduling environment, the production scheduling is again performed, the optimal production route is reselected. The manufacturing task is sent to the target production resource, and the production service $S_{i,k+1}$ is accepted until the manufacturing task is completed according to the production sequence S_i. The two sections will be separately elaborated below.

When the manufacturing task J_i is released, the system first searches for the production resource set of a specific production service in the cloud manufacturing platform according to J_i's production sequence, and then arranges them according to the production order to generate a J_i production scheduling network model as shown in Fig. 1. By reading and processing the status information of the production resources, the weight of each edge in the network is calculated.

Next, according to the scheduling network model, the algorithm will find the optimal production path under the current conditions. Since the production path is all paths from the start node to the end node, this problem can be converted into a single source shortest path solution problem.

The classic algorithms for single source shortest path problem are Bellman-Ford algorithm and Dijkstra algorithm [18, 19]. If the number of nodes in the scheduling network model is $|V|$ and the number of directed edges is $|E|$, it can be proved that the total running time of the Bellman-Ford algorithm is $O(|V||E|)$, and the total running time of the Dijkstra algorithm is $O(|V|^2)$ [20]. Both algorithms have high time complexity and are not suitable for large-scale production scheduling in a cloud manufacturing environment.

The scheduling model in the cloud manufacturing environment has two features. One of the characteristics is that the network is a directed acyclic network. The second feature is that the nodes of the network are arranged strictly in the order of topological expansion. The directional edges in the network are all forward edges. Since the manufacturing task has a strict sequence of production processes, in the production process sequence, if the production services S_{i,k_1} and S_{i,k_2} are adjacent and S_{i,k_1} are before S_{i,k_2}, for any node v_1 which is formed by any production resource of production resource set $O(S_{i,k_1})$ and any node v_2 which is formed by any production resource of production resource set $O(S_{i,k_2})$, there must be a direction edge (v_1, v_2), and there must be no directional edge (v_2, v_1).

Therefore, the scheduling model under the cloud manufacturing environment can be adjusted to facilitate analysis. The adjusted model is shown in Fig. 2.

According to the new production scheduling model, the Bellman-Ford algorithm was modified to reduce the time complexity of the algorithm. The revised algorithm flow is as follows:

(1) Initialize all node attributes in node set V. For source node s, there are $s.p = NULL$ and $s.d = 0$. For all nodes except the source node $v \in \complement_V\{s\}$, there are $v.p = NULL$ and $v.d = \infty$. Go to process (2).
(2) According to the topology of the nodes, find the first node v that is not processed. If all the nodes in node set V are processed, go to flow (4).

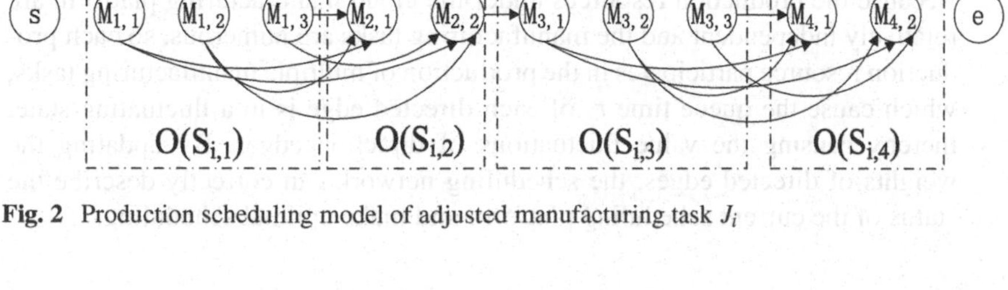

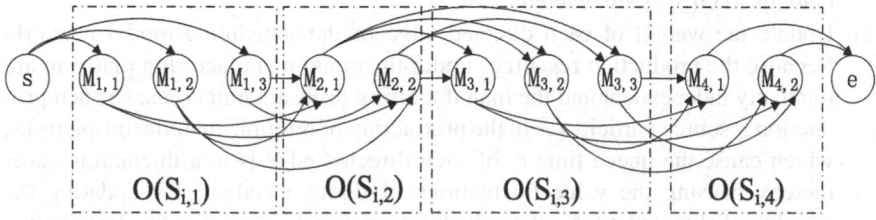

Fig. 2 Production scheduling model of adjusted manufacturing task J_i

(3) Relax all edges starting from node v. After the processing is completed, the node v is marked as processed and the flow returns to step (2).
(4) Starting from the ending node e, searching for the node's predecessor node $e.p$ and the node $e.p$'s predecessor node $(e.p).p$. Iterate to the source node s in this way. The nodes passing through the iteration process and the directed edges connecting these nodes form the optimal production path L_{se} from the source node s to the end node e in the current production scheduling model. Go to process (5).
(5) In L_{se}, look for the node a whose precursor node is source node s, that is, $a.p = \text{s}$. The dispatching system will issue a Manufacturing task J_i to the production resource $M_{1,a}$ represented by node a, and J_i will enter the waiting queue of $M_{1,a}$ for production service $S_{i,1}$. The algorithm flow ends [21, 20].

According to the analysis, the above algorithm is the first part of the scheduling algorithm, and the time complexity is O(|V| + |E|). This complexity is linear and the operation is simple and easy. It is suitable for dealing with the production dynamic scheduling of large-scale scheduling models. The first part of the scheduling algorithm occurs when the new task arrives.

The second part of the scheduling algorithm occurs when the manufacturing task J_i completes the current production service $S_{i,k}(k = 1, 2, \ldots, t)$. At this time, due to the uncertainty of the service time in the production process and the change of the state of the production resources, the original optimal production path L_{se} is no longer applicable to the current scheduling environment, and the production needs to be re-dispatched. The specific algorithm flow of rescheduling is as follows:

(1) Assume that the node formed in the scheduling model by the production resource which has manufacturing task J_i currently is v, and the set of nodes formed in the scheduling model by all production resources of the production resource set $\mathrm{O}(S_{i,k})$ is V_k. Since the manufacturing task J_i has already accepted the production service $S_{i,k}$ at this time, the scheduling system removes all the nodes $\text{v}' \in s \cup \complement_{V_k}\{v\}$ from the node set V to form a new scheduling model network. In the new scheduling model network, node v will become the source node s in the new scheduling model network. Taking the model shown in Fig. 2 as an example, if the manufacturing task J_i accepts the production service $S_{i,1}$ on the production resource $M_{1,2}$, the updated scheduling model network is shown in Fig. 3. At this point, the node of the production resource $M_{1,2}$ in the model becomes the source node s. The original source node and other nodes which belongs to $\mathrm{O}(S_{i,1})$ are deleted.
(2) Update the weight of each directed edge of the scheduling model network. Because the production resources under the cloud manufacturing platform are relatively independent and the manufacturing tasks are numerous, so each production resource participates in the production of multiple manufacturing tasks, which cause the queue time t_q of each directed edge is in a fluctuating state, thereby causing the value fluctuations of direction edges. By updating the weights of directed edges, the scheduling network can correctly describe the status of the current scheduling platform and make optimal scheduling.

Table 1 Manufacturing task type information

Manufacturing task type name	Production process sequence			
	1	2	3	4
TaskType1	Service1	Service2	Service3	Service1
TaskType2	Service4	Service2	Service1	Service3
TaskType3	Service3	Service1	Service2	

(3) Repeat the process (1)–(4) of the first part of the scheduling algorithm to find the optimal production path L_{se} from the source node s to the end node e in the current production scheduling model. The optimal production path at this time is a production route searched by the scheduling algorithm based on the current scheduling platform status.

4 Simulation

Based on the above algorithm, this paper sets up a specific scheduling scenario, performs simulation experiments on the scheduling scenario, and verifies the effectiveness of the scheduling. First, the scheduling of production of the above algorithm under the scene determined by the service time is checked.

In this simulation scheduling scenario, the scheduling platform can provide production services for three types of manufacturing tasks. Each type of manufacturing task has its own sequence of production processes. The specific information of every manufacturing task's type is shown in Table 1.

The production resource pool of the dispatching platform has a large number of production resources and constitutes the total production resource set M of the platform. The basic information, the available production services, and the corresponding service time of each production resource are shown in Table 2.

In the current scheduling scenario, there are six manufacturing tasks that need to be scheduled. Specific information on production tasks is shown in Table 3.

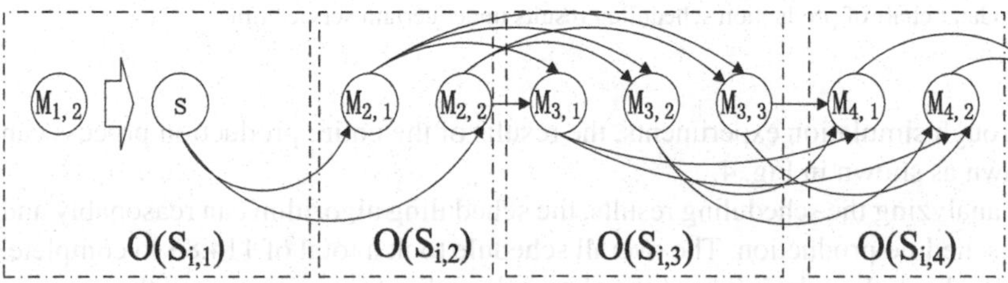

Fig. 3 Production scheduling model of manufacturing task after updating

Table 2 Production resource information under certain service time

Production resource name	Production resource location	Available production service types	Service time
Res1	(2,10)	Service1	20
		Service2	14
Res2	(4,18)	Service3	16
Res3	(10,6)	Service1	12
		Service4	22
Res4	(12,14)	Service4	22
		Service2	10

Table 3 Manufacturing task information

Production task name	Manufacturing task type name	Manufacturing task number
Task1	TaskType1	1
Task2	TaskType1	2
Task3	TaskType2	3
Task4	TaskType2	4
Task5	TaskType3	5
Task6	TaskType3	6

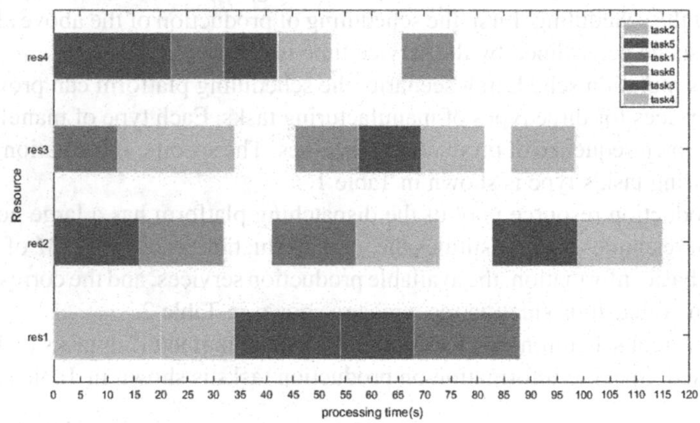

Fig. 4 Gantt chart of production scheduling results under certain service time

Through simulation experiments, the results of the entire production process can be drawn as shown in Fig. 4.

By analyzing the scheduling results, the scheduling algorithm can reasonably and safely schedule production. The overall schedule took a total of 114.83 to complete.

Next, check the scheduling of the above algorithm for production in the scenario where the service time is indefinite. Change the service time in Table 2 to the interval

Table 4 Production resource information under uncertain service time

Production resource name	Production resource location	Available production service types	Service time
Res1	(2,10)	Service1	[11,29]
		Service2	[5,23]
Res2	(4,18)	Service3	[7,25]
Res3	(10,6)	Service1	[3,21]
		Service4	[13,31]
Res4	(12,14)	Service4	[13,31]
		Service2	[1,19]

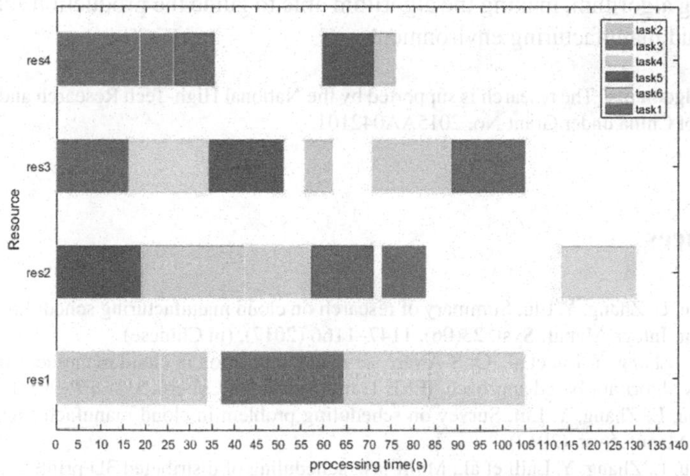

Fig. 5 Gantt chart of production scheduling results under uncertain service time

number. Let the current service time be t, then make the interval number lower boundary $t_u = t - 9$ and the upper boundary $t_h = t + 9$, making the average of the service time t. The reconfigured production resource information is shown in Table 4.

Through simulation experiments, the scheduling results for the entire production process can be shown in Fig. 5.

Through the analysis and comparison of the two scheduling results, it can be seen that the order of the manufacturing tasks receiving production services in production resources has changed, and the time consumed by each production resource to provide corresponding production services also fluctuates. This is because the uncertainty of service time has affected the entire scheduling environment. The algorithm can correctly handle the impact of uncertain service time, so that production can still be carried out smoothly. The overall time spent is 130.36 and the scheduling time is short.

5 Conclusion

This paper proposes an algorithm for dynamic production scheduling in an environment with uncertain service time. The theoretical analysis and simulation experiments show that the method is reasonable. The cloud manufacturing production environment is a typical production environment with uncertain service time. At the same time, the cloud manufacturing production environment has a large scale and the production resources are dispersed, which brings new difficulties for production scheduling. The algorithm proposed in this paper integrates the characteristics of service time uncertainty and other characteristics in the cloud manufacturing environment into the scheduling algorithm system, and reduces the time complexity of the scheduling algorithm, making the algorithm able to guide the production scheduling in the cloud manufacturing environment.

Acknowledgements The research is supported by the National High-Tech Research and Development Plan of China under Grant No. 2015AA042101.

References

1. L. Zhou, L. Zhang, Y. Liu, Summary of research on cloud manufacturing scheduling problem. Comput. Integr. Manuf. Syst. **23**(06), 1147–1166 (2017). (in Chinese)
2. F. Li, L. Zhang, Y. Liu et al., QoS-Aware service composition in cloud manufacturing: a gale-shapley algorithm-based approach. IEEE Trans. Syst. Man Cybern. Syst. **PP**(99), 1–12 (2018)
3. L. Zhou, L. Zhang, Y. Liu, Survey on scheduling problem in cloud manufacturing. Comput. Integr. Manuf. Syst. (2017)
4. L. Zhou, L. Zhang, Y. Laili et al., Multi-task scheduling of distributed 3D printing services in cloud manufacturing. Int. J. Adv. Manuf. Technol. **2**, 1–15 (2018)
5. L. Zhou, L. Zhang, B.R. Sarker et al., An event-triggered dynamic scheduling method for randomly arriving tasks in cloud manufacturing. Int. J. Comput. Integr. Manuf. **31**(3), 1–16 (2017)
6. L. Zhou, L. Zhang, Dynamic task scheduling method based on simulation in cloud manufacturing, in *Theory, Methodology, Tools and Applications for Modeling and Simulation of Complex Systems* (Springer Singapore, 2016), pp. 20–24
7. Y. Liu, X. Xu, L. Zhang et al., An extensible model for multi-task service composition and scheduling in a cloud manufacturing system. J. Comput. Inf. Sci. Eng. **16**(4) (2016)
8. J. Ehm, T. Hildebrandt, M. Freitag et al., Potential of data-driven simulation-based optimization for adaptive scheduling and control of dynamic manufacturing systems, in *Winter Simulation Conference* (IEEE, 2017), pp. 2820–2831
9. N. Keller, X. Hu, Data driven simulation modeling for mobile agent-based systems, in *Theory of Modeling and Simulation* (IEEE, 2017), p. 24
10. Y. Liu, X. Xu, L. Zhang et al., Workload-based multi-task scheduling in cloud manufacturing. Rob. Comput. Integr. Manuf. **45**(C), 3–20 (2016)
11. C.C. Huang, C.L. Huang, Development of cloud computing based scheduling system using optimized layout method for manufacturing quality, in *International Symposium on Computer, Consumer and Control* (IEEE, 2012), pp. 444–447
12. L. Zhou, L. Zhang, C. Zhao et al., Diverse task scheduling for individualized requirements in cloud manufacturing. Enterp. Inf. Syst. **1**, 1–19 (2017)

13. Y. Cheng, D. Zhao, F. Tao et al., Complex networks based manufacturing service and task management in cloud environment, in *Industrial Electronics and Applications* (IEEE, 2015), pp. 242–247
14. Y. Cheng, F. Tao, D Zhao et al., Modeling of manufacturing service supply–demand matching hypernetwork in service-oriented manufacturing systems. Rob. Comput. Integr. Manuf. **45** (2016)
15. H. Yang, Z. Wang, Y. Lv, Z. Xi, H. Wang, Interval number solution method for job shop scheduling problem under uncertain process processing time. Comput. Integr. Manuf. Syst. **23**(06), 1147–1166 (2017). (in Chinese)
16. Y. Yadekar, E. Shehab, J. Mehnen, Uncertainties in Cloud Manufacturing (2014)
17. H. Guo, L. Zhang, F. Tao, A framework for correlation relationship mining of cloud service in cloud manufacturing system. Adv. Mater. Res. **314–316**, 2259–2262 (2011)
18. E.W. Dijkstra, A note on two problems in connexion with graphs. Numer. Math. **1**(1), 269–271 (1959)
19. R. Bellman, On a routing problem. Q. Appl. Math. **16**(1), 87–90 (1958)
20. T.T. Cormen, C.E. Leiserson, R.L. Rivest, Introduction to Algorithms (Higher Education Press, 2002)
21. E.F. Moore, The shortest path through a maze, in *Proceeding of the International Symposium on the Theory of Switching* (1959), pp. 285–292

Static and Dynamic Performance Modeling and Simulation of a Microturbine

Yahao Ren, Jiqiang Wang, Qiangang Zheng, Ying Liu and Zhongzhi Hu

Abstract The paper deals with the development of a thermodynamic simulation model of T100P microturbine from generalized maps. The microturbine unit consists of a compressor, a combustion chamber, a turbine, a recuperator and a diffuser, noted that the compressor and the turbine shared a single shaft with a high—speed generator. Finally, the steady-state and dynamic model have been matched with the experimental data and a good result is obtained. The model is to be used in the research on the development of control systems, verification on dynamic performance and control system hardware verification.

Keywords Simulation · Modelling · Microturbine · Performance analysis

1 Introduction

The overall requirements of Chinese power development are a rational energy consumption structure, efficient energy conversion and use, and lower pollution emissions. Therefore, more environmental-friendly and higher efficient power generation technologies are gradually gaining attention. The knowledge and experience from the mobile microturbines led to the development of a new technology, stationary combined heat and power (CHP) plants, which are widely used in distributed generation [1, 2], can meet the needs of special occasions, but also meet the users with high requirements for safety and stability of power supply. In this paper, the T100 microturbine has taken into analysis for its long history and wide range of use.

From the development of industries in various countries, we have gradually realized that replacing some real gas turbine tests with digital simulation in use of computer technology can effectively reduce the development cost and improve the development efficiency. To design the gas turbine control system and verify the reliability

Y. Ren · J. Wang · Q. Zheng · Y. Liu · Z. Hu (✉)
Jiangsu Province Key Laboratory of Aerospace Power System,
Nanjing University of Aeronautics and Astronautics, Nanjing 210016, China
e-mail: huzhongzhi@nuaa.edu.com

Y. Jia et al. (eds.), *Proceedings of 2018 Chinese Intelligent Systems Conference*, Lecture Notes in Electrical Engineering 529,
https://doi.org/10.1007/978-981-13-2291-4_31

of the hardware, process knowledge of the object is needed. This paper adopts an easy method to develop the model, which can relatively accurately reflect the dynamic process. Many researches have been carried out on the establishment of dynamic models at home and abroad, and considerable results have been achieved [3–5]. In this paper, a component-level modeling methods has been introduced and the result of simulation has been matched with empirical data, which proves that the dynamic performance of the model are consistent with reality.

2 Model Architecture

The microturbine T100 from Turbec AB is a combined heat and power (CHP) generation system, which can produce 100 kW of electric power [6]. T100 microturbine was divided into nine thermodynamic part, as showing in Fig. 1.

Referring to the thermodynamic components of the T100, the architecture of the model has been divided into compressor, burner, turbine, recuperator, exhaust diffuser and a power generator, with a single shaft connecting the turbine, compressor and power generator, as showing in Fig. 2.

The station of the model was set as follows, station 0 is the external environment, station 1 is the outlet of inlet duct, station 2 is the outlet of compressor, station 3

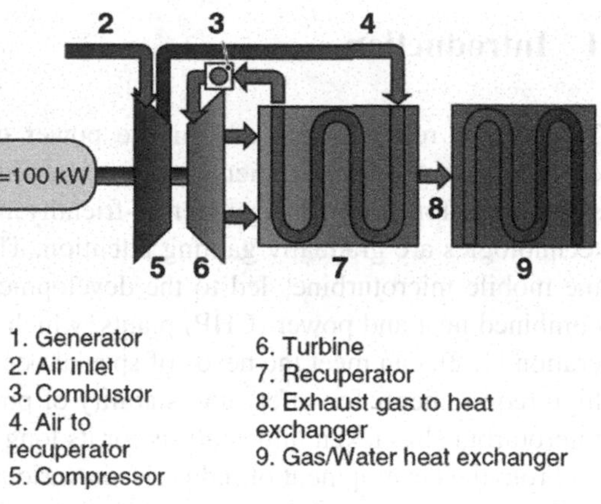

Fig. 1 Scheme of thermodynamic parts of the T100 microturbine

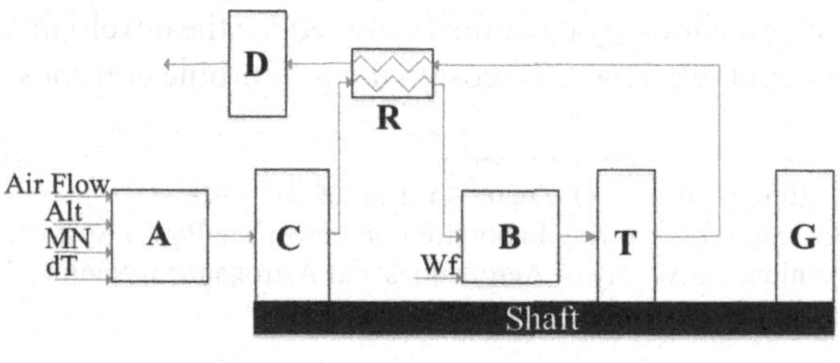

Fig. 2 Architecture of T100 microturbine: Ambient (A), Compressor (C), Turbine (T), Burner (B), Recuperator (R), Generator (G), Diffuser (D)

is the outlet of burner, station 4 is the outlet of turbine and station 5 is the outlet of diffuser. What's more, T_{2R} was defined as gas outlet temperature of recuperator which will get into burner.

3 Microturbine Component Level Modeling

In this paper, microturbine model consists of total system thermodynamics and the component level balance equations. System thermodynamics modeling is based on the Brayton cycle, and the constant pressure specific heat and gas adiabatic index are calculated by temperature and *FAR* (Fuel to air ratio). Enthalpy and entropy are calculated by pressure, temperature and *FAR*. The temperature is calculated by enthalpy and *FAR* or entropy and *FAR*. Component models are built according to the performance maps [7]. Then the operating points on maps will be worked out by using a Newton-Raphson solver.

The emphasis has been on the functionality and performance of the static and dynamic system model. The model is built in Simulink and components from the common turbomachinery block of TMATS.

A. Ambient and inlet modeling

In this block, the properties of mass flow can be calculated by the known values as generally mass flow, altitude, *MN*, and *dT*. Then the enthalpy, entropy, total pressure (P_{t2}) and total temperature (T_{t2}) of the mass flow will go to the compressor for the next calculated.

B. Compressor modeling

The modelling of gas turbine will be very difficult due to the lack of basic design data and characteristics of the compressor. In this paper, generalized maps were adopted when calculating the performance of compressor [8] and the map data came from the GasTurb 11. The empirical performance map showed as Fig. 3. The parameters present pressure ration *PR*, corrected shaft speed N_c and corrected mass flow W_c, in the compressor map are defined as follows:

$$PR = P_{t2}/P_{t1} \tag{1}$$

$$N_c = N/\sqrt{T_{t1d}} \tag{2}$$

$$W_c = W\sqrt{T_{t1d}}/P_{t1d} \tag{3}$$

where, T_{t1}—compressor air inlet total temperature *K*; T_{t1d} —compressor air inlet total temperature in design point, *K*; N_L—rotational speed of the shaft, rpm; the index "*d*" refers to design point.

Combined with the initial guess value, the gas mass flow through the compressor W_{c2}, the compressor pressure ratio *PR* and the compressor efficiency $_C$ can be calculated from the map, and the thermodynamic parameters of the compressor outlet gas are further calculated by the data readout from the map. After completing the above thermal calculation, the thermal parameters of the compressor outlet cross section can be obtained.

C. Recuperator modeling

The regenerator uses waste heat from the exhaust of the gas turbine to heat the compressed air to improve cycle efficiency. The recuperator of the T100 is a counter flow heat exchanger, even though there is some crossflow in the beginning and in the end. This paper used NTU method to do the steady state simulation [9]. The NTU method has good applicability in the case of unknown export temperature. When the recuperator in a dynamic model, use the blow method to do the simulation.

The dynamic mathematical model of the recuperator is reduced from a set of partial differential equations to a set of ordinary differential equations (Fig. 4).

Mass balance equation:

$$W_{in} = W_{out} \tag{4}$$

where, W_{in} and W_{out}—air mass flow rate, kg/s.

Gas energy balance equation:

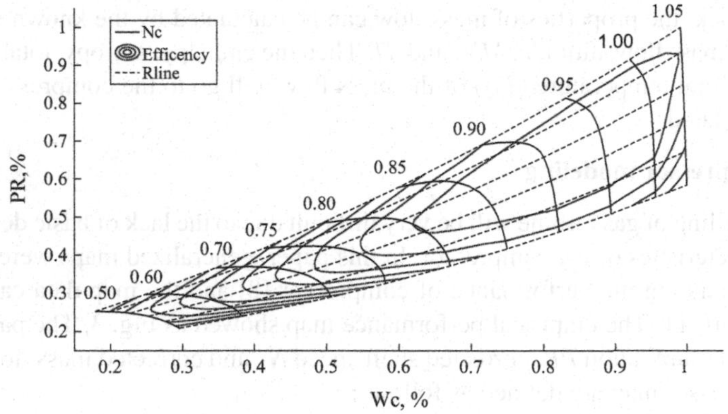

Fig. 3 Comperssor performance map

Fig. 4 Schematic diagram of recuperator

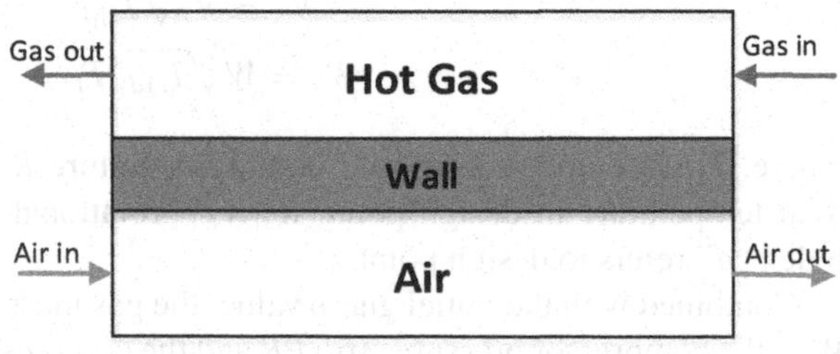

$$W_{out}h_{out} - W_{in}h_{in} = \alpha A_w\left(T_w - \bar{T}\right) \tag{5}$$

where, A_w—Heat exchange area, m^2; h_{in} and h_{out}—inlet and outlet gas specific enthalpy, kJ/kg; p—pressure, Pa; $\bar{T}$—gas average temperature, K; T_w—recuperator wall temperature, K; α—gas heat transfer coefficient, kJ/(m^2 s K).

The energy balance equation of the recuperator:

$$M_W C_W dT_w/dt = \alpha A_w\left(T_w - \bar{T}_g\right) - \alpha A_w\left(T_w - \bar{T}_a\right) \tag{6}$$

where, M_w—surface quality of recuperator, kg; C_w—recuperator metal wall surface specific heat capacity, kJ/(kg K). And dT_w/d_t is equal to zero during steady-state operation and it will cause response delay during dynamic operation.

D. **Burner modeling**

The outlet gas mass flow and the total temperature balance of the combustion chamber are as follows:

$$W_3 = W_2 + W_f \tag{7}$$

$$P_{t3} = \sigma_B P_{t2} \tag{8}$$

where, W_f—fuel mass flow, kg/s; σ_B—burner pressure loss.

Combustion chamber efficiency is constant when the gas turbine is in working condition and the enthalpy of the outlet of the combustion chamber is:

$$\mathrm{H}_3 = \left(W_f \cdot LHV \cdot \eta_B + W_2 H_2\right)/W_2 \tag{9}$$

where, η_B—the efficiency of the combustion chamber; H_2—inlet gas enthalpy of combustion chamber. *LHV*—fuel lower heating value, *BTU/lb*.

Then the *FAR* and the enthalpy value are used to calculate the outlet temperature of the combustion chamber.

E. **Turbine modeling**

Similar to the compressor, the characteristics of the turbine outlet mass flow can be calculated by a performance map, which correlates the relationships between pressure ratio (*PR*), mass flow (W_c), corrected shaft speed (N_c) and efficiency. W_c and efficiency can be calculated by *PR* and N_c, and the calculation of characteristic is same as the compressor model. Cooling flow is not considered and maps are shown in Fig. 5.

F. **Shaft modeling**

Shaft dynamics has no need to be taken into account when developing a steady-state model. In this paper, inertia terms is added into torque balance equations:

$$\frac{dN}{dt} = \frac{Trq_{compressor} + Trq_{turbine} + Trq_{generator}}{2 \cdot \pi \cdot I} \tag{10}$$

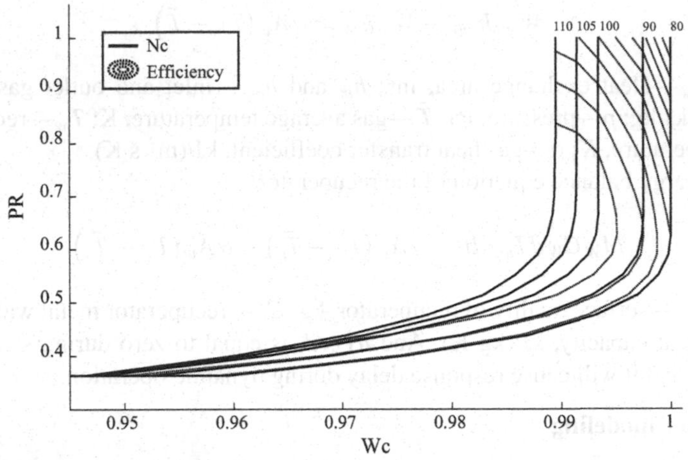

Fig. 5 Turbine performance map

$$\mathrm{N} = \int \frac{dN}{dt} \tag{11}$$

where, *Trq*—component torque, N m; *I*—shaft inertia, kg m^2. Similar to temperature inertia, *dN/dt* is equal to zero during steady-state operation and it will cause response delay during dynamic operation.

4 Simulation and Verification

The performance model of T100 microturbine was created by above modules. When we take design point into account, pressure ratio, efficiencies, speeds, and mass flows are adequate for a specified operating condition analyzing. To build the design point model, known values and assumptions are used to define performance parameters. Adjustments should be made to assumptions to meet specified design point, and obtain Steady-state point by solving co-operating equations. Steady-state co-operating equations are as follows:

(1) Compressor characteristics interpolated air mass flow and imported conversion air mass flow deviation equation:

$$\left(Wc_2 - Wc_{2map}\right)/Wc_2 = \varepsilon_1 \tag{12}$$

(2) Turbine characteristics of air mass flow and input air mass flow deviation equation:

Table 1 Disgn data

Performances	Design point data	Model data	Error
Electric power (kW)	100 ± 3	100.8	0.8%
Electric efficiency (%)	30 ± 1	30.3	1%
Compressor pressure ration	4.5	4.5	–
Shaft speed (rpm)	70,000	70000 rpm	–
Exhaust gas mass flow (kg/s)	0.8	0.795	0.63%
Turbine inlet temperature (K)	1223	1225.1	0.17%
Turbine outlet temperature (K)	918	922.6	0.50%
Exhaust gas temperature (K)	543	547.4	0.80%

$$\left(Wc_4 - Wc_{4map}\right)/Wc_4 = \varepsilon_2 \tag{13}$$

(3) Input and output air mass flow deviation equation of the exhaust diffuser:

$$(Wc_{5in} - Wc_{5out})/Wc_{5in} = \varepsilon_3 \tag{14}$$

(4) Shaft power deviation equation:

$$N_T - N_C - N_P - \left(\frac{\pi}{30}\right)^2 Jn \cdot \frac{dn}{dt} = \varepsilon_4 \tag{15}$$

(5) Recuperator temperature balance equation:

$$(T_4 - T_{4IC})/T_4 = \varepsilon_5 \tag{16}$$

where, the index "map" refers to the value that calculate from the generalized maps; the index "IC" refers to initial guess value;

When the engine is at a steady-state point, solve the above equations by using the Newton-Raphson iteration method, and make sure these deviation of residual $|\varepsilon_i| \leq 10^{-5}$ (i = 1, 2, 3, 4, 5), which we consider a convergence of the model.

In this paper, a method of using a generic map to complete unknown portions is used and map scaling was performed on the generic map to resize it for the operation point data [10]. And data taken from literature are showed in Table 1. Steady-state model from design point data has been verified with good results.

B. **Dynamic Verification**

In this paper, the dynamic model is modified by using the Genetic algorithm to optimize map characteristic, and the experimental data is used to match the operation of the model at part load [11].

Table 2 Result of the verification

	Pressure	Temperature (K)	Air mass flow (kg/s)	Shaft speed (rpm)
Experimental	4.2	862.15	0.803	69,912.5
Numerical	4.215	859.4	0.8056	70,386
Err.	0.36%	−0.32%	0.32%	0.7%
Experimental	3.484	862.15	0.6710	63,350
Numerical	3.48	866.11	0.6775	62,843
Err.	−0.12%	0.46%	0.97%	−0.8%
Experimental	2.973	876.15	0.5750	58,362
Numerical	2.88	871	0.5684	54,382.6
Err.	−3.12%	−0.6%	−1.15%	−6.82%

In this paper, the experimental data of 100, 70 and 50 kW operating point is compared whit the simulation results, and the fuel mass flow is used as input. The fuel mass flow corresponding to each working condition is 0.00707, 0.005086 and 0.003913 kg/s. The result of the data matching is showed in Table 2.

From the results of the static verification above, we see that the largest error is in the exhaust gas temperature. The probable cause of the error is the efficiency of the recuperator. Dynamic verification in the part load experiments has been done and have a good match with the measured data. When we emphasize the complete microturbine system, the errors of 2–4% can be acceptable. When operation point at 50% load, the shaft speed has a relative error of 6.82%, and the errors are probably caused by scaling curves of the generic compressor map and the generic turbine map. What's more an constant value of $\varepsilon_{rec} = (T_{2R} - T_2)/(T_4 - T_2)$ of 0.865 is used at all operating points, but the recuperator efficiency changes with variable shaft speed [12, 13].

5 Conclusion and Possible Application

A simulation method has been presented to analyze microturbine static and dynamic performance. A model of Turbec T100 recuperated MGT was developed and validate the model with experimental data. Steady-state model has high simulation accuracy. Although the speed error of the dynamic model is relatively large, the basic parameters match with the experimental data. A numerical simulation is never perfect. We should make a trade-off between time spend on a component's precision in the model and the control system design for the model. A functional model has been developed and further works such as control system design can be done. What's more, with the hardware-in-the-loop plant, it can be used to initiate hardware failures and develop fault tolerant control system.

References

1. H.B. Puttgen, P.R. Macgregor, F.C. Lambert, Distributed generation: semantic hype or the dawn of a new era. IEEE Power Energ. Mag. **99**(1), 22–29 (2003)
2. H.B. Puttgen, D.R. Volzka, M.I. Olken, Restructuring and reregulation of the US electric utility industry. IEEE Power Eng. Rev. **21**(2), 8–10 (2001)
3. W.I. Rowen, Simplified mathematical representations of heavy-duty gas turbines. J. Eng. Power **105**(4), 865–869 (1983)
4. P.D. Fairchild, *Experimental and Theoretical Study of Microturbine-Based BCHP System* (Oak Ridge National Lab, TN (US), 2001)
5. A. Kumar, K.S. Sandhu, S.P. Jain et al., Modeling and control of micro-turbine based distributed generation system. Int. J. Circuits Syst. Signal Process. **3**(2), 65–72 (2009)
6. A. Malmquist, Analysis of a Gas Turbine Driven Hybrid Drive System for Heavy Vehicles (Institutionen för elkraftteknik, 1999)
7. J.W. Chapman, T.M. Lavelle, J.S. Litt, Practical techniques for modeling gas turbine engine performance, in *52nd AIAA/SAE/ASEE Joint Propulsion Conference* (2016), p. 4527
8. M. Plis, H. Rusinowski, Mathematical modeling of an axial compressor in a gas turbine cycle. J. Power Technol. **96**(3), 194–199 (2016)
9. S. Haugwitz, Modelling of microturbine systems, in *European Control Conference (ECC), 2003* (IEEE, 2003), pp. 1234–1239
10. Turbec T100 CHP system Technical description version 5.0, Turbec AB, 2009
11. R. Calabria, F. Chiariello, P. Massoli et al., CFD Analysis of Turbec T100 combustor at part load by varying fuels, in *ASME Turbo Expo 2015: Turbine Technical Conference and Exposition* (American Society of Mechanical Engineers, 2015), pp. V008T23A020–V008T23A020
12. M. Henke, N. Klempp, M. Hohloch et al., Validation of a T100 micro gas turbine steady-state simulation tool, in *ASME Turbo Expo 2015: Turbine Technical Conference and Exposition* (American Society of Mechanical Engineers, 2015), pp. V003T06A003–V003T06A003
13. A. di Gaeta, F. Reale, F. Chiariello et al., A dynamic model of a 100 kW micro gas turbine fuelled with natural gas and hydrogen blends and its application in a hybrid energy grid. Energy **129**, 299–320 (2017)

An Image Dehazing Algorithm Based on Binocular Disparity

Zhong-Yi Hu and Mian-Lu Zou

Abstract This paper discusses the principle of the common dehazing algorithms, points out their shortcomings, and puts forward a novel image defogging algorithm based on binocular disparity. This method is consisted of two identical cameras. (1) We can get two images on a flat, one is disparity image based on SAD feature matching, the other is depth image corresponding to foggy image by using the inverse principle between disparity and depth. (2) Luminance component and depth images are respectively regarded as guided image with guided filter and the input images, to estimate atmospheric optical transmission diagram. Therefore, (3) we introduce atmospheric optical transmission corresponding to sky corrected by brightness difference mechanism. The experimental results show that the proposed algorithm can achieve better recovery results and eliminate the halo phenomenon caused by defogging single image.

Keywords Binocular disparity · Dehazing · Guided filter

1 Introduction

Discovered by scientists, aerosols is a major cause of formation of foggy weather, whose influence is wide and far beyond the scope of wind and air, making more and more areas affected [1]. In recent successive year, with the occurrence of degeneration of air quality and frequent fog weather, the imaging system is affected by atmospheric light scattering effect generated by haze and smog, leading to the missing of contrast and color reduction of decline quality images. Fortunately dehazing has made important progress in recent years, methods include using two or more images of different polarization directions to dehaze, or successfully depending on stronger prior conditions or assumptions.

Z.-Y. Hu (✉) · M.-L. Zou
Intelligent Information Systems Institute, Wenzhou University, Wenzhou 325035, Zhejiang, China
e-mail: hujunyi@163.com

Y. Jia et al. (eds.), *Proceedings of 2018 Chinese Intelligent Systems Conference*, Lecture Notes in Electrical Engineering 529,
https://doi.org/10.1007/978-981-13-2291-4_32

The methods based on polarization [2–4], which use two or more images of different polarization directions to defog. In the papers [5–8], more constraints can be obtained from many images of same scene in different weather. Tan [9] notices that no fog image compared with original fog image must have a higher contrast, therefore, to defog by maximizing local contrast of recover image. The result is credible on the vision, but not necessarily true in physics. Fattal [10] estimates the reflectivity of object, and speculates the propagation of media, based on the assumption that the propagation and the shadow on the surface of the scene are locally unrelated. The method proposed by Fattal is reasonable physically and gets favorable result. However, this method can't deal with heavy fog image, and it may invalidate if above assumption is not established. He proposes a dehazing algorithm based on dark channel prior [11]. In the no fog images, most of the local area without sky region, where a few pixels (named as black pixels) have very low brightness in at least one color channels (RGB) [12], while in the fog images, the brightness of black channel becomes higher in heavy fog. This method directly use dark channel prior to estimate atmospheric optical transmission diagram, moreover, use the method of image soft matting repair to smooth atmospheric optical transmission diagram, which can be used to obtain clear pictures of foggy images. This method possesses physical validity, but dark channel prior statistics may invalidate, when a big scene of target area is similar with atmospheric light in nature. Furthermore, repairing atmospheric optical transmission diagram not only has limitations, but also its algorithms are extremely complicated. Small repair parameters lead to obvious increase of local error, or excessive repair parameters lead to local details hidden. The above reasons make atmospheric optical transmission depth images lose level, leading to effective distinction of object instance.

Although image clarification processing research has made significant achievements, in turn, it puts forward new requirements for results clarity and realism. At the same time, the current image clarification methods also have some deficiency, and generally only are applied to the scene on the land. In view of the advantages and disadvantages of the existing defogging algorithms, this paper proposes a new method. Firstly, we use binocular disparity to obtain depth map. Then filter original brightness figure to generate atmospheric optical transmission diagram, by restoring degraded images to get a clear picture. This method does not need to calibrate binocular camera, is not dependent on the prior knowledge of inputting fog image, also does not rely on the larger contrast in the fog image or target surface shade, and is able to handle the fog image with different levels of depth of field.

2 Foggy Weather Imaging Model

With the reflected light of target object to light path of the camera, the existence of atmospheric particles scattering, ambient light around light path will act on fog particles, which leads to scattering phenomenon. Then it will deviate from the original direction of light propagation and is merged into the field of imaging light path, along the light path toward camera. Furthermore, it will participate in imaging, with the reflection of target object. Above process we call foggy weather imaging [13]. That is what McCartney model describes in [14], foggy scene target imaging is existed together by the incident light attenuation model and atmospheric scattering model exist together, and plays a leading role. It is the interaction that leads to mass reduction of target image. The more proportion of atmospheric scattering light, the poorer of the image quality; on the contrary, the more clear of images. Therefore, overall strength of light source received by foggy outdoor visual sensor, which can be equivalent to reflective intensity of sensor after incident light experiences atmospheric attenuation, and linear superposition of the surrounding environment of scattered light after entering the visual system. We describe it as formula (1):

$$I(x, v) = I_r(x, v) + A(x, v) = I_0(x, v)e^{-\gamma(v)x} + A_\infty(1 - e^{-\gamma(v)x}) \tag{1}$$

In the research field of video and image clarification, it is necessary to consider the transmission properties of visible light in the atmosphere, and the researchers can't consider the frequency of light waves. The formula (1) can be simplified as formula (2).

$$I(x) = I_0(x)e^{-\gamma x} + A_\infty(1 - e^{-\gamma x}) \tag{2}$$

For being convenient to study, we set $t = e^{-\gamma x}$, and we can further simplify formula (2) as formula (3):

$$I(x) = I_0(x)t + A_\infty(1 - t) \tag{3}$$

It denotes common foggy imaging model in computer vision and image processing.

3 Generate Depth Maps by Using Binocular Disparity Images

3.1 Disparity Estimation Based on SAD Similar Characteristics

The most basic image acquisition device of binocular disparity is consisted of two identical cameras. Their imaging plane is located on a plane, their axes are parallel to each other, and the horizontal axis is overlapped. In this collection device, the image position of same target in the scene on the two camera image plane is different. Where the same target in the scene in two different images, we call it conjugate imaging points, the one of imaging points is correspondence of the other imaging point. The solution of conjugate pairs is to solve corresponding problem. Difference of the conjugate pairs at the overlap of the two images, i.e., the distance between the conjugate pairs, we call it the disparity.

Sum of absolute differences (SAD) is a common and simple measure of similarity in 3D matching feature algorithm, two images collected by the horizontal binocular camera are compared with the absolute difference of each pixel block, and the image block difference value is used to create a simple similarity matrix. We assume that the two images collected by binocular camera respectively are $I_l(x, y, z)$ and $I_r(x, y, z)$. The disparity estimation based on SAD similarity feature in the position of (x, y, z) can be presented with formula (4):

$$d_{SAD} = \arg(\min_{d \in D_B}(D_{SAD}(x, y, z, d))) \tag{4}$$

d denotes disparity trace, d_x, d_y, d_z denote the disparities of the disparities of x, y and z.

$$D_{SAD} = \sum_{(u,v,w) \in B(x,y,z)} \left| I_l(u + d_x, v + d_y, w + d_z) - I_r(u, v, w) \right| \tag{5}$$

u, v, *and* w respectively denote pixel coordinates of x, y, *and* z.

3.2 Depth Maps Estimation

Depth estimation is significant factor of image recover processing [15, 16]. Stereo vision presents a direct way to infer depth information, by using two images, i.e., stereo pair, which are respectively denote left eye and right eye. It is well known that disparity is inversely proportional to the depth. The smaller of object disparity, the greater of distance between object and camera or eyes. On the contrary, the greater of object disparity, the smaller of distance between object and camera or eyes. Therefore, in this paper, we can simply reverse the disparity, and extrude in the interval range [0, 255]. The estimation formula of depth maps is as shown in (6):

$$D = 255 \frac{d_{SAD} - \min(d_{SAD})}{\max(d_{SAD}) - \min(d_{SAD})}. \tag{6}$$

4 Optimization of Atmospheric Light Transmission in Sky Region

Extensive experimental results statistics demonstrate that it is obvious for large area sky defogging to generate color distortion and halo phenomenon. By analyzing reason, we find that when we take a binocular image, the sky is furthest away from the camera, and the same depth reflected disparity change slowly compared with foreground. Therefore, in this paper, it is difficult to obtain the accurate of atmospheric optical transmission by using binocular parallax. To deal with this shortcoming, this paper starts with the detailed atmospheric optical transmission diagram to improve the defogging algorithm and increase the robustness of the algorithm. Our idea is to optimize the atmospheric optical transmission diagram in the sky region through the relationship between the brightness of atmospheric light intensity value and the atmospheric optical transmission diagram after refinement, and we introduce brightness difference dA to address halo phenomenon generated by large area sky defogging. We compare gray value I_g of foggy images with atmospheric light intensity A_∞. If $A_\infty - I_g < dA$, then we regarded the existing region as sky region. Then we need to optimize corresponding atmospheric optical transmission diagram, which is named as lightness difference mechanism. The atmospheric light intensity value A_∞ is the estimated mean. Simultaneous it may appear the situation $A_\infty < I_g$, therefore when optimize atmospheric optical transmission diagram, we also calibrate the error of atmospheric light intensity estimation. Reducing the intensity of defogging is an effective and simple method, i.e., increasing the value of the atmospheric optical transmission map in the corresponding region to achieve optimization and correction. In this paper, the optimization formula based on lightness difference mechanism is defined as formula (7).

$$t = \min(1, \max(1, \frac{dA}{|A_\infty - I_g|})\tilde{t}) \tag{7}$$

The optimization on formula (7) can make the brightness and the atmospheric light intensity value close to corresponding value of sky atmospheric optical transmission. This optimization not only reduces color distortion and halo phenomenon generated by excessively defogging, but also prevents the change of the atmospheric optical transmission diagram apart from sky region. In this paper, seeking the maximum operation is cleverly applied. Where if $\frac{dA}{|A_\infty - I_g|} < 1$, then We replace $\frac{dA}{|A_\infty - I_g|} < 1$ with constant 1. Simultaneously, in order to avoid excessively magnify sky atmospheric optical transmission figure value, we use the minimum operation, i.e., if the optimized value is greater than 1, we replace the value of optimized sky atmospheric optical transmission with constant 1. During the processing of experiment, we find that ideal value region of dA is 10–80, and in our experiment, we set $dA = 50$.

5 Experiments

In order to verify the method validity, in this paper, we compare our experimental results with that of mainstream algorithms [11, 17], and complete subjective effect and quantitative analysis. The experimental platform is MATLAB 2016a, the operating environment is Windows 8 for 64-bit, and computer configuration is Intel Core CPU i7-4500U@1.80 GHz (TM), for the 8 GB memory.

To verify the validity of the algorithm, we use phenix SDC-821 camera to collect foggy images, part images and depth maps, as shown in Fig. 1. For fog images, we implement defogging test with our algorithm and mainstream algorithm, the test results are as shown in Fig. 2. The test results demonstrate that the sharpness and overall contrast are largely improved with our algorithm, and the color of recover images is very realistic.

(a) left (b) right (c) disparity

Fig. 1 Binocular fog images and disparity maps

(a) He's result (b) Berma's result (c) ours result

Fig. 2 Experimental results comparison of different algorithms

6 Summary and Development

Due to the effect of atmosphere scattering, foggy degraded images have the characteristics which include low contrast and indistinct scene, bringing inconveniences to the application. Thus, to recovery the foggy weather degraded image, achieve clear imaging is of great significance. Aiming at the shortcomings of single image fog algorithms, in this paper, we can get fog depth image through binocular disparity images. At the same time, we can simply quickly optimize atmospheric optical transmission based on brightness difference mechanism, to eliminate common halo phenomenon caused by dehazing. Experimental results show that the proposed dehazing algorithm based on binocular parallax algorithm can achieve good recovery results. Researching the influence of maximum disparity on the accuracy of atmospheric optical transmission diagram and calculation, and searching for better binocular disparity algorithm will be one of the future directions of our team.

Acknowledgements The authors acknowledge the financial supported by Zhejiang Provincial Natural Science Foundation of China (project No.: LZ15F030002, LY16F020022). The author is grateful to the anonymous referee for the careful checking of the details of this paper and for helpful comments and constructive criticism.

References

1. R.A. Kerr, Pollutant hazes extend their climate-changing reach. Science **315**(5816), 1217 (2007)
2. S. Shwartz, E. Namer, Y.Y. Schechner, Blind haze separation, in *2006 IEEE Computer Society Conference on Proceedings of the Computer Vision and Pattern Recognition* (2006), pp. 1984–1991
3. Y.Y. Schechner, S.G. Narasimhan, S.K. Nayar, Instant dehazing of images using polarization, in *Proceedings of the 2001 IEEE Computer Society Conference on Computer Vision and Pattern Recognition, CVPR 2001*, pp. 325–332
4. Y.Y. Schechner, S.G. Narasimhan, S.K. Nayar, Polarization-based vision through haze. Appl. Opt. **42**(3), 511–525 (2003)
5. S.G. Narasimhan, S.K. Nayar, Chromatic framework for vision in bad weather, in *IEEE Conference on Proceedings of the Computer Vision and Pattern Recognition* (2000), pp. 598–605
6. S.G. Narasimhan, S.K. Nayar, Contrast restoration of weather degraded images. IEEE Trans. Pattern Anal. Mach. Intell. **25**(6), 713–724 (2003)
7. S.G. Narasimhan, S.K. Nayar, Interactive (de) weathering of an image using physical models, in *Proceedings of the IEEE Workshop on Color and Photometric Methods in Computer Vision* (2003), p. 1
8. S.G. Narasimhan, S.K. Nayar, Vision and the atmosphere. Int. J. Comput. Vision **48**(3), 233–254 (2002)
9. R.T. Tan, Visibility in bad weather from a single image, in *2008 IEEE Conference on Proceedings of the Computer Vision and Pattern Recognition, CVPR 2008*, pp. 1–8
10. R. Fattal, Single image dehazing, in *Proceedings of the ACM Transactions on Graphics (TOG)* (2008), pp. 72–80
11. H. Kaiming, S. Jian, T. Xiaoou, Single image haze removal using dark channel prior. IEEE Trans. Pattern Anal. Mach. Intell. **33**(12), 2341–2353 (2011)
12. P.S. Chavez Jr., An improved dark-object subtraction technique for atmospheric scattering correction of multispectral data. Remote Sens. Environ. **24**(3), 459–479 (1988)

13. Yi-xin Zhang, *Study on the Distortion of Laser Beam in Turbulent Atmosphere* (Nanjing University of Science and Technology, Nanjing, 2005)
14. E.J. Mccartney, Optics of the Atmosphere: Scattering by Molecules and Particles, vol. 1, no. 1 (New York, John Wiley and Sons, Inc., 1976. 421 p, 1976) pp. 123–129
15. S.T. Barnard, W.B. Thompson, Disparity analysis of images. IEEE Trans. Pattern Anal. Mach. Intell. **4**, 333–340 (1980)
16. R. Szeliski, Computer Vision: Algorithms and Applications (Springer, 2010)
17. D. Berman, T. Treibitz, S. Avidan, Non-local Image Dehazing, Computer Vision and Pattern Recognition (IEEE, 2016), pp. 1674–1682

State Prediction Based on ARIMA Model for Aerial Target

Tongle Zhou, Qingxian Wu and Mou Chen

Abstract In order to predict the air combat data accurately and quickly, a prediction method is developed for the aerial target based on autoregressive integrated moving average (ARIMA) model in this paper. The air combat situation data mainly consists of the velocity, altitude of aerial target and the angle between the target line of sight and target velocity. Finally, with an example, the simulation results indicate that the developed method can accurately and efficiently predict the air combat state data.

Keywords Air combat state data · State prediction · ARIMA

1 Introduction

With the increasing complexity of air combat and the development of advanced military technology, predicting target intention in advance can help for making adequate preparations of the autonomous attack and defense decision-making of Unmanned Combat Aerial Vehicles (UCAV) [1]. On the other hand, intention prediction of aerial target also has a number of benefits, such as improving the operational effectiveness of weapon system and reducing the cost of the war. As the first step of intention prediction, the air combat state data prediction also has great contributions on air combat decision-making system.

T. Zhou · Q. Wu (✉) · M. Chen
College of Automation Engineering,
Nanjing University of Aeronautics and Astronautics, Nanjing 211106, China
e-mail: wuqingxian@nuaa.edu.cn

T. Zhou
e-mail: zhoutongleok@hotmail.com

M. Chen
e-mail: chenmou@nuaa.edu.cn

Y. Jia et al. (eds.), *Proceedings of 2018 Chinese Intelligent Systems Conference*, Lecture Notes in Electrical Engineering 529,
https://doi.org/10.1007/978-981-13-2291-4_33

Due to the uncertainty and incompleteness of the modern battle information environment, prediction of air combat data has become well-concerned recent years. Some works can be found in the literature. Dempster-Shafer (D-S) evidence theory was used to predict air combat situation in [2]. In order to predict target intention, a method based on game theory was developed in [3]. A method was studied based on Bayesian network for target intention prediction in [4]. In [5], an adaptive neuro-fuzzy inference system was introduced for the air combat data. Furthermore, the grey incidence analysis method was discussed to predict the aerial target intention in [6].

In statistics and econometrics, and in particular in time series analysis, an ARIMA model is a generalization of an autoregressive moving average (ARMA) model. Both of these models are suited to time series data either to predict future data in the series [7]. In addition, ARIMA model only needs endogenous variable, so the rapidity of air combat data prediction can be guaranteed. Thus, ARIMA model has been widely used in different fields such as PM2.5 concentrations forecasting [8], road accidents prediction [9] and price estimation of commodity [10]. As is well-known, air combat data is a kind of time series, which obtained by different sensors in different moments. Hence, ARIMA model is used to predict the air combat data in this paper.

This paper is organized as follows. The air combat scenario is given in Sect. 2. In Sect. 3, the ARIMA model is presented and the air combat state data prediction algorithm is proposed. And Sect. 4 presents the simulation results. The conclusion is drawn in the last section.

2 Problem Statements and Preliminaries

In general, aerial target intention prediction is mainly obtained by air combat situation and the information of target. In the air combat environment, these air combat data is measured by various kinds of sensors. According to the real time data analysis, the future data can be predicted, and the intention of target can be further obtained. The air combat situation between UCAV i and the target j is shown in Fig. 1. In Fig. 1, we define the line between two fighters is the target line of sight; v is the target velocity, h is the altitude of target and a is the angle between the target line of sight and target velocity.

Fig. 1 Air combat situation

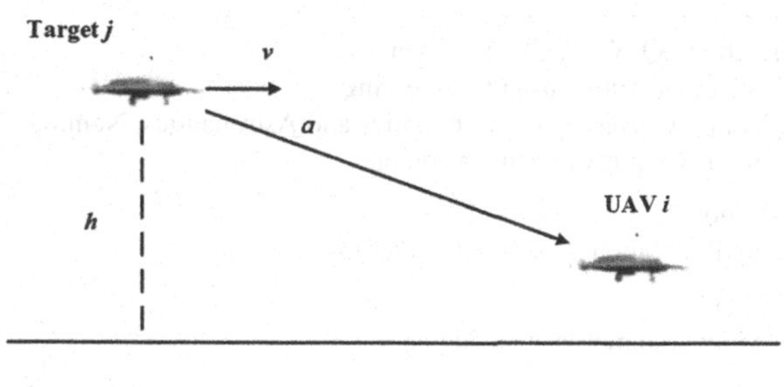

Due to the regularity of aerial target intention, different states of target can reflect the various results. For instance, the high speed targets are more aggressive, besides, if the target flight direction towards the UCAV, the target have greater possibility of attack intention. Thus, in this paper, because of the universality, the velocity, altitude and the angle between the target line of sight and target velocity are mainly considered. Moreover, other factors can be obtained by these three factors. Another benefit is that the state of UAV *i* is not required. Hence, air combat state data prediction is developed to predict the target velocity, target altitude and the angle between the target line of sight and target velocity.

3 ARIMA Model Based State Prediction Algorithm for Aerial Target

In air combat decision-making system, sampling periods of different sensors are different. Sometimes, the sampling periods is even not uniform. Thus, the time alignment should be considered before state prediction. Time alignment is the basis of information fusion, of which would greatly influence situation evaluation and decision generation. In this paper, a approach based on least square data fitting and cubic spline interpolation is used [11]. Discrete observations are imitated to continuous curve. So that the data value at any moment can be calculated based on the continuous curve.

As stated before, an ARIMA model is a generalization of an ARMA model in time series analysis. Both of these models are suited to predict future data in the series [7]. In the most cases, in order to predict the future data, the time series requires being stationary. However, the data of air combat is non-stationary sometimes. ARIMA models are applied in some cases where data show evidence of non-stationarity. In that case, an initial differencing step can be applied one or more times to eliminate the non-stationarity. Moreover, for stationary data, ARMA models are suitable.

The flow diagram of air combat state data prediction system is shown in Fig. 2.

The first step is data preprocessing, assume that the observations are $(t_0, f(t_0))$, $(t_1, f(t_1))$, ..., $(t_n, f(t_n))$, where t_i is the sampling time. $M = \{\varphi_0, \varphi_1, \ldots, \varphi_m\}$ is a function of time. φ_i is the polynomial function and the highest power of polynomial less than n. The fitted curve is [12]

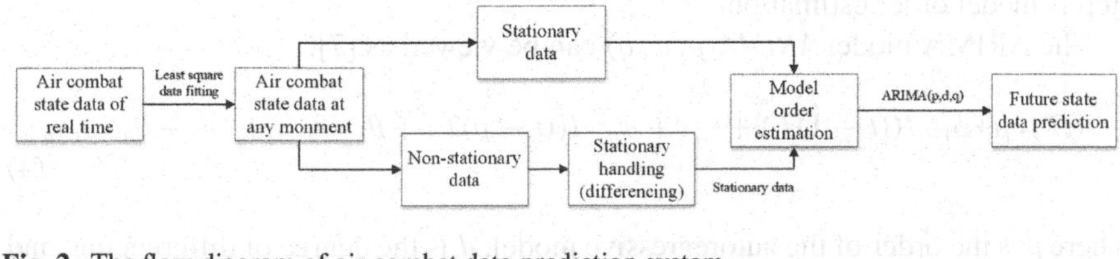

Fig. 2 The flow diagram of air combat data prediction system

$$L(t) = \sum_{j=0}^{n} a_j^* \varphi_j(t) \tag{1}$$

where a_j^* are the polynomial coefficients that lead $\sum_{i=0}^{n}[L(t_i) - f(t_i)]$ getting the minimum value.

After getting the fitted curve $L(t)$, the desired information at any moment can be obtained.

For a time series, if the mean value and variance are both invariant, and there is no periodic change. We call this time series is stationary. In other words, the basic idea of stationarity is that the behavior of the process do not change over time [13]. Hence, the stationarity should be judged before future data prediction.

Generally, the sequence diagram reflects the stationarity of time series. For an air combat state data sequence diagram, if the data is go up and down by a fixed value, the time series is stationary. Otherwise, if the data shows upward or downward trend, it is non-stationary.

The second step of ARIMA model based state prediction algorithm for aerial target is stationary handling. In order to deal with the non-stationary air combat state time series, differencing step is used in ARIMA. Assume that the sampling period is T, the preprocessed data from fitted curve are $L(T)$, $L(2T)$, …, $L(tT)$, for purpose of data differencing, the difference between the observations is computed. Mathematically, this is shown as [7]

$$l(tT) = L(tT) - L((t-1)T) \tag{2}$$

where $l(tT)$ is the difference.

Sometimes, data differencing might be necessary to a second time to obtain a stationary time series, which is referred to as second order differencing [7]:

$$\begin{aligned} l(tT) &= (L(tT) - L((t-1)T)) - (L((t-1)T) - L((t-2)T)) \\ &= L(tT) - 2L((t-1)T) + L((t-2)T) \end{aligned} \tag{3}$$

Given the above, the degree of differencing is depend on the regularity of the air combat state data. The purpose of stationary handing step is to obtain the stationary data on the basis of iterative differencing.

After stationary handling, the stationary data has been obtained. Then, the next step is model order estimation.

The ARIMA model $ARIMA(p, d, q)$ can be viewed as [7]:

$$\hat{l}(tT) = \mu + \phi_1 \cdot l((t-1)T) + \cdots + \phi_p \cdot l((t-p)T) + \theta_1 \cdot \varepsilon_{t-1} + \cdots + \theta_q \cdot \varepsilon_{t-q} \tag{4}$$

where p is the order of the autoregressive model, d is the degree of differencing, and q is the order of the moving-average model. μ is the constant, ϕ_i are the parameters of the autoregressive part of the model, θ_i are the parameters of the moving average part and the ε_t are error terms. The error terms ε_t are generally assumed to be independent, identically distributed variables sampled from a normal distribution with zero mean value.

Furthermore, the purpose of model order estimation is to determine the value of parameters p and q.

In this paper, the Akaike Information Criterion (AIC) used as the evaluation index. The AIC is defined as [14]:

$$AIC = -2\log(\Theta) + 2(p + q + k) \quad (5)$$

Fig. 3 The flow diagram of air combat data prediction algorithm

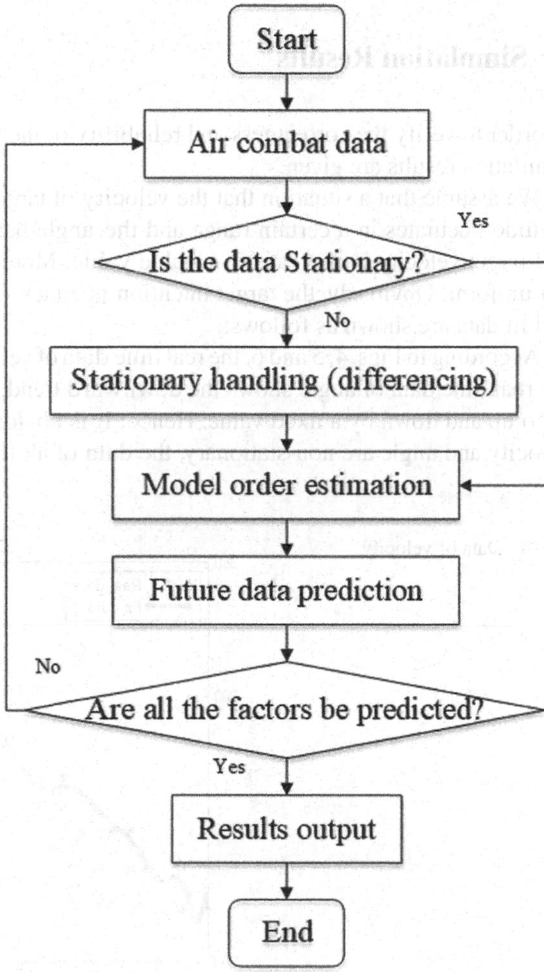

where Θ is the likelihood of the data, k represents the intercept of the ARIMA model. For AIC, if $k = 1$ then there is an intercept in the ARIMA model. Otherwise, if $k = 0$ then there is no intercept in the ARIMA model. Generally, the smaller AIC indicates the better model fitting degree.

Finally, once the parameters of ARIMA model $ARIMA(p, d, q)$ are determined, the model can be used for air combat state data prediction.

The input is the observations $f(t_0), f(t_1), \ldots, f(t_n)$, and the output is the air combat state data of next time $f((n+1)T)$. Moreover, note that difference restoration is needed to be considered.

From what has been discussed above, the flow diagram of air combat state data prediction algorithm is shown in Fig. 3.

4 Simulation Results

In order to verify the correctness and reliability of the above developed method, the simulation results are given.

We assume that a situation that the velocity of target is gradually increasing, the altitude fluctuates in a certain range and the angle between the target line of sight and target velocity is decrease to a stable value. Moreover, the sampling periods is not uniform. Obviously, the target intention is attack. The real time air combat data and fit data are shown as follows:

According to Figs. 4, 5 and 6, the real time data of velocity shows upward trend and the real time data of angle shows the downward trend, the real time data of altitude is go up and down by a fixed value. Hence, It is obviously that the real time data of velocity and angle are non-stationary, the data of altitude is stationary. Namely, the

Fig. 4 Data of velocity

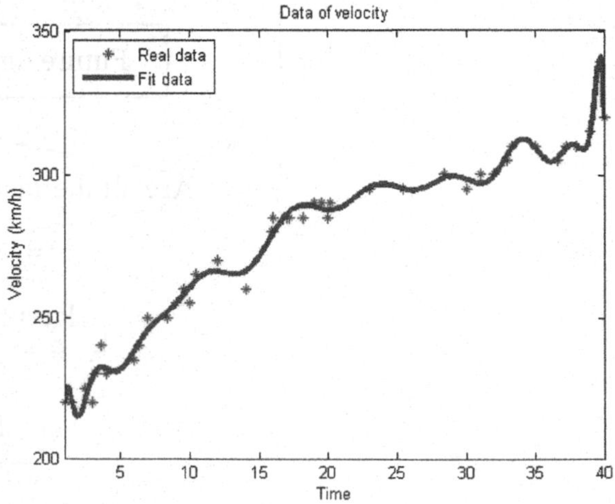

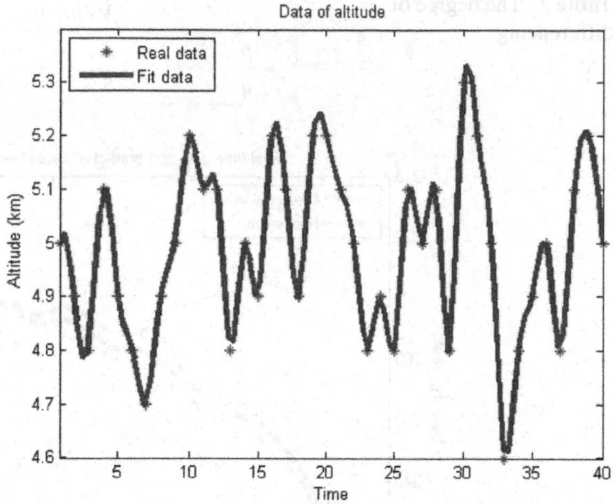

Fig. 5 Data of altitude

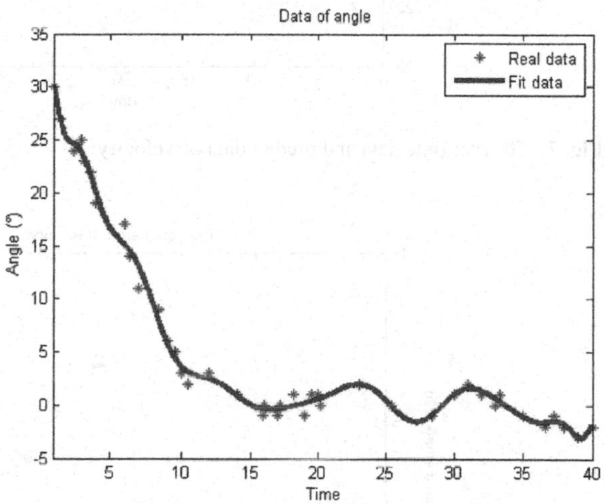

Fig. 6 Data of angle

differencing is needed for velocity data and angle data, and the ARMA model can be used directly for altitude data.

Choosing the time interval $T = 1s$, the degree of differencing d is shown in Table 1.

According to model order estimation, the order of the autoregressive model p and the order of the moving-average model q are shown in Table 2.

Table 1 The degree of differencing

	Velocity	*Altitude*	*Angle*
d	1	0	2

Table 2 The degree of differencing

	Velocity	*Altitude*	*Angle*
p	3	5	4
q	5	2	4

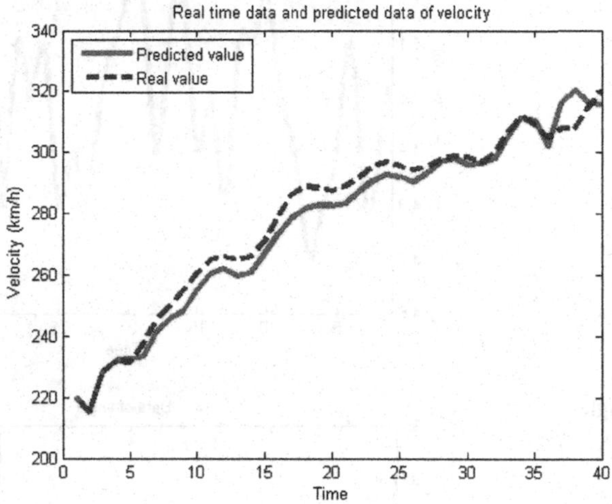

Fig. 7 The real time data and predict data of velocity

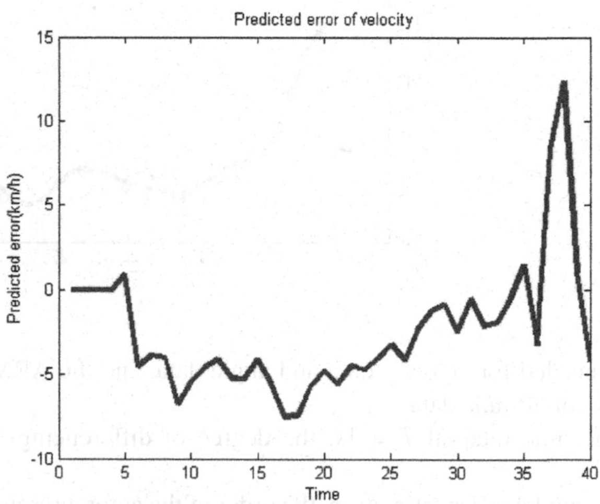

Fig. 8 The predicted error of velocity

The prediction results and error of velocity are shown in Figs. 7 and 8. The altitude prediction result and error are shown in Figs. 9 and 10. The prediction result and error of angle are shown in Figs. 11 and 12.

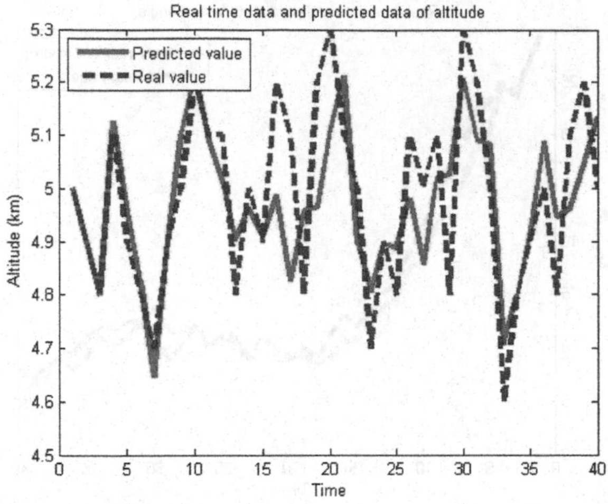

Fig. 9 The real time data and predict data of altitude

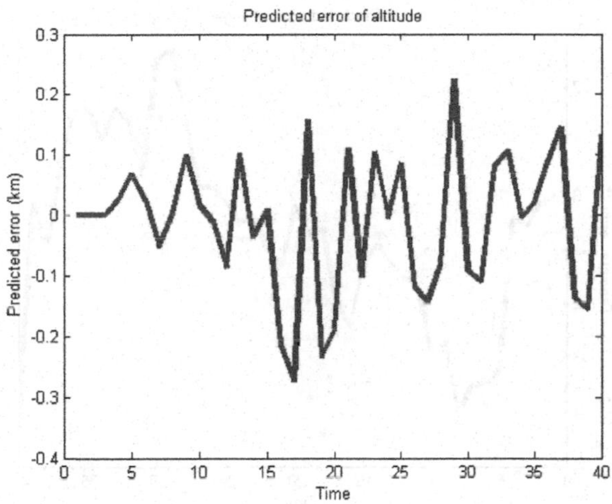

Fig. 10 The predicted error of altitude

Figures 7 and 8 show that ARIMA model can predict non-stationary data well, the error of velocity is small enough. Figures 9 and 10 show that ARIMA model is also suit to stationary data prediction, the minor error of altitude expresses the accuracy of algorithm. Figure 12 shows that the angle prediction has little error, the predict data and real time data have no significant difference in Fig. 11, which indicates the predict effect of angle is also satisfactory. The simulation results show that the prediction method is effective in air combat state data prediction.

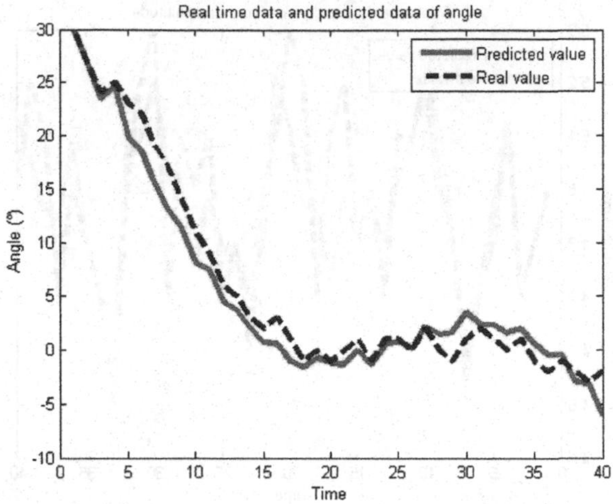

Fig. 11 The real time data and predict data of angle

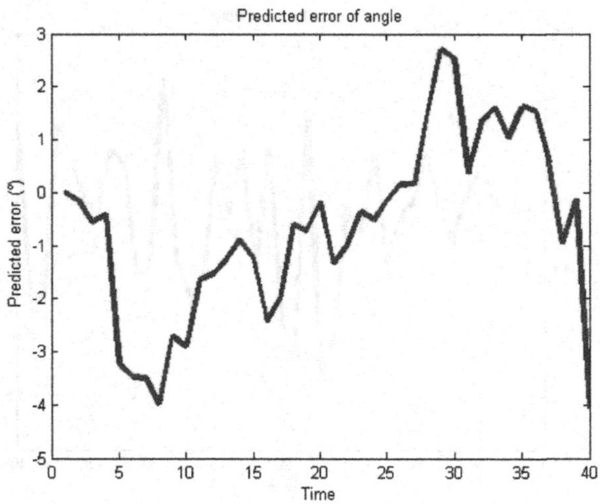

Fig. 12 The predicted error of angle

5 Conclusions

As the first step of intention prediction, the air combat state data prediction has great significance on air combat decision-making system. In this paper, due to the rapidity and accuracy of ARIMA model, a prediction method based on ARIMA is introduced. The simulation results have showed that the deserved prediction method can predict the air combat state data effectively. It can lay a foundation for future air combat decision-making.

Acknowledgements This work is partially supported by Equipment Pre-research Foundation of Laboratory (No. 61425040104) and Science and Technology on Electron-Optic Control Laboratory. The authors also gratefully acknowledge the helpful comments and suggestions of the reviewers, which have improved the presentation.

References

1. A. Pongpunwattana, R. Rysdyk, Real-time planning for multiple autonomous vehicles in dynamic uncertain environments. J. Aerosp. Comput. Inf. Commun. **12**(1), 580–604 (2004)
2. L. Sun, L. Yu, W. Huang, Improved weighted D-S evidence theory in the application of target intention prediction. J. Air Force Eng. Univ.: Nat. Sci. **10**(1), 17–22 (2009)
3. Z. Yuan, The air target sequential game combat intention prediction model. Syst. Eng. Theor. Pract. **7**(7), 70–76 (1997)
4. Y. Song, X. Zhang, Z. Wang, Target Intention Inference Model Based on Variable Structure Bayesian Network. Comput. Intell. Softw. Eng. pp. 1–4 (2009)
5. Y. Cui, Q. Wu, M. Chen, Aerial target intention prediction based on adaptive neuro-fuzzy inference system, in *The 15th Chinese Conference on System Simulation Technology and Application*, pp. 277–281 (2014)
6. T. Zhou, M. Chen, S. Chen, J. Zou, Intention prediction of aerial target under incomplete information. ICIC Express Lett. **8**(3), 623–631 (2017)
7. S. Makridakis, M. Hibon, ARIMA models and the Box-Jenkins methodology. Appl. Econ. pp. 265–286 (2011)
8. P. Wang, H. Zhang, Z. Qin, G. Zhang, A novel hybrid-Garch model based on ARIMA and SVM for PM2.5 concentrations forecasting. Atmos. Pollut. Res. **1**(8), 850–860 (2017)
9. Y. Yang, A study of prediction based on ARIMA model of road accidents. Stat. Appl. **2**(6), 268–275 (2017)
10. J. Liu, H. Li, Price estimation of Pseudo-Ginseng based on ARIMA model and BP neural network. Comput. Sci. Appl. **7**(7), 696–710 (2017)
11. K. Liang, Q. Pan, G. Song, X. Zhang, Z. Zhang, The study of multi-sensor time registration method. J. Shaanxi Univ. Sci. Technol. **24**(6), 111–114 (2006)
12. Z. Pan, W. Dong, Z. Wang, D. Zhao, Study on PRS/IRS time registration based on curve fitting. J. Air Force Radar Acad. **25**(5), 343–346 (2011)
13. Z.H. Munim, H. Schramm, Forecasting container shipping freight rates for the Far East C Northern Europe trade lane. Marit. Econ. Logistics **19**(1), 106–125 (2017)
14. H. Akaike, A new look at the statistical model identification. IEEE Trans. Autom. Control **19**(6), 716–723 (1974)

Evolution of Function Modules in Complex Networks

Lizhen Zhang, Yuan Zhang, Lan Xiang and Jin Zhou

Abstract It is generally accepted that the function module plays an important role for the understanding of structure and function of complex networks. This brief studies the issue of the evolution of function modules in complex networks from the view of automaton state evolution. The definitions of two kinds of function modules, including the fixed module and the period module, are presented to characterize the nature of complex networks, and the corresponding evolutions of complex networks are further addressed. Furthermore, the developed concepts of function module are applied to the typical Boolean network and the representative cyclical group respectively, and then the obtained results can reveal the practical features of these networks.

Keywords Function module · Complex network · Automaton · Boolean network · Cyclic group

1 Introduction

The function module, which is usually regarded as the fundamental unit of complex networks, consists of a large number of vertices that are closely connected each other,

---

L. Zhang · Y. Zhang · J. Zhou (✉)
Shanghai Institute of Applied Mathematics and Mechanics,
Shanghai University, Shanghai 200072, China
e-mail: jzhou@shu.edu.cn

L. Zhang
Department of Mathematics, Tianjin Polytechnic University, Tianjin 300387, China

L. Xiang
Department of Physics, School of Science,
Shanghai University, Shanghai 200444, China

Y. Jia et al. (eds.), *Proceedings of 2018 Chinese Intelligent Systems Conference*, Lecture Notes in Electrical Engineering 529,
https://doi.org/10.1007/978-981-13-2291-4_34

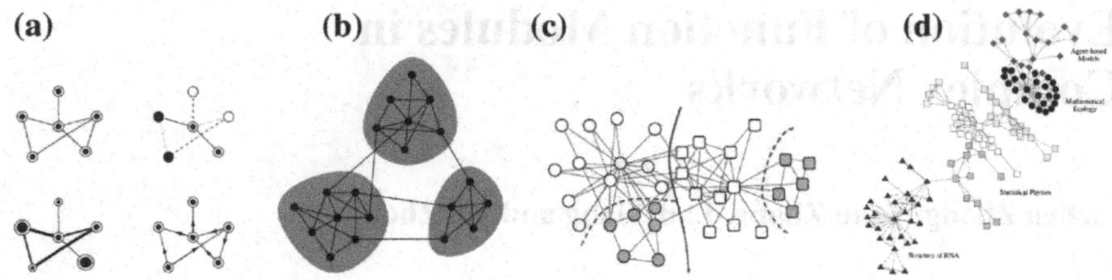

Fig. 1 Complex networks and their modular architecture: **a** Different types complex networks [3]. **b** Complex Network with typical Modular structure [2]. **c** Community structure of a complex network [11]. **d** Communities structure of a social network [4]

and so it plays an important role in the structure and function of complex networks. Because of the centrality of function module in determining the topology properties of complex networks, more and more attention has been paid to the functions of some nodes or edges in complex networks [1, 2]. In the last decades, many studies have indicated that the networks often exhibit the hierarchical organization, which is similar to the blocky structure of the soil [3]. Specifically, the vertices of the networks can divide into several groups that can further subdivide into groups of groups, such as ecological niches of the food webs, communities in social networks [4–6] and a lot of modules in biochemical networks [3, 7–10]. In that light, a lot of successful cases in study of the function module and the complex networks have been obtained, including the exploration of the statistical properties of complex networks, the interaction behaviors with other systems [11–13] (see Fig. 1c, d).

Although a wide variety of topological properties for complex networks have been extensively studied, the understanding of the evolution mechanism of complex network still remains limited [14]. Moreover, because the vertices [15] and the edges are very complex (see Fig. 1a, b), the effective analysis approaches and tools to deal with these complex evolution situations for complex networks have been lacking [16]. These findings motivate us to further explore the evolution of function modules in complex networks.

For our investigation, we here use the automata theory and symbolic dynamical system to explore the evolution mechanism of complex networks. Specifically, the evolution and functional modules of complex networks are discussed.

2 Preliminaries

In general, a dynamical system is characterized by its elements and the other time-dependent development states, in fact, all of which are not enough. In these processes, lots of elements interact with causal feedback fixed points and cycles, resulting in unstable states, other kinds of attractors, and even chaos. In the context, automaton have been regarded as a powerful tool for describing the relationships between the complexity and the complex networks. More importantly, automata can perform a lot of tasks relying on formal language that is the soul of the automata [17]. Therefore, we sought to describe these processes by automaton, and accordingly, some concepts are given as below:

Definition 1 (1) Alphabet (or called vocabulary): An *alphabet* (*or vocabulary*) is a finite nonempty set of symbols actually, denoted by the symbol Σ. For example, $\Sigma = \{0, 1\}$ is a binary alphabet and $\Sigma = \{a, b, c, \ldots, z\}$ is English Language alphabet.

(2) String: A *string* (sometimes called word) is a finite sequence of symbols from the alphabet Σ, we use small letter, x,y and so on to express. For example, "1010101" is a string (or word) of the binary alphabet $\Sigma = \{0, 1\}$ and "shanghai" is a word (or string) of the English Language alphabet $\Sigma = \{a, b, c, \ldots, z\}$.

(3) Length of a string: Length of a string is the number of the symbols in the string, here, we use the notation $|x|$ to express the length of a string x.

(4) The empty string(or Null string): The empty string is with the zero length of a string, denoted by λ.

(5) Formal language : If Σ^* expresses the set of all words (including the empty string) over Σ, then a *language* over Σ is a subset of Σ^*.

In addition, the concepts related automata are given as following:

Definition 2 (1) Finite state automaton: A finite-state automaton M is a quintuple $< S, I, R, \delta, F >$ (see Fig. 2), in which $S = \{s_1, s_2, \ldots, s_n\}$ is a finite states set; $I = \{i_1, i_2, \ldots, i_m\}$ is input symbolic set; R is output symbolic set and

$$\delta : S \times I \rightarrow S$$

is a states transition function; F is a subset of the set S, which is also called the terminating states set—the elements of this set are finally arrived the states from any state of the state space S.

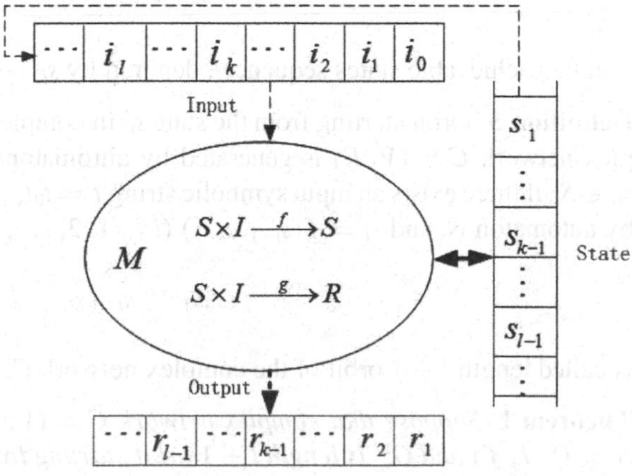

Fig. 2 Diagram of a finite state automaton

(2) The string x is accepted by automaton M. String x acceptable for automaton: Let M be an automaton and $x \in I$ be an input string, if an automaton starts to read the first symbolic of the string x, and then in turn reads each character of string x, when it reads the last character, the state steps into some state of the terminating states set.

3 Function Modules of Complex networks

Firstly, a definition of evolution of complex networks is proposed from the view of automaton below [17]:

Definition 3 The evolution definition of complex networks: Let $N = (S, I, f)$ be an automaton, and the state space S be finite nonempty set, I be a input symbolic set, f be a state transition function defined on S from $S \times I$ to I. If $\forall s_i, s_j (i = 0, 1, \ldots, n-1) \in S, i \neq j$, there exists an acceptable string t_k in I^* s.t. $s_j = f(s_i, t_k)$,
we can get the graph $C = (V, E)$, here $V = S$,

$$E = \{e|e = < s_i, s_j >, exists\ t_k \in I,\ s.t.\ y = f(x, t_k)\},$$

then the graph $C = (V, E)$ is called complex network which is generated by state transition function f and the acceptable string t based on the automaton N.

Definition 4 (1) State sequence starting from s_0 : suppose that complex network $C = (V, E)$ is generated by automaton $N = (S, I, f)$. For a state $s_0 \in S$ and an input acceptable string $t = t_0 t_1 \ldots t_{l-1}$ by automaton N, if there exists a state sequence $s = s_0 s_1 \ldots s_l$ which is formed by the function f and the input string t, then s is called state sequence starting s_0. (2) Achievable states sequence formed from s_i to s_j: for two states $s_i, s_j \in S$, if there exists an input symbolic string $t = t_0 t_1 \ldots t_m \in I^*$ satisfying $s_{i+1} = f(s_i, t_0), s_{i+2} = f(s_{i+1}, t_1), \ldots, s_j = f(s_j - 1, t_m)$, then the state sequence

$$s = s_i s_{i+1} s_{i+2} \ldots s_{j-1} s_j$$

is called achievable states sequence , denoted by $s_i \xrightarrow{t} s_j$.

Definition 5 Orbit starting from the state s_0 in complex network: suppose that complex network $C = (V, E)$ is generated by automaton $N = (S, I, f)$. For any state $s_0 \in S$, if there exists an input symbolic string $t = t_0 t_1 \ldots t_{l-1} \in I^*$ which is accepted by automaton N, and $s_i = f(s_{i-1}, t_{i-1})$ $(i = 1, 2, \ldots, l)$, then the states sequence

$$s_0 s_1 \ldots s_{l-1} s_l$$

is called length $l + 1$ orbit of the complex network C, denoted by $O^t_{s_0}$.

Theorem 1 *Suppose that complex network* $C = (V, E)$ *is generated by automata* $N = (S, I, f)$ *and* $O^t_{s_0}$ *is length* $l + 1$ *orbit starting from* s_0,

(1) for $x, y \in V$, if there exist $s_0 \in S$ and the input strings $t(i), t(j) \in I^*$ s.t. $s_0 \xrightarrow{t(i)} x$ and $s_0 \xrightarrow{t(j)} y$ hold, then x, y have the equivalence relation, denoted $x \sim y$.

(2) for the equivalence relation "$\sim$", we have

$$S = \bigcup_{\lambda \in \Lambda, t(\lambda) \in T} O_{s_\lambda}^{t(\lambda)},$$

here $T = \{t(\lambda) | \lambda \in \Lambda\}$.

Proof (1) For any $x \in O_{s_0}^t$, using the definition of orbits, we have $x \sim x$. For $x, y \in S$, and $s_0 \in S$, if there exist strings $t(i), t(j) \in I^*$ satisfying $s_0 \xrightarrow{t(i)} x$ and $s_0 \xrightarrow{t(j)} y$, we can get $s_0 \xrightarrow{t(j)} y$ and $s_0 \xrightarrow{t(i)} x$, that is $y \sim x$. For $z \in S$, if there exists string $t(k)$, we have $s_0 \xrightarrow{t(k)} z$, and further get $x \sim z$. Therefore, the relation "$\sim$" is equivalence relation. Moreover, from the properties of the equivalence relation, we can get the result (2) in the theorem.

Definition 6 Suppose that complex network $C = (V, E)$ is generated by automaton $N = (S, I, f)$. Let $O_{s_0}^t$ be the length $l + 1$ orbit starting s_0, we have:

(1) fixed module: for the orbit $O_{s_0}^t = s_0 s_1 \ldots s_{l-1} \ldots$ in complex networks, if there exists s_i, we have

$$s_0 s_1 \ldots s_{l-1} \ldots = s_0 s_1 \ldots s_{i-1} \ldots s_i s_i s_i \ldots = s_0 s_1 \ldots s_{i-1} (s_i)^\infty,$$

and then the fragment "$s_i \ldots s_i = (s_i)^\infty$" of the orbit $O_{s_0}^t$ is called fixed module about the state s_0, denoted by : $(s_i)^\infty (or\ F(s_i))$;

(2) cycle (or period) states module: for some orbit $O_{s_0}^t = s_0 s_1 \ldots s_{l-1} \ldots$ of complex networks, if there exist the length n states sequence $s_{i_0} s_{i_1} \ldots s_{i_{k-1}}$ of S satisfying $s_0 s_1 \ldots s_{l-1} \ldots = s_0 \ldots (s_{i_0} s_{i_1} \ldots s_{i_{k-1}})^\infty$, then the fragment "$(s_{i_0} s_{i_1} \ldots s_{i_{k-1}})^\infty$" of the orbit $O_{s_0}^t$ is called cycle (or period) states module about the state s_0, denoted by:
$(s_{i_0} s_{i_1} \ldots s_{i_{k-1}})^\infty (or P(s_{i_0} s_{i_1} \ldots s_{i_{k-1}}))$.

Remark 1 (1) From the definition of the stationary states modules and the cycle (or period) states modules, we can observe that the stationary states module is the cycle (or period) states modules with period "1".

(2) A lot of successful cases of study in the various fields using the structure and properties of the sequence have been studied. For example, the study of the homoclinic orbits in some symbolic dynamical systems and the kneading sequence of the study of the coarse grain chaos and so on [18].

4 Applications to Boolean Networks and Cyclical Groups

4.1 Modules Structure of Boolean Networks

Boolean Networks (BNs) which are firstly proposed by the Kauffman, etc. [16], are a directed graph based discrete system, having great application potential for cell differentiation, immune response, evolutionary biology, neural network and gene regulation, etc. [19]. From a dynamical system perspective, the state evolution of BNs is global convergence, that is to say, it can always convergence to a fixed circle or a fixed point regardless of how changes of the initial state are, both of which are called attractor. In addition, all of the states could evolve into the same set of attractor state that is called the attraction domain of the attractors (or called basin of the attractor).

For a general BN, $\Sigma = (B, F)$ is a directed graph which consists of n-variational nodes $\{b_1, b_2, \ldots, b_n\}$ whose in-degree is $k_i (i = 1, 2, \ldots, n)$, and the corresponding Boolean function sequence is $\{f_1, f_2, \ldots, f_n\}$. Without loss of generality, we directly use the characters of these nodes to express the corresponding variables. Accordingly, we have the following equations.

$$\begin{cases} b_1(t+1) = f(b_{1_1}, b_{1_2}, \ldots, b_{1_{k_1}}), \\ b_2(t+1) = f(b_{2_1}, b_{2_2}, \ldots, b_{2_{k_2}}), \\ \quad\vdots \\ b_n(t+1) = f(b_{n_1}, b_{n_2}, \ldots, b_{n_{k_n}}), \end{cases} \tag{1}$$

Thus, the states of the nodes of BNs are determined by the nodes connectivity and their Boolean functions. As an example, next, we shall consider a Boolean network Σ_1 which consists of three nodes with three Boolean function $\{and, or, or\}$(as shown in Fig. 3a). According to the equations of the Boolean Network Σ_1 and its Boolean functions of every nodes, we can get the table of the state transition (see Fig. 3b) and the complex network Fig. 3c, which are as followings. From the complex network of the BNs Σ_1, we can get the state space of the BN Σ_1 $S = \{000, 001, 010, 011, 100, 101, 110, 111\}$, and can obtain the following results:

(1) The orbit which starts the state 000 converges the state 000 forever. Here, we sign it by $(000)^\infty(orF(000))$, which is a fixed state module.
(2) The orbit which starts the states 100, 101, 110 converges to the state 011 and further converges to the state 111 and finally fixed the state 111. Thus, $F(111)$ is also a fixed state module.
(3) The orbits which start 010 and 001 are the same sequence

$$010001010001\ldots = (010001)^\infty$$

that is a period attractor $(010001)^\infty$, signed by $P(010001)$.

(a)

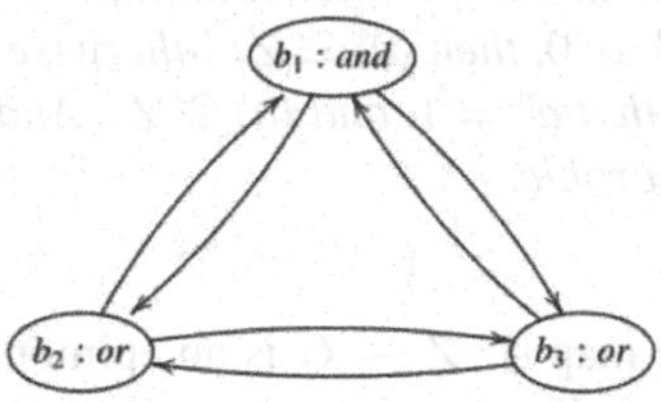

(b)

T			$T+1$		
b_1	b_2	b_3	b_1	b_2	b_3
0	0	0	0	0	0
0	0	1	0	1	0
0	1	0	0	0	1
0	1	1	1	1	1
1	0	0	0	1	1
1	0	1	0	1	1
1	1	0	0	1	1
1	1	1	1	1	1

(c)

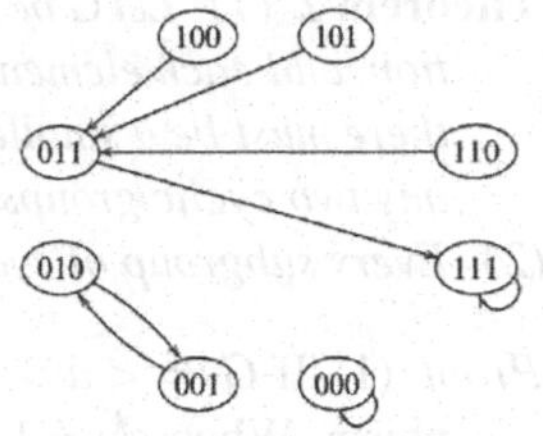

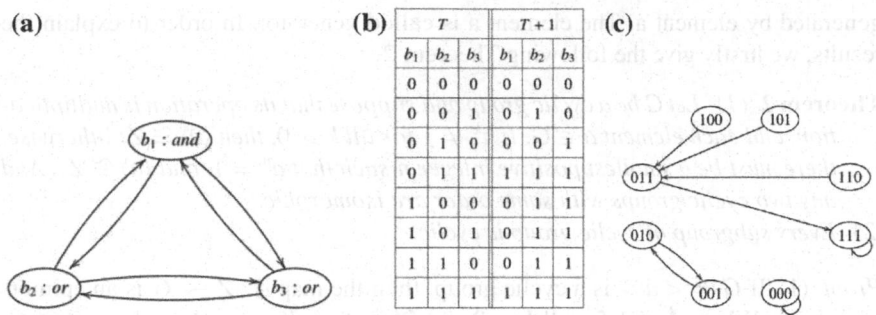

Fig. 3 **a** The Boolean network BN_1 with the nodes functions {and, or, or}. **b** The state transition table of BN_1. **c** The complex network of BN_1

(a)

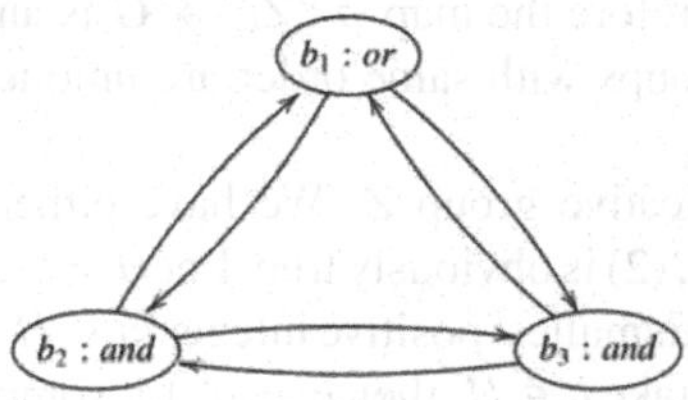

(b)

T			$T+1$		
b_1	b_2	b_3	b_1	b_2	b_3
0	0	0	0	0	0
0	0	1	0	1	0
0	1	0	0	0	1
0	1	1	1	1	1
1	0	0	0	0	1
1	0	1	0	1	1
1	1	0	0	1	1
1	1	1	1	1	1

(c)

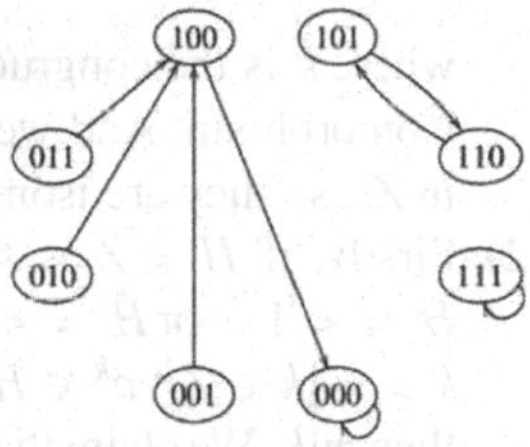

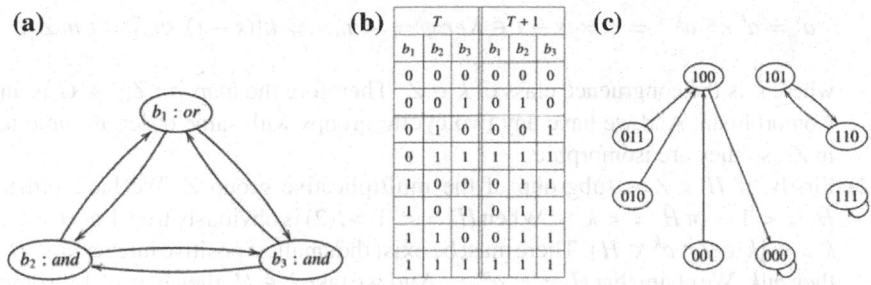

Fig. 4 **a** The Boolean network BN_2 with the nodes functions {or, and, and}. **b** The state transition table of BN_2. **c** The complex network of BN_2

If we change the Boolean functions $\{and, or, or\}$ of nodes (b_1, b_2, b_3) to the functions $\{or, and, and\}$, the shape of the BN dose not change, but the structure of its complex network is changing qualitatively Fig. 4a. This complex network is denoted by Σ_2. So the state transition table of Boolean network Σ_2 and its complex network turn to the state transition table depicted in Fig. 4b and the module structure depicted in Fig. 4c, respectively.

From the complex networks Σ_1 and Σ_2, we can get the following results: modules $F(111)$ and $F(000)$ are the fixed modules and $P(101110)$ is a period module.

Remark 2 By making a comparison for the above two complex networks, it can be known that if we want to change the state structure of the BNs, then we need to change the Boolean functions of the BNs.

4.2 Fundamental Units of Cyclical Group

In algebra, the cyclic group, is very important group, that has been solved completely. If G can be written $\langle a \rangle = \{a^n | n \in Z\}$, then the group G is called a cyclic group generated by element a. The element a is called generator. In order to explain the results, we firstly give the following Theorem 2.

Theorem 2 *(1) Let G be a cyclic group and suppose that its operation is multiplication, and each element $a \in G$. If $a^k \neq 1$ for all $k > 0$, then $\langle a \rangle \cong Z$; otherwise, there must be a smallest positive integer n such that $a^n = 1$, and $\langle a \rangle \cong Z_n$. And any two cyclic groups with same order are isomorphic.*

(2) Every subgroup of cyclic group is cyclic.

Proof (1) If $G = < a >$ is a cyclic group, then the map $\phi : Z \rightarrow G$ is an epimorphism. When $a^k \neq 1$ for all $k > 0$, $\langle a \rangle \cong Z$, then $Ker\phi = 0$, we have $Z \cong G$. Otherwise $Ker\phi$ is a nontrivial subgroup of Z and hence $Ker\phi = < k >$, where k is the least positive integer such that $a^k = e$. For all $s, t \in Z$,

$$a^s = a^t \Leftrightarrow a^{s-t} = e \Leftrightarrow s - t \in Ker\ \phi = < n > \Leftrightarrow k|(s-t) \Leftrightarrow \bar{s} = \bar{t}\ in\ Z_k,$$

where $\bar{k}$ is the congruence class of $k \in Z$. Therefore the map $\varphi : Z_k \rightarrow G$ is an isomorphism. And we have any two cyclic groups with same order are unique in Z_k, so they are isomorphic.

(2) Firstly, if $H < Z$ is subgroup of the multiplicative group Z. We have either $H = < 1 >$ or $H = < k >$. When $H = < 1 >$, (2) is obviously true. Let $H = < k >= \{k \in Z : a^k \in H\}$. There must be exist the smallest positive integer $m \in H$, then $m|k$. We claim that $H = < a^m >$. And we take $h \in H$, then $h = a^k$ for some k, so that $k \in H$ and $k = ms$ for some $s \in Z$. Therefore, $h = a^k = a^{ms} \in < a^m >$. So we have $\langle a \rangle \cong Z_n$.

So we can obtain the results:

Theorem 3 *Suppose that the state space S and the input symbolic set I are the set Z_n; and f = "+", i.e. the automata $M = < Z_n, Z_n, \text{"+"} >$, then the complex network which is generated by the automata M is a n order complete graph (Fig. 5a).*

Proof According to the knowledge of number theory, and defining the addition in n module residue class set $Z_n = \{(Z_n, +), Z_n = \{\bar{0}, \bar{1}, \ldots, \bar{n}\}$, for $\forall \bar{a}, \bar{b} \in Z_n$, we have $\bar{a} + \bar{b} = \overline{a + b} \in Z_n$, then Z_n is closed for "+" and (Z_n, "+") is an Abelian group. In $Z_n \times Z_n$, the map $f = +$ is a surjection, namely, $Z_n \times Z_n \rightarrow Z_n$ is a surjection. Therefore, all of the sequences in the input symbolic set are acceptable sequences of the automata and any two points in the state space are achievable. Collectively, the complex network is a n order complete graph.

For an example, if both of the state set and input symbolic set are taken as Z_3; and f = "+", then one can have that the automaton is $M = (Z_3, Z_3, \text{"+"})$, which leads to the following complex network as shown in Fig. 5a that is a 3 order complete graph. Furthermore, we can obtain three fixed state module $F(\bar{0}), F(\bar{1}), F(\bar{2})$ and two period modules $P(\bar{0}\bar{1}\bar{2})$ (Fig. 5b) and $P(\bar{0}\bar{2}\bar{1})$ (Fig. 5c).

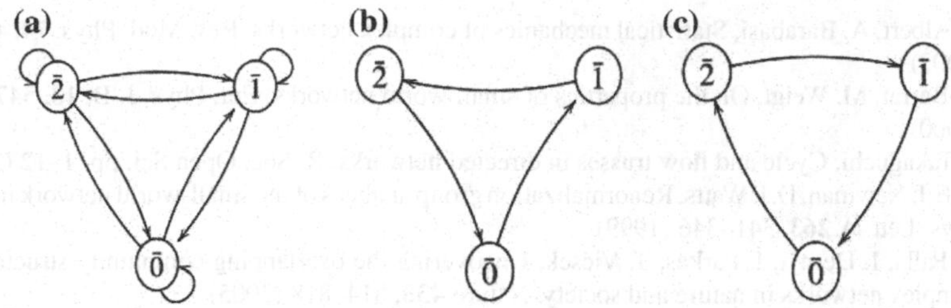

Fig. 5 The complete network of the cyclic group Z_3 and its module structures: **a** The complete network G_3 of the cyclic group Z_3. **b** The period module $P(\bar{0}\bar{1}\bar{2})$ of the complete network G_3. **c** The period module $P(\bar{0}\bar{2}\bar{1})$ of the complete network G_3

5 Conclusions

Although the function module of complex networks has been extensively studied, the understanding of its evolution mechanism still remains limited. Herein, a definition of complex networks is presented based on the point of view from automaton state evolution. Moreover, three important features of complex networks—the trajectory, the fixed module, and the period module—are defined. In particular, these definitions are rigorous mathematic analysis for the behaviour of the complex networks. Finally, the concepts of our definitions are applied in the Boolean network and the cyclical group. Obtained results verify the validity of the definitions proposed in the work.

Acknowledgements The authors wish to thank the editor and the reviewers for their insightful and constructive comments, which improved this paper significantly. This work is supported by the National Science Foundation of China (Grant Nos. 11672169 and 51875331).

References

1. D.J. Watts, S.H. Strogatz, Collective dynamics of small-world networks. Nature **393**(6684), 440–442 (1998)
2. M.E.J. Newman, Modularity and community structure in networks. PNAS **103**(23), 8577–8582 (2006)
3. M.E.J. Newman, The structure and function of complex networks. SIAM Rev. **45**, 167–256 (2003)
4. E. Ravasz, A.L. Somera, D.A. Mongru, Z.N. Oltvai, A. Barabasi, Hierarchical organization of modularity in metabolic networks. Science **30**, 1551–1555 (2002)
5. J. Kim, Robustness analysis of network modularity. IEEE Trans. Control Network Syst. **3**(4), 348–357 (2016)
6. A.L. Albert, Emergence of scaling in random networks. Science **386**(5439), 509–512 (1999)
7. J. Wang, On motifs and functional modules in complex networks. Sci. Technol. Humanity pp. 78–82 (2014)
8. J. Andreas, Neural module networks, in *IEEE Conference on Computer Vision and Pattern Recognition*, pp. 39–48 (2016)
9. S. Wasserman, K. Faust, *Social Network Analysis* (Cambridge Univ. Press, Cambridge, 1994)

10. R. Albert, A. Barabasi, Statistical mechanics of complex networks. Rev. Mod. Phys. **74**, 47–97 (2002)
11. A. Barrat, M. Weigt, On the properties of small world networks. Eur. Phys. J. B. **13**, 547–560 (2000)
12. T. Takaguchi, Cycle and flow trusses in directed networks. R. Soc. Open Sci. pp. 1–12 (2016)
13. M.E.J. Newman, D.J. Watts, Renormalization group analysis of the small-world network model. Phys. Lett. A **263**, 341–346 (1999)
14. G. Palla, I. Dernyi, I. Farkas, T. Vicsek, Uncovering the overlapping community structure of complex networks in nature and society. Nature **435**, 814–818 (2005)
15. A.L. Barabasi, N. Gulbahce, J. Loscalz, Network medicine: a network-based approach to human disease. Nat. Rev. Genet. **12**, 56–68 (2011)
16. S.A. Kauffman, Antichaos and Adaptation (Scientific American, 1991), pp. 78–84
17. P. Linz, *An Introduction to Formal Languages and Automata*, 3rd edn. (Jones and Bartlett Publishers Inc, Burlington, 2001)
18. Z.L. Zhou, *Symbolic Dynamics* (Shanghai Scientific and Technological Education, Shanghai, 1997)
19. C. Gros, *Complex and Adaptive dynamical systems* (Springer, Berlin, 2008)

A NLOS Error Mitigation Algorithm Based on ELM and EKF in Indoor Tracking

Wendong Xiao and Dongliang Zhao

Abstract Extended Kalman Filter (EKF) has a good performance in positioning and tracking systems due to its simple algorithm, low computational complexity, suitable for weak nonlinear systems, and good tracking performance in Gaussian environments. The error caused by non-line-of-sight (NLOS) propagation in the indoor environment on measured value has a great influence on EKF tracking performance and may even be divergent. The method of using neural network to correct NLOS has good adaptability and robustness, which can effectively mitigate errors and improve accuracy. This paper proposes an NLOS error mitigation based on Extreme Learning Machine (ELM) and EKF. ELM uses the state information of EKF to classify the measured values, determines the propagation path affected by NLOS error, and then corrects the measured values using the ELM trained for the path. Experiments show that compared with the traditional method, this algorithm can effectively mitigate the influence of NLOS and further improve tracking accuracy.

Keywords NLOS · Indoor tracking · ELM · EKF

1 Introduction

Indoor wireless positioning technology is one of the key technologies for the application of the Internet of Things. Compared with outdoor positioning technology, due to more complex indoor communication environment, there are different degrees of signal attenuation, reflection, multipath effect and non-line-of-sight error in wireless signal transmission process, which affects the positioning accuracy. Indoor positioning system based Time Difference of Arrival (TDOA) on Time

W. Xiao · D. Zhao (✉)
School of Automation and Electrical Engineering, University of Science and Technology Beijing, Beijing, China
e-mail: 1748787587@qq.com

W. Xiao
e-mail: wendongxiao68@163.com

Y. Jia et al. (eds.), *Proceedings of 2018 Chinese Intelligent Systems Conference*, Lecture Notes in Electrical Engineering 529,
https://doi.org/10.1007/978-981-13-2291-4_35

or of Arrival (TOA), has a certain mitigation on signal attenuation, reflection, and multipath effects, which has high precision and is widely used, but it is difficult to overcome NLOS error.

Paper [1] proposed an improved distance smoothing method which can mitigate the NLOS error significantly, and use variance estimation and online distance mean to identify LOS and NLOS propagation, thereby updating the Kalman filter effectively. Paper [2] proposed a novel algorithm based on semidefinite programming (SDP) which can reduce TOA error in NLOS environment. In paper [3], a NLOS mitigation algorithm based on Kalman filter and channel classification is proposed to improve TOA accuracy in NLOS and multipath environment. By using two classes of non-parametric regressors to form an estimate of the ranging error, paper [4] proposed an approach to reduce NLOS errors in physical layer directly. Paper [5, 6] use neural network to mitigate NLOS and achieve better results. This paper proposes an algorithm that combines ELM and EKF, using the predicted EKF state and the external measured value as the input to the ELM to classify measured values and determine the propagation path affected by NLOS error, then select the appropriate ELM network to correct the measured values so as to effectively mitigate the NLOS error.

2 TOA Positioning and EKF

In indoor positioning based on TOA, the distance between tag and anchor point is measured by symmetrical double-sided two-way ranging (SDS-TWR) algorithm, which can effectively solve the problem of clock synchronization between nodes and obtain high-precision distance. After the distances between tag and anchor points are obtained by distance measuring algorithm, if the distances from tag to at least three anchor points is known, the position of the tag can be determined by trilateration, as shown in Fig. 1.

Kalman Filter is only applicable to linear Gaussian state space model. For nonlinear models, the basic idea of EKF is to use Taylor expansion to linearize the state equations and observation equations, and use the first term in expansion to compare with the nonlinear functions in the model. The linearization process is performed and the Jacobian matrix is calculated so that the Gaussian distribution is used to approximate the posterior distribution of the state and then Kalman Filter is used for estimation.

The equation of state of the moving target can be expressed as:

$$X_k = AX_{k-1} + \omega_{k-1} \tag{1}$$

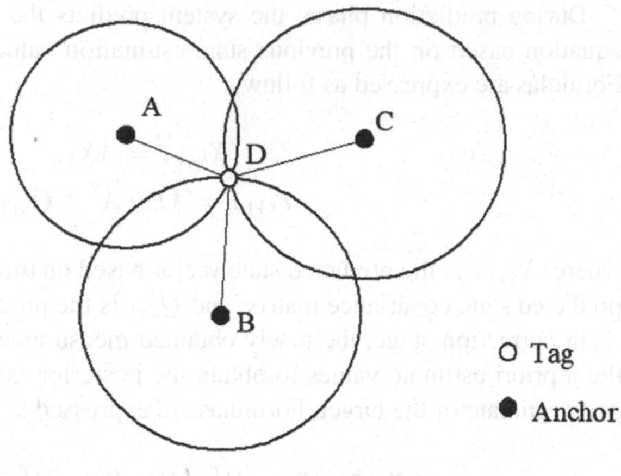

Fig. 1 Positioning algorithm of trilateration

where: $X_k = \begin{bmatrix} x_k & v_x & y_k & v_y \end{bmatrix}^T$ is the state variable and $A = \begin{bmatrix} 1 & T & 0 & 0 \\ 0 & 1 & 0 & 0 \\ 0 & 0 & 1 & T \\ 0 & 0 & 0 & 1 \end{bmatrix}$ is the state matrix of the system. ω_k is Gaussian noise, and its variance is Q.

Then establish an observation model based on the principle of trilateral positioning. The coordinates of the n-th anchor node are known to be (x_n, y_n), The value of the distance from tag to n anchors at time k is $(d_{1k}, d_{2k}, \ldots, d_{nk})$, The formula of distance between anchor node and target is as follows:

$$d_{nk} = \sqrt{(x_k - x_n)^2 + (y_k - y_n)^2} \tag{2}$$

The observation equation is:

$$Z_k = h_k(X_k) + \upsilon_k \tag{3}$$

$$h_k(X_k) = \begin{bmatrix} \sqrt{(x_k - x_1)^2 + (y_k - y_1)^2} \\ \sqrt{(x_k - x_2)^2 + (y_k - y_2)^2} \\ \vdots \\ \sqrt{(x_k - x_n)^2 + (y_k - y_n)^2} \end{bmatrix} \tag{4}$$

where:

$Z_k = \begin{bmatrix} d_{1k} & d_{2k} & \cdots & d_{nk} \end{bmatrix}^T$, and υ_k is the Gaussian noise of the measurement and the variance is R. Since the observation model is nonlinear, the Jacobian matrix of $h_k(X_k)$ H_k is required.

The EKF is mainly divided into two phases: time update (prediction) and measurement update (correction).

During prediction phase, the system predicts the current state using the state equation based on the previous state estimation value to obtain a priori estimate. Formulas are expressed as follows:

$$X_{k+1,k} = AX_k \tag{5}$$

$$P_{k+1,k} = AP_{k,k}A^T + Q_{k+1} \tag{6}$$

where: $X_{k+1,k}$ is the predicted state vector based on time k at time $k+1$, $P_{k+1,k}$ is the predicted state covariance matrix, and Q_{k+1} is the process noise variance matrix.

In correction stage, the newly obtained measurement values are combined with the a priori estimate values to obtain the posterior estimate values and correct the prior estimate of the target. Formulas are expressed as follows:

$$K_{k+1} = P_{k+1,k}H_{k+1}^T\left(H_{k+1}P_{k+1,k}H_{k+1}^T + R_{k+1}\right)^{-1} \tag{7}$$

$$X_{k+1,k+1} = X_{k+1,k} + K_{k+1}\left(Z_{k+1} - H_{k+1}X_{k+1,k}\right) \tag{8}$$

$$P_{k+1,k+1} = (I - K_{k+1}H_{k+1})P_{k+1,k} \tag{9}$$

where: K_{k+1} is Kalman gain at time $k+1$, and observation matrix is Z_{k+1}, observed noise variance matrix is P_{k+1}.

3 NLOS Error

In many scenarios, the propagation of radio signals is non-line-of-sight propagation, which will cause a positive additional excess delay for the TOA measurement, resulting in a large deviation in the position estimate of unknown node. The NLOS error obeys exponential distribution or uniform distribution in different channel environments.

In the NLOS environment, the known TOA of the anchor (x_i, y_i) measured with the tag (x, y) is τ_i, converted to distance:

$$d_i = c\tau_i = l_i + l_i^{NLOS} + \upsilon_i \tag{10}$$

where: c is the speed of light, l_i is the line-of-sight distance from tag to i-th anchor, l_i^{NLOS} is NLOS error, and υ_i is the measurement error.

4 Extreme Learning Machine

Extreme learning machine is a new learning algorithm model applied to a single hidden layer feed-forward neural network. Its characteristic lies in that before network training, only appropriate number of hidden layer nodes is set. The weights of input nodes and hidden layer and thresholds of hidden layer are assigned randomly. Hidden layer output weights can be obtained through analytical operations. The network training process is completed once without complicated iterations. Therefore, the extreme learning machine algorithm model has excellent characteristics of fast learning and good generalization performance [7].

The training process of ELM can be described as follows:

Given N different training samples (x_i, t_i), where $x_i = \left[x_{i1}\ x_{i2}\ \cdots\ x_{in} \right]^T \in R^n$ is the sample input value, $t_i = \left[t_{i1}\ t_{i2}\ \cdots\ t_{im} \right]^T \in R^m$ is the expected output value of sample. The single hidden layer forward neural network model using ELM is shown in Fig. 2.

Assuming the network has n input layer nodes, L hidden layer nodes, and m output layer nodes. The activation function of hidden layer neurons is $g(x)$, and the bias of hidden layer neurons is b_i.

Then the expression of the ELM model is

$$\sum_{i=1}^{L} \beta_i g\left(\omega_i \cdot x_j + b_i\right) = y_j \tag{11}$$

where: $j = 1, 2, \ldots, N$, $\omega_i = \left[\omega_{1i}\ \omega_{2i}\ \cdots\ \omega_{ni} \right]$ represents the connection weight vector for all input layer nodes and the i-th hidden layer node, $\beta_i = \left[\beta_{1i}\ \beta_{2i}\ \cdots\ \beta_{ni} \right]^T$ represents the connection weight vector of the i-th hidden layer node and network output layer node, $y_j = \left[y_{j1}\ y_{j2}\ \cdots\ y_{jm} \right]^T$ represents network output value.

The error approximating these N samples of standard SLFNs, which have L hidden layer nodes, can be infinitely close to zero. That is

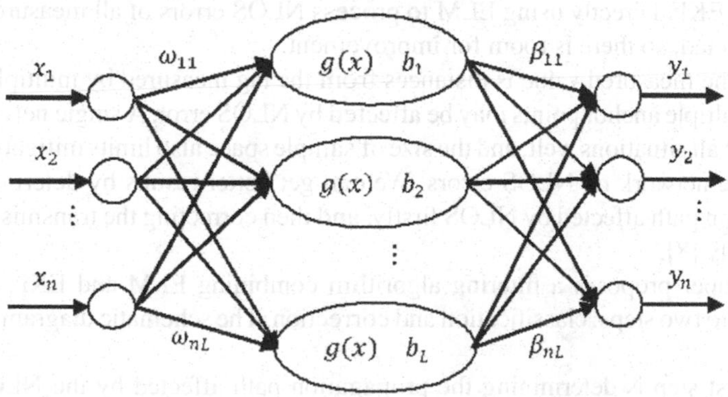

Fig. 2 ELM model structure

$$\sum_{j=1}^{L} \left\| y_j - t_j \right\| = 0 \tag{12}$$

The existence of β_i,ω_i and b_i makes the following expression established

$$\sum_{i=1}^{L} \beta_i g\left(\omega_i \cdot x_j + b_i\right) = t_j \tag{13}$$

Written in matrix form is

$$H\beta = T \tag{14}$$

where: H is output matrix of single hidden layer.

Assuming Moore-Penrose generalized inverse of matrix H is $H^{\dagger}$, the smallest norm least-squares solution of above linear system is

$$\hat{\beta} = H^{\dagger} T \tag{15}$$

5 NLOS Error Mitigation Algorithm Based on ELM and EKF

In order to mitigating NLOS error, the proposed algorithm in this paper combines ELM and EKF. Using the measured value $Z_k = \begin{bmatrix} d_{1k} \ d_{2k} \cdots d_{nk} \end{bmatrix}^T$ as input, the output target is true value of the distance between nodes, and train the ELM network for correcting NLOS error. This method has certain effect on the mitigation of NLOS error, but the relevant state of EKF can not be fully utilized by simply combining ELM and EKF. Directly using ELM to process NLOS errors of all measured values is not targeted, so there is room for improvement.

Since the measured value is distances from the tag measured by multiple anchor points, multiple anchor points may be affected by NLOS error. A single network does not handle all situations well, and the size of sample space also limits mitigation effect of a single network on NLOS errors. We can get better results by determining the propagation path affected by NLOS firstly, and then correcting the transmission path accordingly [8].

This paper proposes a filtering algorithm combining ELM and EKF, which is divided into two steps: classification and correction. The schematic diagram is shown in Fig. 3.

The first step is determining the propagation path affected by the NLOS error. In EKF prediction stage, error of prediction point coordinate with respect to the real coordinate obeys Gaussian distribution. Therefore, although there is an error in distances from prediction point to anchor points, there is no NLOS effect. Therefore,

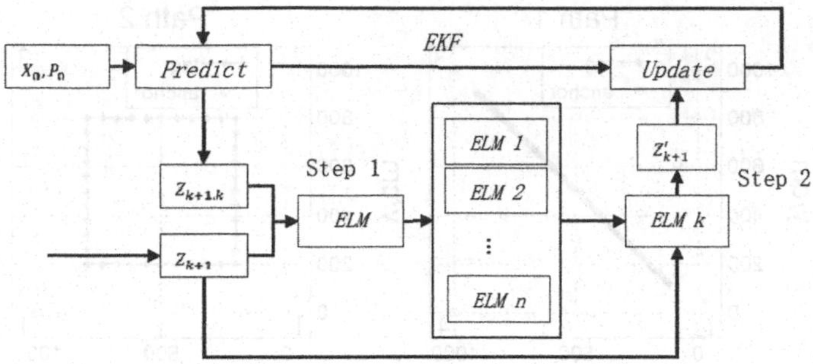

Fig. 3 NLOS error mitigation based on ELM and EKF

it can be used as a reference to determine whether the propagation path is affected by NLOS error. The distance from prediction point to each anchor point is $Z_{k+1,k}$. Here, $Z_{k+1,k}$ and take measured value Z_{k+1} as ELM input, and measured values are classified to determine the propagation path affected by NLOS error in measured value.

The second step is correcting the NLOS error. For NLOS error of different propagation paths, the ELM network for correcting the NLOS error is separately trained, with measured value as input and true value as target. Since only one case is concerned, for mitigation of a particular propagation path NLOS error, the effect of the network is better than a single network that trains all paths. According to the propagation path affected by the NLOS error determined in the first step, the corresponding ELM network is selected to mitigating NLOS error more effectively.

6 Simulation

For the purpose of verifying mitigation of ELM on NLOS error in location tracking, and the effect of the algorithm proposed in this paper, the same experiment data was compared with the following three methods as a control for simulation experiments in NLOS environment.

Method 1: This method uses EKF without any processing of measured values.
Method 2: This method correct measured value Z_{k+1} with a single trained ELM. The corrected measured value is used for updating EKF measurement.
Method 3: This method firstly classify the measured value with measured value Z_{k+1} and predicted value $Z_{k+1,k}$, and then selects the trained ELM network to correct NLOS error. The output is the corrected measured value, which is used as the input for updating the EKF measurement.

Simulation conditions: There are four anchor nodes in a square area of 1000 cm × 1000 cm. Their coordinates are (0, 0), (0, 1000), (1000, 0) and (1000, 1000) respectively. The movement paths of the tag adopt four conditions as shown in Fig. 4, a

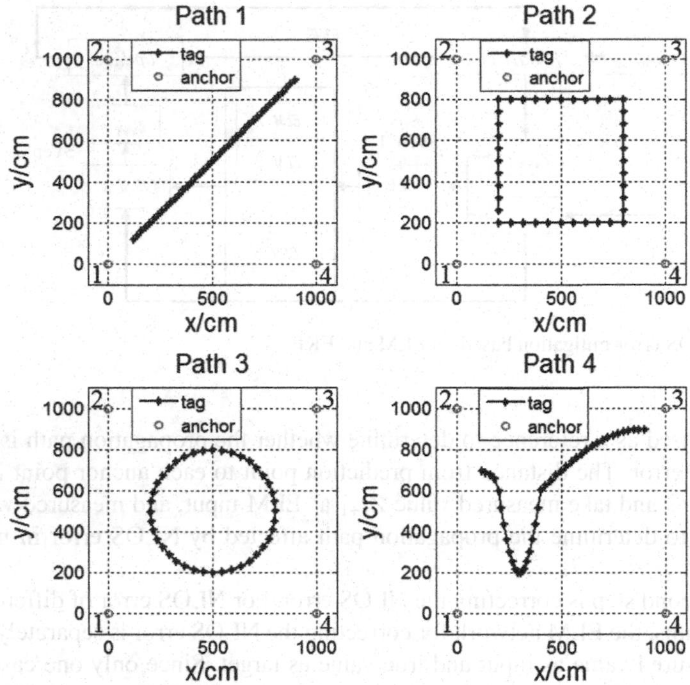

Fig. 4 The moving paths of the tag

Table 1 RMSE in different situations

RMSE/cm	Path 1	Path 2	Path 3	Path 4
Method 1	25.72	24.33	26.85	29.73
Method 2	21.18	20.35	13.05	21.70
Method 3	13.77	9.19	7.49	14.07

straight line, a circle, a square, and an irregular curve. 40 points are sampled for each path, and the distance error from tag to the four anchors obeys Gaussian distribution of the variance 100 cm^2.

The 40 points of each path sample are averagely divided into four segments. The NLOS error is added to the measured distance from anchor 1 to tag in the first segment. In second segment, The NLOS error is added to measured distance from anchor 2 to the tag, and so on. And the NLOS error are uniformly distributed following [10, 90] cm.

Figure 5 shows the simulation results, and Table 1 shows the RMSE of different conditions.

It can be seen from Fig. 5 that EKF algorithm of method 1 has a large error due to none processing of NLOS. The measured values in Methods 2 and 3 are corrected by ELM, so there is a certain improvement relative to Method 1. However, because method 2 uses a single ELM network to correct NLOS error of all propagation path,

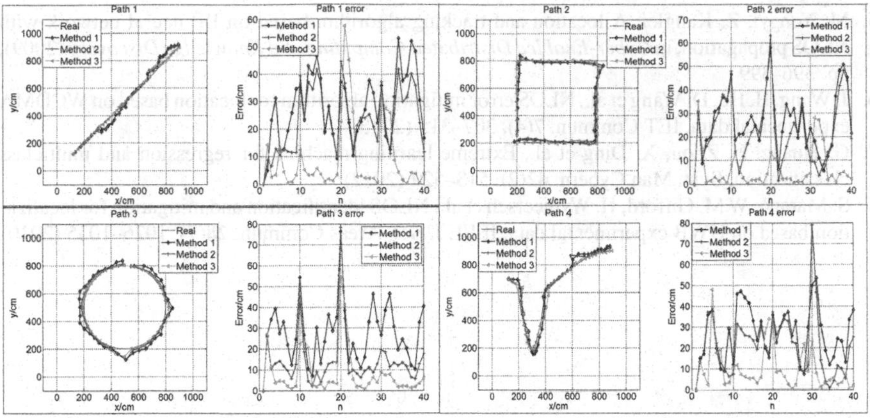

Fig. 5 Experimental results

the effect of mitigating NLOS error is limited. As it can be seen from the four curves and data in the experiment, the promotion of method 2 is not very obvious. In most cases, method 3 can well determine the propagation path affected by the NLOS error, so that the targeted ELM network can be used to correct the NLOS effectively.

7 Conclusion

For the problem of NLOS error in indoor tracking, a solution combining ELM and EKF is proposed in this paper. The EKF predicted value and the measured value are used to classify measured value first, and then the corresponding ELM network is used to correct NLOS error of different propagation paths. As we can see from experimental results, this algorithm can effectively mitigate the NLOS error and improve the positioning and tracking accuracy.

References

1. K. Yu, E. Dutkiewicz, Improved Kalman filtering algorithms for mobile tracking in NLOS scenarios, in *Wireless Communications and Networking Conference* (2012), pp. 2390–2394
2. R.M. Vaghefi, J. Schloemann, R.M. Buehrer et al., NLOS mitigation in TOA-based localization using semidefinite programming, in *Workshop on Positioning Navigation and Communication* (2013), pp. 1–6
3. J. He, Y. Geng, F. Liu et al., CC-KF: enhanced TOA performance in multipath and NLOS indoor extreme environment. IEEE Sens. J. **14**(11), 3766–3774 (2014)
4. H. Wymeersch, S. Marano, W.M. Gifford et al., A machine learning approach to ranging error mitigation for UWB localization. IEEE Trans. Commun. **60**(6), 1719–1728 (2012)

5. M. Yongyi, Z. Kanglei, A location and tracking algorithm based on BP neural network with NLOS propagation, in *Cyber-Enabled Distributed Computing and Knowledge Discovery* (2009), pp. 396–399
6. J. Wang, H. Hu, D. Wang et al., NLOS error mitigation algorithm for location based on WCDMA experimental data. IET Commun. **7**(4), 307–315 (2013)
7. G. Huang, H. Zhou, X. Ding et al., Extreme learning machine for regression and multiclass classification. Syst. Man Cybern. **42**(2), 513–529 (2012)
8. S. Maranò, W.M. Gifford, H. Wymeersch et al., NLOS identification and mitigation for localization based on UWB experimental data. IEEE J. Sel. Areas Commun. **28**(7), 1026–1035 (2010)

Forecasting Short-Term Residential Electricity Consumption Using a Deep Fusion Model

Ming Lei, Liyang Tang, Mingxing Li, Zhenyu Ye and Liwei Pan

Abstract Electricity consumption forecasting is practically significant for either detecting abnormal power usage pattern or resource-conserving purpose. Indeed, it is a non-trivial task since electricity consumption is related to multiple complex factors, including historical amount of consumption, calendar dates and holidays, as well as residential power consumption habits. To this end, we propose an end-to-end structure to collectively forecast short-term power consumption of private households, called RCFNet (Residual Conventional Fusion Network). Specifically, our RCFNet uses (1) three branches of residual convolutional units to model the temporal proximity, periodicity and tendency properties of electricity consumption, (2) one fully connected neural network to model the weekday or weekend property, and (3) a residual convolution network to fuse the above output to produce short-term prediction. All the convolutions used here are one-dimensional. Through experimental studies on residential electricity consumption dataset in Australia, it is validated that the proposed RCFNet outperforms several well-known methods. Besides, we demonstrate that residential power consumption is closely related to the living characteristics of residents.

M. Lei · L. Tang · Z. Ye · L. Pan (✉)
Key Laboratory of Public Safety Emergency Information Technology of Anhui Province, China Electronic Technology Group Corporation 38th Research Institute, Hefei 230088, China
e-mail: panliwei0813@163.com

M. Lei
e-mail: buaalei001@163.com

L. Tang
e-mail: tangliyang921@gmail.com

Z. Ye
e-mail: yyyu200@163.com

M. Li
Seventh Research Division and the Center for Information and Control, School of Automation Science and Electrical Engineering, Beihang University (BUAA), Beijing 100191, China
e-mail: lmx196@126.com

Y. Jia et al. (eds.), *Proceedings of 2018 Chinese Intelligent Systems Conference*, Lecture Notes in Electrical Engineering 529,
https://doi.org/10.1007/978-981-13-2291-4_36

Keywords Residential electricity · Short-term load forecasting
Residual convolutional neural network

1 Introduction

With the development of the economic, consumption pattern of electricity becomes the result of varying environmental parameters, and flexibility in balancing supply and demand is increasingly required. And the stability of the modern power system is undergoing unprecedented challenges. Accurate load forecasting can guarantee stable and secure power grid, reduce the cost of power generation, and facilitate the power system operations [1]. With the massive deployment of smart meters, residential electricity consumption data can be easily collected, and thus create opportunities for short-term load forecasting for individual customers. On the one hand, abnormal electricity consumption can be detected through historical electricity consumption data mining. On the other hand, household electricity consumption could be profiled considering tiered pricing strategy and meters readings, so that electricity consumption could be adjusted in time, which can reduce the waste of electricity consumption and achieve the purpose of energy conservation and environmental protection [2].

In this paper, we aim to make contributions to short-term residential load forecasting using residential electricity consumption data collected by smart meters. It is very challenging since household-level consumption is much more irregular than the transmission or distribution levels. Specifically, the power consumption of individual households is affected by many factors such as:

Temporal dependencies. Residential power usage may be affected by recent time intervals. For example, power consumption at 10:00 a.m. is likely to be affected by that of 9:00 a.m. Intuitively, residential electricity consumption shows periodicity. Those people who are regular office workers should have similar consumption on consecutive workdays, repeating every 24 h. However, the law of electricity consumption may also have certain differences. For example, temperature and sunrises differs among different seasons, say summer and winter, which affects the daily routines of residents and therefore affects their power consumption characteristics.

External factors. For example, weather conditions such as sunny or rainy, calendar days such as workdays, weekends and holidays, as well as other events may influence power consumption characteristics.

To tackle these challenges, we propose a deep network called RCFNet (Residual Conventional Fusion Network) to collectively predict short-term residential power consumption. Specifically, our RCFNet uses three branches of residual convolutional units to model the temporal proximity, periodicity and tendency properties of electricity consumption, one fully connected neural network to model the weekday or weekend property, and a residual convolution network to fuse the above output to produce short-term prediction. Besides, we evaluate our approach using smart meter data from Australia.[1] Experimental results demonstrate the advantages of our approach compared with 4 baselines: History Average, Seasonal Autoregressive Integrated Moving Average Model, Multi-Layer Perceptron and Long Short-Term Memory.

The remainder of this paper is structured as follows. Section 2 reviews related works. In Sect. 3 we present our methodology. In Sect. 4 we describe the data set used in our study, and present our experimental settings, evaluation measures, and the empirical results with discussion. Finally, Sect. 5 concludes the paper and gives an outlook on future work.

2 Related Works

Many approaches had been reported in the literature on short-term load forecasting. However, most existing works deal with group behavior of power consumption instead of individual residents, since residential forecasts were conventionally considered trivial due to the volatile nature of individual loads. Therefore, the issues on short-term household load forecasting remain open [3–10]. In literature, methods for energy load forecasting include: time-series analysis (e.g. ARIMA [11]), regression analysis method [12], neural networks [13], support vector machine [14], fuzzy theory [15], etc.

Cao et al. [11] adopted ARIMA considering meteorological conditions for intraday load forecasting, so as to enhance the forecasting accuracy and robustness. Pang et al. [16] proposed a neural network which takes only the load values of current and previous time steps as the input to predict the load value at next time step. The result is further compensated by rough set to increase the forecasting accuracy. Chaouch et al. [17] proposed a functional time series for short term forecasting of household-level intra-day electricity load curve. Ghofrani et al. [18] used spectral analysis and Kalman filter on short-term load forecasting for residential customers.

As an improved method of Recurrent Neural Network (RNN), the Long Short-Term Memory (LSTM) [19] has made great progress in the field of sequence learning. Marino et al. [20] validated two variants of the LSTM: standard LSTM and LSTM-based Sequence to Sequence (S2S) architecture for single-meter residential load forecasting issue. Kong et al. [21] also proposed a LSTM based framework in the same task.

Deeper and deeper CNNs (Convolutional Neural Network) were proposed by researchers [22, 23], which reveals the importance of network depth, that is, deeper networks typically lead to superior results. However, with the increase of network depth, the gradient vanishing problem emerges in the process of neural network back propagation. By using shortcut connections, Residual Networks (ResNets) [24] perform residual mapping fitted by stacked nonlinear layers. It solves the gradient vanishing problem along with the increase of network depth. Combining these two approaches, Zhang et al. [25] proposed a deep-learning based approach, called ST-

[1] http://data.gov.au/dataset/electricity-consumption-benchmarks.

ResNet, to forecast the inflow and outflow of crowds in each and every region of a city. Inspired by above existing works, we use one-dimensional convolution to extract features from power consumption data sequence collected through smart meters and optimize the structure of fusion network.

3 Methodology

Smart meters will periodically collect residential consumption data, and time interval is typically half an hour. Therefore, our task is formulated as following: given meters reading data series of observations $\{S_t | t = 0, 1, \ldots, n-1\}$, predict future electricity consumption S_n. Unlike electric load at the system level, the complexity of household load forecasting lies in the significant volatility and uncertainty. As mentioned in [21], the diversity in the aggregated level smooths the daily load profiles, while the electricity consumption of a single resident is more dependent on the underlying human behavior. Daily routines and lifestyle of individual residents, together with various types of appliances, may have a more direct impact on the load profile. In order to improve the prediction performance, we propose to abstract hidden knowledge from previous observation data and establish a learning algorithm to model the correlation between abstracted knowledge and forecasting targets.

3.1 Residual Networks

The general idea of the Residual Networks [24] is that, by using shortcut connections, residual networks perform residual mapping fitted by stacked nonlinear layers, which is easier to be optimized than the original mapping. Identity Mapping ResNet [26] simplifies the residual networks and solves gradient vanishing problem along with the increase of the network depth, and thus greatly improves the training effect of the network. A residual unit [26] with an identity mapping is defined as:

$$X^{(l+1)} = X^{(l)} + F(X^{(l)}) \tag{1}$$

where $X^{(l)}$ and $X^{(l+1)}$ are the input and output serial of the lth residual unit respectively, and F is a residual function [24].

3.2 Residual Conventional Fusion Network (RCFNet)

Figure 1 shows the structure of the model used in this paper.

As described earlier, the time series data of power consumption is obtained by half-an-hour interval. Here we divide the temporal correlation of data into three types: (a) Adjacent data, we use the most recent 24 h (half an hour interval, 48 data points) for proximity structure; (b) Period data, such as electricity consumption data at the same time every day, and the last 28 days (28 data points) are selected; (c) Trend data, we choose the last ten weeks of data (10 data points) for tendency structure. In this way, time series data of three characteristics is constructed as input of the model. The same residual convolution network structure is employed to extract above three types of features, which is shown by Fig. 1 (details of ResConvNet in Fig. 1 will be discussed in Sect. 3.3).

As for external factors, we manually extract some features. Considering the difficulties in external information acquisition, here we simply differentiate weekdays and weekends. Specifically, consumption data during weekdays or weekends is encoded into a one-hot vector, as the input to a two-layer fully connected network. The first three modules (i.e. tendency, periodicity, proximity) output a feature vector with a length of fixed size (e.g. 10), and the external module also outputs a vector with the same length, and then the four outputs are spliced into one vector. Then, the vector is used as the input of the FusionNet (detail shown in Fig. 2c). Finally, output from FusionNet is fed to a ReLU (Rectified Linear Unit) activation function and the results are mapped to [0, 1].

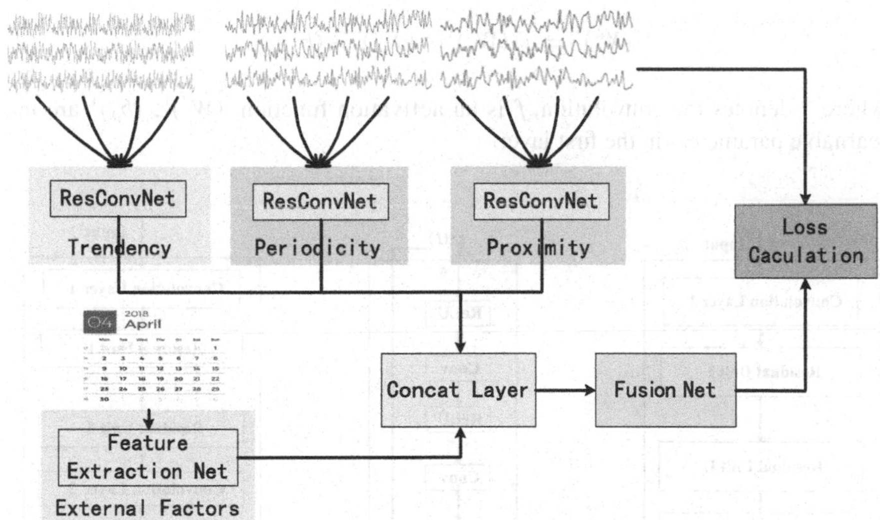

Fig. 1 RCFNet architecture

3.3 Time Serial Feature Extraction Network

The first three components (tendency, periodicity, proximity) share the same network structure, as shown in Fig. 2a. Details of each component are explained in the following.

(1) **Convolution**. A good model should be able to capture time series features. Convolution Neural Network has a strong ability to grasp the spatial structure information characteristics [27], in which two-dimensional convolution kernel is used to extract features of adjacent areas of image. Note that power consumption data is one-dimensional time series data. Thus we consider using one-dimensional convolution kernel to perform one-dimensional convolution, so that the adjacent features in time series can be extracted. By stacking multiple layers of convolution network layer, we can obtain larger receptive field. Considering the loss of resolution caused by subsampling while preserving distant dependencies [28], we do not use subsampling, but only convolutions.

The proximity module in the structure showed in Fig. 1 samples the power consumption data over a recent period to model the proximity characteristics of data. Here $X_t^a = (X_t^a)^0 = [S_{t-l_a}, S_{t-(l_a-1)}, \ldots, S_{t-1}]$ represents the proximity dependency sequence as an input to the entire network at time t, and proceeding to the next layer through convolution operations as:

$$(X_t^a)^1 = f((W_a)^1 * (X_t^a)^0 + (b_a)^1) \quad (2)$$

where $*$ denotes the convolution, f is an activation function, $(W_a)^1$, $(b_a)^1$ are the learnable parameters in the first layer.

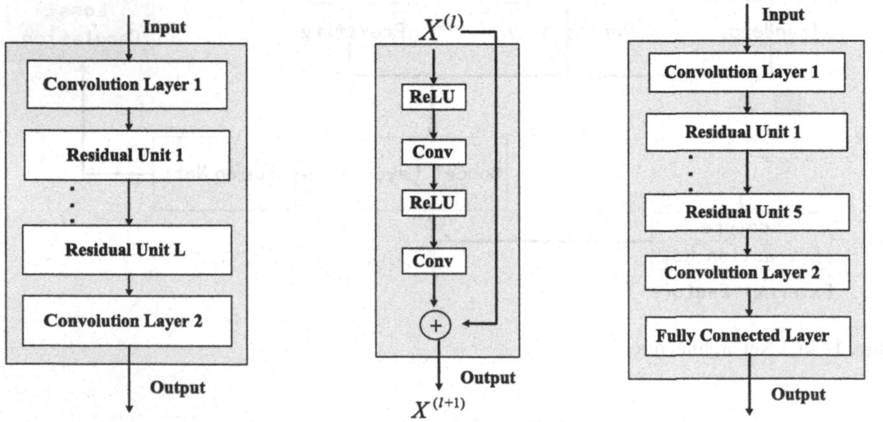

(a) structure of ResConvNet **(b) structure of Residual Unit** **(c) structure of FusionNet**

Fig. 2 Time serial feature extraction network

(2) **Residual Unit**. We use the residual learning method [26] to extract features of the sequence from deep network. In our residual convolution network as in Fig. 2a, b, we stack L residual units on the convolution output

$$(X_t^a)^{l+1} = (X_t^a)^l + F((X_t^a)^l; (\theta_a)^l), l = 1, \ldots, L \quad (3)$$

Here, F represents a residual function (e.g. a combination of ReLU and convolution, see Fig. 2b), and $(\theta_a)^l$ contains all the learnable parameters of layer L. We use batch normalization [29] before ReLU layer. At the top of all residual units, we add a convolution layer to build a network structure composed of two convolution layers and one residual unit, which is used to extract the adjacent features.

Similarly, extraction of periodic and trend characteristics are achieved through the same structure (see Fig. 1). Here $X_t^p = (X_t^p)^0 = [S_{t-l_p \cdot p}, S_{t-(l_p-1) \cdot p}, \ldots, S_{t-p}]$ and $X_t^q = (X_t^q)^0 = [S_{t-l_q \cdot q}, S_{t-(l_q-1) \cdot q}, \ldots, S_{t-q}]$ respectively represent that extraction of periodic and trend sampling data, l_p and l_q respectively represent the length of the corresponding sequence. p is the sampling span representing the periodicity of the sequence, which is equal to one-day that describes daily periodicity. q is equal to one-week representing the trend of the sequence. The number of data sampling points is 48 per day, and the corresponding settings for periodicity interval p and trend interval q are respectively $p = 48$, $q = 7 \times 48$.

(3) **External Component**. Power consumption largely depends on residential behavior, and there are a lot of influential factors, such as weather conditions, holidays, etc. Power consumption behavior differs among weekdays and weekends. Limited by data availability, here we simply consider the influence of weekdays and weekends on residential power consumption. We encode it as a one-hot vector connected to the two layers of the full connection layer, where the external output is defined as E_t, and the output is mapped to a vector length of fixed size (e.g. 10) to ensure consistent outputs of previous three components.

(4) **Fusion**. We concatenate four modules' output (i.e. tendency, periodicity, proximity and external factor) above to one vector. We feed the vector into the FusionNet (see Fig. 2c),which consist of one convolution layer, five stacked residual units, one convolution layer and one fully connected layer. Finally, output from FusionNet is fed to a ReLU (Rectified Linear Unit) activation function and the results are mapped to [0, 1].

The network trains predictions $\widehat{S}_t$ by inputting three data sequences and extrinsic features, and then, minimizes the mean squared error between the predicted power consumption and the actual power consumption:

$$MSE(\theta) = \left\| S_t - \widehat{S}_t \right\|_2^2 \quad (4)$$

Here θ denotes learnable parameters, and S_t is observed value at time t.

3.4 RCFNet Training

The program design is demonstrated with pseudo code in Program 1. First, training instances are constructed from original sequence data (lines ①–②). Then, RCFNet is trained via back-propagation and Adam [30] optimizer (lines ③–⑧) (Here we use Adam optimizer because [31] indicates it outperforms other candidates, including the stochastic gradient descent (SGD), Adagrad [32], Adadelta [33] and RMSProp [34]). Finally, we validate the model on the test set (⑨–⑩).

Program 1:RCFNet Training Program

Input:

length of adjacent, periodicity, trend sequences: l_a, l_p, l_q;
period interval: p; trend interval: q;
historical observations: $S = \{ S_0, \ldots, S_{n-1} \}$;
external features: $E = \{ E_0, \ldots, E_{n-1} \}$;

Data preparing:

① for all available time interval $t(1 \leq t \leq n-1)$, add series data $X_t^a = [S_{t-l_a}, S_{t-(l_a-1)}, \ldots, S_{t-1}]$, $X_t^p = [S_{t-l_p \cdot p}, S_{t-(l_p-1)\cdot p}, \ldots, S_{t-p}]$, $X_t^q = [S_{t-l_q \cdot q}, S_{t-(l_q-1)\cdot q}, \ldots, S_{t-q}]$, E_t, $S_t$to set $D = (\{X_t^a, X_t^p, X_t^q, E_t\}, S_t)$ (S_t and E_t are the actual power consumption and external factors at time t, respectively)
② divide D into training set D_{tr} and test set D_{ts}

Model Training:

③ build network on Keras [35] with Tensorflow [36] backend
④ initialize all learnable parameters θ in RCFNet
⑤ repeat:
⑥ randomly select a batch training data set from that set D_{tr}
⑦ evaluate performance by mean squared error to find trainable parameter θ
⑧ until the convergence condition is satisfied

Output:

Learned RCFNet model

Model Evaluation:

⑨ fetch all the samples from test set D_{ts} as the input to the learned RCFNet model
⑩ evaluate performance by using Root Mean Square Error (RMSE)

4 Experiments

In this section, we perform experiments on power consumption dataset from Australia to validate the efficiency of the proposed RCFNet. The dataset contains power consumption from 25 households in Victoria, published by Australian Government, Department of Industry, Innovation and Science, with a two-year coverage from April 1, 2012 to March 31, 2014. The sampling interval is half an hour. To eliminate data missing issue, we selected 8 households with full data coverage in this work.

4.1 Baselines

We compare our RCFNet with the following 4 baselines:

(1) **HA** (History Average). It is the most straightforward way for prediction. The basic assumption is that power consumption during each period over days exhibits strong regularity. The amount of electricity consumed is predicted by average the historical consumption during the same period on previous days.
(2) **SARIMA** (Seasonal ARIMA). It takes seasonal period difference into consideration. The seasonal parameter is set to days (48 sampling intervals) as the periodic characteristics of residential electricity usage are repeated on a daily basis.
(3) **MLP** (Multi-Layer Perceptron). A three-layer artificial neural network is used to predict the amount of electricity used. The model uses one day's data before the point (48 time periods, half an hour apart) as input.
(4) **LSTM** (Long Short-Term Memory). The model can connect previous time information to the task of the current time to complete the memory of the power consumption habits. A three-layer LSTM network is used here to predict the amount of power used the day before the point in time (48 time periods, half an hour apart) as input.

4.2 Preprocessing

All models are built on a desktop PC with a 3.4 GHz Intel i7-6700 processor and 16 GB of memory, using Python 3.6.4 and Keras 2.1.5 with TensorFlow 1.3.0 backend as model training tools.

In our RCFNet model, we first use the min-max regularization method to normalized training data to values between [0, 1]. In the output part of RCFNet, we use the ReLU activation function, which has better training convergence speed and better effect. During the validation phase, we convert the predicted values into normal values by reversing the operation. The external factors here, such as weekdays or weekends, are encoded as a one hot vector.

Convolutions use filters of size 3×1. Each residual unit uses two convolution layers. Here we use fixed layer size of residual units. The training set is divided into two parts, where the first 90% is used for training and the last 10% for validation. To prevent excessive training iterations, we implement early stopping technique to find optimal number of training iterations. After obtaining the best model, we continue to do fixed epoch training on the best model to get better results. Here we continue to train 50 times as the final result of model training. In the LSTM and MLP models, the sequence data with the length of 48 is used as input, while the final fusion model adopts the following parameters: the length of adjacent sequence is 48, the length of periodic sequence is 28, and the length of trend sequence is 10.

In order to maintain the consistency of the results, we use the last 60 days for model validation and the rest of the data as training data sets.

We validate the prediction performance using Root Mean Square Error (RMSE), which is defined as:

$$RMSE = \sqrt{\frac{1}{n}\sum_{t=1}^{n}(observed_t - predicted_t)^2} \tag{5}$$

where n is the total number of predictions.

4.3 Results

We validate our model against the other four models in Table 1, where the performance of each model is listed by each household ID. Moreover, in order to testify the influence of different parameter factors on RCFNet, we design six different model variants on RCFNet, RCF (network considering adjacent, periodicity, tendency and external factors), RCF-NoE (RCF without considering external factors), RCF-NoP (RCF without considering periodicity), RCF-NoQ (RCF without considering tendency), RCF-NoPQ (RCF without considering periodicity and tendency) and RCF-NoEPQ (RCF without considering periodicity, tendency and external factors). The best result in each column is shown by the bold underlined numbers.

From Table 1, we can see that RCFNet outperforms the other four baselines, which indicates that RCFNet captures electricity consumption characteristics well. In addition, we can also find that the impacts of proximity, periodicity, tendency, and external factors differ among households. This reflects that different residents' electricity consumption characteristics are quite different. Some residents have stronger periodicity, while others show more obvious trend characteristics. It can be seen that if some feature branches are not taken into account, some results of RCFNet may be worse than certain baseline models, which shows that these features have great influence on the accuracy of RCFNet. Here, we only simply consider the settings $l_a = 48$, $l_p = 28$, $l_q = 10$ for the length of adjacent l_a, periodic l_p and trend sequence l_q, and the influence of weekdays or weekends for external factor. If we consider the influence of these factors more comprehensively, our model may achieve better results. On the whole, the network which comprehensively considers the proximity, periodicity, tendency and external factors has better prediction performance.

Table 1 Comparison among different methods on Australia Set

Model	ID-1098	ID-17625	ID-3117	ID-3494	ID-3762	ID-4192	ID-8927	ID-9918
HA	259.0383	143.8904	387.866	103.2604	2997.6932	326.3549	181.6399	100.9354
SARIMA	175.1828	107.4669	280.309	91.3854	2106.2326	221.5403	138.7098	78.6782
MLP	196.7551	132.1633	306.0178	119.6558	2892.7752	300.1101	170.6765	101.0996
LSTM	185.5381	112.8852	284.9665	91.2207	2169.6357	229.701	142.3328	81.8849
RCFNet (ours)								
RCF	<u>**168.9863**</u>	<u>**107.2604**</u>	<u>**280.037**</u>	<u>**86.3546**</u>	<u>**2087.2318**</u>	<u>**215.0213**</u>	<u>**137.0944**</u>	<u>**77.0784**</u>
RCF-NoE	175.0846	109.1304	280.7952	88.4304	2099.0213	219.0942	139.6173	77.2698
RCF-NoP	171.8832	108.7702	281.3781	86.974	2138.8793	219.5323	138.2551	78.2808
RCF-NoQ	172.3574	108.7478	280.7664	87.3681	2156.0119	216.7504	138.9258	77.4132
RCF-NoPQ	175.8988	109.157	282.172	87.6574	2284.9714	220.4727	140.4179	78.978
RCF-NoEPQ	178.489	109.4951	282.8467	89.7897	2309.8326	224.6844	140.9597	79.9351

5 Conclusion and Future Work

In this paper, we propose a deep network architecture, called RCFNet, based on the historical data of residential electricity consumption to achieve accurate short-term electricity consumption forecast. The network structure takes into account the proximity, periodicity and tendency property of electricity consumption, as well as the influence of external factors. Through an extensive case study on residential power consumption dataset in Australia, the forecasting performance of proposed RCFNet is validated compared with other four typical forecasting models (i.e. History Average, Seasonal ARIMA, Multi-Layer Perceptron and Long Short-Term Memory). Besides, RCFNet has good scalability and can be widely used in other types of time series prediction. In future, we will further explore the characteristics of power consumption, and consider to optimize the RCFNet structure using GAN (Generative Adversarial Networks).

Acknowledgements This research work was partly supported by National Key Research and Development Program of China (Grant No. 2016YFC0800100), Major Research Program of the National Natural Science Foundation of China (Grant No. 91546103), and Anhui Provincial Natural Science Foundation (Grant No. 1708085QG162).

References

1. W. Kong, Z.Y. Dong, Y. Jia et al., Short-term residential load forecasting based on LSTM recurrent neural network. IEEE Trans. Smart Grid **PP**(99), 1–1 (2017)
2. O.I. Asensio, M.A. Delmas, Nonprice incentives and energy conservation. Proc. Natl. Acad. Sci. **112**(6), 510–515 (2015)
3. A. Marinescu, C. Harris, I. Dusparic et al., Residential electrical demand forecasting in very small scale: An evaluation of forecasting methods, in *International Workshop on Software Engineering Challenges for the Smart Grid* (IEEE, 2013), pp. 25–32
4. E. Mocanu, P.H. Nguyen, M. Gibescu et al., Deep learning for estimating building energy consumption. Sustain. Energy Grids Netw. **6**, 91–99 (2016)
5. A. Veit, C. Goebel et al., Household electricity demand forecasting: benchmarking state-of-the-art methods (2014)
6. B. Stephen, X. Tang, P.R. Harvey et al., Incorporating practice theory in sub-profile models for short term aggregated residential load forecasting. IEEE Trans. Smart Grid **8**(4), 1591–1598 (2017)
7. S. Humeau, T.K. Wijaya, M. Vasirani et al., Electricity load forecasting for residential customers: exploiting aggregation and correlation between households, in *Sustainable Internet and ICT for Sustainability* (IEEE, 2013), pp. 1–6
8. M.D. Wagy, J.C. Bongard, J.P. Bagrow et al., Crowdsourcing predictors of residential electric energy usage. IEEE Syst. J. **PP**(99), 1–10 (2017)
9. Y.H. Hsiao, Household electricity demand forecast based on context information and user daily schedule analysis from meter data. IEEE Trans. Industr. Inf. **11**(1), 33–43 (2017)
10. Y. Wang, Q. Xia, C. Kang, Secondary forecasting based on deviation analysis for short-term load forecasting. IEEE Trans. Power Syst. **26**(2), 500–507 (2011)
11. X. Cao, S. Dong, Z. Wu et al., A data-driven hybrid optimization model for short-term residential load forecasting, *in IEEE International Conference on Computer and Information Technology;*

Ubiquitous Computing and Communications; Dependable, Autonomic and Secure Computing; Pervasive Intelligence and Computing (IEEE, 2015), pp. 283–287
12. Y. Li, D. Niu, Application of principal component regression analysis in power load forecasting for medium and long term, in *International Conference on Advanced Computer Theory and Engineering* (IEEE, 2010), pp. V3-201–V3-203
13. Z.H. Osman, M.L. Awad, T.K. Mahmoud, Neural network based approach for short-term load forecasting. Int. J. Sci. Environ. Technol. **1**(5), 1–8 (2012)
14. N. Ye, Y. Liu, Y. Wang, Short-term power load forecasting based on SVM, in *World Automation Congress* (IEEE, 2012), pp. 47–51
15. C.L. Zhang, Power system short-term load forecasting based on fuzzy clustering analysis and rough sets. J. North China Electric Power Univ. (2008)
16. Q. Pang, M. Zhang, Very short-term load forecasting based on neural network and rough set, in *International Conference on Intelligent Computation Technology and Automation* (IEEE, 2010), pp. 1132–1135
17. M. Chaouch, Clustering-based improvement of nonparametric functional time series forecasting: application to intra-day household-level load curves. IEEE Trans. Smart Grid **5**(1), 411–419 (2014)
18. M. Ghofrani, M. Hassanzadeh, M. Etezadi-Amoli et al., Smart meter based short-term load forecasting for residential customers, in *North American Power Symposium* (IEEE, 2011), pp. 1–5
19. S. Hochreiter, J. Schmidhuber, Long short-term memory. Neural Comput. **9**(8), 1735–1780 (1997)
20. D.L. Marino, K. Amarasinghe, M. Manic, Building Energy Load Forecasting Using Deep Neural Networks (2016)
21. W. Kong, Z.Y. Dong, Y. Jia et al., Short-term residential load forecasting based on LSTM recurrent neural network. IEEE Trans. Smart Grid **PP**(99), 1–1 (2017)
22. K. Simonyan, A. Zisserman, Very deep convolutional networks for large-scale image recognition. Comput. Sci. (2014)
23. C. Szegedy, W. Liu, Y. Jia, P. Sermanet, S. Reed, D. Anguelov, D. Erhan, V. Vanhoucke, A. Rabinovich, Going deeper with convolutions, in *Proceedings of the IEEE Conference on Computer Vision and Pattern Recognition* (2015), pp. 1–9
24. K. He, X. Zhang, S. Ren et al., Deep Residual Learning for Image Recognition (2015), pp. 770–778
25. J. Zhang, Y. Zheng, D. Qi, Deep Spatio-Temporal Residual Networks for Citywide Crowd Flows Prediction (2016)
26. K. He, X. Zhang, S. Ren et al., Identity mappings in deep residual networks, in *European Conference on Computer Vision* (Springer, Cham, 2016), pp. 630–645
27. Y. Lecun, L. Bottou, Y. Bengio et al., Gradient-based learning applied to document recognition. Proc. IEEE **86**(11), 2278–2324 (1998)
28. J. Long, E. Shelhamer, T. Darrell, Fully convolutional networks for semantic segmentation, in *Computer Vision and Pattern Recognition* (IEEE, 2015), pp. 3431–3440
29. S. Ioffe, C. Szegedy, Batch normalization: accelerating deep network training by reducing internal covariate shift, in *ICML*, pp. 448–456
30. D.P. Kingma, J. Ba, Adam: a method for stochastic optimization. Comput. Sci. (2014)
31. S. Ruder, An Overview of Gradient Descent Optimization Algorithms (2016)
32. J. Duchi, E. Hazan, Y. Singer, Adaptive subgradient methods for online learning and stochastic optimization. J. Mach. Learn. Res. **12**, 2121–2159 (2011)
33. M.D. Zeiler, ADADELTA: an adaptive learning rate method. Comput. Sci. (2012)
34. G. Hinton, N. Srivastava, K. Swersky, RMSProp: divide the gradient by a running average of its recent magnitude, in *Neural Networks for Machine Learning, Coursera lecture 6e* (2012)
35. F. Chollet, Keras (2015). https://github.com/fchollet/keras
36. M. Abadi, A. Agarwal, P. Barham et al., TensorFlow: Large-Scale Machine Learning on Heterogeneous Distributed Systems (2016)

Prediction of Permeability Index of Blast Furnace Based on Online Sequential Extreme Learning Machine

Yan Di, Sen Zhang, Xiaoli Su, Yixin Yin and Baoyong Zhao

Abstract Permeability index of the blast furnace is one of the vital monitoring parameters to reflect the operation status of the blast furnace. At present, there are few prediction models for the permeability index at home and abroad. Therefore, this paper proposes to establish a prediction model of the permeability index by using online sequential extreme learning machine (OS-ELM) combined with wavelet analysis, and this paper compares it with the prediction models established by extreme learning machine (ELM), support vector machine (SVM) and BP neural network algorithm. The simulation results show that the prediction model based on OS-ELM has better accuracy than others.

Keywords Blast furnace · Permeability index · Prediction model
Online sequential extreme learning machine

1 Introduction

Blast furnace ironmaking is a highly complex and multivariable production process [1]. The permeability index is a characteristic quantity of the permeability change of the blast furnace column, when permeability index is poor, it will cause the static pressure of the furnace shell to rise. If the permeability is too high, the gas utilization rate will be reduced. The permeability index is one of the significant indicators

Y. Di · X. Su · Y. Yin · B. Zhao
University of Science and Technology Beijing, Beijing 100083, China

S. Zhang (✉)
School of Electrical and Electronic Engineering,
University of Science and Technology Beijing, Beijing, China
e-mail: zhangsen@ustb.edu.cn

Y. Di · X. Su · Y. Yin · B. Zhao
Key Laboratory of Knowledge Automation for Industrial Processes
of Ministry of Education, School of Automation and Electrical Engineering,
University of Science and Technology Beijing, Beijing 100083, China

Y. Jia et al. (eds.), *Proceedings of 2018 Chinese Intelligent Systems Conference*, Lecture Notes in Electrical Engineering 529,
https://doi.org/10.1007/978-981-13-2291-4_37

of measuring the anterograde state of blast furnace. According to the permeability index, the blast furnace operator can find and avoid the abnormal state of the blast furnace, such as hanging and slip, so as to judge blast furnace in time and to operate reasonably according to the change of the blast furnace. From this it can be seen that accurate prediction of permeability index of blast furnace is crucial for blast furnace operators.

The permeability index characterizes the change in the permeability of the blast furnace [2]. The index of the cold air flow and the pressure difference are used to predict the permeability index in the conventional model [3]. However, there are many parameters in the process of blast furnace ironmaking, and there are many non-linear relationships between parameters. It is difficult to accurately characterize the permeability index only by using cold air flow and pressure difference. And this will affect the accuracy of prediction and cannot meet actual production requirements.

Online Sequential Extreme Learning Machine (OS-ELM) is a fast single hidden layer neural network algorithm [4]. In the initial part, the weights from the hidden layer to the output layer are learned through a small number of samples by using ELM. The second part is online learning. The output weights learned in the initial part are updated through a single sample or data block sequentially in this phase. Compared with the traditional learning algorithm, it not only considers the timeliness of the data, but also has the features of fast computing speed, strong generalization ability, and does not fall into the local minimum. Therefore, it has been widely used in various fields [5, 6].

This paper uses the measured data to establish the model of permeability index based on the OS-ELM algorithm. Firstly, in order to improve the accuracy of the model, this paper analyses the factors affecting the permeability index according to the relevant mechanism of the blast furnace, and given to blast furnace production data contain noise, this paper uses wavelet transform to process the production data. Then the model is compared with the model established by ELM, SVM and BP [7, 8]. The experimental results show that the prediction model established by using OS- ELM algorithm can accurately predict the permeability index in real time. This model is more suitable for solving the problems in this paper and provides support for the subsequent operation of the blast furnace.

2 Selection of Input Parameters of Model and Data Processing

2.1 The Permeability Index Factors Analysis

The parameters of the blast furnace ironmaking process are numerous and interact with each other. Furnace foremen who have experience not only use air volume, wind pressure, and top pressure to predict the permeability index, but also commonly

Table 1 Mutual information between parameters of the blast furnace and permeability index

Operating parameters	Mutual information
Blast volume	0.6831
Blast pressure	0.7621
Blast velocity	0.7815
Blast temperature	0.7346
Top pressure	0.6975
Top temperature	0.8089
Permeability index	2.8527

combine other operating parameters of the blast furnace to jointly forecast in actual production.

The higher the blast temperature is, the more the heat is brought into the blast furnace, which in turn creates thermal fluctuations in the lower part of the blast furnace. The blast velocity can affect the activity of the hearth of the blast furnace and affect the distribution of the gas flow in the lower part of the blast furnace, thus affecting the lower part of the blast furnace. The top temperature can indirectly reflect the heat exchange status inside the blast furnace when the temperature of the blast temperature changes little. Therefore, the blast temperature, blast velocity, and top temperature are added in the parameter input of the model.

In addition, the permeability index data is a non-linear time series, data trends in the early can predict future trends. Therefore, the historical data of permeability index before prediction point is also considered as input parameters during the modeling process.

From the above analysis, this paper selects blast volume blast temperature, blast pressure, blast velocity, top temperature, top pressure the historical data of permeability index as input parameters. Moreover, this paper analysis the correlation between input parameters and the permeability index individually combined with the mutual information.

For the whole variable X, the entropy of the reduced uncertainty is called mutual information entropy (mutual information) due to the occurrence of the variable Y and the correlation between the two. And it can be formulated as

$$I(X, Y) = H(X) - H(X/Y) = H(Y) - H(Y/X) = I(Y, X) \tag{1}$$

If $I(X, Y) > \delta I(Y, Y)$, it is considered that there is a strong correlation between the parameter X and Y. Generally $\delta = 0.2$. It can be seen from Table 1 that these 6 parameters and have a strong correlation with the permeability index, and it can be seen from Table 2 the correlation between permeability index at t time and the previous data decreases with time, so the first three time points of data are also selected as input parameters.

Table 2 Mutual information between permeability index before *t* time and permeability index at *t* time

Time	*t*-6	*t*-5	*t*-4	*t*-3	*t*-2	*t*-1
Mutual information	0.3259	0.3704	0.4786	0.6380	0.8099	0.7425

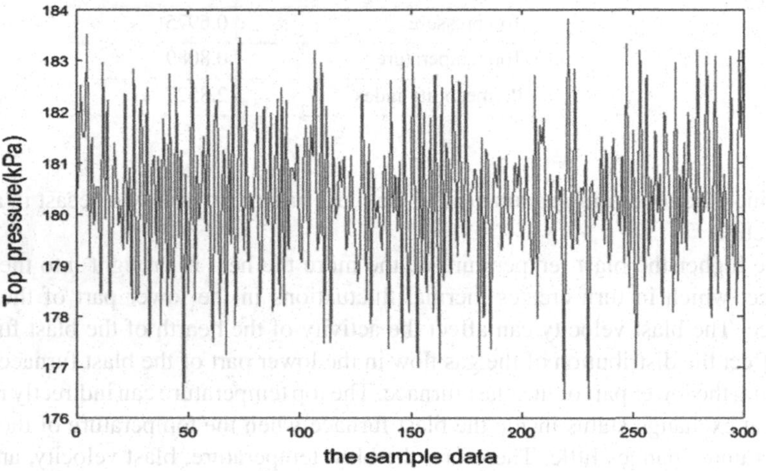

Fig. 1 The row data of top pressure

2.2 Data Processing

Because the operation environment of the blast furnace is complex, there are a lot of noises of the production data [9]. The use of these blast furnace data to set up a prediction model will make the prediction effect worse, so the production data of the blast furnace will be processed first.

Wavelet transform can decompose the signal into sub-signals with different frequency and scale, and effectively distinguish the abrupt part and noise in the signal, so as to achieve the non-stationary signals de-noising [10]. Take top pressure data as an example in this paper, using wden() function in MATLAB to process the data.

After many experiments, the soft threshold method is used to de-noising, and demy is selected as wavelet function, at the same time the de-noising effect is not obvious when the decomposition level is greater than 4 layers, so the decomposition level selects 4 layers. Figure 1 shows the top pressure data before data processed, Fig. 2 shows the top pressure data after data processed.

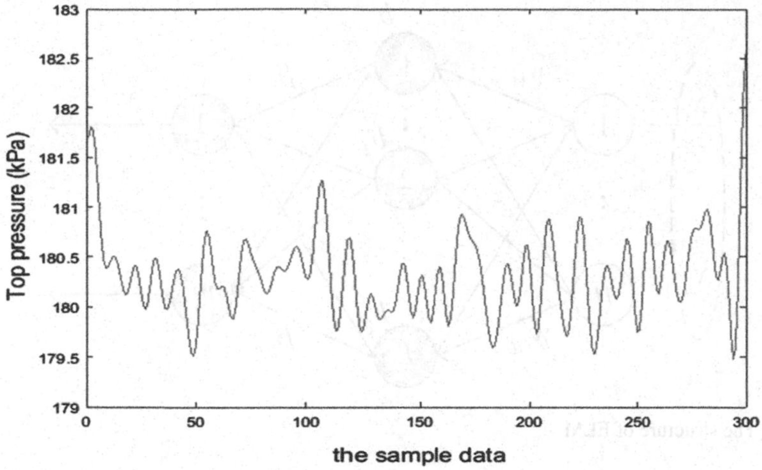

Fig. 2 The de-noised data of top pressure

3 Prediction of Permeability Index of Blast Furnace Based on Online Sequential Extreme Learning Machine

3.1 Online Sequential Extreme Learning Machine (OS-ELM)

Online sequential extreme learning machine algorithm is an online learning model of ELM algorithm. The network structure of extreme learning machine can be shown in Fig. 3. In OS-ELM, with the acquisition of new data, the output weights are updated sequentially, which involves two aspects: (1) All the data are involved in the training; (2) In the online learning process, regardless of the data acquisition time, the data is treated equally.

The OS-ELM algorithm is divided into the following two steps:

(1) Initialization phase: Suppose given N_0 distinct samples (x_i, t_i), $x_i = [x_{i1}, x_{i2}, \ldots, x_{in}]^T \in R^n$, $t_i = [t_{i1}, t_{i2}, \ldots, t_{im}]^T \in R^m$. Do not require complex iterations and optimization during the training. The input weights and biases of the hidden layers are randomly generated. The output weight $\beta^{(0)}$ is calculated by ELM in order to $\|H\hat{\beta}\|_{\beta^{(0)}} = \min_{\beta^{(0)}}\|H\beta^{(0)} - T\| = 0$.

$$Where\ \ H_0 = \begin{bmatrix} g(\omega_1 x_1 + b_1) & \vdots & g(\omega_{N_0} x_1 + b_{N_0}) \\ \vdots & \ddots & \vdots \\ g(\omega_1 x_{N_0}) + b_1) & \vdots & g(\omega_{N_0} x_{N_0} + b_{N_0}) \end{bmatrix} \tag{2}$$

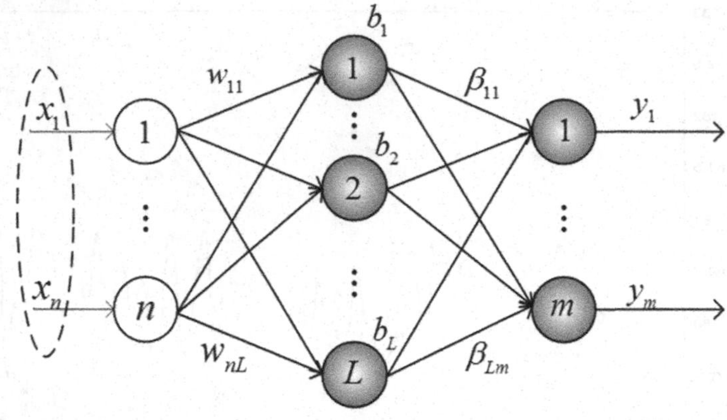

Fig. 3 The structure of ELM

This paper use SVD to solve the pseudo-inverse of the matrix.

$$\beta^{(0)} = K_0^{-1} H_0^T T_0,\ K_0 = H_0^T H_0 \tag{3}$$

(2) Phase of Online Learning: When N_1 distinct samples are added into the model, the output weight $\beta^{(1)}$ is calculated by ELM. Considering both training data sets N_1 and N_0, the output weight becomes

$$\beta^{(1)} = K_1^{-1} \begin{bmatrix} H_0 \\ H_1 \end{bmatrix}^T \begin{bmatrix} T_0 \\ T_1 \end{bmatrix},\ K_1 = \begin{bmatrix} H_0 \\ H_1 \end{bmatrix}^T \begin{bmatrix} H_0 \\ H_1 \end{bmatrix} \tag{4}$$

For sequential learning, we have to express $\beta^{(1)}$ as a function of $\beta^{(0)}$, and K_1 is not a function of the data set N_0, now $\beta^{(1)}$ and K_1 can be written as

$$\begin{aligned} \beta^{(1)} &= \beta^{(0)} + K_1^{-1} H_1^T (T_1 - H_1 \beta^{(0)}) \\ K_1 &= K_0 + H_1^T H_1 \end{aligned} \tag{5}$$

Thus we have

$$\begin{aligned} \beta^{(K+1)} &= \beta^{(K)} + K_{K+1}^{-1} H_{K+1}^T (T_{K+1} - H_{K+1} \beta^{(K)}) \\ K_{K+1} &= K_K + H_{K+1}^T H_{K+1} \end{aligned} \tag{6}$$

Let $P_{K+1} = K_{K+1}^{-1}$, then the equations for updating $\beta^{(K+1)}$ can be written as

$$\begin{aligned} \beta^{(K+1)} &= \beta^{(K)} + P_{K+1} H_{K+1}^T (T_{K+1} - H_{K+1} \beta^{(K)}) \\ P_{K+1}^{-1} &= P_K^{-1} + H_{K+1}^T H_{K+1} \end{aligned} \tag{7}$$

3.2 *Prediction Model Based on OS-ELM*

The specific steps of established the model are as follows:

(1) Make sure the input layer has 9 nodes and the output layer has an output node.

Combined with the analysis of this paper, nine important blast furnace parameters are chosen as input nodes, they are the blast volume, the blast pressure, the blast temperature, the top pressure, the top temperature, the blast velocity and 3 historical permeability index, the permeability index is the output node.

(2) Establish OS-ELM network structure.

  (1) Select the activation function from several common activation functions such as sigmoid, sin, hardlim, radbas.
  (2) Set the number of hidden layer nodes, Randomly generate the weight matrix ω and the bias vector b;
  (3) Using the training samples, calculate the hidden layer output matrix H_0 and the output weight $\beta^{(0)}$;
  (4) According to Eqs. (3)–(6), updating the hidden layer output matrix and the output weight and find out the output vector of the model.

(3) Verify OS-ELM network structure.
Enter the testing samples, calculate the predicted output value by using this model, if the mean square error between the predicted output value and the actual measured value obtained meets the requirements, then the model is established. If not, return to step (2) to reset the number of the hidden layer nodes.

4 Simulation Experiment and Analysis

Select 1496 actual measurement data, of which 1196 data are used to train the model and 300 data are used to be tested. It will cause great fluctuations in the prediction results and affect the accuracy of the prediction model because the unit of the input parameters is different and the numerical differences are large. Therefore, the data processed by wavelet de-noising is normalized. Then do Simulation experiment by using OS-ELM, ELM, SVM, BP model.

Figure 4 shows a comparison of the predictive results of the blast furnace permeability index with the OS-ELM prediction model and the actual data. Figure 5 shows the relative error of the predicted permeability index of OS-ELM.

After repeated adjustments and analysis, this paper found that when the number of nodes in the hidden layer is 30, the size of N_0 is 360 and the size of $N_{K+1} = 1$, the performance of the established prediction model achieves the best goal. So the number of nodes is determined as 30, N_0 is 360 and $N_{K+1} = 1$.

In order to further compare the prediction effect of each prediction model, Fig. 6 shows the comparison results between the prediction results of the blast furnace

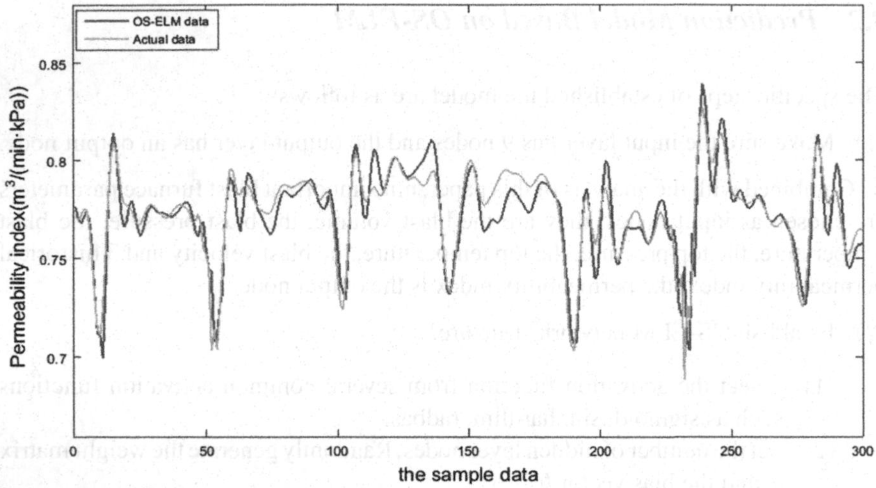

Fig. 4 Prediction results of permeability index based on OS-ELM

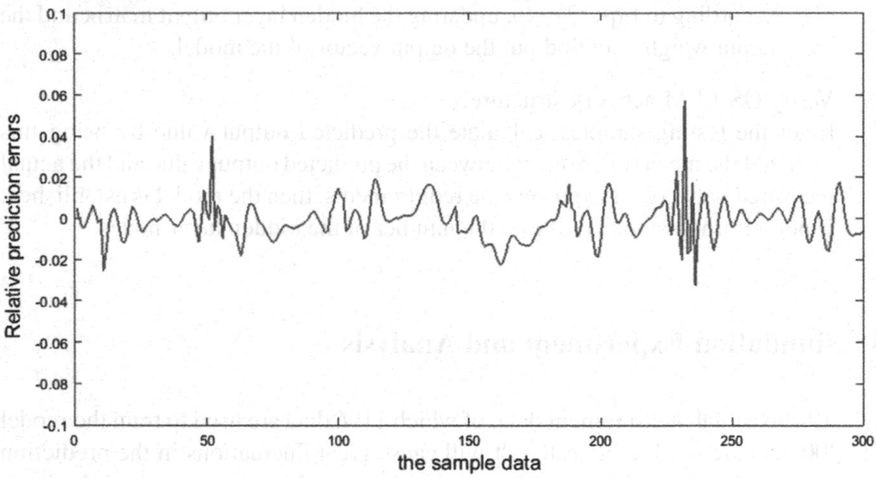

Fig. 5 Relative prediction errors of permeability index based on OS-ELM

permeability index based on OS-ELM, ELM, SVM, and BP models and the actual data. Figure 7 shows relative prediction errors of permeability index based on SVM, BP, ELM and OS-ELM. It can be seen that OS-ELM model has a better prediction performance compared with ELM, BP and SVM.

The root mean square error RMSE, training time, and testing time are used as evaluation indexes for each prediction model in this paper. The RMSE expression is:

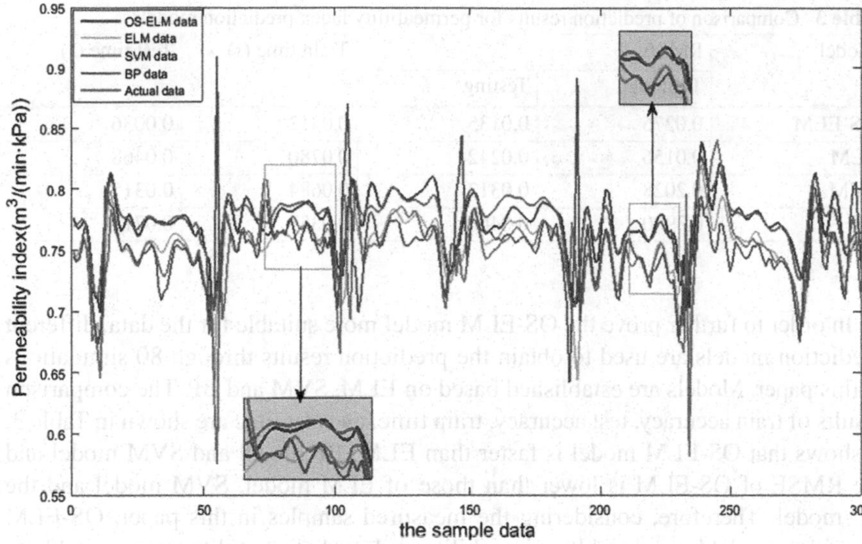

Fig. 6 Prediction results of permeability index based on different algorithms

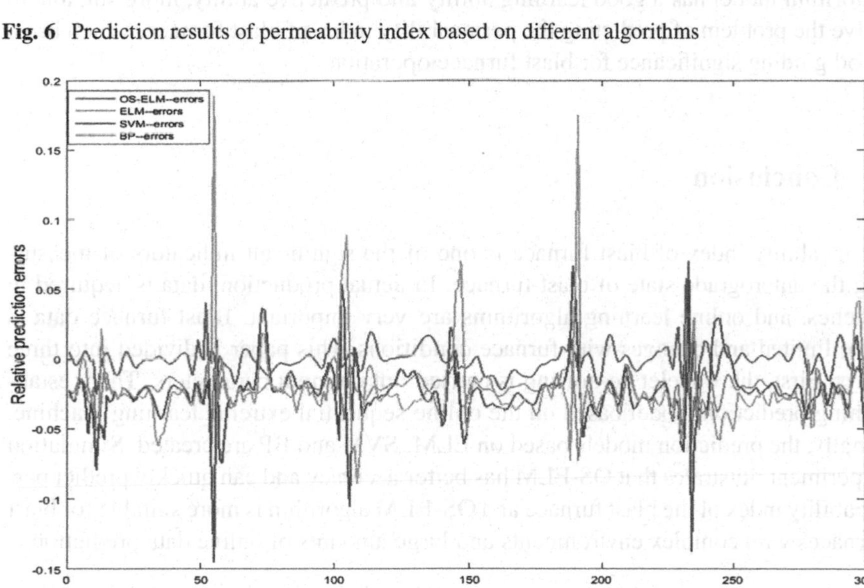

Fig. 7 Relative prediction errors of permeability index based on different algorithms

$$RMSE = \sqrt{\frac{\sum_{i=1}^{n} (y_{ob,i} - y_{\text{model},i})^2}{n}} \tag{8}$$

Table 3 Comparison of prediction results for permeability index prediction models

Model	RMSE		Train time (s)	Test time (s)
	Training	Testing		
OS-ELM	0.0276	0.0135	0.0312	0.0036
ELM	0.0156	0.0242	0.0780	0.0468
SVM	0.2028	0.0312	0.0684	0.0315
BP	0.3900	0.0468	0.0357	0.0712

In order to further prove the OS-ELM model more suitable for the data, different prediction models are used to obtain the prediction results through 80 simulations in this paper. Models are established based on ELM, SVM and BP. The comparison results of train accuracy, test accuracy, train time, and test time are shown in Table 3. It shows that OS-ELM model is faster than ELM, BP model and SVM model and the RMSE of OS-ELM is lower than those of ELM model, SVM model and the BP model. Therefore, considering the measured samples in this paper, OS-ELM algorithm model has a good learning ability and predictive ability, more suitable to solve the problem of predicting the permeability index of blast furnace, which has a good guiding significance for blast furnace operation.

5 Conclusion

Permeability index of blast furnace is one of the significant indicators of measuring the anterograde state of blast furnace. In actual production, data is acquired in batches, and online learning algorithms are very important. Blast furnace data is time limited and changes with furnace conditions. This paper is divided into three parts. First, the wavelet de-noising is carried out on input parameters. Then, establishing prediction model based on the online sequential extreme learning machine. Finally, the prediction models based on ELM, SVM and BP are created. Simulation experiment illustrates that OS-ELM has better accuracy and can quickly predict permeability index of the blast furnace and OS-ELM algorithm is more suitable for blast furnaces with complex environments and large amounts of online data prediction.

Acknowledgements This work was supported by the Key Program of National Nature Science Foundation of China under grant No. 61,333,002, the National Nature Science Foundation of China under grants No. 61,673,056, the Beijing Natural Science Foundation under grant No. 4,182,039, and the Beijing Key Discipline Construction Project (XK100080537).

References

1. X.G. Liu, S.H. Luo, Y.H. Liu et al., Nonlinear mixed control for silicon contents of hot metal in BF ironmaking processes. Control Theor. Appl. **23**, 391–396 (2006)
2. H. Liu, P.R. Li, Z.J. BAO et al., Intelligent predictive modeling of blast furnace system. J. Cent. South Univ. (Science and Technology) **43**, 1787–1794 (2012)
3. C.D. Zhou, Blast Furnace Ironmaking Production of Technical Manual. Metallurgical Industry Press (2002)
4. H.G. Zhang, S. Zhang, Y.X. Yin, Online sequential ELM algorithm with forgetting factor for real applications. Neurocomputing **261**, 144–152 (2017)
5. G.B Huang, C.W. Deng, J. Xu, J.X. Tang, Extreme learning machines: new trends and application. Sci. China (information Sciences) **02**, 5–20 (2015)
6. X.L. Su, Y.X. Yin, S. Zhang, Prediction model of improved multi-layer extreme learning machine for permeability index of blast furnace. Control Theor. Appl. **33**(12), 1674–1684 (2016)
7. Nello Cristanini, John Shawe Taylor. An Introduction to support vector machines and other kernel-based learning methods.[M] China Machine Press,2005
8. K. Kim, K. Tk, J. Kittler, Locally linear discriminant analysis for multimodally distributed classes for face recognition with a single model image. IEEE Trans. Pattern Anal. Mach. Intell. **27**, 318–327 (2005)
9. C.H. Gao, L. Jian, J.M. Chen et al., Data-driven modeling and predictive algorithm for complex blast furnace ironmaking process. Acta Autom. Ainica **35**(6), 725–730 (2009)
10. Z. Madadi, G.V. Anand, A.B. Premkumar, Signal detection in generalized gaussian noise by nonlinear wavelet de-noising [J]. IEEE Trans. Circuits Syst. I Regul. Pap. **60**(11), 2973–2986 (2013)

3D Printing Fault Detection Based on Process Data

Bing Li, Lin Zhang, Lei Ren and Xiao Luo

Abstract 3D printing technology is a kind of rapid prototyping technology. In the 3D printing process, several common faults often happen, resulting in interruption of the printing process or poor-quality of the printed product. In order to maintain normal function of the 3D printer, users need to manually check the scene all the time. In order to perform real-time detection of faults in the printing process of the 3D printer, multiple sets of experiments were conducted. We use sensors to obtain multiple parameters of the 3D printer. The machine learning method is used to classify and detect whether the printing process is in a fault state. This method can effectively detect the fault condition that occurs during real-time 3D printing process and can be promoted in more 3D printers.

Keywords 3D printing fault prediction · Logistic regression Classification model

1 Introduction

3D printing technology [1] is a highly digitized, automated and highly personalized manufacturing technology. 3D printing technology is different traditional machining and other "reducing material manufacturing" technologies [2]. It is based on the theory of discrete deposition, using layered processing and superposition molding,

B. Li · L. Zhang (✉) · L. Ren · X. Luo
School of Automation Science and Electrical Engineering, Beihang University, Beijing 100191, China
e-mail: johnlin9999@163.com

B. Li
e-mail: icearl@qq.com

L. Ren
e-mail: lei_ren@126.com

X. Luo
e-mail: 243219055@qq.com

Y. Jia et al. (eds.), *Proceedings of 2018 Chinese Intelligent Systems Conference*, Lecture Notes in Electrical Engineering 529,
https://doi.org/10.1007/978-981-13-2291-4_38

through the layer-by-layer accumulation of materials, achieving the manufacture process. It uses a computer to cut the 3D model of the formed part into a series of "thin" pieces of a certain thickness. The 3D printing device manufactures each layer "sheet" from the bottom up to finally form a three-dimensional solid part. This manufacturing technology eliminates the need for conventional tools or molds, enabling the manufacture of complex structures that are difficult or impossible in conventional method, and can effectively simplify the production process and shorten the manufacturing time cycle [3]. However, not matching with the rapid development of 3D printing technology, the fault detection and product quality inspection of 3D printing are still in a very imperfect state and lacking systematic detection methods [4–6].

In the 3D printing process, several common failures are often encountered, resulting in interruption of the printing process or poor-quality of the printed product. Specifically, it mainly includes two types, one is that the printing material is exhausted or broken, resulting in the uncompleted product (Fig. 1), and the other is that at the start of printing the material does not adhere to the printing base firmly, resulting in the poor-quality product (Fig. 2).

For current 3D printing technology, one single 3D printed product takes at least a few hours to completely print out a usable product. If there some faults in the printing process, the product will become unable to use. Simultaneously, if the user fails to notice when the printing faults, hours (including materials) will be wasted. If users want to keep it operating properly and discover the above-mentioned possible faults in time, they need to manually check 3D printer all the time. This costs people and time unnecessarily. So, this reduces the overall efficiency of successful printing, especially when users manage multiple 3D printers at the same time with cloud environment and distribution collaboration, manual supervision of the printer becomes even more infeasible. So, the efficiency is quite low, and human resource is not worthwhile for this. What we want is a detection device that can monitor the status of the current 3D printer in real time and detect whether it is in a fault state by monitoring the status parameter and provide an alarm service for the users [7], thereby saving a lot of time and printing materials.

Fig. 1 The uncompleted product

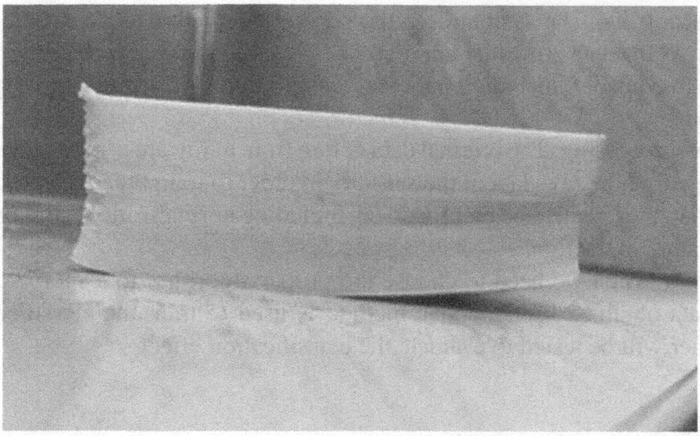

Fig. 2 The poor-quality product

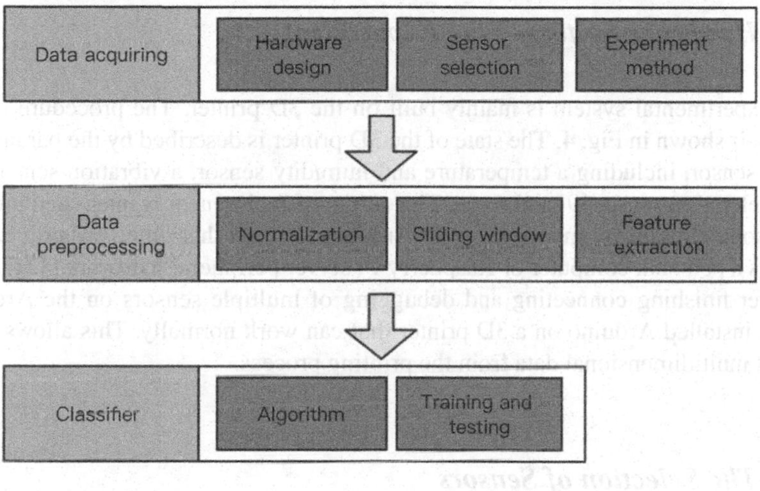

Fig. 3 Procedure of the system

This article takes the 3D printer as the research object, and obtains several important parameters in the printing process through various sensors, and use these parameters to predict which state the current printing is in.

2 Architecture Overview

The purpose of this system is to detect the printing status using the values of multiple sensors. Specifically, the experiment includes the following aspects, shown in Fig. 3.

1. Data acquiring: The system uses the sensor parameters to detect the printing status. So, the data is the first import process we need to deal with. There are three things to consider, including hardware design, sensor selection and experiment method.
2. Data preprocessing: The original data gotten from hardware is not usable directly. After getting the raw data of the sensors, in order to train the data needed by the classifier, they need to be preprocessed, including normalization, sliding window and feature extraction.
3. Classifier: Then we need to choose the proper algorithm to train the classifier. The data obtained by the above method is used to train the classifier and the classifier will be tested to evaluate the classification effect.

3 Hardware System and Data Acquiring

3.1 Hardware System of the Experiment

The Experimental system is mainly built on the 3D printer. The procedure of the system is shown in Fig. 4. The state of the 3D printer is described by the parameters of the sensor, including a temperature and humidity sensor, a vibration sensor [8], an acceleration sensor [9] and so on. The output of each sensor is integrated into the open-source hardware microcontroller Arduido [10]. At this time, Arduido can be used as a personal computer or Raspberry Pi server peripheral hardware [11].

After finishing connecting and debugging of multiple sensors on the Arduino board, installed Arduino on a 3D printer that can work normally. This allows us to collect multidimensional data from the printing process.

3.2 The Selection of Sensors

Choosing proper sensors in order to have a good classification is also a challenge. Here we use the combination of prior knowledge and experimentation. Firstly, according to the composition and characteristics of 3D printers, relevant literature research results and certain printing experience, select the several possible key state parameters, in order to obtain the most relevant parameters with the fault. When relevant parameters and corresponding sensors are configured, then the data can be analyzed based on these sensors. Finally, several sensors with high correlation of fault classification is left for future use, for model training and verification.

In the experiment, current, voltage, vibration and acceleration has been tested, and the test shows that the most related feature is vibration and acceleration signal.

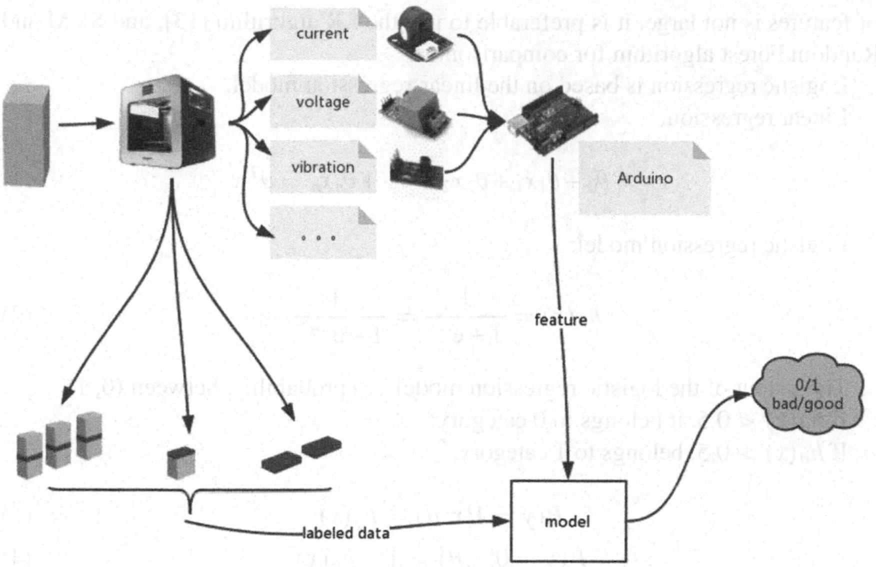

Fig. 4 Procedure of the system

3.3 Experiment Method

For data analysis, first we need to get a large set of labeled samples [12]. Each sample consists of values of multiple sensors as its characteristic values on whether it is a fault as a label. For this reason, the experiment method is to control the cross-section of the printed product during the experiment to be the same, so that a certain time can be deducted from the height, so that we can know what the label of a certain sample is from the time stamp certain moment corresponds to a certain height. Based on this method, the experiment was printed multiple times for various cross sections, resulting in multiple sets of data. These data will be used as the sample set for training and testing.

4 Algorithm

Based on the above algorithm, this paper sets up a specific scheduling scenario, performs simulation experiments on the scheduling scenario, and verifies the effectiveness of the scheduling. First, the scheduling of production of the above algorithm under the scene determined by the service time is checked.

There are two kinds of faults mentioned above, and each kind of fault can be regarded as a two-class problem (whether it is in a fault state). Besides, the number of features is not large, it is preferable to use the LR algorithm [13], and SVM and Random Forest algorithm for comparison.

Logistic regression is based on the linear regression model.

Linear regression:

$$z = \theta_0 + \theta_1 x_1 + \theta_2 x_2 + \cdots + \theta_n x_n = \theta^T x \tag{1}$$

Logistic regression model:

$$h_\theta(x) = \frac{1}{1 + \mathrm{e}^{-z}} = \frac{1}{1 + \mathrm{e}^{-\theta^T x}} \tag{2}$$

The output of the logistic regression model is a probability, between (0, 1).

If $h_\theta(x) < 0.5$, it belongs to 0 category;

If $h_\theta(x) > 0.5$, belongs to 1 category.

$$P(y = 1|x;\theta) = h_\theta(x) \tag{3}$$

$$P(y = 0|x;\theta) = 1 - h_\theta(x) \tag{4}$$

The corresponding probability function is:

$$P(y|x;\theta) = (h_\theta(x))^y * (1 - h_\theta(x))^{1-y} \tag{5}$$

The likelihood function corresponding to m samples is:

$$L(\theta) = \prod_{i=1}^{m} \left(h_\theta\left(x^{(i)}\right)\right)^{y^{(i)}} * \left(1 - h_\theta\left(x^{(i)}\right)\right)^{1-y^{(i)}} \tag{6}$$

The log likelihood function is:

$$l(\theta) = \log(L(\theta)) = \sum_{i=1}^{m} y^{(i)} \log\left(h_\theta\left(x^{(i)}\right)\right) + \left(1 - y^{(i)}\right) \log\left(1 - h_\theta\left(x^{(i)}\right)\right) \tag{7}$$

Finally:

$$J(\theta) = -\frac{1}{m} l(\theta) \tag{8}$$

Then the gradient descent method can then be used to find the optimal θ parameter.

Random forests [14] are based on decision trees and use bagging methods to integrate learning.

Support vector machine [15] use the theory of maximizing the interval.

5 Experiment Process

5.1 *Data Preprocessing*

After getting the data set using the hardware and experiment method, now start the analysis.

The original data is shown in Fig. 5.

There are several following problems about the original data:

1. The sensitivity to noise is relatively high. This is because that 3D printers produce relatively large vibrations at the edges of the 3D product, and this is not necessary for data analysis. Besides, it is not easy for visualization and data understanding when they exist. Therefore, they can be treated as noises;
2. Different products are printed, and the sensor will produce signals in different numerical ranges. We need to control them within the same numerical range. Otherwise, different prints will not be comparable and the predicted result is not reasonable;
3. We actually need to know whether the printer is in an abnormal state in a small period of time (an abnormal state at a certain moment may encounter noise interference), so the data should be smoothed then the prediction effect will be better.

Based on above problems, we need to do some data preprocessing.

1. Sliding windows [16]

Make each n data points as a sliding window. This can reduce the sensitivity to noise points. The exact value of n needs to be selected during the experiment. If the value of n is too low, it will be sensitive to noise. But if the value of n is too high, the

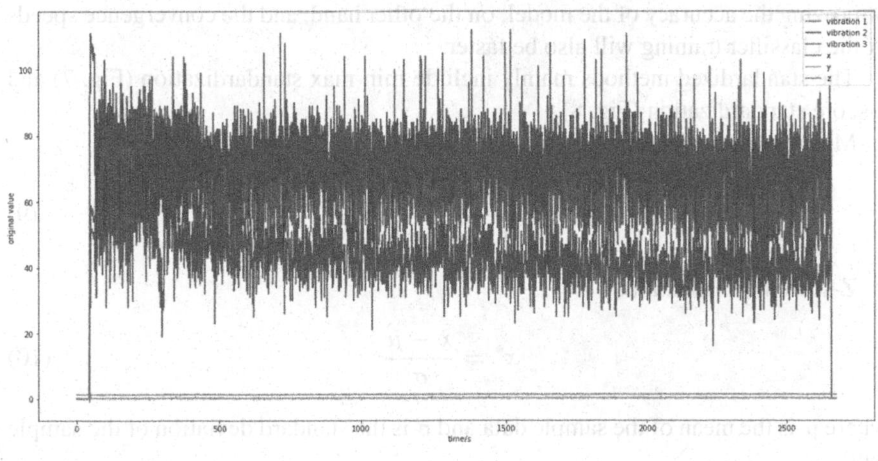

Fig. 5 Origin data

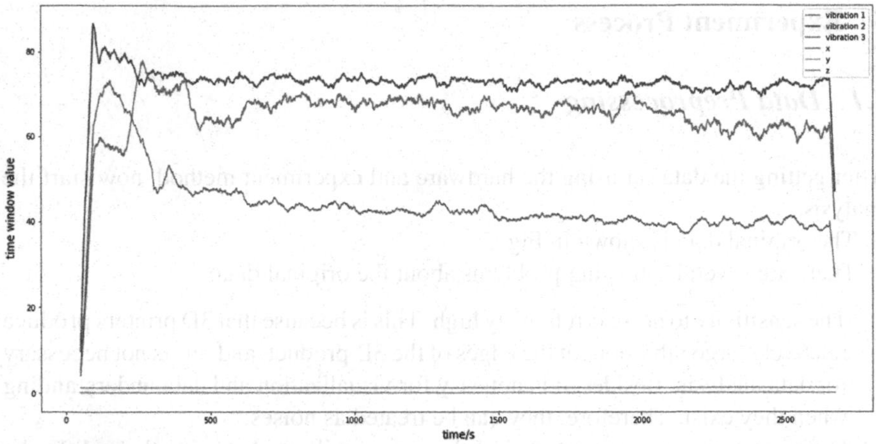

Fig. 6 Data after preprocess of time window

result will lose meaning because the ultimate goal of the experiment is to determine whether the current parameter state is in an abnormal state. In the course of this experiment, n = 50 will lead to a good classification effect. Figure 6 shows the data after the preprocess of sliding window.

2. Normalization

The normalization of data is to scale the data so that it falls into a small specific range.

This makes the numeric range of printing data to be controlled within a specific range. On the one hand, the randomness brought about by different printing products and printing parameter settings can be eliminated, making different batches of printing data comparable, maintaining them the same distribution, thereby greatly improving the accuracy of the model; on the other hand, and the convergence speeds of the classifier training will also be faster.

The standardized methods mainly include min-max standardization (Fig. 7) and z-score standardization (Fig. 8).

Min-max normalization:

$$x^* = \frac{x - x\_min}{x\_max - x\_min} \tag{9}$$

Z-score normalization:

$$x^* = \frac{x - \mu}{\sigma} \tag{10}$$

where μ is the mean of the sample data and σ is the standard deviation of the sample data.

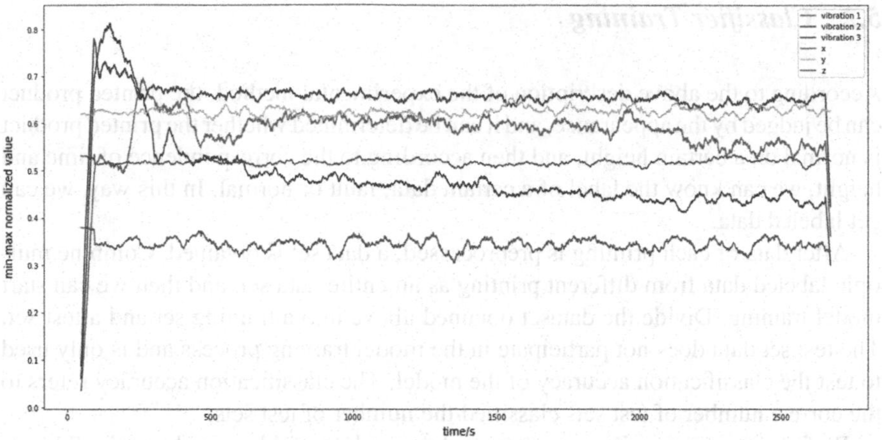

Fig. 7 Data after min-max normalization

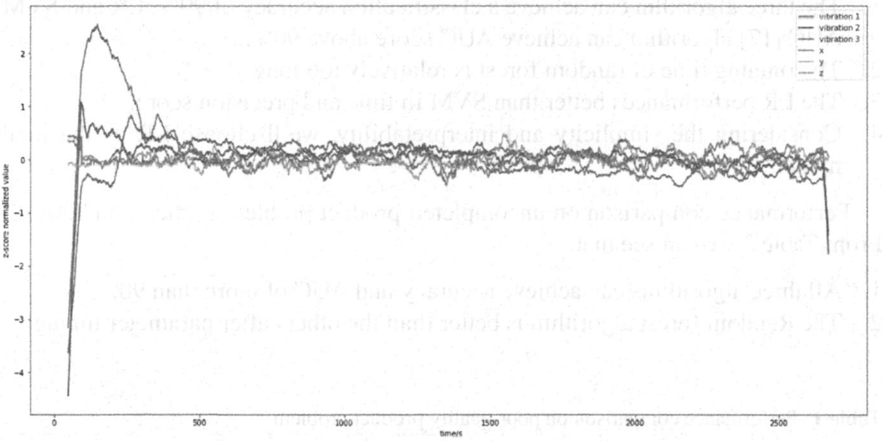

Fig. 8 Data after z-score normalization

In the experiment, after testing, using z-score can greatly improve class accuracy of model. So we'll use z-score normalization for data preprocess.

Moreover, before Normalization, add the sum of three vibration signal and three acceleration signal values separately for two more feature. This will benefit to the accuracy of the model.

5.2 *Classifier Training*

According to the above description of the experimental method, the printed product can be judged by the appearance, and it can be determined whether the printed product is normal on a certain height, and then according to the correspondence of time and height, we can know the label of a certain data, fault or normal. In this way, we can get labeled data.

After data of each printing is preprocessed, a data set is obtained. Combine multiple labeled data from different printing as an entire data set, and then we can start model training. Divide the dataset obtained above into a training set and a test set. The test set data does not participate in the model training process and is only used to test the classification accuracy of the model. The classification accuracy refers to the correct number of test sets classified/the number of test sets.

Performance comparison on poor-quality product problem is shown in Table 1. From Table 1 we can see that:

1. The three algorithm can achieve a classification accuracy of 90%. LR and SVM AUC [17] algorithm can achieve AUC score above 90%.
2. The running time of random forest is relatively too long.
3. The LR performances better than SVM in time and precision score.
4. Considering the simplicity and interpretability, we'll choose LR as the final model.

Performance comparison on uncompleted product problem is shown in Table 2. From Table 2 we can see that:

1. All three algorithms can achieve accuracy and AUC of more than 90.
2. The Random forest algorithm is better than the others after parameter tuning

Table 1 Performance comparison on poor-quality product problem

	LR	SVM	Random forest
Accuracy/%	91.76	91.56	90.54
AUC score/%	90.00	90.78	87.58
Recall score/%	87.09	89.49	82.59
Precision score/%	75.40	72.56	73.39
Time/s	55.00	55.44	238.66

Table 2 Performance comparison on uncompleted product problem

	LR	SVM	Random forest
Accuracy/%	92.5	94.93	95.43
AUC score/%	96.5	95.93	96.61
Recall score/%	91.35	92.79	93.64
Precision score/%	93.67	96.96	97.52
Time/s	28.31	32.03	36.83

3. The Random forest algorithm is slightly insufficient in time.
4. For a comprehensive assessment, the Random forest can be used as a model in the final system.

6 Conclusion

Based the experiment and data analysis methods, the status of the 3D printer can be monitored by several sensors, and the machine learning method can be used to predict whether the printer is in a fault state.

This experiment can be extended to more 3D printers, especially in cloud manufacturing [18, 19]. Based on the idea of this paper, a real-time fault detection solution for 3D printers can be developed. Specifically, after experimental verification, the vibration of the stepping motor in the 3D printer is most important for the failure detection. Therefore, the 3D printer provider can add the built-in vibration sensor in the stepping motor at the factory, so that the real-time fault can be monitored.

Acknowledgements The research is supported by the National High-Tech Research and Development Plan of China under Grant No. 2015AA042101.

References

1. E.L. Nyberg, A.L. Farris, B.P. Hung et al., 3D-printing technologies for craniofacial rehabilitation, reconstruction, and regeneration. Ann. Biomed. Eng. **45**(1), 45 (2017)
2. C.W. Foster, M.P. Down, Y. Zhang et al., 3D printed graphene based energy storage devices. Sci. Rep. **7**, 42233 (2017)
3. N. Bhattacharjee, A. Urrios, S. Kang et al., The upcoming 3D-printing revolution in microfluidics. Lab Chip **16**(10), 1720 (2016)
4. J. Straub, Automated testing and quality assurance of 3D printing/3D printed hardware: Assessment for quality assurance and cybersecurity purposes. IEEE Autotestcon, pp. 1–5. IEEE, (2016)
5. W. Ruinan, W. Yao, Design of an intelligent monitoring system for the fluid-feeding system of the car assembly line. Electrical Automation (2017)
6. M. Ataş, Y. Yardimci, A. Temizel, A new approach to aflatoxin detection in chili pepper by machine vision. Comput. Electron. Agric. **87**(9), 129–141 (2012)
7. X. Liu, Research on Establishment of Internet-based Product Quality Tracking System in Mobile Internet Age. Standard Science (2016)
8. G. Galdos, P. Tamigniaux, J.P. Morel et al. Vibration Sensor (2017)
9. M. Li, L. Zhao, The classification of human lower limb motion based on acceleration sensor, in *Guidance, Navigation and Control Conference* (IEEE, 2017), pp. 2210–2214
10. L. Guerriero, G. Guerriero, G. Grelle et al., Brief communication: a low-cost Ar-duino®-based wire extensometer for earth flow monitoring. Nat. Hazards Earth Syst. Sci. **17**(6), 881–885 (2017)
11. A. Samourkasidis, I. Athanasiadis, A miniature data repository on a raspberry Pi. Electronics **6**(1), 1 (2017)
12. R. Devore, G. Kerkyacharian, D. Picard et al., Mathematical methods for supervised learn-ing. Found. Comput. Math, **6**(1), 3–58 (2017)

13. S. Edition, Applied logistic regression analysis. Technometrics **38**(2), 184–186 (2017)
14. S.J. Rigatti, Random Forest. J. Insur. Med. **47**, 31–39 (2017)
15. R. Fu, B. Li, Y. Gao et al., Content-based image retrieval based on CNN and SVM, in *IEEE International Conference on Computer and Communications* (IEEE, 2017), pp. 638–642
16. L. Zhang, J. Lin, R. Karim, Sliding window-based fault detection from high-dimensional data streams. IEEE Trans. Syst. Man Cybern. Syst. **47**(2), 289–303 (2017)
17. N.T. Khajavi, A. Kuh, The goodness of covariance selection problem from AUC bounds. Communication, Control, and Computing (IEEE, 2017), pp. 1252–1258
18. J. Mai, L. Zhang, F. Tao et al., Customized production based on distributed 3D printing services in cloud manufacturing. Int. J. Adv. Manuf. Technol. **84**(1–4), 71–83 (2016)
19. J. Mai, L. Zhang, F. Tao et al., Architecture of hybrid cloud for manufacturing enterprise. System Simulation and Scientific Computing (Springer Berlin, Heidelberg, 2012), pp. 365–372

Finite Element Analysis of the Bearing Assembly of Motion Simulator in Wide Temperature Range

Qingsheng Xiao, Yongling Fu, Linjie Li and Xiao Han

Abstract In this paper, a bearing assembly which is the part of a motion simulator operating at ambient temperature −173 °C and vacuum of 1×10^{-5} Pa is studied. ANSYS Workbench finite element software is used to simulate the stress and strain of the bearing assembly which contains shaft, sliding bearing and bearing seat in different temperature fields. To study the change of clearance between sliding bearing and bearing seat and the equivalent stress on shaft subjected to maximum load, It can guide the design and machining of the motion simulator, also can prevent the occurrence of stuck or loose sliding fault.

Keywords Motion simulator · Wide temperature range
Steady state temperature field · Deformation

1 Introduction

It is necessary to develop a motion simulator [1, 2] which can simulate the motion state of spacecraft because the spacecraft needs to carry out space environment simulation test under the motion state. The motion simulator developed in this paper is used to provide the load with two degrees of freedom of spin and pitch in high vacuum and ultra-low temperature environment. The space environment simulation test of the load is realized. The operating environment temperature range is 100–333 K, and the operating environment vacuum requirement is 1×10^{-5} Pa. In order to study that gap variation of the sliding bear and bearing seat in a wide temperature range, it is important to ensure the normal operation of the simulator.

Q. Xiao (✉) · X. Han
Beijing Institute of Spacecraft Environment Engineering, Beijing 100094, China
e-mail: xqs.hit@163.com

Y. Fu · L. Li
School of Mechanical Engineering and Automation, Beihang University, Beijing 100191, China

Y. Jia et al. (eds.), *Proceedings of 2018 Chinese Intelligent Systems Conference*, Lecture Notes in Electrical Engineering 529,
https://doi.org/10.1007/978-981-13-2291-4_39

In the transmission system of the motion simulator, the shaft and bearing seat are made of 316L, while the sliding bearing fitted with them is made of copper alloy. In a wide temperature range, the physical properties of copper alloy and stainless steel will change significantly with the change of temperature, which will cause the interference fit between sliding bearing and bearing housing to be small. When the interference fit is reduced to a certain extent, the sliding between sliding bearing and bearing housing may occur, which will affect the accuracy of the system. The change of the clearance between the shaft and the sliding bearing will also affect the transmission characteristics of the system, even cause jamming phenomenon, leading to the damage of the drive motor. Therefore, it is very necessary to study the reasonable setting of the fit between sliding bearing and shaft and bearing seat.

Based on ANSYS Workbench, the thermal-structural analysis model of shaft, bearing housing and sliding bearing is established, and the steady-state temperature field of the model under different temperature changes is analyzed. ANSYS is one of the most advanced large-scale general finite element analysis software in the world. It has powerful calculation function and extensive simulation function, especially it can simulate nonlinear problems such as multi-disciplinary simulation [3, 4]. The finite element software ANSYS is used to simulate the steady-state temperature field of shaft, sliding bearing and bearing seat under different temperature fields. Through the data obtained, the design dimension of the model is improved and the design of the motion simulator is optimized.

2 Finite Element Theory for Steady-State Temperature Field Analysis

In order to unify that thermal analysis model and the structural analysis model, a finite element three-dimensional model can be used to carry out the thermal analysis and the structural analysis, and there is no need for data conversion between the thermal model and the structural model, The calculation precision is also improved. The theoretical solution process of this method is as follows:

(1) The interpolation form of unit temperature field is as follows:

$$T^e(x, y, z, t) = \sum_{i=1}^{r} N_i(x, y, z)T_i(t) = [N(x, y, z)]\{T(t)\}^e \tag{1}$$

In which, $N_i(x, y, z)$—interpolating function, $T_i(t)$—unit node temperature.

(2) The temperature field control equation can be established according to heat transfer:

$$\rho c\frac{\partial T}{\partial t} = \frac{\partial}{\partial x}\left(k_x\frac{\partial T}{\partial x}\right) + \frac{\partial}{\partial y}\left(k_y\frac{\partial T}{\partial y}\right) + \frac{\partial}{\partial z}\left(k_z\frac{\partial T}{\partial z}\right) \tag{2}$$

The boundary conditions are determined as follows:

$$-\left(k_x\frac{\partial T}{\partial x}n_x + k_y\frac{\partial T}{\partial y}n_y + k_z\frac{\partial T}{\partial z}n_z\right) = \sigma\varepsilon T^4 - aq_r \tag{3}$$

The initial conditions are as follows:

$$T(x, y, z, t = 0) = T_0 \tag{4}$$

(3) According to the Galerkin method, according to the following formula (1), (2) and (3), it can be derived:

$$\frac{\partial J^e}{\partial T_i} = [C]^e\frac{\partial}{\partial T}\{T(t)\}^e + \left([K_k]^e + [K_r]^e\right)\{T(t)\}^e - \{F_r\}^e \tag{5}$$

In that time domain, the Eq. (5) is discrete by difference, and then the finite element equations of each element are synthesize to obtain the calculation equation of the total temperature field:

$$\left(\frac{[C]}{\Delta t}[K_k + K_r]\right)\{T(t)\}_t = \{F_r\} + \frac{[C]}{\Delta t}\{T(t)\}_{t-\Delta t} \tag{6}$$

3 Steady-State Thermal-Structural Finite Element Analysis

The main flow of thermal-structural coupled steady-state temperature field analysis is shown in Fig. 1.

3.1 The Establishment of Simulation Model

Proper simulation of sliding bearing and shaft deformation requires consideration of the state of stress under both thermal and force loads, that is, the state of stress under thermal-structural coupling. See Fig. 2 for the three-dimensional model of shaft, bearing housing and sliding bearing. The main structural parameters are: Diameter of outer circle of shaft 70 mm; Bearing housing inner hole diameter 80 mm, length 94 mm; Sliding bearing outer ring inner diameter 70 mm, outer diameter 80 mm, length 94 mm. The material of the shaft and bearing housing is S316L, and the material of the sliding bearing is copper alloy, and their physical characteristic parameters at different temperatures are shown in Table 1. Sliding bearing and bearing housing interference fit together and, for initial analysis, assume that the interference on the sliding bearing radius is 0.05 mm at room temperature (25 °C).

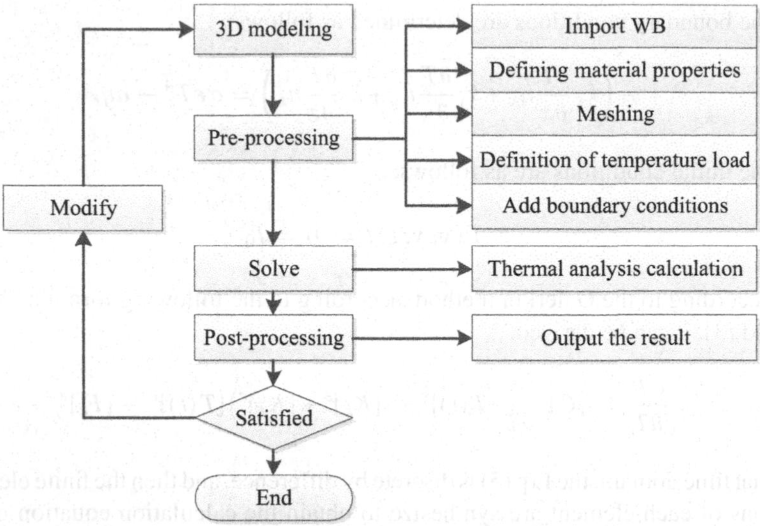

Fig. 1 Thermal analysis steps

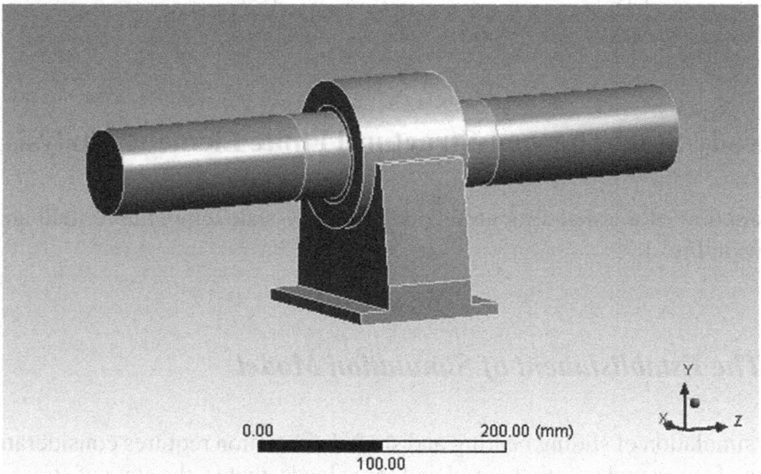

Fig. 2 3D model of shaft, bearing seat and sliding bearing

3.2 *Pre-processing of Finite Element Calculations*

Pre-processing includes defining material parameters, selecting thermal-structural units, dividing grids, and defining boundary conditions.

After building the model in the workbench software, set the material parameters according to Table 1, define the friction coefficient of the contact surface, set to 0.1.

Table 1 Physical properties of each material [5]

Parameters	Density kg/m^3	Temperature °C	Modulus of elasticity GPa	Poisson ratio	Coefficient of linear expansion 10^{-6} °C^{-1}	Thermal conductivity W/(m * k)
316L	7860	25	195	0.294	16	16.86
		−100	250	0.287	14.5	
		−196	259	0.283	8.31	
Copper alloy	8230	25	125.6	0.35	16.6	83.7
		−100	122.7		15.1	
		−196	129.6		9.7	

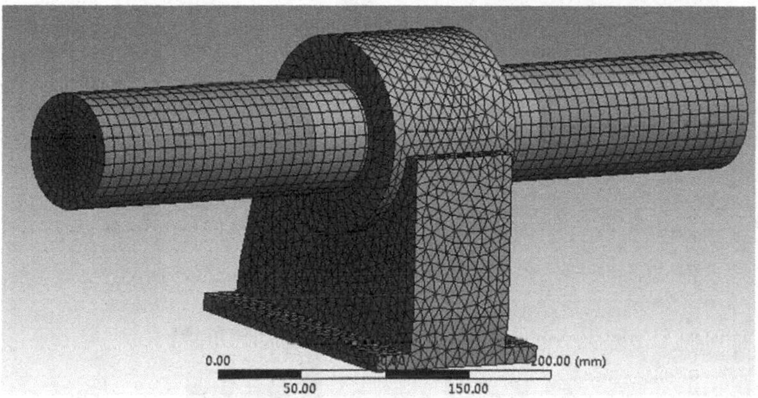

Fig. 3 Meshing

Enter the Mesh function module to divide the mesh, and use the three-dimensional solid coupling element SOLID5 coupling to divide the whole assembly into 2564 elements and 3641 nodes. The finite element model is shown in Fig. 3.

Add a fixed constraint to both ends of the shaft and the self-weight of the part to the shaft according to the calculation result, and then use the temperature load as the boundary condition of the subsequent thermal-structure coupling analysis to conduct the thermal-structure coupling analysis on the assembly, as shown in Fig. 4.

3.3 *Solution and Post Processing*

Through the above flow, the thermal structure coupling analysis is carried out by using workbench solver, the whole model is defined as Set-All, and the initial temperature field (25 °C) is defined for the whole model. Through solving the calculation, the stress distribution map and displacement cloud picture of the assembly under the

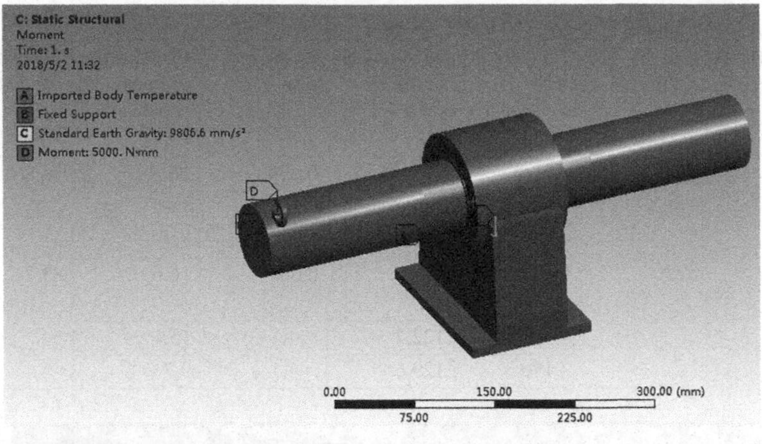

Fig. 4 Add boundary conditions

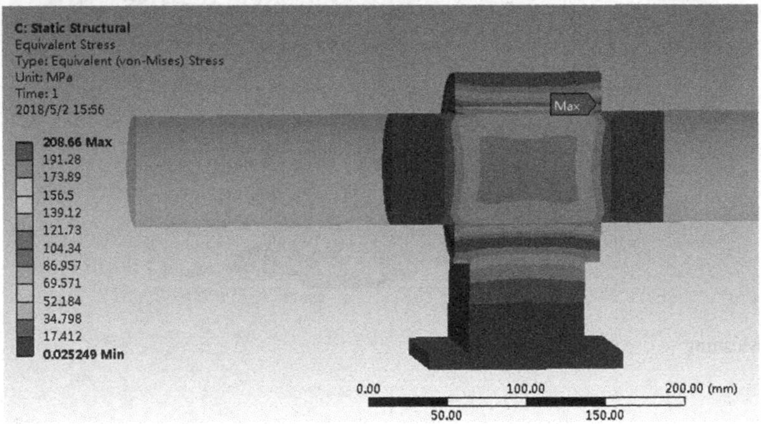

Fig. 5 Stress cloud map of assembly at 25 °C

action of thermal load and structural load are analyzed. Due to the stress concentration of the shaft at the fixed constraint, only the area close to the sliding bearing part is extracted when the result is viewed, as shown in Figs. 5 and 6.

As shown in Fig. 7, the maximum pressure between the sliding bearing and the bearing housing is 69.242 MPa, which indicates that the three parts are in contact state at 25 °C.

Then set the boundary conditions under −173 °C temperature field. In this analysis, the boundary conditions, loads, constraint relations and field variables are the same as those of the basic model. Only an initial temperature load needs to be edited to simulate the state of the model at −173 °C. After the restart analysis is completed, the stress cloud images of the sliding bearing contact surface are shown in Fig. 8, it shows that the contact pressure between the sliding bearing and the bearing housing

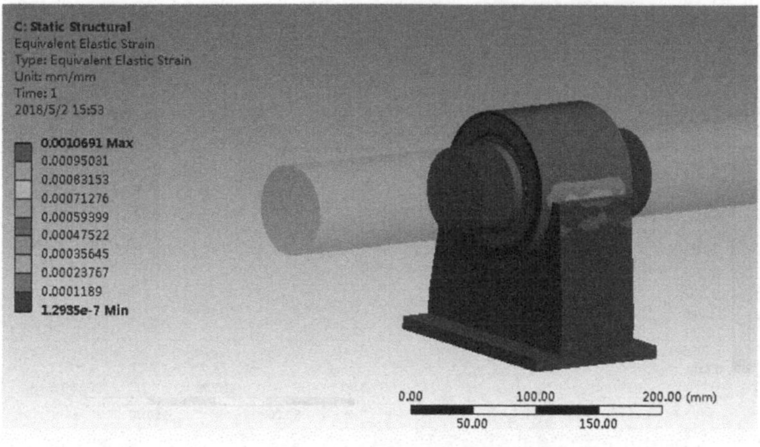

Fig. 6 Strain cloud map of assembly at 25 °C

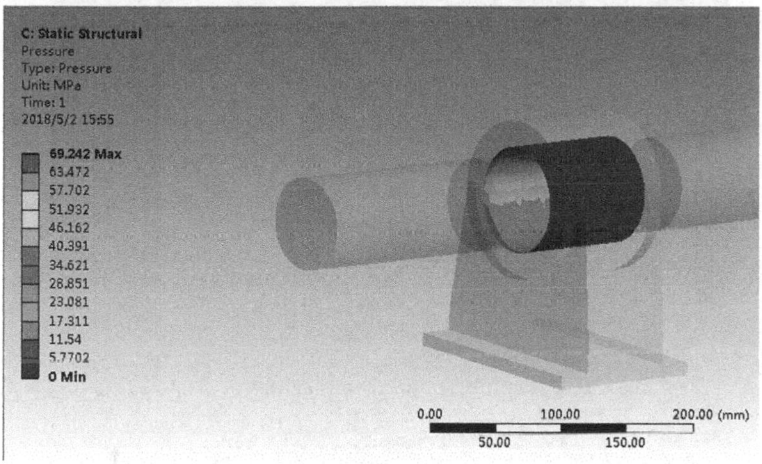

Fig. 7 Pressure distribution between sliding bearing and bearing seat at 25 °C

becomes 45.595 MPa. This indicates that at −173 °C the amount of interference fit between the sliding bearing and the shaft as well as the bearing housing becomes small, but still in contact, so the design of the model should be improved to reduce the initial interference fit.

Change the assembly clearance between the sliding bearing and the bearing housing to reduce the interference between them. Finally, through several simulations, when the interference on the radius of the sliding bearing is 0.025 mm, the maximum contact pressure between the sliding bearing and the bearing housing is 6.4 MPa at the temperature of −173 °C, as shown in Fig. 9, which is a satisfactory assembled state.

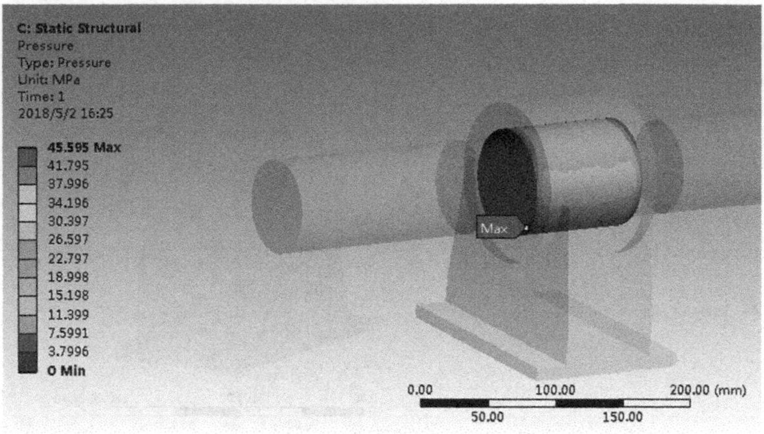

Fig. 8 Pressure distribution between sliding bearing and bearing seat at −173 °C

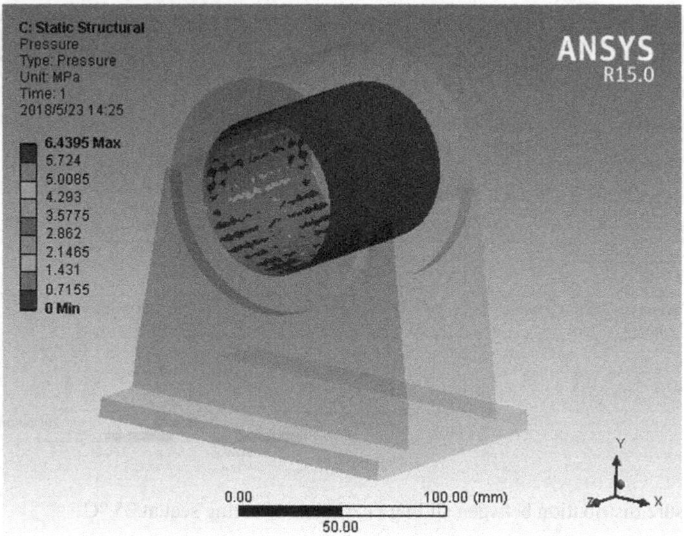

Fig. 9 Pressure distribution between sliding bearing and bearing seat at −173 °C

Then, by adding pressure load to the shaft, the maximum equivalent stress of the shaft at −173 °C is 455.94 MPa, which is less than the yield stress of 316L at this temperature (539.2 MPa). As shown in Fig. 10, the strength of the bearing assembly meets the design requirements.

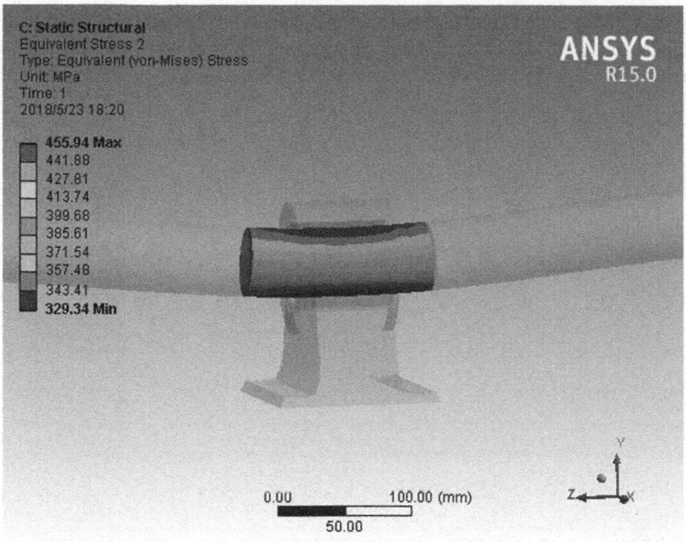

Fig. 10 The equivalent stress on a shaft subjected to maximum load at −173 °C

4 Conclusion

Through the analysis of various boundary conditions mentioned above, it can be seen that the variation of assembly clearance between sliding bearing and shaft and between sliding bearing and bearing seat reaches 35% in a wide temperature range. If not properly assembled, the sliding bearing will cause serious deformation, which will affect the transmission characteristics of the system. The simulation results show that when the initial assembly clearance between the sliding bearing and the bearing seat is 0.025 mm at room temperature, the assembly clearance between the sliding bearing and the bearing seat becomes almost zero when the ambient temperature drops to 100 K and the shaft can also withstand maximum loads at minimum temperatures. This simulation analysis can guide the conceptual design of the subsequent motion simulator.

References

1. Z. Xinbang, The research on application of hardware in the loop simulation for space-craft. Aerosp. Control **33**(1), 77–83 (2015)
2. Z. XinBang, L.D. LIU, S.Z. LIU, Motion simulators for rendezvous simulation test. Aerosp. Control Appl. **35**(2), 51–55 (2009)
3. X.U.E. Mingde, D.I.N.G. Hongwei, W.A.N.G. Lihua, 3-D finite element analysis of the temperature field, thermal deformation and stress of pistons in diesel engines. Acta Armamentarii **22**(1), 11 (2001)

4. W. Shuogui, X. Yuanming, Effect of interference fit and axial preload in the stiffness of the high speed angular contact ball bearing. J. Univ. Sci. Technol. China **36**(12), 1314–1320 (2006)
5. Chen Guobang, *Cryogenic Engineering Materials* (Zhejiang University Press, Hangzhou, 1998)

State-Feedback Stabilization of Stochastic Non-holonomic Systems (SNSs) Under Arbitrary Switchings with Time-Varying Delays

Huining Wu, Yushan Jiang, Dongkai Zhang and Zengxiao Guo

Abstract This result deals with the state-feedback stabilization of SNSs with time-varying delays and arbitrary switchings with. The backstepping state feedback stabilizing controllers are given.

Keywords Stabilization · Arbitrary switching · Time-varying delays

1 Introduction

Recently, the stabilization of SNSs was considered [1, 2]. State-feedback stabilization for SNSs with time-varying delays was discussed [3, 4]. State-feedback stabilization controllers of switching SNSs were given [5, 6]. Output feedback stabilization for SNSs under arbitrary switching was studied [7]. stochastic non-linear systems' stabilizing controller with arbitrary switchings was designed [8, 9].

Based on above analysis, we tries to discuss the stabilization of SNSs with time-varying delays and arbitrary switchings. The main idea is to design the backstepping state feedback stabilizing controllers which guarantee all sates is asymptotical stabilization in probability.

H. Wu (✉) · D. Zhang · Z. Guo
School of Science, Shijiazhuang University, Shijiazhuang 050035, China
e-mail: wuhuining9999@163.com

D. Zhang
e-mail: zdkmailhot@126.com

Z. Guo
e-mail: gdmath@163.com

Y. Jiang
School of Mathematics and Statistics, Northeastern University
at Qinhuangdao, Qinhuangdao 066004, China
e-mail: soblev@126.com

Y. Jia et al. (eds.), *Proceedings of 2018 Chinese Intelligent Systems Conference*, Lecture Notes in Electrical Engineering 529,
https://doi.org/10.1007/978-981-13-2291-4_40

2 Controllers Design

The systems are given as:

$$dx_0(t) = u_0(t)dt + f_{[\sigma(t),0]}dt, \tag{1.1}$$

$$\left.\begin{aligned} dx_1(t) &= x_2(t)u_0(t)dt + f_{[\sigma(t),1]}dt + g^T_{[\sigma(t),1]}d\omega, \\ dx_2(t) &= x_3(t)u_0(t)dt + f_{[\sigma(t),2]}dt + g^T_{[\sigma(t),2]}d\omega, \\ dx_3(t) &= u(t)dt + f_{[\sigma(t),3]}dt + g^T_{[\sigma(t),3]}d\omega, \end{aligned}\right\} \tag{1.2}$$

where $x_0 \in \mathbb{R}$, $x = (x_1, x_2, x_3)^T \in \mathbb{R}^3$, u_0, u are control inputs,

$$\begin{aligned} f_{[\sigma(t),0]} &\triangleq f_{[\sigma(t),0]}(t, x_0(t)), \\ f_{[\sigma(t),i]} &\triangleq f_{[\sigma(t),i]}(t, x_0(t), x(t), x(t-d(t))), \\ g^T_{[\sigma(t),i]} &\triangleq g^T_{[\sigma(t),i]}(t, x_0(t), x(t), x(t-d(t))) \end{aligned}$$

are locally Lipschitz continuous functions with $f_{[\sigma(t),0]}(t, 0) = 0$, $f_{[\sigma(t),i]}(t, 0) = 0$, $g^T_{[\sigma(t),i]}(t, 0) = 0$, $i = 1, 2, 3$; ω is an r-dimensional independent standard Wiener process; $\sigma(t) : [0, +\infty) \to M = \{1, 2, \ldots, m\}$ is a piecewise constant switching signal; $d(t) : \mathbb{R}_+ \to [0, d]$ is the time-varying delay with $\dot{d}(t) \le \eta < 1$.

Assumption 1 For a given positive constant c_0, one has

$$\left|f_{[\sigma(t),0]}\right| \le c_0|x_0(t)|.$$

Assumption 2 For given non-negative smooth functions $a_{[\sigma(t),i]}(\cdot)$ and $b_{[\sigma(t),i]}(\cdot)$, we have

$$\left|f_{[\sigma(t),i]}\right| \le a_{[\sigma(t),i]}(x_0(t))\left(\sum_{k=1}^{i}|x_k(t)| + \sum_{k=1}^{i}|x_k(t-d(t))|\right),$$

$$\left\|g_{[\sigma(t),i]}\right\| \le b_{[\sigma(t),i]}(x_0(t))\left(\sum_{k=1}^{i}|x_k(t)| + \sum_{k=1}^{i}|x_k(t-d(t))|\right).$$

Theorem 1 *For any $x_0(t_0)$, $t_0 > 0$, there exist positive constant λ_0, such that (1.1) with*

$$u_0 = -\lambda_0 x_0, \lambda_0 > 0 \tag{2}$$

is asymptotically stable.

In the following, to design the controller u, (3) is needed.

$$z_i = \begin{cases} \frac{x_i}{u_0^{n-i}}, t \geq t_0 \\ x_i, t_0 - d \leq t \leq t_0 \end{cases}, i = 1, 2, 3. \tag{3}$$

By (1.2) and (3), one has

$$\begin{cases} \mathrm{d}z_i(t) = z_{i+1}(t)\mathrm{d}t + \bar{f}_{[\sigma(t),i]}\mathrm{d}t + \bar{g}^T_{[\sigma(t),i]}\mathrm{d}\omega, i = 1, 2 \\ \mathrm{d}z_3(t) = u(t)\mathrm{d}t + \bar{f}_{[\sigma(t),3]}\mathrm{d}t + \bar{g}^T_{[\sigma(t),3]}\mathrm{d}\omega, \end{cases}$$

and

$$\bar{f}_{[\sigma(t),i]} \triangleq \frac{f_{[\sigma(t),i]}}{u_0^{3-i}} + (3-i)\lambda_0 z_i \frac{u_0 + f_{[\sigma(t),0]}}{u_0}, \bar{g}_{[\sigma(t),i]} \triangleq \frac{g_{[\sigma(t),i]}}{u_0^{3-i}}.$$

Lemma 1 *For $i = 1, 2, 3$, one has*

$$\left|\bar{f}_{[\sigma(t),i]}\right| \leq \varphi_{[\sigma(t),i]}(x_0(t))\left(\sum_{k=1}^{i}|z_k(t)| + \sum_{k=1}^{i}|z_k(t-d(t))|\right),$$

$$\left\|\bar{g}_{[\sigma(t),i]}\right\| \leq \psi_{[\sigma(t),i]}(x_0(t))\left(\sum_{k=1}^{i}|z_k(t)| + \sum_{k=1}^{i}|z_k(t-d(t))|\right),$$

and $\varphi_{[\sigma(t),i]}(\cdot)$, $\psi_{[\sigma(t),i]}(\cdot)$ are non-negative smooth functions and

$\varphi_{[\sigma(t),i]} \triangleq \varphi_{[\sigma(t),i]}(x_0(t)) = max\{a_{[\sigma(t),i]}, \bar{a}_{[\sigma(t),i]}, \bar{\bar{a}}_{[\sigma(t),i]}\}$,

$\bar{a}_{[\sigma(t),i]} \triangleq a_{[\sigma(t),i]}\lambda_0^{i-k}|x_0(t)|^{i-k} + (n-i)(\lambda_0 + c_0)$,

$\bar{\bar{a}}_{[\sigma(t),i]} \triangleq a_{[\sigma(t),i]}\lambda_0^{3-k}\exp\{(n-i)[d(\lambda_0 - c_0) + 2c_0(t-t_0)]\}$,

$\psi_{[\sigma(t),i]} \triangleq \psi_{[\sigma(t),i]}(x_0(t)) = b_{[\sigma(t),i]}max\left\{1, \lambda_0^{3-k}\exp[(3-i)(d(\lambda_0 - c_0) + 2c_0(t-t_0))]\right\}$.

Remark 1 The similar assumptions and Lemma 1 can be found in [4], but the main difference is arbitrary switching in this paper.

Letting

$$z_1^* = 0, \quad \xi_1 = z_1 - z_1^*, \quad z_2^* = -\alpha_1\xi_1, \quad \xi_2 = z_2 - z_2^*, \quad z_3^* = -\alpha_2\xi_2, \quad \xi_3 = z_3 - z_3^*,$$

then

$$\begin{cases} \mathrm{d}\xi_1 = z_2\mathrm{d}t + \bar{\bar{f}}_{[\sigma(t),1]}\mathrm{d}t + \bar{\bar{g}}^T_{[\sigma(t),1]}\mathrm{d}\omega, \\ \mathrm{d}\xi_2 = z_3\mathrm{d}t + \bar{\bar{f}}_{[\sigma(t),2]}\mathrm{d}t + \bar{\bar{g}}^T_{[\sigma(t),2]}\mathrm{d}\omega, \\ \mathrm{d}\xi_3 = u_1\mathrm{d}t + \bar{\bar{f}}_{[\sigma(t),3]}\mathrm{d}t + \bar{\bar{g}}^T_{[\sigma(t),3]}\mathrm{d}\omega, \end{cases}$$

where

$$\bar{f}_{[\sigma(t),1]} = \bar{\bar{f}}_{[\sigma(t),1]}, \bar{g}_{[\sigma(t),1]} = \bar{\bar{g}}_{[\sigma(t),1]},$$

$$
\begin{aligned}
\bar{\bar{f}}_{[\sigma(t),2]} =& \bar{f}_{[\sigma(t),2]} + \alpha_1 z_2 + \alpha_1 \bar{f}_{[\sigma(t),1]} + \xi_1 \frac{\partial \alpha_1}{\partial x_0}\left(u_0 + f_{[\sigma(t),0]}\right),\\
\bar{\bar{g}}_{[\sigma(t),2]} =& \alpha_1 \bar{g}_{[\sigma(t),1]} + \bar{g}_{[\sigma(t),2]},\\
\bar{\bar{f}}_{[\sigma(t),3]} =& \bar{f}_{[\sigma(t),3]} + \alpha_2 z_3 + \alpha_2 \bar{f}_{[\sigma(t),2]} + \alpha_1\alpha_2 \bar{f}_{[\sigma(t),21]}\\
&+ \left(\xi_1 \alpha_2 \frac{\partial \alpha_1}{\partial x_0} + \xi_2 \frac{\partial \alpha_2}{\partial x_0}\right)\left(u_0 + f_{[\sigma(t),0]}\right),\\
\bar{\bar{g}}_{[\sigma(t),3]} =& \bar{g}_{[\sigma(t),3]} + \alpha_1\alpha_2 \bar{g}_{[\sigma(t),1]} + \alpha_2 \bar{g}_{[\sigma(t),2]}.
\end{aligned}
$$

Step 1 Define the 1st Lyapunov function

$$
V_1 = \frac{1}{4} z_1^4 + \frac{3}{1-\eta} \int_{t-d(t)}^{t} z_1^4(s)\mathrm{d}s.
$$

By Itô formula, one has

$$
\begin{aligned}
\mathcal{L}V_1 \leq& z_1^3\left(z_2 + \bar{f}_{[\sigma(t),1]}\right) + \frac{3}{2} z_1^2 \mathrm{Tr}\left\{\bar{g}_1^T \bar{g}_1\right\} + \frac{3}{1-\eta}\left\{z_1^4 - z_1^4(t-d(t))\left(1-\dot{d}(t)\right)\right\}\\
\leq& z_1^3 z_2 + \left\{\varphi_{[\sigma(t),1]} + \frac{3}{4}\left(\frac{1}{4}\right)^{\frac{1}{3}} 2^{\frac{1}{3}} \varphi^{\frac{4}{3}}_{[\sigma(t),1]} + 3\psi^2_{[\sigma(t),1]} + \frac{3}{1-\eta} + \frac{3}{2}\psi^4_{[\sigma(t),1]}\right\} z_1^4\\
&- 2z_1^4(t-d(t))\\
\leq& z_1^3 z_2 + \left\{\varphi_{[\sigma(t),1]} + 3\psi^2_{[\sigma(t),1]} + \frac{3}{1-\eta} + l_{11} + l_{22}\right\} z_1^4 - 2z_1^4(t-d(t))\\
\leq& -3z_1^4 - 2z_1^4(t-d(t)) + z_1^3\left(z_2 - z_2^*\right),
\end{aligned}
$$

where

$$
\begin{aligned}
z_2^* =& -\left(3 + \frac{3}{1-\eta} + \varphi_{[\sigma(t),1]} + 3\psi^2_{[\sigma(t),1]} + l_{11} + l_{12}\right) z_1 \triangleq -\alpha_1(x_0) z_1,\\
l_{11} =& \frac{3}{4} \times \left(\frac{1}{4}\right)^3 \times \left(\frac{1}{2}\right)^{\frac{1}{3}} \times \varphi^{\frac{4}{3}}_{[\sigma(t),1]},\ l_{12} = \frac{9}{2}\psi^4_{[\sigma(t),1]}.
\end{aligned}
$$

Step 2 Define the 2nd Lyapunov function

$$
V_2 = V_1 + \frac{1}{4}\xi_2^4 + \frac{2}{1-\eta} \int_{t-d(t)}^{t} \xi_2^4(s)\mathrm{d}s.
$$

So, one can obtain

$$
\mathcal{L}V_2 \leq -3z_1^4 - 2z_1^4(t-d(t)) + z_1^3\left(z_2 - z_2^*\right)
$$

$$
\begin{aligned}
&+\xi_2^3\left(z_3+\bar{f}_{[\sigma(t),2]}+\alpha_1\bar{f}_{[\sigma(t),1]}+\xi_1\frac{\partial\alpha_1}{\partial x_0}(u_0+f_0)\right)\\
&+\frac{3}{2}\xi_2^2 Tr\{(\alpha_1\bar{g}^T_{[\sigma(t),1]}+\bar{g}^T_{[\sigma(t),2]})(\alpha_1\bar{g}_{[\sigma(t),1]}+\bar{g}_{[\sigma(t),2]})\}\\
&+\frac{2}{1-\eta}\xi_2^4-\frac{2}{1-\eta}\xi_2^4(t-d(t))(1-\dot{d}(t))\\
&\le -2\xi_1^4-2\xi_2^4-\xi_1^4(t-d(t))-\xi_2^4(t-d(t))+\xi_2^3(z_3-z_3^*),
\end{aligned}
$$

where

$$
\begin{aligned}
z_3^*=-\Bigg\{&2+\frac{2}{1-\eta}+\frac{1}{4}\left(\frac{4}{27}\right)^{-3}+\frac{3}{4}\left(\left(\frac{4}{9}\right)^{-\frac{1}{3}}+\left(\frac{2}{3}\right)^{-\frac{1}{3}}\right)(\alpha_1\varphi_{[\sigma(t),1]})^{\frac{4}{3}}+\frac{3}{4}\left(\frac{4}{9}\right)^{-\frac{1}{3}}\varphi^{\frac{4}{3}}_{[\sigma(t),2]}\\
&+\varphi_{[\sigma(t),2]}+\frac{3}{4}\left(\left(\frac{4}{9}\right)^{-\frac{1}{3}}+\left(\frac{2}{3}\right)^{-\frac{1}{3}}\right)(\alpha_1\varphi_{[\sigma(t),2]})^{\frac{4}{3}}+\alpha_1+\frac{3}{4}\left(\frac{4}{9}\right)^{-\frac{1}{3}}\alpha_1^{\frac{8}{3}}+828\psi^4_{[\sigma(t),2]}\\
&+\frac{3}{4}\left(\frac{4}{9}\right)^{-\frac{1}{3}}\left[\frac{\partial\alpha_1}{\partial x_0}(\lambda_0x_0-c_0x_0)\right]^{\frac{4}{3}}+24\psi^2_{[\sigma(t),2]}+2160(\alpha_1\psi_{[\sigma(t),2]})^4\\
&+135(\alpha_1\psi_{[\sigma(t),1]})^4\Bigg\}\xi_2\triangleq-\alpha_2\xi_2.
\end{aligned}
$$

Step 3 Define the 3rd Lyapunov function

$$
V_3=V_2+\frac{1}{4}\xi_3^4+\frac{1}{1-\eta}\int_{t-d(t)}^{t}\xi_3^4(s)\mathrm{d}s.
$$

So, one can obtain

$$
\begin{aligned}
\mathcal{L}V_3\le&-2\xi_1^4-2\xi_2^4-\xi_1^4(t-d(t))-\xi_2^4(t-d(t))+\xi_2^3(z_3-z_3^*)\\
&+\xi_3^3\{u+\bar{f}_{[\sigma(t),3]}+\alpha_2z_3+\alpha_2\bar{f}_{[\sigma(t),2]}+\alpha_1\alpha_2z_2+\alpha_1\alpha_2\bar{f}_{[\sigma(t),1]}\\
&+\xi_1\alpha_2\frac{\partial\alpha_1}{\partial x_0}(u_0+f_0)+\xi_2\frac{\partial\alpha_2}{\partial x_0}(u_0+f_0)\Bigg\}\\
&+\frac{3}{2}\xi_3^2(\alpha_1\alpha_2\bar{g}^T_{[\sigma(t),1]}+\alpha_2\bar{g}^T_{[\sigma(t),2]}+\bar{g}^T_{[\sigma(t),3]})^2\\
&+\frac{1}{1-\eta}\xi_3^4-\frac{1}{1-\eta}\xi_3^4(t-d(t))(1-\dot{d}(t))\\
&\le-\xi_1^4-\xi_2^4+\xi_3^3u-\xi_3^3\xi_4^*-\xi_3^4\\
&\le-\xi_1^4-\xi_2^4-\xi_3^4,
\end{aligned}
$$

where

$$
u=\xi_4^*=-\alpha_3\xi_3 \tag{4}
$$

$$
\begin{aligned}
\alpha_3 = {} & 1 + \frac{1}{1-\eta} + \frac{1}{4}\left(\frac{2}{15}\right)^{-3} + \frac{3}{4\sqrt[3]{4}}\left(\left(\frac{1}{12}\right)^{-\frac{1}{3}} + 3\left(\frac{1}{10}\right)^{-\frac{1}{3}} + \left(\frac{1}{6}\right)^{-\frac{1}{3}} + \left(\frac{1}{2}\right)^{-\frac{1}{3}}\right)\varphi_{[\sigma(t),3]}^{\frac{4}{3}} \\
& + \varphi_{[\sigma(t),3]} + \frac{3}{4\sqrt[3]{4}}\left(\left(\frac{1}{12}\right)^{-\frac{1}{3}} + \left(\frac{1}{10}\right)^{-\frac{1}{3}}\right)(\alpha_1\varphi_{[\sigma(t),3]})^{\frac{4}{3}} \\
& + \frac{3}{4\sqrt[3]{4}}\left(\left(\frac{1}{6}\right)^{-\frac{1}{3}} + \left(\frac{1}{10}\right)^{-\frac{1}{3}}\right)(\alpha_2\varphi_{[\sigma(t),3]})^{\frac{4}{3}} + \frac{3}{4\sqrt[3]{4}}\left(\frac{1}{12}\right)^{-\frac{1}{3}}\left[\alpha_2\frac{\partial\alpha_1}{\partial x_0}(\lambda_0 - c_0)x_0\right] \\
& + \alpha_2 + \frac{3}{4\sqrt[3]{4}}\left(\frac{1}{10}\right)^{-\frac{1}{3}}\alpha_2^{\frac{8}{3}} + \frac{3}{4\sqrt[3]{4}}\left(\frac{1}{12}\right)^{-\frac{1}{3}}(\alpha_2\varphi_{[\sigma(t),1]})^{\frac{4}{3}} + \frac{3}{4\sqrt[3]{4}}\left(\frac{1}{12}\right)^{-\frac{1}{3}}(\alpha_1\alpha_2\varphi_{[\sigma(t),2]})^{\frac{4}{3}} \\
& + \frac{3}{4\sqrt[3]{4}}\left(\left(\frac{1}{10}\right)^{-\frac{1}{3}} + \left(\frac{1}{6}\right)^{-\frac{1}{3}}\right)(\alpha_2\varphi_{[\sigma(t),2]})^{\frac{4}{3}} + \frac{3}{4\sqrt[3]{4}}\left(\frac{1}{10}\right)^{-\frac{1}{3}}(\alpha_1\alpha_2)^{\frac{4}{3}} \\
& + \frac{3}{4\sqrt[3]{4}}\left(\frac{1}{12}\right)^{-\frac{1}{3}}\left(\alpha_1^2\alpha_2\right)^{\frac{4}{3}} + \frac{3}{4\sqrt[3]{4}}\left(\left(\frac{1}{12}\right)^{-\frac{1}{3}} + \left(\frac{1}{10}\right)^{-\frac{1}{3}}\right)(\alpha_1\alpha_2\varphi_{[\sigma(t),1]})^{\frac{4}{3}} \\
& + \frac{3}{4\sqrt[3]{4}}\left(\frac{1}{10}\right)^{-\frac{1}{3}}\left[\frac{\partial\alpha_2}{\partial x_0}(\lambda_0 - c_0)x_0\right]^{\frac{4}{3}} + \frac{1}{4}\left(\left(\frac{1}{12}\right)^{-1} + \left(\frac{1}{10}\right)^{-1}\right)(3\alpha_1\alpha_2\psi_{[\sigma(t),1]})^4 \\
& + \frac{1}{4}\left(\left(\frac{1}{12}\right)^{-1} + \left(\frac{1}{10}\right)^{-1}\right)\left(3\sqrt{2}\alpha_2\psi_{[\sigma(t),2]}\right)^4 + \frac{1}{4}\left(\left(\frac{1}{12}\right)^{-1} + \left(\frac{1}{10}\right)^{-1}\right)(6\alpha_1\alpha_2\psi_{[\sigma(t),2]})^4 \\
& + \frac{1}{4}\left(\left(\frac{1}{10}\right)^{-1} + \left(\frac{1}{6}\right)^{-1}\right)(6\alpha_2\psi_{[\sigma(t),2]})^4 + \frac{1}{4}\left(\left(\frac{1}{12}\right)^{-1} + \left(\frac{1}{10}\right)^{-1}\right)\left(3\sqrt{3}\psi_{[\sigma(t),3]}\right)^4 \\
& + \frac{1}{4}\left(\left(\frac{1}{10}\right)^{-1} + \left(\frac{1}{6}\right)^{-1}\right)\left(3\sqrt{6}\psi_{[\sigma(t),3]}\right)^4 + \frac{1}{4}\left(\left(\frac{1}{12}\right)^{-1} + \left(\frac{1}{10}\right)^{-1}\right)\left(3\sqrt{6}\alpha_1\psi_{[\sigma(t),3]}\right)^4 \\
& + \frac{1}{4}\left(\left(\frac{1}{10}\right)^{-1} + \left(\frac{1}{6}\right)^{-1}\right)\left(3\sqrt{6}\alpha_2\psi_{[\sigma(t),3]}\right)^4 + \frac{1}{2}\left(3\sqrt{6}\psi_{[\sigma(t),3]}\right)^4 + 54\psi_{[\sigma(t),3]}^2
\end{aligned}
$$

Choosing

$$V = V_0 + V_3,$$

one has

$$\mathcal{L}V \le -\lambda_0 x_0^2 - \xi_1^4 - \xi_2^4 - \xi_3^4.$$

Theorem 2 *For systems (1.1) and (1.2), letting u_0 and u as (2) and (4), then all states are asymptotically stabilized to origin in probability.*

In above analysis, for $x_0(t_0) \neq 0$, we give the controllers u_0 and u. But if $x_0(t_0) = 0$, one can apply the same switching control in [4] to (1.1) and (1.2). So the following result is correct.

Theorem 3 *For systems (1.1) and (1.2) and any $x_0(t_0)$, applying the same switching control in [4], all states are asymptotically stabilized to origin in probability.*

3 Conclusions

For SNSs with arbitrary switchings and time-varying delays, the backstepping state feedback stabilizing controllers are given.

Acknowledgements This work has been supported by National Natural Science Foundation of China under Grant 61,503,262, Natural Science Foundation of Hebei Province under Grant A2014106035, Foundation for High-level Talents of Hebei Province under grant number A2016001144.

References

1. D. Zhang, C. Wang, G. Wei, H. Chen, Output feedback stabilization for stochastic nonholonomic systems with nonlinear drifts and Markovian switching. Asian J. Control **16**(6), 1679–1692 (2014)
2. Y. Zhao, C. Wang, J. Yu, Partial-state feedback stabilization for a class of generalized nonholonomic systems with ISS dynamic uncertainties. Int. J. Control Autom. Syst. **16**(1), 79–86 (2018)
3. X. Qin, H. Min, Further results on adaptive state-feedback stabilization for a class of stochastic nonholonomic systems with time delays. Int. J. Control Autom. Syst. **16**(2), 640–648 (2018)
4. F. Gao, F. Yuan, Y. Wu, State-feedback stabilisation for stochastic non-holonomic systems with time-varying delays. IET Control Theor. Appl. **6**(17), 2593–2600 (2012)
5. D. Zhang, H. Zhang, H. Wu, Y. Yue, Q. Du, State feedback stabilization of stochastic nonholonomic mobile robots under arbitrary switchings. Proc. 2017 Chin. Intell. Syst. Conf. **459**, 713–725 (2017)
6. D. Zhang, C. Wang, J. Qiu, H. Chen, State-feedback stabilisation for stochastic non-holonomic systems with Markovian switching. Int. J. Model. Ident. Control **16**(3), 221–228 (2012)
7. X. Liu, Y. Zhao, Y. Yue, D. Zhang, Output feedback stabilization for stochastic nonholonomic systems under arbitrary switching. Discrete Dyn. Nat. Soc. **2017**, 1–8 (2017)
8. W. Si, X. Dong, Adaptive neural DSC for stochastic nonlinear constrained systems under arbitrary switchings. Nonlinear Dyn. **90**(4), 2531–2544 (2017)
9. F. Wang, B. Chen, Y. Sun, C. Lin, Finite time control of switched stochastic nonlinear systems. Fuzzy Sets and Systems (2018), in press, https://doi.org/10.1016/j.fss.2018.04.016

Online RPCA on Background Modeling

Huini Fu, Benzhang Wang and HengZhu Liu

Abstract Fast RPCA method for background modeling has to load video sequence into memory to conduct matrix decomposition. It is efficient with small-size video or limited image sequences, but not for online update or big data processing. So to satisfy the need for dynamic subspace, this work provides an option for processing one sample per time instance with an online optimization scheme to recover static camera background scene from video sequence. Experiments are conducted via LRSLibrary benchmarks. Results show that our proposed online fast RPCA can be a supplement to other online approaches.

Keywords Online RPCA · LRSLibrary · BRP · ADM · Matrix restoration

1 Introduction

Video surveillance and monitoring system has much more efficient applications with deep research of computer vision and machine learning algorithm, like abnormal activity detection and danger pre-detection. Just as what was proposed in [1], recovering background and foreground from video sequence is important in video analysis. Even today, background model generation is an important image process. Specific background model generation methods are supposed to handle specific issues including light changes, highlighted regions or shadows, period jittering, and etc. Background model generation method constantly lead to poor estimations constrained by these trouble issues and many other difficulties [2–7].

Casted by static camera, the background of video sequence is presumed to be the same. Under this circumstance, the target is to obtain a pure background image, then subtract the background from video sequence to acquire foregrounds. Considering

H. Fu (✉) · B. Wang · H. Liu
School of Computer Science and Technology, National University of Defense Technology, Changsha 410000, China
e-mail: 798768548@qq.com

Y. Jia et al. (eds.), *Proceedings of 2018 Chinese Intelligent Systems Conference*, Lecture Notes in Electrical Engineering 529,
https://doi.org/10.1007/978-981-13-2291-4_41

more realistic and complex video applications, it is difficult to get a satisfying background for disturbance caused by changing lights and dynamic background objects.

Subspace learning algorithms are proposed to solve this problem.

The Robust Principal Component Analysis (RPCA) is a traditional algorithm for subspace learning, it showed great potential in sampling and reconstructing signals which is very important in compressed sensing technology. In 2009, John [8] proposed an algorithm using RPCA in dimension reduction and achieved great improvements in face recognition.

Robust PCA algorithm is generally used in subspace projection and dimension reduction, and it achieved a great performance. Oliver [9] proposed a background model building method with Principal Component Analysis (PCA) technique. Casted by static camera, backgrounds of the video sequence basically stay the same. Then we could use RPCA in background model generation problem. Assuming a matrix could represent image sequences, with one image represented by a column vector, then the background of the sequence could be reconstructed by computing the low rank matrix through convex optimization. Through experiments, this method shows great advantage over state-of-the-art background modeling methods in Wallflower dataset in reference [9].

In the family of RPCA [10, 11], only a few online optimization algorithms were proposed and have achieved better performance [12], which replaced batch computation with stochastic optimization. It points out a way for big data decomposition.

This work is based on the online stochastic optimization algorithm, and. A refined online RPCA method is proposed to improve the initialization and optimization process of online RPCA. Our approach put focus on the equipment of the refined optimization process on existing RPCA framework. Experiments show improvements in building the background, and in segmenting foreground and background, and it is also efficient in time complexity.

The organization of this paper is illustrated as follows. Section 2 gives specification of the matrix norms used in this paper. Section 3 briefly demonstrates the RPCA problem for background model building. In Sect. 4, the improved optimization process is introduced. In Sect. 5, experiments and analyses are presented.

2 Notation

In this work, bold letter $\boldsymbol{x} \in R^p$ denote data vector, while capital letter $\boldsymbol{x} \in R^{p\times n}$ denote matrices. For an arbitrary real matrix E, let

$\|E\|_F = \left(\sum_{i=1}^{m}\sum_{j=1}^{n}|a_{i,j}|^2\right)^{\frac{1}{2}}$: the Frobenius norm

$\|E\|_0$: ℓ_0—norm, the numbers of none-zero elements in E

$\|E\|_1 = \sum_{i,j} |E_{i,j}|$: ℓ_1—norm, the sum of the absolute values of the elements in E

$\|E\|_{2,1} = \sum_{i=1}^{n} \|e_i\|$: $\ell_{2,1}$—norm where e_i is the i-th column of matrix E.

$\|E\|_* = \sum_i \sigma_i(E)$: nuclear norm, the sum of E's singular values.

3 The RPCA Problem

Robust PCA is a popular method to estimate the underlying subspace from an observed one with corrupted sparse noise. Let $X = [\boldsymbol{x}_1, \boldsymbol{x}_2, \ldots, \boldsymbol{x}_n] \in R^{d\times n}$ be a rectangular matrix, RPCA is proposed to decompose the observed matrix X into a low rank matrix Z plus a sparse matrix E. Under mild condition, RPCA is supposed to recover z and E from X by solving the following problem:

$$\min_{Z,E} rank(Z) + \lambda\|E\|_0 \quad s.t., X = Z + E \tag{3.1}$$

where $X = Z + E$, and λ is a tradeoff parameter. While the $rank(L)$ and ℓ_0-norm are NP hard problems, then the target function can be reformulated to compute nuclear norm and ℓ_1-norm separately, which are convex and could be optimized. (ℓ_0-norm is to compute the number of none-zero elements in a vector. ℓ_1-norm is to compute the sum of the absolute values of the vector, which is also called sparse rule operator—lasso regularization)

$$\begin{aligned} &\min_{Z,E} \|Z\|_* + \lambda\|E\|_1 \\ &s.t., \ X = Z + E \end{aligned} \tag{3.2}$$

ℓ_1-norm considers only the minimization of sparse matrix. while with ℓ_2-norm, overfitting could be prevented and generalization ability could be improved. Just as described in Sect. 2, $\ell_{2,1}$-norm is defined based on ℓ_2-norm, and it shares similar sparse patterns from multiple different tasks.

Regularization of $\ell_{2,1}$-norm is a critical computation in foreground detection. After $\ell_{2,1}$-norm regularization, each column vector of a matrix shares the same pattern.

Then the target function of RPCA can be formulated as,

$$\begin{aligned} &\min_{Z,E} \|Z\|_* + \lambda\|E\|_{2,1} \\ &s.t., \ X = Z + E \end{aligned} \tag{3.3}$$

Removing the equation constraint, the problem 1.3.3 can be reformulated as

$$\min_{Z,E} \|Z\|_* + \lambda\|E\|_{2,1} + \frac{\mu}{2}\|X - Z - E\|_F^2 \tag{3.4}$$

The nuclear norm is a batch computation and it couples samples tightly, which makes one sample per time difficult. Here we use a substitution equivalent for the nuclear norm of Z, whose rank is denoted as r, as follows [13],

$$\|Z\|_* = \inf_{L \in R^{d\times r}, R \in R^{n\times r}} \left\{ \frac{1}{2}\|L\|_F^2 + \frac{1}{2}\|R\|_F^2 : Z = LR^T \right\}$$

Let r denote the rank of X. In this decomposition, $L \in R^{d\times r}$ is the basis of the subspace and $R \in R^{r\times n}$ denotes the coefficients of the samples w.r.t. $L \in R^{d\times r}$ the basis. Thus the RPCA problem in (3.4) can be reformulated as

$$\min_{Z, L \in R^{d\times r}, R \in R^{n\times r}, E} \frac{1}{2}(\|L\|_F^2 + \|R\|_F^2) + \lambda\|E\|_{2,1} + \frac{\mu}{2}\|X - Z - E\|_F^2 \tag{3.5}$$

where μ is penalty parameter. Substituting Z with LR^T, the above problem can be rewritten as,

$$\min_{Z, L \in R^{d\times r}, R \in R^{n\times r}, E} \frac{1}{2}(\|L\|_F^2 + \|R\|_F^2) + \lambda\|E\|_{2,1} + \frac{\mu}{2}\left\|X - LR^T - E\right\|_F^2 \tag{3.6}$$

Given a set of samples $X = [\boldsymbol{x}_1, \boldsymbol{x}_2, \ldots, \boldsymbol{x}_n] \in R^{d\times n}$, solving problem (3.6) indeed minimizes the following empirical cost function,

$$f_n(L) \overset{\Delta}{=} \frac{1}{2n}\sum_{i=1}^{n} \ell(x_i, L) + \frac{1}{2n}\|L\|_F^2 \tag{3.7}$$

wherein the loss function of each sample is defined as

$$\ell(\boldsymbol{x}_i, L) \overset{\Delta}{=} \min_{r,e} \frac{\mu}{2}\|\boldsymbol{x}_i - Lr - e\|_F^2 + \frac{1}{2}\|r\|_2^2 + \lambda\|e\| \tag{3.8}$$

the empirical cost function is optimized to minimize the loss with each sample loss function were minimized.

4 Optimization

Compared with APG used in [14], ALM can provide a higher recovering accuracy. In the optimization of RPCA family, ALM is widely applied. While ALM treats the decomposition problem as a general convex optimization problem, it did not take the separable structure of Z and E into consideration. Under this circumstance, ADM is suitable for recovering low rank matrix and sparse matrix. ADM is a refinement of quadratic penalty method, which ensures that the algorithm would converge fast meanwhile increase its adaptability. In order to solve this problem, ADM (Alternating Direction Method) is adopted.

The corresponding augmented Lagrangian of (3.8) is

$$\ell(r, e, L, Y) \triangleq \frac{1}{2}\|r\|_2^2 + \lambda\|e\| + \langle Y, x - Lr - e\rangle + \frac{\mu}{2}\|\boldsymbol{x} - Lr - e\|_F^2 \tag{4.1}$$

where Y is the lagrangian multiplier.

In traditional ADM method, r and e are updated iteratively.

4.1 Basic Initialization with BRP

BRP was proposed in Godec algorithm for low rank & sparse matrix decomposition in [15]. Given a dense matrix X, BRP can generate an approximation of X through,

$$Y_1 = XA_1; \quad Y_2 = X^T A_2; \quad L = Y_1(A_2^T Y_1)^{-1} Y_2^T$$

wherein L is a fast rank-r approximation of X.

4.2 Updating r

The sub-problem $\min_r \ell(r, e, L)$ for updating r is given by,

$$\begin{aligned}\min_r \ell(r, e, L, Y) &= \frac{1}{2}\|r\|_2^2 + \langle Y, x - Lr - e\rangle + \frac{\mu}{2}\|\boldsymbol{x} - Lr - e\|_F^2 \\ &= \frac{1}{2}\|r\|_2^2 + \frac{\mu}{2}\left\|\boldsymbol{x} - Lr - e + \frac{Y}{\mu}\right\|_F^2\end{aligned} \tag{4.2}$$

4.3 Updating e

The subproblem $\min_e \ell(r, e, L)$ for updating e is given by,

$$\begin{aligned}\min_r \ell(r, e, L, Y) &= \lambda\|e\| + \langle Y, x - Lr - e\rangle + \frac{\mu}{2}\|\boldsymbol{x} - Lr - e\|_F^2 \\ &= \lambda\|e\| + \frac{\mu}{2}\left\|\boldsymbol{x} - Lr - e + \frac{Y}{\mu}\right\|_F^2\end{aligned} \tag{4.3}$$

Algorithm 1 Online RPCA

Input $X = [x_1, x_2, ..., x_n] \in R^{d\times n}$

Initialize $A = 0 \in R^{r\times r}, B = 0 \in R^{n\times r}, L_0 = Y_1(A_2^T Y_1)^{-1} Y_2^T \in R^{d\times r}, A_1 \in R^{1\times r}, A_2 \in R^{d\times r}$

Parameters $\lambda = 1/sqrt(size(x)), \mu, \gamma$

for $t = 1$ to T do

1. $r_t = \text{argmin} \min_r \ell(r, e, L_{t-1}, Y)$
2. $e_t = \text{argmin} \min_e \ell(r, e, L_{t-1}, Y)$
3. $Y_t = Y + \mu(x_t - L_{t-1} r_t - e_t)$
4. $\mu = \min(\gamma * \mu, \mu_{\max})$
5. $A_t \leftarrow A_{t-1} + rr^T$
6. $B_t \leftarrow B_{t-1} + (x - e) * r^T$
7. update the basis L_t using algorithm 2

end for

Output: LR^T, E

Algorithm 2 Basis Update

Input : $L = [l_1, ..., l_r] \in R^{d\times r}, AB = [a_1, ..., a_r] \in R^{r\times r}, and\ B = [b_1, ..., b_r] \in R^{d\times r}$.

Initialize $\tilde{A} \leftarrow A + \lambda I$

for $j = 1$ to r **do**

$$l_j \leftarrow \frac{1}{\tilde{A}_{j,j}}(b_j - L\tilde{a}_j) + l_j$$

end for

Output L.

5 Experiments

λ, γ, μ are three parameters used in online RPCA algorithm. γ is growth parameter and μ is penalty parameters in ADM, which could monitor the convergence speed. Empirically set $\gamma = 1.01$, $\mu = 0.5$, For setting these parameters in an appropriate range makes little difference to the result.

We test our online RPCA algorithm on **LRSLibrary** [16] with matlabR2011 (windows Intel Core i5-4570 CPU 3.2 GHz 12 GB RAM, 64bit). The result is aspiring.

Table 1 describes of the information of datasets.

Table 1 the brief introduction of datasets

DataSets	Size	Number of frames
Car	48 * 48	51
Escalator	130 * 160	197
Highway	240 * 320	1698
Shop	144 * 192	156

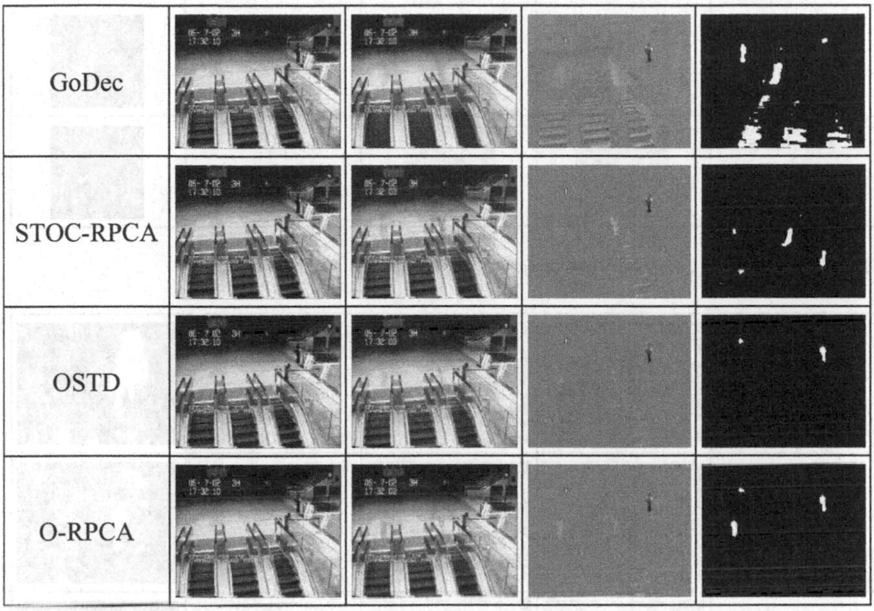

Fig. 2 Performance on escalator video sequence

Figure 1, 2, 3, 4 compared the performance of different RPCA variants on different video sequences.

Experiments show that our online RPCA shares the same decomposition performance but less computer memory occupation, which improved the performance of other online algorithms. It illustrates the proposed optimization process is effective and efficient. The results show our method with updating procedure does not take up too much computation memory and is good enough to deal with big data.

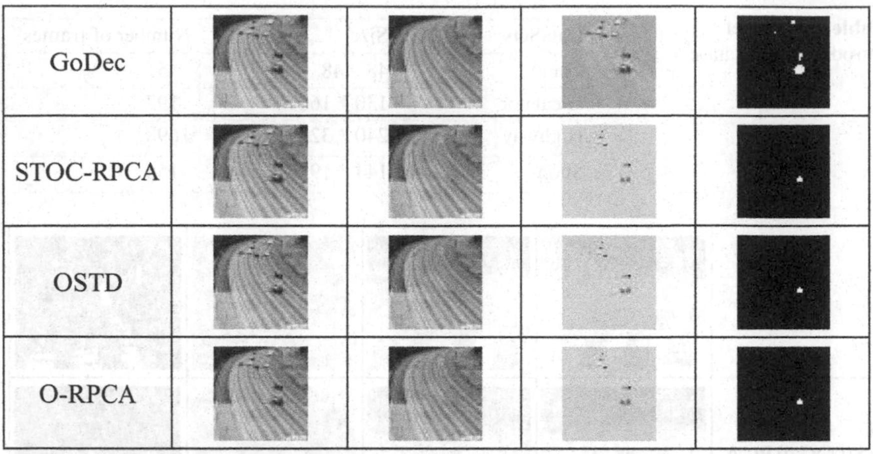

Fig. 1 Performance on car video sequence

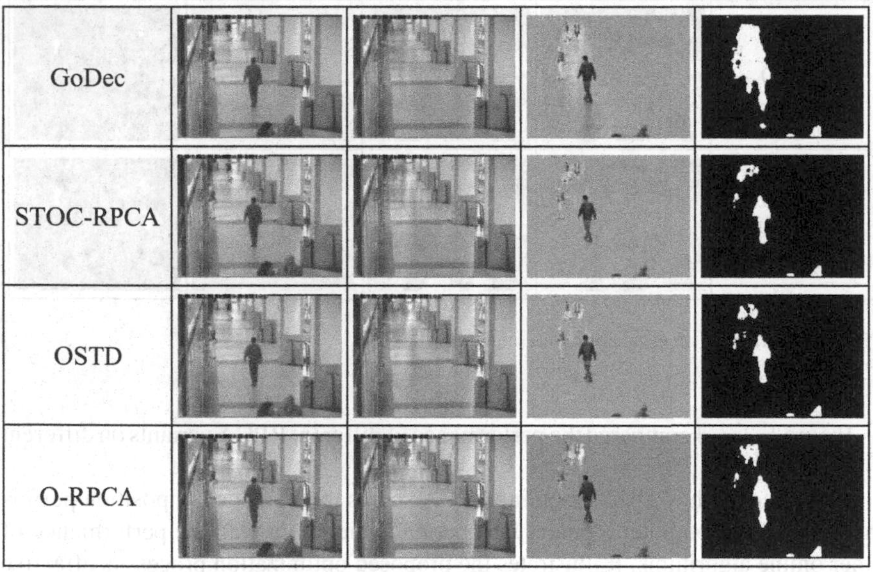

Fig. 3 Performance on shop video sequence

Fig. 4 Performance on highway video sequence

References

1. F. Huini, G. Zhihui, L. Hengzhu, Fast Robust PCA on background modeling, in *Proceedings of 2017 Chinese Intelligent Systems Conference* (2017)
2. H. Dengyuan et al., Reliable moving vehicle detection based on the filtering of swinging tree leaves and raindrops. J. Vis. Commun. Image Represent. **23**(4), 648–664 (2012)
3. C.R. Wren, Real-time tracking of the human body. IEEE Trans. Pattern Anal. Mach. Intell. **19**(7), 780–785 (1997)
4. E. Ahmed, Non-parametric model for background subtraction [J]. ECCV **1843**, 751–767 (2000)
5. S. Chris, Adaptive Background mixture models for real-time tracking, in *IEEE Computer Society Conference on Computer Vision and Pattern Recognition*, p. 252 (1999)
6. B. Oliver, Van D. Marc, ViBe: a universal background subtraction algorithm for video sequences. IEEE Trans. Image Process. **20**(6), 1709–1724 (2011)
7. K. Tomasz, K. Mateusz, M. Gorgon, *Real-time Background Generation and Foreground Object Segmentation for High-definition Colour Video Stream in FPGA Device*. (Springer, New York, Inc. 2014)
8. W. John et al., Robust Principal Component analysis: exact recovery of corrupted low-rank matrices. neural networks for signal processing X, 2000, in *Proceedings of the 2000 IEEE Signal Processing Society Workshop*, vol. 1 (IEEE, 2009), pp. 289–298
9. B. Thierry, F.E. Baf, B. Vachon, Statistical background modeling for foreground detection: a survey. Handbook Of Pattern Recognition And Computer Vision, pp. 181–199 (2009)
10. De La T. Fernando, M.J. Black, Robust principal component analysis for computer vision, in *Proceedings. Eighth IEEE International Conference on Computer Vision, 2001. ICCV 2001*, vol. 1(IEEE, 2001), pp. 362–369
11. X. Shijie et al., *FaLRR: A Fast Low Rank Representation Solver* (CVPR, IEEE, 2015), pp. 4612–4620
12. X. Feng Jiashi, Y.S. Huan, Online robust PCA via stochastic optimization. Adv. Neural. Inf. Process. Syst. **26**, 404–412 (2013)
13. P. Sprechmann, A.M. Bronstein, G. Sapiro, *Learning Efficient Sparse and Low Rank Models*. arXiv preprint arXiv:1212.3631 (2012)
14. S. Andrews et al, Online Stochastic Tensor decomposition for background subtraction in multispectral video sequences, in *IEEE International Conference on Computer Vision Workshop* (IEEE, 2016), pp. 946–953
15. Z. Tianyi, T. Dacheng. GoDec: Randomized Lowrank & Sparse Matrix Decomposition in Noisy Case. ICML. DBLP, pp. 33–40 (2011)
16. S. Andrews, B. Thierry, E.H. Zahzah, LRSLibrary: Low-Rank and Sparse tools for Background Modeling and Subtraction in Videos. Handbook on “Robust Low-Rank and Sparse Matrix Decomposition: Applications in Image and Video Processing” (2016)

Bearing Fault Diagnosis Based on Generalized S Transform Denoising and Convolutional Neural Network

Wei Liu, Minghong Han and Lihua Chen

Abstract This paper utilizes convolutional neural network (CNN) combining generalized S transform denoising (GSTD) method to complete noisy bearing fault diagnosis. After GSTD, images with more obvious failure information can be obtained. Then these feature images are trained by convolutional neural network. The recognition accuracy of the proposed method on testing dataset achieves as high as 99.25%. Finally, the proposed method is compared with other diagnosis methods to prove its effectiveness in processing noise signal.

Keywords Deep learning · CNN · Fault diagnosis
Generalized S transform denoising · Bearing

1 Introduction

About 30% of the mechanical problems in rotating machinery are caused by roller bearings failure, so the fault diagnosis of roller bearings is of great significance.

The core of the algorithm of bearing fault diagnosis is signal feature extraction and pattern classification. Time-frequency analysis is a traditional signal processing technology. Common pattern classification algorithms include support vector machine [1, 2], BP neural network, nearest neighbor classifier, etc. Through signal processing, feature extraction, and state diagnosis, the traditional method relies on expert fault diagnosis experience and signal processing techniques, which is easily disturbed by human factors.

The traditional shadow models is inferior to Deep Learning (DL) methods. When it comes to fault diagnosis, the input data is the one-dimensional time signals. CNN used to deal with it has been investigated to achieve motor fault diagnosis [3]. More common is to convert one-dimensional signals into two-dimensional images. 2-D

W. Liu · M. Han (✉) · L. Chen
School of Reliability and Systems Engineering, Beihang University,
Beijing 100191, China
e-mail: hanminghong@buaa.edu.cn

Y. Jia et al. (eds.), *Proceedings of 2018 Chinese Intelligent Systems Conference*, Lecture Notes in Electrical Engineering 529,
https://doi.org/10.1007/978-981-13-2291-4_42

gray-level images was obtained from 1-D vibration signals [4], the proposed approach has succeeded in diagnosing induction motor fault.

Stockwell et al. [5] proposed S-transform in 2002. It has good performance and is superior to other conventional time-frequency methods. In order to change the time-frequency resolution, it came into being generalized S transform.

In this paper, the generalized S transform is used to transform one-dimensional signal into 2-dimensional time frequency image, using frequency domain denoising technology to deal with the image. The noise interference can be reduced effectively, and the robustness of the feature is increased, combing convolutional neural network for fault classification of rolling bearings.

2 Convolutional Neural Network and Generalized S Transform Denoising

CNN has got widely successful applications, especially in image recognition tasks. LeNet-5 [6], AlexNet [7], VGGNet [8] and GoogLeNet [9] are common models in CNN. Convolutional layer, pooling layer, and softmax layer et al. are included in a typical convolutional neural network.

2.1 *Convolutional and Feature Pooling Layer*

The input layer of a CNN is generally an image, denoted by X_0. The operation of the convolutional layer is expressed as:

$$X_{i\,j\,k} = X_{i-1} \otimes W_i + b_i \tag{1}$$

Where $x_{i\,j\,k}$ is the (i, j) component of the kth feature map. W_i is the weight matrix of the convolution kernel of the i-th layer, $\otimes$ is the convolution operation, and b_i is the offset vector of the i-th layer.

Get the feature mapping from the previous layer to the next layer through a nonlinear function, denoted as $X_i = F(X_{i-1})$. Among them: F is a non-linear function, the commonly used nonlinear functions are ReLU functions [6].

The most common use of these is maximum pooling. Max-pooling is often used in recent models. The expression is:

$$y_{i\,j\,k} = \max\left(y_{i_1 j_1 k} : i \le i_1 \le i + p, j \le j_1 \le j + q\right) \tag{2}$$

where p is the length of the pooling window and q is the stride length.

2.2 Softmax Layer

The Softmax model is a generalization of logistic regression models for multi-classification problems. For training set $\left\{\left(x^{(1)}, y^{(1)}\right), \ldots, \left(x^{(m)}, y^{(m)}\right)\right\}$, we have $y^{(i)} \in \{1, 2, \ldots, k\}$. The class label y can take k different values.

The cost function of is defined as:

$$J(\theta) = -\frac{1}{m}\left[\sum_{i=1}^{m}\sum_{j=1}^{m} 1\left\{y^{(i)} = j\right\}\log\frac{e^{\theta_j^T x(i)}}{\sum_{l=1}^{k} e^{\theta_l^T x(i)}}\right] \quad (3)$$

where $j = 1, 2, \ldots, k$. $\theta = \left[\theta_1^{(T)}, \theta_2^{(T)}, \ldots \theta_k^{(T)}\right]$ is the parameter of the softmax regression model. $1\{\cdot\}$ is the indicator function. Update the parameters in process of minimizing the loss function.

2.3 Generalized S Transform Denoising (GSTD)

The continuous S-transform $S(\tau, f)$ of signal $x(t)$ can refer to [5]. The original signal is one-dimensional time signal. It becomes two-dimensional images after S-transform. The expression of the generalized S transform (GST) is:

$$s(\tau, f) = \int_{-\infty}^{\infty} x(t)\frac{|f|^p|\lambda|}{\sqrt{2\pi}} \cdot \exp\left(\frac{-(\tau - t)^2 f^{2p}\lambda^2}{2}\right) \cdot \exp(-i2\pi f t)dt \quad (4)$$

Where: τ is the time shift factor; f is the frequency. The regulatory factors λ, p. An image $S_p(u, v)$ having size M * N is obtained by generalized S-transformation.

The filter function is $H(u, v)$, and its size is the same as $S_p(u, v)$. We dot-multiply $S_p(u, v)$ and $H(u, v)$, then the filtered image is obtained.

Where:

$$H(u, v) = e^{\frac{-D^2(u,v)}{2D_0^2}} \quad (5)$$

D_0 is the cut-off radius of the filter. The larger D_0 is, the wider the filter band is, and the smoother the better. And the $D(u, v)$ is the distance of a point to the filter center.

3 CNN-Based Fault Diagnosis

3.1 Create a Training Database

This experiment uses the United States Case Western Reserve University bearing fault data. We used the drive end acceleration data collected at the sampling frequency of 12 kHz, including raw data for operating conditions including normal (NR), inner race fault (IRF), ball fault (BF), and outer race fault (ORF), shown as in Table 1. All samples have a signal-to-noise ratio of 5 white noise. The time domain maps are shown in Fig. 1. The ratio of training set and testing set is 3:1 for each working condition. The total training sample is 1800 and the total test sample is 600.

3.2 Create a CNN

The convolution kernel size is 5 ∗ 5, and the step size is 1 ∗ 1. We have used zero padding before performing convolution on the feature map, zero padding size is 2. The max pooling layer is with 2 ∗ 2 filters and stride 2, reducing the size of the feature map by one-half. The convolutional and max pooling layer are called Conv-Maxpool layer. Dropouts are generally used in fully connected layers. In this paper,

Table 1 Four working conditions bearing data

Working condition	Fault diameter (mil)	Rotating speed (rpm)	Number of sample	Sample length
Normal	0	1750	600	512
Inner race fault	21	1750	600	512
Ball fault	21	1750	600	512
Outer race fault	21	1750	600	512

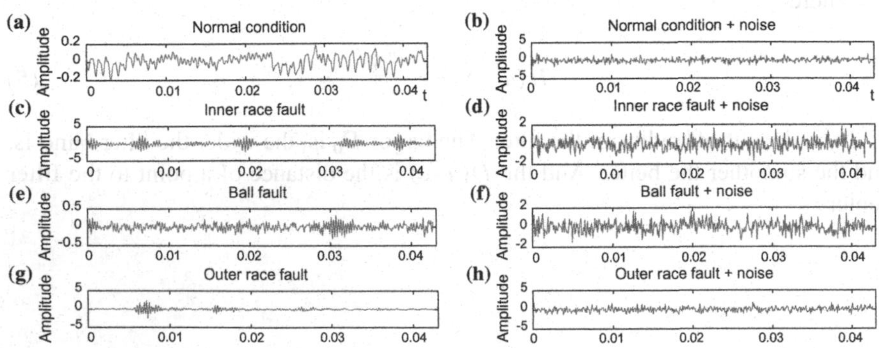

Fig. 1 Vibration signals of different working conditions

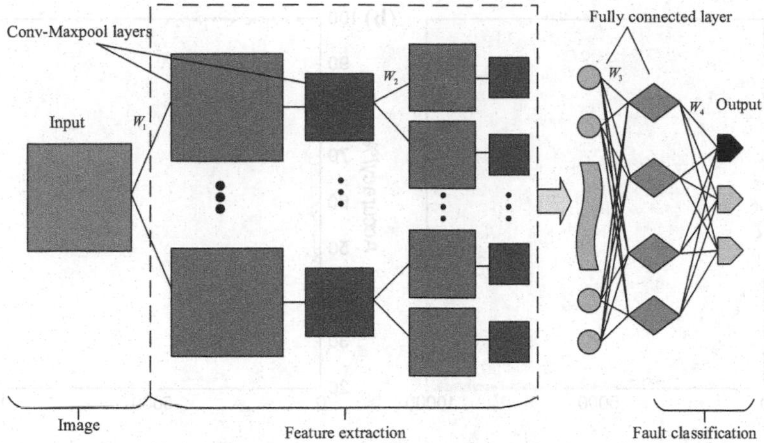

Fig. 2 The structure of CNN

the convolutional neural network has two Conv-Maxpool layers. The number of fully connected layers is 1, number of nodes is 256. The network structure is shown in Fig. 2.

3.3 CNN Training

The images obtained by GSTD are used as training database. The hyperparameters selected for training the rolling bearing neural network are as follows: Mini-batch size is 50, learning rate is 0.0001, and the number of iterations is 10,000. The optimization algorithm is Root Mean Square Propagation (RMSProp). RMSProp [10] is an optimizer that normalizes the gradients utilizing the magnitude of recent gradients. The RMSProp algorithm is more suitable to train deep networks, converges much faster than stochastic gradient decent (SGD).

3.4 Result Analysis

After 10,000 iterations, the CNN loss values and total recognition accuracy on the testing dataset were presented on Fig. 3. The CNN diagnosis results show that GSTD combined with convolutional neural networks achieves a total recognition accuracy rate 99.25% for the four conditions of the bearing.

Comparison of different denoising methods. Method 1: SVD denoising. A is a matrix of $m \times n$ composed of signals and noise generated by GST. $A = \mathrm{UDV}^T$,

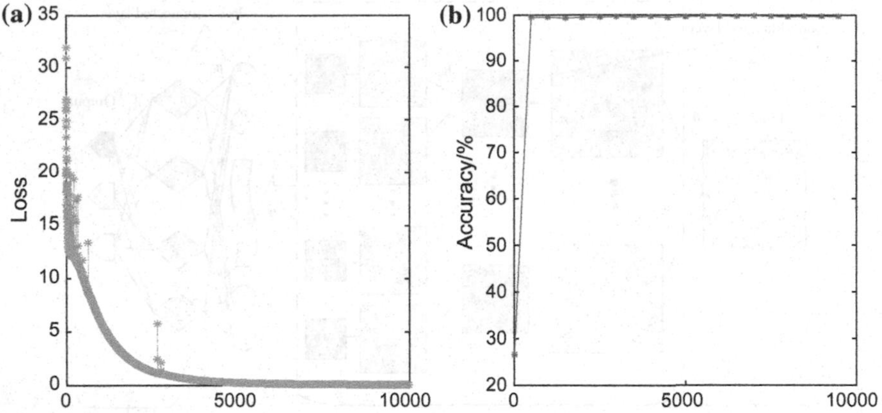

Fig. 3 The loss curve and prediction accuracy on testing dataset

$D = \begin{bmatrix} S & 0 \\ 0 & 0 \end{bmatrix}$, $S = diag(\sigma_1, \sigma_2, \ldots, \sigma_r)$. The smaller singular values react noise. This part of the smaller singular value is set to 0 in order to remove the noise. Zhao et al. proposes a difference spectrum of singular value (DSOFV) theory [11], using the maximum peak value of DSOFV to determine the threshold σ_{th} between the ideal signal and the noise.

Method 2: Threshold filtering. When filtering is performed using global threshold filtering, getting filter $C(\tau, f) = \begin{cases} 1 & S \geq \lambda S_{max} \\ 0 & S < \lambda S_{max} \end{cases}$. S_{max} is the maximum value of the time-frequency matrix after the generalized S transform. λ is the threshold factor, $0 \leq \lambda \leq 1$.

Method 3: In order to improve the SVD local filtering ability, the S transform spectrum is divided into multiple blocks, and SVD is used in each block.

Method 4: The method proposed in part 1.2.3.

The results of GSTD on four working conditions are shown in Fig. 4, among them λ=1, $p = 0.9$. The identification accuracy rate on the testing dataset of the four methods through the above established convolutional neural network is presented on the Table 2, method 4 has the highest recognition accuracy 99.25%. Method 3 has the worst recognition accuracy 73.36%. Method 1 has a higher accuracy 85.74% among method 1–4. The results prove the effectiveness of the GSTD method.

In order to show the good performance of the proposed method, we compared CNN with other traditional recognition models. In this part, we combined CNN, Multi-Layer Perceptron (MLP), SVM with Generalized S transform respectively for bearing fault diagnosis. Multi-Layer Perceptron is a feedforward neural network that includes at least one hidden layer, which was given two hidden layers with 512, and 512 nodes, respectively. The learning rate and mini-batch size was set 0.01, 80 respectively, choosing RMSPro optimization. The number of iteration is

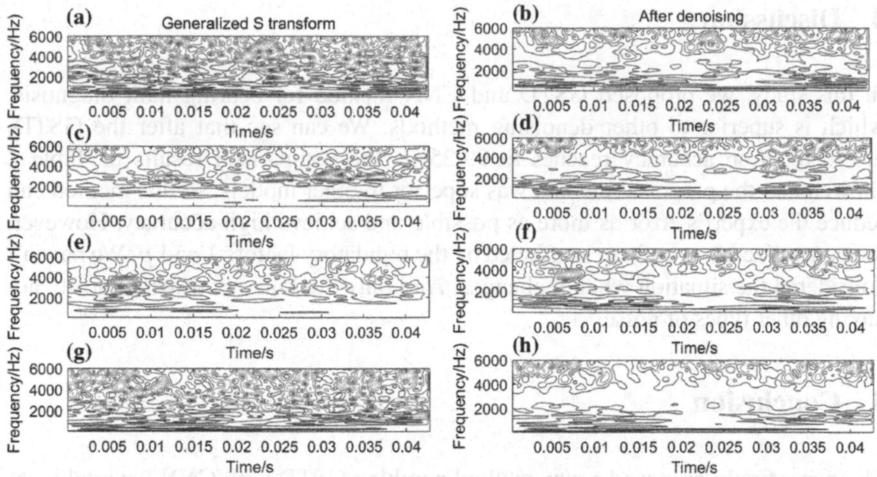

Fig. 4 Generalized S-Transform denoising of four states of roller bearing: **a, b** ball fault, D_0= 80 **c, d** inner race fault, D_0= 70 **e, f** outer race fault, D_0= 72 and **g, h** normal, D_0= 68

Table 2 Comparison of different denoising methods

Method		Total accuracy (%)	Error
Method 1		85.74	0.1426
Method 2	λ=0.2	79.6	0.204
	λ=0.8	79.3	0.207
Method 3 (Divided into 8 * 8)		73.36	0.2664
Method 4 (GSTD)		99.25	0.0075

Table 3 Identification accuracy of different methods

Methods	Normal (%)	Inner fault (%)	Ball fault (%)	Outer fault (%)	Total accuracy (%)
CNN+GST	68.44	100	56.08	97.44	81.44
MLP+GST	56.76	100	68.32	96	80.58
SVM+GST	72.67	100	97.33	72	85.5
Proposed method	99.44	100	98.24	99.32	99.25

1000. SVM, which is an excellent classifier. Specifically, SVM with a radial basis function (SVM.R) was included in the performance evaluation. From the result of Table 3, it was obvious that the proposed method had a good effect compared with other methods. The classification results of CNN+GST, MLP+GST, SVM+GST are 81.44, 80.58, 85.5%. The results showed that the GSTD and CNN method has better diagnosis effect for noisy bearing signals.

4 Discussion

In this study, we proposed GSTD and CNN method for bearing fault diagnosis, which is superior to other denoising methods. We can see that after the GSTD, the recognition accuracy reaches to 99.25%. The comparison results of Table 2 showed that the proposed method was superior to other models, so this method can reduce the expert's error as more as possible and achieve high accuracy. However, the generalized S-transform is affected by the regulatory factors λ and p. We haven't considered the situation when it changes, And only thinking about white noise, not having other types of noise.

5 Conclusion

The paper firstly proposed a new method combing GSTD with CNN for intelligent fault diagnosis. GSTD could not only reduce noise interference but enhance image useful features. The proposed theory in the paper could achieve a high recognition accuracy under noise conditions. It provides a new idea for dealing with noise. The recognition accuracy of the model is as high as 99.25%, which is superior to other diagnosis methods. Indicate the potential of this method on the intelligent diagnosis field.

References

1. A. Tabrizi, L. Garibaldi, A. Fasana, et al., Early damage detection of roller bearings using wavelet packet decomposition, ensemble empirical mode decomposition and support vector machine. Meccanica **50**(3), 865–874 (2015)
2. X. Hongbo, C. Guohua, An intelligent fault identification method of rolling bearings based on LSSVM optimized by improved PSO. Mech. Syst. Signal Process. **1–2**, 167–175 (2013)
3. I. Turker et al., Real-time motor fault detection by 1-D convolutional neural networks. IEEE Trans. Ind. Electron. **63**, 7067–7075 (2016)
4. D. Van Tan, C. Ui-Pil, Signal model-based fault detection and diagnosis for induction motors using features of vibration signal in two-dimension domain. Strojniski vestnik **57**, 655–666 (2011)
5. R.G. Stockwell, L. Mansinha, R.P. Lowe, Localization of the complex spectrum: the S transform. IEEE Trans. Signal Process. **4**, 998–1001 (2002)
6. Y. LeCun, LeNet-5, Convolutional Neural Networks [Online]. (2015), Available: http://yann.lecun.com/exdb/lenet/
7. K. Alex, I. Sutskever, G.E. Hinton, Imagenet classification with deep convolutional neural networks, in *International Conference on Neural Information Processing Systems*, vol. 60, pp. 1097–1105 (2012)
8. S. Karen, A. Zisserman, Very Deep Convolutional Networks for Large-Scale Image Recognition. Computer Science, 2014
9. S. Christian, et al., Going deeper with convolutions, in *2015 IEEE Conference on Computer Vision and Pattern Recognition*, pp. 1–9 (2015)
10. T. Tieleman, G. Hinton, Lecture 6.5-rmsprop: divide the gradient by a running average of its recent magnitude, in *Technical Report: Neural Networks of Machine Learning*, pp. 26–30 (2012)
11. X. Zhao et al., Difference spectrum theory of singular value and its application to the fault diagnosis of headstock of lathe. J. Mech. Eng. **1** (2010)

An Assessment Method of Vacuum Circuit Breaker Based on Variable Weight and Cloud Model

Yumei Wang, Yin Zhang and Di Lu

Abstract This paper uses assessment method of vacuum circuit breaker (VCB) conditions based on variable weight and cloud model. The state indicators are determined by the test data and relevant literature. The improved analytic hierarchy process method is used to optimize the subjectivity of the weights, and the variable weight is introduced to correct the constant weight, which according to the actual data. The cloud model is used to dividing the condition level of vacuum circuit breaker and calculating the membership degree of each grade cloud. The condition of the vacuum circuit breaker is determined by the fuzzy synthesis of the weights and the condition matrix. Taking the ZN63A-12 vacuum circuit breaker as an example, the results based on the method are closer to the actual condition, which can provide a reference to the condition maintenance.

Keywords Vacuum circuit breakers · Conditional assessment · Variable weight
Cloud model

1 Introduction

With the construction and development of smart grids, circuit breakers are one of the important equipment for the safe operation of power systems, and their maintenance methods are also an inevitable trend from the periodical maintenance to the condition-based maintenance. Therefore, an objective and accurate assessment method is needed to evaluate the equipment status. At present, there are many methods for assessing the status of power equipment, such as fuzzy theory, grey correlation, association rules, Bayesian networks and other methods [1–4]. The fuzzy comprehensive evaluation method is mostly used to evaluate the operating status of circuit breakers in complex operating environments and complex conditions. In

Y. Wang (✉) · Y. Zhang · D. Lu
School of Electrical Engineering and Automation,
Henan Polytechnic University, 454000 Jiaozuo, China
e-mail: 1799223255@qq.com

Y. Jia et al. (eds.), *Proceedings of 2018 Chinese Intelligent Systems Conference*, Lecture Notes in Electrical Engineering 529,
https://doi.org/10.1007/978-981-13-2291-4_43

applying fuzzy comprehensive evaluation, index weights reflect the importance of the evaluation index in the operation of vacuum circuit breakers. Reference [5, 6] uses analytic hierarchy process of evaluation index empowerment, but over-rely on their own experience of experts in the weight coefficient established rights, leading to weight coefficient is too strong on subjectivity. For this, reference [7] introduces the balance function to the weight of the comprehensive index modified. The reference [8] uses the triangular fuzzy number to correct the subjective weight.

To set the weight coefficient more objectively, this paper uses an improved analytic hierarchy process to determine the index weights, introduces variable weight formulas to adjust the constant weights. Combining the cloud model with the multi-level structure of fuzzy mathematics, a state evaluation method of vacuum circuit breaker based on variable weight and cloud model is proposed.

2 Condition Evaluation Index of Vacuum Circuit Breaker

Vacuum circuit breaker condition assessment involves a number of factors, and it is necessary to comprehensively consider various influencing factors and conduct objective evaluations to truly reflect the operating status of vacuum circuit breakers. The vacuum circuit breaker condition assessment is divided into four parts, namely vacuum characteristics, mechanical characteristics, electrical characteristics and other items. The mechanical and electrical characteristics are subdivided into specific sub-items, and other items are composed of objective conditions such as work environment and maintenance, as shown in Table 1.

3 Index Weight

The commonly used weighting methods often obtain the constant weights of the indicators and cannot flexibly reflect the index status. This paper uses the analytic hierarchy process of the interval judgment matrix to determine the constant weights and obtains variable weights through variable weight formulas to objectively reflect the importance of the indicators.

3.1 Determination of Constant Weight

Analytic Hierarchy Process (AHP) establishes the comparison of the importance of two elements with respect to the upper level by experts, and quantifies the relative importance to $a_{ij} = a_i / a_j$, forms a judgment matrix for this item, and then obtains the weights. However, in the actual evaluation, taking into account the experts' subjec-

Table 1 Evaluation index system of vacuum circuit breaker condition

Target level	Project level	Subproject level	Indicator level
Vacuum circuit breaker condition	Vacuum characteristics	Vacuum degree of arc chamber	
	Mechanical characteristics	Time parameter	Non-synchronous closing/breaking and other
		Speed parameter	Closing/Breaking velocity and other
		Current parameter	Current of closing/breaking circuit
	Electrical characteristics	Contact electrical wear	
		Working years	
		Accumulative used times	
	Other items	Environment, appearance manufacturers, etc.	

tivity and other complex factors, this article uses the interval number $a_{ij} = [l_{ij}, u_{ij}]$ to express the relative degree, and constructs the interval judgment matrix as follows:

$$A = \begin{bmatrix} [l_{11}, u_{11}] & [l_{12}, u_{12}] & \ldots & [l_{1n}, u_{1n}] \\ [l_{21}, u_{21}] & [l_{11}, u_{11}] & \ldots & [l_{2n}, u_{2n}] \\ \vdots & \vdots & \ldots & \vdots \\ [l_{n1}, u_{n1}] & [l_{n2}, u_{n2}] & \ldots & [l_{11}, u_{11}] \end{bmatrix} \tag{1}$$

Each element of the matrix is derived using the 1–9 scale method, where $\frac{1}{9} \leq l_{ij} = \frac{1}{u_{ji}} \leq 9, l_{ii} = u_{ii} = 1$.

After the interval judgment matrix A is constructed, there are formulas:

$$b_{ij} = \sqrt[2n]{\prod_{k=1}^{n} \frac{l_{ik} u_{ik}}{l_{jk} u_{jk}}} \tag{2}$$

Get the point approximation matrix B of the consistency of A; calculate the maximum eigenvalue and the corresponding vector for matrix B. Then have:

$$\lambda\text{max} = \sum_{i=1}^{n} \frac{(BW)_i}{nW_i} \tag{3}$$

$$W_i = \sqrt[n]{\prod_{j=1}^{n} b_{ij}} \Big/ \sum_{i=1}^{n} \sqrt[n]{\prod_{j=1}^{n} b_{ij}} \tag{4}$$

Perform a random consistency check on the matrix:

$$CR = (\lambda_{\max} - n)/(RI \times (\mathrm{n} - 1)) \tag{5}$$

The RI values in the formula are shown in Table 2:

When the random consistency indicator satisfies the requirement $CR < 0.10$, then the eigenvector sought is the constant weight.

3.2 *Determination of Variable Weight*

When the value of certain indicators of the circuit breaker seriously deviates from the normal value, and the weight is relatively small, it often results in a misjudgment in which the state evaluation is still normal when it is necessary to pay attention. Therefore, the variable weight theory was introduced to improve the weight coefficient. The variable weight formula [9] is:

$$W_i^v = \frac{W_i x_i^{\alpha-1}}{\sum_{i=1}^{n} W_i x_i^{\alpha-1}} \quad i = 1, 2, \cdots, n \tag{6}$$

The W_i^v is the variable weight, x_i is the score value of the state quantity; n is the number of state quantities; W_i is the constant weight of the state quantity. The α is an equilibrium function which value depends on the relative importance of each indicator. This paper takes $\alpha = 0$ for the quantitative indicator and takes $\alpha = 1$ for the qualitative indicator.

Table 2 RI value of n module estimate matrix

n	1	2	3	4	5	6	7	8
RI	0	0	0.58	0.90	1.12	1.24	1.32	1.41

4 Vacuum Circuit Breaker Condition Assessment

4.1 Creating Factor Sets and Comment Sets

The factor set of vacuum circuit breaker named U is composed of a vacuum degree characteristic U1, a mechanical characteristic U2, an electrical characteristic U3, and other factors U4. The factor set U is U = {U1, U2, U3, U4}.

According to the maintenance requirements of the vacuum circuit breaker, the operating status of the circuit breaker is divided into four types: good (V1), moderate (V2), attention (V3) and alert (V4). The comment set V is V = {V1, V2, V3, V4}.

4.2 Membership Function Based on Cloud Model

In the fuzzy comprehensive evaluation, the commonly used interval division method only considers the fuzziness. To solve the problem of subjective uncertainty, such as randomness and fuzziness, the cloud theory founded by Li Deyi was used widely in data mining and effectiveness evaluation of complicated system [10]. The cloud model is an uncertainty model that realizes the conversion to qualitative concepts and quantitative values. The cloud has three digital features: Expected (Ex), Entropy (En), and Hyperentropy (He), which will integrate the fuzziness and randomness of spatial concepts in a unified way. The C(x) cloud's expectation-entropy curves is:

$$C(x) = \exp\left[-(x - Ex)^2 / (2En)^2\right] \tag{7}$$

According to the comment set, four state intervals are divided to generate four clouds corresponding to each state, which are denoted as C1, C2, C3, and C4. Among them, C1 and C4 are semi-normal clouds, and C2 and C3 are normal clouds. The digital features of the cloud model Ci (Exi, Eni, Hei) are determined as follows:

$$\begin{cases} Exi = (Ci_{\max} + Ci_{\min})/2 \\ Eni = (Ci_{\max} - Ci_{\min})/6 \\ Hei = 0.05 \end{cases} \tag{8}$$

The $Ci_{\max}$ and $Ci_{\min}$ represent the upper and lower bounds of the cloud interval, i mean the four grades, i = 1, 2, 3, 4. The resulting cloud membership is shown in Fig. 1.

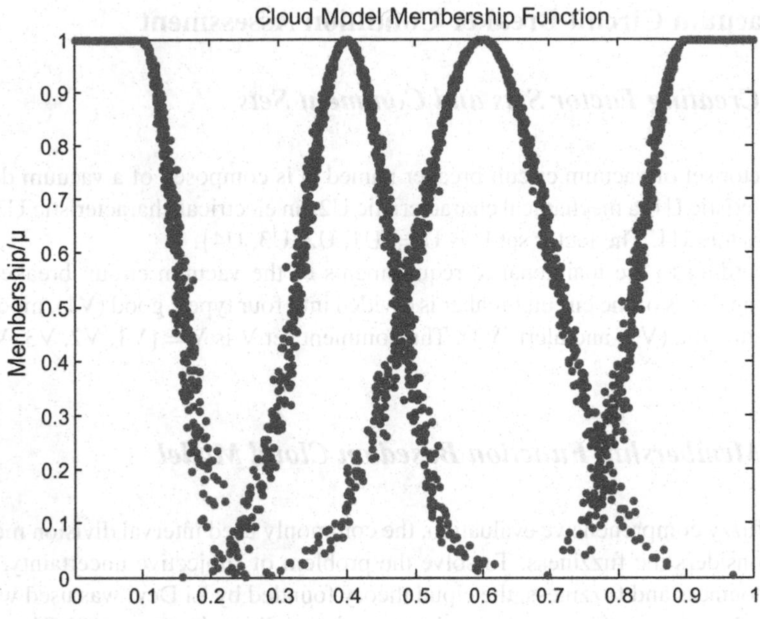

Fig. 1 Cloud model membership function

4.3 Fuzzy Comprehensive Evaluation Results

For the evaluation of vacuum circuit breakers, it is necessary to consider the influence of various and to retain all information about individual factors. Therefore, the weighted average type operator $M(\bullet, \oplus)$ is selected to finally obtain the state matrix D:

$$d_j = \sum_{i=1}^{n} w_i' \bullet r_{ij} \ (j = 1, 2, \ldots, n) \tag{9}$$

The w_i' is variable weight of the indicator, r_{ij} is the corresponding degree of membership, d_j is the element of the state matrix D and n is the number of element in the state matrix.

5 Experimental Verification

Taking the ZN63A-12 vacuum circuit breaker as the condition assessment object, the operating status of the vacuum circuit breaker was evaluated using the method of this paper. Experimental data and scores are shown in Table 3.

Table 3 Experimental data and scores

Test parameters	Parameter value	Score
Non-synchronous time of breaking (ms)	1.26	0.729
Non-synchronous time of closing (ms)	0.73	0.691
Breaking velocity (m/s)	1.65	0.939
Closing velocity (m/s)	0.96	0.779
Currents peak of breaking circuit (A)	1.22	0.742
Currents peak of closing circuit (A)	1.17	0.667

5.1 Determination of Constant Weight and Variable Weight

According to Table 1, each judgment matrix is constructed and the corresponding constant weights are obtained. Taking the mechanical characteristics as an example, the state of mechanical characteristics of vacuum circuit breakers is mainly analyzed from three aspects: time parameters, speed parameters, and current parameters. The interval judgment matrix of mechanical characteristics is constructed.

$$A_m = \begin{bmatrix} [1,1] & [2,3] & [3,4] \\ [1/3,1/2] & [1,1] & [1,2] \\ [1/4,1/3] & [1/2,1] & [1,1] \end{bmatrix}$$

The judgment matrix B of mechanical characteristics can be obtained by formula (2):

$$B_m = \begin{bmatrix} 1 & 2.450 & 3.464 \\ 0.408 & 1 & 1.414 \\ 0.289 & 0.707 & 1 \end{bmatrix}$$

The maximum available eigenvalues and eigenvectors are calculated by formula (3) and (4), $\lambda_{\max} = 3$, $W = (0.8944, 0.3651, 0.2582)^T$. And the random consistency verification by formula (5), $CR = CI/RI = 0 < 0.1$, shows that the consistency of the judgment matrix meets the requirements. Therefore, the constant weight of mechanical characteristics is $W_m = [0.589, 0.241, 0.170]$.

Variable weight can be obtained by formula (6). The constant weights and variable weights of various indicators of mechanical properties are shown in Table 4.

Table 4 Constant weights and variable weights

Project	Constant weight W	Variable weights W′
Mechanical characteristics	[0.589, 0.241, 0.170]	
Time parameter	[0.309, 0.582, 0.109]	[0.308, 0.612, 0.080]
Speed parameter	[0.297, 0.540, 0.163]	[0.270, 0.591, 0.139]
Current parameter	[0.724, 0.276]	[0.708, 0.292]

5.2 Determination of Fuzzy Relationship Matrix

Taking the current parameters as an example, the membership functions based on the cloud model are constructed by formula (7) and (8). The results are shown in Table 5:

The current condition matrix named $R_{currents}$ is:

$$R_{currents} = \begin{bmatrix} 0.180\ 0.803\ 0\ 0 \\ 0.722\ 0.374\ 0\ 0 \end{bmatrix}$$

Through the respective cloud model membership function, the remaining two condition matrices R_{time} and R_{speed} can be obtained as:

$$R_{time} = \begin{bmatrix} 0.581 & 0.642 & 0 & 0 \\ 0.277 & 0.802 & 0 & 0 \\ 0.300 & 0.500 & 0.200 & 0 \end{bmatrix}$$

$$R_{speed} = \begin{bmatrix} 1 & 0 & 0 & 0 \\ 0.952 & 0.030 & 0 & 0 \\ 0.500 & 0.300 & 0.200 & 0 \end{bmatrix}$$

Table 5 Membership functions based on the cloud model

Evaluation level	Cloud model	Membership	
		Breaking	Closing
V1	C1 (0.80, 0.067, 0.005)	0.722	0.180
V2	C2 (0.60, 0.100, 0.005)	0.374	0.803
V3	C3 (0.30, 0.100, 0.005)	0	0
V4	C4 (0.10, 0.033, 0.005)	0	0

5.3 Fuzzy Comprehensive Evaluation

From formula (9), the evaluation matrix of mechanical characteristics can be obtained:

$$D_m = W_m \circ R_m$$
$$= W_m \circ \begin{bmatrix} W'_{time} \circ R_{time} \\ W'_{speed} \circ R_{speed} \\ W'_{current} \circ R_{current} \end{bmatrix} = \begin{bmatrix} 0.491\ 0.559\ 0.016\ 0 \end{bmatrix}$$

According to the above method, the condition evaluation matrix of each project is shown in Table 6.

According to the principle of the maximum degree of membership, it can be concluded from Table 6 that the vacuum characteristics, mechanical characteristics and electrical characteristics are in a moderate state, while the other items are in a good state.

The weight of circuit breakers is $W_{VCB} = [0.468, 0.312, 0.173, 0.048]$ and circuit breaker condition matrix composed by the evaluations matrix of four projects, such as $R_{VCB} = \begin{bmatrix} D_v\ D_m\ D_e\ D_o \end{bmatrix}^T$. The evaluation matrix of the circuit breaker is obtained by formula (9):

$$D_{VCB} = W_{VCB} \circ R_{VCB}$$
$$= \begin{bmatrix} 0.383\ 0.586\ 0.042\ 0.003 \end{bmatrix}$$

According to the principle of maximum degree of membership, the largest item in the evaluation matrix is the second item, corresponding to the V2 item in the evaluation sets, and the condition of the circuit breaker is in a moderate state. In fact, compared with the factory data, it was found that some of the data had deviated from the factory data, and the operating status of the circuit breaker had declined from the optimal state. It is credible that the circuit breaker is in moderate condition in the evaluation result.

Table 6 Evaluation matrix of various indicators

Project	Evaluation matrix
Vacuum Characteristics (v)	[0.293 0.653 0 0]
Mechanical characteristics (m)	[0.491 0.559 0.016 0]
Electrical characteristics (e)	[0.418 0.526 0.170 0]
Other items (o)	[0.426 0.319 0.169 0.072]

6 Conclusion

Vacuum circuit breaker status assessment is a complex process. This paper comprehensively considers various factors that affect the vacuum circuit breaker and classifies the factors and establishes a hierarchical comprehensive evaluation index system for vacuum circuit breakers. Using the interval numbers analytic hierarchy process to obtain the constant weights, introducing variable weights to avoid the impact on the individual indicator values seriously deviating from the normal value of the vacuum circuit breaker evaluation rating, and improve the accuracy of the evaluation results. Use the cloud model membership function to calculate the index evaluation level. This method improves the accuracy of vacuum circuit breaker condition evaluation and provides a reliable reference to vacuum circuit breaker condition-based maintenance.

References

1. C. Pengyong, S. Xiaofei, C. Yunfei et al., Fuzzy comprehensive evaluation on the operation conditions of SF6 high voltage circuit breaker. High Voltage Apparatus **52**(3), 171–176 (2016)
2. G. Lianyu, L. Kejun, L. Yongliang et al., HV circuit breaker state assessment based on gray-fuzzy comprehensive evaluation. Electr. Power Autom. Equip. **34**(11), 161–167 (2014)
3. H. Xuyong, S. Peng, G. Sujie et al., Assessment method of SF6 high-voltage circuit breaker based on association rules and variable weight coefficients. Power Syst. Prot. Control **46**(2), 50–56 (2018)
4. Y. Wenhui, W. Zhan, Z. Xiangjun et al., Reliability tracing analysis for multi-state power transformers using Bayesian network. Power Syst. Prot. Control **43**(6), 78–85 (2015)
5. H. Lingjie, W. Wei, W. Zhensheng et al., Diagnosis model of HV SF6 circuit breaker based on fuzzy theory. High Voltage Apparatus **44**(3), 246–249 (2008)
6. L. Yu, Z. Guogang, G. Yingsan et al., The condition assessment method for HV circuit breakers based on fuzzy theory. High Voltage Apparatus **43**(4), 274–277 (2007)
7. C. Weigen, W. Yanqin, L. Ruijin, Variable weight fuzzy comprehensive evaluation method for operation condition of high voltage circuit breaker. High Voltage Apparatus **45**(3), 73–77 (2009)
8. W. Junxia, Z. Bide, Z. Jinlong et al., The condition assessment method for SF6 circuit breaker based on improved fuzzy AHP. Sichuan Electric Power Technol. **36**(3), 5-10+23 (2013)
9. Z. Jinchun, Q. Hangping, Q. Jichuan, The construction of balance function in variable weight evaluation. Fire Control Command Control **32**(7), 107–110 (2007)
10. L. Deyi, M. Haijun, S. Xuemei, Membership cloud and membership cloud generator. Comput. Res. Dev. **6**, 15–20 (1995)

Composite-Rotating Consensus of Leaderless Multi-agent Systems with Time-Delay

Lipo Mo, Yixuan Jiang, Hong Liu, Xinyue Yang and Zhiyuan Ouyang

Abstract This paper addresses the composite-rotating consensus problem of a class of second order multi-agent systems with time-delay. In order to solve the composite-rotating consensus problems, a distributed control protocol is introduced. Then the stability analysis is completed by using the method of frequency domain analysis and the maximum upper bound of time-delay is also obtained. Finally, the effectiveness of the theoretical results are verified by simulations.

Keywords Multi-agent systems · Composite-rotation · Consensus · Time-delay

1 Introduction

During the past decades, the consensus problem of multi-agent systems is one of the hotspot problems in the field of control, and many meaningful results have been reported, for example, Atkins et al. provided an overview of information consensus in multi-agent cooperative control [1]. Olfati-Saber et al. provided a theoretical framework for analyzing the consensus algorithm of multi-agent network systems[2], and these enlighten us to study the effect of time-delay. The consensus of multi-agent systems means that the state variables will tend to be identical under some conditions. Vicsek et al. proposed a novel simple dynamics model to simulate the particle's auto-ordering phenomenon and obtained the conclusion that there is consistency among the particles through simulation [3]. Jadbabaie et al. analyzed the Vicsek's model and applied the theory of graph to illustrate that if the graph is connected, the consensus can be guaranteed [4]. Ren et al. proved that for the dynamically changing interaction topology, if the combination of directed interaction diagraphs at certain time intervals has a spanning tree, the asymptotically consensus can be obtained when the system evolves [5]. The problem of rotation

L. Mo (✉) · Y. Jiang · H. Liu and X. Yang
School of Science, Beijing Technology and Business University,
Beijing 100048, People's Republic of China
e-mail: beihangmlp@126.com

Y. Jia et al. (eds.), *Proceedings of 2018 Chinese Intelligent Systems Conference*, Lecture Notes in Electrical Engineering 529,
https://doi.org/10.1007/978-981-13-2291-4_44

consensus of multi-agent systems is a new problem which has been proposed to describe the celestial bodies motion in recent years. Rotation consensus mainly solves the state consensus of the agents in the process of rotating motion. The role issue of rotation consensus is to design proper controller to make the movement of agents around the same center at the same angular velocity. This problem was first proposed and solved in Refs. [6, 7]. Due to there exist many composite-rotating motions in nature, such as the motion of the moon relative to the sun, the composite-rotation consensus problem was proposed based on the rotation consensus. The agents move around the center of the circle, the center of the circle rotates around the other center simultaneously. Recently, Refs. [9, 13] have studied the collective composite-rotating consensus and mean-square composite-rotating consensus of multi-agent systems, respectively. Furthermore, many good results were obtained in Refs. [10–12].

In real engineering systems, due to the uncertainty of environment, there must exist time-delays during the information exchanges process among agents, which may results in the instability of the closed-loop system. To solve the rotation consensus problem of multi-agent systems with time-delays, numbers of excellent results have been obtained [13–16]. However, during the process of exploring composite-rotation consensus problems, most of the scholars assumed that there isn't time-delay phenomenon when the information exchange occurs. Hence, these obtained results cannot be applied directly to real engineering systems and the effect of time-delay must be considered.

In this paper, we study the composite-rotation consensus problem of second-order multi-agent systems with time-delay. First, a distributed controller is introduced to make all agents reach composite-rotation consensus moments. Then some sufficient conditions are obtained for composite-rotation consensus via the method of frequency domain analysis. In addition, the maximum upper bound of time-delay also be obtained. Finally, numerical simulations are given to verify the reliability of the conclusion.

Notations: R, C represent the set of real number and complex number, respectively; $\otimes$ represents the Kronecker product, j represents the imaginary unit; I_n represents the n-dimensional unit matrix and $\mathbf{1}_n$ represents the vector that all elements are 1; A^* represents the conjugate transpose of matrix A; For complex vector $x \in C^n$, $||x|| = x^*x$; $\mathrm{tr}(P)$ represents the trace of P.

2 Graph Theory

The theory of Graph is effective for solving the coordination problems. Given a multi-agent system is made by N agents, which can be seen as N nodes and each agent can exchange information with its adjacent agent, we use the undirected connected graph as the communication topology for the multi-agent system. Denotes $V = \{v_i, i = 1, 2, \ldots, N\}$ as the set of the N nodes and the undirected graph $G = \{V, \varepsilon\}$ as the communication topology of the N agents, where $\varepsilon \subset V \times V$ is the set of edges of the graph. v_i can be called the neighbor of v_j, if $(v_i, v_j) \in \varepsilon$. $N_i = \{j|(v_i, v_j) \in \varepsilon\}$ represents the set of labels whose neighbor is agent i. Denote $A = [a_{ij}] \in R^{N\times N}$ as the weighted adjacency matrix of graph G, in which $a_{ii} = 0$ and $a_{ij} > 0$, if $(i, j) \in$

ε, otherwise $a_{ij} = 0, i \neq j$. Let $D = diag\{d_1, d_2, \ldots, d_N\} \in R^{N\times N}$ be the degree matrix of graph G, where $d_i = \sum_{j=1}^{N} a_{ij}$. $L = D - A$ is defined as the Laplacian of the weighted graph, which is symmetric. The path connects v_i and v_j in the graph G is a sequence of distinct vertices $v_{i_0}, \ldots, v_{i_m}$, where $v_{i_0} = v_i$, $v_{i_m} = v_j$ and $(v_{i_r}, v_{i_{r+1}}) \in \varepsilon, 0 \leq r \leq m-1$. If a path exists between any two vertices v_i and v_j $(i \neq j)$, then the undirected graph can be seen as connected.

Lemma 1 *[17] Let G be a graph on N nodes with Laplacian L, and $\lambda_1, \ldots, \lambda_N$ be the eigenvalues of L satisfying $\lambda_1 \leq \cdots \leq \lambda_N$. Then $\lambda_1 = 0$ and $\mathbf{1} = [1, 1, \ldots, 1]^T \in \mathcal{R}^N$ is the eigenvector associated with zero eigenvalue. Moreover, $\lambda_2 > 0$ if G is connected.*

3 Problem Statement

Consider a multi-agent system with n agents. The dynamics of the ith agent are given as follows:

$$\begin{aligned} \dot{r}_i(t) &= v_i(t), \\ \dot{v}_i(t) &= u_i(t), \quad i \in I, \end{aligned} \tag{1}$$

where $I = \{1, 2, \ldots, n\}$, $r_i(t), v_i(t) \in C$ represents the the state of position and information of velocity of the ith agent, respectively. $u_i(t) \in C$ is the control input of the ith agent. In real applications, there are many systems are described by this kind of dynamics, for example, vehicles always have this kind of dynamics [18].

Definition 1 The multi-agent system (1) is said to achieve composite-rotating consensus if

$$\lim_{t\to\infty} |v_i(t) - v_k(t)| = 0, \tag{2}$$

$$\lim_{t\to\infty} |r_i(t) - r_k(t)| = 0, \tag{3}$$

and

$$\lim_{t\to\infty} |\dot{c}_i(t) + j\omega_2 c_i(t)| = 0, \quad i \in I, \tag{4}$$

where $c_i(t) = r_i(t) + j\omega_1^{-1} v_i(t)$, ω_1, ω_2 are positive constants.

Remark 1 In Definition 1, $c_i(t)$ represents the rotation center of the ith agent and ω_1 is the corresponding angular velocity. (2) and (3) imply that all agents can reach an agreement on position and velocity as time tends to infinity. (4) implies that the rotation center $c_i(t)$ would finally rotates around original point with angular velocity ω_2 asymptotically.

4 Main Results

The purpose of this paper is to design proper control protocol for each agent to assure all agent achieve composite-rotating consensus under the existence of time-delay and meanwhile estimate the upper bound of the maximum time-delay.

Before we make an analysis for the stability of the system, we make the following assumption:

Assumption 1 The communication graph G considered in this paper is undirected and connected.

In this paper, the following control protocol is used:

$$u_i(t) = u_{i1}(t) + u_{i2}(t), \tag{5}$$

where $u_{i1}(t) = j\omega_1 v_i(t) - \omega_1\omega_2 c_i(t)$ and $u_{i2}(t) = -\sum_{k\in N_i} a_{ik}(v_i(t-\tau) - v_k(t-\tau)) - \sum_{k\in N_i} a_{ik}(c_i(t-\tau) - c_k(t-\tau))$.

Let $\xi(t) = [r_1(t), c_1(t), \ldots, r_n(t), c_n(t)]^T$, then

$$\dot{\xi}(t) = (I_n \otimes A)\xi(t) - (L \otimes B)\xi(t-\tau), \tag{6}$$

where

$$A = \begin{bmatrix} j\omega_1 & -j\omega_1 \\ 0 & -j\omega_2 \end{bmatrix}, B = \begin{bmatrix} 0 & 0 \\ -1 & 1 + j\omega_1^{-1} \end{bmatrix}.$$

Denote $\alpha(t) = \frac{1}{n}\sum_{i=1}^{n} r_i(t)$, $\beta(t) = \frac{1}{n}\sum_{i=1}^{n} c_i(t)$, then we have $\dot{\alpha}(t) = j\omega_1\alpha(t) - j\omega_1\beta(t)$ and $\dot{\beta}(t) = -j\omega_2\beta(t)$.

Let

$$e(t) = \xi(t) - \mathbf{1}_n \otimes [\alpha(t), \beta(t)]^T.$$

Noted that $(L \otimes B) \cdot \left(\mathbf{1}_n \otimes [\alpha(t), \beta(t)]^T\right) = 0$, we have

$$\begin{aligned}
\dot{e}(t) &= \dot{\xi}(t) - \mathbf{1}_n \otimes [\dot{\alpha}(t), \dot{\beta}(t)]^T \\
&= (I_n \otimes A)\xi(t) - (L \otimes B)\xi(t-\tau) - \mathbf{1}_n \otimes \begin{bmatrix} j\omega_1\alpha(t) - j\omega_1\beta(t) \\ -j\omega_2\beta(t) \end{bmatrix} \\
&= (I_n \otimes A)\xi(t) - (L \otimes B)\xi(t-\tau) - (I_n \otimes A)\left(\mathbf{1}_n \otimes [\alpha(t), \beta(t)]^T\right) \\
&= (I_n \otimes A)e(t) - (L \otimes B)e(t-\tau).
\end{aligned}$$

From Lemma 1, there exists an orthogonal matrix $U = \left[\frac{1}{\sqrt{n}}\mathbf{1}_n, \overline{U}\right]$, $\overline{U} \in R^{n\times(n-1)}$, such that $U^T L U = diag\{0, \lambda_2, \ldots, \lambda_n\}$, where $0 < \lambda_2 \le \lambda_3 \le \ldots \le \lambda_n$ are the eigenvalues of L.

Let $\delta(t) = (U^T \otimes I_2)e(t)$, then

$$\begin{aligned}\dot{\delta}(t) &= (U^T \otimes I_2)\dot{e}(t)\\ &= (U^T \otimes I_2)[(I_n \otimes A)e(t) - (L \otimes B)e(t-\tau)]\\ &= (U^T \otimes I_2)[(I_n \otimes A)(U \otimes I_2)\delta(t) - (L \otimes B)(U \otimes I_2)\delta(t-\tau)]\\ &= (I_n \otimes A)\delta(t) - (diag\{0, \lambda_2, \ldots, \lambda_n\} \otimes B)\, \delta(t-\tau).\end{aligned}$$

Noted that $(\mathbf{1}_n^T \otimes I_2)e(t) = 0$. Let $\delta(t) = [0, \overline{\delta}(t)]^T$, where $\overline{\delta}(t) \in C^{2n-2}$, then

$$\dot{\overline{\delta}}(t) = (I_{n-1} \otimes A)\overline{\delta}(t) - (diag\{\lambda_2, \ldots, \lambda_n\} \otimes B)\, \overline{\delta}(t-\tau). \quad (7)$$

Remark 2 When there exists no time-delay, system (7) can be changed into the following form:

$$\dot{\overline{\delta}}(t) = (I_{n-1} \otimes A - diag\{\lambda_2, \ldots, \lambda_n\} \otimes B)\overline{\delta}(t).$$

Lemma 2 [8] *If*

$$3\omega_1\omega_2\|\frac{1}{n}\sum_{i=1}^{n} c_i(0)\| < (\omega_1 + \omega_2)\|\frac{1}{n}\sum_{i=1}^{n} v_i(0)\|.$$

then $I_{n-1} \otimes A - diag\{\lambda_2, \ldots, \lambda_n\} \otimes B$ is Hurwitz.

Theorem 1 *Consider the multi-agent system* (1). *Under Assumption 1, if*

$$\begin{aligned}&3\omega_1\omega_2\|\frac{1}{n}\sum_{i=1}^{n} c_i(0)\| < (\omega_1 + \omega_2)\|\frac{1}{n}\sum_{i=1}^{n} v_i(0)\|,\\ &\omega_1\omega_2 < \lambda_2,\\ &\tau < \tau^*,\end{aligned}$$

where $\tau^* = [\arctan\left(\frac{\omega_1(\omega_1\omega_2+\lambda_2)}{\omega_1(\omega_1+\omega_2)+\lambda_2}\right)]/[\omega_2 + \frac{\sqrt{\omega_1^2+1}}{\omega_1}\lambda_n]$. *Then the multi-agent system* (1) *can achieve composite-rotating consensus under control protocol* (5).

Proof Let us consider system (7) in the complex domain. By taking the Laplace transform of (7), we have

$$sE(s) - \overline{\delta}(0) = (I_{n-1} \otimes A)E(s) - diag\{\lambda_2, \ldots, \lambda_n\} \otimes BE(s)e^{-\tau s},$$

where $E(s) = L[\overline{\delta}(t)]$ is the Laplacian transformation of $\overline{\delta}(t)$. Hence,

$$E(s) = G_\tau(s)^{-1}\overline{\delta}(0),$$

where

$$G_\tau(s) = sI_{2n-2} - I_n \otimes A + diag\{\lambda_2, \ldots, \lambda_n\} \otimes Be^{-\tau s}.$$

To guarantee the stability of system (6), we only need to assure $G_\tau(s)$ has no zero point on the imaginary axis for time-delay. Suppose that $s = j\omega$ is the zero point of $G_\tau(s)$ and $u \in C^{2n-2}$ is the corresponding eigenvector, i.e.,

$$[j\omega I_{2n-2} - I_n \otimes A + diag\{\lambda_2, \ldots, \lambda_n\} \otimes Be^{-j\tau\omega}]u = 0. \tag{8}$$

Let $u = [u_2^T, \ldots, u_n^T]^T$, $u_i = [u_{i1}, u_{i2}]^T$, $i = 2, \ldots, n$. Clearly, $u_i \neq 0$. It follows from (8) that for each $i = 2, 3, \ldots, n$, we have

$$[j\omega I_2 - A + \lambda_i Be^{-j\tau\omega}]u_i = 0. \tag{9}$$

From the first row of (9), we have $j\omega u_{i1} - j\omega_1 u_{i1} + j\omega_1 u_{i2} = 0$, i.e.,

$$u_{i2} = \frac{\omega_1 - \omega}{\omega_1} u_{i1}. \tag{10}$$

It is clear that $\omega \neq \omega_1$, if not, $\omega = \omega_1$, then $u_{i2} = 0$. Besides, from (9), $\lambda_i u_{i1} e^{-j\tau\omega} = 0$ which implies $u_{i1} = 0$ and hence $u_1 = 0$, this is a contradiction.

Multiplying both sides of (9) by u_i^*, we have

$$u_i^*[j\omega I_2 - A + \lambda_i Be^{-j\tau\omega}]u_i = 0. \tag{11}$$

Calculating the real part and imaginary part of (11) respectively, we have

$$-\frac{\omega}{\omega_1 - \omega}\cos(\tau\omega) + \frac{1}{\omega_1}\sin(\tau\omega) = 0, \tag{12}$$

$$\frac{\omega}{\omega_1 - \omega}\sin(\tau\omega) + \frac{1}{\omega_1}\cos(\tau\omega) = -\frac{\omega + \omega_2}{\lambda_i}. \tag{13}$$

Take τ sufficiently small to guarantee $|\tau\omega| < \frac{\pi}{2}$. We claim that $\omega < 0$. If not, $\omega > 0$, then it follows from (12) that $\omega < \omega_1$, under which equation (13) does not hold anymore. Taking the square of (12) and adds the square of (13), we have

$$\left(\frac{\omega}{\omega_1 - \omega}\right)^2 + \left(\frac{1}{\omega_1}\right)^2 = \frac{(\omega + \omega_2)^2}{\lambda_i^2}. \tag{14}$$

Since $\omega < 0$ and $\omega_1 > 0$, we have $\left(\frac{\omega}{\omega_1 - \omega}\right) < 1$, thus

$$\left(\frac{\omega + \omega_2}{\lambda_i}\right)^2 \leq 1 + \frac{1}{\omega_1^2} = \frac{\omega_1^2 + 1}{\omega_1^2},$$

which implies that

$$\omega \geq -\omega_2 - \frac{\sqrt{\omega_1^2+1}}{\omega_1}\lambda_i \geq -\omega_2 - \frac{\sqrt{\omega_1^2+1}}{\omega_1}\lambda_n.$$

Besides

$$\left(\frac{\omega+\omega_2}{\lambda_i}\right)^2 \geq \frac{1}{\omega_1^2},$$

which implies $\omega > -\omega_2 + \frac{\lambda_i}{\omega_1}$ or $\omega < -\omega_2 - \frac{\lambda_i}{\omega_1}$.

Noted that $\omega_1\omega_2 < \lambda_2$, we have $\omega_2 < \frac{\lambda_2}{\omega_1} \leq \frac{\lambda_i}{\omega_1}$ and $-\omega_2 + \frac{\lambda_i}{\omega_1} > 0$. Hence, $\omega <- \omega_2 - \frac{\lambda_i}{\omega_1}$. From (12), we have

$$\tan(\tau\omega) = \frac{\omega\omega_1}{\omega_1 - \omega}.$$

Let $g(\omega) = \frac{\omega\omega_1}{\omega_1-\omega}$. Then $g'(\omega) = \frac{\omega_1^2}{(\omega_1-\omega)^2} > 0$. Hence $g(\omega)$ is an increasing function about ω. So

$$\tan(\tau\omega) \leq -\frac{\omega_1(\omega_1\omega_2+\lambda_2)}{\omega_1(\omega_1+\omega_2)+\lambda_2},$$

then

$$\tau\omega \leq -\arctan\left(\frac{\omega_1(\omega_1\omega_2+\lambda_2)}{\omega_1(\omega_1+\omega_2)+\lambda_2}\right).$$

Noted that $\omega < 0$, we have

$$\tau \geq \frac{-\arctan\left(\frac{\omega_1(\omega_1\omega_2+\lambda_2)}{\omega_1(\omega_1+\omega_2)+\lambda_2}\right)}{\omega} \geq \frac{\arctan\left(\frac{\omega_1(\omega_1\omega_2+\lambda_2)}{\omega_1(\omega_1+\omega_2)+\lambda_2}\right)}{\omega_2 + \frac{\sqrt{\omega_1^2+1}}{\omega_1}\lambda_n} = \tau^*.$$

If $\tau < \tau^*$, then Eqs. (12) and (13) don't hold anymore, i.e., $G_\tau(s)$ has no zero point on the imaginary axis. From Lemma 2, when $\tau = 0$, all the zero point of $G_0(s)$ has negative real parts. Because the zero points are the continuous function of time-delay τ, all the zero points of $G_\tau(s)$ have negative real parts when $\tau < \tau^*$, this is to say, system (7) is Hurtwiz. Therefore, $\lim_{t\to\infty} \overline{\delta}(t) = 0$ and $\lim_{t\to\infty} e(t) = 0$, i.e., $\lim_{t\to\infty}(r_i(t) - r_k(t)) = 0$ and $\lim_{t\to\infty}(c_i(t) - c_k(t)) = 0$, $i, k \in I$ and then $\lim_{t\to\infty}(v_i(t) - v_k(t)) = 0$. From the expression of $\dot{c}_i(t)$, we have $\lim_{t\to\infty}[\dot{c}_i(t) + j\omega_2 c_i(t)] = 0$. Therefore, the multi-agent system (1) achieves composite-rotating consensus under control algorithm (5). □

5 Simulation

In this section, a numerical simulation is given to illustrate the theoretical results obtained in the previous sections. The communication topology graph G that considered in this paper is given as in Fig. 1. The initial values of this system are chosen as $[r_1(0), r_2(0), r_3(0), r_4(0), r_5(0), r_6(0)]^T = [0.1, 0.15, 0.2, 0.25, 0.3, 0.35]^T$, $[v_1(0), v_2(0), v_3(0), v_4(0), v_5(0), v_6(0)]^T = [0.3, 0.3, 0.3, 0.3, 0.3, 0.3]^T$. Moreover, the time-delay in this system is set as 0.03 and the upper bound of the maximum time delay we estimated is 0.0871.

From Fig. 2 we could see that all agents achieve composite-rotating consensus under the existence of time-delay in the leaderless multi-agent system asymptotically.

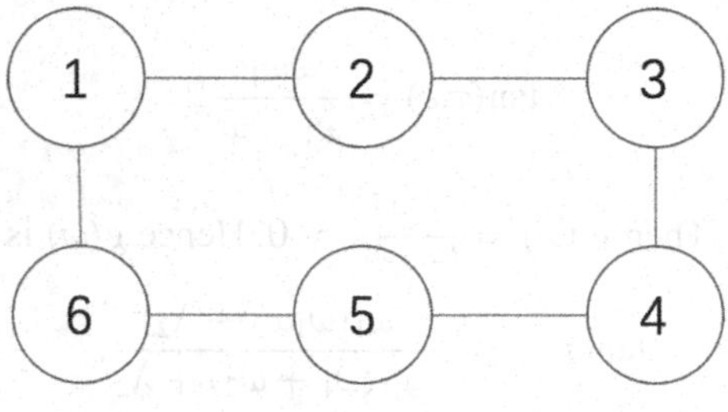

Fig. 1 Communication topology

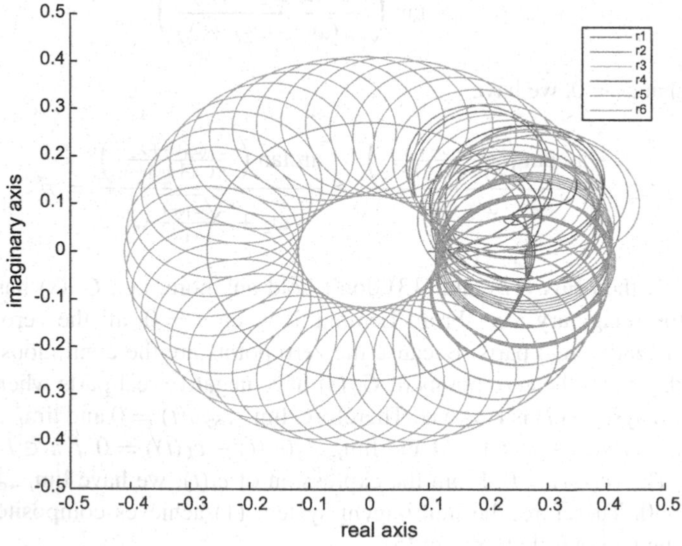

Fig. 2 Trajectories of all agents

6 Conclusions

This paper studied the composite-rotation consensus problem of multi-agent system with time-delay under fixed topology. First, a novel distributed control protocol was given. And then some stability conditions and the maximum time-delay margin were obtained via the method of frequency domain analysis. Finally, simulation results were given to demonstrate the validity of our theoretical results. In the future, we will consider the composite-rotation consensus problem of multi-agent systems with switching topologies.

Acknowledgements This work is supported by National Natural Science Foundation (NNSF) of China (Grant No. 61304155) and the College Student Research and Career-creation Program of Beijing City(Grant No. 201810011022)

References

1. R. Wei, R.W. Beard, E.M. Atkins, A survey of consensus problems in multi-agent coordination, in *2005 American Control Conference*, vol. 3, pp. 1859–1864, 2005
2. R. Olfati-Saber, J.A. Fax, R.M. Murray, Consensus and cooperation in networked multi-agent systems. Proc. IEEE **95**(1), 215–233 (2007)
3. T. Vicsek, A. Czirók, E. Benjacob, I.I. Cohen, O. Shochet, Novel type of phase transition in a system of self-driven particles. Phys. Rev. Lett. **75**(6), 1226 (1995)
4. A. Jadbabaie, J. Lin, A.S. Morse, Coordination of groups of mobile autonomous agents using nearest neighbor rules. IEEE Trans. Autom. Control **48**(6), 988–1001 (2003)
5. R. Wei, R.W. Beard, Consensus seeking in multiagent systems under dynamically changing interaction topologies. IEEE Trans. Autom. Control **50**(5), 655–661 (2005)
6. Peng Lin, Yingmin Jia, Distributed rotating formation control of multi-agent systems. Syst. Control Lett. **59**(10), 587–595 (2010)
7. Peng Lin, Kaiyu Qin, Zhongkui Li, Wei Ren, Collective rotating motions of second-order multi-agent systems in three-dimensional space. Syst. Control Lett. **60**(6), 365–372 (2011)
8. Peng Lin, Lu Wanting, Yongrui Song, Collective composite-rotating consensus of multi-agent systems. Chin. Phys. B **23**(4), 178–182 (2014)
9. Lipo Mo, Shaoyan Guo, Yongguang Yu, Mean-square composite-rotating consensus of second-order systems with communication noises. Chin. Phys. B **27**(7), 070504 (2018)
10. Lipo Mo, Xiaolin Yuan, Yongguang Yu, Target-encirclement control of fractional-order multi-agent systems with a leader. Physica A **509**, 479–491 (2018)
11. Lina Jin, Shuanghe Yu, Dongxu Ren, Circular formation control of multiagent systems with any preset phase arrangement. J. Control Sci. Eng. **2018**, 1–11 (2018)
12. Youcheng Lou, Yiguang Hong, Distributed surrounding design of target region with complex adjacency matrices. IEEE Trans. Autom. Control **60**(1), 283–288 (2014)
13. P. Lin, Y. Jia, J. Du, S. Yuan, Distributed consensus control for second-order agents with fixed topology and time-delay, in *Chinese Control Conference*, pp. 577-581, 2007
14. Peng Lin, Kaiyu Qin, Hongmei Zhao, Man Sun, A new approach to average consensus problems with multiple time-delays and jointly-connected topologies. J. Franklin Inst. **349**(1), 293–304 (2012)
15. Peng Lin, Mingxiang Dai, Yongduan Song, Consensus stability of a class of second-order multi-agent systems with nonuniform time-delays. J. Franklin Inst. **351**(3), 1571–1576 (2014)

16. Peng Lin, Yingmin Jia, Consensus of a class of second-order multi-agent systems with time-delay and jointly-connected topologies. IEEE Trans. Autom. Control **55**(3), 778–784 (2010)
17. C. Godsil, G.F. Royle, *Algebraic graph theory* (New York, Springer, 2001)
18. Yingmin Jia, Robust control with decoupling performance for steering and traction of 4WS vehicles under velocity-varying motion. IEEE Trans. Control Syst. Technol. **8**(3), 554–568 (2000)

Finite-Time Consensus Problem of Second-Order Multi-agent Systems with External Disturbances

Yan Cui and Danni Qiao

Abstract Finite-time consensus for second-order multi-agent systems with external disturbances is investigated in this article. By turning the original system into an equivalent system, which can be represented by disagreement vector, sufficient conditions that guarantee all agents reach finite-time average consensus are derived. Finally, simulations are supplied to show the validity of the gotten theoretic results.

Keywords Finite-time consensus · Disagreement vector · External disturbances
Second-order multi-agent systems

1 Introduction

In recent years, more and more concern has been got to the consensus issues of multi-agent systems. Part of the reason is that it is widely used in solving multi-domain coordination and cooperation in product design, production and manufacturing, and even throughout the product life cycle. Multi-agent systems provide more effective means for system integration, parallel design, and intelligent manufacturing. Moreover, the speed of convergence is an serious index to evaluate the consistency algorithm. The researchers found that improving the second minimum eigenvalue for the Laplacian matrix can increase the convergence speed of the system. Compared with asymptotic convergence, the finite time consensus control method has obvious advantages, which not only guarantees the faster convergence speed of the system, but also shows better robustness in the presence of external interference. Therefore, it has great practical significance to research the above issues.

The consensus theory has become a challenging in some aspects. It is an important research topic in the control discipline. The finite-time consensus of second-order multi-agent systems has been discussed and studied by many scholars [1–5]. More-

Y. Cui · D. Qiao (✉)
College of Physics and Information Engineering, Shanxi Normal University,
Linfen 041000, Shanxi, China
e-mail: 897304019@qq.com

Y. Jia et al. (eds.), *Proceedings of 2018 Chinese Intelligent Systems Conference*, Lecture Notes in Electrical Engineering 529,
https://doi.org/10.1007/978-981-13-2291-4_45

over, how to attain a consensus within a limited period of time is a crucial issue in these studies. In [4], based on the homogeneous theory, Zhang and Yang attempted to establish some sufficient conditions for reaching an agreement of the finite-time that has leader. Bhat [5] has done more detailed and precise research. In some practical applications, the finite-time consensus needs to be within the limited control accuracy range. Some researchers apply full state feedback to propose some control algorithms to attain the finite-time consensus of second-order multi-agent systems [6–8]. Recently, consensus problem for the first-order multi-agent systems with disturbances has been researched in some literatures [9–13]. Cao et al. [9] proposed the consensus algorithm by applying extended state observer based on state and disturbances estimations. In [10], for multi-agent systems that has ordinary Lipschitz-type nonlinear network and external disturbances, a distributed consensus problem has been researched.

Most studies in the current literature have little consideration of the causes of instability in the environment. However, the real network is often in an unstable communication environment, such as noise interference, weather instability and so on. Moreover, there are less external interference in the study of second-order multi-agent systems. Therefore, it would be very significant to study the relevant issues. In order to handle these issues, the finite-time consensus for the second-order multi-agent systems with external disturbances is investigated in this article.

The rest of this article is arranged as follows. The graph theoretic knowledge and lemma needed are be written-out in Sect. 2. The finite-time consensus problem for second-order multi-agent systems with external disturbances is researched in Sect. 3. In Sect. 4, an example and some simulation results are supplied to illustrate the validity of the gotten theoretic results. The last conclusion is in Sect. 5.

2 Preliminaries

Note $G = (V, E)$ as a directed graph, where $V = \{1, 2, \ldots, N\}$ indicate the set of nodes, and $E \subseteq V \times V$ are the edges, respectively. The collection of neighborhoods of vertex i are denoted by $N_i = \{ j | (i, j) \in E\}$. The nodes of an undirected graph do not have a starting point or an ending point. Otherwise, it is a directed graph. An undirected graph is a special case of digraph. $A = \left[a_{ij}\right]$ is called the weighted adjacency matrix. It is defined as $a_{ij} > 0$ when $(i, j) \in E$, otherwise, $a_{ij} = 0$. Specifically, we can assume that $a_{ij} = a_{ji}$ for the undirected graph. Then the adjacency matrix is symmetric. The Laplacian matrix $L = \left[l_{ij}\right]$ of G is defined as $l_{ii} = \sum_{j=1, j\neq i}^{M} a_{ij}$ and $l_{ij} = -a_{ij}, \forall i \neq j$.

Definition 1 [14] Finite-time consensus can be achieved, if for any starting state, there exists a determined value $x^* < \infty$, $v^* < \infty$ and make systems (1) satisfies

$$\lim_{t\to T} \left|x_i(t) - x^*\right| = 0, \quad i = 1, 2, \ldots, n, \quad ; \quad \lim_{t\to T} \left|v_i(t) - v^*\right| = 0, \quad i = 1, 2, \ldots, n,$$

With $x^* = \frac{1}{n}\sum_{i=1}^{n} x_i(t)$, $v^* = \frac{1}{n}\sum_{i=1}^{n} v_i(t)$, $T \in (0, \infty)$ is called the settling time.

Lemma 1 [15] *Let $a_1, a_2, a_M, \ldots, \geq 0$ and let $0 < \beta \leq 1$. Then $\sum_{j=1}^{M} a_j^{\beta} \geq (\sum_{j=1}^{M} a_j)^{\beta}$.*

Lemma 2 [16] *Let $L = [l_{ij}] \in R^{M \times M}$ represent the Laplacian matrix a of graph G, which is defined as $l_{ij} = \begin{cases} \sum_{\mu=1,\mu\neq i}^{n} a_{i\mu} & i = j \\ -a_{ij} & i \neq j \end{cases}$, L(A) has the some characteristics:*

*(a) zero is an eigenvalue of L and **1** is the associated eigenvector, i.e. $L(A)\mathbf{1} = 0$;*
(b) $y^T L y = \frac{1}{2}\sum_{i,j=1}^{n} a_{ij}(y_j - y_i)^2$, and the positive semi-definite of L(A) reveals that all eigenvalues of L(A) are real and greater than or equal to zero;
(c) If G is connected, the second smallest eigenvalue of L(A), which is called the algebraic connectivity of G and represented by λ_2, is greater than zero;
(d) λ_2 of G is equal to $min_{\zeta\neq 0, 1^T\zeta=0} \frac{\zeta^T L(A)\zeta}{\zeta^T\zeta}$, and thus, if $1^T\zeta = 0$, then $\zeta^T L(A)\zeta \geq \lambda_2\zeta^T\zeta$.

Lemma 3 [17] *Assume that the topology of system (1) is an undirected and connected graph. Provided that there is a function $V(\delta(t))$ that satisfy the following terms, then the protocol (2) that has the disagreement vector $\delta(t) = (\delta_1(t), \delta_2(t), \ldots, \delta_n(t))^T$ can solves a finite-time agreement problem.*

(a) When $\delta(t)$ is equal to 0, $V(\delta(t)) = 0$; and when $\delta(t)$ is not equal to 0, $V(\delta(t)) > 0$.
(b) $V(\delta(t))$ is Lipschitz continuous on R^n that has the Lipschitz invariant quantity $N_1 > 0$.
(c) There are real numbers $\gamma > 0$ and $\alpha \in (0, 1/2)$ and make $\dot{V}(\delta(t)) + \gamma(V(\delta(t)))^{\alpha} \leq 0$, Then there is an open neighborhood $\Xi \subseteq R^n$ of any $N_2 \geq 0$ and make $\|\Psi(t, \delta)\| \leq N_2\|\delta\|$ for any continuous function $\Psi : [0, \infty) \times R^n \to R^n$. In addition, the solution of $T(\delta(t)) \leq \frac{1}{\gamma(1-\alpha)}(V(\delta(0)))^{1-\alpha}$ with T is continuous is defined on the definition domain $[0, \infty)$ that satisfies $\delta(t) \in \Xi$ while $t \in [0, \infty)$, and $\delta(t) = 0$ while $t \geq T$.

$$\dot{x}_i(t) = u_i(t) + f_i^d(t, x(t)), i = 1, 2, \ldots, n, \tag{1}$$

$$u_i(t) = f_i(x(t)), i = 1, 2, \ldots, n \tag{2}$$

For the protocol (2), it needs to be satisfied that $\sum_{i=1}^{n} f_i(x(t)) = 0$.

3 Main Results

Assume that the second-order multi-agent system is made up of n agents. Each agent is considered as a vertex in a undirected graph *G*. For the following second-order multi-agent system:

$$\dot{x}_i(t) = v_i(t)$$
$$\dot{v}_i(t) = u_i(t) + f_i^d(t, x(t), v(t)), \quad \mathrm{i} = 1, 2, \ldots, n \tag{3}$$

where $x_i(t) \in R$ and $v_i(t) \in R$ represents position and speed states, as well as $u_i(t) \in R$ represents the control input. $f_i^d(t, x(t), v(t))$ is the external disturbance with regard to t, $x(t)$ and $v(t)$.

Lemma 4 *Assume that the topology of system (3) is an undirected and connected graph. Provided that there is a function $V(\xi(t))$ that satisfy the following terms, then the protocol (4) that has the disagreement vector $\delta(t) = (\delta_1(t), \delta_2(t), \ldots, \delta_n(t))^T$ and $\xi(t) = (\xi_1(t), \xi_2(t), \ldots, \xi_n(t))^T$ can solves a finite-time agreement problem.*

(a) When $\xi(t) = 0$, $V(\xi(t)) = 0$; and when $\xi(t) \neq 0$, $V(\xi(t)) > 0$.
(b) $V(\xi(t))$ is Lipschitz continuous on R^n that has the Lipschitz invariant quantity $M_1 > 0$.
(c) There are real numbers $\gamma > 0$ and $\alpha \in (0, \frac{1}{2})$ and make $\dot{V}(\xi(t)) + \gamma(V(\xi(t)))^\alpha \leq 0$, Then there is an open neighborhood $\Omega \subseteq R^n$ of for any $N_2 \geq 0$ and make $\|Z(t, \xi)\| \leq N_2\|\xi\|$ for any continuous function $Z : [0, \infty) \times R^n \to R^n$. In addition, the solution of $T(\xi(t)) \leq \frac{1}{\gamma(1-\alpha)}(\xi(\delta(0)))^{1-\alpha}$ with T is continuous is defined on the definition domain $[0, \infty)$ that satisfies $\xi(t) \in \Omega$ when $t \in [0, \infty)$, and $\xi(t) = 0$ when $t \geq T$.

$$u_i(t) = \sum_{i=1}^{n}\sum_{j=1}^{n} a_{ij}[f_i(x(t)) + g_i(v(t))] \tag{4}$$

For the protocol (4), it needs to be satisfied that $\sum_{i=1}^{n} f_i(x(t)) = 0$, $\sum_{i=1}^{n} g_i(v(t)) = 0$.

Proof Since $a_{ij} = a_{ji}$ for an undirected and connected graph, we can know that $\sum_{i=1}^{n} \dot{x}_i(t) = 0$ and $\sum_{i=1}^{n} \dot{v}_i(t) = 0$, thus x^* and v^* is time-invariant.

Because of invariance of x^* and v^*, x(t) and v(t) can be decomposed into

$$x(t) = x^*1 + \delta(t) \tag{5}$$
$$v(t) = v^*1 + \xi(t) \tag{6}$$

Where $\delta(\mathrm{t}) = (\delta_1(t), \delta_2(t), \ldots, \delta_n(t))^T$, $\xi(\mathrm{t}) = (\xi_1(t), \xi_2(t), \ldots, \xi_n(t))^T$. In [18], δ and ξ are called the group disagreement vector, they are both orthogonal to **1**.

Therefore, rewriting (4), we have

$$u_i(t) = \sum_{i=1}^{n}\sum_{j=1}^{n} a_{ij}\left[\tilde{f}_i(\delta(t)) + \tilde{g}_i(\xi(t))\right] \tag{7}$$

by Lemma 3, under protocol (7) with the disagreement vector the δ and ξ, finite-time average consensus is attained. This completes the proof.

Because of $sig(y)^{\alpha} = |y|^{\alpha} sig(y)$, then $|x_i - x_j|^{\alpha} sig(x_i - x_j) = -|x_j - x_i|^{\alpha} sig(x_j - x_i)$, $|v_i - v_j|^{\alpha} sig(v_i - v_j) = -|v_j - v_i|^{\alpha} sig(v_j - v_i)$, for adjacency matrix, $a_{ij} = a_{ji}$, therefore, from (7) and the condition of lemma 4, the consensus protocol is described by

$$u_i = \sum_{i=1}^{n}\sum_{j=1}^{n} a_{ij}\left[sig(x_j - x_i)^{\alpha} + sig(v_j - v_i)^{\beta}\right] \tag{8}$$

where $0 < \alpha < 1$, $\beta = \frac{\alpha}{1+\alpha}$.

Assumption 1 Because $f_i^d(t, x(t), v(t))$ is the external disturbance with respect to t, $x(t)$ and $v(t)$, it can be represented by disagreement vector δ and ξ by employing (5) and (6). Suppose $f_i^d(t, x(t), v(t)) = k\xi_i$.

Theorem 1 *Suppose that the topology of second-order system (3) is an undirected and connected graph, as well as Assumption 1 is true. Given the protocol (8), system (3) achieve the finite-time average consensus, if there is a continuous function $V(\xi(t))$ that meet the conditions (a)–(c) of Lemma 4.*

Proof By (5) and (6), differentiating the function $x(t)$ and $v(t)$, we have

$$\dot{x}(t) = \dot{\delta}(t) \tag{9}$$

$$\dot{v}(t) = \dot{\xi}(t) \tag{10}$$

what should be paid attention to is

$$\dot{\xi}_i(t) = \sum_{i=1}^{n}\sum_{j=1}^{n} a_{ij}\left[sig(x_j - x_i)^{\alpha} + sig(v_j - v_i)^{\beta}\right] + k\xi_i \tag{11}$$

Select the alternate Lyapunov function

$$V = \sum_{i=1}^{n}\sum_{j=1}^{n} \int_{0}^{x_i - x_j} a_{ij} sig(s)^{\alpha} ds + \frac{1}{2}\sum_{i=1}^{n} \xi_i^2 \tag{12}$$

The time derivative of the Lyapunov function is

$$\begin{aligned}
\dot{V} &= \sum_{i=1}^{n}\sum_{j=1}^{n} a_{ij} sig(x_i - x_j)^{\alpha} v_i + \sum_{i=1}^{n} v_i \dot{v}_i \\
&= \sum_{i=1}^{n}\sum_{j=1}^{n} a_{ij} sig(x_i - x_j)^{\alpha} v_i + \sum_{i=1}^{n} v_i \left\{ \sum_{j=1}^{n} a_{ij}\left[sig(x_j - x_i)^{\alpha} + sig(v_j - v_i)^{\beta}\right] + k\xi_i \right\} \\
&= \sum_{i=1}^{n}\sum_{j=1}^{n} v_i \left[a_{ij} sig(v_j - v_i)^{\beta} + k\xi_i\right]
\end{aligned}$$

$$= \frac{1}{2}\sum_{i=1}^{n}\sum_{j=1}^{n}\left[a_{ij}(v_i - v_j)sig(v_j - v_i)^{\beta} + 2kv_i\xi_i\right]$$

$$= \frac{1}{2}\sum_{i=1}^{n}\sum_{j=1}^{n}\left[a_{ij}(\xi_i - \xi_j)sig(\xi_j - \xi_i)^{\beta} + 2k\xi_i^2\right]$$

$$= -\frac{1}{2}\sum_{i=1}^{n}\sum_{j=1}^{n}\left(a_{ij}^{\frac{2}{1+\beta}}(\xi_j - \xi_i)^2\right)^{\frac{1+\beta}{2}} + \sum_{i=1}^{n}\sum_{j=1}^{n}\sigma\xi_i^2$$

In the above formula, $\sigma = 2k$. By Lemma 1, we have

$$-\frac{1}{2}\sum_{i=1}^{n}\sum_{j=1}^{n}\left(a_{ij}^{\frac{2}{1+\beta}}(\xi_j - \xi_i)^2\right)^{\frac{1+\beta}{2}} \le -\frac{1}{2}\left(\sum_{i=1}^{n}\sum_{j=1}^{n}a_{ij}^{\frac{2}{1+\beta}}(\xi_i - \xi_j)^2\right)^{\frac{1+\beta}{2}}$$

Since $1^T\delta = 0$ and $1^T\xi = 0$, by Lemma 2,

$$-\frac{1}{2}\sum_{i=1}^{n}\sum_{j=1}^{n}\left(a_{ij}^{\frac{2}{1+\beta}}(\xi_j - \xi_i)^2\right)^{\frac{1+\beta}{2}} \le -\frac{1}{2}(4\lambda_2(L)V)^{\frac{1+\beta}{2}} = -2^{\beta}(\lambda_2(L)V)^{\frac{1+\beta}{2}}$$

Let $\mu = 2^{\frac{1+\beta}{2}}\lambda_2(L)^{\frac{1+\beta}{2}}$, then

$$-\frac{1}{2}\sum_{i=1}^{n}\sum_{j=1}^{n}\left(a_{ij}^{\frac{2}{1+\beta}}(\xi_j - \xi_i)^2\right)^{\frac{1+\beta}{2}} \le -2^{\frac{\beta-1}{2}}\mu V^{\frac{1+\beta}{2}} \tag{13}$$

Since 0<β < 1/2, then $\left(\|\xi^2\|\right)^{\frac{1+\beta}{2}} \ge \|\xi\|^2$,
Hence, we have

$$\sum_{i=1}^{n}\sum_{j=1}^{n}\sigma\xi_i^2 \le \sigma 2^{\frac{1+\beta}{2}}V^{\frac{1+\beta}{2}} \tag{14}$$

combine (13) with (14), we have

$$\dot{V} \le (2^{\frac{1+\beta}{2}}\sigma - 2^{\frac{\beta-1}{2}}\mu)V^{\frac{1+\beta}{2}} \le 2^{\frac{1+\beta}{2}}(\sigma - \frac{1}{2}\mu)V^{\frac{1+\beta}{2}}$$

Note that $\sigma = 0.01$ implies that $\sigma - \frac{1}{2}\mu < 0$, denote $\sigma - \frac{1}{2}\mu < -K, K > 0$, therefore $\dot{V} \le -KV^{\frac{1+\beta}{2}}$.

By Lemma 4, $V = 0$ for $t \ge T = \frac{2}{K(1-\beta)}V(\xi(0))^{\frac{1-\beta}{2}}$.

The second-order multi-agent systems (1) solves the finite-time average consensus problem.

4 Example and Simulations

Consider the Second-order multi-agent system with 4 agents represented by the vertex i, i = 1, …, 4. The starting states of the agents: $x(0) = (10, -0.4, 5.2, -8.5)^T$, $v(0) = (3.0, 0.6, 1.5, -0.5)^T$

Consider adjacency matrix of the network G is A, as follows:

$$A = \begin{bmatrix} 0\,1\,1\,0 \\ 1\,0\,0\,0 \\ 1\,0\,0\,1 \\ 0\,0\,1\,0 \end{bmatrix}, \quad \text{therefore,} \quad L = \begin{bmatrix} 2 & -1 & -1 & 0 \\ -1 & 1 & 0 & 0 \\ -1 & 0 & 2 & -1 \\ 0 & 0 & -1 & 1 \end{bmatrix}$$

Note that network G is connected. It can be calculated that $\lambda_2(L(A)) = 0.5858$, $\mu = 0.8819$. From $\kappa = 0.01,\ \mu = 0.8819$, we can get $\kappa - \frac{1}{2}\mu < 0$.

Under protocol (2) with $\alpha = 0.5,\ \beta = \frac{\alpha}{1+\alpha} = 0.333$, Figs. 1 and 2 show thc dynamic tracks of the agents' position and velocity states, respectively.

5 Conclusions

The finite-time consensus problem of second-order multi-agent system with external disturbance is investigated in this article. By transferring the original system into

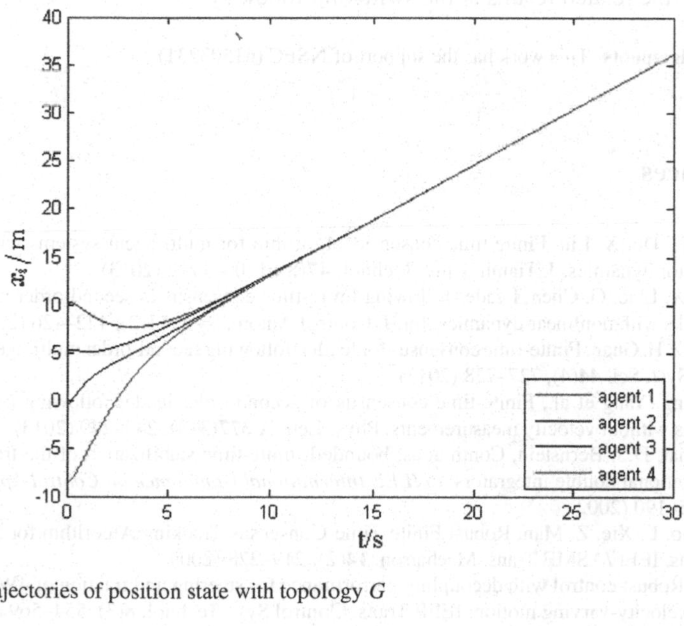

Fig. 1 Trajectories of position state with topology G

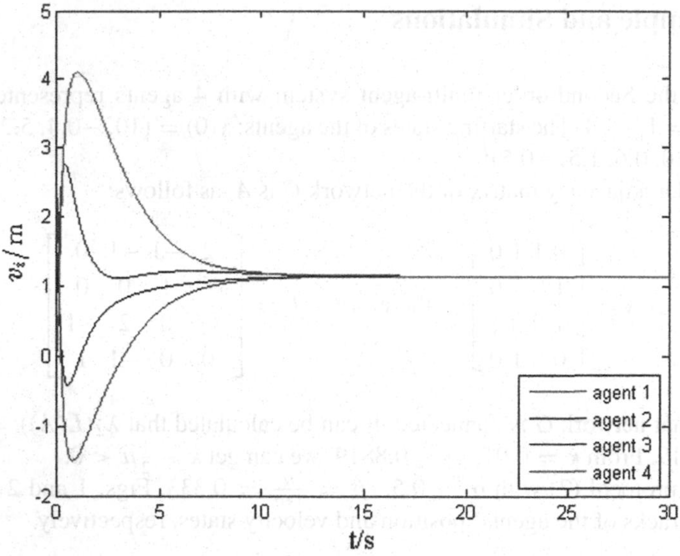

Fig. 2 Trajectories of velocity state with topology G

an equivalent system, which can be represented by disagreement vector, the sufficient conditions for solving the above problem are obtained. Some simulations are supplied to show the validity of the gotten theoretic results. Future works include the researching and discussing of consensus for discrete multi-agent systems and expanding the related results to the switching topology.

Acknowledgements This work has the support of NSFC (61503231).

References

1. S. Li, H. Du, X. Lin, Finite-time consensus algorithm for multi-agent systems with double-integrator dynamics. J. Tianjin Univ. Technol. **47**(8), 1706–1712 (2013)
2. H. Li, X. Liao, G. Chen, Leader-following finite-time consensus in second-order multi-agent networks with nonlinear dynamics. Int. J. Control Autom. Syst. **11**(2), 422–426 (2013)
3. F. Sun, Z.H. Guan, Finite-time consensus for leader-following second-order multi-agent system. Int. J. Syst. Sci. **44**(4), 727–738 (2013)
4. Y. Zhang, Yang et al., Finite-time consensus of second-order leader-following multi-agent; systems without velocity measurements. Phys. Lett. A **377**(3–4), 243–249 (2013)
5. S.P. Bhat, D.S. Bernstein, Continuous, bounded, finite-time stabilization of the translational and rotational double integrators. in *IEEE International Conference on Control Applications*, pp. 185–190 (2002)
6. S. Khoo, L. Xie, Z. Man, Robust Finite-Time Consensus Tracking Algorithm for Multirobot Systems. IEEE/ASME Trans. Mechatron. **14**(2), 219–228 (2009)
7. Y. Jia, Robust control with decoupling performance for steering and traction of 4WS vehicles under velocity-varying motion. IEEE Trans. Control Syst. Technol. **8**(3), 554–569 (2000)

8. Y. Jia, Alternative proofs for improved LMI representations for the analysis and the design of continuous-time systems with polytopic type uncertainty: a predictive approach. IEEE Trans. Autom. Control **48**(8), 1413–1416 (2003)
9. W. Cao, J. Zhang, W. Ren, Leader–follower consensus of linear multi-agent systems with unknown external disturbances. Syst. Control Lett. **82**, 64–70 (2015)
10. D. Ouyang, Z. Yu, H. Jiang et al., Consensus for general multi-agent networks with external disturbances. Neurocomputing **198**(C), 100–108 (2016)
11. G. Ren, Y. Yu, *Robust Consensus of Fractional Multi-agent Systems with External Disturbances*. Elsevier Science Publishers B. V. (2016)
12. C. Ma, H. Qiao, Distributed asynchronous event-triggered consensus of nonlinear multi-agent systems with disturbances: An extended dissipative approach. Neurocomputing **243**, 177–186 (2017)
13. M.C. Fan, Y. Wu, Global leader-following consensus of nonlinear multi-agent systems with unknown control directions and unknown external disturbances. Appl. Math. Comput. **331**, 274–286 (2018)
14. Q. Song, J. Cao, W. Yu, Second-order leader-following consensus of nonlinear multi-agent systems via pinning control. Syst. Control Lett. **59**(9), 553–562 (2010)
15. F. Xiao, L. Wang, Y. Jia, Fast information sharing in networks of autonomous agents, in *IEEE American Control Conference*, pp. 4388–4393 (2008)
16. L. Wang, S. Sun, C. Xia, Finite-time stability of multi-agent system in disturbed environment. Nonlinear Dyn. **67**(3), 2009–2016 (2012)
17. F. Sun, W. Zhu, Y. Li et al., Finite-time consensus problem of multi-agent systems with disturbance. J. Franklin Inst. **353**(12), 2576–2587 (2016)
18. R. Olfati-Saber, R.M. Murray, Consensus problems in networks of agents with switching topology and time-delays. IEEE Trans. Autom. Control **49**(9), 1520–1533 (2004)

Prediction of Gas Utilization Ratio Based on the Kernel Extreme Learning Machine

Xuejiao Huang, Sen Zhang and Yixin Yin

Abstract Gas utilization ratio (GUR) is a significant indicator to measure the operation status and energy consumption of blast furnaces (BFs). Accurately predicting the GUR can reflect the actual operating status of the BFs and the consumption of the charge in real time. Kernel extreme learning machine (KELM) algorithm not only has the characteristics of fast computation speed of extreme learning machine (ELM), but also has better stability and generalization ability. This study applies KELM to investigate the relationship between GUR and some significant factors which affect GUR. An improved fruit fly optimization algorithm (IFOA) is used to optimize the parameters in the KELM model. The experimental results demonstrate that the prediction model based on the KELM has better prediction effect in forecasting accuracy and modeling time needed.

Keywords Gas utilization ratio · Blast furnace · Kernel extreme learning machine
Improved fruit fly optimization algorithm

1 Introduction

The steel industry plays an important role in the national economy in China. Blast furnace (BF) ironmaking as an essential part of the steel industry is a highly energy-intensive and polluting production process. According to statistics, the national ironmaking energy consumption accounts for about 70% of the total energy consumption of the steel industry [1]. In the face of increasingly serious environmental pollution

X. Huang · S. Zhang (✉) · Y. Yin
School of Automation and Electrical Engineering, University of Science and Technology Beijing, Beijing 100083, People's Republic of China
e-mail: zhangsen@ustb.edu.cn

X. Huang · S. Zhang · Y. Yin
Key Laboratory of Knowledge Automation for Industrial Processes of Ministry of Education, School of Automation and Electrical, Beijing 100083, People's Republic of China

Y. Jia et al. (eds.), *Proceedings of 2018 Chinese Intelligent Systems Conference*, Lecture Notes in Electrical Engineering 529,
https://doi.org/10.1007/978-981-13-2291-4_46

and shortage of resources, the BF ironmaking process needs to undertake the task of energy conservation and emissions reduction [2]. At present, because the coke ratio in blast furnace process is an important economic and technical indicator, domestic and foreign scholars mainly carries on the related research. However, the coke ratio is an energy consumption statistics in days, and its prediction and control are mainly judged and evaluated by experience of operators in the factory. The long time span may result in the lag of scheduling and the waste of resources.

The Gas utilization ratio (GUR) is characterized by the ratio of carbon monoxide into carbon dioxide in the BF. It is a key parameter to measure the degree of gas-solid reduction reaction in blast furnace ironmaking process. It essentially represents the reduction and utilization rate of carbon, which is the main raw material of blast furnace, and directly affects the energy consumption of ironmaking [3]. It can well evaluate the energy utilization of BF ironmaking. The GUR can be real-time reflected by the composition of the top gas of the BF, which can timely characterize the current operation level and energy consumption of the BF. Xiang et al. proposed the GUR should also be used as a criteria to determine reasonable gas flow distribution apart from smooth running, for choosing charging system and distribution system [4]. Hu introduces a few effective measures to raise the GUR by 1% [5]. Zhang presented the recovery and utilization circumstances of blast furnace gas and showed the relationship of sufficient utilization of comprehensive efficiency and energy consumption of the ironmaking process [6]. However, research on the GUR prediction nowadays is only concentrated at the mechanism level [7].

Actually BF smelting is a highly complicated production processing, the mechanism model has many assumptions, complex parameters and low precision. The mechanism model is difficult to reflect the strong nonlinear relationship between the energy consumption and operation of BF, which results in that we cannot get operation optimization strategy based on the model to improve the GUR.

The prediction method based on the artificial neural networks (ANNs) has been widely used. The back propagation (BP) neural network model and the support vector machine (SVM) model are in the guiding position [8]. However, traditional neural network models and support vector machine models still have the following problems: (1) learning speed is slow; (2) manual intervention is too tedious; (3) easy to produce local optimal solution. The extreme learning machine (ELM) is a novel single hidden layer feed forward neural network proposed by Huang. As a new forecasting method, this problem has been overcome by more and more scholars [9, 10]. However, the initial value of the model is randomly set, resulting in the instability of the algorithm. Huang et al. applied the idea of the kernel function in the support vector machine model to the ELM, absorbing the advantages of the support vector machine and putting forward the model of the kernel extreme learning machine (KELM) [11, 12].

In this paper, some control parameters of BF bottom are taken as inputs and the GUR of the top gas mixture is taken as the output, and the KELM prediction model is established. For the uncertain parameters in the KELM, improved fruit fly optimization algorithm (IFOA) is used to select the optimal value.

The rest of paper has four sections: Sect. 2 makes first-phase preparations, concluding the selection of input parameters of the model, the introduction of the kernel extreme learning machine model, and the introduction of the improved fruit fly optimization algorithm; Sect. 3 is the establishment of the IFOA-KELM model; Sect. 4 is the result and analysis of the experiment; Sect. 5 is the conclusion of the paper.

2 Preliminaries

2.1 Analysis of the Relevant Factors of the GUR

GUR can be calculated by the content of the gas and in the top gas of the BF; the calculation formula is shown in Eq. (1):

$$\eta_{co} = \frac{(\mathrm{co}_2)}{(\mathrm{co}_2) + (\mathrm{co})} \tag{1}$$

The inside of the blast furnace is basically a large counter current heat exchanger and a chemical reactor, which includes reactants, including iron ore, coke and limestone, and feed by layer by layer. In addition, the preheated air and extra oxygen occasionally blow into the bottom through the mouth of the wind. Due to the downward movement of raw materials and the upward flow of hot gas, a series of complex chemical reactions and heat transfer phenomena occur in different areas of blast furnace at different temperatures [13], which may generate a lot of energy, and the temperature of blast furnace may reach 2000°. When production continues, the produced iron and slag will flow out from the bottom, and the entire iron making cycle will take 6–8 h.

Because of the high complexity of BF ironmaking, there are many parameters that may affect the GUR. It is worth noting that choosing too many parameters as input variables may increase the complexity of the model, and too few parameters will reduce the performance of the model. Therefore, the parameter selection of the prediction model of the GUR plays an important role in improving the precision of the model. Based on expert knowledge and iron production process mechanism, we select air temperature, the air pressure, the air volume, the air speed, the top pressure, the permeability index the oxygen enrichment, and the top temperature as inputs parameters.

2.2 The Theory of the Kernel Extreme Learning Machine

Extreme Learning Machine (ELM) based on the study of single hidden layer feed-forward neural network is a simple and effective learning algorithm proposed by Huang in the early 2004. The structure of the ELM is shown as Fig. 1.

It is assumed that the model input sample set contains Q group sample, and the corresponding input matrix and the output matrix can be expressed as (2):

$$\boldsymbol{X} = \begin{bmatrix} x_{11} & x_{12} & \cdots & x_{1Q} \\ x_{21} & x_{22} & \cdots & x_{2Q} \\ \vdots & \vdots & \ddots & \vdots \\ x_{n1} & x_{n2} & \cdots & x_{nQ} \end{bmatrix}_{n\times Q}, \quad \boldsymbol{Y} = \begin{bmatrix} y_{11} & y_{12} & \cdots & y_{mQ} \\ y_{21} & y_{22} & \cdots & y_{mQ} \\ \vdots & \vdots & \ddots & \vdots \\ y_{m1} & y_{m2} & \cdots & y_{mQ} \end{bmatrix}_{m\times Q} \tag{2}$$

If the activation function of the hidden layer is $g(x)$, the output of the network can be obtained by operation.

$$T = [t_1, t_2, \ldots, t_Q]_{m\times Q}, \quad t_j = \begin{bmatrix} t_{1j} \\ t_{2j} \\ \vdots \\ t_{mj} \end{bmatrix}_{m\times 1} = \begin{bmatrix} \sum_{i=1}^{L} \beta_{i1} g(w_i x_j + b_i) \\ \sum_{i=1}^{L} \beta_{i2} g(w_i x_j + b_i) \\ \vdots \\ \sum_{i=1}^{L} \beta_{im} g(w_i x_j + b_i) \end{bmatrix}_{m\times 1} \tag{3}$$

w_i represents the weight values between the neuron of the input layer and the neuron of the hidden layer. β_i represents the weight values between the neuron of the hidden layer and the neuron of the output layer. b_i represents the bias vector of the hidden layer.

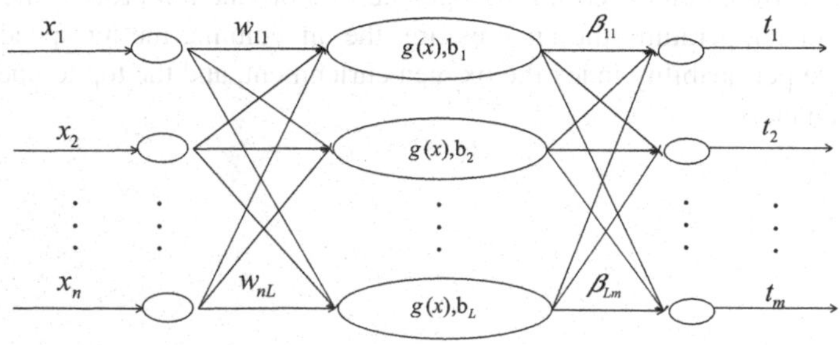

Fig. 1 The structure of ELM

The equation can be simply written as (4):

$$\boldsymbol{H}\beta = \boldsymbol{T} \tag{4}$$

The $\boldsymbol{H}$ is the output matrix of hidden layer network.

$$\boldsymbol{H} = \begin{bmatrix} g(w_1x_1+b_1) & g(w_2x_1+b_2) & \cdots & g(w_Lx_1+b_L) \\ g(w_1x_2+b_1) & g(w_2x_2+b_2) & \cdots & g(w_Lx_2+b_L) \\ \vdots & \vdots & \ddots & \vdots \\ g(w_1x_Q+b_1) & g(w_2x_Q+b_2) & \cdots & g(w_Lx_Q+b_L) \end{bmatrix} \tag{5}$$

Traditional neural network, the input weight and the bias vector of the hidden node are generated by random generation in ELM. Therefore, the elm model is equivalent to finding the least square solution β of the linear function: $\boldsymbol{H}\beta\boldsymbol{=}\boldsymbol{T}$.

$$\left\|\mathbf{H}\hat{\beta} - \mathbf{T}\right\| = \min_{\beta}\|\mathbf{H}\beta - \mathbf{T}\| \tag{6}$$

The least square solution is $\hat{\beta} = \boldsymbol{H}^{-1}\boldsymbol{T}$, the matrix $\boldsymbol{H}^{-1}$ the inverse matrix of the matrix $\boldsymbol{H}$.

In order to solve the possible complex linear problems in the sample data samples in various applications, Huang et al. introduced a regular parameter, and the output expression of the model is as follows:

$$f(x) = g(x)\boldsymbol{\beta} = g(x)\boldsymbol{H}^T\left(\frac{\boldsymbol{I}}{C} + \boldsymbol{H}\boldsymbol{H}^T\right)^{-1} T \tag{7}$$

When $g(x)$ is an unknown feature map, the KELM model can be constructed and the kernel matrix is defined as follows:

$$\boldsymbol{H}\boldsymbol{H}^T(i,j) = K(x_i, x_j),\ \boldsymbol{H}\boldsymbol{H}^T = \Omega_{ELM} = \begin{bmatrix} K(x_1,x_1) & \cdots & K(x_1,x_j) \\ \vdots & \ddots & \vdots \\ K(x_i,x_1) & \cdots & K(x_i,x_j) \end{bmatrix} = K(\boldsymbol{x}_i, \boldsymbol{x}_j) \tag{8}$$

The output function of KELM can be expressed as:

$$f(x) = \begin{bmatrix} K(x,x_1) \\ \vdots \\ K(x,x_N) \end{bmatrix}^T \left(\frac{\boldsymbol{I}}{C} + \boldsymbol{\Omega}_{ELM}\right)^{-1} T \tag{9}$$

In the process of modeling, it is not necessary to know the specific form of the activation function of the hidden layer, and the value of the output function can be obtained by constructing the kernel function.The number of the hidden layer nodes is not set in the solution process, and the bias matrix of the weight matrix and the hidden layer node need not be set.

2.3 The Theory of the Improved Fruit Fly Optimization Algorithm

In 2011, Professor Pan, inspired by the fruit fly foraging to simulate the process of the fruit fly population searching for food through a strong sense of smell and visual sense and the complex relationship between the fruit flies, proposed a swarm intelligence algorithm-the fruit fly optimization(FOA) algorithm [14]. The algorithm flow is shown as follows:

(1) Random initial fruit fly swarm location: (X_axis, Y_axis)
(2) Give the random direction and distance for the search of food to fruit fly individually: $\begin{cases} X_i = X\_axis + (\lambda - 0.5) \\ Y_i = Y\_axis + (\lambda - 0.5) \end{cases}$
(3) Calculate the distance to the origin: $Dist_i = \sqrt{(X_i^2 + Y_i^2)}$
Calculate the smell concentration judgment value: $S_i = \frac{1}{Dist_i}$
(4) Based on the smell concentration judgment function, find the smell concentration of the individual location of the fruit fly: $Smell_i = Function(S_i)$
(5) Find out the fruit fly with maximal smell concentration: $[bestSmell \quad bestIndex] = max(Smell_i)$
(6) If the smell concentration is better than the previous values, implement step (7), otherwise repeat execution steps (1)–(6).
(7) Keep the best smell concentration value and the location, meanwhile, the fruit fly swarm will fly towards that location using vision: $\begin{cases} Smellbest = bestSmell \\ X\_axis = X(bestSmell) \\ Y\_axis = Y(bestSmell) \end{cases}$

It is found that the range of the search food for each individual is limited by the initial set of r, and the selection of r will lead to the loss of the optimal solution in the optimization process. In order to solve this problem, this paper uses IFOA: in order to improve the local optimization ability in the iterative process, the search radius of large search r_0 is fixed, and the search radius r_1 in small range search decreases with the increase of the number of iterations. $r_1 = \left(1 - \frac{g-1}{g_{max}}\right) r_0$, The improved algorithm ensures the global search ability, and avoids the local optimal solution in the optimization process, and realizes the balance of the global search capability and the local optimization ability.

3 The Establishment of Prediction Model Based on the IFOA-KELM

(1) Normalize the data and divide the processed data into test set and training set.
(2) Use the IFOA to optimize the two parameters (C, σ) of the KELM model. The kernel function selected is shown: $K(x, x_i) = \exp\left(-\frac{x-x_i^2}{\sigma^2}\right)$
(3) Set the size of the swarm *SizePop*, the number of iterations *Maxgen* and the initial location $(x_1, y_1), (x_2, y_2)$. Give the random direction and distance for the search of food to fruit fly individually. Calculate the distance to the origin and the smell concentration judgment value.
(4) Based on the smell concentration judgment function, find the smell concentration of the individual location of the fruit fly.
(5) Repeat step (3) and step (4) until the number of iterations reaches the initial setting value. The optimal solution of the iteration process is obtained, and the best parameters are obtained.
(6) According to the optimal parameters obtained by optimization, the KELM model used the optimal parameters calculate the output of test data.

4 Simulation and Result Analysis

The data are divided into training samples and test samples. The first 1620 sets of data are selected as training samples, and the latter 180 sets of data are taken as test samples.

The BP, LSSVM, ELM, KELM and IFOA-KELM algorithms are used to establish the prediction model of GUR respectively. The actual values of the 180 sets of the test samples are compared with the predicted values as shown in Fig. 2. Figure 2a is the prediction result of BP model. Figure 2b is the prediction result of LSSVM model. Figure 2c is the prediction result of ELM model. Figure 2d is the prediction result of KELM model. Figure 2e is the prediction result of IFOA-KELM model.

We use the mean square error(MSE) and the mean absolute error(MAE) to measure the prediction effect of the gas utilization prediction model:

$$MSE = \frac{1}{n}\sum_{i=1}^{n}(y_i' - y_i)^2 \tag{10}$$

$$MAE = \frac{1}{n}\sum_{i=1}^{n}\left|y_i' - y_i\right| \tag{11}$$

where n is the number of samples, yi is the actual value of the GUR, and yi' is the prediction value of the output of the model.

Table 1 Comparison results of different algorithms

	MSE	MAE	Training time/s	Testing time/s
IFOA-KELM	0.0854	0.2014	43.7854	0.0146
KELM	0.0950	0.2291	0.1991	0.0119
ELM	0.1577	0.3013	0.0780	0.0156
LSSVM	0.1379	0.2636	0.5616	0.0468
BP	0.1678	0.3182	1.4664	0.0312

In this paper, 10-fold cross validation is used, and the initial samples are divided into ten subsets whose sizes are 180. Each subset is used for testing and others are used for training. The average of the ten test sets error is the test error of the model.

It can be seen from the Table 1 that the MSE of the KELM model is 0.0950, the BP model is 0.1678, and the LSSVM model is 0.1379 for the test samples of GUR. According to the data, the MSE of the KELM model is 0.0950, and the prediction time is faster than the BP and the LSSVM. Using IFOA algorithm to optimize uncertain parameters in KELM model, the optimization process takes some time, but the KELM model after optimization has higher prediction accuracy. The MSE and MAE of the IFOA-KELM model are the lowest compared with BP, LSSVM, ELM and KELM model.

From the above analysis, compared with the BP and the LSSVM, the KELM is better than the comparison model in the time of training and the prediction accuracy. KELM has better overall prediction performance. Only considering the prediction accuracy, it is better to use the IFOA-KELM model.

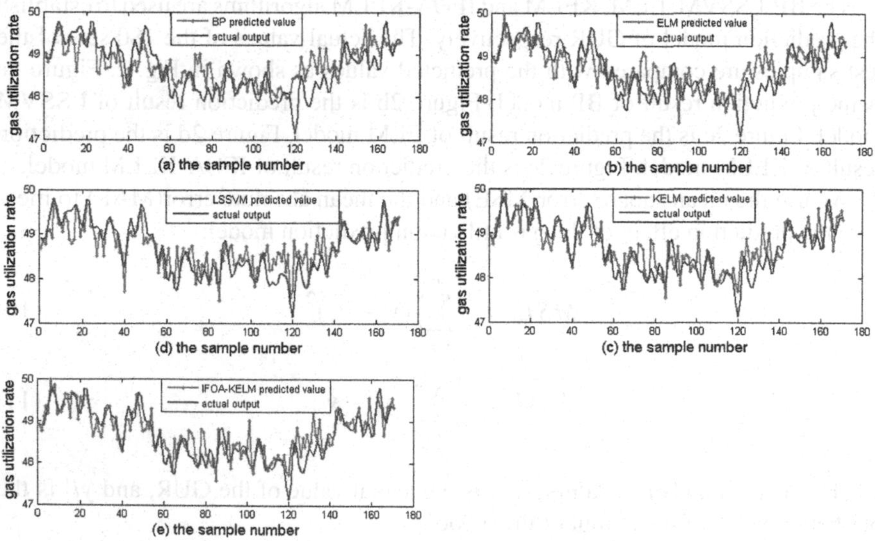

Fig. 2 The predicted results of models

5 Conclusion

GUR is an important index to measure the running state and energy consumption of the blast furnace. It has great theoretical significance and application value to establish a model to predict the GUR. Compared with the BP and the LSSVM, the GUR prediction model based on KELM improves the prediction effect, and greatly reduces the time required for the modeling. Optimize the parameters in the KELM model by IFOA algorithm, and further improves the prediction precision of the model.

Acknowledgements This work was supported by the Key Program of National Nature Science Foundation of China under grant No. 61333002, the National Nature Science Foundation of China under grants No. 61673056, the Beijing Natural Science Foundation under grant No. 4182039, and the Beijing Key Discipline Construction Project (XK100080537).

References

1. T.J. Yang, H.B. Zuo, Scientific Development Strategy of Blast Furnace Ironmaking Technology in China. Iron & Steel (2008)
2. S. Lin, Y. Wen, G. Zhao et al., Recognition of blast furnace gas flow center distribution based on infrared image processing. J. Iron. Steel Res. Int. **23**(3), 203–209 (2016)
3. D. Xiao, J. An, M. Wu et al., Chaotic prediction of carbon-monoxide utilization ratio in the blast furnace, in Control Conference, pp. 9499–9503, IEEE, 2016
4. Z.Y. Xiang, X.L. Wang, H. Yin, More discussion on evaluation method for productive efficiency of ironmaking blast furnace. Iron and Steel/ Gangtie **48**(3), 86–91 (2013)
5. Z.G. Hu, Strategy for improvement in utilization rate of No. 5 BF gas in WISCO and its practice[J]. Wugang jishu- WISCO Technol. **50**(2), 8–11 (2012)
6. J. Zhang, Increasing utilization rate of blast furnace gas to decrease energy consumption of ironmaking process. Metallurgical Power (2005)
7. C. Zhou, Blast Furnace Ironmaking Production of Technical Manual (Metallurgical Industry Press: Beijing, China, 2005), pp. 174–177. [Google Scholar]
8. J. Liu, Y.L. Huang, Nonlinear network traffic prediction based on BP neural network. Jisuanji Yingyong/ J. Comput. Appl. **27**(7), 1770–1772 (2007)
9. G.B. Huang, Q.Y. Zhu, C.K. Siew, Extreme learning machine: a new learning scheme of feedforward neural networks, in Proceedings. 2004 IEEE International Joint Conference on Neural Networks, vol. 2, pp. 985–990, IEEE, 2004
10. G.B. Huang, H. Zhou, X. Ding et al., Extreme learning machine for regression and multiclass classification. IEEE Trans. Syst. Man Cybern. Part B (Cybern.) **42**(2), 513–529 (2014)
11. W.Y. Deng, Q.H. Zheng, Z.M. Wang, Cross-person activity recognition using reduced kernel extreme learning machine. Neural Netw. **53**, 1–7 (2014)
12. X. Liu, L. Wang, G.B. Huang et al., Multiple kernel extreme learning machine. Neurocomputing **149**, 253–264 (2015)
13. J. Yang, B. Liu, Life cycle inventory of steel products in China. Acta Sci. Circum. **22**(4), 519–522 (2002)
14. W.T. Pan, A new fruit fly optimization algorithm: taking the financial distress model as an example. Knowl.-Based Syst. **26**, 69–74 (2012)
15. H.Z. Li, S. Guo, C.J. Li et al., A hybrid annual power load forecasting model based on generalized regression neural network with fruit fly optimization algorithm. Knowl.-Based Syst. **37**, 378–387 (2013)

Adaptive Synchronization Control for Dual-Manipulator System Using Finite Time Parameter Estimation

Miaomiao Gao, Qiang Chen, Liang Tao and Yurong Nan

Abstract In this paper, an adaptive synchronization control method is proposed for dual-manipulator systems to guarantee the satisfactory synchronization and tracking performance on the basis of effective finite time parameter estimation. The mean-coupling synchronization scheme is employed to obtain the mean-coupling error, and a fast finite time sliding mode surface and a auxiliary control variable are presented based on the position tracking error and mean-coupling error to facilitate the control design. Then, the fast terminal sliding mode synchronization controller is proposed to ensure that both the tracking error and the synchronization error can converge to zero in finite time. Moreover, an adaptive parameter estimation law is developed by the extracted parameter error information to accurately identify the unknown system parameters. Comparative simulations are provided to validate the effectiveness of the proposed method.

Keywords Adaptive control · Dual-manipulator system · Mean-coupling synchronization · Parameter estimation

1 Introduction

Over the past decades, robotic manipulator technology has applied in many fields, such as medical, military, industry, service and so on. Compared with the single manipulator system, two or more manipulators systems can accomplish more complex production tasks [1]. Therefore, more researches on the multiple manipulators systems have been developed [2, 3]. Lots of synchronous control schemes have been proposed for multiple manipulators such as master-slave control [4], leader-follower control [5], cross-coupling control [6], etc.

M. Gao · Q. Chen (✉) · L. Tao · Y. Nan
College of Information Engineering, Zhejiang University of Technology, Hangzhou 310023, China
e-mail: sdnjchq@zjut.edu.cn

Y. Jia et al. (eds.), *Proceedings of 2018 Chinese Intelligent Systems Conference*, Lecture Notes in Electrical Engineering 529,
https://doi.org/10.1007/978-981-13-2291-4_47

Recently, the mean deviation coupling control was proposed in [7] for multiple motors, which can improve the synchronization performance and reduce the complexity of control structure with the increasing number of motors. Compared to multiple motor systems, multiple manipulators systems have more complex structures and uncertainties, and the superiorities of mean-coupling synchronization scheme with respect to complexity reduction and synchronization performance improvement are exactly what the multiple manipulators requires. Thus, the mean-coupling scheme is suitable for the design of multiple manipulators system, so as to further improve the system control performance.

To further improve the position tracking control performance, numerous control schemes combined with synchronous control have been proposed for multiple manipulators. A feedback controller with master-slave scheme for two manipulators system was designed in [8] to make the tracking errors eventually uniformly bounded. Among the methods, sliding mode control (SMC) is widely used due to its good robustness and anti-disturbance ability [9, 10]. An adaptive terminal sliding mode controller with cross-coupling scheme for multiple manipulator systems was proposed in [10] to ensure the tracking errors and synchronization errors converge to zero in finite time. However, the parameters of multiple manipulators systems are usually unknown or uncertain, which will bring difficulties to the control system. Adaptive estimation has been widely studied when the systems have unknown or immeasurable parameters [11–14]. In [13], a composite adaptive law was introduced by utilizing tracking and prediction errors. However, this method requires the construction of an extra predictor, and the parameters are difficult to converge to the true values. Recently, an adaptive parameter estimation scheme was proposed in [14], which is driving by the parameter estimation error. The advantage of the design is that the adaptation law is independent of any observer or predictor, and the parameters can converge to the true values.

In this paper, an adaptive synchronization control based on finite time parameter estimation is proposed for dual-manipulator system. A mean-coupling error is defined to construct the fast finite time sliding mode surface and auxiliary control variable, which can simplify the control design. Then, the fast terminal sliding mode control based on mean-coupling synchronization scheme is proposed to ensure the position tracking errors and synchronization errors can converge to zero in finite time. Besides, several auxiliary filtered variables are introduced to obtain the expression of parameter estimation errors, and an adaptive parameter estimation law is developed with the estimated errors to achieve the accurate identification of the unknown system parameters.

2 Problem Formulation and Preliminaries

The dynamics of the two-link dual-manipulator system is expressed as

$$M\left(q\right)\ddot{q}+C\left(q,\dot{q}\right)\dot{q}+G\left(q\right)=\tau \tag{1}$$

where $q = \left[q_1^T \ q_2^T\right]^T \in R^{4\times1}$, $\dot{q} = \left[\dot{q}_1^T \ \dot{q}_2^T\right]^T \in R^{4\times1}$, $\ddot{q} = \left[\ddot{q}_1^T \ \ddot{q}_2^T\right]^T \in R^{4\times1}$ are the vectors of manipulator joint position, velocity and acceleration, respectively, $M(q) = diag([M_1(q) \ M_2(q)]) \in R^{4\times4}$ is the inertia matrix, $C(q, \dot{q}) = diag\left(\left[C_1(q, \dot{q}) \ C_2(q, \dot{q})\right]\right) \in R^{4\times4}$ is the centripetal and Coriolis torques, $G(q) = \left[G_1^T(q) \ G_2^T(q)\right]^T \in R^{4\times1}$ is the gravity torques and $\tau = \left[\tau_1^T \ \tau_2^T\right]^T \in R^{4\times1}$ represents the control torque vector.

Before the controller design, some useful definitions and properties are provided as follows:

Definition 1 [15]: A vector or matrix function Φ is persistently excited (PE) if there exist $T > 0$ and $\sigma > 0$ such that $\int_t^{t+T} \Phi^T(r)\Phi(r)dr \geq \sigma I$ holds for any $t \geq 0$.

Property 1 $\dot{M}(q) - 2C(q, \dot{q})$ is a skew symmetric matrix

$$x^T\left[\dot{M}(q) - 2C(q, \dot{q})\right]x = 0, \forall x \in R^{4\times1} \tag{2}$$

Property 2 The left side of the Eq. (1) can be represented in the following linearly parametric form

$$M(q)\ddot{q} + C(q, \dot{q})\dot{q} + G(q) = \Phi(q, \dot{q}, \ddot{q})\theta \tag{3}$$

where θ is a constant parameter vector, composed of the parameters to be estimated, such as mass, inertia moment, length and so on, $\Phi(q, \dot{q}, \ddot{q})$ is a known regression matrix.

3 Fast Finite-Time Sliding Mode Surface

The position tracking error e and synchronization error ε are defined as

$$e = q_d - q \tag{4}$$

and

$$\varepsilon = Te \tag{5}$$

where $e = \left[e_1^T \ e_2^T\right]^T \in R^{4\times1}$ is the position tracking error, $q_d = \left[q_{d1}^T \ q_{d2}^T\right]^T \in R^{4\times1}$ is the desired trajectory, T is the synchronization transformation matrix in the form of

$$T = \begin{bmatrix} \frac{I}{2} & -\frac{I}{2} \\ -\frac{I}{2} & \frac{I}{2} \end{bmatrix} \tag{6}$$

In order to ensure the position tracking error e and synchronization error ε convergence to zero simultaneously, a mean-coupling error is given by

$$E = e + \beta\varepsilon = (I + \beta T)\, e \tag{7}$$

where $E = \left[E_1^T \; E_2^T \right]^T \in R^{4\times 1}$, I is the unit matrix, β respects the control gain that is the positive definite diagonal matrix. As $(I + \beta T)$ is symmetric and with full rank, it is concluded that when the mean-coupling error $E = 0$, the position tracking error $e = 0$ and synchronization error $\varepsilon = 0$ are both achieved at the same time.

A fast terminal sliding mode (FTSM) is constructed as

$$\begin{aligned} S &= \dot{E} + \lambda_1 E + \lambda_2 sign^{\gamma} E \\ &= (I + \beta T)(\dot{q}_r - \dot{q}) \end{aligned} \tag{8}$$

where λ_1 and λ_2 are positive definite diagonal matrices, $\gamma = q/p$, p, q are positive odd numbers satisfying $q < p$, $sign^{\gamma} E = \left[|E_1|^{\gamma} sign(E_1) \; \cdots \; |E_n|^{\gamma} sign(E_n) \right]^T$, $\dot{q}_r$ is a auxiliary variable satisfying

$$\dot{q}_r = \dot{q}_d + (I + \beta T)^{-1} \lambda_1 E + (I + \beta T)^{-1} \lambda_2 sign^{\gamma} E \tag{9}$$

Then, the derivative of S is

$$S = \ddot{E} + \lambda_1 \dot{E} + \gamma \lambda_2 |E|^{\gamma-1} \dot{E} = (I + \beta T)(\ddot{q}_r - \ddot{q}) \tag{10}$$

with

$$\ddot{q}_r = \ddot{q}_d + (I + \beta T)^{-1} \lambda_1 \dot{E} + \gamma (I + \beta T)^{-1} \lambda_2 |E|^{\gamma-1} \dot{E} \tag{11}$$

As $\ddot{q}_r$ contains a negative fractional power $\gamma - 1$, there exists the singularity problem if $E = 0$. Motivated by the work of [16], the singularity problem is avoided by modifying $\dot{q}_r$ as

$$\dot{q}_r = \dot{q}_d + (I + \beta T)^{-1} \lambda_1 E + (I + \beta T)^{-1} \lambda_2 \varphi(E) \tag{12}$$

where

$$\varphi(E) = \begin{cases} sign^{\gamma} E, & S = 0 \; or \; S \neq 0, |E| > \mu \\ l_1 E + l_2 sign^2 E & S \neq 0, |E| \leq \mu \end{cases} \tag{13}$$

with $l_1 = (2 - \gamma)\, \mu^{\gamma-1}$, $l_2 = (\gamma - 1)\, \mu^{\gamma-2}$, $\mu > 0$ is a small positive constant.

Then, the derivative of $\dot{q}_r$ is addressed as

$$\ddot{q}_r = \begin{cases} \ddot{q}_d + (I + \beta T)^{-1} \lambda_1 \dot{E} + \gamma (I + \beta T)^{-1} \lambda_2 |E|^{\gamma-1} \dot{E}, & S = 0 \; or \; S \neq 0, |E| > \mu \\ \ddot{q}_d + (I + \beta T)^{-1} \lambda_1 \dot{E} + l_1 (I + \beta T)^{-1} \lambda_2 \dot{E} + 2 l_2 \lambda_2 \,|E|\, \dot{E}, & S \neq 0, |E| \leq \mu \end{cases} \tag{14}$$

4 Adaptive Controller Design

According to (10), the dual-manipulator system (1) is rewritten as

$$M(q)(I+\beta T)^{-1}\dot{S}+C(q,\dot{q})(I+\beta T)^{-1}S-R(q,\dot{q})=-\tau \tag{15}$$

where auxiliary function $R(q,\dot{q})$ is

$$R(q,\dot{q})=M(q)\ddot{q}_r+C(q,\dot{q})\dot{q}_r+G(q) \tag{16}$$

Based on the Property 2, the auxiliary function $R(q,\dot{q})$ is presented in linearly parametric form

$$R(q,\dot{q})=M(q)\ddot{q}_r+C(q,\dot{q})\dot{q}_r+G(q)=\Phi_R(q,\dot{q})\theta \tag{17}$$

where $\Phi_R(q,\dot{q})$ is the known regression matrix, θ is the unknown parameter to be estimated.

In (15), the joint acceleration $\ddot{q}$ is contained in $\dot{S}$, which is sensitive to noises. In order to avoid the requirement of $\dot{S}$, the new auxiliary functions $F(q,\dot{q})=M(q)(I+\beta T)^{-1}S$ and $H(q,\dot{q})=-\dot{M}(q)(I+\beta T)^{-1}S+C(q,\dot{q})(I+\beta T)^{-1}S$ are introduced as

$$\begin{cases} F(q,\dot{q})=M(q)(I+\beta T)^{-1}S=\Phi_F(q,\dot{q})\theta \\ H(q,\dot{q})=-\dot{M}(q)(I+\beta T)^{-1}S+C(q,\dot{q})(I+\beta T)^{-1}S=\Phi_H(q,\dot{q})\theta \end{cases} \tag{18}$$

where $\Phi_F(q,\dot{q})$ and $\Phi_H(q,\dot{q})$ are new regressor matrices without the joint acceleration $\ddot{q}$.

Then, the system (15) is reformulated as

$$\dot{F}(q,\dot{q})+H(q,\dot{q})-R(q,\dot{q})=\Phi(q,\dot{q},\ddot{q})\theta=-\tau \tag{19}$$

where $\dot{F}(q,\dot{q})=\frac{d}{dt}\left[M(q)(I+\beta T)^{-1}S\right]=\dot{\Phi}_F(q,\dot{q})\theta$, $\Phi(q,\dot{q},\ddot{q})=\dot{\Phi}_F(q,\dot{q})+\Phi_H(q,\dot{q})-\Phi_R(q,\dot{q})$ is the regressor matrix. However, $\dot{\Phi}_F(q,\dot{q})$ includes the joint acceleration $\ddot{q}$, and to avoid using $\ddot{q}$, the following linear filter operation on both sides of (19) is given by

$$\begin{cases} k\dot{\Phi}_{Ff}(q,\dot{q})+\Phi_{Ff}(q,\dot{q})=\Phi_F(q,\dot{q}), & \Phi_{Ff}(q,\dot{q})\mid_{t=0}=0 \\ k\dot{\Phi}_{Hf}(q,\dot{q})+\Phi_{Hf}(q,\dot{q})=\Phi_H(q,\dot{q}), & \Phi_{Hf}(q,\dot{q})\mid_{t=0}=0 \\ k\dot{\Phi}_{Rf}(q,\dot{q})+\Phi_{Rf}(q,\dot{q})=\Phi_R(q,\dot{q}), & \Phi_{Rf}(q,\dot{q})\mid_{t=0}=0 \\ k\dot{\tau}_f+\tau_f=\tau, & \tau_f\mid_{t=0}=0 \end{cases} \tag{20}$$

where k is the filtered tuning parameter, $\Phi_{Ff}(q,\dot{q})$, $\Phi_{Hf}(q,\dot{q})$, $\Phi_{Rf}(q,\dot{q})$ and τ_f are the filtered form of $\Phi_F(q,\dot{q})$, $\Phi_H(q,\dot{q})$, $\Phi_R(q,\dot{q})$ and τ, respectively.

From (19) and (20), it is obtained as

$$\left[\frac{\Phi_F(q,\dot{q}) - \Phi_{Ff}(q,\dot{q})}{k} + \Phi_{Hf}(q,\dot{q}) - \Phi_{Rf}(q,\dot{q})\right]\theta = \Phi_f(q,\dot{q})\,\theta = -\tau_f \tag{21}$$

where $\Phi_f(q,\dot{q}) = \left[\frac{\Phi_F(q,\dot{q}) - \Phi_{Ff}(q,\dot{q})}{k} + \Phi_{Hf}(q,\dot{q}) - \Phi_{Rf}(q,\dot{q})\right]$ is the new regressor matrix. It is clearly shown in (21) that the joint acceleration $\ddot{q}$ is avoided through the introduction of the filter operation (20).

In the following, the auxiliary matrix P and the vector Q in [14] are employed in the form of

$$\begin{cases} \dot{P} = -lP + \Phi_f^T \Phi_f, P(0) = 0 \\ \dot{Q} = -lQ + \Phi_f^T \tau_f, \; Q(0) = 0 \end{cases} \tag{22}$$

where l is the tuning parameter, $P(0)$ and $Q(0)$ are the initial values of P and Q, respectively.

The solution of (22) is given by

$$\begin{cases} P = \int_0^t e^{-l(t-r)} \Phi_f^T(r) \Phi_f(r) dr \\ Q = \int_0^t e^{-l(t-r)} \Phi_f^T(r) \tau_f(r) dr \end{cases} \tag{23}$$

According to (21) and (23), it can be verified that $Q = -P\theta$. In order to obtain the information of parameter errors, the auxiliary vector W is defined as

$$W = P\hat{\theta} + Q = P\hat{\theta} - P\theta = -P\tilde{\theta} \tag{24}$$

where $\tilde{\theta} = \theta - \hat{\theta}$ is the parameter estimation error.

Then, the adaptive control law in the system (1) is designed as

$$\tau = \Phi_R \hat{\theta} + K_1 S + K_2 sign^{\rho} S \tag{25}$$

where $K_1 > 0$, $K_2 > 0$ are the positive gain matrices, $0 < \rho < 1$ is a positive constant, and the adaptive updating law of $\hat{\theta}$ is given by

$$\dot{\hat{\theta}} = \Gamma \left(\Phi_R^T S - \kappa_1 W - \kappa_2 P^T sign^{\rho} W\right) \tag{26}$$

where $\Gamma > 0$ and $\kappa_1, \kappa_2 > 0$ are the adaptive gain matrices.

5 Stability Analysis

Lemma 1 [17]: *For a continuous nonlinear system $\dot{x}(t) = g(x(t))$, $g(0) = 0$, $x(t) \in R^N$, if there exists a continuously positive-definite function $V(x)$ satisfying $\dot{V}(x) + c_1 V(x) + c_2 V^{\gamma}(x) \leq 0$, $c_1 > 0$, $c_2 > 0$, $0 < \gamma < 1$, $V(x)$ will converge to zero in a finite time and the settling time can be given by*

$$T_1 \leq \frac{1}{c_1(1-\gamma)} \ln \frac{c_1 V^{1-\gamma}(x_0) + c_2}{c_2} \tag{27}$$

where $V(x_0)$ is the initial value of $V(x)$.

Lemma 2 [14]: *According to Definition 1 the regressor matrix $\Phi_f(q, \dot{q})$ in (21) is persistently excited (PE), then the matrix P in (23) is positive definite and satisfies $\lambda_{\min}(P) > \delta > 0$.*

Theorem 1 *Considering the dual-manipulator system (1) with adaptive controller (25) and adaptive law (26), if the regressor matrix $\Phi_f(q, \dot{q})$ is persistently excited (PE), the parameter estimation error $\tilde{\theta}$ and the fast terminal sliding-mode manifold S converge to zero in finite time.*

Proof Design a Lyapunov function as

$$V = \frac{1}{2} S^T M (I + \beta T)^{-1} S + \frac{1}{2} \tilde{\theta}^T \Gamma^{-1} \tilde{\theta} \tag{28}$$

Differentiating (28) with respect to time yields

$$\dot{V} = S^T M (I + \beta T)^{-1} \dot{S} + \frac{1}{2} S^T \dot{M} (I + \beta T)^{-1} S + \tilde{\theta}^T \Gamma^{-1} \dot{\tilde{\theta}} \tag{29}$$

Substituting (15), (25) and (26) to (29) leads to

$$\begin{aligned}
\dot{V} &= S^T(-\tau - C(q, \dot{q})(I + \beta T)^{-1} S - \Phi(q, \dot{q})\theta) + \frac{1}{2} S^T \dot{M} (I + \beta T)^{-1} S - \tilde{\theta}^T \Gamma^{-1} \dot{\hat{\theta}} \\
&= S^T(\Phi_R \hat{\theta} + K_1 S + K_2 sign^{\rho} S - C(q, \dot{q})(I + \beta T)^{-1} S - \Phi(q, \dot{q})\theta) \\
&\quad + \frac{1}{2} S^T \dot{M} (I + \beta T)^{-1} S - \tilde{\theta}^T \Gamma^{-1} \Gamma (\Phi_R^T S - \kappa_1 W - \kappa_2 P^T sign^{\rho} W) \\
&= S^T \Phi_R \tilde{\theta} - S^T K_1 S - S^T K_2 sign^{\rho} S - S^T C(q, \dot{q})(I + \beta T)^{-1} S \\
&\quad + \frac{1}{2} S^T \dot{M} (I + \beta T)^{-1} S - \tilde{\theta}^T \Phi_R^T S + \tilde{\theta}^T \kappa_1 W + \tilde{\theta}^T \kappa_2 P^T sign^{\rho} W
\end{aligned} \tag{30}$$

According to Property 1, (30) is further written as

$$\begin{aligned}\dot{V} &= -S^T K_1 S - S^T K_2 \|S\|^{\rho} - \tilde{\theta}^T \kappa_1 P \tilde{\theta} - \tilde{\theta}^T \kappa_2 P^{\rho+1} \left\|\tilde{\theta}\right\|^{\rho} \\ &\le -\mu_1 V - \mu_2 V^{\frac{\rho+1}{2}}\end{aligned} \tag{31}$$

where $\mu_1 = min\{2\lambda_{\min}(K_1)/\lambda_{\max}(M(I+\beta T)^{-1}), 2\kappa_1\lambda_{\min}(P)/\lambda_{\max}(\Gamma^{-1})\}$, $\mu_2 = min\{\lambda_{\min}(K_2)\cdot(2/\lambda_{\max}(M(I+\beta T)^{-1}))^{\frac{\rho+1}{2}}, \kappa_2\lambda_{\min}^{\rho+1}(P)\cdot(2/\lambda_{\max}(\Gamma^{-1}))^{\frac{\rho+1}{2}}\}$ are positive constants, $\lambda_{\max}(\cdot)$ and $\lambda_{\min}(\cdot)$ are the maximum and minimum eigenvalues of the corresponding matrices.

According to Lemma 1, it is concluded that $\tilde{\theta}$ and S converge to zero in finite time. According to [17], when S converge to zero, the mean-coupling error E converge to zero in finite time. According to (7), as $(I+\beta T)$ is a full rank matrix, the mean-coupling error E converge to zero in finite time, such that the position tracking error e and synchronization error ε are both converge to zero in finite time. The theorem is proved.

6 Simulation

In this section, a dual-manipulator system consisting of two same manipulators is considered, and the dynamic equation is given as:

$$\begin{bmatrix} M_{i_{11}}(q) & M_{i_{12}}(q) \\ M_{i_{21}}(q) & M_{i_{22}}(q) \end{bmatrix}\begin{bmatrix} \ddot{q}_{i1} \\ \ddot{q}_{i2} \end{bmatrix} + \begin{bmatrix} C_{i_{11}}(q,\dot{q}) & C_{i_{12}}(q,\dot{q}) \\ C_{i_{21}}(q,\dot{q}) & C_{i_{22}}(q,\dot{q}) \end{bmatrix}\begin{bmatrix} \dot{q}_{i1} \\ \dot{q}_{i2} \end{bmatrix} + \begin{bmatrix} G_{i1}(q) \\ G_{i2}(q) \end{bmatrix} = \tau$$

with

$M_{i_{11}}(q) = (m_{i1}+m_{i2})r_{i1}^2 + m_{i2}r_{i2}^2 + 2m_{i2}r_{i1}r_{i2}\cos(q_{i2}) + J_{i1}$,
$M_{i_{12}}(q) = M_{i_{21}}(q) = m_{i2}r_{i2}^2 + m_{i2}r_{i1}r_{i2}\cos(q_{i2})$, $M_{i_{22}}(q) = m_{i2}r_{i2}^2 + J_{i2}$,
$C_{i_{11}}(q,\dot{q}) = -m_{i2}r_{i1}r_{i2}\sin(q_{i2})\dot{q}_{i1}$, $C_{i_{12}}(q,\dot{q}) = -2m_{i2}r_{i1}r_{i2}\sin(q_{i2})\dot{q}_{i1}$,
$C_{i_{21}}(q,\dot{q}) = 0$, $C_{i_{22}}(q,\dot{q}) = m_{i2}r_{i1}r_{i2}\sin(q_{i2})\dot{q}_{i1}$,
$G_{i1}(q) = (m_{i1}+m_{i2})gr_{i1}\cos(q_{i2}) + m_{i2}gr_{i2}\cos(q_{i1}+q_{i2})$,
$G_{i2}(q) = m_{i2}gr_{i2}\cos(q_{i1}+q_{i2})$.

where $i = 1, 2$, m_{i1}, m_{i2} are the mess of manipulator, r_{i1}, r_{i2} are the length of each link, J_{i1}, J_{i2} are the inertia of each link, g is the gravity constant. In this study, the unknown parameters of each manipulator to be estimated is $\theta = \begin{bmatrix} m_{i1} & m_{i2} & J_{i1} & J_{i2} \end{bmatrix}^T$. In order to verify the effectiveness and superior of the proposed approach, two different control schemes are considered: (S1) proposed adaptive synchronization control with finite time parameter estimation (ASCFTPE); (S2) traditional adaptive synchronization control (TASC), which the sliding mode surface, the controller and adaptive law are given by

$$S_1 = \dot{E} + \lambda_1^* E \tag{32}$$

$$\tau_1 = \Phi_R \hat{\theta} + K_{11}^* S \tag{33}$$

$$\dot{\hat{\theta}}_1 = \Gamma^* \Phi_R^T S \tag{34}$$

For fair comparison, the system parameters, most of identification parameters, control parameters and initial conditions are selected the same. The parameter values of each manipulator are set as $m_{i1} = 0.3\,\text{kg}$, $m_{i2} = 0.5\,\text{kg}$, $r_{i1} = 0.2\,\text{m}$, $r_{i2} = 0.3\,\text{m}$, $J_{i1} = 0.05\,\text{kg}\cdot\text{m}$, $J_{i2} = 0.1\,\text{kg}\cdot\text{m}$, $g = 9.81\,\text{m/s}^2$. The identification parameters and the control parameters of two schemes are set the same, i.e., $\lambda_1 = \lambda_1^* = 3I$, $K_{11} = K_{11}^* = 8I$, $\Gamma = \begin{bmatrix} 2\ 2\ 2\ 2\ 10\ 10\ 10\ 10 \end{bmatrix}$. The other parameters of A1 are set as $l = 1$, $k = 0.001$, $\kappa_1 = 2$, $\kappa_2 = 1$, $\beta = 0.8$, $\lambda_2 = 2I$, $K_{12} = 2I$, $\gamma = 7/9$, $\rho = 9/11$, $\mu = 0.02$, $\zeta = 0.05$.The initial conditions are given as $q_1(0) = [0.12 \quad 0.25]^T$, $q_2(0) = [0.18 \quad 0.22]^T$, $\dot{q}_1(0) = \dot{q}_2(0) = [0 \quad 0]^T$, $\Phi_{Rf}(0) = 0$, $\Phi_{Ff}(0) = 0$, $\Phi_{Hf}(0) = 0$, $P(0) = 0$, $Q(0) = 0$, $\theta(0) = 0$.

The reference trajectory of each manipulator is specified by $q_d = 0.5\sin(t)$. The position tracking performance and tracking error are shown in Figs. 1 and 2, respectively. From Figs. 1 and 2, it can be seen that proposed method has faster convergence rate and higher convergence precision for dual-manipulator than the TASC. The synchronization error are shown in Fig. 3. The figure shows that the proposed method has the superior performance of high synchronization precision and faster synchronization rate in comparison with TASC. The estimation results of the dual-manipulator parameters $m_{11}, m_{12}, m_{21}, m_{22}, J_{11}, J_{12}, J_{21}, J_{22}$ is shown in Figs. 4 and 5. From Figs. 4 and 5, we can see that the parameter estimation of proposed scheme can converge to the true value, while TASC can not converge to the true value. Figure 6 shows the control torque of the proposed method.

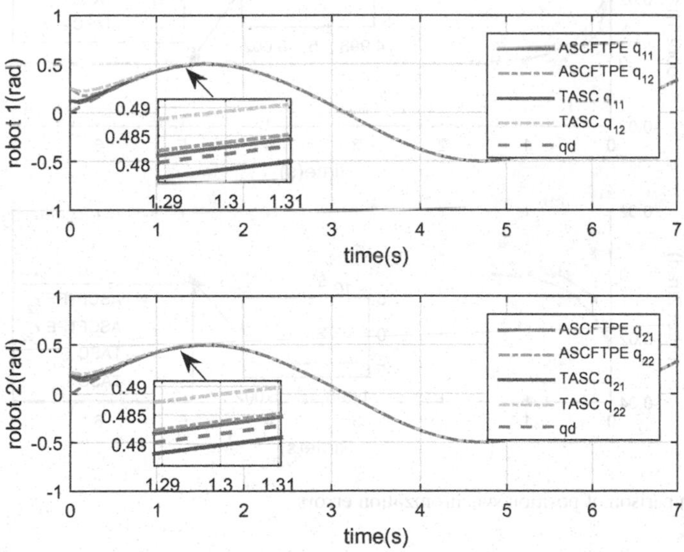

Fig. 1 Comparison of position tracking trajectory

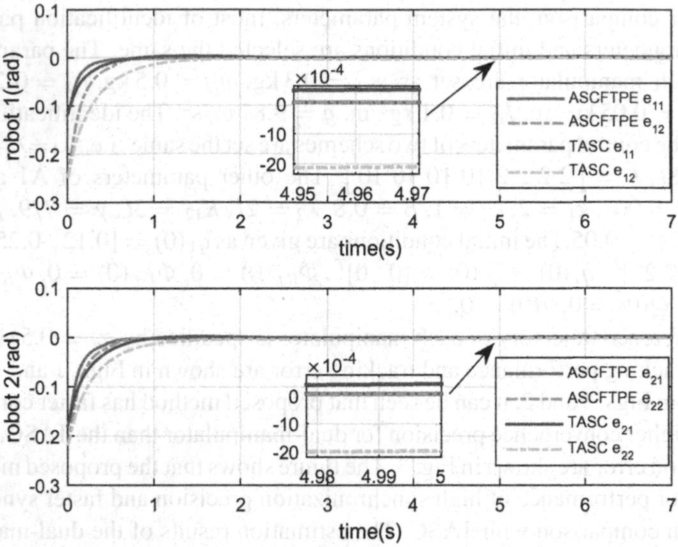

Fig. 2 Comparison of position tracking errors

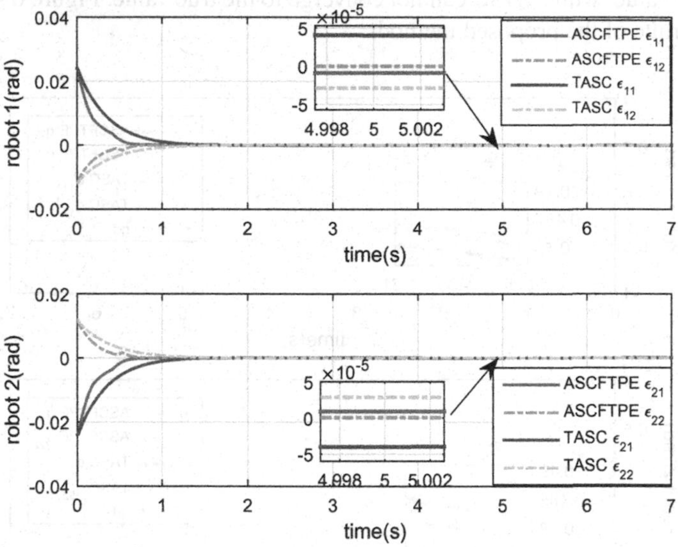

Fig. 3 Comparison of position synchronization errors

From Figs. 1, 2, 3, 4 and 5, it can be clearly demonstrated that the proposed method can achieve the superior performance with faster and higher precision position tracking and synchronization performance. Moreover, the proposed method has the higher estimation accuracy.

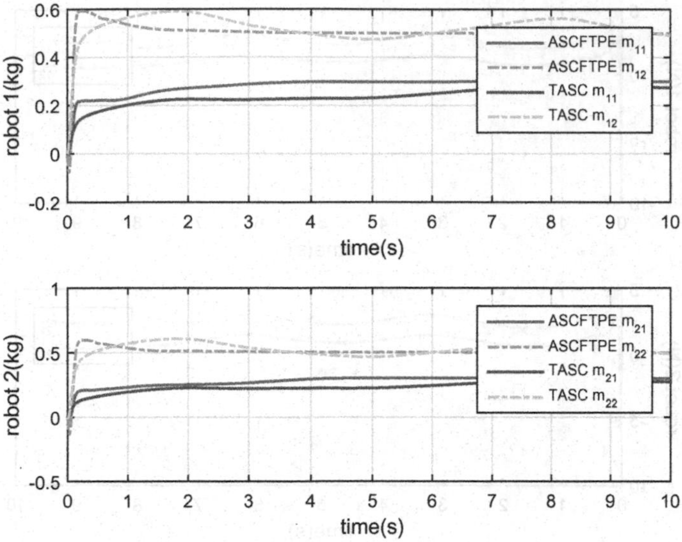

Fig. 4 Comparison of parameters estimation

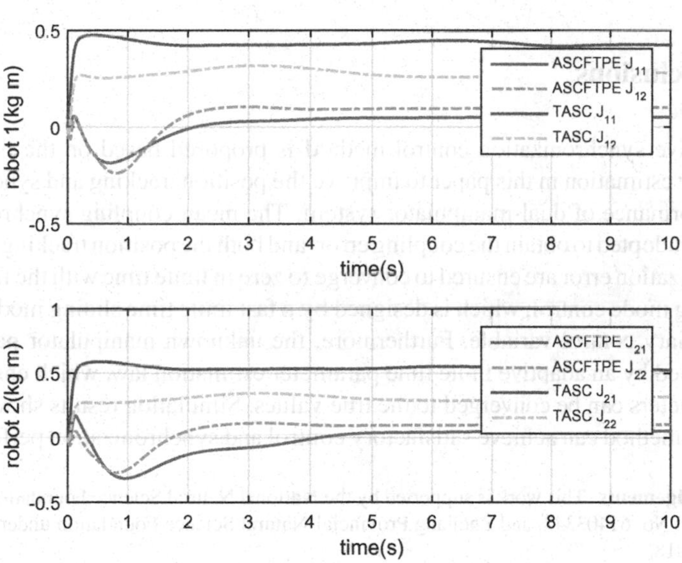

Fig. 5 Comparison of parameters estimation

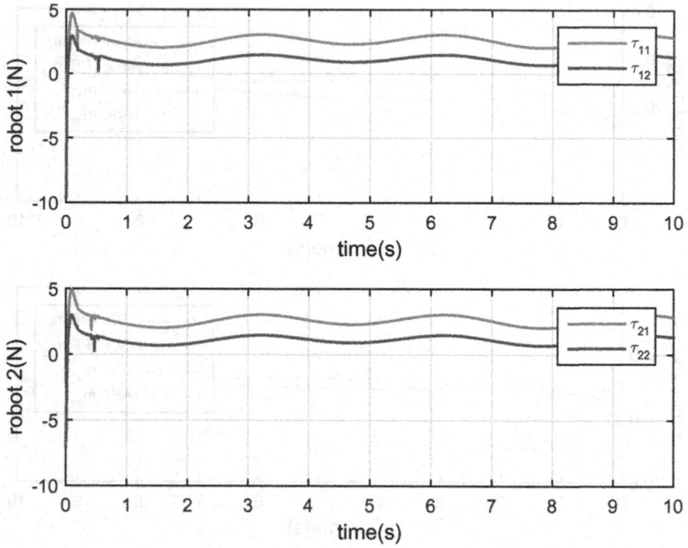

Fig. 6 Control inputs

7 Conclusions

An adaptive synchronization control method is proposed based on the finite time parameter estimation in this paper to improve the position tracking and synchronization performance of dual-manipulator system. The mean-coupling synchronization scheme is adopted to obtain the coupling error, and both the position tracking error and synchronization error are ensured to converge to zero in finite time with the fast terminal sliding mode control, which is designed by a fast finite time sliding mode surface and auxiliary control variable. Furthermore, the unknown manipulator parameters are obtained by an adaptive finite time parameter estimation law, which ensures that the parameters can be converged to the true values. Simulation results show that the proposed method can achieve satisfactory control and synchronization performance.

Acknowledgements This work is supported by the National Natural Science Foundation of China under Grant No. 61403343, and Zhejiang Provincial Natural Science Foundation under Grant No. LY17F030018.

References

1. L.E. Pfeffer, The design and control of a two-armed, cooperating, flexible-drivetrain robot system (1993)
2. P. Zhang, Y.C. Li, Simulations and trajectory tracking of two manipulators manipulating a flexible payload, in *2008 IEEE Conference on Robotics, Automation and Mechatronics*, IEEE, pp. 72–77 (2008)

3. D. Sun, Y.H. Liu, Position and force tracking of a two-manipulator system manipulating a flexible beam payload, in *Proceedings IEEE International Conference on Robotics and Automation, 2001* IEEE, pp. 3483–3488 (2001)
4. A.K. Bondhus, K.Y. Pettersen, H. Nijmeijer, Master-slave synchronization of robot manipulators, in *Proceedings IFAC Symposium on Nonlinear Control Systems*, pp. 91–596 (2004)
5. G. Zhao, D. Zhao. Robust consensus control for multiple robotic manipulators, in *Control Conference*. IEEE, pp. 2229–2233 (2014)
6. D. Sun, J.K. Mills, Adaptive synchronized control for coordination of multirobot assembly tasks. IEEE Trans. Rob. Autom. **18**(4), 498–510 (2002)
7. L. Li, L. Sun, S. Zhang, Mean deviation coupling synchronous control for multiple motors via second-order adaptive sliding mode control. Isa Trans **62**, 222–235 (2016)
8. A. Rodriguez-Angeles, H. Nijmeijer, Coordination of two robot manipulators based on position measurements only. Int. J. Control **74**(13), 1311–1323 (2001)
9. D. Zhao, C. Li, Q. Zhu, Low-pass-filter-based position synchronization sliding mode control for multiple robotic manipulator systems. Proc. Inst. Mech. Eng. Part I: J. Syst. Control Eng. **225**(8), 1136–1148 (2011)
10. D. Zhao, S. Li, F. Gao, Robust adaptive terminal sliding mode-based synchronised position control for multiple motion axes systems. IET Control Theory & Appl. **3**(1), 136–150 (2009)
11. V. Adetola, M. Guay, Finite-time parameter estimation in adaptive control of nonlinear systems. IEEE Trans. Autom. Control **53**(3), 807–811 (2008)
12. C.Y. Su, T.P. Leung, A sliding mode controller with bound estimation for robot manipulators. IEEE Trans. Rob. Autom. **11**(1), 165–166 (1993)
13. J.J.E. Slotine, W. Li, Composite adaptive control of robot manipulators. Automatica **25**(4), 509–519 (1989)
14. J. Na, M.N. Mahyuddin, G. Herrmann, Robust adaptive finite-time parameter estimation and control for robotic systems. Int. J. Rob. Nonlinear Control **25**(16), 3045–3071 (2015)
15. S. Sastry, M. Bodson, *Adaptive Control: Stability, Convergence, and Robustness* (Prentice Hall, New Jersey, 1989)
16. A.M. Zou, K.D. Kumar, Z.G. Hou, X. Liu, Finite-time attitude tracking control for spacecraft using terminal sliding mode and Chebyshev neural network. IEEE Trans. Syst. Man Cybern B Cybern. **41**(4), 950–963 (2011)
17. S. Yu, X. Yu, B. Shirinzadeh, Continuous finite-time control for robotic manipulators with terminal sliding mode. Automatica **41**(11), 1957–1964 (2005)

Link Selection in Radio Tomographic Imaging with Backprojection Transformation

Jiaju Tan, Xuemei Guo and Guoli Wang

Abstract Multi-path interference in Radio Tomographic Imaging(RTI), often brings unpredictable degeneration to the reconstructed image and degrades the accuracy of Device-Free Localization(DFL). By analyzing the reconstruction process of RTI, this paper certifies that the shadow fading can be transformed as a linear combination of the contribution of RF links. This transformation named backprojection indicates that the selection of informative RF links is helpful to resist the multi-path noise. Then a method based on Bayesian Compressive Sensing(BCS) and backprojection is proposed to figure out the contributive RF links and reconstruct the image. Besides, by transforming the reconstruction issue of high-dimensional image into the analysis problem of low-dimensional measured data, the proposed method also decreases the time complexity of BCS without reducing the accuracy. The experimental results show the effectiveness and practicability of the method in RTI and DFL.

Keywords Radio tomographic imaging · Backprojection · Bayesian compressive sensing · Multi-path interference

J. Tan
Institute of Robotics and Automatic Information System,
Nankai University, Tianjin, China

J. Tan
Tianjin Key Laboratory of Intelligent Robotics, Tianjin, China

X. Guo · G. Wang
School of Data and Computer Science, Sun Yat-sen University, Guangzhou, China

X. Guo · G. Wang (✉)
Key Laboratory of Machine Intelligence and Advanced Computing,
Ministry of Education, Guangzhou, China
e-mail: isswgl@mail.sysu.edu.cn

Y. Jia et al. (eds.), *Proceedings of 2018 Chinese Intelligent Systems Conference*, Lecture Notes in Electrical Engineering 529,
https://doi.org/10.1007/978-981-13-2291-4_48

1 Introduction

Narrow-band Radio Tomographic Imaging (RTI) is a novel wireless technology of computational imaging to perceive the surrounding environment [1]. As the Radio Frequency (RF) waves can penetrate through opaque obstructions like walls and keep stable from illumination influence, RTI can recognize the position variation of target in the ambient environment without requiring target to wear or carry any electronic devices [2]. And RTI protects the target privacy as it only captures the location information of target so that it is becoming a helpful intelligent technology in many Device-Free Localization(DFL) scenarios [3], like the roadside surveillance [4] and the assisted healthcare for the elderly [5].

However, unpredictable multi-path interference exists in the sensing network and leads to great uncertainty in RTI [6]. And because of the bandwidth limits of the RF signal, the received RSS data includes not only the target-induced shadow fading but also the multi-path fading [7]. The multi-path fading causes degradation of the reconstruction image and becomes a tough challenge [8]. To obtain a better estimate of the shadow fading in the multi-path environment, bayesian compressive sensing (BCS) [9, 10] has been applied to RTI by treating the target in the RF network as sparse [11]. And by adding apriori information of the noise distribution, BCS can obtain better reconstruction performance of RTI [12]. While the computation load to inference the high-dimensional image is high as the accuracy has positive correlation with the resolution of RTI [13]. Thus a method to analyze the low-dimensional RSS directly and then indirectly reconstruct the high-dimensional image is practical in actual RTI application.

This papers proposes a method of RF-link selection to efficiently resist the multi-path interference and improve the performance of RTI. This proposed method certificates that the RTI reconstruction issue can be transformed into a support vector regression (SVR) problem [14]. This transformation is named the back-projection imaging because it indicates that the shading image is a weighted combination of the selected informative RF links. It explains the feasibility of the RF-link selection in some researches base on fade level [15, 16], mutual information criterion [17] or energy criterion [18]. Besides, the target-affected RF links are only a small portion of total RF links then can be also regarded as sparse. Then the bayesian compressive sensing (BCS) can be applied to estimate these critical RF links. Therefore a method based on the BCS and backprojection to inference the target-affected RF links and reconstruct the image is proposed. It can not only recognize the contributive RF links but also infer their contribution ratio to the target-induced shadow fading. In addition, the proposed method can eliminate the irrelevant RF links to resist the redundant RSS measurement which induces interference in the imaging. Furthermore, as the proposed method only analyzes the RSS data in the low-dimensional space rather than the total image in high-dimensional space in most of RTI reconstruction methods, the time complexity of the proposed method can be greatly reduced.

The paper is organized as follows. In Sect. 2, a brief description of RTI is given. In Sect. 3, a detailed principle of the backprojection transformation of RTI is formulated, then a method of link selection based on BCS is described. In Sect. 4, the RTI experiments are deployed to validate the effectiveness of the proposed method.

2 Problem Statement

In general, considering a 2D area of which L sensors which are deployed along the perimeter, then an equipotent sensing network of $M = L(L-1)/2$ links is constructed. And this monitored region can be divided into N virtual pixels then the vector $\boldsymbol{x} \in \mathbb{R}^N$ represents the shadow fading of this area. The RSS of each link forms the measurement vector $\boldsymbol{y} \in \mathbb{R}^M$. It is computed as $\boldsymbol{y} = \boldsymbol{R}_0 - \boldsymbol{R}_1$, where $\boldsymbol{R}_1$ is the received RSS at RF nodes when the target enters the sensing network and $\boldsymbol{R}_0$ is the baseline RSS when the monitor area is empty of targets. Actually, $\boldsymbol{y}$ is the weighted sum of the shadow fading at pixels which are across by the link so that the measuring model of RSS can be represented as

$$y_i = \sum\nolimits_{j=1}^{N} \Psi_{i,j} x_j + e_i, \tag{1}$$

where $\boldsymbol{e} \in \mathbb{R}^M$ is the measuring noise and $\Psi = [\psi_1, \ldots, \psi_M]^T \in \mathbb{R}^{M \times N}$ is the projection matrix where $\psi_i^T \in \mathbb{R}^N$ is the projection weight of link i to fading image $\boldsymbol{x}$. The element $\Psi_{i,j}$ represents the impact of the shadow fading at pixel j to the RSS of link i. And the simplest projection matrix Ψ can be spatially analyzed as an ellipse model,

$$\Psi_{i,j} = \frac{1}{\sqrt{d_i}} = \begin{cases} 1, & \text{if } d_{ij}^{\text{tran}} + d_{ij}^{\text{rec}} < d_i + \lambda \\ 0, & \text{otherwise,} \end{cases} \tag{2}$$

where d_i is the distance between the transmitter and receiver of link i, d_{ij}^{tran} and d_{ij}^{rec} are the distance from the centroid of pixel j to the transmitter and receiver of link i. And $\lambda = 0.01$ is a threshold parameter to control the size of ellipse model. Thus RTI is to estimate the image $\boldsymbol{x}$ from the obtained RSS $\boldsymbol{y}$ and locate the target at the centroid of pixel where the maximal fading locates in $\boldsymbol{x}$.

3 Methods

3.1 Backprojection Transformation

The measurement process of link i in (1) can be reformulated as

$$y_i = \sum\nolimits_{j=1}^{N} \Psi_{i,j} x_j + e_i = \psi_i^T \boldsymbol{x} + e_i \tag{3}$$

meaning that the inverse problem of solving $\boldsymbol{x}$ from Eq. (1) can be regarded as a regression problem, where $\{(\psi_i^T, y_i)\}_{i=1}^M$ can be treated as the training set. While the fading image $\boldsymbol{x}$ is sparse as the target only covers a small region in the sensing network. Then this regression problem can be transformed as a convex optimization problem named support vector regression(SVR) [14] to estimate the endogenous variable $\boldsymbol{x}$ by defining the lost function as the sum of the sparsity of $\boldsymbol{x}$ and the fitting error, as

$$\min_{\boldsymbol{x}} \frac{1}{2}\|\boldsymbol{x}\|_2^2 + C\sum\nolimits_{i=1}^M \ell_\epsilon(y_i - \psi_i^T \boldsymbol{x}), \tag{4}$$

where ϵ is the maximal fitting error of all the samples in the set $\{(\psi_i^T, y_i)\}_{i=1}^M$. And ℓ_ϵ is the ϵ non-sensitive loss function,

$$\ell_\epsilon(\delta) = \begin{cases} 0, & \text{if } |\delta| \le \epsilon, \\ |\delta| - \epsilon, & \text{otherwise}, \end{cases} \tag{5}$$

and controls that only the outliers with the fitting cost exceeding ϵ may be penalized. In order to make the model more feasible to tolerate the outliers for generalization, the slack variables are introduced into the (4) for soft margin as,

$$\begin{aligned} &\min_{\boldsymbol{x}} \frac{1}{2}\|\boldsymbol{x}\|_2^2 + C\sum\nolimits_{i=1}^M (\eta_i + \eta_i^*) \\ &\text{s.t.} \begin{cases} y_i - \psi_i^T \boldsymbol{x} \le \epsilon + \eta_i \\ \psi_i^T \boldsymbol{x} - y_i \le \epsilon + \eta_i^* \\ \eta_i \quad\quad\;\; \ge 0, \\ \eta_i^* \quad\quad\;\; \ge 0, \end{cases} \end{aligned} \tag{6}$$

where $C > 0$ is a regularization constant to balance the regression performance and the tolerance to outliers. Equation (6) can be solved by the Lagrange multipliers, as

$$\begin{aligned} \mathcal{L}(\boldsymbol{x}, \eta_i, \eta_i^*, \zeta_i, \zeta_i^*, \tau_i, \tau_i^*) = & \frac{1}{2}\|\boldsymbol{x}\|_2^2 + C\sum\nolimits_{i=1}^M (\eta_i + \eta_i^*) - \sum\nolimits_{i=1}^M (\tau_i\eta_i + \tau_i^*\eta_i^*) \\ & + \sum\nolimits_{i=1}^M \zeta_i(y_i - \psi_i^T \boldsymbol{x} - \epsilon - \eta_i) \\ & + \sum\nolimits_{i=1}^M \zeta_i^*(\psi_i^T \boldsymbol{x} - y_i - \epsilon - \eta_i^*), \end{aligned} \tag{7}$$

where $\mathcal{L}$ is the Lagrange function, and $\zeta_i, \zeta_i^*, \tau_i, \tau_i^*$ are the according Lagrange Multipliers of (6) with positive constraint. The optimal solution of $\boldsymbol{x}$ can be estimated by setting the partial derivatives of $\mathcal{L}$ with respect to $\boldsymbol{x}$ to $\boldsymbol{0}$, as

$$\frac{\partial \mathcal{L}}{\partial \boldsymbol{x}} = \boldsymbol{x} - \sum\nolimits_{i=1}^M (\zeta_i - \zeta_i^*)\psi_i = \boldsymbol{0}. \tag{8}$$

Then the optimal $\boldsymbol{x}$ can be obtained as

$$\hat{x} = \sum\nolimits_{i=1}^{M} (\zeta_i - \zeta_i^*)\psi_i \triangleq \sum\nolimits_{i=1}^{M} z_i\psi_i = \Psi^T z, \tag{9}$$

where $z = [z_1, \ldots, z_M]^T$.

Besides, to ensure the optimality of x in (9), the Karush-Kuhn-Tucker(KKT) condition and the dual problem of (7) should be analyzed. By setting the partial derivatives of $\mathcal{L}$ with respect to η_i, η_i^* to 0 as,

$$\frac{\partial \mathcal{L}}{\partial \eta_i} = C - \zeta_i - \tau_i = 0, \quad \frac{\partial \mathcal{L}}{\partial \eta_i^*} = C - \zeta_i^* - \tau_i^* = 0, \tag{10}$$

the dual problem of (7) can be obtained by substituting (8) and (10) into (7),

$$\begin{aligned} \max_{\zeta_i, \zeta_i^*} & \sum\nolimits_{i=1}^{M} y_i(\zeta_i - \zeta_i^*) - \epsilon(\zeta_i + \zeta_i^*) \\ & - \frac{1}{2} \sum\nolimits_{i=1}^{M} \sum\nolimits_{j=1}^{M} (\zeta_i - \zeta_i^*)(\zeta_j - \zeta_j^*)\psi_i^T \psi_j \\ \text{s.t.} & \sum\nolimits_{i=1}^{M} (\zeta_i - \zeta_i^*) = 0, 0 \le \zeta_i, \zeta_i^* \le C \end{aligned} \tag{11}$$

Then the according KKT condition is

$$\begin{cases} \zeta_i(y_i - \psi_i^T x - \epsilon - \eta_i) = 0 \\ \zeta_i^*(\psi_i^T x - y_i - \epsilon - \eta_i^*) = 0 \\ \zeta_i \zeta_i^* = 0, \tau_i \tau_i^* = 0, \\ (C - \zeta_i)\tau_i = 0, (C - \zeta_i^*)\tau_i^* = 0. \end{cases} \tag{12}$$

The expression (9) shows that the shadow fading x can be regarded as a linear combination of the detectivity of each link with the weight z, indicating that z is the contribution ratio of the links to the shadow fading x. Thus z can reflect that which link is target-affected or keeps unaffected. Then the informative links which is crucial to reconstruct x can be found out by estimating z. And by substituting (9) into (1), the measurement model can be transformed into

$$y = \Psi x + e = \Psi \Psi^T z + e \triangleq \Omega z + e, \tag{13}$$

Then when z in (13) has been estimated, the x can be obtained by the backprojection computation (9).

Besides, the value of z in (9) depends on ζ and ζ^*. The KKT condition (12) limits that $\zeta_i \neq 0$ and $\zeta_i^* \neq 0$ can not be both satisfied because $\zeta_i \zeta_i^* = 0$. And $\zeta_i \neq 0$ if and only if $y_i - \psi_i^T x - \epsilon - \eta_i = 0$, $\zeta_i^* \neq 0$ if and only if $\psi_i^T x - y_i - \epsilon - \eta_i^* = 0$. It indicates that only when the sample (ψ_i^T, y_i) is called the support vector by being outside the ϵ margin, the according ζ_i and ζ_i^* can be non-zero so that $z_i = \zeta_i - \zeta_i^* \neq 0$. But only a small part of all samples satisfy that this condition so that the solution z can be regarded as sparse.

3.2 Link Selection Based on BCS

In order to obtain a better estimate of the sparse weight z in (13), BCS [9] can be used to obtain the sparse solution.

The measuring noise e_i of link i in (13) is assumed as the Gaussian white noise with the same variance σ as $\boldsymbol{e} \sim \mathcal{N}(\boldsymbol{0}, \sigma I)$, thus the measurement $\boldsymbol{y}$ also approximately follows a Gaussian distribution $\boldsymbol{y} \sim \mathcal{N}(\Psi z, \sigma I)$.

And each z_i of link i is regarded as independent and follows the Gaussian distribution with a precision α_i as $z_i \sim \mathcal{N}(0, \alpha_i^{-1})$. Then $z \sim \mathcal{N}(\boldsymbol{0}, \boldsymbol{\alpha}^{-1} I)$ by setting $\boldsymbol{\alpha} = [\alpha_1^{-1}, \ldots, \alpha_n^{-1}]^T$ and $P(z|\boldsymbol{\alpha}) = \prod_{i=1}^{M} P(z_i|\alpha_i) = \prod_{i=1}^{M} \mathcal{N}(z_i|0, \alpha_i^{-1})$.

To estimate z and the hyperparameters $\boldsymbol{\alpha}$ and β, by applying the Bayesian law and type-II maximum likelihood method [10], z can be estimated as

$$\hat{z} = \hat{\beta}\hat{\Sigma}\Omega^T \boldsymbol{y}. \tag{14}$$

where $\hat{\Sigma} = [\hat{\beta}\Psi^T\Psi + \hat{A}]^{-1}$ and $\hat{A} = \mathrm{diag}(\hat{\boldsymbol{\alpha}})$. And $\hat{\boldsymbol{\alpha}}$ and $\hat{\beta}$ can be inferenced as

$$\hat{\alpha}_i = \frac{\hat{\gamma}_i}{\hat{z}_i^2}, \hat{\beta} = \frac{\|\boldsymbol{y} - \Omega\hat{z}\|_2^2}{M - \sum_i \hat{\gamma}_i}, \tag{15}$$

where $\hat{\gamma}_i = 1 - \alpha\hat{\Sigma}_{ii}$ and $\hat{\Sigma}_{ii}$ is the i-th diagonal element of Σ.

Totally, when the measured RSS $\boldsymbol{y}$ is obtained, the $\hat{z}$ in (13) can be estimated by (14) through updating the $\hat{\boldsymbol{\alpha}}$ and $\hat{\beta}$ by (15) iteratively. Then the fading image $\boldsymbol{x}$ can be obtained from the updated $\hat{z}$ by the backprojection process (9).

4 Experiments and Results

Two different experimental scenes are deployed to evaluate the proposed method, as shown in Fig. 1. The RF nodes along the boundary are MICAz devices made by Crossbow, which communicated with a 2.4 GHz IEEE 802.15.4 standard. The period for token transmission is about 120 ms. Before the experiment, the average RSS of 5 min in the empty monitor area is measured as the baseline RSS $\boldsymbol{R}_0$. And during the localization experiment, to each position, the RSS is the average value of 5 measurement and each measurement lasts 30 s. And the size of the virtual pixels in the sensing area is 0.1 m × 0.1 m.

- Indoor Experiment: A cluttered indoor area 6 m × 4 m is deployed as Fig. 1a and b shown. The number of pixels and links is $N = 2400$ and $M = 190$. The experiment includes 36 single target and 8 multiple targets.
- Outdoor Experiment: A outdoor square area of 6 m×6 m is tested as Fig. 1c and Fig. 1d shown. The number of pixels and links is $N = 3600$ and $M = 276$. This experiment includes 36 single target and 19 multiple targets.

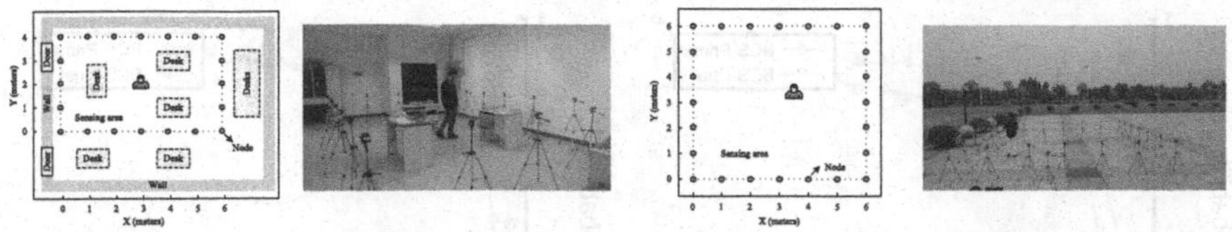

(a) Indoor Layout (b) Indoor Scenario (c) Outdoor Layout (d) Outdoor Scenario

Fig. 1 Experimental deployment

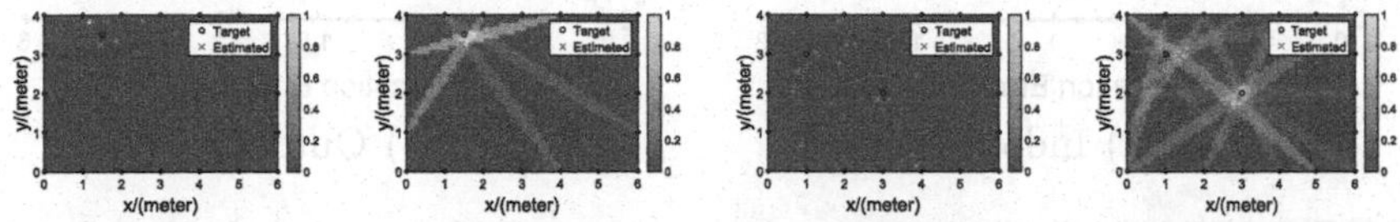

(a) Indoor Single Target BCS-P (b) Indoor Single Target BCS-B (c) Indoor Multiple Target BCS-P (d) Indoor Multiple Target BCS-B

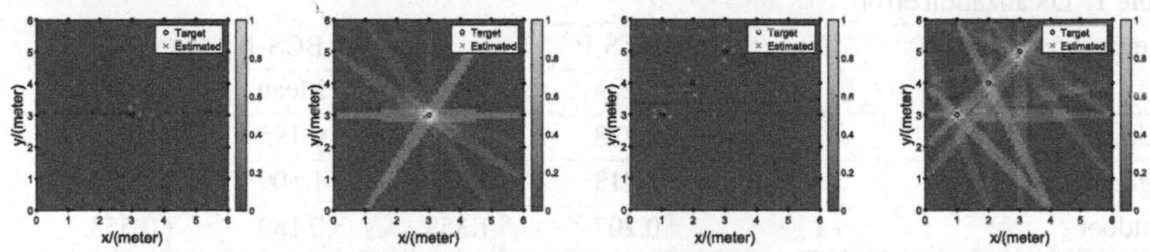

(e) Outdoor Single Target BCS-P (f) Outdoor Single Target BCS-B (g) Outdoor Multiple Target BCS-P (h) Outdoor Multiple Target BCS-B

Fig. 2 Reconstruction image

Four reconstruction examples by prime BCS(BCS-P) [9] and the proposed Backprojection BCS(BCS-B) are shown in Fig. 2. From these reconstruction images, the informative RF links which are truly affected by target appearance can be figured out by the proposed BCS-B, as the Fig. 2b, d, f and h shown. Then the shadow image can be obtained by the linear combination of these chosen RF links. Besides, the artifact by BCS-P in Fig. 2a, c, e and g can be also reduced by the proposed BCS-B as it effectively eliminates the uninformative RF links. Then the quality of RTI is improved by resisting the multi-path interference and the position of target is accurately estimated.

The error is calculated as the 2-order OMAT distance [19], which is the root mean square error of the best assignment of estimation to the true positions, as

$$d_O(\boldsymbol{P}, \hat{\boldsymbol{P}}) = \left(\frac{1}{T} \min_{\pi \in \boldsymbol{\Pi}} \sum_{t=1}^{T} d(\boldsymbol{p}_t, \hat{\boldsymbol{p}}_{\pi(t)})^2\right)^{\frac{1}{2}}, \tag{16}$$

where T is the number of target, $\boldsymbol{P} = \{\boldsymbol{p}_1, \ldots, \boldsymbol{p}_T\}$ and $\hat{\boldsymbol{P}} = \{\hat{\boldsymbol{p}}_1, \ldots, \hat{\boldsymbol{p}}_T\}$ are the true and estimated position of target, $\boldsymbol{\Pi}$ is the set of all possible permutations of $\{1, \ldots, T\}$. And the error comparison of 44 indoor and 55 outdoor measurement is

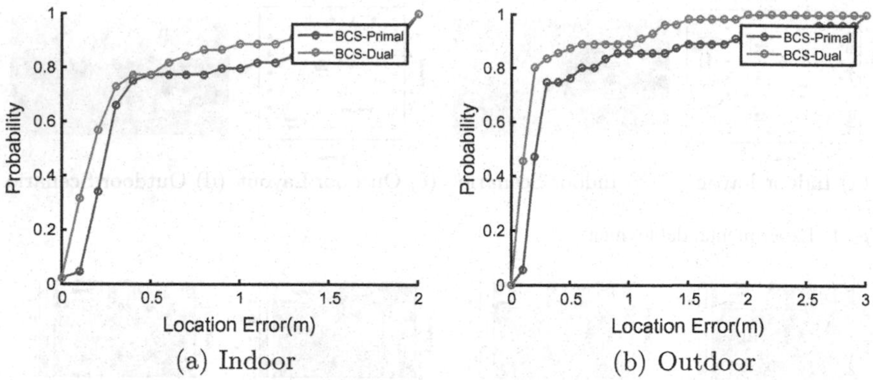

Fig. 3 Error cumulative distribution

Table 1 Localization error

Scenario (error/m)	Target	BCS-P		BCS-B	
		Mean	Max	Mean	Max
Indoor	1	0.228	1.321	0.195	0.454
Figure 1a, b	2–3	1.713	3.359	1.509	3.090
Outdoor	1	0.207	1.358	0.189	0.553
Figure 1c, d	2–5	1.205	3.420	0.627	2.186

Table 2 Localization time

Scenario (time/s)	Target	BCS-P	BCS-B
		Mean	Mean
Indoor	1	6.032	0.504
Figure 1a, b	2–3	4.926	0.679
Outdoor	1	9.611	1.584
Figure 1c, d	2–5	11.419	1.874

shown in Fig. 3 and Table 1. The lower error show that the BCS-B refines the primal BCS-P by eliminating the uninformative links and multi-path noise.

Moreover, the consuming time of two methods are reported in Table 2. It shows that the complexity of the proposed BCS-B is greatly decreased from that of BCS-P by an order of magnitude. These results indicate that compared to directly estimating the high-dimensional image, the combination of selecting the low-dimensional RF links by BCS and then obtaining the high-dimensional image only by linear combination, can realize fast reconstruction of RTI without decreasing the accuracy. Thus the backprojection property makes it possible to transform the computation of the high-dimensional shadow-image reconstruction into the low-dimensional informative-link recognition.

5 Conclusions

The paper demonstrates that the shadow fading in RTI can be formulated as a linear weighted combination of the contribution of RF links. This backprojection transformation indicates that the selection of informative RF links is useful to resist the multipath interference and improve the performance of RTI. Then a method combined of RF-link selection by BCS and RTI reconstruction by backprojection is developed. The proposed method modifies the quality of primal BCS and decrease the computation load by transforming the direct reconstruction issue of high-dimensional shading image into the accurate recognition problem of low-dimensional RF links. And the experiments show that the proposed method can not only accurately recognize the contributive RF links and reconstruct the shadow fading but also greatly decrease the time complexity of direct RTI.

Acknowledgements This work was supported by the National Natural Science Foundation of P.R. China under Grant Nos. 61772574 and 61375080, the Key Program of Natural Science Foundation of Guangdong, China under Grant No. 2015A030311049.

References

1. J. Wilson, N. Patwari, Radio tomographic imaging with wireless networks. IEEE Trans. Mobile Comput. **9**(5), 621–632 (2010)
2. C. Alippi, M. Bocca, G. Boracchi, N. Patwari, M. Roveri, RTI goes wild: radio Ttomographic imaging for outdoor people detection and localization. IEEE Trans. Mobile Comput. **15**, 2585–2598
3. N. Patwari, J. Wilson, RF sensor networks for device-free localization: measurements, models, and algorithms. Proc. IEEE **98**(11), 1961–1973 (2010)
4. C. Anderson, R. Martin, T. Walker, R. Thomas, Radio tomography for roadside surveillance. IEEE J. Selected top. Sig. Proc. **8**(1), 66–79 (2014)
5. M. Bocca, O. Kaltiokallio, N. Patwari, *Radio Tomographic Imaging for Ambient Assisted Living*, vol. 362 (Springer, Berlin Heidelberg, 2012), pp. 108–130
6. Y. Guo, K. Huang, N. Jiang, X. Guo, Y. Li, G. Wang, An exponential-rayleigh model for RSS-based device-free localization and tracking. IEEE Trans. Mobile Comput. **14**(3), 484–494 (2015)
7. Z. Yang, K. Huang, X. Guo, G. Wang, A real-time device-free localization system using correlated RSS measurements. Eurasip J. Wireless Commun. Networking, pp. 1–12 (2013)
8. Y. Luo, K. Huang, X. Guo, G. Wang, A hierarchical RSS model for RF-based device-free localization. Pervasive Mobile Comput. **31**, 124–136 (2016)
9. S. Ji, Y. Xue, L. Carin, Bayesian compressive sensing. IEEE Trans. Signal Proc. **56**(6), 2346–2356 (2008)
10. M. Tipping, A. Smola, Sparse bayesian learning and the relevance vector machine. J. Mach. Learn. Res. **1**(3), 211–244 (2001)
11. K. Huang, Y. Guo, L. Yang, X. Guo, G. Wang, Optimal information based adaptive compressed radio tomographic imaging, in *Proceedings of the 32th Chinese Intelligent Systems Conference*, vol. 12, no. 7, pp. 7438–7444 (2013)
12. K. Huang, Y. Guo, X. Guo, G. Wang, Heterogeneous Bayesian compressive sensing for sparse signal recovery. IET Proc. Iet **8**(9), 1009–1017 (2014)

13. K. Huang, S. Tan, Y. Luo, X. Guo, G. Wang, Enhanced radio tomographic imaging with heterogeneous Bayesian compressive sensing. Pervasive Mobile Comput. **40**(9), 450–463 (2017)
14. A. Smola, B. Scolkopf, A Tutorial on Support Vector Regression. Kluwer Academic Publishers, **14**(3), pp. 199–222 (2004)
15. J. Wilson, N. Patwari, A fade-level skew-laplace signal strength model for device-free localization with wireless networks. IEEE Trans. Mobile Comput. **11**(6), 947–958 (2012)
16. O. Kaltiokallio, M. Bocca, N. Patwari, A fade level-based spatial model for radio tomographic imaging. IEEE Trans. Mobile Comput. **13**(6), 1159–1172 (2014)
17. K. Huang, Y. Luo, X. Guo, G. Wang, Data-efficient radio tomographic imaging with adaptive Bayesian compressive sensing, in *IEEE International Conference on Information and Automation*, pp. 1859–1864 (2015)
18. M. Khaledi, SK. Kasera, N. Patwari, M. Bocca, Energy efficient radio tomographic imaging, In *Eleventh IEEE International Conference on Sensing*, pp. 609–617 (2014)
19. D. Schuhmacher, B.T. Vo, B.N. Vo, A consistent metric for performance evaluation of multi-object filters. IEEE Trans. Signal Process **56**, 3447–3457 (2008)

Indoor Navigation for Quadrotor Using RGB-D Camera

Peng Zhang, Rui Li, Yingjing Shi and Liang He

Abstract In this paper, we present hardware design and software architecture of a navigation system for quadrotor. By getting data from RGB-D camera and processing it on the onboard computer with visual simultaneous localization and mapping (SLAM) algorithm that we proposed, we can obtain real-time pose of the quadrotor and a 3D dense map of the surroundings. At the same time, we use an improved rapidly exploring random tree algorithm (RRT*) to get a safe global path with obstacle avoidance in Octomap (the map is stored in Octotree format), and using the path to perform navigation. The experimental results demonstrate the navigation and 3D SLAM capabilities of the quadrotor in our system.

Keywords Quadrotor · 3D SLAM · RRT* · Navigation

1 Introduction

In recent years, navigation and control of automation vehicles have attracted considerable attention due to the requirement of the engineering applications [1–3]. Especially, the quadrotor has gained enormous commercial potential during the last years because of the Low-cost, small-size and the hovering ability. Among multiple utilities of quadrotors, navigation of quadrotors is a hot research subject. Outdoor navigation mainly relies on the Global Positioning System (GPS), which is not almost

P. Zhang · R. Li (✉) · Y. Shi · L. He
School of Automation Engineering, University of Electronic Science and Technology of China, Chengdu 611731, China
e-mail: lirui@uestc.edu.cn

L. He
Shanghai Aerospace Control Technology Institute, Shanghai 201109, People's Republic of China

L. He
Shanghai Key Laboratory of Aerospace Intelligent Control Technology, Shanghai 201109, People's Republic of China

Y. Jia et al. (eds.), *Proceedings of 2018 Chinese Intelligent Systems Conference*, Lecture Notes in Electrical Engineering 529,
https://doi.org/10.1007/978-981-13-2291-4_49

reliable indoors because of its unacceptable precision and weak reception of the signal. Therefore, indoor navigation relies on two kinds of sensors: laser scanners and cameras. Laser scanners provide range information with high precision and odometry can be derived by scan matching technique. Previously [4] applies 2D laser for tiny quadrotors performing navigation and SLAM indoors. These systems use laser-scan matching algorithms and combine the data with inertial measurement unit (IMU) measurements in an EKF filter to provide sate estimation. Girish C. et al. present their self-contained quadrotors navigation system in [5], which uses scanning laser rangefinder and an extension of EKF based on SLAM algorithms to provide position and heading estimation.

However, laser scanners are not an ideal choice because of their heavy weight and high-power consumption in comparison with cameras. There are also lots of research progress of the camera SLAM. Achtelick et al. [6] control a quadrotor for both indoor and outdoor flight by using monocular vision, in which depth information is recovered by fusing a pressure sensor and an accelerometer through an Extended Kalman Filter. In Yi Lin et al. [7] construct a monocular (VINS) and achieve safe navigation through unknown cluttered environment. Park et al. apply a stereo camera to control a quadrotor and avoid collision in [8].

For special indoor environments, RGB-D camera is becoming popular in the quadrotor community owing to its limited size, low weight and affordable cost. RGB-D cameras, such as Microsoft Kinect and Asus Xtion, are widely adopted because of their ability to provide the visual information of the cameras to create accurate 3D environment maps. Furthermore, there are also multiple related works based on RGB-D cameras. For instance, in [9] the authors present an autonomous and real-time navigation system for flight of quadrotor in unknown indoor settings with an onboard RGB-D camera. In [10] they show a new approach to perform indoor waypoint MAV navigation using an RGB-D system.

In this paper, an autonomous flight and navigation system for a low-cost quadrotor is presented. We get the real-time 3D position of the quadrotor and 3D dense map of the environment though the SLAM algorithm, even without any GPU acceleration. By using the RRT* path planning algorithm the quadrotor flies freely in three-dimensional space. The remainder of the paper is organized as follows: Sect. 2 presents the hardware design and the software architectures of our quadrotor experimental testbed. Section 3 describes the SLAM algorithm for the precise position and the reconstructed 3D map. The safe and collision-free trajectory is produced for autonomous flight of the robot by using RRT* path planning method which is described in Sect. 4. Then, Sect. 5 presents experimental results, showing the availability of the proposed methods. Finally, conclusions are discussed in Sect. 6.

2 System Architecture

2.1 *Hardware Architecture*

The purpose of this paper is to design a compact and constructible aerial robot platform consisting of the following main components: (1) ASUS Xtion Pro Live RGB-D Camera with RGB image, a depth image, a raw IR image (all of resolution 640×480) for 3D SLAM and navigation; (2) Pixhawk as a flying controller with MCU, gyroscope and accelerometer; (3) JFRC U2814 motors and propellers are used to power quadrotor flight, providing approximately 10 min of flight with the 4S Li-Po battery; (4) a radio controller; (5) the onboard mini-computer, which is a dual-core Intel NUC i7-5557u processor with ROS-Kinect operating system whose software runs under Ubuntu 16.04 system. The advantage of NUC is that it is only $25 \times 105 \times 108$ mm in size and weighs 180 g. In addition, the frame of quadrotor is made of carbon fiber tubes and aluminum components which makes it light and durable. The platform is shown in Fig. 1.

2.2 *Software Architecture*

The software architecture of our system is shown in Fig. 2. The entire quadrotor experimental platform mainly consists of three parts: a flight control unit, an onboard computer and a ground station. (1) The flight control unit includes a powerful micro controller unit (MCU), which makes it possible to run high-level tasks such as attitude control and position control of quadrotor. (2) The onboard computer is usually used for image feature extraction and matching, inter frame pose estimation, 3D SLAM as well as path planning. (3) The ground station is applied for visualization, communication and navigation of quadrotor. In this paper, ROS (Robot Operating System) is

Fig. 1 The RGB-D-based quadrotor platform. (1) ASUS Xtion Pro Live RGB-D Camera (2) Pixhawk (3) JFRC U2814 motors (4) radio controller (5) Intel NUC mini computer

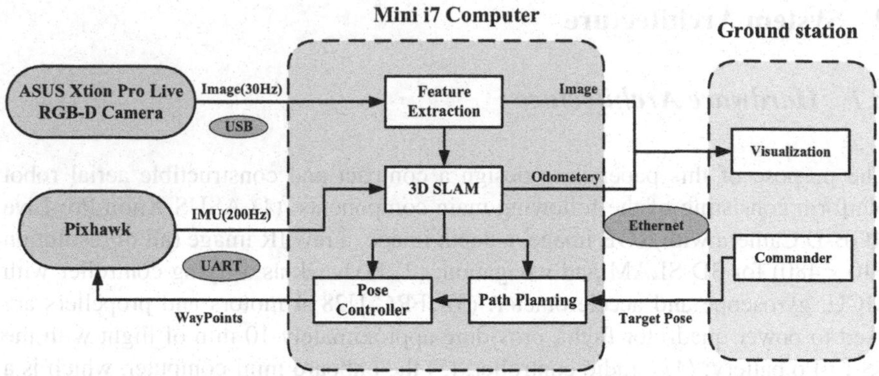

Fig. 2 Software architecture

viewed as a middleware in software framework, and it allows communication among different systems, platforms, even different architectures. Thus we run ROS-based several nodes (an implementation of a program in ROS) to implement the 3D SLAM function. Furthermore, the ground station and onboard computer exchange information via 5.8 GHz WIFI, by the way, onboard computer sends the planning flight path to the flight control unit for quadrotor position controlling by USB interfaces.

3 Real-Time Visual SLAM

3.1 Feature Extraction and Matching

Due to the large amount of information in image, we usually take some representative points (feature points) of the image instead of the whole one to do image processing. Meanwhile, SIFT, SURF and ORB are some extraction methods that widely used in image feature extraction. We use different methods to extract 1000 feature points simultaneously from a same image, SURF takes 271.3 ms, SIFT takes 5228.7 ms, and ORB only costs 15.3 ms. Because ORB utilizes improved FAST (Features from Accelerated Segment Test) Algorithms for feature points extraction and BRIEF (Binary Robust Independent Elementary Features) for feature points description, which greatly improves the extraction speed of features. Thus, ORB algorithm is applied for feature extraction in this paper. We select the two feature points with the smallest Hamming distance as the similarity pairs because ORB uses Hamming space to represent binary string descriptors. After getting the initial matching points, RANSAC (Random Sample Consensus) algorithm is applied to eliminate wrong points. Figure 3 shows the matching result.

Fig. 3 Matching result

3.2 Pose Estimation

After the ORB feature extraction and matching, we can easily get the corresponding relation, such as rotation and translation, between the two key frame image feature points belonging to respective depth image from the RGB-D camera. Then ICP (Iterative Closest Point) can be used to get rotation matrix R and translation vector t. Suppose that P and P' are two piles of 3D matched points. Here,

$$P = \{p_1, \ldots, p_n\}\ P' = \{p'_1, \ldots, p'_n\}\ \forall i, p_i = Rp'_i + t \tag{1}$$

Since the 3D point matching pair is calculated by 2D matched point, the existence of noise and the projection matching relationship cannot be fully balanced. So, we define $e_i = p_i - (Rp'_i + t)$. Our goal is to minimize this bias for N feature points. The error model is described as follows:

$$\min_{R,t} J = \frac{1}{2}\sum_{i=1}^{n} \left\|(p_i - (Rp'_i + t)\right\|_2^2$$
$$\min_{R,t} J = \frac{1}{2}\sum_{i=1}^{n} \left\|(p_i - p - R(p'_i - p')\right\|^2 + \left\|p - Rp' - t\right\|^2 \tag{2}$$

Suppose p is the center of mass of P and p' is the center of mass of P'. Through (2), it is clear that the left term is only related to R, but the right term is related R and t. As R is known, correspondingly, t can be solved from this formula.

The procedure of ICP can be divided into three steps:

Step 1: Calculate the position of the center of mass of two sets of 3D points, then calculate the decentralized coordinates for each point,

$$q'_i = p'_i - p'\ q_i = p_i - p \tag{3}$$

Step 2: Calculate the rotation matrix according to the following optimization problem

$$R = \arg\min_{R} \frac{1}{2}\sum_{i=1}^{n} \|q_i - Rq_i'\|^2 = \frac{1}{2}\sum_{i=1}^{n} q_i^T q_i - q_i'^T R^T R q_i' - 2q_i^T R q_i' \tag{4}$$

Step 3: Calculate t according to R obtained in step 2:

$$t = p - Rp' \tag{5}$$

3.3 Local Map

If only the previous and the next frame are used for matching, the number of matching points are not enough when the reference frame is with weak-quality due to illumination or occlusion. This case will lead to a deviation of pose estimation of the camera and the deviation will accumulate causing a large deviation for pose estimate after long periods. Therefore, we apply a different method where a local map is maintained that describes the feature point information close to the current frame, then the local map is matched with the current frame, and subsequently the pose of the current frame is estimated. To some extent, this method can solve the above problem and avoid the accumulation of deviations caused by inferior quality frames. At the same time, we also have to discard the feature points which are far away from the current frame or outside the view and control the scale of the local map to meet the requirements of the arithmetic speed. Once the local map is built, the back-end seeks to solve a nonlinear-least square problem to derive the optimal configuration of poses and landmarks. We use the open source framework g2o to optimize the constructed pose.

3.4 Loop Closing

We detect the similarity among images with Bag-of-Words (BoW) in this paper. The main idea of the Bow is that the feature points in the image are clustered by K-means clustering method and then a vocabulary tree and the word description vector of the image can be obtained. After that we calculate the acquaintance of the word description vector and use a certain loop detection strategy to complete the loop detection. This method converts the continuously changing features into separated words, which can not only classify similar features and reduce the storage space, but also speed up the search by using reversed-index method to directly determine the scene which contains the feature word. Moreover, closed-loop detection compensates the shortcomings of the visual odometer's long-term drift. Finally, we get a dense map, but it can't be used for navigation. Therefore, we also need to convert the dense map to Octomap.

4 Path Planning

After the previous steps, the maps are ready to be used for autonomous flight indoors. For this purpose, the path planning methods, such as potential field and RRT* algorithms, should be applied.

Although the RRT algorithm has many good characteristics, the overall cost of the path is not considered in the course of search. The randomness of the node expansion causes the planned path to be random, leading to the unsmooth path. Thus, the better trajectory is needed. Due to the shortcomings of RRT algorithm, improvement is proposed to make our algorithm suitable for autonomous navigation of UAV.

Improvement 1: Bidirectional Expansion
Two random trees are defined in the searching space. One starts at the starting point and the other starts at the target point. Two random trees grow the leaf nodes alternately. When a new leaf node is created, a test should be conducted to detect whether it is close enough from the other tree (a certain preset value which is more than the Euclidean distance of the two nodes). If the above conditions are satisfied, two random trees are connected.

Improvement 2: Increasing the Gravitational Component
In order to reduce the path randomness, the idea of gravity in the artificial potential field method can be introduced into the RRT algorithm so as to guide the random tree to grow in the direction of the target, avoiding the generation of local minimum values. The core idea is to introduce the target gravitational function $G(n)$ at each node n in the path. In this function, x-rand and x-goal denote respectively a random node and the end of the path, and p represents the length of the search step. The node nth represents the nth x-new node extended out from the starting point x-init which can be expressed as formula (6):

$$F(n) = R(N) + G(n) \tag{6}$$

Here $F(n)$ represents the growth guidance function from node n to the new node, $R(n)$ represents the random growth function of node n, so $G(n)$ is the target gravity function as formula (7) and generation of new nodes after introducing gravity is formula (8).

$$G = k_p * \left\| x_{\text{goal}} - x_{\text{near}} \right\| \tag{7}$$

$$x_{\text{new}} = x_{\text{near}} + p * \left(\frac{x_{\text{rand}} - x_{\text{near}}}{\| x_{\text{rand}} - x_{\text{near}} \|} \right) + k_{\text{p}} * \frac{x_{\text{goal}} - x_{\text{near}}}{\left\| x_{\text{goal}} - x_{\text{near}} \right\|} \tag{8}$$

Here p is the length of growth step of the random tree, k_p is the gravitational coefficient and x_{goal} is the target position vector of the UAV.

Improvement 3: Smoothing Based On B-spline Curve
First, greedy algorithm is used to filter out invalid path points, and then further smooth processing based on B-spline curves is conducted. Planning a path with a B-spline

curve can meet the requirement to modify the local path without changing the entire path shape.

The *K*-order *B*-spline curve is expressed as:

$$C(u) = \sum_{i=0}^{n} N_{i,k}(u) * P_i \tag{9}$$

Here P_i is the control point. The improved RRT* algorithm based path planning algorithm for UAV is as follows:

Step 1: Initialize two search trees with the current position of the UAV as the planning starting point. One starts from the starting point and the other starts from the target point.
Step 2: Find a tree node of the existing expansion tree, which is closest to the sampling point, through the sample points generated by random sampling. Then the new node is obtained by a certain distance from the gravity function.
Step 3: Determine whether the Euclidean distance between the new node and another random tree is less than the present value. If condition is not satisfied, turning to step 2, otherwise proceed to the next step.
Step 4: Connect two random trees and filter out invalid path points by greedy criterion after obtaining the path point from the starting point to q-goal.
Step 5: Use a cubic B-spline curve to smooth the path to get the final planned path.

5 Experiments and Result

To evaluate the performance of our system, we conduct several experiments in autonomous flight. Our visual SLAM system can achieve real-time performance and the rate of visual odometry is up to 20 Hz by letting the quadrotor fly in a corridor size 2.5×20 m in Fig. 4. (a) is the real corridor image, (b) is the result of 3D reconstruction.

It is easy to get Octomap from dense point cloud. Figure 5 shows the Octomap, the color grid in the figure indicates that it has been occupied and the color of the grid is determined by the height. The resolution of Octomap is 0.1 m.

From Fig. 6, it can be seen that the RRT* algorithm can plan a route to avoid all the obstacles detected in Octomap. The red arrow is the current position of the quadrotor, and the green line is the planned safe routes last, In the experiment of Fig. 7, the quadrotor is in position control mode after sending a sequence of waypoints. Both experiments prove the effectiveness of the state estimation and automation control with a maximum error of 10 cm or less.

(a) (b)

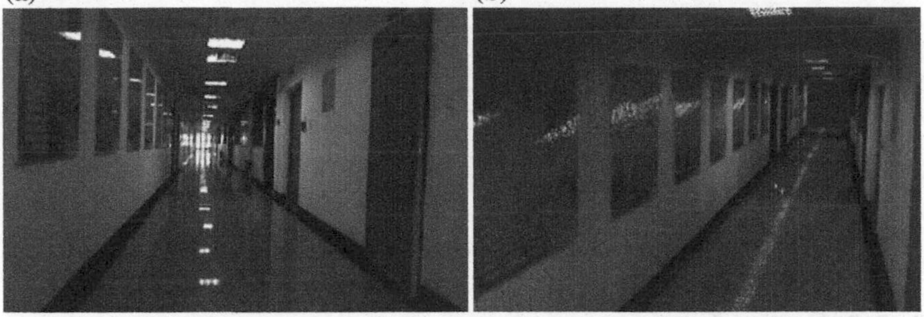

Fig. 4 **a** Real corridor, **b** 3D SLAM result

Fig. 5 Octomap result

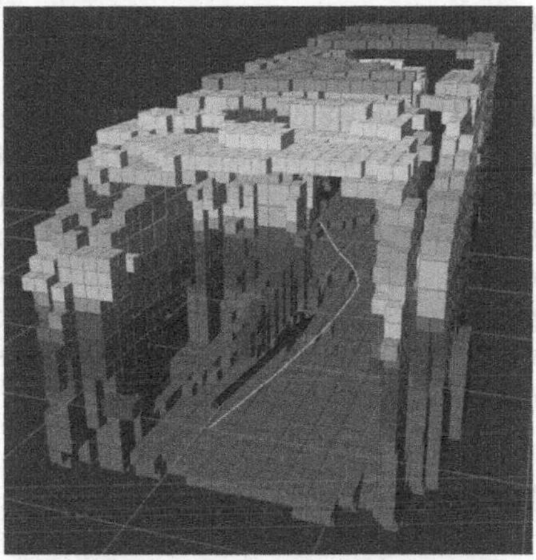

Fig. 6 RRT* path

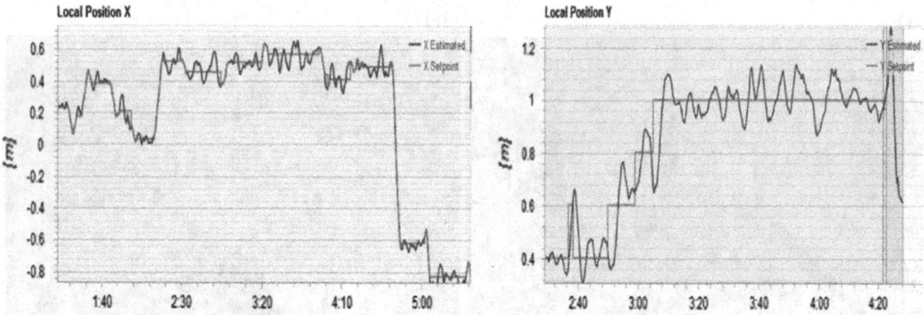

Fig. 7 Position control results

6 Conclusions

In this paper, we first introduced our vision SLAM algorithm based on the RGB-D camera, and then proposed the RRT* path planning algorithm. Finally the proposed methods were implemented on our quadrotor platform. As the experimental results show, our system can perform precise navigation autonomously and successfully map its surroundings in complete unknown environment without any other auxiliary position systems. Moreover, the developed system has the potential to be widely used in complex environment such as disaster area and tunnels with high accuracy of navigation and control ability.

References

1. Y. Jia, Robust control with decoupling performance for steering and traction of 4WS vehicles under velocity-varying motion. IEEE Trans. Control Syst. Technol. 554–569 (2000)
2. A. Katriniok, D. Abel, Adaptive EKF-based vehicle state estimation with online assessment of local observability. IEEE Trans. Control Syst. Technol. 1368–1381 (2016)
3. C. Shen, Y. Shi, B. Buckham, Trajectory tracking control of an autonomous underwater vehicle using lyapunov-based model predictive control. IEEE Trans. Ind Electron. 5796–5805 (2018)
4. A. Bachrach, A. de Winter, Range—robust autonomous navigation in gps-denied environments, in *Proceeding. of the IEEE international Conference on Robotics* and Automation *(ICRA)*, pp. 1096–1097 (2010)
5. G. Chowdhary, D.M. Sobers, Self-contained autonomous indoor flight with ranging sensor navigation. AIAA J. Guid. Control Dyn. 1843–1854 (2012)
6. M. Achtelik, M. Achtelik, S. Weiss, R. Siegwart, Onboard IMU and monocular vision based control for MAVs in unknown in- and outdoor environments, in *IEEE International Conference On Robotics and Automation (ICRA)* (2011)
7. T. Qin, S. Shen, Robust initialization of monocular visual-inertial estimation on aerial robots, in *Proceedings IEEE/RSJ International Conference* on · Intelligent *Robots Systems*, pp. 4225–4232, 2017
8. J. Park, Y. Kim, Stereo vision based collision avoidance of quadrotor UAV. Control, Automation and Systems (ICCAS), 173–178, 2012
9. L. Tang, S. Yang. Toward autonomous navigation using an RGB-D camera for flight in unknown indoor environments, in *Guidance, Navigation and Control Conference IEEE*, pp. 2007–2012 (2015)
10. M.C.P. Santos, M. Sarcinelli-Filho, R. Carelli, Indoor waypoint UAV navigation using a RGB-D system, in *The Workshop on Research IEEE*, pp. 84–91 (2015)

Target-Enclosing for Multi-agent Systems at Same Height

Xueyan Wang, Yingjing Shi, Rui Li and Yimin Bao

Abstract In this paper, we consider the target-enclosing problem of multi-agent systems, where two algorithms are provided to achieve enclosure for static or moving targets at same height. We first establish the center estimator for targets. Then two control protocols are designed. The first protocol is designed with distance-only measurement for static targets. The second protocol is designed with local position information of neighbor targets. Some simulation results are provided to show the effectiveness of the obtained theoretical results.

Keywords Enclosing control · Multi-agent systems
Lasalle's invariance principle · Lyapunov theory · Distributed control

1 Introduction

With the demand of engineering applications, various control methods have been developed for single system in the last decades [1–4]. However, the multi-agent system can play more important role than single system due to the reduced cost, improved efficiency and so on. Thus, the distributed cooperative control problems of multi-agent systems have attracted extensive attention over the past decades, including the following areas: formation control of multi-agent systems [5, 6], distributed optimization [7, 8], distributed estimation control [9] and so on. Target enclosure can be considered as a special problem which combines formation control with consensus problem. At present, the target enclosure theory can be applied to unmanned aerial vehicles, such as reconnaissance missions, search and rescue missions, and residential security.

Target enclosure has been extensively investigated in literature. Deghat and Shames [10] proposes a control approach to localize and circumnavigate a slow

X. Wang · Y. Shi · R. Li (✉) · Y. Bao
School of Automation Engineering, University of Electronic Science
and Technology of China, Chengdu 611731, China
e-mail: lirui@uestc.edu.cn

Y. Jia et al. (eds.), *Proceedings of 2018 Chinese Intelligent Systems Conference*, Lecture Notes in Electrical Engineering 529,
https://doi.org/10.1007/978-981-13-2291-4_50

drift target with bearing measurement. Shames and Dasgupta [11] considers a similar problem to it, where the circumnavigation problem is solved with distance-only measurement. There are also much research results for an agent circumnavigates a group of targets. For example [12] considers the problem of localization and circumnavigation of a group of targets, which are either stationary or moving slowly with unknown speed, by a single agent. Deghat et al. propose an estimator initially for the stationary targets case to localize the targets and the center of mass of them. Guo et al. [13] studies the moving target-enclosing problem for a group of autonomous mobile robots, which can be considered as the demand of achieving a formation surrounding a moving target whose movement is not known as a priori. Another situation that a group of agents enclose a group of targets has also been studied extensively. Marasco et al. [14] solves the problem of creating a dynamic circular formation around a target and proposes a Decentralized Model Predictive Control (DMPC) policy which is used when a single UAV encircles a stationary target, a single UAV encircles a moving target, and a group of UAVs encircle a stationary target. Chen et al. [15] uses a decentralized estimation and control framework to solve the surrounding problem which is about a team of leaders and a team of followers. Shi et al. [16] poses a distributed control algorithm for a group of agents and a group of targets with local information. However, in many cases, the assumption of knowing the position of the targets is not always practical.

In this paper, we propose control algorithms by which a group of agents achieve enclosure for a group of targets at same height. These algorithms can be applied to many multiagent joint actions, such as joint monitoring and united strike. We first propose a targets' center estimator with distance-only measurement which play a critical role for the design of enclosing control algorithm. Then, we present a control law to achieve target-enclosing at same height based on the target estimator. Further, another control law with local position information for moving targets is given. The main novelty in our work is that the proposed algorithms can achieve target-enclosure at same height by multi-agent systems with distance-only measurement.

The rest of the paper is organized as follows: In Sect. 2, the target-enclosing control at same height is formally defined. In Sect. 3, we propose the target-enclosing control of stationary targets with distance-only measurement firstly. Then we provide another algorithm with local position information when targets are dynamic. Simulation results are shown in Sect. 4 to demonstrate the effectiveness of the proposed algorithm. Finally, Sect. 5 summarizes the main conclusions.

2 Problem Statement

Consider n stationary targets with unknown positions $r_i(t) \in R^m (m \in \{2, 3\})$ and n agents with known positions $x_i(t) \in R^m (m \in \{2, 3\})$ at time t, for $i = 1, 2, \ldots, n$. The communication network of agents can be modeled as an undirected connected graph G. $A = [a_{ij}] \in R^{n \times n}$ is the adjacency matrix associated with the graph G.

The measurement $D_i(t) = \|x_i(t) - r_i(t)\|$ means the distance between agent i and its' neighbor target i. The kinetic model of each agent is

$$\begin{cases} \dot{x}_i(t) = v_i(t) \\ \dot{v}_i(t) = u_i(t). \end{cases} \tag{1}$$

Here $V_i(t) \in R^m (m \in \{2, 3\})$ denotes the velocity. $u_i(t) \in R^m (m \in \{2, 3\})$ denotes the control input,$i = 1, 2, \ldots, n$. In the rest of this paper, $\|\cdot\|$ denotes 2-norm.

In this paper, we consider the enclosure problem of multi-agent systems for multiple targets at same height keeping formation formed by the targets. We will propose two algorithms for the target-enclosing problem. In the first situation, it is assumed that all the agents can only obtain the distance information among agents and neighbor targets. In the second situation, all the local information of neighbor targets is known. In order to achieve the enclosure task, we need to perform the following two sub-problems:

(1) Construct center estimator;
(2) Propose an algorithm that make the agent system asymptotically form a formation enclosing all the targets at same height.

Definition 2.1 Denote $\bar{r}(t)$ as the center of targets, and build a virtual point $c(t) = \bar{r}(t) + z_0$, $z_0 = (0, 0, -a)^T$, $a > 0$. We say that the target-enclosing control at same height is achieved, if

$$\lim_{t\to\infty} \{[x_i(t) - c(t)] - k[r_i(t) - c(t)]\} = 0, \tag{2}$$

for $i = 1, 2, \ldots, n$. Here k is a constant and $k > 1$.

Remark 2.1 In this equation, $x_i(t) - c(t)$ denotes the vector from the virtual point to the agent i, and $r_i(t) - c(t)$ denotes the vector from the virtual point to the target i.

3 Main Results

As mentioned earlier, we have two goals to achieve. Section 3.1 describes the control algorithm for stationary targets with distance-only measurement. Section 3.2 describes the control algorithm for moving targets with local position information.

3.1 Target-Enclosing with Distance-Only Measurement

Since each agent only can obtain the distance information from the neighbor target, we first need to estimate the position of the neighbor target, then estimate the targets' center.

The localization algorithm for estimating is given below, which is borrowed from [8]. For $\rho > 0$, let

$$\begin{aligned} \eta_i(t) &= \dot{z}_1(t) = -\rho z_1(t) + \tfrac{1}{2}D_i^2(t); \\ m_i(t) &= \dot{z}_2(t) = -\rho z_2(t) + \tfrac{1}{2}x_i^T(t)x_i(t); \\ f_i(t) &= \dot{z}_3(t) = -\rho z_3(t) + x_i(t). \end{aligned} \tag{3}$$

Here, $\eta_i(t)$, $m_i(t)$, $f_i(t)$ are respectively the state variable filtered versions of $\frac{1}{2}D_i^2(t)$, $\frac{1}{2}x_i^T(t)x_i(t)$, $x_i(t)$. With the measurement $D_i(t)$ and the knowledge of agents' own position,$\eta_i(t)$, $m_i(t)$, $f_i(t)$ can be obtained without explicit differentation.

The localization algorithm can be defined as follows:

$$\dot{\hat{r}}_i(t) = -\gamma f_i(t)(\eta_i(t) - m_i(t) + f_i^T(t)\hat{r}_i(t)). \tag{4}$$

Here $\hat{r}_i(t)$ denotes the estimator of targets position.

The next thing to do is to design the targets' center estimator $\hat{x}_i(t)$. The targets' center estimator should satisfy three conditions:

(1) $\sum_{i=1}^{n} \hat{x}_i(t) = \sum_{i=1}^{n} \hat{r}_i(t)$;
(2) $\lim_{t\to\infty} \left\|\hat{x}_j(t) - \hat{x}_i(t)\right\| = 0$;
(3) there is $T > 0$, such that $\left\|\hat{x}_j(t) - \hat{x}_i(t)\right\| = 0$ when $t > T$.

Conditions (2) and (3) ensure that all $\hat{x}_i(t)$ will converge to the same position. Then, the target center estimating algorithm used in this paper is defined as:

$$\hat{x}_i(t) = \phi_i(t) + \hat{r}_i(t) \tag{5}$$

for all $i = 1, 2, \ldots, n$. Here

$$\dot{\phi}_i(t) = \begin{cases} \alpha \sum_{j=1}^{n} a_{ij} \frac{\hat{x}_j(t) - \hat{x}_i(t)}{\|\hat{x}_j(t) - \hat{x}_i(t)\|} & , \hat{x}_j(t) \neq \hat{x}_i(t) \\ 0 & , \hat{x}_j(t) = \hat{x}_i(t) \end{cases}. \tag{6}$$

In Eq. (6), $\phi_i(t)$ denotes the dynamic estimating state and $\hat{x}_i(t)$ denotes the ith agent's estimator, $\phi_i(0) = 0$, $\alpha > 0$.

In the following, we will show that the proposed targets' center estimator satisfies conditions (1)–(3). That is to say the estimator can represent the target center.

Theorem 3.1. *Consider $\eta_i(t)$, $m_i(t)$, $f_i(t)$ defined in (3), with $\rho > 0$. For all $t \geq 0$, we have*

$$-\beta \leq \dot{\hat{r}}_i(t) \leq \beta, \tag{7}$$

if and only if there exists $\alpha_1 > 0, \alpha_2 > 0, T > 0$, such that for all $t \geq 0$

$$\alpha_1 I \leq \int_t^{t+T} f_i(\tau) f_i^T(\tau) d\tau \leq \alpha_2 I. \tag{8}$$

Proof As shown in [17], we denote p as the derivate operator. We can easily get the relationship $\dot{\eta}_i(t) + \eta_i(t) = \frac{d}{dt}\{\frac{1}{2}D_i^2(t)\}$. Then as $\rho > 0$, in operator notation, $\eta_i(t) \approx \frac{p}{p+\rho}\{\frac{1}{2}D_i^2(t)\}$. Similarly, $m_i(t) \approx \frac{p}{p+\rho}\{\frac{1}{2}x_i^T(t)x_i(t)\}$ and $f_i(t) \approx \frac{p}{p+\rho}x_i(t)$. We get

$$\eta_i(t) \approx m_i(t) - f_i^T(t) r_i(t). \tag{9}$$

Define $\tilde{r}_i(t) = \hat{r}_i(t) - r_i(t)$. Because of (4) and (9), we obtain

$$\dot{\tilde{r}}_i(t) = -\gamma f_i(t) f_i^T(t) \tilde{r}_i(t). \tag{10}$$

It's well known that the linear time varying system with $\gamma > 0$,

$$\dot{z}(t) = -\gamma f_i(t) f_i^T(t) z(t) \tag{11}$$

is exponentially asymptotically stable iff (8) holds. Notice,

$$\ddot{\tilde{r}}_i(t) = -2\alpha \dot{\tilde{r}}_i(t) - \gamma \dot{x}_i(t) f_i^T(t) \tilde{r}_i(t) + \| f_i(t) \|^2 f_i(t) f_i^T(t) \tilde{r}_i(t) - \gamma f(t) \dot{x}_i^T(t) \dot{\tilde{r}}_i(t). \tag{12}$$

One can observe that $\dot{\tilde{r}}_i(t)$ is differential, then $\dot{\tilde{r}}_i(t)$ is continuous. Thus $\dot{\tilde{r}}_i(t)$ is bounded, and the result follows. ■

Theorem 3.2. *If the graph G is undirected and connected, and $\alpha > (n-1)\beta$, then $\hat{x}_i(t)$ given in (5) satisfies the conditions (1)–(3)*

Proof From (6), we can see $\sum_{i=1}^{n} \phi_i(t) \equiv 0$. It's obvious that $\sum_{i=1}^{n} \hat{x}_i(t) = \sum_{i=1}^{n} \hat{r}_i(t)$. In order to prove condition (2), define a Lyapunov function:

$$V(t) = \frac{1}{2} \sum_{i=1}^{n} [\hat{x}_i(t) - \frac{1}{n} \sum_{k=1}^{n} \hat{x}_k(t)]^2. \tag{13}$$

Calculating $\dot{V}(t)$

$$\dot{V}(t) = \sum_{i=1}^{n} \{[\hat{x}_i(t) - \frac{1}{n}\sum_{k=1}^{n} \hat{x}_k(t)] \times [\dot{\phi}_i(t) + \dot{\hat{r}}_i(t) - \frac{1}{n}\sum_{k=1}^{n} \dot{\hat{x}}_k(t)]\}.$$

According to Eq. (6), we have

$$\sum_{i=1}^{n} \{[\hat{x}_i(t) - \frac{1}{n}\sum_{i=1}^{n} \hat{x}_k(t)]\dot{\phi}_i(t)\} = -\frac{\alpha}{2}\sum_{i=1}^{n}\sum_{j=1}^{n} [a_{ij}|\hat{x}_i(t) - \hat{x}_j(t)|] \tag{14}$$

From Theorem 3.1, we know $-\beta \le \dot{\hat{r}}_i(t) \le \beta$. Since the graph G is undirected and connected, we have

$$\sum_{i=1}^{n} \{[\hat{x}_i(t) - \frac{1}{n}\sum_{k=1}^{n} \hat{x}_k(t)]\dot{\hat{r}}_i(t)\}$$

$$\begin{aligned} &\le \sum_{i=1}^{n} \left|[\hat{x}_i(t) - \tfrac{1}{n}\sum_{k=1}^{n} \hat{x}_k(t)]\dot{\hat{r}}_i(t)\right| \le \beta \sum_{i=1}^{n} \left|\hat{x}_i(t) - \tfrac{1}{n}\sum_{k=1}^{n} \hat{x}_k(t)\right| \\ &\le \tfrac{\beta}{n}\sum_{i=1}^{n}\sum_{k=1}^{n} \left|\hat{x}_i(t) - \hat{x}_k(t)\right| \le (n-1)\beta \max_{i,k=1,2,\ldots,n} \left|\hat{x}_i(t) - \hat{x}_k(t)\right| \\ &\le \tfrac{(n-1)\beta}{2}\sum_{i=1}^{n}\sum_{k=1}^{n} [a_{ij}\left|\hat{x}_i(t) - \hat{x}_k(t)\right|]. \end{aligned} \tag{15}$$

Moreover, we know

$$\sum_{i=1}^{n} \{[\hat{x}_i(t) - \frac{1}{n}\sum_{k=1}^{n} \hat{x}_k(t)][-\frac{1}{n}\sum_{k=1}^{n} \dot{\hat{x}}_k(t)]\} = -\frac{1}{n}\sum_{k=1}^{n} \dot{\hat{x}}_k(t)\sum_{i=1}^{n} [\hat{x}_i(t) - \frac{1}{n}\sum_{k=1}^{n} \hat{x}_k(t)] = 0. \tag{16}$$

According to Eqs. (14)–(16), we get

$$\dot{V}(t) \le \frac{(n-1)\beta - \alpha}{2}\sum_{i=1}^{n}\sum_{j=1}^{n} [a_{ij}|\hat{x}_i(t) - \hat{x}_j(t)|]. \tag{17}$$

Since $\alpha > (n-1)\beta$, we can obtain $\dot{V}(t) \le 0$. Based on Lasalle's invariance principle, condition (2) holds.

To demonstrate condition (3), we have

$$\tfrac{1}{n}\sum_{j=1}^{n} \left|\hat{x}_i(t) - \hat{x}_k(t)\right| \le \left|\hat{x}_{i_0}(t) - \hat{x}_{j_0}(t)\right| \le \textstyle\sum_{i=1}^{n}\sum_{j=1}^{n} \left|\hat{x}_i(t) - \hat{x}_j(t)\right|$$

$$\left|\hat{x}_{i_0}(t) - \hat{x}_{j_0}(t)\right| = \max\left|\hat{x}_i(t) - \hat{x}_j(t)\right|.$$

We obtain

$$\frac{\dot{V}(t)}{\sqrt{V(t)}} \leq \frac{-\frac{\alpha}{2}\sum_{i=1}^{n}\sum_{j=1}^{n}[a_{ij}|\hat{x}_i(t)-\hat{x}_j(t)|]}{\frac{\sqrt{n}}{\sqrt{2}}|\hat{x}_{i_0}(t)-\hat{x}_{j_0}(t)|}.$$

Since $\alpha > (n-1)\beta$, we have

$$\sqrt{V(t)} \leq \sqrt{V(0)} - \frac{\beta(n-1)-\alpha}{\sqrt{2n}} \times t. \tag{18}$$

Equation (18) shows that there is $T > 0$, such that $\|\hat{x}_j(t) - \hat{x}_i(t)\| = 0$ when $t > T$.

Theorem 3.3. *Denote $c_i(t) = \hat{x}_i(t) + z_0$, $z_0 = (0, 0, -a)^T$, $a > 0$. In keeping with our outlined strategy, the controller for the multi-agent systems can be written as:*

$$\begin{aligned} u_i(t) = \dot{\hat{v}}_i(t) &- \{[x_i(t) - c_i(t)] - \tfrac{h+a}{a}[\hat{r}_i(t) - c_i(t)]\} \\ &- \{[v_i(t) - \hat{v}_i(t)] - \tfrac{h+a}{a}[\dot{\hat{r}}_i(t) - \hat{v}_i(t)]\}. \end{aligned} \tag{19}$$

Here $\hat{v}_i(t) = \dot{\phi}_i(t) + \dot{\hat{r}}_i(t)$, h is the objective height. Then the multi-agent systems can achieve the target-enclosing problem with protocol (19).

Proof Define

$$\begin{cases} \xi_i(t) = [x_i(t) - c_i(t)] - \frac{h+a}{a}[\hat{r}_i(t) - c_i(t)]; \\ \lambda_i(t) = [v_i(t) - \hat{v}_i(t)] - \frac{h+a}{a}[\dot{\hat{r}}_i(t) - \hat{v}_i(t)]. \end{cases}$$

Thus (19) can be written as

$$u_i(t) = \dot{\hat{v}}_i(t) - \xi_i(t) - \lambda_i(t).$$

Define the Lyapunov function

$$V_1(t) = \frac{1}{2}\sum_{i=1}^{n}[\xi_i(t)]^2 + \frac{1}{2}\sum_{i=1}^{n}[\xi_i(t) + \lambda_i(t)]^2. \tag{20}$$

Calculating $\dot{V}_1(t)$,

$$\begin{aligned} \dot{V}_1(t) &= \sum_{i=1}^{n}[\xi_i(t)\lambda_i(t)] - \sum_{i=1}^{n}\{[\xi_i(t) + \lambda_i(t)] \times [\xi_i(t) + \frac{h+a}{a}\ddot{\hat{r}}_i(t) - \frac{h+a}{a}\dot{\hat{v}}_i(t)]\} \\ &= -\sum_{i=1}^{n}[\xi_i(t)]^2 - \frac{h+a}{a}\sum_{i=1}^{n}\{[\xi_i(t) + \lambda_i(t)][\ddot{\hat{r}}_i(t) - \dot{\hat{v}}_i(t)]\}. \end{aligned} \tag{21}$$

According to Theorem 3.2, there exists $T > 0$, such that $\|\hat{x}_i(t) - \hat{x}_j(t)\| = 0$, thus we have $\|c_i(t) - c_j(t)\| = 0$. From Eq. (6), we have $\dot{\phi}_i(t) = 0$ at the time when $\|\hat{x}_i(t) - \hat{x}_j(t)\| = 0$. Hence, $\ddot{r}_i(t) - \dot{\hat{v}}_i(t) = \dot{\phi}_i(t) = 0$ when $t \to \infty$, for all $i = 1, 2, \ldots, n$. It is followed that $\xi_i(t) = 0$. Therefore, Eq. (2) holds. This control law realizes the desired effect. ■

3.2 *Target-Enclosing with Local Position Information*

When the local position information of the neighbor targets can be obtained, we can achieve the enclosing control for moving targets at same height by replacing the estimated value of targets in Sect. 3.1 with the actual position value.

The estimator can be defined as follows:

$$\hat{x}_i(t) = \phi_i(t) + r_i(t).$$

Suppose that all the targets are dynamic and

$$-\beta \leq |D^+ r_i(t)| = \left| \lim \sup_{h \to 0^+} \frac{1}{h}[r_i(t+h) - r_i(t)] \right| \leq \beta.$$

According to Theorem 3.2, the estimator satisfies conditions (1)–(3).

Define the controller

$$\begin{aligned} u_i(t) = \dot{\hat{v}}_i(t) - \{[x_i(t) - c_i(t)] - \frac{h+a}{a}[r_i(t) - c_i(t)]\} \\ - \{[v_i(t) - \hat{v}_i(t)] - \frac{h+a}{a}[s_i(t) - \hat{v}_i(t)]\}. \end{aligned} \quad (22)$$

Here $s_i(t) = \dot{r}_i(t)$. Then the multi-agent systems can achieve enclosure control for moving targets with controller (22). The proof of validity is similar to that of Theorem 3.3, and is omitted hence.

4 Simulations

In this section, we testify the proposed algorithm with numerical simulations.

Consider a networked multi-agent system with $n = 4$ members. Each agent and target belongs to R^3. The adjacency matrix A is given below.

Fig. 1 Graph G

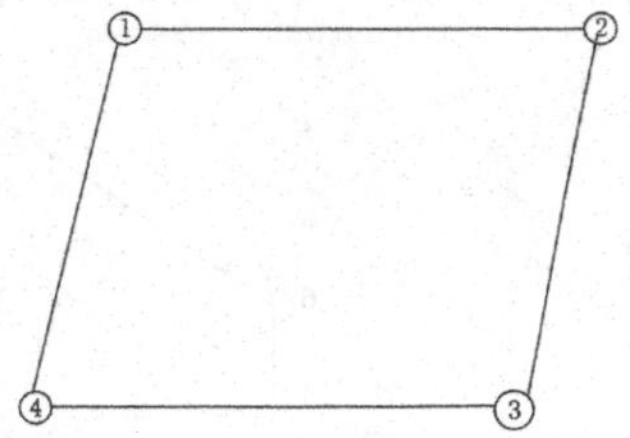

$$A = \left[a_{ij}\right] = \begin{bmatrix} 0\,1\,0\,1 \\ 1\,0\,1\,0 \\ 0\,1\,0\,1 \\ 1\,0\,1\,0 \end{bmatrix}$$

Topology structure described by graph G is shown in Fig. 1.

4.1 *Stationary Targets Enclosing with Distance-Only Measurement*

The control protocol is given by Eq. (19). We take $\alpha = \frac{3}{5}$, $a = 2$, $h = 4$. The initial positions $x_{ai}(0)$ and the initial velocities $v_{ai}(0)$ of the agents are chosen as

$$x_{a1}(0) = \begin{bmatrix} -2 \\ 4 \\ 2.2 \end{bmatrix}, x_{a2}(0) = \begin{bmatrix} -1.5 \\ 1 \\ 1.3 \end{bmatrix}, x_{a3}(0) = \begin{bmatrix} -1.5 \\ -2 \\ 5 \end{bmatrix}, x_{a4}(0) = \begin{bmatrix} -2 \\ -3 \\ -1 \end{bmatrix}.$$

$$v_{a1} = \begin{bmatrix} 2 \\ 2 \\ 3 \end{bmatrix}, v_{a2} = \begin{bmatrix} 3 \\ 3 \\ 0 \end{bmatrix}, v_{a3} = \begin{bmatrix} 1 \\ 1 \\ 2 \end{bmatrix}, v_{a4} = \begin{bmatrix} 0.5 \\ 5 \\ 1.5 \end{bmatrix}.$$

The simulation results are shown in Fig. 2. In Fig. 2, the black spots show the position of targets. The pink X marks denote the initial positions of agents. The pink spots denote the terminal position of targets. The dotted line denotes $\hat{x}_i(t)$, which is the estimated value of target center. The solid lines show the trajectory of the agents. From these simulation results, we can find that with the estimator algorithms (5) and (6) and enclosing protocol (19), the agents finally achieve a formation at same height around the targets with the same geometry.

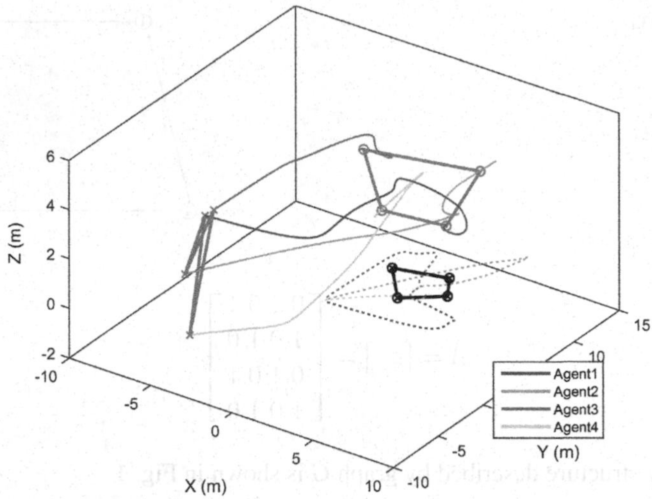

Fig. 2 Simulation results of protocol (19)

4.2 Moving Targets Enclosing with Local Information

The control protocol is given in Eq. (22). We take $\alpha=\frac{3}{5}$, $a=2$, $h=4$. The initial positions $x_{ai}(0)$ and the initial velocities $v_{ai}(0)$ of the agents are chosen as

$$x_{a1}(0)=\begin{bmatrix}-2\\4\\2.2\end{bmatrix},\ x_{a2}(0)=\begin{bmatrix}-1.5\\1\\1.3\end{bmatrix},\ x_{a3}(0)=\begin{bmatrix}-1.5\\-2\\5\end{bmatrix},\ x_{a4}(0)=\begin{bmatrix}-2\\-3\\-1\end{bmatrix}.$$

$$v_{a1}(0)=\begin{bmatrix}2\\2\\3\end{bmatrix},\ v_{a2}(0)=\begin{bmatrix}3\\3\\0\end{bmatrix},\ v_{a3}(0)=\begin{bmatrix}1\\1\\2\end{bmatrix},\ v_{a4}(0)=\begin{bmatrix}0.5\\5\\1.5\end{bmatrix}.$$

The initial positions $x_{ti}(0)$ and the initial velocities $v_{ti}(0)$ of the targets are chosen as

$$x_{t1}(0)=\begin{bmatrix}5\\5\\0\end{bmatrix},\ x_{t2}(0)=\begin{bmatrix}4\\7\\0\end{bmatrix},\ x_{t3}(0)=\begin{bmatrix}1\\6\\0\end{bmatrix},\ x_{t4}(0)=\begin{bmatrix}3\\3\\0\end{bmatrix}.$$

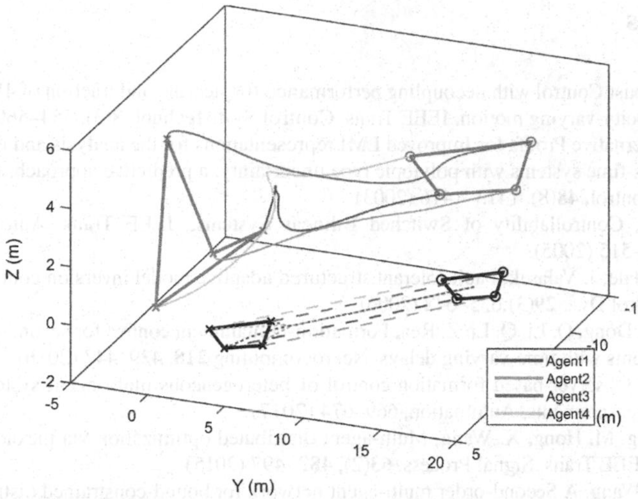

Fig. 3 Simulation results of protocol (22)

$$v_{t1}(0) = \begin{bmatrix} -1 \\ 1 \\ 0 \end{bmatrix}, v_{t2}(0) = \begin{bmatrix} -1 \\ 1 \\ 0 \end{bmatrix}, v_{t3}(0) = \begin{bmatrix} -1 \\ 1 \\ 0 \end{bmatrix}, v_{t4}(0) = \begin{bmatrix} -1 \\ 1 \\ 0 \end{bmatrix}.$$

The simulation results are shown in Fig. 3. In Fig. 3, the black spots show the positions of targets. The pink X mark denotes the initial positions of agents. The pink spots denote the terminal positions of targets. The dotted line denotes, which is the estimated value of target center. Another 4 broken lines are the trajectory of the targets. The solid lines show the trajectory of the agents. It can be seen from Fig. 3 that multi-agent systems achieve a formation at same height around the targets with the same geometry.

5 Conclusions

In this paper, we propose an estimator based control method for target-enclosing problem which can be used in two situations. The first situation is that only the distance information from neighbor targets can be known. The second situation is the local position information of neighbor targets can be obtained. It's worth noticing that the targets of the first situation only can be static, and the targets of the second situation can be dynamic. The simulations are provided to demonstrate the effectiveness of the proposed control methods.

References

1. Y. Jia, Robust Control with decoupling performance for steering and traction of 4WS vehicles under velocity-varying motion. IEEE Trans. Control Syst. Technol. **8**(3), 554–569 (2000)
2. Y. Jia, Alternative Proofs for Improved LMI representations for the analysis and the design of continuous-time systems with polytopic type uncertainty: a predictive approach. IEEE Trans. Autom. Control, **48**(8), 1413–1416 (2003)
3. D. Cheng, Controllability of Switched Bilinear Systems., IEEE Trans. Autom. Control, **50**(4):511–515 (2005)
4. M.D. Tandale, J. Valasek, Fault-tolerant structured adaptive model inversion control. J. Guidance, Control Dyn. **29**(3):635–642 (2006)
5. L. Han, X. Dong, Q. Li, Q. Li, Z. Ren, Formation-containment control for second-order multi-agent systems with time-varying delays. Neurocomputing **218**, 439–447 (2016)
6. K. Zhang, Observer-based formation control of heterogeneous multi-agent systems without velocity measurements, Automation, 669–674 (2017)
7. T.H. Chang, M. Hong, X. Wang, Multi-agent distributed optimization via inexact consensus ADMM. IEEE Trans. Signal Process. **63**(2), 482–497 (2015)
8. Q. Liu, J. Wang, A Second-order multi-agent network for bound-constrained distributed optimization. IEEE Trans. on Autom. Control **60**(12), 3310–3315 (2015)
9. R. Olfati-Saber, P. Jalalkamali, Coupled distributed estimation and control for mobile sensor networks. IEEE Trans. on Autom. Control **57**(10), 2609–2614 (2012)
10. M. Deghat, I. Shames, B.D.O. Anderson, C. Yu, Localization and circumnavigation of a slowly moving target using bearing measurements. IEEE Trans. Autom. Control, **59**(8), 2182–2188 (2014)
11. I. Shames, S. Dasgupta, B¸ Fidan, B.D.O. Anderson, Circumnavigation using distance measurements under slow drift. IEEE Trans. Autom. Control, **57**(4), 889–902 (2012)
12. M. Deghat, L. Xia, B.D.O. Anderson, Y.G. Hong, Multi-target localization and circumnavigation by a single agent using bearing measurements. Int. J. Robust Nonlinear Control **25**, 2362–2374 (2015)
13. J. Guo, G. Yan, Z. Lin, Local control strategy for moving-target-enclosing under dynamically changing network topology. Syst. Control Lett., **59**(10):654–661 (2010)
14. A.J. Marasco, S.N. Givigi, C.A. Rabbath, Model predictive control for the dynamic encirclement of a target. Am. Control Conference **50**(6), 2004–2009 (2012)
15. F. Chen, W. Ren, Y. Cao, Surrounding control in cooperative agent networks. Syst. Control Lett., **59**(11), 704–712 (2010)
16. Y.J. Shi, R. Li, T.T. Wei, Target-enclosing control for second-order multi-agent systems. Int. J. Systems Science **46**(12), 2279–2286 (2015)
17. S.H. Dandach, B. Fidan, S. Dasgupta, B.D.O. Anderson, A continuous time linear adaptive source localization algorithm robust to persistent drift. Syst. Control Lett., **58**(1):7–16 (2009)

Sliding Window Based Monocular SLAM Using Nonlinear Optimization

Jingyun Duo, Long Zhao and Jianing Mao

Abstract In this paper, a sliding window based real-time monocular SLAM is proposed. In our method, latest multiple states are estimated in a sliding window by using nonlinear optimization, and the other states are marginalized out from the sliding window. Meanwhile, we convert measurements corresponding to marginalized states into prior, so as to bound the computational complexity and improve the accuracy of state estimation without loop detection. Two experiments are designed to evaluate the accuracy and effectiveness of our method. The results show that the performance of our method is much better than the monocular ORB-SLAM, and our method can effectively estimate the sparse point cloud of map structure and camera motion with unknown scale.

Keywords Visual SLAM · Monocular · Sliding window · Nonlinear optimization

1 Introduction

Visual SLAM (Simultaneous Localization and Mapping) is a computational problem of constructing or updating a map of an unknown environment by using visual sensors while simultaneously keeping track of an agent's location within it [1]. In recent years, with the development of computer vision technology and the increase ability of the computer information processing, visual SLAM has become a hot research topic, and shows great application value in many fields, such as self-driving cars, unmanned aerial vehicles, newly emerging domestic robots, virtual reality and augmented reality.

J. Duo · L. Zhao (✉) · J. Mao
School of Automation Science and Electrical Engineering,
Beihang University, Beijing 100191, China
e-mail: flylong@buaa.edu.cn

L. Zhao
Science and Technology on Aircraft Control Laboratory,
Beihang University, Beijing 100191, China

Y. Jia et al. (eds.), *Proceedings of 2018 Chinese Intelligent Systems Conference*, Lecture Notes in Electrical Engineering 529,
https://doi.org/10.1007/978-981-13-2291-4_51

Visual SLAM needs to solve both localization and mapping tasks, and location information and map information are highly correlated. Therefore, visual SLAM is a very challenging technology, which appears to be a chicken-and-egg problem. The visual sensors used in the research of SLAM mainly include monocular camera, binocular camera and depth camera. The binocular camera can calculate the pixel depth, but it suffers from large amounts of calculation and often needs to be accelerated by the GPU (Graphic Processing Unit). The depth camera is seriously affected by the illumination condition, which is not suitable for the outdoor scene. Although the monocular camera cannot recover the scale information, monocular SLAM still has received a great many attentions due to its low cost, small volume, low energy consumption and easy hardware setup [2].

Monocular SLAM can be divided into two categories: feature based methods and direct methods, and the above two kinds of methods both have advantages and disadvantages in practical application. Feature based methods match features between successive frames, recover both camera motion and map structure using multiple view geometry, then refine the state variables using bundle adjustment. Owing to the use of robust feature matching, feature based methods are more robust and accurate than direct methods, and they can work normally even under large inter-frame movement and fierce illumination change. However, feature based methods only consider the features in images, ignoring the rest of pixels, so these methods are unable to build dense or semi-dense map. The representative feature based monocular SLAM are probably PTAM [3] and ORB-SLAM [4]. PTAM is the first system to segment the tracking and mapping step in parallel threads, so as to realize real-time states estimation. However, PTAM is only designed for applications in small scenes, and multiple modifications are needed if used in large-scale outdoor environments [5]. ORB-SLAM is built on the main ideas of PTAM and uses ORB features for all the tasks, which makes ORB-SLAM more efficient, simple and reliable. The direct methods only consider the intensity value of pixels, and the camera motion and map structure are recovered by minimizing the photometric error. Direct methods are more efficient, because the time for features matching can be saved. However, direct methods have poor robustness, because these methods are based on the assumption of intensity value invariant. The most representative direct monocular SLAM are probably LSD-SLAM [6]. LSD-SLAM directly operates on image intensities both for tracking and mapping, and the semi-dense map constructed by LSD-SLAM gets better visualization than the sparse map built by featured based SLAM.

This paper aims at designing a high precision and strong robustness monocular SLAM system, so we select feature based methods as the research direction. ORB-SLAM is considered as one of the most perfect and easy-to-use real-time monocular SLAM systems. However, the tracking module of monocular ORB-SLAM only considers the current camera states, ignoring the constraints of past states to current, therefore the precision of states estimation is limited, and the state variables need to be refined by the module of loop closing. If the actual motion trajectory does not include the loop path or the loop detection failed, the results of states estimation will drift slowly with the system running time. In this paper, we improve the monocular ORB-SLAM, and propose a sliding window based monocular SLAM. In our method,

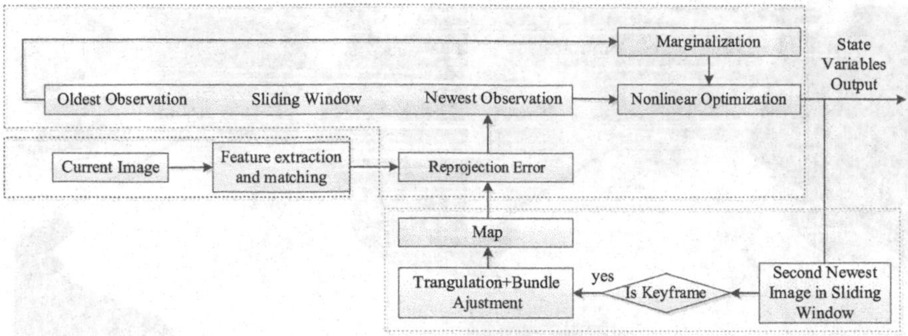

Fig. 1 The flow diagram of our proposed monocular SLAM

latest multiple states are estimated in a sliding window by using nonlinear optimization, and the other status are marginalized. Meanwhile, we convert measurements corresponding to marginalized states into prior, so as to bound the computational complexity and improve the accuracy of state estimation without loop detection.

This paper is organized as follows: In Sect. 1, we discuss the research background and actuality. In Sect. 2, the systematic framework of our method is presented. In Sect. 3, we introduce the task of features extraction and matching. State estimation and local mapping are presented in In Sects. 4 and 5 respectively. Experimental results are presented in Sect. 6. Finally, we draw some conclusions in Sect. 7.

2 Overview

The structure of our proposed monocular SLAM is shown in Fig. 1. The basic framework mainly involves three parallel threads. The thread of features extraction and matching (the red part in Fig. 1) is mainly responsible for extracting features from images and matching them in successive frames. The thread of state estimation (the blue part in Fig. 1) is mainly responsible for calculating the reprojection error of feature points, solving the nonlinear optimization problem and doing marginalization step. The thread of local mapping (the green part in Fig. 1) is mainly responsible for updating the map.

We employ the following notations and definitions throughout this paper. $(\cdot)^w$ denotes the world frame, a feature point f_i represented in the world frame is written as $\boldsymbol{p}^w_{fi}$.$(\cdot)^k$ denotes the camera frame corresponding to k-th image in the sliding window. $\boldsymbol{p}^w_k$, $\boldsymbol{R}_{wk}$, $\boldsymbol{q}_{wk}$ and $\boldsymbol{\theta}_{wk}$ denote translation, rotation matrix, Hamilton quaternion and rotation vector from frame $(\cdot)^k$ to the world frame $(\cdot)^w$ respectively. $\otimes$ denotes the multiplication operation between two quaternions.

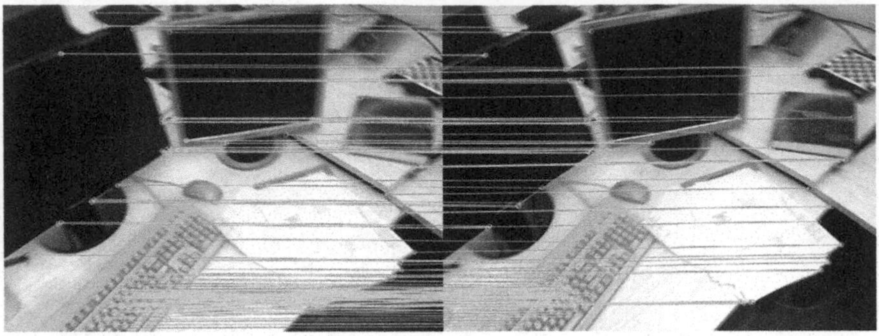

Fig. 2 The matching result among successive frames

3 Features Extraction and Matching

ORB [7] feature is adopted for extracting corners from images and matching them among successive frames. The related steps are as follows. Firstly, we build an 8-layer image pyramid with the scale factor of 1.2, and the FAST [8] corners are extracted from each layer of the image pyramid. Secondly, in order to ensure a uniform distribution of the corners, the original image is divided into several non-overlap grids, we select 5 corners with the largest response in each grid and add them to feature point set. If the corners in the grid are insufficient, we reduce the extraction threshold. Finally, we calculate the direction of each feature point, build the descriptors and match them between successive frames. The matching result among successive frames is shown in Fig. 2.

4 State Estimation

As show in Fig. 3, we select feature points reprojection error based on pinhole camera model to construct cost function, and the Gauss-Newton method is used to optimize state variables. The cost function is defined as

$$J(\boldsymbol{\chi}) = \sum_{k=1}^{n}\sum_{i=1}^{m} \boldsymbol{r}(\hat{\boldsymbol{l}}_{f_i}^{C_k}, \boldsymbol{\chi})^T \boldsymbol{W}_{\boldsymbol{r}(\hat{\boldsymbol{l}}_{f_i}^{c_k}, \boldsymbol{\chi})} \boldsymbol{r}(\hat{\boldsymbol{l}}_{f_i}^{C_k}, \boldsymbol{\chi}) + J_{mar}(\boldsymbol{\chi}) \tag{1}$$

where $\boldsymbol{\chi}$ denotes the set of state variables in the sliding window, $\boldsymbol{\chi} = [\boldsymbol{x}_1, \boldsymbol{x}_2, \cdots, \boldsymbol{x}_n, \boldsymbol{p}_{f_1}^w, \boldsymbol{p}_{f_2}^w, \cdots, \boldsymbol{p}_{f_m}^w]$ and $\boldsymbol{x}_k = [\boldsymbol{p}_k^w, \boldsymbol{R}_{w\,k}]$, n denotes the number of frames in the sliding window, m denotes the number of feature points which can be observed, $\boldsymbol{r}(\hat{\boldsymbol{l}}_{f_i}^{C_k}, \boldsymbol{\chi})$ denotes the reprojection error of f_i to the k-th image in the sliding window, $\boldsymbol{W}_{\boldsymbol{r}(\hat{\boldsymbol{l}}_{f_i}^{c_k}, \boldsymbol{\chi})}$ denotes the information matrix corresponding to $\boldsymbol{r}(\hat{\boldsymbol{l}}_{f_i}^{C_k}, \boldsymbol{\chi})$,

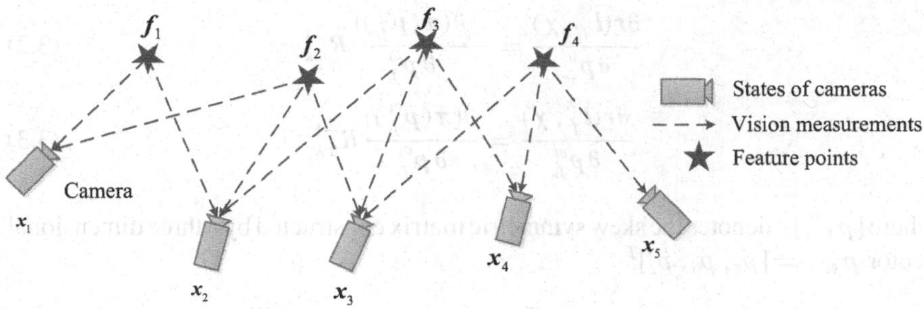

Fig. 3 A schematic diagram of our sliding window based monocular SLAM

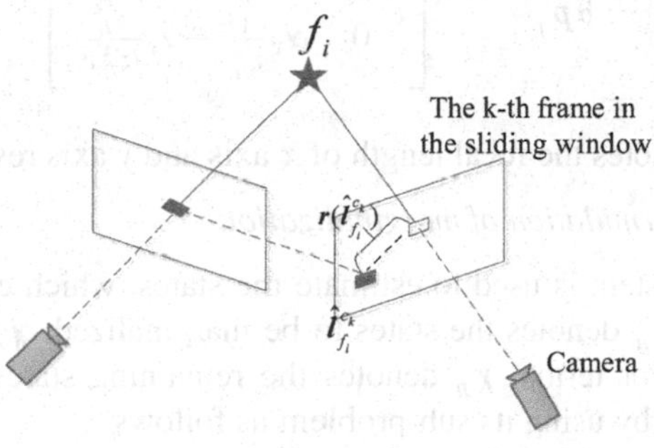

Fig. 4 The reprojection error of feature point f_i to k-th image in sliding window

$J_{mar}(\boldsymbol{\chi})$ denotes the prior information from marginalization. In the following, we present the details of reprojection error and marginalization formulations respectively.

A. *Mathematical formulation of reprojection error*

As show in Fig. 4, the reprojection error of feature point f_i to k-th image in the sliding window can be expressed as

$$\boldsymbol{r}(\hat{\boldsymbol{l}}_{f_i}^{C_k}, \boldsymbol{\chi}) = \boldsymbol{\pi}(\boldsymbol{R}_{wk}^{-1}(\boldsymbol{p}_{f_i}^{w} - \boldsymbol{p}_k^{w})) - \hat{\boldsymbol{l}}_{f_i}^{C_k} \tag{2}$$

where $\hat{\boldsymbol{l}}_{f_i}^{C_k}$ denotes the observation of the feature point f_i in the k-th image, which can be obtained by feature matching in Sect. 3. We define the notion $\boldsymbol{p}_{f_i}^{C_k} = [x_{f_i}^{C_k}, y_{f_i}^{C_k}, z_{f_i}^{C_k}]^T = \boldsymbol{R}_{wk}^{-1}(\boldsymbol{p}_{f_i}^{w} - \boldsymbol{p}_k^{w})$ denotes the position of the feature point f_i in the k-th camera frame, $\boldsymbol{\pi}(\boldsymbol{p}_{f_i}^{C_k})$ denotes the function to transform $\boldsymbol{p}_{f_i}^{C_k}$ from camera coordinate to pixel coordinate. We provide the Jacobians of $\boldsymbol{r}(\hat{\boldsymbol{l}}_{f_i}^{C_k}, \boldsymbol{\chi})$ as follows

$$\frac{\partial \boldsymbol{r}(\hat{\boldsymbol{l}}_{f_i}^{C_k}, \boldsymbol{\chi})}{\partial \boldsymbol{\theta}_{wk}} = \frac{\partial(\boldsymbol{\pi}(\boldsymbol{p}_{f_i}^{C_k}))}{\partial \boldsymbol{p}_{f_i}^{C_k}} [\boldsymbol{R}_{wk}^{-1}(\boldsymbol{p}_{f_i}^{w} - \boldsymbol{p}_k^{w})]_{\times} \tag{3.1}$$

$$\frac{\partial r(\hat{l}_{f_i}^{c_k}, \chi)}{\partial p_k^w} = -\frac{\partial(\pi(p_{f_i}^{c_k}))}{\partial p_{f_i}^{c_k}} R_{wk}^{-1} \tag{3.2}$$

$$\frac{\partial r(\hat{l}_{f_i}^{c_k}, \chi)}{\partial p_{f_i}^w} = \frac{\partial(\pi(p_{f_i}^{c_k}))}{\partial p_{f_i}^{c_k}} R_{wk}^{-1} \tag{3.3}$$

where $[p_{3\times1}]_\times$ denotes the skew symmetric matrix constructed by a three dimensional vector $p_{3\times1} = [p_x, p_y, p_z]^T$

$$\frac{\partial(\pi(p_{f_i}^c))}{\partial p_{f_i}^c} = \begin{bmatrix} \alpha_x \frac{1}{z_{f_i}^{c_k}} & 0 & -\alpha_x \frac{x_{f_i}^{c_k}}{(z_{f_i}^{c_k})^2} \\ 0 & \alpha_y \frac{1}{z_{f_i}^{c_k}} & -\alpha_y \frac{y_{f_i}^{c_k}}{(z_{f_i}^{c_k})^2} \end{bmatrix} \tag{4}$$

where α_x and α_y denotes the focal length of x axis and y axis respectively.

B. *Mathematical formulation of marginalization*

A Gauss-Newton system is used to estimate the states, which can be expressed as $H\delta\chi = b$. Suppose χ_μ denotes the states to be marginalized, χ_λ denotes the states related to χ_μ by error terms, χ_p denotes the remaining states. We can simplify marginalization step by using its sub-problem as follows

$$\begin{bmatrix} H_{\mu\mu} & H_{\mu\lambda} \\ H_{\lambda\mu} & H_{\lambda\lambda} \end{bmatrix} \begin{bmatrix} \delta\chi_\mu \\ \delta\chi_\lambda \end{bmatrix} = \begin{bmatrix} b_\mu \\ b_\lambda \end{bmatrix} \tag{5}$$

Using Schur complement [9] to simplify Eq. (5) yields

$$\begin{bmatrix} H_{\mu\mu} & H_{\mu\lambda} \\ 0 & H_{\lambda\lambda}^* \end{bmatrix} \begin{bmatrix} \delta\chi_\mu \\ \delta\chi_\lambda \end{bmatrix} = \begin{bmatrix} b_\mu \\ b_\lambda^* \end{bmatrix} \tag{6}$$

where $H_{\lambda\lambda}^* = H_{\lambda\lambda} - H_{\lambda\mu} H_{\mu\mu} H_{\mu\lambda}$, $b_\lambda^* = b_\lambda - H_{\lambda\mu} H_{\mu\mu}^{-1} b_\mu$. Solving the equation $H_{\lambda\lambda}^* \delta\chi_\lambda = b_\lambda^*$, and the value of $\delta\chi_\lambda$ can be obtained. Owing that the value of $\delta\chi_\mu$ is not solved by the above steps, we cannot update the states χ_μ, so the "first estimate Jacobians [10]" is used. We fix the linearization point around initial point, that is to say the value of $H_{\lambda\lambda}^*$ stays the same, and the value of b_λ^* updates as follows

$$b_\lambda^* \leftarrow b_\lambda^* + \frac{\partial b_\lambda^*}{\partial \chi_\lambda} \delta\chi_\lambda = b_\lambda^* - H_{\lambda\lambda}^* \delta\chi_\lambda \tag{7}$$

In our monocular SLAM, the number of state variables will increase significantly over time, and the algorithm complexity will increase with the number of states. In order to limit the computational complexity of our method, marginalization strategy is used. As show in Fig. 5, if the second latest frame in sliding window is a key frame, the oldest key frame in sliding window is marginalized, meanwhile, we convert

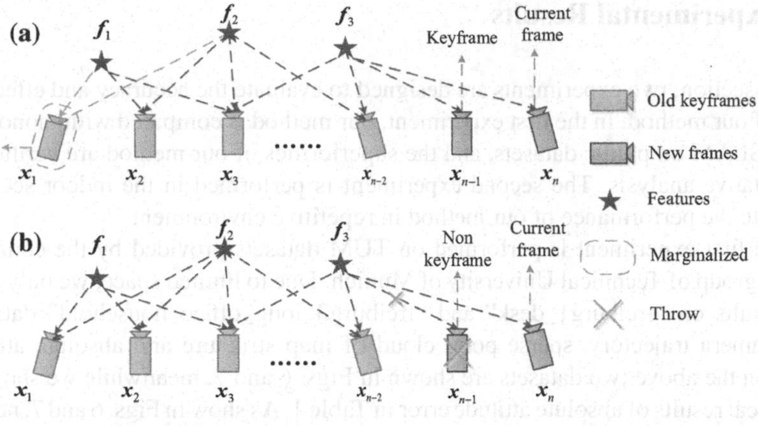

Fig. 5 A schematic diagram of our marginalization strategy **a** the second latest frame is a keyframe **b** the second latest frame is not a keyframe

measurements corresponding to the marginalized states into prior. If the second latest frame in sliding window is not a key frame, we directly throw this frame with its corresponding measurements. In this case, throw strategy is used instead of marginalization, which can maintain the sparsity of the system.

5 Local Mapping

Our method needs to update the map structure according to the new key frame states and the observed feature point information. The local mapping thread only operating on key frames is not limited to the frame rate of the camera. When the second latest frame in sliding window is a key frame, linear triangulation method [11] is used to initialize the position of feature points. Then we use local *BA* [12] to optimize the position of new feature points. In order to maintain a compact map structure, we discard the "bad" map points, which have been initialized for a long time, but can be observed by only few key frames. We also test repeatability of newly created points with the existing points in the map, then merge them if create repeatedly. A frame is considered as a key frame, if meeting one of the following conditions

(1) The number of the feature points tracked by this frame is more than 60, and 50% tracked feature points have not been initialized.
(2) The local mapping thread is idle.

6 Experimental Results

In this section, two experiments are designed to evaluate the accuracy and effectiveness of our method. In the first experiment, our method is compared with monocular ORB-SLAM on public datasets, and the superiorities of our method are verified by quantitative analysis. The second experiment is performed in the indoor scene to evaluate the performance of our method in repetitive environment.

The first experiment is performed on TUM datasets provided by the computer vision group of Technical University of Munich. Due to limited space, we only show the results on "freiburg1_desk" and "freiburg3_long_office_household" datasets. The camera trajectory, sparse point cloud of map structure and absolute attitude error on the above two datasets are shown in Figs. 6 and 7, meanwhile we show the statistical results of absolute attitude error in Table 1. As show in Figs. 6 and 7, neither our method nor monocular ORB-SLAM can recover the scale information, which is indeed an unsurmountable problem for monocular SLAM. Therefore, we only select absolute attitude error as the criterion to evaluate the performance. It can be seen from Table 1 that the attitude root mean square error (RMSE) of monocular ORB-SLAM on "freiburg1_desk" dataset is 2.48 degrees, while our method is only 1.55 degrees. Meanwhile, the corresponding values on "freiburg3_long_office_household" dataset are 1.03 and 0.42 degrees. The performance of our method is obviously better than the monocular ORB-SLAM. The good performance comes from the fact that we estimate the latest multiple states in a sliding window by using nonlinear optimization, and convert measurements corresponding to marginalized states into prior. The absolute attitude error for same algorithm on different datasets are much different. That is because visual SLAM is influenced by scene information and camera movement. Fast camera movement and weak texture environment both can reduce the accuracy of the estimated results. The time consumption of our method is presented in Table 2, our algorithm is programmed by C++ language and run on an Intel(R) Core(TM) i7 CPU of 2.80 GHz. The local mapping thread only operating on key frames is not limited to the frame rate of the camera. Therefore, the real-time performance of our method depends on the state estimation thread. Furthermore, the calculation efficiency of state estimation thread is mainly affected by the number of states in the sliding window, and running time increases with the growing number of states in the sliding window. In order to ensure real-time state estimation, we set the number of frames in the sliding window to 5. In practice, this number can be adjusted according to the calculation ability of the platform.

The second experiment is performed in indoor corridor environment of New Main Building of Beihang University. This environment is mostly weak or repetitive texture regions, which presents many challenges to visual SLAM. In this experiment, the researcher holds the monocular camera (MT9V034) with front-downward view, walks along the square corridor at a normal pace and eventually returns to the origin. It totally takes 129.8 s to finish the test, and the estimated trajectory length is

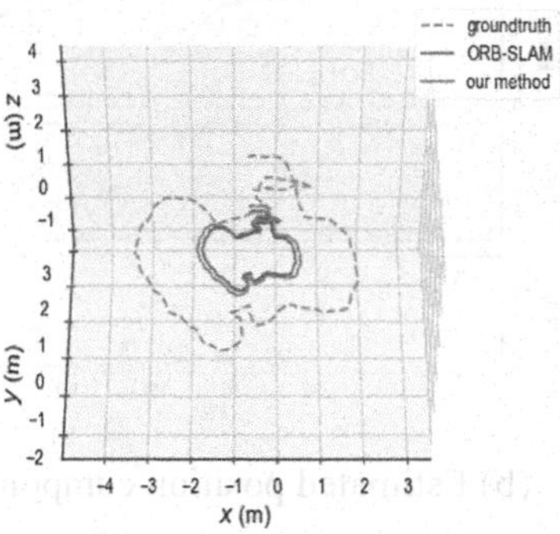

(a) Trajectory

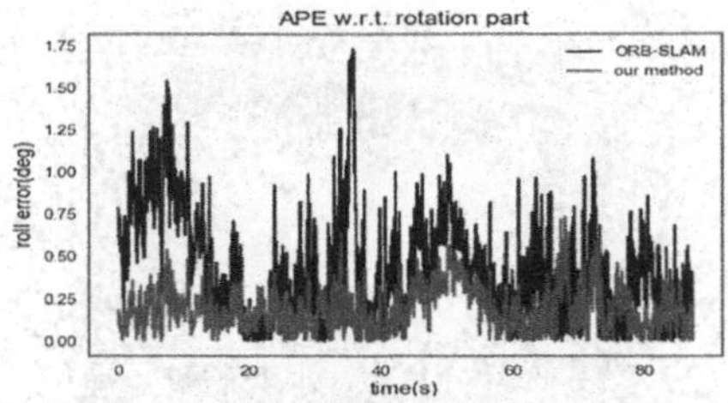

(c) Absolute error of roll

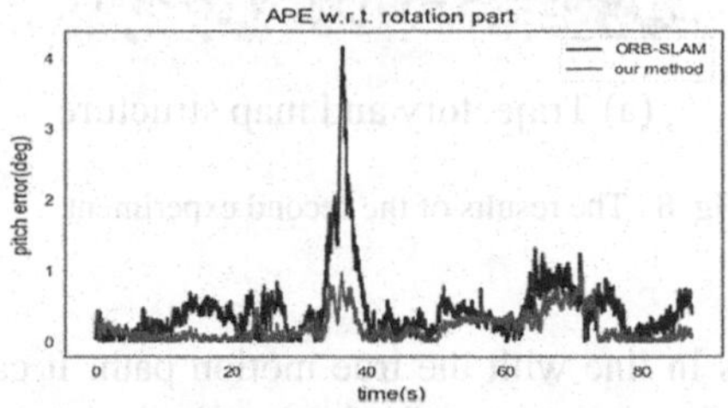

(d) Absolute error of pitch

(b) Sparse point cloud of map structure

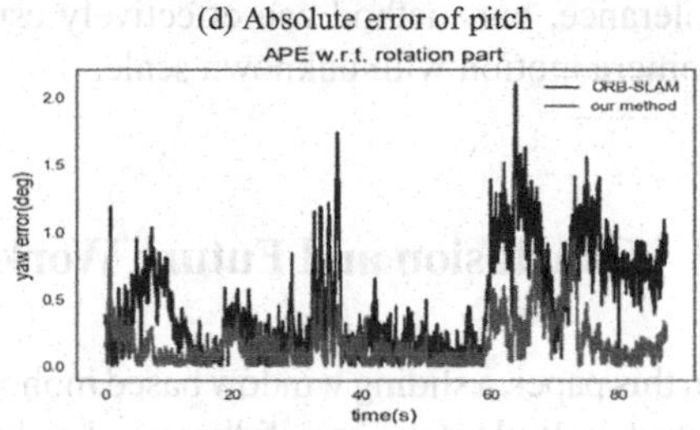

(e) Absolute error of yaw

Fig. 7 Experimental results on "f freiburg3_long_office_household" dataset

Table 1 The statistical results of absolute attitude error

Datasets	Systems	RMSE[°]	MEAN[°]	MEDIAN[°]	STD[°]
freiburg1_desk	ORB-SLAM	2.483761	2.338919	2.192515	0.835777
	Our method	1.549332	1.434604	1.426759	0.585101
freiburg3_long_office_household	ORB-SLAM	1.031849	0.897872	0.793732	0.508466
	Our method	0.425993	0.365495	0.317269	0.218824

Table 2 The average time consumption of our method

Threads	Features tracking	State estimation	Local mapping
Running time [ms]	23	27	117

52.15 m. The final estimated position of camera is (0.21, 0.13, 0.09 m), however, the ideal value should be (0, 0, 0 m). The position absolute error is only 0.51% of the trajectory length. As show in Fig. 8, the estimated motion trajectory of our method

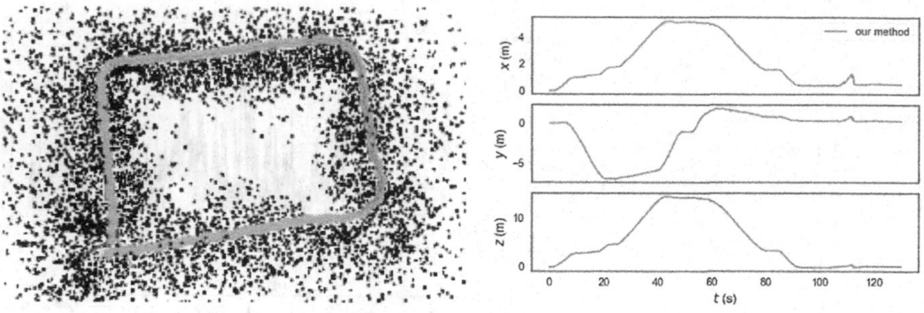

(a) Trajectory and map structure (b) Estimated position components

Fig. 8 The results of the second experiment

is in line with the true motion path. It can be seen that, within the range of error tolerance, our method can effectively estimate the sparse point cloud of map and camera motion with unknown scale.

7 Conclusion and Future Work

In this paper, a sliding window based monocular SLAM is introduced. We estimate the latest multiple states in a sliding window by using nonlinear optimization, meanwhile we use marginalization strategy to limit the computational complexity. Our method is performed on both public datasets and actual real-time test. The results shows that our method can effectively estimate the sparse point cloud of map and camera motion with unknown scale. However, our method still cannot recover the scale information, our further interest lies in how to integrate inertial measurements in this framework.

References

1. C. Xiangkun, J. Min, Z. Liangyu, et al. A feature matching method for simultaneous localization and mapping, in *2017 IEEE Information Technology, Networking, Electronic and Automation Control Conference*, pp. 1091–1094
2. L. Yi, G. Fei, Q. Tong et al., Autonomous aerial navigation using monocular visual-inertial fusion. J. Field Rob. **35**(4), 23–51 (2017)
3. K. Georg, M. David, Parallel tracking and mapping for small AR workspaces, in *2007 IEEE and ACM International Symposium on Mixed and Augmented Reality*, pp. 225–234
4. M.-A. Raul, Montiel JMM, Tardos, JD. ORB-SLAM: a versatile and accurate monocular SLAM system. IEEE Trans. Rob. **31**(5), 1147–1163 (2015)
5. F. Christian, P. Matia, S. Davide, SVO: fast semi-direct monocular visual odometry, in *2014 IEEE International Conference on Robotics and Automation*, pp. 15–22
6. E. Jakob, S. Thomas, Cremers daniel LSD-SLAM: large-scale direct monocular SLAM, in *2014 European Conference on Computer Vison*, pp. 834–849

7. E. Rublee, V. Rabaud, K. Konolige, et al. ORB: an efficient alternative to SIFT or SURF, in *2012 IEEE International Conference on Computer Vision*, pp. 2564–2571
8. R. Edward, Drummond tom Machine learning for high-speed corner detection. in *2006 European Conference on Computer Vision*, pp. 430–443
9. S. Gabe, M. Larry, S. Gaurav, Sliding window filter with application to planetary landing. J. Field Rob. **27**(5), 587–608 (2010)
10. T.C. Dong-Si, Mourikis AI motion tracking with fixed-lag smoothing: Algorithm and consistency analysis, in *2011 IEEE International Conference on Robotics and Automation*, pp. 5655–5662
11. P. Mikael, P. Tommaso, F. Michael, et al. Robust stereo visual odometry from monocular techniques, in *2015 IEEE Intelligent Vehicles Symposium*, pp. 686–691
12. S. Hauke, Montiel J. M. M, Davison Andrew J. Real-time monocular SLAM: why filter?, in *2010 IEEE International Conference on Robotics and Automation*, pp. 2657–2664

Application of the Fuzzy C-Means Clustering Algorithm for the Burden Distribution Matrix of Blast Furnace

Yuanzhe Hui, Sen Zhang, Xiaoli Su and Yixin Yin

Abstract Burden distribution matrix is the key to guarantee the long-term stable production of the blast furnace. The optimization of the burden distribution matrix aims to form a reasonable burden surface. It can help to achieve the goal of smooth, high-quality and low-consumption blast furnace production. This paper uses the blast furnace condition parameter to measure the burden distribution matrix. And these data is characterized by panel data in statistics. The fuzzy c-means algorithm is used to cluster. Finally, evaluation indicators are using to analyze the clustering effect. It has important reference value for the blast furnace actual production.

Keywords Burden distribution matrix · Fuzzy c-means algorithm · Clustering Panel data · Principal component analysis

1 Introduction

The blast furnace ironmaking process is complicated. Optimization of the burden distribution matrix is an important method for blast furnace's upper regulation. Reasonable burden distribution matrix combined with the bottom adjustments to form an ideal material surface shape and stabilize the gas flow distribution, so as to improve the gas utilization rate and ensure blast furnace operation stable [1]. Therefore, exploring how to optimize burden distribution matrix is extremely important for blast furnace operation.

Y. Hui · S. Zhang (✉) · X. Su · Y. Yin
School of Automation and Electrical Engineering, University of Science and Technology Beijing, Beijing 100083, People's Republic of China
e-mail: zhangsen@ustb.edu.cn

Y. Hui · S. Zhang · X. Su · Y. Yin
Key Laboratory of Knowledge Automation for Industrial Processes of Ministry of Education, School of Automation and Electrical Engineering, University of Science and Technology Beijing, Beijing 100083, People's Republic of China

Y. Jia et al. (eds.), *Proceedings of 2018 Chinese Intelligent Systems Conference*, Lecture Notes in Electrical Engineering 529,
https://doi.org/10.1007/978-981-13-2291-4_52

This paper applies real production data and chooses reasonable clustering method to cluster the burden distribution matrix. Each burden distribution matrix has its number. It is aims to find the best effect of the burden distribution matrix's clustering.

Since the interior of the blast furnace is a constantly changing dynamic environment, the description of the burden distribution matrix is not easy. However, the panel data can describe the dynamic changes of individuals. The data is based on seven furnace condition indicators as the cross-section data to form the multi-index panel data. The panel data is normalized and dimensionality reduced by SPSS software to construct the evaluation function sequence matrix.

Due to the diversity of the burden distribution matrix, the furnace condition parameters are not same. For analyzing, the burden distribution matrices are required to cluster. The FCM is an improvement of the ordinary C-means algorithm. The common C-means algorithm is hard division for the data, but the FCM is flexible fuzzy division. It uses the concept of membership and has the high clustering accuracy. It can satisfy industrial requirements generally. This paper applies the FCM for clustering, and chooses Dunn and Silhouette indicators to analyze the clustering results.

2 Clustering Algorithm Introduction and Panel Data

2.1 Fuzzy C-Means Algorithm

Dunn proposed the fuzzy clustering algorithm in 1974. He applied fuzzy theory to cluster analysis [2]. Bezdek improved the algorithm and proposed a clustering optimization algorithm for fuzzy objective functions, namely fuzzy C-means clustering algorithm, and proved the convergence of this algorithm. The principle of fuzzy C-means algorithm is easy to understand, it can solve a wide range of problems. So it is attracted more and more attention [3].

The core idea of FCM is to obtain the membership degree matrix and transform the clustering of the original data into the division of the membership degree matrix. There are two unknown parameters that need to be determined in the FCM algorithm: the number of clusters c and the smoothing parameter m, m is called the fuzzy weighted index. $m \in (0, +\infty)$. Usually, the value of m is 1.5–2.5. If m is too large, the clustering result is not good. If m is too small, the FCM algorithm will be close to the hard clustering algorithm [4].

If you use x_{jk} that represents the value of kth dimensionality of the sample, $X = \{x_1, x_2, \ldots, x_n\}$ is the data set, N is the number of sample, n is each sample's dimension, $x_j = \{x_{j1}, x_{j2}, \ldots, x_{jn}\}$. The membership of the data X set is $U = [u_{ij}]$, u_{ij} is the probability or degree of membership of the j-th sample x_j of X in the i-th category. c indicates that the data set X is divided into c classes. The degree of membership satisfies: $\sum_{i=1}^{c} u_{ij} = 1, \forall j = 1 \ldots N, 0 \le u_{ij} \le 1; 1 \le i \le c; 1 \le j \le N$.

Set of c cluster center: $V = [v_1, v_2 \ldots v_c]^T$, which $v_i = [v_{i1}, v_{i2} \ldots v_{in}]$ are n dimensional vectors, FCM algorithm through the continuous iteration to make the objective function reaches the minimum. General definition of the objective function J is:

$$\mathrm{J(U}, v_1 \ldots v_c) = \sum_{i=1}^{c} J_i = \sum_{i=1}^{c}\sum_{j=1}^{n} u_{ij}^{m} d_{ij}^{2} \tag{1}$$

where $d_{ij} = \sqrt{\sum_{q=1}^{n} \left(x_{jq} - v_{iq}\right)^2}$, indicates the Euclidean distance from the first sample j to the center of the i-th cluster.

In order to obtain the necessary conditions for minimizing the objective function, the Lagrangian multiplier method is used to reconstruct the objective function $\overline{J}$. Treat $\sum_{i=1}^{c} u_{ij} = 1$ as a constraint, then:

$$\overline{J}(U, v_1, \ldots, v_c, \lambda_1'', \ldots, \lambda_n) = J(U, v_1, \ldots, v_c)$$

$$+\sum_{j=1}^{n} \lambda_j \left(\sum_{i=1}^{c} u_{ij} - 1\right) = \sum_{i=1}^{c}\left(\sum_{j=1}^{n} u_{ij}^{m} d_{ij}^{2}\right) + \sum_{j=1}^{n} \lambda_j \left(\sum_{i=1}^{c} u_{ij} - 1\right) \tag{2}$$

where $\lambda_j (j = 1, 2 \ldots n)$ is the Lagrangian multiplier. Then find the partial derivative of the input function for each input parameter and establish the system of equation. The necessary condition for obtaining the minimum value of the objective function is:

$$v_i = \frac{\sum_{j=1}^{n} u_{ij}^{m} x_j}{\sum_{j=1}^{n} u_{ij}^{m}}, i = 1, 2 \ldots c \tag{3}$$

$$u_{ij} = \frac{1}{\sum_{k=1}^{c} \left(\frac{d_{ij}}{d_{kj}}\right)^{2/(m-1)}}, i = 1, 2 \ldots c; j = 1, 2 \ldots n \tag{4}$$

Repeat the above two equations until the algorithm converges.

2.2 *Statistical Description of Panel Data*

In statistics, there are various forms of data. In many data formats, panel data has its unique advantages. The panel data includes both cross-sectional data and time series, which can better represent the development status of the sample and the dynamic development trend from a time and space perspective. Bonzo introduced multivariate statistical methods into panel data analysis firstly, and improved cluster analysis methods using probabilistic link functions [5]. Xiao Zelei constructed a similar index of the comprehensive evaluation function sequence matrix, and implemented dimensionality reduction and system cluster analysis [6].

Time / Sample	1	...	t	...	T
	$X_1 \ldots X_t \ldots X_p$		$X_1 \ldots X_t \ldots X_p$		$X_1 \ldots X_t \ldots X_p$
1	$X_{11}(1) \ldots X_{1j}(1) \ldots X_{1p}(1)$		$X_{11}(t) \ldots X_{1j}(t) \ldots X_{1p}(t)$		$X_{11}(T) \ldots X_{1j}(T) \ldots X_{1p}(T)$
i	$X_{i1}(1) \ldots X_{ij}(1) \ldots X_{ip}(1)$		$X_{i1}(t) \ldots X_{ij}(t) \ldots X_{ip}(t)$		$X_{i1}(T) \ldots X_{ij}(T) \ldots X_{ip}(T)$
N	$X_{N1}(1) \ldots X_{Nj}(1) \ldots X_{Np}(1)$		$X_{N1}(t) \ldots X_{Nj}(t) \ldots X_{Np}(t)$		$X_{N1}(T) \ldots X_{Nj}(T) \ldots X_{Np}(T)$

Fig. 1 Multiple indicators panel data

With the statistical description of the panel data, it is possible to use the obtained data information to construct a measurement model of panel data and perform cluster analysis based on the problems to be solved [7].

The descriptive statistics of the multi-indicator panel data can be represented in the form of a two-dimensional table. There are N samples. The features of each sample are represented by P indicators and the length of time is T. As shown in Fig. 1.

The following shows the statistics required for several multi-indicator panel data in cluster analysis.

Where $i \in [1, N]; j \in [1, p]; t \in [1, T]$.

The mean of the j-th indicator at t-time:

$$\overline{X}_j(t) = \frac{1}{N}\sum_{i=1}^{N} x_{ij}(t) \tag{5}$$

The mean of the j-th indicator:

$$\overline{X}_j(t) = \frac{1}{T}\frac{1}{N}\sum_{t=1}^{T}\sum_{i=1}^{N} x_{ij}(t) \tag{6}$$

The variance of the j-th indicator at time t is:

$$\text{var}x_j(t) = \frac{1}{N-1}\sum_{i=1}^{N}\left[x_{ij}(t) - \overline{x_j}(t)\right]^2 \tag{7}$$

The variance of the j-th indicator is:

$$\mathrm{varx_j} = \frac{1}{T}\frac{1}{N-1}\sum_{i=1}^{N}\left[\mathrm{x_{ij}(t)} - \overline{\mathrm{x_j}}(t)\right]^2 \tag{8}$$

After the multi-indicator panel data format is defined and several basic statistics are defined, multi-indicator panel data can be studied using quantitative modeling analysis and multivariate statistical analysis for different practical problems [8].

2.3 Cluster Effect Evaluation Index

The clustering effect evaluation index is a variable that measures the performance of the clustering algorithm. This section describes two clustering effect evaluation indicators [9].

Dataset with n data objects $D = \{x_1, x_2, \ldots, x_n\}$, $x_i (i = 1, 2, \ldots, n)$ represents the i-th p-dimensional data object. Divide the data set D into a set of NC subsets $D = \{C_1, C_2, \ldots, C_{NC}\}$ by using a corresponding clustering algorithm. c_j represents the cluster center of subset C_j. n_j represents the number of data in C_j. $d(x_i, x_j)$ represents the Euclidean distance between the i-th and j-th objects.

(1) Dunn index:

$$\mathrm{D(NC)} = \min\left\{\min\frac{\min_{x\in C_i, y\in C_j} d(x, y)}{\max\left[\max_{x,y\in C_k} d(x, y)\right]}\right\} \tag{9}$$

The D indicator is obtained by dividing the degree of separation between classes and compactness within the class. Between-class separation refers to the smallest two-point distance between a class and another class. Within-class tightness refers to the largest class diameter in all classes. The larger the D index, the farther the gap between the class and the class is, and the more satisfactory the clustering effect is.

(2) Silhouette index:

$$\mathrm{S(NC)} = \frac{1}{NC}\sum_{i=1}^{NC}\left\{\frac{1}{n_i}\sum_{x\in C_i}\frac{b(x) - a(x)}{\max[b(x), a(x)]}\right\} \tag{10}$$

In this formula: $a(x) = \frac{1}{n_i - 1}\sum_{x,y\in C_i, x\neq y} d(x, y)$, $b(x) = \min_{j,j\neq i}\left[\frac{1}{n_j}\sum_{x\in C_i, y\in C_j} d(x, y)\right]$.

In the S indicator, a(x) denotes the distance between two objects in the class, and b(x) denotes the distance between each object between the class and the class. This index also indicates that the clustering result is optimal when the maximum value is taken.

3 Data Processing

3.1 Data Selection

The data in this paper uses blast furnace production data from a steel mill. The index be used is blast temperature (BT), blast pressure (BP), top pressure (TP), blast volume of oxygen(BVO), blast volume (BV), and permeability index, gas utilization rate.

3.2 Constructing Comprehensive Evaluation Matrix with Factor Analysis

In this paper, 36 different numbered burden distribution matrices are selected as samples, and three hours average divided into 36 time series. A 36-by-36 data matrix is constructed. Each number in the data matrix represents a comprehensive evaluation of all 7 indicators of the current sample at the current time (Fig. 2).

From the perspective of each column in the matrix, value made up of several burden distribution matrices at a certain moment in time. From each row, it is an observed value of the burden distribution matrix at different times. Each burden distribution matrix is described by several blast furnace indicators. Assume that there are K burden distribution matrix, each burden distribution matrix has P periods, there are K*P data. Among them, each row of data can be expanded as shown in the Fig. 3.

The steps of using SPSS software to establish a comprehensive evaluation function matrix sequence are as follows:

	Time 1	Time 2	Time 3	···	Time p
burden distribution matrix 1	Blast furnace condition indicator	Blast furnace condition indicator		...	Blast furnace condition indicator
burden distribution matrix 2	Blast furnace condition indicator	Blast furnace condition indicator		...	Blast furnace condition indicator
burden distribution matrix 3	..	..			..
..	.	.			.
burden distribution matrix k	Blast furnace condition indicator	Blast furnace condition indicator		...	Blast furnace condition indicator

Fig. 2 Data format

burden distribution matrix 1

	Index 1	Index 2	...	Index m
Time 1	X11	X12		
Time 2	X21	X22		
.				
.				
.				
Time p	Xp1	Xp2		Xpm

Fig. 3 The structure of the burden distribution matrix 1

(1) Standardized raw data processing

$$x_{ij}^{*}(t_n) = \frac{x_{ij}(t_n) - \overline{x_{.s}}(t_n)}{\sqrt{\mathrm{var}(x_{.s}(t_n))}}, s = 1, 2 \ldots m, n = 1, 2 \ldots p \quad (11)$$

s is the index component, total m; t_n is the time, total p; i is the study sample, total k; $x_{.s}(t_n)$ represents the average value of the current indicator, $\mathrm{var}(x_{.s}(t_n))$ indicating the standard deviation of the indicator.

(2) dimension reduction, factor analysis

Use the dimension reduction and factor analysis tools of SPSS software to select standardized variables for factor analysis and obtain results.

For example, the no.1 burden distribution matrix, eigenvalues of the first three common factors are greater than one, and the cumulative contribution rate of the three common factors is 84.377%, which can explain most of the information of the original variables. Therefore, the original seven indicators can extract three common factors fac1-1, fac1-2, fac1-3.

(3) Obtain comprehensive score index

Save the scores of the three common factors as variables, denoted as f_1, f_2, f_3. Let the eigenvalues of the three common factors be λ_1, λ_2, λ_3 respectively. Calculate the comprehensive score indicator by the following formula.

$$\mathrm{F} = \frac{\lambda_1}{\lambda_1 + \lambda_2 + \lambda_3} * f_1 + \frac{\lambda_2}{\lambda_1 + \lambda_2 + \lambda_3} * f_2 + \frac{\lambda_3}{\lambda_1 + \lambda_2 + \lambda_3} * f_3 \quad (12)$$

According to the above steps, the comprehensive evaluation score index of each burden distribution matrix at each moment is separately obtained, Finally a 36*36 data matrix as shown in Fig. 2 is obtained. Next, clustering analysis is performed on Matlab software.

Table 1 FCM evaluation results

	Three categories	Four categories	Five categories
Dunn	0.04048	0.09590	0.08421
Silhouette	0.1542	0.2290	0.1586

Table 2 Comparison of FCM and kmeans under four categories

	kmeans	FCM
Dunn	0.05704	0.09590
Silhouette	0.1301	0.2290

4 Clustering and Results

Because there are few samples in the dataset and the dimensions are high, the sample in the high-dimensional space is sparse. It is easy to cause overfitting. To make the data fit the model better and avoid overfitting. Use the principal component analysis (PCA) to reduce the data dimension. On the other hand, dimension reduction can make the clustering effect of data show in a picture clearly.

The result of dimension reduction by PCA is that the first three principal components greater than 1 can represent 74% of the original data, so the high dimensional data set is reduced to three dimensions. Then use FCM algorithm to analyze the data after dimension reduction. The clustering results are presented in the three-dimensional scatter plot. Samples of different clusters are represented by different color.

In the FCM algorithm program, the maximum number of iterations is 100, the threshold is 0.000001, and the value of the fuzzy weighted index m is set to 2.

This paper tests the different clustering effects produced by different cluster numbers, and uses cluster evaluation indicators to judge the quality of the clustering. We know that the higher the cluster evaluation index, the better the clustering effect. Through the Table 1, four cluster centers have the best clustering effect. Figure 4 shows the FCM three-dimensional scatter plot clustered into four categories.

In order to reflect the clustering effect of FCM algorithm better, this paper compares it with the kmeans algorithm. The number of clusters is set to four, the maximum number of iterations is 100, the threshold is 0.000001. According to Table 2, FCM has a better effect when the number of clusters is four. Figure 5 shows the kmeans three-dimensional scatter plot clustered into four categories.

Through analysis, it is concluded that FCM introduces the concept of membership compared to the traditional kmeans algorithm and has higher precision to satisfy industrial requirements. When the number of clusters is four, the clustering effect is the best.

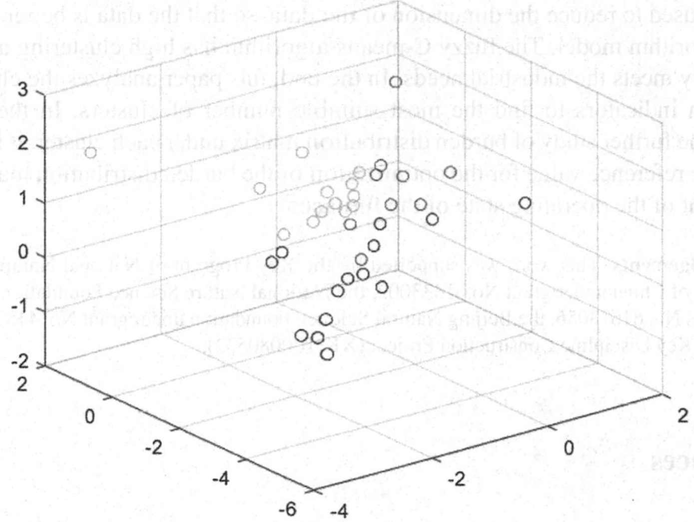

Fig. 4 FCM clustering results

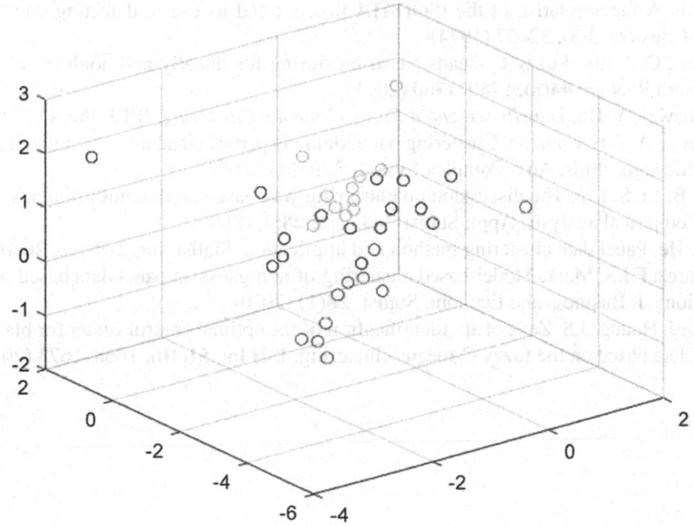

Fig. 5 Kmeans clustering results

5 Conclusions

This paper describes the burden distribution matrix through the parameters of furnace conditions, and applies the form of the panel data. It is well reflects the dynamic process of the blast furnace operation. Besides, the principal component analysis

(PCA) is used to reduce the dimension of the data, so that the data is better adapted to the algorithm model. The fuzzy C-means algorithm has high clustering accuracy and largely meets the industrial needs. In the end, this paper analyzes the clustering evaluation indicators to find the most suitable number of clusters. In the future, through the further study of burden distribution matrix under each cluster, it is hoped to provide reference value for the optimization of the burden distribution matrix and adjustment of the operating state of the furnace.

Acknowledgements This work was supported by the Key Program of National Nature Science Foundation of China under grant No.61333002, the National Nature Science Foundation of China under grants No. 61673056, the Beijing Natural Science Foundation under grant No. 4182039, and the Beijing Key Discipline Construction Project (XK 100080537).

References

1. C. Shaoyong, L. Jueming, Optimization and practice of burden distribution matrix for No.3 BF of Wuhu Xinxing ductile iron pipe limited company. Appl. Practice of New Technol. pp. 193–196 (2017)
2. J.C. Dunn, A fuzzy relative of the ISODATA process and its use in detecting compact well-separated clusters. **3**(3), 32–57 (1974)
3. D.M. Tsai, C.C. Lin, Fuzzy C-means based clustering for linearly and nonlinearly separable data. Pattern Recogn. **44**(8), 1750–1760 (2011)
4. R.J. Hathaway, Y. Hu, *Density-weighted Fuzzy C-means Clustering*. IEEE Press (2009)
5. D.C. Bonzo, A.Y. Hermosilla, Clustering panel data via perturbed adaptive simulated annealing and genetic algorithms. Adv. Complex Syst. **4**, 339–360 (2002)
6. Z. Xiao, B. Li, S. Liu, The discussion on clustering way based on the multi-dimensional panel data and empirical analysis. Appl. Statist. Manage. **28**(5) (2009)
7. Y. Li, X. He, Panel data clustering method and application. Statist. Inf. Tribune, **26**(10) (2010)
8. M.A. Juarez, F.J.S. Mark, Model-based clustering of non-gaussian panel data based on skew-T distributions. J. Business and Econom. Statist. **28**(1) (2010)
9. S.H. Luo, J. Huang, J.S. Zeng et al., Identification of the optimal control center for blast furnace thermal state based on the fuzzy C-means clustering. ISIJ Int. **51**(10), 1668–1673 (2011)

Adaptive Terminal Sliding Mode Trajectory Tracking Control of Mobile Robot Based on Disturbance Observer

Junxiong Yan and Wuxi Shi

Abstract This paper proposes an adaptive terminal sliding mode trajectory tracking control scheme for the wheel mobile robot in the presence of wheel skidding and slipping and unknown center of mass. An auxiliary kinematics controller is designed to make the auxiliary velocity of the robot asymptotically converge to the desired velocity, and a torque controller is designed to make the velocity of the robot converge to the desired velocity within a limited time. The disturbance observer is used to estimate the lumped disturbance. It is proved that all the signals in the closed-loop system are bounded and that the tracking error converges to zero. Simulation results demonstrate the effectiveness of the proposed scheme.

Keywords Wheel mobile robot · Trajectory tracking · Skidding and slipping
Disturbance observer · Terminal sliding mode control

1 Introduction

In recent years, trajectory tracking control of wheeled mobile robot (WMR) has attracted considerable attention, and the research results are reported in [1–6]. Among them, a trajectory tracking control scheme was proposed for two-wheeled mobile robots using sliding mode in [1], an adaptive controller was designed to achieve trajectory tracking in [2], and in [3], a backstepping method was used to design a trajectory tracking controller. Although these control schemes have achieved good results, however, all of them assumed that the robot was under nonholonomic constraints of pure rolling without skidding and slipping. When the robot performs a curve motion on a smooth road surface, it will produce skidding and slipping, which can destroy the nonholonomic of the robot system. In [4], by GPS and other auxiliary sensors on the skidding and slipping disturbance, a tracking controller was devel-

J. Yan · W. Shi (✉)
School of Electrical Engineering and Automation,
Tianjin Polytechnic University, Tianjin 300387, China
e-mail: shiwuxi@163.com

Y. Jia et al. (eds.), *Proceedings of 2018 Chinese Intelligent Systems Conference*, Lecture Notes in Electrical Engineering 529,
https://doi.org/10.1007/978-981-13-2291-4_53

Fig. 1 Model of wheeled mobile robot model

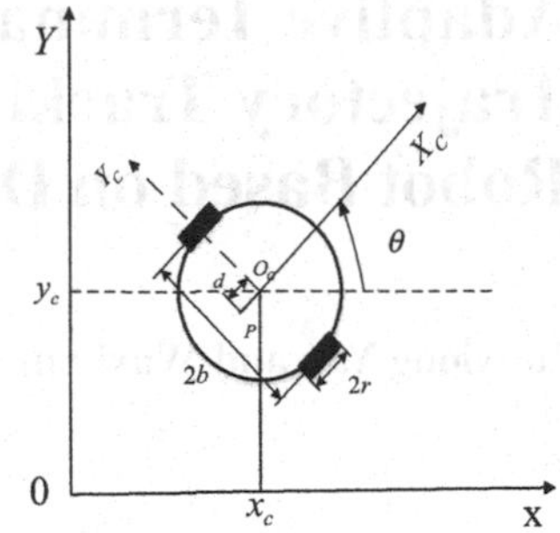

oped for a car-like WMR with skidding and slipping, while in [5], a polar coordinate transformation is used to compensate the robot's skidding and slipping disturbances. Although the above results can effectively overcome the skidding and slipping disturbance in the trajectory tracking, however, the proposed schemes can not consider the influence of internal parameter uncertainty. In [6], by using a disturbance observer to estimate the internal parameters uncertainty, however, the skidding and slipping were not considered.

In this paper, an adaptive terminal sliding mode trajectory tracking controller is proposed for the wheel mobile robot with the wheel skidding and slipping and unknown center of mass. Based on the kinematics model, an auxiliary kinematics controller is designed to make the auxiliary velocity of the robot asymptotically converge to the desired velocity, and an adaptive terminal sliding mode torque controller is developed. The disturbance observer is used to estimate the skidding and slipping disturbance. The proposed scheme ensures that all the signals in the closed-loop system are bounded, and that the tracking errors converge to zero. Simulation results demonstrate the effectiveness of the proposed scheme.

2 Wheeled Mobile Robot Model

In this paper, a two wheeled mobile robot model is considered in Fig. 1. X_c is the moving direction of the WMR, the distance between the two wheels is $2b$, $2r$ is the wheel's diameter. O_c denotes the geometrical center of the WMR, p denotes the mass center of WMR, d represents the distance between the mass and geometrical center. The heading direction θ is defined as the angle between axis X_c and axis X.

The WMR's model is as follows [4]:

$$M(q)\ddot{q} + W(q,\dot{q})\dot{q} + G(q) + \zeta = B(q)\tau - A^T(q)\rho \tag{1}$$

where $q = \begin{bmatrix} x, & y, & \theta, & \varphi_r, & \varphi_l \end{bmatrix}^T$ denotes the position and orientation of WMR, $M(q)$ is the system inertia matrix, $W(q,\dot{q})$ is the Coriolis matrix and centrifugal force,

$G(q)$ is the gravitational vector, ζ is the unknown disturbance, $B(q)$ is the input transformation matrix, $\tau = (\tau_1, \tau_2)^T$ is the two control torques provided by the machine, $A^T(q)$ is the nonholonomic constraint matrix, ρ is a constraint force vector.

With pure rolling and non-slipping nonholonomic constraints, the WMR's model is expressed as

$$A(q)\dot{q} = 0 \tag{2}$$

where $A(q) = \begin{bmatrix} -\sin\theta & \cos\theta & -d & 0 & 0 \\ \cos\theta & \sin\theta & b & -r & 0 \\ \cos\theta & \sin\theta & -b & 0 & -r \end{bmatrix}$.

The kinematics of WMR is obtained as

$$\dot{q} = S(q)\dot{\eta} \tag{3}$$

$S(q)$ is given as follows:

$$S(q) = \begin{bmatrix} \cos\theta & \sin\theta & 0 & 1/r & 1/r \\ -d\,\sin\theta & d\,\cos\theta & 1 & b/r & -b/r \end{bmatrix}^T,$$

where $\eta = \begin{bmatrix} v & w \end{bmatrix}^T$, $v = r(\dot{\varphi}_r + \dot{\varphi}_l)/2$ is linear velocity, $w = r(\dot{\varphi}_r - \dot{\varphi}_l)/(2b)$ is the angular velocity, $[\varphi_r, \varphi_l]^T$ specify the angular positions for the right and left driving wheels. With the skidding and slipping, the WMR's model is expressed as

$$A(q)\dot{q} = \Lambda \tag{4}$$

where $\Lambda = [\phi \ -r\dot{\psi}_r \ -r\dot{\psi}_l]^T$, with $\psi = \begin{bmatrix} \psi_r & \psi_l \end{bmatrix}^T$ being the slipping speed of the left and right wheels, and ϕ being the lateral skidding velocity in WMR.

Consider skidding and slipping, the kinematics of WMR's is [8]

$$\dot{q} = S(q)(\eta - u) + z(q, \phi) \tag{5}$$

where $u = \begin{bmatrix} u_v & u_w \end{bmatrix}^T$, with $u_v = r(\psi_r + \psi_l)/2$ being the longitudinal slip velocity, $u_w = r(\psi_r - \psi_l)/(2b)$ being the yaw rate perturbation due to the slippage of the wheels, and $z(q, \phi) = \begin{bmatrix} -\phi\,\sin\theta & \phi\,\cos\theta & 0 & \psi_r & \psi_l \end{bmatrix}^T$ being the unmatched disturbance vector induced from the perturbed nonholonomic constraints.

Substituting (4) and (5) into (1), we have

$$\dot{\eta} = N(q)\tau + E(q, \dot{q})\eta + D \tag{6}$$

where $N(q) = -((S^T(q)B(q))^{-1}S^T(q)M(q)S(q))^{-1}$, D is disturbance, $E(q, \dot{q}) = N(q)(S^T(q)B(q))^{-1}S^T(q)(M(q)\dot{S}(q) + W(q, \dot{q})S(q))$.

From the expression of matrices $M(q)$, $B(q)$, $W(q)$, $S(q)$, we obtain $E(q, \dot{q}) = 0$, then (6) can become as

$$H\dot{\eta} = C\tau + \delta \tag{7}$$

where $H = \begin{bmatrix} h_1 & 0 \\ 0 & h_2 \end{bmatrix}$, $C = \begin{bmatrix} 1 & 1 \\ 1 & -1 \end{bmatrix}$, $h_1 = \frac{3br^2}{-md^2r^2+Ir^2+2b^2I_w}$, $h_2 = \frac{2r^2}{mr^2+2I_w}$, $\delta = HD$.

3 Wheeled Mobile Robot Controller Design

In this section, an auxiliary kinematics controller is designed to make the auxiliary speed of the robot asymptotically converge to the desired speed. Then, a torque controller is designed to generate real linear and angular velocities to track the desired one. The disturbance observer is used to estimate the lumped disturbance. Assume that the kinematic model of reference trajectory of the mobile robot is as follows:

$$\begin{cases} \dot{x}_r = v_r \cos\theta_r \\ \dot{y}_r = v_r \sin\theta_r \\ \dot{\theta}_r = w_r \end{cases} \tag{8}$$

where v_r is the desired linear velocity, w_r is the desired angular velocity, and (x_r, y_r, θ_r) is the desired reference coordinate. The tracking error is defined as

$$\begin{bmatrix} x_e \\ y_e \\ \theta_e \end{bmatrix} = \begin{bmatrix} \cos\theta & \sin\theta & 0 \\ -\sin\theta & \cos\theta & 0 \\ 0 & 0 & 1 \end{bmatrix} \begin{bmatrix} x_r - x \\ y_r - y \\ \theta_r - \theta \end{bmatrix} \tag{9}$$

By some manipulation, from (9) we have

$$\begin{bmatrix} \dot{x}_e \\ \dot{y}_e \\ \dot{\theta}_e \end{bmatrix} = \begin{bmatrix} -1 & y_e \\ 0 & -x_e - d \\ 0 & -1 \end{bmatrix} \begin{bmatrix} v \\ w \end{bmatrix} + \begin{bmatrix} v_r \cos e_3 - w_r d \sin e_3 \\ v_r \sin e_3 + w_r d \cos e_3 \\ w_r \end{bmatrix} \tag{10}$$

The kinematics auxiliary controller is designed as follows [9]:

$$\eta_c = \begin{bmatrix} v_c \\ w_c \end{bmatrix} = \begin{bmatrix} v_r \cos\theta_e + k_1 x_e \\ w_r + k_2 \sin\theta_e + k_3 v_r y_e \end{bmatrix} \tag{11}$$

where $k_1, k_2, k_3 > 0$ are design parameters, v_c is auxiliary linear velocity and w_c is auxiliary angular velocity. Substituting (11) into (10) yields

$$\begin{bmatrix} \dot{x}_e \\ \dot{y}_e \\ \dot{\theta}_e \end{bmatrix} = \begin{bmatrix} -1 & y_e \\ 0 & -x_e - d \\ 0 & -1 \end{bmatrix} \begin{bmatrix} v_r \cos\theta_e + k_1 x_e \\ w_r + k_2 \sin\theta_e + k_3 v_r y_e \end{bmatrix} + \begin{bmatrix} v_r \cos e_3 - w_r d \sin e_3 \\ v_r \sin e_3 + w_r d \cos e_3 \\ w_r \end{bmatrix} \tag{12}$$

Choose a Lyapunov function as

$$V_1 = \frac{1}{2}(x_e + d - d\cos\theta_e)^2 + \frac{1}{2}(y_e - d\sin\theta_e)^2 + \frac{(1-\cos\theta_e)}{k_3} \tag{13}$$

The first order derivative of V_1 is

$$\dot{V}_1 = -k_1 x_e^2 - \frac{k_2}{k_3}(\sin\theta_e)^2 - d v_r(\sin\theta_e)^2 - k_1 d x_e(1-\cos\theta_e) \tag{14}$$

For unknown constant d, we design the following adaptive law

$$\dot{\hat{d}} = -v_r(\sin\theta_e)^2 - k_1 x_e(1-\cos\theta_e) \tag{15}$$

where $\hat{d}$ is the estimated value of d. Using the estimated value $\hat{d}$, (7) can be rewritten as

$$\bar{H}\dot{\eta} = C\tau + \xi \tag{16}$$

where $\bar{H} = \begin{bmatrix} \hat{h}_1 & 0 \\ 0 & h_2 \end{bmatrix}$, $\hat{h}_1 = \frac{3br^2}{-m\hat{d}^2r^2+Ir^2+2b^2I_w}$, $\xi = \delta - (h_1 - \hat{h}_1)\dot{v}$ is the lumped disturbance. To compensate the lumped disturbance ξ, the disturbance observer is designed. Motivated by Dawei et al. [6], we construct the following disturbance observer

$$\begin{aligned} \dot{z} &= -L\bar{H}^{-1}z - L(\bar{H}^{-1}C\tau + \bar{H}^{-1}L\eta) \\ \hat{\xi} &= z + L\eta \end{aligned} \tag{17}$$

where $\hat{\xi}$ is the estimate of ξ, L represents the gain matrix of the observer, $z = \begin{bmatrix} z_1 & z_2 \end{bmatrix}^T$ is the internal state variables of the observer.

Assumption 1 ξ and its derivatives are bounded, and $\lim_{t\to\infty}\|\dot{\xi}\| = 0$.

Define the estimate error as $\tilde{\xi} = \xi - \hat{\xi}$. According to (16) and (17), one has

$$\dot{\tilde{\xi}} = \dot{\xi} - \dot{\hat{\xi}} = -\dot{z} - L\dot{\eta} = -L(\xi - \hat{\xi}) = -L\tilde{\xi} \tag{18}$$

By adjusting the gain matrix L, $\hat{\xi}$ can exponentially converge to ξ.

According to (16), one has

$$\bar{H}\dot{\eta}_e = C\tau - \bar{H}\dot{\eta}_c + \xi \tag{19}$$

where $\eta_e = \eta - \eta_c = \begin{bmatrix} v - v_c \\ w - w_c \end{bmatrix} = \begin{bmatrix} v_e \\ w_e \end{bmatrix}$.

To make the actual velocity of the robot converges to the desired velocity within a limited time, define a sliding variable s as [7]

$$s = \sigma + \lambda sig(\dot{\sigma})^{\gamma} \tag{20}$$

where $\sigma = \int_0^t \eta_e(\varsigma)d\varsigma$, $1 < \gamma < 2$, $\lambda = diag\{\lambda_1, \lambda_2\} > 0$, $sig(\dot{\sigma})^{\gamma} = |\dot{\sigma}|^{\gamma}\text{sgn}(\dot{\sigma})$.

The time derivative of s is

$$\dot{s} = \dot{\sigma} + \lambda\gamma|\dot{\sigma}|^{\gamma-1}\ddot{\sigma} = \eta_e + \lambda\gamma|\eta_e|^{\gamma-1}\dot{\eta}_e \tag{21}$$

Design the torque controller as follows

$$\tau = C^{-1}\bar{H}[-\alpha s - \beta sig(s)^{\varepsilon} - j\,\text{sgn}(s) - \bar{H}^{-1}\hat{\xi} - \frac{|\eta_e|^{2-\gamma}}{\lambda\gamma} + \dot{\eta}_c + \frac{|\eta_e|^{1-\gamma}s^{-1}Y\hat{d}}{\lambda\gamma}] \tag{22}$$

where $0 < \varepsilon < 1$, $Y = v_r(\sin\theta_e)^2 + k_1x_e(1 - \cos\theta_e)$, $\alpha > 0$, $\beta > 0$, $\delta > 0$.

Theorem 1 *For the WMR system, if the kinematics controller (11) and the torque controller (22) with the disturbance observer (17) are applied, then all the signals in the closed-loop system are bounded, and the tracking errors x_e, y_e, θ_e converge to zero.*

Proof Choose the Lyapunov function as

$$V_2 = V_1 + \frac{1}{2}s^T s + \frac{1}{2}\tilde{d}^2 \tag{23}$$

Where $\tilde{d} = d - \hat{d}$. The first order derivative of V_2 is

$$\dot{V}_2 = V_1 + s[\eta_e + \lambda\gamma|\eta_e|^{\gamma-1}\bar{H}^{-1}(C\tau - \bar{H}\dot{\eta}_c + \xi)] - \tilde{d}\dot{\hat{d}} \tag{24}$$

Denote $\phi(\eta_e) = \lambda\gamma|\eta_e|^{\gamma-1}$, substituting (22) into (24), we obtain

$$\dot{V}_2 \le -k_1x_e^2 - \frac{k_2}{k_3}(\sin\theta_e)^2 - \phi(\eta_e)[(j - \bar{H}^{-1}\tilde{\xi})\|s\| + \alpha s^2 + \beta|s|^{\varepsilon+1}] \tag{25}$$

From (25), we can select $j - \bar{H}^{-1}\tilde{\xi} > 0$ to ensure that $\dot{V}_2 \le 0$. Therefore, all the signals in the closed-loop system are bounded.

Choose the Lyapunov function as $V_3 = \frac{1}{2}s^2$, we have

$$\dot{V}_3 \leq -\phi(\eta_e)((\alpha - \phi(\eta_e)^{-1}Y\hat{d}|s|^{-2})s^2 + \beta s^{\varepsilon+1}) \tag{26}$$

If there exists α_0 and $\alpha - \phi(\eta_e)^{-1}Y\hat{d}|s|^{-2} > \alpha_0 > 0, \sigma \neq 0, \dot{\sigma} \neq 0$, then there exists $\vartheta_1 > 0, \vartheta_2 > 0, \alpha_0\phi(\eta_e) \geq \vartheta_1, \beta\phi(\eta_e) \geq \vartheta_2$ such that

$$\dot{V}_3 \leq -\vartheta_1 s^2 - \vartheta_2|s|^{\varepsilon+1} = -2\vartheta_1 V_3 - 2^{\frac{\varepsilon+1}{2}}\vartheta_2 V_3^{\frac{\varepsilon+1}{2}} \tag{27}$$

According to [7], from (27), we can know that V_3 convergence to the origin in a limited time, which follows that s also convergence to the origin in a limited time t_r, where $t_r \leq \frac{1}{(1-\varepsilon)} In[1 + \frac{\vartheta_1}{\vartheta_2}(2V)^{\frac{\varepsilon+1}{2}}]$.

If $\dot{V}_2 = 0$, one has

$$x_e^2 = 0, \sin\theta_e = 0, -k_1 d x_e(1 - \cos\theta_e) = 0, s = 0 \tag{28}$$

which means that

$$\theta_e = 0, x_e = 0, v = v_c, w = w_c \tag{29}$$

From (10) and (11), one has

$$k_2 \sin\theta_e + k_3 v_r y_e = w_c - w_r, \dot{\theta}_e = w_r - w \tag{30}$$

Substituting (29) into (30) yields $y_e = 0$, therefore, only $(x_e, y_e, \theta_e) = (0, 0, 0)$, $\dot{V}_2 = 0$. From the above discussion, the closed-loop system is asymptotically stable around $(x_e, y_e, \theta_e) = (0, 0, 0)$. Thus the tracking errors x_e, y_e, θ_e converge to zero.

4 Simulation

The main physical parameters of the WMR as follows:

$r = 0.045\text{m}, I_\omega = 1.05\,\text{kg}\cdot\text{m}^2, m = 3.79\,\text{kg}, b = 0.118\,\text{m}, \; I = 2.4\,\text{kg}\cdot\text{m}^2.$

The WMR start point is $(x, y, \theta) = (0, 0, 0)$, the reference velocities are set $v_r = 1$ m/s,$w_r = 1$ rad/s, the desired trajectory is $x_r = \sin t$,$y_r = -\cos t$, $\theta_r = t$. In simulation, the disturbance will be added between 8–10 s, the disturbance is set as $\delta = \begin{bmatrix} (5\sin(t) + 3)f(t) \\ -(10\sin(t) + 3)f(t) \end{bmatrix}$, and $f(t) = u(t-8) - u(t-10)$ is a step function. The

Fig. 2 The trajectory diagram of the mobile robot

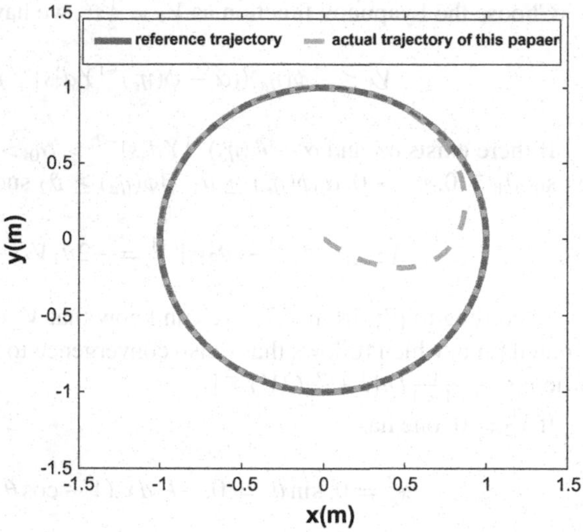

Fig. 3 The output torque of WMR

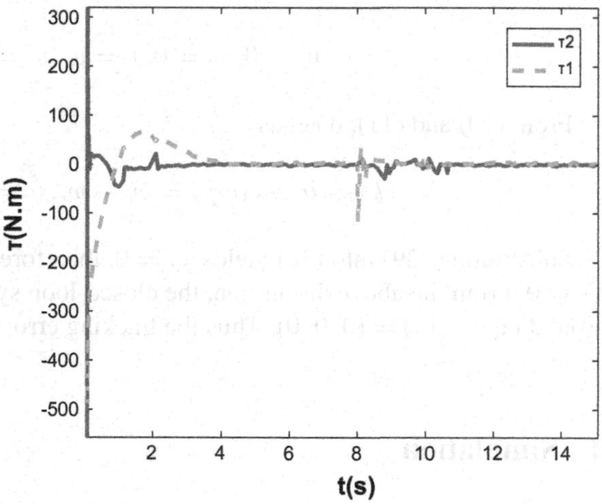

parameters of the observer are chosen as $L = \begin{bmatrix} l_1 & 0 \\ 0 & l_2 \end{bmatrix}, l_1 = 20, l_2 = 15, z_1(0) = z_2(0) = 1$.

The parameters of the kinematics controller are chosen as $k_1 = 1.5, k_2 = 77, k_3 = 64$ the parameters of the dynamic controller are chosen as $\alpha = diag\{8, 21\}, \delta = diag\{9, 1.1\}, \varepsilon = diag\{0.06, 0.06\}, \lambda = diag\{0.0056, 0.004\}, \gamma = diag\{13/11, 13/11\}, \beta = diag\{80, 100\}$.

The simulation results are shown in Figs. 2, 3, and 4. From the simulation results, it can be observed that the designed disturbance observer can quickly track the

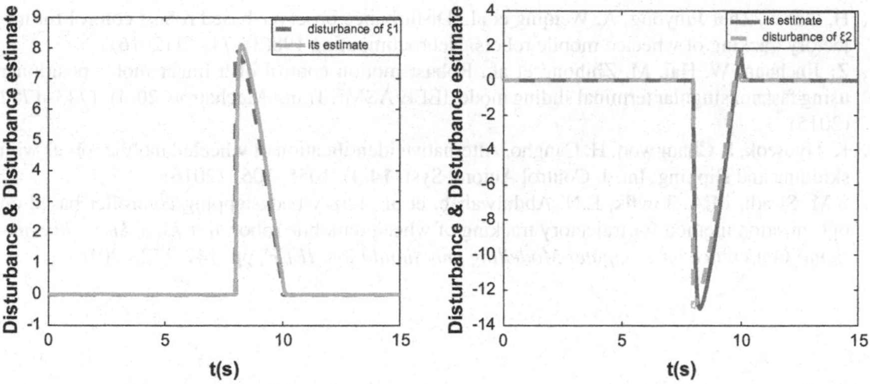

Fig. 4 Disturbance and its estimate

disturbances. Moreover, all the signals in the closed-loop system are bounded, and the tracking errors x_e, y_e, θ_e converge to zero.

5 Conclusion

In this paper, an adaptive terminal sliding mode trajectory tracking control based on disturbance observer has been proposed to resolve the skidding and slipping disturbances. The scheme ensures that all signals in the closed-loop system are bounded, and that the velocity of the robot converges to the desired velocity within a limited time. It has been proved that all the signals in the closed-loop system are bounded, and that the tracking errors converge to zero. Simulation results demonstrate the effectiveness of the proposed scheme.

References

1. M. Jianqiu, Y. Xinggang, S.K. Spurgeon, et al., Trajectory tracking control of a two-wheeled mobile robot using sliding mode techniques, in *Control Conference. IEEE*, pp. 3307–3312 (2015)
2. T. Fukao, H. Nakagawa, N. Adachi, Adaptive tracking control of a nonholonomic mobile robot. Rob. Autom. IEEE Trans. **16**(5), 609–615 (2004)
3. W.M.E. Mahgoub, I.M.H. Sanhoury, Back stepping tracking controller for wheeled mobile robot. in *International Conference on Communication, Control, Computing and Electronics Engineering. IEEE*, pp. 1–5 (2017)
4. C.B. Low, W. Danwei, GPS-based tracking control for a car-like wheeled mobile robot with skidding and slipping. IEEE/ASME Trans. Mechatronics, **13**(4), 480-484 (2009)
5. S.J. Yoo, Adaptive tracking control for a class of wheeled mobile robots with unknown skidding and slipping. IET Control Theory Appl. **4**(10), 2109–2119 (2010)

6. H. Dawei, Zhai Junyong, A. Weiqing et al., Disturbance observer-based robust control for trajectory tracking of wheeled mobile robots. Neurocomputing **198**(3), 74–79 (2016)
7. Z. Jinchuan, W. Hai, M. Zhihong et al., Robust motion control of a linear motor positioner using fast nonsingular terminal sliding mode. IEEE/ASME Trans. Mechatron. **20**(4), 1743–1752 (2015)
8. K. Hyoseok, P. Changwoo, H. Cangho, Alternative identification of wheeled mobile robots with skidding and slipping. Int. J. Control Autom. Syst. **14**(4), 1055–1062 (2016)
9. S.M. Swadi, M.A. Tawfik, E.N. Abdulwahab, et al., Fuzzy-backstepping controller based on optimization method for trajectory tracking of wheeled mobile robot, in *Uksim-Amss, International Conference on Computer Modelling and Simulation. IEEE*, pp. 147–152, (2016)

Combining STFT and Random Forest Algorithm for Epileptic Detection

Xiashuang Wang, Guanghong Gong and Ni Li

Abstract In the automatic detection of epileptic seizures, time varying electroencephalography (EEG) signals monitoring of critically ill patients is an essential procedure in intensive care units. There is increasing interest in using seizure detection algorithms, such as random forest, for seizures EEG analysis, but a better understanding of how to design and train random forest for EEG decoding and how to visualize the informative EEG time and frequency features the dimensionality reduction of PCA is still needed. Here, we studied seizure detection algorithms designed for recognizing diseased signals from raw seizures EEG. Our results show the recognizing performance of random forest algorithm reaching at mean recognizing accuracies 96%. It can exploit and might help doctors better diagnose the extent of epilepsy.

Keywords Epileptic detection · RF · Time-frequency analysis · Simulation model

1 Introduction

Epilepsy is the clinical manifestation of hyperpolarizing electrical activity in paroxysmal neurons in the brain, which is characterized by recurrent, sudden and transient [1]. Continuous electroencephalography (cEEG) is a significant tool for the diagnosis and treatment of epilepsy diseases of ill patient in intensive care units (ICU). Other physiological detection methods cannot reflect the seizure information in real time as compared with cEEG [2]. However, the epileptic seizures signals can't be monitored in the short term, so it is indispensable to raw EEG continuously over a long period, which is typically 20 min recording of the patient's brain waves. It is a tedious and time-consuming task to detect epileptic seizures through visual examination of experienced doctors. Therefore, an accurate and efficient automatic detection algorithm included time information, frequency information, time-frequency information and

---

X. Wang · G. Gong · N. Li (✉)
School of Automation and Electrical Engineering,
Beihang University, Beijing, China
e-mail: lini@buaa.edu.cn

Y. Jia et al. (eds.), *Proceedings of 2018 Chinese Intelligent Systems Conference*, Lecture Notes in Electrical Engineering 529,
https://doi.org/10.1007/978-981-13-2291-4_54

nonlinearity for epilepsy is very important and urgent in the long time monitoring and detection.

In past years, some studies have solved the problem of non-stationary EEG activity of automatic monitoring and detection of epileptic. Tzallas et al. had used to the time domain method through computerized EEG analysis during epileptic seizures. Additionally, a popular Fourier-based technique of spectral analysis method is commonly employed for EEG signals of analysis in the frequency domain. Manish Sharmaa et al. proposed a time-frequency method for detection of active seizure EEG in long-term no active EEG using flexible wavelet transform and fractal dimension [3]. Lately, based multiscale radial basis functions algorithm have shown promising results in epileptic seizure EEG decoding by Yang et al. [4]. Because of epileptic EEG signals is refers to the qualitative and the authenticity of multi-component. The following content briefly discuss the widely used TFA as short time fourier transform (STFT), WT and power spectral density (PSD) of the parametric method to analyze EEG and has been widely applied to the automatic detection of epileptic [5, 6, 7].

At the same time, medical workers also want to understand how machine learning models extract information from brain signals in many cases. To detect time interval variations EEG input data of epileptic seizures have been used by Machine learning techniques including three categories: the Generative Model, the Linear Classifiers, and the Non- Linear Classifiers. In past years, numerous machining learning models of seizure detection have been developed [8–10].

Although the above algorithm classifies two types of data, it has achieved good results. For different degree conditions, only relying on two or two classifications of EEG signals are not only troublesome, but also time consuming, which is totally unfavorable for practical applications. From the actual point of view, this paper believes that three types of detection and classification of cEEG signals directly including healthy people, severe epilepsy, and intermittent epilepsy. This detection method not only saves time, but also solves the problems that reading doctors are tired to dealing with diagnosis and resulting in misdiagnosis or missed diagnosis. Make medical care more accurate.

The paper proposes an automatic detection method of epileptic seizures based on the time-frequency image using combined short-time fourier transform (STFT) and random forest (RF) algorithm. The statistical features including mean, variance, in the histogram of segmented gray scale time-frequency image have been extracted. A hypothesis testing with lower p-values indicates that the four waveforms (theta, alpha, beta and gamma) with features of mean, variance and skewness are highly determinant.

The paper adopts the most effective PCA dimension reduction feature processing method for cEEG signals [11], which is easier to process and use through the reduced-dimension epilepsy EEG signals, thereby reducing the overhead of the algorithm.

The optimal features of theta, alpha, beta and gamma waves are fed into the RF for classification of seizure and non-seizure EEG signals. Experiments are conducted in ten cross validation to test the performance of the proposed method. The results show that the proposed method can achieve the best average three type's accuracy of 96% and gain better recognizing accuracy.

The rest of the paper is organized as follows: the methodology of whole experiment including STFT method, feature extraction and RF classifier are introduced in Sect. 3. The results of his experiment for the classification of seizure and non-seizure EEG are given in Sects. 2, 3, 4, 5. Finally, Sect. 6 concludes the paper.

2 Simulation EEG Data Model

First, we use the fourier transform, multitaper spectrum analysis, PACF and (STFT) to describe the time-frequency characteristics of EEG simulation model.

$$y(t) = \begin{cases} 2|t|^{\omega}\sin(2\pi f_{\theta}t), t \in [0,2); \\ 2|t|^{\omega}\sin\left(2\pi f_{\beta}t\right), t \in [2,4); \\ 2|t|^{\sigma}\sin(2\pi f_{\alpha}t), t \in [4,6); \\ 2|t|^{\sigma}\sin\left(2\pi f_{\gamma}t\right), t \in [6,8); \\ 0, otherwise \end{cases} \tag{1}$$

The frequency components mean to emulate the four sub-bands theta (4–8 Hz), alpha (8–16 Hz), beta (16–32 Hz), and gamma (32–60 Hz) bands of EEG recording respectively. The above signals is sampled with a sampling frequency 100 Hz, and corrupted by a Gaussian white noise sequence of variance 0.04, and thus a total of 800 observations are obtained. According to the EEG simulation data expression, we see clear four peak and lines on the spectrum that can detect four frequency components including 7, 15, 25, 40 Hz by FFT, STFT time frequency analysis as shown following figures in the first row. This indicates that the original time domain signals mainly contains four frequency signals. They are also the point of energy concentration, which is consistent with the simulating signals. There are some small peaks except these four peaks by using the FFT, that is, there is a spectral leakage, which may cause the spectrum to be blurred and distorted.

- Actual clinical EEG dataset analysis and verification

Twenty-five sampled patients with medically intractable partial epilepsy noninvasive EEG data are derived from the German epilepsy laboratory in Bonn. The patients reported are consecutive patients who are selected according to the following inclusion criteria.

The datasets are divided into five groups of ictal scalp EEG signals: O, Z, F, N and S. Each group of data contained 100 samples and the subjects are 5 people. Each sample contains 4097 sample points with time of 23.6 s, and the signals record is based on a standard 10–20 system with a sampling frequency of 173. 61 Hz. The age of subjects ranged from 19 to 60 years old with right-handed and the location of epileptogenic foci for each subject are determined by skilled epileptologists.

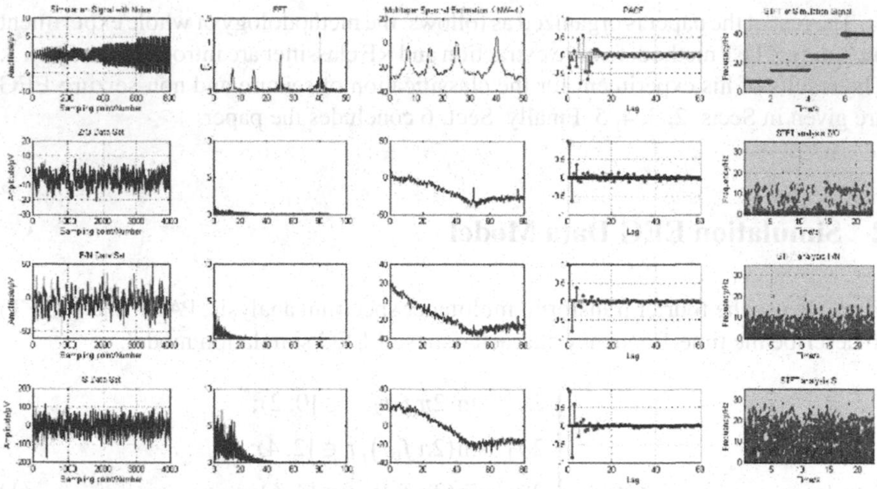

Fig. 1 Simulation and actual clinical EEG data analysis

This paper uses O/Z, F/N and S datasets to classify EEG signals. The O/Z dataset are healthy people under a state of alert only on the surface of the scalp EEG signals. The dataset F/N did not attack for epilepsy patients in the period between lesions within the area of intracranial EEG signals. The S dataset are ill patients with epilepsy in episodes cause lesions within the area of intracranial EEG signals respectively to stage between the normal and attacks and attacks the brain electrical signals. The cEEG signals are shown in Fig. 1.

- Discussion

EEG is mainly distributed in the frequency range of 0–40 Hz. Therefore, in the experiment, a butterworth digital low-pass filter with cut off frequency of 40 Hz and an order of 49 stages is used to filter the EEG and extract the effective frequency band. Then use Fourier transform, multitaper spectral analysis, PACF, STFT to describe the time-frequency characteristics of the signals. In order to avoid negative energy, the absolute value of the transformed data is taken in the experiment. STFT treats every non-stationary signals as a stationary signals, superimposes a series of small signals, and performs windowing to obtain a two-dimensional function of the signals on the time-frequency plane. After the optimal order fractional Fourier transform of the EEG signals, the absolute value of the obtained transform result is taken. Then, we use the Hamming window with a window width of 256 points, STFT is performed on the transform results to extract the time-frequency characteristics of three types of brain electrical signals: Z/O, F/N, and S. Their corresponding time-frequency energy distributions. After the above transformation, each sample is decomposed into a 129*69 time-frequency matrix. The obtained time-frequency matrix is vectorized on this basis, and the original time-frequency characteristic matrix is converted into a 4907*1 one-dimensional vector. The Z/O, F/N and S are processed in the above manner to obtain the time-frequency feature vectors of the three types of samples. Each type of sample corresponds to a one-dimensional matrix [12, 13]. For convenience of description, Z/O 4907*200, F/N 4907*200, and S4907*100 are used to represent the time-frequency characteristics of normal EEG signals, intermittent seizures, and EEG signals during seizures. In order to achieve the credibility of the test results, the three groups of data need be arithmetic averaged processing, compressed into a single column of 4097*1 matrixes. The above Fig. 1. illustrates that the fourier transform analysis the frequency of the EEG. It reflects the frequency feature of the entire signals. However, smooth signals using FFT is possible. For EEG signals, STFT comes into being. The window is added to the signals, and then the window function is moved. Assuming that the windowed signals is a stationary signals in different time widths, the power spectrum at different moments can be calculated.

Compared with the above results, the spectra of the three data sets that Z/O, F/N and S are significantly different and the analysis results of STFT are clearer. Compared with the Z/O and F/N data sets, thc EEG signals in the S dataset have very higher energy. There are wave peaks in the alpha band and the theta band, especially the S dataset with high energy near 0–3 and 7 Hz of frequencies. The overall energy of the brain of F/N data sets are lower that slow wave activity with low amplitude mainly and the wave peak is not particularly obvious. The brain power of Z/O is the lowest in the five datasets. Therefore, it can be inferred that the brain power of S dataset is severe and persistent epileptic. The brain powcr of the F/N dataset is a mild epileptic person with latent epilepsy. In the Z/O dataset, the spectrum of the EEG showed a little energy concentration, and the two data sets are the EEG data of healthy pcople.

3 STFT and PCA Based Feature Analysis

The cEEG signals scalp at time t can be defined as a vector:

$$\mathrm{S(t)} = (\mathrm{s}(t_1), \mathrm{s}(t_2), \cdots, \mathrm{s}(t_n)) \tag{2}$$

S(t) (n = 1, 2, … , 4097) is the cEEG. For each segment of EEG, a MVAR model with order p can be constituted as follows:

$$\mathrm{S(t)} = \sum_{i=1}^{p} \omega_i S(t-i) + \varepsilon(t) \tag{3}$$

ω_i is a matrix of coefficients and ε(t) is the estimation error which is time sequence of Gaussian white noise. The order p can be depended by the Schwarz's Bayesian Criterion. Covariance inflation criterion The 4 order of MVAR model can describe the feature of cEEG signals. Therefore, the STFT of coefficient ω_i can be calculated by:

$$\omega(\mathrm{f}) = \sum_{i=1}^{p} \omega_i e^{-2\pi i r f} \tag{4}$$

The average frequency, standard deviation, and average amplitude of the EEG signal are extracted as features for identification to form a three-dimensional observation vector. Among them, the average frequency is:

$$\mathrm{P} = \frac{\sum f_j M_j}{\sum M_j} \tag{5}$$

where M_j is the power spectrum value of frequency. It is very important to choose the right model type and order for power spectrum estimation based on the parametric model; otherwise, it may cause large errors in the results.

The time-frequency analysis reveals the frequency distribution of the signal and the regularity of each frequency component over time. The principle of STFT is to use a window function $k(\tau - t)$ to extract a section of EEG centered on a certain moment and make a Fourier

The leaf transformation then moves the window function to repeat the above calculations for different moments. The expression is as follows:

$$T_s(t, f) = \int_{-\infty}^{+\infty} s(\tau)k^*(\tau - t)e^{-jf\tau}d\tau \tag{6}$$

In this paper, the PCA model is used to reduce the dimension of EEG signals, which makes it easier to process and use the reduced-dimension epileptic EEG, thereby reducing the overhead of the algorithm. Harikumar etc. introduce 14 kinds of dimensionality reduction techniques1 that be used for epileptic EEG processing. The relevant important features are more easily revealed from the data. In the PCA algorithm, the EEG signal data is converted from the original coordinate system to the new coordinate system, and the selection of the new coordinate system is determined by the data itself. Because the maximum variance of the data gives the most important information of the data. When converting the coordinate system, the direction with the largest variance is used as the coordinate axis direction. The first new coordinate axis selects the method with the largest variance S_i' in the original data, and the second new coordinate axis selects the orthogonality with the first new coordinate axis and the direction of the second largest variance S_j'. Repeat this process, the number of times is the feature dimension of the original data.

$$S_i' * S_j' = 0, (\mathrm{i} \neq \mathrm{j}) \tag{7}$$

In the new coordinate system obtained in this way, most of the variances are contained in the first several axes, and the subsequent axes contain a variance of almost zero. In fact, this method retains dimensional features that contain most of the variance and ignores feature dimensions that contain variances of almost zero, so achieves dimensionality reduction of data features.

4 Random Forest Design Method

At present, the literature is divided into two groups for this dataset. In this paper, the dataset is directly divided into three categories including healthy people, intermittent epilepsy, and continuous epilepsy, reducing unnecessary troubles in one-step. We use random forest algorithm for classification, which combines the fully grown classification and regression tree (C&RT) decision tree into a bagging form, resulting in a large decision model. There are three main advantages of the random forest algorithm. Parallel training of different hosts can generate first, different decision trees with high efficiency. Second, the random forest algorithm inherits the advantages of C&RT. Third, all decision trees are combined by bagging to avoid a single decision tree. Caused a problem with fitting.

5 Analysis of Experimental Results

The classification performance of the RF model is determined on the statistical analysis results of all subjects to avoid probable deviations from the testing results of epileptic seizures in single sample of trial run. Evaluating indicators of performance of the proposed algorithm include accuracy, selectivity, sensitivity [14], specificity, average detection rate, which are defined as shown in Eqs:

1. Accuracy (ACC): this evaluation indicator is used to detect the interictal and ictal cEEG.

$$\mathrm{ACC} = \frac{TP + TN}{TN + FP + TP + FN} 100\% \tag{8}$$

2. Selectivity (SEL): this evaluation indicator is used to reject false detections of ictal.

$$\mathrm{SEL} = \frac{TP}{TP + FP} 100\% \tag{9}$$

3. Sensitivity (SEN): this evaluation indicator is used to detect ictal.

$$\mathrm{SEN} = \frac{TP}{TP + FN} 100\% \tag{10}$$

4. Specificity (SPE): this evaluation indicator is used to interictal segments.

Table 1 Evaluation of statistical indicators

Index	ACC	SEL	SEN	SPE	ADR
Percentage	96.7	96.2	95.6	96.4	96.0

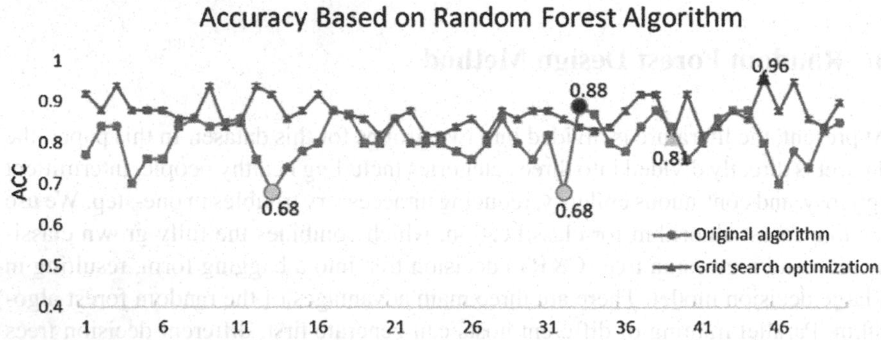

Fig. 2 Accuracy of classification of actual clinical EEG

$$\mathrm{SPE} = \frac{TN}{TN + FP} 100\% \tag{11}$$

5. Average detection rate (ADR): this evaluation indicator is used to compute the average of SEN and SPE.

$$\mathrm{ADR} = \frac{SEN + SPE}{2} 100\% \tag{12}$$

where a true positive (TP) identifies the algorithm of recognizing the total number of true normal events correctly. A true negative (TN) denotes the total number of true events of epileptic seizure period correctly. False positive (FP) and false negative (FN) are the detection algorithm identifies the total number of false normal events false events of epileptic seizure period incorrectly by the epileptologists. We evaluate the statistical indicators such as ACC, SEL, SEN, SPE, ADR using detection algorithm to EEG, as seen from (Table 1).

In this study, 10 fold cross-validation is used to obtain more reliable and stable algorithm performance results. The cEEG datasets are divided into ten subsets randomly, which only one subset is used for test set. While the other residual subsets are considered train set to classify cEEG of different level of epileptic seizure patient. At last, the 10 fold across-validation results are acquire by computing the average indicators of ten trials to evaluate the recognizing algorithm. At the same time, only the train and test subsets cause the over-fitting phenomenon to increases the credibility of classifying data. The correct classification accuracy is shown in Fig. 2.

6 Conclusion

In this paper, the simulation experiment shows that the model of EEG with the CPU simulation takes a long time to adopted time-frequency analysis method of STFT to detect the frequency of epileptic seizure EEG. Then, we introduce the random forest algorithm to classify the three type's epilepsy EEG data. From experiments results, we can see that it is very suitable for epilepsy EEG, but when the number of EEG is more, its calculating time will be longer. In the actual clinical application, the scale of the detection is not too large. In the future, we intend to implement the optimized model of machine learning to realize classification of different types of EEG data to assist doctors in diagnosis.

References

1. S. Patidar, T. Panigrahi, Detection of epileptic seizure using Kraskov entropy applied on tunable-Q wavelet transform of EEG signals. Biomed. Signal Process. Control **34**, 74–80 (2017)
2. D. Wang, D. Ren, K. Li, Y. Feng, D. Ma, X. Yan, et al., Epileptic seizure detection in long-term eeg recordings by using wavelet-based directed transfer function. IEEE Trans. Biomed. Eng. **6**:1–1 (2018)
3. M. Sharma, R.B. Pachori, Acharya U. Rajendra, A new approach to characterize epileptic seizures using analytic time-frequency flexible wavelet transform and fractal dimension. Pattern Recogn. Lett. **94**, 11–12 (2017)
4. L. Yang, W. Xu, L. Lin, L. Ke, Y. Xiao, G. Qi, Epileptic seizure classification of EEGs using time-frequency analysis based multiscale radial basis functions. IEEE J. Biomed. Health Inf. **25**, 1–7 (2017)
5. K. Fu, J. Qu, Y. Chai, Y. Dong, Classification of seizure based on the time-frequency image of EEG signals using HHT and SVM. Biomed. Signal Process. Control **13**, 15–22 (2014)
6. A.T. Tzallas, M.G. Tsipouras, D.I. Fotiadis, Epileptic seizure detection in EEGs using time-frequency analysis. IEEE Trans. Inf. Technol. Biomed. A Publication of the IEEE Eng. Med. Biol. Soc. **13**, 703–10 (2009)
7. B. Boashash, L. Boubchir, G. Azemi, A methodology for time-frequency image processing applied to the classification of non-stationary multichannel signals using instantaneous frequency descriptors with application to newborn EEG signals. EURASIP J. Adv. Signal Process. **12**, 117–119 (2012)
8. R.B. Pachori, R. Sharma, S. Patidar, Classification of normal and epileptic seizure EEG signals based on empirical mode decomposition: springer international publishing **66**, 34–36 (2015)
9. C. Guerrero-Mosquera, A.M. Trigueros, J.I. Franco, Á. Navia-Vázquez, New feature extraction approach for epileptic EEG signal detection using time-frequency distributions. Med. Biol. Eng. Compu. **48**, 321–324 (2010)
10. R.R. Sharma, R.B. Pachori, Time–frequency representation using IEVDHM–HT with application to classification of epileptic EEG signals. IET Sci. Meas. Technol. **12**, 72–82 (2018)
11. J.G. Bogaarts, D.M. Hilkman, E.D. Gommer, K.M. Van, J.P. Reulen, Improved epileptic seizure detection combining dynamic feature normalization with EEG novelty detection. Med. Biol. Eng. Compu. **54**, 1–10 (2016)
12. Y. Jia, Alternative proofs for improved LMI representations for the analysis and the design of continuous-time systems with polytopic type uncertainty: a predictive approach. Autom. Control IEEE Trans. **48**, 1413–1416 (2003)
13. Y. Jia, Robust control with decoupling performance for steering and traction of 4WS vehicles under velocity-varying motion. IEEE Trans. Control Syst. Technol. **8**, 554–69 (2000)
14. H.A. Haider, R. Esteller, C.D. Hahn, M.B. Westover, J.J. Halford, J.W. Lee, et al., Sensitivity of quantitative EEG for seizure identification in theintensive care unit. Neurology 87–7 (2016)

Practical Distributed Cooperative Control of Multiple Nonholonomic Unicycle Robots

Baoli Ma and Wenjing Xie

Abstract In this work, we study the distributed cooperative control problem of multiple nonholonomic unicycle robots with a time-varying reference trajectory. Under the mild assumptions that the communication topology is bidirectional connected, the reference trajectory is bounded and known for at least one robot and the velocity of the reference trajectory is bounded but unknown for all robots, a novel distributed cooperative control protocol is proposed guaranteeing that all the robots follow the reference trajectory with an arbitrarily small ultimate tracking errors. Simulation examples are given to verify the proposed distributed cooperative scheme.

Keywords Nonholonomic unicycle robots · Distributed cooperative control · Time-varying reference trajectory · Unified controller

1 Introduction

The distributed cooperative control (DCC) of multiple agent systems has earned great interests of many researchers for more than ten years [1]. This work concerns with the DCC problems of multiple nonholonomic unicycle agent systems with a time-varying reference trajectory. Previous studies on such systems can be divided into two catalogues: DCC with no leader [2–12] and DCC with a leader [4, 13]. For the latter case, the schemes in [4, 13] require that: (I) the leader is always moving such that the trajectory satisfies persistently exciting conditions; (II) the leader and the followers have the same dynamics such that the reference trajectory is feasible for the followers; (III) the reference trajectory and its derivative generated by the

B. Ma (✉)
The Seventh Research Division, School of Automation Science and Electrical Engineering, Beihang University, Beijing 100191, China
e-mail: mabaoli@buaa.edu.cn

W. Xie
School of Computer and Information Science, Southwest University, Chongqing 400715, China
e-mail: yongchao@mail.shufe.edu.cn

Y. Jia et al. (eds.), *Proceedings of 2018 Chinese Intelligent Systems Conference*, Lecture Notes in Electrical Engineering 529,
https://doi.org/10.1007/978-981-13-2291-4_55

leader are known for all followers. It is well known that the linear tools [14, 15] can not be applied directly to the control design of nonhlonomic systems. Regarding to practical considerations, it is expected to develop a novel DCC scheme that can solve the DCC problems of multiple nonholonomic/underactuated agent systems under the more relaxed assumptions such that the above requirements are removed.

In this contribution, we study the DCC problem of multiple nonholonomic *unicycle* robots with a time-varying reference trajectory generated by a leader, which may have different dynamics from the followers. Under the mild assumptions that the communication graph of the followers is bidirectional connected, the leader's trajectory is bounded and known by at least one follower, and the leader's velocity is bounded but unknown to all followers, we develop a unified DCC scheme ensuring that all the followers track the leader's trajectory with arbitrarily small ultimate tracking errors. The controller design is divided into two steps: (I) convert the dynamics of each unicycle robot into three separate first-order linear integrators by making several state/input transformations and by introducing auxiliary variables and additional input; (II) design a DCC law for the converted linear agent systems such that all the followers follow the leader with arbitrarily small tracking errors.

The rest of the paper is organized as follows. The DCC problem of nonholonomic unicycle agent systems is formulated in Sect. 2, and solved in Sect. 3. Simulation examples are provided in Sect. 4. Section 5 concludes the work.

2 Problem Formulation

Consider a group of unicycle robots:

$$\dot{x}_i = v_i \cos\theta_i, \dot{y}_i = v_i \sin\theta_i, \dot{\theta}_i = r_i, \tag{1}$$

where $i \in \{1, 2, \cdots, n\}$ denotes the index of agents, (x_i, y_i) and θ_i are the position and the orientation of agent i, and (v_i, r_i) the linear and angular velocity control inputs of agent i, respectively. The reference trajectory $(x_0(t), y_0(t), \theta_0(t))$ is generated via some external dynamics as follows

$$\dot{x}_0 = v_{10}, \dot{y}_0 = v_{20}, \dot{\theta}_0 = v_{30}, \tag{2}$$

where (v_{10}, v_{20}, v_{30}) are the velocity of the reference trajectory.

For the n multiple nonholonomic unicycle systems (1), we suppose that there is a bidirectional information interchange during systems through sensors or wireless communication. Regarding each system as a node, the information interchange during systems is described by an undirected graph $G = (\mathcal{V}, \mathcal{E})$, where $\mathcal{V}$ is the set of agent nodes, $\mathcal{E}$ is the set of edges indicating information flow during the systems. If the state of system j is available to system i, system j is said to be a neighbor of system i. The index numbers of all neighbors of system i form a set denoted by N_i. For the undirected graph, $j \in N_i$ means that $i \in N_j$. It is noted that a system does not consider itself as a neighbor, i.e., $i \notin N_i$. A graph is called connected if there exist a set of edges connecting any two nodes.

Control problem: Given a reference trajectory $(x_0(t), y_0(t), \theta_0(t))$ generated by (2), find a bounded DCC law $v_i(\cdot), r_i(\cdot)$ for nonholonomic agent i with dynamics (1) such that the state tracking errors $(x_i - x_0, y_i - y_0, \theta_i - \theta_0), i \in \{1, 2, \cdots, n\}$ between the leader and the followers are globally uniformly ultimately bounded (GUUB) with arbitrarily small ultimate bounds.

Assumption 1 The reference trajectory (x_0, y_0, θ_0) and its velocity (v_{10}, v_{20}, v_{30}) are bounded for all time.

Assumption 2 Each follower agent i knows the states of itself and its neighbors; the communication graph G of the follower agents is bidirectional connected; at least one (but not all) follower knows the reference state (x_0, y_0, θ_0), and the reference velocity (v_{10}, v_{20}, v_{30}) is unknown for all followers.

3 Controller Design

3.1 Coordinate and Input Transformations

To facilitate the controller design, we first convert the dynamics of the follower and the leader (1)–(2) to some advantageous forms.

Define new state variables for the follower agents as

$$\begin{aligned} &\bar{x}_i = x_i \cos\theta_i + y_i \sin\theta_i,\ \ \bar{y}_i = y_i \cos\theta_i - x_i \sin\theta_i,\ \ \theta_i = \theta_i, \\ &z_{1i} = \theta_i,\ \ z_{2i} = \bar{x}_i,\ \ z_{3i} = -(2\bar{y}_i + \bar{x}_i\theta_i) \end{aligned} \tag{3}$$

then the dynamics (1) becomes

$$\dot{z}_{1i} = u_{1i},\ \ \dot{z}_{2i} = u_{2i},\ \ \dot{z}_{3i} = z_{2i}u_{1i} - z_{1i}u_{2i}, \tag{4}$$

where $(u_{1i} = r_i, u_{2i} = v_i + \bar{y}_i r_i)$ denote the new inputs.

For the leader agent system (2), we construct the coordinate transformations:

$$\begin{aligned} &\bar{x}_0 = x_0 \cos\theta_0 + y_0 \sin\theta_0,\ \bar{y}_0 = y_0 \cos\theta_0 - x_0 \sin\theta_0,\ \theta_0 = \theta_0, \\ &e_{10} = \theta_0,\ \ e_{20} = \bar{x}_0,\ \ e_{30} = -(2\bar{y}_0 + \bar{x}_0\theta_0), \end{aligned} \tag{5}$$

which convert the leader's dynamics (2) to the following form

$$\dot{e}_{10} = w_{10},\ \ \dot{e}_{20} = w_{20},\ \ \dot{e}_{30} = w_{30}, \tag{6}$$

with (w_{10}, w_{20}, w_{30}) denoting the following new reference velocity variables

$$w_{10} = v_{30}, \quad w_{20} = v_{10}\cos\theta_0 - v_{30}x_0\sin\theta_0 + v_{20}\sin\theta_0 + v_{30}y_0\cos\theta_0,$$
$$w_{30} = -2(v_{20}\cos\theta_0 - v_{30}y_0\sin\theta_0 - v_{10}\sin\theta_0 - v_{30}x_0\cos\theta_0) - w_{20}\theta_0 - \bar{x}_0 v_{30}. \tag{7}$$

To solve the DCC problem, we introduce an additional dynamics with auxiliary input η_i and auxiliary state (ξ_{1i}, ξ_{2i}) for each follower system (4) as

$$\dot{\xi}_{1i} = -k_1\frac{\rho_i-\varepsilon^2}{\rho_i}\xi_{1i} + \frac{\xi_{2i}}{\rho_i}\eta_i, \qquad \dot{\xi}_{2i} = -k_1\frac{\rho_i-\varepsilon^2}{\rho_i}\xi_{2i} - \frac{\xi_{1i}}{\rho_i}\eta_i, \tag{8}$$

where $\rho_i := \xi_{1i}^2 + \xi_{2i}^2$, the initial auxiliary state $(\xi_1(0), \xi_2(0))$ is selected such that $\rho_i(0) = (\xi_{1i}(0))^2 + (\xi_{2i}(0))^2 \geq \varepsilon^2$, and (k_1, ε) are two positive constants.

With the auxiliary state (ξ_{1i}, ξ_{2i}), the final state transformation is

$$e_{1i} = z_{1i} - \xi_{1i}, \quad e_{2i} = z_{2i} - \xi_{2i}, \quad e_{3i} = z_{3i} - e_{2i}\xi_{1i} + e_{1i}\xi_{2i}. \tag{9}$$

The dynamics of the new states defined in (9) can be calculated as

$$\dot{e}_{1i} = u_{1i} - \dot{\xi}_{1i} := w_{1i}, \qquad \dot{e}_{2i} = u_{2i} - \dot{\xi}_{2i} := w_{2i},$$
$$\dot{e}_{3i} = \eta_i + (e_{2i} + 2\xi_{2i})w_{1i} - (e_{1i} + 2\xi_{1i})w_{2i} := w_{3i},$$

where (w_{1i}, w_{2i}, w_{3i}) are the new inputs:

$$w_{1i} = u_{1i} + k_1\frac{\rho_i-\varepsilon^2}{\rho_i}\xi_{1i} - \frac{\xi_{2i}}{\rho_i}\eta_i, \quad w_{2i} = u_{2i} + k_1\frac{\rho_i-\varepsilon^2}{\rho_i}\xi_{2i} + \frac{\xi_{1i}}{\rho_i}\eta_i,$$
$$w_{3i} = \eta_i + (e_{2i} + 2\xi_{2i})w_{1i} - (e_{1i} + 2\xi_{1i})w_{2i}. \tag{10}$$

Therefore, the dynamics of the follower agents can be written as

$$\dot{e}_{1i} = w_{1i}, \dot{e}_{2i} = w_{2i}, \dot{e}_{3i} = w_{3i}, i \in \{1, 2, \cdots, n\}. \tag{11}$$

Lemma 1 *Under Assumption 1 and* $(k_1 > 0, \rho_i(0) \geq \varepsilon^2 > 0)$*, any DCC law* $w_{li}(\cdot)$ *that makes the converted tracking errors* $e_{li} - e_{l0}$ $(l \in \{1, 2, 3\}; i \in \{1, 2, \cdots, n\})$ *GUUB with arbitrarily small ultimate bounds also makes the original state tracking errors* $(x_i - x_0, y_i - y_0, \theta_i - \theta_0)$ *GUUB with arbitrarily small ultimate bounds.*

Proof The proof is omitted here due to the limited space. □

According to Lemma 1, we know that the remaining task to achieve the control goal is to find a DCC law $w_{li}(\cdot)$ such that the state of each follower agent (11) tracks the state of the leader agent (6) with arbitrarily small tracking errors.

3.2 Controller Design

Let $L \in R^{n\times n}$ represent the Laplacian matrix corresponding to G (the communication graph of the follower agents (1) or (11)), and its (i, j)th element L_{ij} satisfies $L_{ij} = -1$, if $j \in N_i$; $L_{ij} = 0$, if $j \notin N_i$ and $j \neq i$; and $L_{ii} = -\sum_{j=1,\ j\neq i}^{n} L_{ij}$.

Denote G_A as the augmented communication graph with the nodes including both the follower agents (11) and the leader agent (6), where the leader is numbered by $n+1$. Let $L_A \in R^{(n+1)\times(n+1)}$ represent the Laplacian matrix of G_A, then we have $L_A = \begin{bmatrix} \bar{L} & -b \\ 0 & 0 \end{bmatrix}$, where $\bar{L} = L + \text{diag}\{b_1, b_2, \cdots, b_n\}$, $b = [b_1, b_2, \cdots, b_n]^{\mathrm{T}} \in R^n$, and $b_i = 1$ if agent i knows the state of the leader, otherwise $b_i = 0$. As G is undirected, $L = L^{\mathrm{T}}$ and $\bar{L} = \bar{L}^{\mathrm{T}}$.

As the follower agents are bidirectional connected and at least one follower knows the state of the leader (Assumption 2), we conclude that the augmented graph G_A is directional connected. This means that L_A has a sole zero eigenvalue, and all other $n-1$ eigenvalues of L_A have positive real parts, thus $\bar{L}$ in L_A is symmetric positive definite due to the last zero row of L_A.

We propose the following DCC protocol

$$w_{li} = -k_2 s_{li} - k_{3l}\frac{s_{li}}{\sqrt{s_{li}^2 + \rho_i^2}}, (l = 1, 2, 3; i = 1, 2, \cdots, n), \tag{12}$$

where $s_{li} = \sum_{j\in N_i} (e_{li} - e_{lj}) + b_i(e_{li} - e_{l0}) = \sum_{j\in N_i} (e_{li} - e_{lj}) + b_i\bar{e}_{li}$, the coefficients (k_2, k_{3l}) are positive, and k_{3l} is sufficiently large such that $k_{3l} \geq |w_{l0}|_{\max}$ with $|w_{l0}|_{\max}$ the upper bound of $|w_{l0}(t)|$. Then, the following results can be obtained.

Lemma 2 *The control law (12) with ($k_2 > 0, k_{3l} \geq |w_{l0}|_{\max}$) guarantees that the tracking errors $\bar{e}_{li}(l \in \{1, 2, 3\}, i \in \{1, 2, \cdots, n\})$ are GUUB with arbitrarily small ultimate bounds, or say more precisely that, there exists a sufficiently large positive constant T such that*

$$\|\bar{e}_l\|_2 \leq \frac{\gamma\varepsilon\sqrt{n}\,|w_{l0}|_{\max}}{\lambda_{\min}(\bar{L})\sqrt{k_2}}, \quad \forall t \geq t_0 + T, \tag{13}$$

where $\gamma > 1$, $\bar{e}_l = [\bar{e}_{l1}, \bar{e}_{l2}, \cdots, \bar{e}_{ln}]^{\mathrm{T}}$, and $\lambda_{\min}(\cdot)$ is the minimum eigenvalue of the symmetric positive definite matrix '$\cdot$'.

Proof Denote $s_l := [s_{l1}, s_{l2}, \cdots s_{ln}]^{\mathrm{T}}$, and note that

$$s_{li} = \sum_{j\in N_i} \left((e_{li} - e_{l0}) - (e_{lj} - e_{l0})\right) + b_i\bar{e}_{li} = \sum_{j\in N_i} \left(\bar{e}_{li} - \bar{e}_{lj}\right) + b_i\bar{e}_{li} = (\bar{L}\bar{e}_l)_i,$$
$$s_l = \bar{L}\bar{e}_l,$$

where $(\bar{L}\bar{e}_l)_i$ is the ith element of $\bar{L}\bar{e}_l$. Then control law (12) can be rewritten as

$$w_l := [w_{l1}, w_{l2}, \cdots, w_{ln}]^{\mathrm{T}} = -k_2 s_l - k_{3l} P_l s_l = -k_2 \bar{L}\bar{e}_l - k_{3l} P_l \bar{L}\bar{e}_l,$$

with $P_l = \mathrm{diag}\left\{\frac{1}{\sqrt{s_{l1}^2+\rho_1^2}}, \frac{1}{\sqrt{s_{l2}^2+\rho_2^2}}, \cdots, \frac{1}{\sqrt{s_{ln}^2+\rho_n^2}}\right\}$. The closed-loop dynamics is

$$\dot{\bar{e}}_l = \dot{e}_l - \dot{e}_{l0}\mathbf{1} = w_l - w_{l0}\mathbf{1} = -k_2 \bar{L}\bar{e}_l - k_{3l} P_l \bar{L}\bar{e}_l - w_{l0}\mathbf{1}, \tag{14}$$

where $e_l = [e_{l1}, e_{l2}, \cdots, e_{ln}]^{\mathrm{T}}, \mathbf{1} = [1, 1, \cdots, 1]^{\mathrm{T}} \in R^n$.

Consider the Lyapunov-like function $V_l = 0.5\bar{e}_l^{\mathrm{T}}\bar{L}\bar{e}_l$, and its derivative is

$$\dot{V}_l \le -k_2 \|s_l\|_2^2 + |w_{l0}| \sum_{i=1}^{n} \rho_i,$$

where $k_{3l} \ge |w_{l0}|_{\max}$ is used to ensure $k_{3l} - |w_{l0}| \ge 0$.
As $V_l = 0.5\bar{e}_l^{\mathrm{T}}\bar{L}\bar{e}_l = 0.5(\bar{L}\bar{e}_l)^{\mathrm{T}}\bar{L}^{-1}(\bar{L}\bar{e}_l) = 0.5 s_l^{\mathrm{T}}\bar{L}^{-1}s_l$, one obtains $0.5\lambda_{\min}(\bar{L}) \|\bar{e}_l\|_2^2 \le V_l \le 0.5\lambda_{\max}(\bar{L}^{-1}) \|s_l\|_2^2$, where $\lambda_{\max}(\cdot)$ is the minimum eigenvalue of the symmetric positive definite matrix '$\cdot$'. Therefore,

$$\dot{V}_l \le -\frac{2k_2}{\lambda_{\max}(\bar{L}^{-1})} V_l + |w_{l0}| \sum_{i=1}^{n} \rho_i. \tag{15}$$

Since ρ_i is GUUB with the upper bound $\rho_i(0)$ and the ultimate bound $\gamma^2\varepsilon^2 (\gamma > 1)$, we conclude from (15) and comparison principle [35] that V_l is GUUB with an upper bound $\max\left\{\frac{1}{2k_2}\lambda_{\max}(\bar{L}^{-1}) |w_{l0}|_{\max} \sum_{i=1}^{n} \rho_i(0), V_l(0)\right\} := \bar{V}_l$ and an ultimate bound $\frac{n}{2k_2}\lambda_{\max}(\bar{L}^{-1}) |w_{l0}|_{\max} \gamma^2\varepsilon^2$. From $\|\bar{e}_l\|_2^2 \le \frac{2V_l}{\lambda_{\min}(\bar{L})}$, we can further infer that $\bar{e}_l$ is GUUB with an upper bound $\frac{2\bar{V}_l}{\lambda_{\min}(\bar{L})}$ and an arbitrarily small ultimate bound $\left(\sqrt{\frac{n\lambda_{\max}(\bar{L}^{-1})|w_{l0}|_{\max}}{k_2\lambda_{\min}(\bar{L})}}\right)\gamma\varepsilon = \frac{\gamma\varepsilon\sqrt{n|w_{l0}|_{\max}}}{\lambda_{\min}(\bar{L})\sqrt{k_2}}$, where $\lambda_{\max}(\bar{L}^{-1}) = \left(\lambda_{\min}(\bar{L})\right)^{-1}$ is used. We finally conclude that, there exists a sufficiently large constant $T > 0$ such that

$$\|\bar{e}_l\|_2 \le \frac{\gamma\varepsilon\sqrt{n\,|w_{l0}|_{\max}}}{\lambda_{\min}(\bar{L})\sqrt{k_2}}, \quad \forall t \ge t_0 + T. \tag{16}$$

□

Theorem 1 *Under Assumptions 1–2, the distributed cooperative control law (12) with* ($k_1 > 0, k_2 > 0, \varepsilon > 0, \rho_i(0) \ge \varepsilon^2, k_{3l} \ge |w_{l0}|_{\max}$) *guarantees that the state tracking errors* ($x_i - x_0, y_i - y_0, \theta_i - \theta_0$) *are GUUB with arbitrarily small ultimate bounds, where* $l \in \{1, 2, 3\}, i \in \{1, 2, \cdots, n\}$.

Proof Theorem 1 is a direct consequence of Lemmas 1 and 2. □

Corollary 1 *Under Assumptions 1–2, ($v_{10} = 0$, $v_{20} = 0$, $v_{30} = 0$) and ($k_1 > 0$, $k_2 > 0$, $\varepsilon > 0$, $\rho_i(0) \geq \varepsilon^2$, $k_{3l} \geq |w_{l0}|_{\max}$), the distributed cooperative control law (12) guarantees that the tracking errors ($x_i - x_0, y_i - y_0, \theta_i - \theta_0$)($i \in \{1, 2, \cdots, n\}$) globally exponentially converge to arbitrarily small constants, and the original control inputs (v_i, r_i) are bounded. Furthermore, by setting $k_{3l} = 0$, $\varepsilon = 0$, $\rho_i(0) > 0$ and k_2 sufficiently large such that $k_2 > k_1\lambda_{\max}(\bar{L}^{-1})$, the tracking errors ($x_i - x_0, y_i - y_0, \theta_i - \theta_0$) can be made converging to zero exponentially with vanishing control inputs (v_i, r_i).*

Proof The proof is omitted here due to the limited space. □

Now we consider another special case that the reference trajectory does not exist, or equivalently the reference trajectory is unknown for all follower robots ($b = 0$). In this case, Assumption 1 is no longer needed, and Assumption 2 is required changed to the following Assumption 3.

Assumption 3 Each follower agent i knows the states of itself and its neighbors, and the communication graph G of the follower agents is bidirectional connected.

Corollary 2 *Under Assumption 3, the distributed cooperative control (12) achieves the practical consensus, guaranteeing the consensus errors ($x_i - x_j, y_i - y_j, \theta_i - \theta_j$) globally exponentially convergent to arbitrary small constants and ensuring the original control inputs (v_i, r_i) bounded. Furthermore, by setting $k_{3l} = 0$, $\varepsilon = 0$, $\rho_i(0) > 0$ and k_2 sufficiently large such that $k_2\lambda_2(L) > k_1$, the consensus errors ($x_i - x_j, y_i - y_j, \theta_i - \theta_j$) can be made converging to zero exponentially with vanishing control inputs (v_i, r_i), where $\lambda_2(L) > 0$ is the second small eigenvalue of L.*

Proof The proof is omitted here due to the limited space.

4 Simulation Examples

In this section, we present several simulation examples to verify the effectiveness of the proposed control laws. Suppose that there are three unicycle agents connected by the undirected communication graph shown in Fig. 1. The reference trajectory (x_0, y_0, θ_0) is generated by

$$\begin{aligned} x_0 &= x_d + h\cos\theta_d, y_0 = y_d + h\sin\theta_d, \theta_0 = \theta_d, \\ \dot{x}_d &= v_d\cos\theta_d, \dot{y}_d = v_d\sin\theta_d, \dot{\theta}_d = r_d, \end{aligned} \tag{17}$$

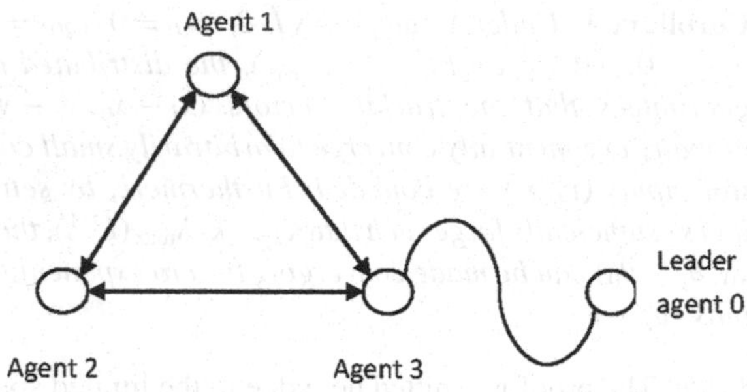

Fig. 1 A communication graph of three agents with agent 3 knowing/unknowing the leader's state (reference trajectory)

where h is a constant, and the bounded (v_d, r_d) can be chosen freely. This reference trajectory (x_0, y_0, θ_0) can be regarded as the one generated by a gripper fixed on the front point of the robot with a distance h from the center point of the wheel's axis. It follows from $\dot{x}_0 = v_d \cos\theta_0 - hr_d \sin\theta_0 := v_{01}, \dot{y}_0 = v_d \sin\theta_0 + hr_d \cos\theta_0 := v_{02}, \dot{\theta}_0 = r_d := v_{03}$ that the reference trajectory (x_0, y_0, θ_0) is feasible for $h = 0$ or $r_d = 0$, and unfeasible for $(h \neq 0, r_d \neq 0)$.

Case1 *(feasible circular reference trajectory)*. Agent 3 knows the reference trajectory generated via (17) with $(x_d, y_d.\theta_d)(0) = (0, 0, 0), h = 0, v_d = \frac{1}{2}\pi \sin(0.1t)$, $r_d = 0.2v_d$.

Case 2 *(unfeasible circular reference trajectory)*. Agent 3 knows the reference trajectory generated via (17) with $(x_d, y_d.\theta_d)(0) = (0, 0, 0)$, $h = 0.1$, $v_d = \frac{1}{2}\pi \sin(0.1\, t)$, $r_d = 0.2\, v_d$.

Case 3 *(fixed reference trajectory)*. Agent 3 knows the reference trajectory generated via (17) with $(x_d, y_d.\theta_d)(0) = (0, 0, 0)$, $h = 0$, and $v_d = 0, r_d = 0$.

Case 4 *(no reference trajectory)*. There is no reference trajectory, or no agent knows the reference trajectory, that is, $b_j \equiv 0 (j = 1, 2, 3)$.

In all the four cases, the control parameters are selected as $k_1 = 0.1, k_2 = 2, k_{31} = k_{32} = k_{33} = 8, \varepsilon = 0.02, \xi_{11}(0) = -3, \xi_{21}(0) = 0, \xi_{12}(0) = -3, \xi_{22}(0) = 0, \xi_{13}(0) = -3, \xi_{23}(0) = 0$, and the initial states are $(x_1, y_1, \theta_1)(0) = (-5, -2, 0), (x_2, y_2, \theta_2)(0) = (0, 6, 0)$, $(x_3, y_3, \theta_3)(0) = (6, -2, 0)$. The simulation results corresponding to the four cases are plotted in Fig. 2, illustrating the geometry paths generated by both the leader and the follower agents.

One can observe from Fig. 2a–c that the states of all the follower agents can track the reference trajectory with small tracking errors no matter whether it is feasible (Fig. 2a) or unfeasible (Fig. 2b), and even just a fixed point (Fig. 2c). One can also see that the three agents automatically achieve consensus with small ultimate agreement errors if the reference trajectory does not exist or is unknown for any agent (Fig. 2d).

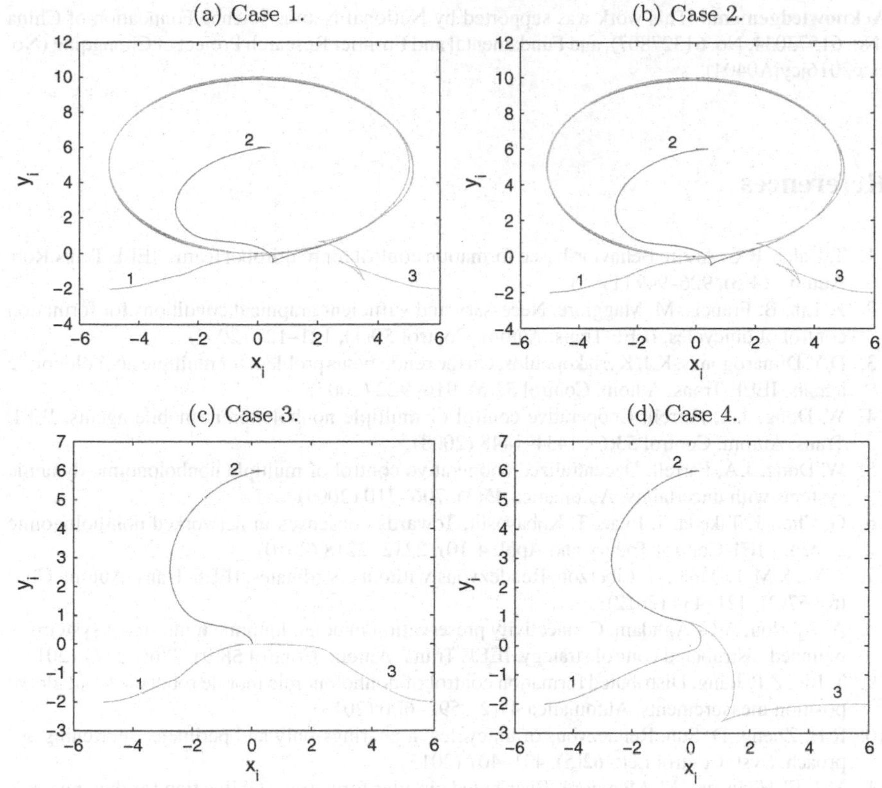

Fig. 2 Simulation results

5 Conclusion

In this work, we developed a distributed cooperative control scheme of multiple nonholonomic unicycle agent systems with a reference trajectory. At the cost of non-vanishing tracking or rendezvous errors that can be made arbitrarily small at will, the proposed scheme is advantageous in two aspects: (1) it is universal in the sense that it is applicable no matter whether the reference trajectory is feasible, unfeasible, a fixed point, or even absent; (2) the reference trajectory is only required known for partial (at least one) agents, and its velocity is only required bounded, not necessarily known for any agent, greatly reducing the sensing or communication costs. We are currently focusing on extending the results to more complicated nonholonomic/underactuated agent systems such as underactuated ships, hovercrafts, underwater vehicles, spacecrafts and so on, expecting to find universal solutions for the distributed cooperative control of such systems.

Acknowledgements This work was supported by National Nature Science Foundation of China (No. 61573034, No. 61327807), and Fundamental and Frontier Research Project of Chongqing (No. cstc2016jcyjA0404).

References

1. T. Balch, R.C. Arkin, Behavior-based formation control for multirobot teams. IEEE Trans.Rob. Autom. **14**(6), 926–949 (1998)
2. Z. Lin, B. Francis, M. Maggiore, Necessary and sufficient graphical conditions for formation control of unicycles. IEEE Trans. Autom. Control **50**(1), 121–127 (2005)
3. D.V. Dimarogonas, K.J. Kyriakopoulos, On the rendezvous problem for multiple nonholonomic agents. IEEE Trans. Autom. Control **52**(5), 916–922 (2007)
4. W. Dong, J.A. Farrell, Cooperative control of multiple nonholonomic mobile agents. IEEE Trans. Autom. Control **53**(6), 1434–1448 (2008)
5. W. Dong, J.A. Farrell, Decentralized cooperative control of multiple nonholonomic dynamic systems with uncertainty. Automatica **45**(3), 706–710 (2009)
6. G. Zhai, J. Takeda, J. Imae, T. Kobayashi, Towards consensus in networked non-holonomic systems. IET Control Theory and Appl. **4**(10), 2212–2218 (2010)
7. J. Yu, S.M. LaValle, D. Liberzon, Rendezvous without coordinates. IEEE Trans. Autom. Control **57**(2), 421–434 (2012)
8. A. Ajorlou, A.G. Aghdam, Connectivity preservation in nonholonomic multi-agent systems: a bounded distrubuted control strategy. IEEE Trans. Autom. Control **58**(9), 2366–2371 (2013)
9. T. Liu, Z.P. Jiang, Distributed formation control of nonholonomic mobile robots without global position measurements. Automatica **49**(2), 592–600 (2013)
10. R.H. Zheng, D. Sun, Rendezvous of unicycles: a bearings-only and perimeter shortening approach. Syst. Control Lett. **62**(5), 401–407 (2013)
11. M.I. El-Hawwary, M. Maggiore, Distributed circular formation stabilization for dynamic unicycles. IEEE Trans. Autom. Control **58**(1), 149–162 (2013)
12. W. Dong, J.A. Farrell, Formation control of multiple underactuated surface vessels. IET Control Theory Appll. **2**(12), 1077–1085 2008
13. J. Ghommam, H. Mehrjerdi, F. Mnif, M. Saad, Cascade design for formation control of non-holonomic systems in chained form. J. Franklin Inst. **348**(6), 973–998 (2011)
14. Y. Jia, Robust control with decoupling performance for steering and traction of 4WS vehicles under velocity-varying motion. IEEE Trans. Control Syst. Technol. **8**(3), 554–569 (2000)
15. Y. Jia, Alternative proofs for improved LMI representations for the analysis and the design of continuous-time systems with polytopic type uncertainty: a predictive approach. IEEE Trans. Autom. Control **48**(8), 1413–1416 (2003)

A Novel Fuzzy Logic System with Consequents as Fuzzy Weighted Averages of Antecedents

Qiye Zhang, Yuqing Liu and Xiao Tian

Abstract Fuzzy logic system is an intelligent system based on IF-THEN rules, which can handle uncertainties effectively, and has been applied to various fields. The design of rules is a key step when a fuzzy logic system is modelled in a practical situation. In this paper, a novel fuzzy logic system named FWA with novel rules is proposed, in which the consequents are fuzzy weighted averages of antecedents. The proposed rules establish some relationship between consequents and antecedents in advance, so that the proposed FWA fuzzy logic system will reduce training time, improve training efficiency, and optimize parameters faster.

Keywords Fuzzy logic system · Trapezoidal fuzzy number · Fuzzy weighted average · Error back-propagation · Steepest descent algorithm

1 Introduction

Fuzzy sets were introduced by L. A. Zadeh in 1965 [1] to describe the ambiguity and fuzziness that often appeared in our daily life. Since then the theory of fuzzy sets has been developed rapidly [2]. Fuzzy logic systems (FLSs) were first proposed by Mamdani in 1974 [3] for controlling of simple dynamic plant, and then have been widely applied to various fields, such as system identification, pattern recognition,

Q. Zhang · Y. Liu (✉) · X. Tian
School of Mathematics and Systems Science and LMIB of the Ministry of Education, Beihang University, Beijing 100191, China
e-mail: 13998127561@163.com

Q. Zhang
e-mail: zhangqiye@buaa.edu.cn

X. Tian
e-mail: 649172415@qq.com

Y. Jia et al. (eds.), *Proceedings of 2018 Chinese Intelligent Systems Conference*, Lecture Notes in Electrical Engineering 529,
https://doi.org/10.1007/978-981-13-2291-4_56

signal processing, communication network and financial investment, etc, in which there are many uncertainties [4]. The two frequently used fuzzy logic systems (FLSs) today are Mamdani [3, 4] and Takagi-Sugeno-Kang (TSK) systems [5, 6]. Both of them are characterized by IF-THEN rules, and have the same antecedent form. They differ in the form of the consequents. The consequent of a Mamdani rule is a fuzzy set, while the consequent of a TSK rule is a function of input variables.

Unfortunately, neither rules of Mamdani FLSs nor those of TSK FLSs establishes the connection between the consequents and antecedents. They take consequents arbitrarily, or only describe the relationship between consequents and inputs. Even in type-2 FLSs developed by Mendel's work group, etc. [7], the rule consequents are also designed arbitrarily, or they are fuzzy affine combinations of inputs. There are also a great number of literatures on FLSs, which are concerned with rules design, e.g. [8–11], but they are all based on (type-2) Mamdani rules or TSK rules, what they cared about were how to get the optimal parameters involved in rule consequents.

In this paper, we suggest a novel FLS named FWA with novel rules, in which the consequents are fuzzy weighted averages of antecedents. The proposed rules establish some relationship between consequents and antecedents in advance, so that our FWA FLS will reduce training time, improve training efficiency, and optimize parameters faster.

The rest of this paper is organized as follows. Section 2 provides some preliminaries of fuzzy numbers and FLS. Section 3 describes the proposed FLS in detail. A numerical simulation of Mackey-Glass time series forecasting is conducted in Sect. 4 to verify the performance of our proposed FLS. Section 5 provides conclusions and some suggestions for future research.

2 Preliminaries

In this section, let us recall some concepts and results about fuzzy numbers [12–14] and fuzzy logic systems [4, 7]. Without loss of generality, let $\mathbb{R}$ denote the set of the real numbers; for a non-empty set X, let $\mathcal{F}(X)$ denote the set of all fuzzy subsets on X. For $A \in \mathcal{F}(X)$, let $\mu_A(x)$ denote the membership function (MF), and $\forall\lambda \in (0, 1]$, $A_\lambda = \{x \in X \mid \mu_A(x) \geq \lambda\}$ is called *λ-cut* (or *λ-level set*) of A.

Definition 1 ([12]) Let $a, b, c, d \in \mathbb{R}$. A *trapezoidal fuzzy number* (TrapFN for short) A is determined by the following membership function $\mu_A(x)$:

$$\mu_A(x) = \begin{cases} \dfrac{x-a}{b-a}, & a \leq x < b, \\ 1, & b \leq x \leq c, \\ \dfrac{x-d}{c-d}, & c < x \leq d, \\ 0, & \text{otherwise}, \end{cases} \tag{1}$$

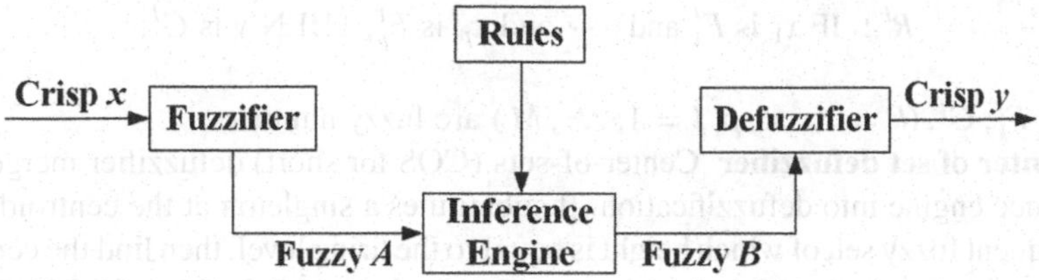

Fig. 1 The structure of a Mamdani fuzzy logic system

which is denoted by $A = (a, b, c, d)$. When $b = c$, a trapezoidal fuzzy number is reduced to a *triangular fuzzy number* (TriFN for short), denoted by $A = (a, b, d)$.

Definition 2 ([12]) Let $A = (a, b, c, d)$ be a TrapFN, L_A^{-1} and R_A^{-1} be the inverse functions of the functions L_A and R_A, respectively. Define

$$P(A) = \frac{\int_0^1 h \frac{L^{-1}(h) + R^{-1}(h)}{2} dh}{\int_0^1 h dh}, \quad (2)$$

it is called the *graded mean integration representation* (GMIR for short) of A.

Remark ([12]) For a TrapFN $A = (a, b, c, d)$, the GMIR of A can be expressed as $P(A) = \dfrac{a + 2b + 2c + d}{6}$. For a TriFN $A = (a, b, d)$, the GMIR of A can be expressed as $P(A) = \dfrac{a + 4b + d}{6}$.

In the following, we introduce the structure of a Mamdani FLS. As we know, it is composed of four modules: fuzzier, IF-THEN rules, inference and defuzzifier, as shown in Fig. 1, see references [4, 7]. They are described in detail next.

Fuzzifier The fuzzifier in a Mamdani FLS maps a crisp point $\mathbf{x}' = (x_1', x_2' \ldots, x_p')$ $\in X$ into a fuzzy set $A_{\mathbf{x}'}$ with membership function $\mu_{A_{\mathbf{x}'}}(\mathbf{x}) = T_{k=1}^{p} \mu_{X_k}(x_k)$, where X_k, $(k = 1, \ldots, p)$ is fuzzy number located at x_k', $\mathbf{x} = (x_1, x_2, \cdots, x_p)$, and T denotes a t-norm, which is usually taken to be minimum operation $\wedge$ or product operation $\cdot$.

The commonly used fuzzifiers include singleton fuzzifier, non-singleton triangular or Gaussian fuzzifier [4, 7]. In this paper, we will use singleton fuzzifier, which is the most widely used fuzzifier, and has the lowest computation complexity. A *singleton fuzzifier* is nothing more than a fuzzy singleton, i.e., $A_{\mathbf{x}'}$ is a *fuzzy singleton* with support $\mathbf{x}' = (x_1', x_2', \ldots, x_p')$ if

$$\mu_{X_k}(x_k) = \begin{cases} 1, & x_k = x_k', \\ 0, & x_k \neq x_k', \end{cases} \qquad k = 1, \ldots, p$$

Rules Suppose that the system has p-input $x_1 \in X_1, \ldots, x_p \in X_p$, 1-output and M rules, so that the lth rule in a Mamdani FLS has the form of (3), we rewrite it here:

$$R^l: \text{ IF } x_1 \text{ is } F_1^l \text{ and } \cdots \text{ and } x_p \text{ is } F_p^l, \text{ THEN } y \text{ is } G^l, \tag{3}$$

where $F_k^l, G^l, (k = 1, \ldots, p, l = 1, \ldots, M)$ are fuzzy numbers.

Center of set defuzzifier Center-of-sets (COS for short) defuzzifier merges the inference engine into defuzzification. It substitutes a singleton at the centroid for a consequent fuzzy set, of which height is equal to the firing level, then find the centroid of fuzzy set consisting of the singletons. The final output of a Mamdani FLS with COS defuzzifier can be computed as:

$$y_{\cos}(\mathbf{x}) = \frac{\sum_{l=1}^{M} c^l T_{k=1}^{p} \mu_{F_k^l}(x_k)}{\sum_{l=1}^{M} T_{k=1}^{p} \mu_{F_k^l}(x_k)} \triangleq \frac{\sum_{l=1}^{M} c^l f^l(\mathbf{x})}{\sum_{l=1}^{M} f^l(\mathbf{x})} \tag{4}$$

where c^l is the centroid of the lth consequent G^l, T stands for t-norm operation, and $f^l(\mathbf{x}) = T_{k=1}^{p} \mu_{F_k^l}(x_k)$ is called *firing level* (or *firing strength*) of the lth rule.

3 The Design of the Proposed Fuzzy Logic System

In this section, we are going to introduce our proposed FWA FLS in detail. Firstly, we describe the structure of the FWA FLS. Then, we provide how to update the parameters involved in the system using steepest descent method.

3.1 The Structure of the Proposed Fuzzy Logic System

The structure of our FWA FLS is the same as that of a Mamdani FLS, as shown in Fig. 1.

Fuzzifier For computational simplicity, we use singleton fuzzifier.

Rules Like a Mamdani FLS, our proposed FLS has p inputs $x_j \in [\alpha_j, \beta_j]$, $(j = 1, 2, \cdots, p)$, and one output $y \in Y$. For each $j = 1, 2, \cdots, p$, we define N_j fuzzy sets $F_j^{l_j}$ $(l_j = 1, 2, \cdots, N_j)$ on $[\alpha_j, \beta_j]$, and design $M = N_1 N_2 \cdots N_p$ rules as follows:

$$\begin{aligned} \mathbf{R}^{l_1 l_2 \ldots l_p}: &\text{ IF } x_1 \text{ is } F_1^{l_1} \text{ and } \cdots \text{ and } x_p \text{ is } F_p^{l_p}, \text{ THEN} \\ &y^{l_1 l_2 \ldots l_p} \text{ is } Y^{l_1 l_2 \ldots l_p} = \frac{\sum_{j=1}^{p} W_j^{l_j} F_j^{l_j}}{\sum_{j=1}^{p} W_j^{l_j}} + B^{l_1 l_2 \ldots l_p} \end{aligned} \tag{5}$$

where $j = 1, 2, \cdots, p$, $l_j = 1, 2, \cdots, N_j$; $W_j^{l_j}$ is a fuzzy set, viewed as the “fuzzy weight”, and $B^{l_1 l_2 \ldots l_p}$ is also a fuzzy set, viewed as the“fuzzy bias”.

From (5), one can see that each consequent fuzzy set $Y^{l_1 l_2 \ldots l_p}$ in our proposed FLS is a fuzzy weighted average of antecedent fuzzy sets $F_j^{l_j}, (j = 1, 2, \cdots, p, l_j = 1, 2, \cdots, N_j)$ with a fuzzy bias $B^{l_1 l_2 \ldots l_p}$. So we will call our proposed FLS "FWA".

Inference and defuzzifier To get a compromise between accuracy and computational complexity, we would like to choose COS defuzzifier. For this, and to obtain the final output of our proposed FWA FLS, we need to calculate each consequent fuzzy set $Y^l = \frac{\sum_{j=1}^p W_j^{l_j} F_j^{l_j}}{\sum_{j=1}^p W_j^{l_j}} + B^l$ at first; then compute the centroid y^l of $Y^l, l = 1, \ldots, M$; finally, we can get the final output through calculating the following numerical weighted average:

$$y_{FWA}(\mathbf{x}) = \frac{\sum_{l_1=1}^{N_1} \sum_{l_2=1}^{N_2} \cdots \sum_{l_p=1}^{N_p} f^{l_1 l_2 \ldots l_p}(\mathbf{x}) y^{l_1 l_2 \ldots l_p}}{\sum_{l_1=1}^{N_1} \sum_{l_2=1}^{N_2} \cdots \sum_{l_p=1}^{N_p} f^{l_1 l_2 \ldots l_p}(\mathbf{x})} \triangleq \frac{\sum_{l=1}^M y^l f^l(\mathbf{x})}{\sum_{l=1}^M f^l(\mathbf{x})} \tag{6}$$

where $f^l(\mathbf{x}) \triangleq f^{l_1 l_2 \ldots l_p}(\mathbf{x}) = \prod_{j=1}^p \mu_{F_j^{l_j}}(x_j)$ is the firing level (or firing strength) of the $l \triangleq l_1 l_2 \ldots l_p$ th rule; $y^l \triangleq y^{l_1 l_2 \ldots l_p}$ is the GMIR (graded mean integration representation) (Eq. 2) of $Y^l \triangleq Y^{l_1 l_2 \ldots l_p}$ (substitute for the centroid of Y^l), and $M = N_1 N_2 \cdots N_p$.

The output of the proposed FWA FLS $y_{FWA}(\mathbf{x})$ can be also written by using FBFs as follows:

$$y_{FWA}(\mathbf{x}) = \sum_{l=1}^{M} y^l \phi^l(\mathbf{x}) \tag{7}$$

where the FBFs $\phi^l(\mathbf{x})$ are calculated by $\phi^l(\mathbf{x}) = \frac{f^l(\mathbf{x})}{\sum_{l=1}^M f^l(\mathbf{x})}, l = 1, \ldots, M$.

In this paper, to simplify computation, we will take trapezoidal fuzzy numbers for both antecedent and consequent fuzzy sets. Let $Y_{FWA}^l \triangleq Y_{FWA}^{l_1 l_2 \ldots l_p}$ denote the fuzzy weighted average term $\frac{\sum_{j=1}^p W_j^{l_j} F_j^{l_j}}{\sum_{j=1}^p W_j^{l_j}}$ in Y^l, which will be calculated through the method introduced in [15]. By this, one first computes the α-cuts of the involved fuzzy sets and then calculates the corresponding interval weighted averages using KM algorithm [16] or EKM algorithm [17].

In order to compute derivatives of $\underline{y}_{FWA}^l(\alpha)$ and $\overline{y}_{FWA}^l(\alpha)$ with respect to MF parameters, we will adopt the procedures suggested in [18].

First, we deduce the analytic expressions of $\underline{y}_{FWA}^l(\alpha)$ and $\overline{y}_{FWA}^l(\alpha)$. For this, we introduce the following notations (see Table 1):

Refer to the method in [18], we can obtain the analytic formulas of $\underline{y}_{FWA}^l(\alpha)$ and $\overline{y}_{FWA}^l(\alpha)$ as follows:

Table 1 Notations and formulas involved in calculating $\underline{y}_{FWA}^l(\alpha)$ and $\overline{y}_{FWA}^l(\alpha)$

Calculation of $\underline{y}_{FWA}^l(\alpha)$	Calculation of $\overline{y}_{FWA}^l(\alpha)$
$\underline{\mathbf{g}}^l = (\underline{g}_1^l, \underline{g}_2^l, \ldots, \underline{g}_p^l) \equiv \mathbf{Q}_L^l(\alpha)\underline{\mathbf{W}}(\alpha)$	$\underline{\mathbf{g}}^l = (\underline{g}_1^l, \underline{g}_2^l, \ldots, \underline{g}_p^l) \equiv \mathbf{Q}_R^l(\alpha)\underline{\mathbf{W}}(\alpha)$
$\overline{\mathbf{g}}^l = (\overline{g}_1^l, \overline{g}_2^l, \ldots, \overline{g}_p^l) \equiv \mathbf{Q}_L^l(\alpha)\overline{\mathbf{W}}(\alpha)$	$\overline{\mathbf{g}}^l = (\overline{g}_1^l, \overline{g}_2^l, \ldots, \overline{g}_p^l) \equiv \mathbf{Q}_R^l(\alpha)\overline{\mathbf{W}}(\alpha)$
$\mathbf{E}_1 \equiv (\mathbf{e}_1\vert\mathbf{e}_2\vert\cdots\vert\mathbf{e}_{K_l}\vert\mathbf{0}\vert\cdots\vert\mathbf{0})^{\mathrm{T}}$ $K_l \times p$	$\mathbf{E}_1 \equiv (\mathbf{e}_1\vert\mathbf{e}_2\vert\cdots\vert\mathbf{e}_{K_r}\vert\mathbf{0}\vert\cdots\vert\mathbf{0})^{\mathrm{T}}$ $K_r \times p$
$\mathbf{e}_i = K_l \times 1$ elementary vector	$\mathbf{e}_i = K_r \times 1$ elementary vector
$\mathbf{E}_2 \equiv (\mathbf{0}\vert\cdots\vert\mathbf{0}\vert\varepsilon_1\vert\varepsilon_2\vert\cdots\vert\varepsilon_{\mathbf{p}-\mathbf{K_l}})^{\mathrm{T}}$ $(p-K_l)\times 1$	$\mathbf{E}_2 \equiv (\mathbf{0}\vert\cdots\vert\mathbf{0}\vert\varepsilon_1\vert\varepsilon_2\vert\cdots\vert\varepsilon_{\mathbf{p}-\mathbf{K_r}})^{\mathrm{T}}$ $(p-K_r)\times 1$
$\varepsilon_\mathbf{i} = (\mathbf{p}-\mathbf{K_l})\times \mathbf{1}$ elementary vector	$\varepsilon_\mathbf{i} = (\mathbf{p}-\mathbf{K_r})\times \mathbf{1}$ elementary vector
$\mathbf{M}_{L1}^l(\alpha) \equiv \mathbf{Q}_L^{l\mathrm{T}}(\alpha)\mathbf{E}_1^{\mathrm{T}}\mathbf{E}_1\mathbf{Q}_L^l(\alpha)$ $p\times p$	$\mathbf{M}_{R1}^l(\alpha) \equiv \mathbf{Q}_R^{l\mathrm{T}}(\alpha)\mathbf{E}_1^{\mathrm{T}}\mathbf{E}_1\mathbf{Q}_R^l(\alpha)$ $p\times p$
$\mathbf{M}_{L2}^l(\alpha) \equiv \mathbf{Q}_L^{l\mathrm{T}}(\alpha)\mathbf{E}_2^{\mathrm{T}}\mathbf{E}_2\mathbf{Q}_L^l(\alpha)$ $p\times p$	$\mathbf{M}_{R2}^l(\alpha) \equiv \mathbf{Q}_R^{l\mathrm{T}}(\alpha)\mathbf{E}_2^{\mathrm{T}}\mathbf{E}_2\mathbf{Q}_R^l(\alpha)$ $p\times p$
$\mathbf{r}_L \equiv (\underbrace{1,1,\ldots,1}_{K_l},0,\ldots,0)^{\mathrm{T}}$ $p\times 1$	$\mathbf{s}_R \equiv (0,0,\ldots,\overbrace{1,\ldots,1}^{K_r})^{\mathrm{T}}$ $p\times 1$
$\mathbf{F}_L^l(\alpha) = (F_{1L}^{l_1}(\alpha), F_{2L}^{l_2}(\alpha), \cdots, F_{pL}^{l_p}(\alpha))^{\mathrm{T}}$	$\mathbf{F}_R^l(\alpha) = (F_{1R}^{l_1}(\alpha), F_{2R}^{l_2}(\alpha), \cdots, F_{pR}^{l_p}(\alpha))^{\mathrm{T}}$
$\underline{\mathbf{W}}(\alpha) = (\underline{W}_1^{l_1}(\alpha), \underline{W}_2^{l_2}(\alpha), \cdots, \underline{W}_p^{l_p}(\alpha))^{\mathrm{T}}$	$\overline{\mathbf{W}}(\alpha) = (\overline{W}_1^{l_1}(\alpha), \overline{W}_2^{l_2}(\alpha), \cdots, \overline{W}_p^{l_p}(\alpha))^{\mathrm{T}}$
$\mathbf{m}_L^l(\alpha) \equiv \mathbf{M}_{L1}^l(\alpha)\mathbf{F}_L^l(\alpha)$ $p\times 1$	$\mathbf{m}_R^l(\alpha) \equiv \mathbf{M}_{R1}^l(\alpha)\mathbf{F}_R^l(\alpha)$ $p\times 1$
$\mathbf{k}_L^l(\alpha) \equiv \mathbf{M}_{L2}^l(\alpha)\mathbf{F}_L^l(\alpha)$ $p\times 1$	$\mathbf{k}_R^l(\alpha) \equiv \mathbf{M}_{R2}^l(\alpha)\mathbf{F}_R^l(\alpha)$ $p\times 1$
$\mathbf{p}_L^l(\alpha) \equiv \mathbf{r}_L^{\mathrm{T}}\mathbf{Q}_L(\alpha)$ $1\times p$	$\mathbf{p}_R^l(\alpha) \equiv \mathbf{r}_R^{\mathrm{T}}\mathbf{Q}_R(\alpha)$ $1\times p$
$\mathbf{q}_L^l(\alpha) \equiv \mathbf{s}_L^{\mathrm{T}}\mathbf{Q}_L(\alpha)$ $1\times p$	$\mathbf{q}_R^l(\alpha) \equiv \mathbf{s}_R^{\mathrm{T}}\mathbf{Q}_R(\alpha)$ $1\times p$

$$\begin{aligned}\underline{y}_{FWA}^l(\alpha) &= \frac{\overline{\mathbf{W}}^{\mathrm{T}}(\alpha)\mathbf{M}_{L1}^l(\alpha)\mathbf{F}_L^l(\alpha) + \underline{\mathbf{W}}^{\mathrm{T}}(\alpha)\mathbf{M}_{L2}^l(\alpha)\mathbf{F}_L^l(\alpha)}{\overline{\mathbf{W}}^{\mathrm{T}}(\alpha)\mathbf{r}_L^{\mathrm{T}}\mathbf{Q}_L(\alpha) + \underline{\mathbf{W}}^{\mathrm{T}}(\alpha)\mathbf{s}_L^{\mathrm{T}}\mathbf{Q}_L(\alpha)} = \frac{\overline{\mathbf{W}}^{\mathrm{T}}(\alpha)\mathbf{m}_L^l(\alpha) + \underline{\mathbf{W}}^{\mathrm{T}}(\alpha)\mathbf{k}_L^l(\alpha)}{\overline{\mathbf{W}}^{\mathrm{T}}(\alpha)\mathbf{p}_L^l(\alpha) + \underline{\mathbf{W}}^{\mathrm{T}}(\alpha)\mathbf{q}_L^l(\alpha)} \\ &= \frac{\sum_{i=1}^p m_{L,i}^l(\alpha)\overline{W}_i^{l_i}(\alpha) + \sum_{j=1}^p k_{L,j}^l(\alpha)\underline{W}_j^{l_j}(\alpha)}{\sum_{i=1}^p p_{L,i}^l(\alpha)\overline{W}_i^{l_i}(\alpha) + \sum_{j=1}^p q_{L,j}^l(\alpha)\underline{W}_j^{l_j}(\alpha)}\end{aligned} \tag{8}$$

$$\begin{aligned}\overline{y}_{FWA}^l(\alpha) &= \frac{\underline{\mathbf{W}}^{\mathrm{T}}(\alpha)\mathbf{M}_{R1}^l(\alpha)\mathbf{F}_R^l(\alpha) + \overline{\mathbf{W}}^{\mathrm{T}}(\alpha)\mathbf{M}_{R2}^l(\alpha)\mathbf{F}_R^l(\alpha)}{\underline{\mathbf{W}}^{\mathrm{T}}(\alpha)\mathbf{r}_L^{\mathrm{T}}\mathbf{Q}_R(\alpha) + \overline{\mathbf{W}}^{\mathrm{T}}(\alpha)\mathbf{s}_L^{\mathrm{T}}\mathbf{Q}_R(\alpha)} = \frac{\underline{\mathbf{W}}^{\mathrm{T}}(\alpha)\mathbf{m}_R^l(\alpha) + \overline{\mathbf{W}}^{\mathrm{T}}(\alpha)\mathbf{k}_R^l(\alpha)}{\underline{\mathbf{W}}^{\mathrm{T}}(\alpha)\mathbf{p}_R^l(\alpha) + \overline{\mathbf{W}}^{\mathrm{T}}(\alpha)\mathbf{q}_R^l(\alpha)} \\ &= \frac{\sum_{i=1}^p m_{R,i}^l(\alpha)\underline{W}_i^{l_i}(\alpha) + \sum_{j=1}^p k_{R,j}^l(\alpha)\overline{W}_j^{l_j}(\alpha)}{\sum_{i=1}^p p_{R,i}^l(\alpha)\underline{W}_i^{l_i}(\alpha) + \sum_{j=1}^p q_{R,j}^l(\alpha)\overline{W}_j^{l_j}(\alpha)}\end{aligned} \tag{9}$$

By taking α in (8) and (9) to be 0 and 1 in turn, we can obtain the analytic expressions of $a_{Y_{FWA}^l}$ immediately:

$$a_{Y_{FWA}^l} = \underline{y}_{FWA}^l(0) = \frac{\overline{\mathbf{W}}^{\mathrm{T}}(0)\mathbf{m}_L^l(0) + \underline{\mathbf{W}}^{\mathrm{T}}(0)\mathbf{k}_L^l(0)}{\overline{\mathbf{W}}^{\mathrm{T}}(0)\mathbf{p}_L^l(0) + \underline{\mathbf{W}}^{\mathrm{T}}(0)\mathbf{q}_L^l(0)} = \frac{\sum_{i=1}^p m_{L,i}^l(0)\overline{W}_i^{l_i}(0) + \sum_{j=1}^p k_{L,j}^l(0)\underline{W}_j^{l_j}(0)}{\sum_{i=1}^p p_{L,i}^l(0)\overline{W}_i^{l_i}(0) + \sum_{j=1}^p q_{L,j}^l(0)\underline{W}_j^{l_j}(0)} \tag{10}$$

we can also get $b_{Y^l_{FWA}}$, $c_{Y^l_{FWA}}$, $d_{Y^l_{FWA}}$ in the same way. Furthermore, the GMIR y^l of the consequent $Y^l, l = 1, \ldots, M$ can be computed by

$$\begin{aligned} y^l &\triangleq y^{l_1 l_2 \ldots l_p} = P(Y^l_{FWA}) + P(B^l) \\ &= \frac{1}{6}\left(a_{Y^l_{FWA}} + 2b_{Y^l_{FWA}} + 2c_{Y^l_{FWA}} + d_{Y^l_{FWA}} + a_{B^l} + 2b_{B^l} + 2c_{B^l} + d_{B^l}\right) \end{aligned} \tag{11}$$

where $P(Y^l_{FWA})$ and $P(B^l)$ denote the GMIR of Y^l_{FWA} and B^l, respectively, and $a_{Y^l_{FWA}}$, $b_{Y^l_{FWA}}$, $c_{Y^l_{FWA}}$, $d_{Y^l_{FWA}}$ can be analytically calculated. To this end, we rewrite the output of the proposed FWA FLS $y_{FWA}(\mathbf{x})$ as follows:

$$y_{FWA}(\mathbf{x}) = \frac{1}{6}\sum_{l=1}^{M}\phi^l(\mathbf{x})\left(a_{Y^l_{FWA}} + 2b_{Y^l_{FWA}} + 2c_{Y^l_{FWA}} + d_{Y^l_{FWA}} + a_{B^l} + 2b_{B^l} + 2c_{B^l} + d_{B^l}\right) \tag{12}$$

3.2 *Parameters Optimization Based on the Error Back-Propagation and Steepest Descent Method*

In our numerical simulations, we will adopt the steepest descent method to optimize all the parameters in our proposed FWA FLS. Let θ denote a generic symbol that represents all parameters to be optimized for the generic objective function E by the following steepest descent algorithm: $\theta_{new} = \theta_{old} - \alpha \frac{\partial E}{\partial \theta}|_{\theta_{old}}$ where $\alpha > 0$.

In the proposed FWA FLS, there are $4Mp$ antecedent parameters and $4Mp + 4M$ consequent parameters to be tuned, they are $a_{F_j^{l_j}}, b_{F_j^{l_j}}, c_{F_j^{l_j}}, d_{F_j^{l_j}}$, $a_{W_j^{l_j}}$, $b_{W_j^{l_j}}$, $c_{W_j^{l_j}}$, $d_{W_j^{l_j}}, (j = 1, 2, \ldots, p, l_1 = 1, \ldots, N_1, l_2 = 1, \ldots, N_2, \cdots, l_p = 1, \ldots, N_p)$, and a_{B^l}, $b_{B^l}, c_{B^l}, d_{B^l}, (l = 1, \ldots, M), M = N_1 N_2 \cdots N_p$. We will optimize all aforementioned parameters based on the following error function:

$$E^{(t)}_{FWA} = \frac{1}{2}\left(y_{FWA}(\mathbf{x}^{(t)}) - y^{(t)}\right)^2 \tag{13}$$

where $(\mathbf{x}^{(t)}, y^{(t)}), (t = 1, \ldots, T)$ are T training data pairs, $y_{FWA}(\mathbf{x}^{(t)})$ is the crisp output of the proposed FWA FLS.

Using chain rule, we can get the updating formulas for antecedent and consequent parameters as follows:

$$a_{F_j^{l_j}}(q+1) = a_{F_j^{l_j}}(q) - \eta \cdot (y_{FWA} - y^{(t)}) \cdot \phi^l \cdot \begin{cases} \left[\frac{1}{6} \frac{\partial d_{Y_{FWA}^l}}{\partial a_{F_j^{l_j}}} + \frac{y^l - y_{FWA}}{\mu_{F_j^{l_j}}(x_j)} \cdot \frac{x_j - b}{(b-a)^2} \right] \Big|_q, & a_{F_j^{l_j}} \le x_j \le b_{F_j^{l_j}} \\ \frac{1}{6} P(W_j^{l_j}) \Big|_q, & \text{others} \end{cases} \tag{14}$$

$$a_{W_j^{l_j}}(q+1) = a_{W_j^{l_j}}(q) - \eta \cdot (y_{FWA} - y^{(t)}) \cdot \frac{1}{6}\phi^l \left[\frac{k_{L,i}^l(0) - q_{L,i}^l(0) a_{Y_{FWA}^l}}{\sum_{j=1}^p p_{L,j}^l(0) \overline{W}_j^{l_j}(0) + \sum_{i=1}^p q_{L,i}^l(0) \underline{W}_i^{l_i}(0)} + \frac{m_{R,i}^l(0) - p_{R,i}^l(0) d_{Y_{FWA}^l}}{\sum_{i=1}^p p_{R,i}^l(0) \underline{W}_i^{l_i}(0) + \sum_{j=1}^p q_{R,j}^l(0) \overline{W}_j^{l_j}(0)} \right] \Bigg|_q \tag{15}$$

We can also get $b_{F_j^{l_j}}(q+1)$, $c_{F_j^{l_j}}(q+1)$, $d_{F_j^{l_j}}(q+1)$ and $b_{W_j^{l_j}}(q+1)$, $c_{W_j^{l_j}}(q+1)$, $d_{W_j^{l_j}}(q+1)$ in the same way.

$$\begin{aligned} a_{B^l}(q+1) &= a_{B^l}(q) - \eta \cdot (y_{FWA} - y^{(t)}) \cdot \frac{1}{6}\phi^l \Big|_q, \\ b_{B^l}(q+1) &= b_{B^l}(q) - \eta \cdot (y_{FWA} - y^{(t)}) \cdot \frac{1}{3}\phi^l \Big|_q, \\ c_{B^l}(q+1) &= c_{B^l}(q) - \eta \cdot (y_{FWA} - y^{(t)}) \cdot \frac{1}{3}\phi^l \Big|_q, \\ d_{B^l}(q+1) &= d_{B^l}(q) - \eta \cdot (y_{FWA} - y^{(t)}) \cdot \frac{1}{6}\phi^l \Big|_q \end{aligned} \tag{16}$$

The detailed calculation for the updating formulas are put in [19]. We summarize the parameters updating steps as follows:

Step 1: Initialization, set the input number p, rule number M, epoch number *epoch*, Monte-Carlo number n, and training and testing data pair numbers;

Step 2: Given the training input-output data pairs $(\mathbf{x}^{(t)}, y^{(t)})$, $(t = 1, \ldots, T_{\text{train}})$ and calculate the FWA FLS output $y_{FWA}(\mathbf{x}^{(t)})$;

Step 3: For each training pair, update antecedent parameters $a_{F_j^{l_j}}$, $b_{F_j^{l_j}}$, $c_{F_j^{l_j}}$, $d_{F_j^{l_j}}$, consequent parameters $a_{W_j^{l_j}}$, $b_{W_j^{l_j}}$, $c_{W_j^{l_j}}$, $d_{W_j^{l_j}}$, and a_{B^l}, b_{B^l}, c_{B^l}, d_{B^l};

Step 4: For each epoch e, execute Step 3, until $e = epoch$, return the finally updated parameters;

Step 5: Given the testing input-output data pairs $(\mathbf{x}^{(t)}, y^{(t)})$, $(t = 1, \ldots, T_{\text{test}})$, and calculate the FWA FLS output $y_{FWA}(\mathbf{x}^{(t)})$ with finally updated parameters.

4 Numerical Simulations

In this section, to compare our proposed FWA FLS with Mamdani and TSK FLSs, we conduct several numerical simulations on the issue of Mackey-Glass chaotic time series forecasting, which is a benchmark problem in both the neural network and fuzzy logic fields [7]. Mackey-Glass chaotic time series is defined by the dynamics of the following first-order nonlinear delay differential equation [20]:

$$\frac{ds(t)}{dt} = \frac{0.2s(t-\tau)}{1+s^{10}(t-\tau)} - 0.1s(t) \tag{17}$$

This delay differential equation models physiological systems that has become known as the equation. For $\tau > 17$, (17) is known to exhibit chaos. We chose $\tau = 30$.

In our simulations, let $s(t) (t = 1, 2, \ldots, T)$ be a time series, and its measured values $y(t)$ are presented as $y(t) = s(t) + n(t), t = 1, \ldots, T$, where $n(t)$ denotes the additive white Gaussian noise. The problem of forecasting a time series is in the following: Given a window of p past measurements of $s(t)$, namely, $y(t-p+1), y(t-p+2), \ldots, y(t)$, determine an estimate of a future value of s, $\hat{s}(t+1)$. The variable p is a fixed positive integer, and in this paper, $p = 4$, namely $y(t-3)$, $y(t-2)$, $y(t-1)$, and $y(t)$, to predict $y(t+1)$. The FLS forecasters are based on 1000 time points, that is, $y(2001), \ldots, y(3000)$. For training, 500 data sets were used, where $t = 2005, \ldots, 2504$, and for testing, another 496 data sets were used, where $t = 2505, \ldots, 3000$.

In the numerical simulations, we all adopt the steepest descent algorithm to optimize the parameters involved in three FLSs based on the following error function:

$$E_{FLS}^{(t)} = \frac{1}{2}\left(y_{FLS}(\mathbf{x}^{(t)}) - y^{(t)}\right)^2, \tag{18}$$

where $(\mathbf{x}^{(t)}, y^{(t)})$, $(t = 1, \ldots, T)$ are T training data pairs, $y_{FLS}(\mathbf{x}^{(t)})$ is the output of Mamdani FLS, or TSK FLS, or the proposed FWA FLS.

Let one epoch be defined as the collection of T training data as in [7]. In this paper, ten epoches were run for each case of the simulations. After each epoch $j, (j = 1, \ldots, 10)$ of training, the performance of each FLS was evaluated by using N testing data pairs $(\mathbf{x}^{(t)}, y^{(t)})$, $(t = T+1, \ldots, T+N)$ and the following root-mean-squared errors (RMSE):

$$RMSE_{FLS}(j) = \sqrt{\frac{1}{N}\sum_{t=T+1}^{T+N} (y_{FLS}(\mathbf{x}^{(t)}|\theta_{FLS}^{j}) - y^{(t)})^2} \tag{19}$$

The learning rate α will decay exponentially with epoch going on, that is, $\alpha(j) = \gamma^{j-1}\alpha_0$, $j = 1, \ldots, 10$, where α_0 is the initial learning rate and γ is the decay rate.

In addition, in our simulations, the following were used: (1) trapezoidal MFs for antecedents and consequents in three FLSs; (2) product t-norm; (3) Mamdani

FLS uses singleton fuzzification and COS defuzzification; (4) Twenty Monte Carlo simulations and designs were performed for each FLS. For each FLS design, its 20 Monte Carlo RMSE values are statistically independent; hence, we were able to use the bootstrap method to get more accurate estimates of the mean and standard deviations (STD for short) of each FLS RMSE.

Detailed designs of the three FLS forecasters are given below. Each antecedent is divided into 2 fuzzy sets, so that the number of rules is $M = 2^4 = 16$. Mamdani FLS forecaster has $4Mp + M = (4p + 1)M = 282$ parameters to be optimized; TSK FLS forecaster has $4Mp + (p + 1)M = (5p + 1)M = 336$ parameters to be optimized; and FWA FLS forecaster has $4Mp + 4Mp + 4M = (8p + 4)M = 576$ parameters to be optimized. The initial values of the antecedent parameters $a_{F_j^{l_j}}, b_{F_j^{l_j}}, c_{F_j^{l_j}}, d_{F_j^{l_j}}$ in the three FLSs were all taken uniformly in the interval $[-2, 4]$. The initial values of consequent parameters in Mamdani and TSK FLSs were chose randomly in the interval [0, 1], and the initial values of consequent parameters $a_{W_j^{l_j}}, b_{W_j^{l_j}}, c_{W_j^{l_j}}, d_{W_j^{l_j}}$ and a_{B^l}, b_{B^l}, c_{B^l}, d_{B^l} in proposed FWA FLS were chose randomly in the interval [0, 0.5].

The simulation results for Mackey-Glass time series forecasting are summarized in Tables 2, 3 and Fig. 2. Table 2 provides the raw values for the mean and standard deviation (STD) of 20 Monte-Carlo simulations for the three FLS forecasters. Table 3 and Fig. 2 provide the comparative results of their mean RMSE.

From the results of Tables 2 and 3, and Fig. 2, one can see that our proposed FWA FLS outperforms Mamdani and TSK FLSs in RMSE for each noise case. From Fig. 2, one also find that RMSE line of the proposed FWA FLS is almost below the other RMSE lines, which means that our FWA FLS can reach to small error earlier than the other two FLSs.

Table 2 Comparisons on mean and STD of RMSE for Mackey-Glass time series forecasting

SNR	Mamdani	TSK	FWA
0dB	0.2921 ± 0.0184	0.2776 ± 0.0219	0.2761 ± 0.0196
5dB	0.2208 ± 0.0124	0.2125 ± 0.0142	0.2117 ± 0.0137
10dB	0.1599 ± 0.0110	0.1607 ± 0.0123	0.1586 ± 0.0109
20dB	0.0837 ± 0.0045	0.0821 ± 0.0041	0.0816 ± 0.0040

Table 3 Percentage improvement of mean RMSE of FWA FLS from that of Mamdani and TSK FLSs Mackey-Glass time series forecasting

SNR	FWA versus Mamdani (%)	FWA versus TSK (%)
0dB	5.48	0.54
5dB	4.12	0.38
10dB	0.81	1.30
20dB	2.51	0.61

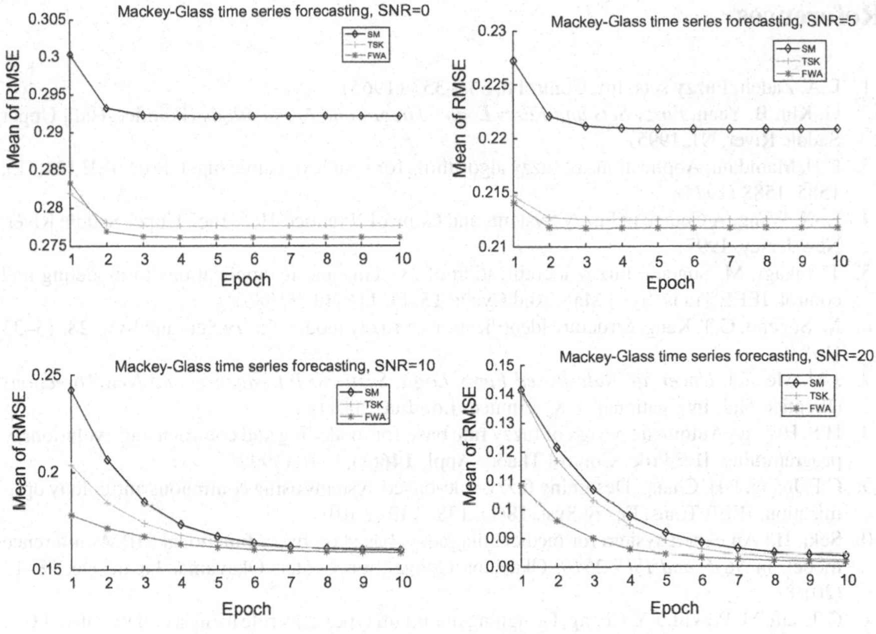

Fig. 2 Mean of RMSE for three FLSs, and SNRs = 0, 5, 10, and 20 dB

5 Conclusions

In this paper, a novel fuzzy logic system named FWA with novel rules was proposed, in which the consequents are fuzzy affine combination of antecedents. The proposed rules establish connection between consequents and antecedents in advance, so that the proposed FWA FLS can reduce the epoches of training, and improve training efficiency. Numerical simulations for Mackey-Glass time-series forecasting have been conducted in the MATLAB environment to compare the proposed FLS with the popular Mamdani FLS and TSK FLS. The simulation results indicate that our proposed FWA FLS outperforms a Mamdani FLS and a TSK FLS in all SNRs noised environments. From the simulations, one can also observe that our proposed FWA FLS is more suitable for the system model in which each input has different influence on the output. In the future work, we would like to apply our FWA FLS to some practical experimentation, e.g. the control systems in [21, 22]. Also, we will try to use other shaped membership functions in our proposed FWA FLS.

Acknowledgements This work is supported by National Natural Science Foundation of China (61403011).

References

1. L.A. Zadeh, Fuzzy sets. Inf. Control **8**, 338–353 (1965)
2. G. Klir, B. Yuan, *Fuzzy Sets and Fuzzy Logic: Theory and Applications* (Prentice-Hall, Upper Saddle River, NJ, 1995)
3. E.H. Mamdani, Applications of fuzzy algorithms for simple dynamic plant. Proc. IEE **121**(12), 1585–1588 (1974)
4. L.-X. Wang, A Course in Fuzzy Systems and Control (Prentice-Hall, Inc., Upper Saddle River, New Jersey, 1997)
5. T. Takagi, M. Sugeno, Fuzzy identification of systems and its applications to modeling and control. IEEE Trans. Syst. Man, And Cyber. **15**(1), 116–132 (1985)
6. M. Sugeno, G.T. Kang, Structure identification of fuzzy model. Fuzzy Sets and Syst. **28**, 15–33 (1988)
7. J.M. Mendel, *Uncertain Rule-Based Fuzzy Logic Systems: Introduction and New Directions* (Prentice-Hall International (UK) Limited, London, 2001)
8. H.S. Hwang, Automatic design of fuzzy rule base for modelling and control using evolutionary programming. IEE Proc.-Control Theory Appl. **146**(1), 9–16 (1999)
9. C.F. Juang, P.H. Chang, Designing fuzzy-rule-based systems using continuous ant-colony optimization. IEEE Trans. Fuzzy Syst. **18**(1), 138–149 (2010)
10. Seki, H.: An expert system for medical diagnosis based on fuzzy functional SIRMs inference model, in *SCIS and ISIS 2010*, Okayama Convention Center, Okayama, Japan, Dec. 8–12 (2010)
11. C.T. Lin, M. Prasad, J.Y. Chang, Designing mamdani type fuzzy rule using a collaborative FCM scheme, in *Proceedings of 2013 International Conference on Fuzzy Theory and Its Application*. (National Taiwan University of Science and Technology, Taipei, Taiwan, Dec. 2013), pp. 279-282
12. S.H. Chen, G.C. Li, Representation, ranking, and distance of fuzzy number with exponential membership function using graded mean integration method. Tamsui Oxford J. Mathe. Sci. **16**(2), 123–131 (2000)
13. C.C. Chou, The canonical representation of multiplication operation on triangular fuzzy numbers. Comput. Mathe. Appl. **45**, 1601–1610 (2003)
14. C.C. Chou, The Representation of Multiplication Operation on Fuzzy numbers and Application to Solving Fuzzy Multiple Criteria Decision Making Problems, in *PRICAI 2006, LNCS(LNAI)* vol. 4099, ed. by Q. Yang and G. Webb (Springer Heidelberg, 2006), pp. 161–169
15. J.M. Mendel, D.R. Wu, *Perceptual computing: aiding people in making subjective judgements* (Wiley, Hoboken, 2010)
16. F.L. Liu, J.M. Mendel, Aggregation using the fuzzy weighted average as computed by the KarnikCMendel algorithms. IEEE Trans. Fuzzy Syst. **16**(1), 1–12 (2008)
17. D.R. Wu, J.M. Mendel, Enhanced karnik-mendel algorithms. IEEE Trans. Fuzzy Syst. **17**(4), 923–934 (2009)
18. J.M. Mendel, Computing derivatives in interval type-2 fuzzy logic systems. Trans. Fuzzy Syst. **12**(1), 84–98 (2004)
19. Liu, Y.-Q., Zhang, Q.-Y., Tian, X.: Computing derivatives in fuzzy logic system with consequents as fuzzy weighted averages of antecedents, in *Submitted to the 14th Chinese Intelligent Systems Conference* (Wenzhou, China, Oct. 13–14, 2018)
20. M.C. Mackey, L. Glass, Oscillation and chaos in physiological control system. Science **197**, 287–289 (1977)
21. Y.M. Jia, Robust control with decoupling performance for steering and traction of 4WS vehicles under velocity-varying motion. IEEE Trans. Control Syst. Technol. **8**(3), 554–569 (2000)
22. Y.M. Jia, Alternative proofs for improved LMI representations for the analysis and the design of continuous-time systems with polytopic type uncertainty: a predictive approach. IEEE Trans. Autom. Control **48**(8), 1413–1416 (2003)

An Information Theory Based Approach for Link Prediction in Complex Networks

Xuecheng Yu, Rui Li and Tianguang Chu

Abstract We consider the link prediction problem in complex networks from the perspective of information theory. An approach is developed to take advantage of different structural features of networks. Specifically, in case only one feature is available, the conditional self-information of the event that there is a link connecting two nodes is used to evaluate the link existence likelihood. In case of multiple available features, we give a linear model to evaluate the existence likelihood of all potential links. Simulation results show that our approach gives satisfying results in synthetic complex networks compared with other methods using typical proximity indices.

Keywords Link prediction · Partial observation · Information theory

1 Introduction

Link prediction aims at estimating the existence of links with observations of social networks. The problem has received extensive attention in disparate fields due to its wide applications [1]. It can also provide useful insight into the mechanism of the growth and evolution of complex networks [2].

There have been many methods for prediction of links in complex networks. A basic one uses similarity-based algorithms based on the assumption that the more similar two nodes are, the higher probability of their connection tends to be. Many

X. Yu · T. Chu (✉)
College of Engineering, Peking University, Beijing 100871, China
e-mail: chutg@pku.edu.cn

X. Yu
e-mail: yuxuecheng@pku.edu.cn

R. Li
School of Mathematical Sciences,
Dalian University of Technology, Liaoning 116024, China
e-mail: rui_li@dlut.edu.cn

Y. Jia et al. (eds.), *Proceedings of 2018 Chinese Intelligent Systems Conference*, Lecture Notes in Electrical Engineering 529,
https://doi.org/10.1007/978-981-13-2291-4_57

similarity-based algorithms use neighboring information and assume that two nodes sharing more common structural features, such as the number of common neighbors and node attributes, are more likely to be connected. In applications, node attributes such as personal information of users in social networks are usually unavailable due to privacy reasons, whereas structural features of a network are often easier to obtain. Therefore, many similarity-based methods focus on structural similarity.

To calculate the probability of potential links, similarity-based methods make use of various indices. For example, Common Neighbors (CN) [3] index is used to count the number of common neighbors but ignore different contributions on the connection likelihood; Adamic-Adar (AA) index [4] and Resource Allocation (RA) index [5] are extensions of the CN index to improve the prediction accuracy, where lower degree node is allocated with higher importance. Note that these indices involve only local structural features. There have been other indices taking into account global similarity measures, e.g., the Katz index [6] and the Propflow index [7]. However, both local and global structural similarity-based methods may have limitations. For example, two unconnected node pairs can have different neighborhoods sharing a common neighbor. In this case, the score of the common neighbor assigned to each node pairs by using the CN index or the RA index is irrespective of respective neighborhoods. In addition, predicting missing links using global structural similarity measures usually takes more time than using local structural similarity measures. To overcome these limitations, Ref. [8] gave a subnetwork similarity measure using the node degree distribution of neighborhoods of two unconnected node pairs and the node degree distribution of the global network, which assigns different scores to the same node for its different neighborhoods, and has a smaller time consumption in dealing with the global structural feature.

So far, studies on link prediction problems are mostly based on likelihood estimation of the event of connection between two nodes. On the other hand, from the perspective of information theory, information measures the uncertainty associated with the outcome of a random variable or an event. Therefore, the connecting likelihood between a node pair can also be evaluated by using the information theory. In view of this, a mutual information method was proposed in [9], which considers the connecting likelihood as the conditional self-information of the connection event between node pairs with given common neighbors of them.

In this paper, we intend to give an information theory based approach to link prediction problem for complex networks. Our approach can take advantage of different structural features of the networks and facilitate prediction missing links. To be specific, we use the conditional self-information to estimate the existence likelihood of a link in case of only one feature available. For the case of multiple available features, we give a linear model incorporating local and global structural features to evaluate the likelihood of a link. We will provide simulation results on different scale networks to compare our approach with other structural similarity based methods.

2 Preliminaries

For a social network $G = (V, E)$, V, E denote the set of nodes and links respectively. Nodes in social networks represent people or users, and links represent connections or communications among nodes. For a set, $|\cdot|$ denotes the cardinality of it, and $|V| = n$ for a set with n nodes.

For two events or random variables X and Y, the conditional probability function is $p(x|y), x \in X, y \in Y$, and the marginal probability functions are $p(x)$ and $p(y)$. The mutual information is the relative entropy between the joint distribution and the product distribution $p(x)p(y)$, calculated by:

$$\begin{aligned} I(x;y) &= \log \frac{p(x,y)}{p(x)p(y)} \\ &= \log \frac{p(x|y)}{p(x)} \\ &= -\log p(x) + \log p(x|y) \\ &= I(x) - I(x|y), \end{aligned}$$

where $I(x|y)$ is the conditional self-information indicating the uncertainty of the occurrence of x given the occurrence of y, and $I(x)$ is the self information of the uncertainty of x. The mutual information is a measure of the amount of information that one event contains about another event. It is actually the reduction in the uncertainty of one event due to the knowledge of the other. Clearly, the mutual information of two independent events is zero.

For convenience of notation, we use N_i to represent the set of neighbors of node i, $N_{ij} = N_i \cup N_j$ to represent the neighborhood of nodes i and j, and d_i to represent the degree of node i.

3 Method

According to structural similarity methods, similarities of node pairs are regarded as the probabilities of potential links between them. From the perspective of information theory, the estimation of connecting likelihood of a potential link between two nodes can be understood as calculating the information of the event that there is a link between them.

In structural similarity methods, all missing links are ranked according to their similarity measure scores, and links connecting more similar nodes are supposed to be of higher existence likelihoods. By the information theory, the existence likelihood is estimated based on the information extracted from neighborhood. Specifically, let L_{ij} be the event of connection between nodes i and j. For an unconnected node pair (i, j), the link likelihood can be estimated by $-I(L_{ij}|N_{ij})$ with the knowledge of neighborhood of nodes i and j [9], and $I(L_{ij}|N_{ij})$ can be calculated as

$$I(L_{ij}|N_{ij}) = I(L_{ij}) - I(L_{ij}; N_{ij}),$$

where $L(L_{ij}; N_{ij})$ is the mutual information between events of connection between nodes i and j and available neighborhood of them.

To calculate $I(L_{ij})$, we introduce a prior probability $p(L_{ij})$ as

$$p(L_{ij}) = \frac{1}{2}\left[\frac{|N_i|}{|V|-1} + \frac{|N_j|}{|V|-1}\right], \tag{1}$$

where $\frac{|N_i|}{|V|-1}$ and $\frac{|N_j|}{|V|-1}$ represent the sociability of nodes i and j respectively.

If elements in N_{ij} are supposed to be independent from each other, it has

$$I(L_{ij}; N_{ij}) = \sum_{z \in N_{ij}} I(L_{ij}; z), \tag{2}$$

with the average mutual information over all node pairs connecting to node z

$$I(L_{ij}; z) = I(L_{ij}) - I(L_{ij}|z),$$

where $I(L_{ij}|z)$ is the conditional self-information of the event of connection of node pair (i, j) with the neighborhood. To calculate $I(L_{ij}|z)$, we need to obtain $p(L_{ij}|z)$ first, which is the probability of a link between node pair (i, j), and can be calculated as the clustering coefficient of node z by

$$p(L_{ij}|z) = C_z = \frac{N_{\Delta z}}{N_{\Delta z} + N_{\wedge z}}, \tag{3}$$

where $N_{\Delta z}$ and $N_{\wedge z}$ are the number of connected and disconnected node pairs sharing the common neighbor z, respectively [9].

Substituting (1) and (3) into (2), we have

$$\begin{aligned} I(L_{ij}; N_{ij}) &= \sum_{z \in N_{ij}} I(L_{ij}; z) \\ &= \sum_{z \in N_{ij}} [I(L_{ij}) - I(L_{ij}|z)] \\ &= \sum_{z \in N_{ij}} [-\log p(L_{ij}) + \log p(L_{ij}|z)] \\ &= \sum_{z \in N_{ij}} \left[-\log \frac{|N_z|}{|V|-1} + \log \frac{N_{\Delta z}}{N_{\Delta z} + N_{\wedge z}}\right]. \end{aligned}$$

If nodes i and j have no common neighbors, $I(L_{ij}; z)$ is zero. Clearly, if $C_i = 1$ for all nodes, $I(L_{ij}; N_{ij})$ degenerates to the common neighbor measure [10]. Therefore, in accordance to the clustering coefficient C_z, different common neighbors make different contributions to the connection likelihood.

Since $I(L_{ij})$ represents the uncertainty of the event of connection between nodes i and j via a link, the existence likelihood of a link can be estimated by the conditional self-information $I(L_{ij}|N_{ij})$ based on the observed neighborhood N_{ij}. And the smaller $I(L_{ij}|N_{ij})$ is, the higher the probability of a link between nodes i and j tends to be.

Therefore, the mutual information $I(L_{ij}|N_{ij})$ gives the reduction in uncertainty of an event L_{ij} when an event N_{ij} is given. For the link prediction task, the likelihood score can be defined as

$$s_{ij} = -I(L_{ij}|N_{ij}). \tag{4}$$

It can be inferred that s_{ij} has positive correlation with p_{ij}.

In practice, different structural features of the network imply different aspects of network properties. For example, the shortest path and clustering, commonly used in link prediction problems, can characterize the small-world property of networks and indicate more links connecting to a node with a dense neighborhood than one with a sparse neighborhood respectively. Generally speaking, more features used in prediction can improve the prediction accuracy. In view of this, we introduce parameters λ_k in evaluation of the contribution of feature k to the connection likelihood as follows

$$s_{ij} = \sum_k \lambda_k s_{ij}^k(N_{ij}) = -\sum_k \lambda_k I_k(L_{ij}|N_{ij}).$$

Based on the above consideration, we give a two-part connection likelihood using local and global structural information. The local information is extracted from the neighborhood of each node pair, which is formulated as a naive Bayes model by

$$p(L_{ij}|\Omega_{ij}) = \frac{p(L_{ij}, \Omega_{ij})}{p(\Omega_{ij})}, \tag{5}$$

where $\Omega_{ij} = \{l_{st} \in E, s \in (N_i \cup \{i\}), t \in (N_j \cup \{j\})\}$ denotes the links across neighborhood of nodes i and j.

Figure 1 gives an illustrate of (5).

In Fig. 1, nodes i and j are disconnected. The neighbor set of i is $N_i = \{a, b, c\}$, and the neighbor set of j is $N_j = \{c, d, e, f\}$. Node c is a common neighbor of nodes i and j, and its neighbor pairs (d, j), (f, j), (b, i) are connected respectively. We then have $N_{\triangle c} = 3$. Due to that (i, d),(i, j), (i, f), (b, d), (b, j), (b, f), (d, f) are disconnected, $N_{\wedge c} = 7$. For nodes i and j, $\Omega_{ij} = |(i, c), (b, c), (b, e), (c, f), (c, j), (c, d)| = 6$.

The conditional probability $p(L_{ij}, \Omega)$ can be estimated as

Fig. 1 A small scale network containing 8 nodes

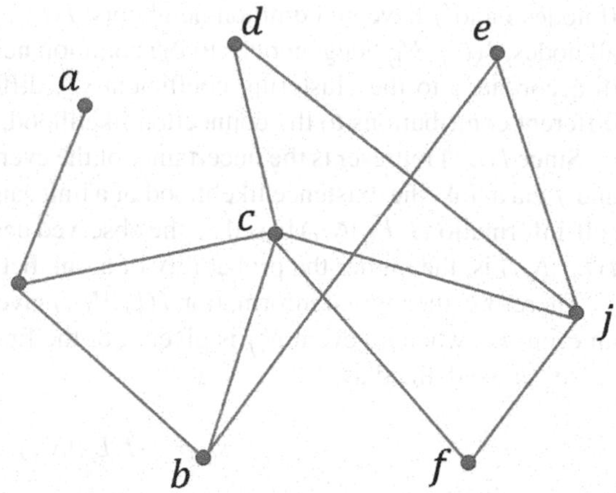

$$p(L_{ij}, \Omega_{ij}) = \frac{|\{l_{is}, s \in CN(i, j)\}| + |\{l_{js}, s \in CN(i, j)\}|}{|\Omega_{ij}|} = \frac{2|CN|}{|\Omega_{ij}|}, \tag{6}$$

where $CN(i, j)$ is the set of common neighbors of nodes i and j, and $|\Omega_{ij}|$ is the number of links across neighborhoods of nodes i and j. In the CN index, two nodes having more common neighbors tend to be more similar. Then, $p(\Omega_{ij})$ can be calculated as the edge density of Ω_{ij}

$$p(\Omega_{ij}) = \frac{2|\Omega_{ij}|}{|N_i \cup N_j|(|N_i \cup N_j| - 1)}. \tag{7}$$

For the global structural feature, we use the Katz index as an example, which is defined as

$$S_{Katz} = (I - \beta A)^{-1} - I, \tag{8}$$

where S_{Katz} is the similarity matrix with the similarity of each node pair as its elements, A is the adjacent matrix of G, and β is a given constant. Note that β should be smaller than the largest eigenvalue of A.

Substituting (6)–(8) into (4), we have

$$\begin{aligned} s_{ij} =& \lambda_1 \left\{\log |CN(i, j)| + \log[|N_i \cup N_j|(|N_i \cup N_j| - 1)] - \log |\Omega_{ij}|\right\} \\ &+ \lambda_2[(I - \beta A)^{-1} - I]_{ij}. \end{aligned}$$

Then, missing links can be predicted according to the rank of s_{ij} for all potential links. Note that the score based on information theory may be positive or negative.

4 Simulations

To verify the effectiveness of our approach, we have performed a number of numerical simulations on different scale of synthetic complex networks, and made comparison with other typical proximity indices such as

- The AA index. Similarity between two nodes is calculated as the sum of inverse of the logarithm of their each common neighbor's degree as

$$AA(a,b) = \sum_{z \in N_a \cap N_b} \ln(d_z)^{-1}.$$

- The RA index. Similarity between two nodes is calculated as the sum of inverse of their each common neighbor's degree as

$$RA(a,b) = \sum_{z \in N_a \cap N_b} d_z^{-1}.$$

- The CN index. Similarity between two nodes is calculated as the number of common neighbors between them as

$$CN(a,b) = |N_a \cap N_b|.$$

- The preferential attachment (PA) index. Similarity between two nodes is calculated as the product of their degrees as

$$PA(a,b) = d_a \cdot d_b.$$

Table 1 shows the average prediction accuracy of 50 independent simulations in each group. The number of nodes and percentage of missing links in each group are same, but the network structure is randomly generated in each round of simulation. The first column from shows the node numbers of networks, and the second column gives the percentage of missing links in simulations. The values in the third to the eighth columns are the average accuracy using our method with single structural feature (ITA-1) and multiple features (ITA-m), AA index, CN index, PA index, and RA index respectively. It can be seen that our method gives better predictions in all cases.

Table 1 Simulation results. Here, all simulations are performed on synthetic complex networks, where the number of nodes varies from 60 to 200

Net size	Deleted links (%)	ITA-1 (%)	ITA-m (%)	AA (%)	CN (%)	PA (%)	RA (%)
60	10	36.703	33.120	23.127	24.481	25.499	23.768
	20	35.879	29.780	20.668	23.187	23.947	21.062
	30	32.692	32.436	20.852	21.850	22.390	20.980
70	10	41.154	32.234	22.861	25.222	25.655	23.081
	20	34.176	28.948	21.484	23.642	23.870	21.491
	30	30.769	32.684	21.546	22.316	23.218	21.546
80	10	39.176	30.440	23.990	25.288	25.089	23.702
	20	34.835	30.096	21.257	22.988	23.523	21.456
	30	36.703	32.102	21.751	22.328	23.977	21.092
90	10	41.648	30.781	23.284	24.359	24.695	23.010
	20	43.022	33.040	21.789	22.827	24.151	22.222
	30	35.330	30.317	22.198	22.631	24.426	21.532
100	10	33.077	29.571	22.434	23.560	24.022	22.110
	20	42.143	33.967	21.115	22.214	23.868	21.659
	30	38.187	29.703	22.022	22.802	24.791	21.731
110	10	36.538	30.859	22.038	23.352	23.352	21.603
	20	36.245	32.850	20.327	21.530	23.679	20.682
	30	36.538	30.200	22.248	23.242	25.035	22.328
120	10	37.527	31.090	21.218	22.230	22.743	20.888
	20	40.659	33.172	20.158	21.064	23.679	20.222
	30	32.088	30.742	21.607	22.770	25.179	21.745
130	10	33.791	31.200	22.244	23.119	23.246	21.822
	20	36.593	32.387	21.071	21.793	24.495	21.071
	30	36.703	30.904	20.976	22.215	24.620	21.048
140	10	39.835	30.746	21.111	22.033	22.806	20.828
	20	30.495	32.452	20.614	21.340	24.170	20.669
	30	29.231	31.994	20.907	22.002	24.015	20.973
150	10	34.451	30.381	21.158	22.319	23.150	21.095
	20	38.901	32.625	20.456	21.386	23.940	20.460
	30	38.901	31.136	20.974	21.982	23.659	21.033
160	10	35.934	30.604	20.642	21.975	22.795	20.584
	20	36.648	32.695	20.352	21.203	23.858	20.256
	30	31.429	31.580	21.058	22.146	23.530	21.106
170	10	40.339	30.067	20.941	22.066	22.699	20.900
	20	34.890	31.528	20.205	21.139	23.851	19.976
	30	41.291	32.265	20.496	21.646	22.962	20.457
180	10	36.154	30.222	20.733	22.137	22.863	20.743
	20	33.846	31.681	20.035	21.298	24.077	19.818
	30	43.791	32.676	20.547	21.799	23.133	20.639

(continued)

Table 1 (continued)

Net size	Deleted links (%)	ITA-1 (%)	ITA-m (%)	AA (%)	CN (%)	PA (%)	RA (%)
190	10	37.802	30.268	20.862	22.389	23.161	20.869
	20	42.308	31.702	20.473	21.785	24.264	20.227
	30	33.077	32.333	20.721	21.704	23.049	20.605
200	10	28.462	30.293	20.550	21.731	22.849	20.710
	20	32.473	31.266	20.765	22.163	24.405	20.493
	30	38.791	32.557	20.718	21.418	22.927	20.534

5 Conclusion

In this paper, we studied the link prediction problem using information theory, and gave an approach to benefit from available structural properties. The information theory based approach has two advantages. The first is that the link existence likelihood can be evaluated via the value of information of different structural features of networks, and the value of information brought by these features are additive. Therefore, our methods can make use of diverse available features. The second advantage is that, when focusing on one feature of the network, the value of information provided by different features are still distinguishable. With multiple available variables, a linear model can synthesize them and make their contributions to predict missing links. In accordance with those available features in an observation of social networks, we used the link existence likelihood to measure different information, and designed a conditional self-information method and a linear combination model. Finally, numerical simulations on synthetic networks verified the effectiveness of our approach.

Acknowledgements This work was supported by National Natural Science Foundation of China under Grant No. 61673027, and partly by National Basic Research Program of China (973 Program, No. 2012CB821203).

References

1. L. Lü, T. Zhou. Link prediction in complex networks: a survey. Phys. Stat. Mech. Its Appl. **390**(6), 1150–1170 (2010)
2. B. Zhu, Y. Xia, An information-theoretic model for link prediction in complex networks. Sci. Rep. **5**, 13707 (2015)
3. M. Newman, Clustering and preferential attachment in growing networks. Phys. Rev. E Stat. Nonlinear Soft Matter Phys. **64**, 025102 (2001)
4. L.A. Adamic, E. Adar, Friends and neighbors on the Web. Social Networks **25**(3), 211–230 (2003)
5. T. Zhou, L. Lü, Y. Zhang, Predicting missing links via local information. Eur. Phys. J. B **71**(4), 623–630 (2009)

6. L. Katz, A new status index derived from sociometric analysis. Psychometrika **18**(1), 39–43 (1953)
7. R. Lichtenwalter, J. Lussier, N. Chawla, New perspectives and methods in link prediction, in *ACM SIGKDD International Conference on Knowledge Discovery and Data Mining*, pp. 243–252 (2010)
8. X. Yu, T. Chu, Link prediction from partial observation in scale-free networks. Lect. Notes, Electr. Eng. **460**(2), 199–206 (2017)
9. F. Tan, Y. Xia, B. Zhu, Link prediction in complex networks: a mutual information perspective. Plos One **9**(9), e107056 (2014)
10. T.M. Cover, J.A. Thomas, *Elements of Information Theory*, 2nd edn. (Wiley, Inc., New-York, 2006)

Event-Triggered Synchronization of Linear Multi-Agent Systems with Time-Varing Communication Delays

Meina Bi and Yang Liu

Abstract In this paper, the event-triggered control protocol of the general linear multi-agent systems with time-varing delays is considered. The state of each agent is sampled when a certain event was triggered, and its state can be transmitted to its neighbors after a time-varing communication delay. The distributed event-triggered protocols, which is consisted of the event-triggered control laws and the triggering functions, are designed according to Riccati matrix equation, under which the consensus problem of multi-agent systems can be solved. Finally, some simulation examples are presented to demonstrate the effectiveness of the theoretical results.

Keywords Multi-agent systems · Consensus · Event-triggered control Time-varing delays

1 Introduction

Many researches of multi-agent systems have been conducted since professor Holland of Michigan University in the United States first proposed the concept of multi-agent systems at the end of the 60s of last century. The researches of the synchronization of multi-agent systems originated from the management science and statistics [1], and have been wildly used in various areas, including flocking, formation control and distributed sensor network.

With the increasing applications of consistency, it is found that time delay is a very important factor that affects the consistency results. In the past few decades, the control problem of systems with time delay has attracted many attentions. On the

M. Bi · Y. Liu
School of Automation Science and Electrical Engineering,
Beihang University, Beijing 100191, China
e-mail: meina_bi@163.com

Y. Liu (✉)
The Seventh Research Division, Beihang University, Beijing 100191, China
e-mail: ylbuaa@163.com

Y. Jia et al. (eds.), *Proceedings of 2018 Chinese Intelligent Systems Conference*, Lecture Notes in Electrical Engineering 529,
https://doi.org/10.1007/978-981-13-2291-4_58

one hand, most of systems include time delay, such as network based control system [2], sampling process control [3] and so on. On the other hand, the analysis of time delay system is very challenging. The control problem of time delay system has been involved in the various branches of the control fields, such as stability analysis [4, 5], H infinite control and prediction [6, 7], filtering [8] and so on.

In these works, each agent need to exchange their state information continuously with their neighbors to achieve the consensus. However, the bandwidth of the real communicating network is inevitably constrained, which will limit the communications. In order to avoid the continuous communications, some researchers have made their efforts on the event-triggered multi-agent consensus problem [9–11]. The event-triggered consensus algorithm for single-integrator multi-agent system was studied in [12], while the reference [13] considered the periodic event-triggered synchronization of double-integrator multi-agent systems. Event-based consensus problem of multi-agent systems with general linear models was studied in [14–20].

In this paper, we consider the distributed event-triggered consensus of general linear multi-agent systems with time-varing communication delays. Based on the event-triggered strategy, communication frequency can be reduced apparently so as to save the energy. The main achievement of this paper is that the protocol designed doesn't need global information, such as the information of nonzero eigenvalue of the Laplacian matrix of the communicating graph.

The rest of this paper is formed as follow. Section 2 provides a brief statement of graph theory and describes the researching problem and protocol. The proof of the protocol in detail is introduced in Sect. 3. Some numerical simulations are displayed in Sect. 4 to demonstrate the consensus algorithm. Finally, some conclusions are given in Sect. 5.

2 Preliminaries

2.1 Graph Theory

Generally, we use $G = \{V, E, W\}$ denote a undirected communication graph among a multi-agent system consisting of N agents. The $V = \{v_1, v_2, \ldots, v_N\}$ is the node set and the $E \subseteq V \times V$ is the edge set. The agent i and j are neighbors if there is an edge between v_i and v_j. An undirected graph is connected if there exists a edge between every pair of distinct nodes, otherwise, it is disconnected. The $W = \{w_{ij}\}$ represents the connection weight matrix between node and node. The $w_{ij} = 1$ if i and j are neighbors and $w_{ij} = 0$ otherwise. The degree matrix of G is denoted by $D = diag\{d_1, d_2, \ldots d_N\}$ while $d_i = \sum_{j=1}^{N} w_{ij}$ represent the degree of agent i. The Laplacian matrix of G is denoted by $L = D - W = [l_{ij}]$. As we can see that L is symmetric and positive semi-definite when G is an undirected graph. An important conclusion is that all the row sums of L are zero so that 1_N is an eigenvector of L corresponding to the zero eigenvalue, which can be written as $1_N^T L = 0$.

2.2 Problem Statement

Consider a group of N agents with general linear dynamics under an undirected and connected communication graph. Each agent's state can be described as follow:

$$\dot{x}_i(t) = Ax_i(t) + Bu_i(t), i = 1, 2, \ldots, N \tag{2.1}$$

where

$$A \in R^{n \times n}, B \in R^{n \times m}, x_i(t) \in R^n, u_i(t) \in R^m$$

A and B are constant matrix with proper dimensions. x_i denotes the state of the i th agent and u_i denotes the controlled input of the ith agent.

The consensus of this kind of multi-agent system is achieved when the following formula is satisfied:

$$\lim_{t\to\infty}\left\|x_i(t) - x_j(t)\right\| = 0 \tag{2.2}$$

Assume 1
The ith agent can be triggered at the time t_k^i, which can be described as follow:

$$t_0 = t_0^i < t_1^i < t_2^i < \cdots < t_k^i < \cdots \tag{2.3}$$

Assume 2
Assume that each agent can get its own state information without communication delay. The ith agent will get the information from its neighbors after a time-varing delay, $\tau(t)$, which satisfied the restraint as follow:

$$0 \le \tau(t) \le \min(t_{k+1}^i - t_k^i) \tag{2.4}$$

Lemma 1 *[21]:*
For an undirected graph G, zero is a simple eigenvalue of L, if and only if G is connected. The smallest nonzero eigenvalue of L satisfies

$$\lambda_2(L) = \min_{x \ne 0, 1_N^T x = 0} \frac{x^T Lx}{x^T x}$$

Lemma 2 *[22]:*
If x and $\dot{x}$ are bounded, and $\int_0^\infty x^T(t)x(t)dt < \infty$, then we have:

$$\lim_{t\to\infty} x(t) = 0$$

Make the definition:

$$x(t) = \left[x_1^T(t), x_2^T(t), \cdots, x_N^T(t)\right]^T \tag{2.5}$$

$$x(t_{k(t)} - \tau(t)) = \left[x_1^T(t^1_{k_1(t)} - \tau_1(t)), x_2^T(t^2_{k_2(t)} - \tau_2(t)), \cdots, x_N^T(t^N_{k_N(t)} - \tau_N(t))\right]^T \tag{2.6}$$

where

$$t^i_{k_i(t)} = \inf\left\{t > t^i_{k_i(t)-1} : f_i(t) \geq 0\right\} \tag{2.7}$$

In order to achieve the consensus of the multi-agent systems, we designed a controlled input as follow:

$$u_i(t) = cK \sum_{j \in N_i} w_{ij}[x_i(t) - x_j(t^j_{k_j(t)} - \tau_j(t))] \tag{2.8}$$

where $K \in R^{m \times n}$ is a parameter matrix can be designed as we need, then we have

$$u(t) = (cD \otimes K)x(t) - (cW \otimes K)x(t_{k(t)} - \tau(t)) \tag{2.9}$$

and

$$\dot{x}(t) = [(I_N \otimes A) + (cD \otimes BK)]x(t) - (cW \otimes BK)x(t_{k(t)} - \tau(t)) \tag{2.10}$$

Make the definition: $\zeta_i(t) = x_i(t) - \frac{1}{N}\sum_{j=1}^N x_j(t)$, $e_i(t) = x_i(t^i_{k_i(t)} - \tau(t)) - x_i(t)$, then we have that $\zeta(t_{k(t)} - \tau(t)) = (M \otimes I_n)x(t_{k(t)} - \tau(t))$, $\zeta(t) = (M \otimes I_n)x(t)$, $(M \otimes I_n)e(t) = \zeta(t_{k(t)} - \tau(t)) - \zeta(t)$ where $M = I_N - \frac{1}{N}1_N \cdot 1_N^T$.

As we can see that $\zeta(t) = 0$ if and only if $x_1(t) = x_2(t) = \cdots = x_N(t)$, so we can regard $\zeta(t)$ as the consensus error of the system, which will solve the consensus problem of the multi-agent when $\lim_{t\to\infty} \zeta(t) = 0$.

Then we can get the derive of $\zeta(t)$:

$$\begin{aligned}\dot{\zeta}(t) &= (M \otimes I_n)\dot{x}(t) \\ &= (M \otimes I_n)\{[(I_N \otimes A) + (cD \otimes BK)]x(t) - (cW \otimes BK)x(t_{k(t)} - \tau(t))\} \\ &= [(I_N \otimes A) + (cD \otimes BK)]\zeta(t) - (cW \otimes BK)\zeta(t_{k(t)} - \tau(t))\end{aligned} \tag{2.11}$$

The triggering function of each agent is designed as follow:

$$f_i(t) = 2d_i\|Ke_i(t)\|^2 - \frac{1}{2}\sum_{j=1}^N w_{ij}\left\|K[x_i(t^i_{k_i(t)} - \tau_i(t)) - x_j(t^j_{k_j(t)} - \tau_j(t))]\right\|^2 - \mu e^{-\upsilon t} \tag{2.12}$$

3 Main Results

Theorem 3.1
Consider the multi-agent system (2.1) under the undirected and connected graph G. Assume the time delay between neighbor agents can be described as (2.4). Under the designed event-triggering condition (2.12) and controlled input protocol (2.8), the consensus problem of the multi-agents systems can be solved if $2c > \frac{1}{\lambda_2(L)}$ and $K = -B^T P$, where $P > 0$ is the positive definite solve of the Riccati matrix equation $PA + A^T P - PBB^T P + Q = 0$.

Proof Choose the Lyapunov function $V(t)$ as follow:

$$V(t) = \zeta^T(t)(I_N \otimes P)\zeta(t) \tag{3.1}$$

Consider the derivative of $V(t)$ along the trajectories of the system (2.1)

$$\begin{aligned}\dot{V}(t) &= 2\zeta^T(t)(I_N \otimes PA + cD \otimes PBK)\zeta(t) - 2\zeta^T(t)(cW \otimes PBK)\zeta(t_k - \tau)\\ &= 2\zeta^T(t)(I_N \otimes PA - cD \otimes PBB^T P)\zeta(t) + 2\zeta^T(t)(cW \otimes PBB^T P)\zeta(t_k - \tau)\end{aligned}$$

where:

$$\begin{aligned}&\zeta^T(t)(cW \otimes PBB^T P)\zeta(t_k - \tau)\\ &= [\frac{1}{2}\zeta^T(t) + \frac{1}{2}\zeta^T(t_k - \tau) - \frac{1}{2}e^T(t)(M \otimes I_n)](cW \otimes PBB^T P)\\ &\quad \times [\frac{1}{2}\zeta(t_k - \tau) + \frac{1}{2}\zeta(t) + \frac{1}{2}(M \otimes I_n)e(t)]\\ &= \frac{1}{4}\zeta^T(t)(cW \otimes PBB^T P)\zeta(t) + \frac{1}{4}\zeta^T(t_k - \tau)(cW \otimes PBB^T P)\zeta(t_k - \tau)\\ &\quad + \frac{1}{2}\zeta^T(t)(cW \otimes PBB^T P)\zeta(t_k - \tau) - \frac{1}{4}e^T(t)(M \otimes I_n)(cW \otimes PBB^T P)(M \otimes I_n)e(t)\end{aligned}$$

Reform the above equation we can get:

$$\begin{aligned}&\zeta^T(t)(cW \otimes PBB^T P)\zeta(t_k - \tau)\\ &= \frac{1}{2}\zeta^T(t)(cW \otimes PBB^T P)\zeta(t) + \frac{1}{2}\zeta^T(t_k - \tau)(cW \otimes PBB^T P)\zeta(t_k - \tau)\\ &\quad - \frac{1}{2}e^T(t)(M \otimes I_n)(cW \otimes PBB^T P)(M \otimes I_n)e(t)\end{aligned}$$

Then we have:

$$
\begin{aligned}
\dot{V}(t) &= 2\zeta^T(t)(I_N \otimes PA - cD \otimes PBB^T P)\zeta(t) \\
&\quad + \zeta^T(t)(cW \otimes PBB^T P)\zeta(t) + \zeta^T(t_k - \tau)(cW \otimes PBB^T P)\zeta(t_k - \tau) \\
&\quad - e^T(t)(M \otimes I_n)(cW \otimes PBB^T P)(M \otimes I_n)e(t) \\
&\le \zeta^T(t)[I_N \otimes (PA + A^T P) - 2cL \otimes PBB^T P]\zeta(t) \\
&\quad + \zeta^T(t_k - \tau)(cW \otimes PBB^T P)\zeta(t_k - \tau) \\
&\quad - e^T(t)(M \otimes I_n)(cW \otimes PBB^T P)(M \otimes I_n)e(t) \\
&= \zeta^T(t)[I_N \otimes (PA + A^T P) - 2cL \otimes PBB^T P]\zeta(t) - e^T(t)(cMWM \otimes PBB^T P)e(t) \\
&\quad + x^T(t_k - \tau)(cMWM \otimes PBB^T P)x(t_k - \tau)
\end{aligned} \tag{3.2}
$$

where:

$$
\begin{aligned}
& e^T(t)(cMWM \otimes PBB^T P)e(t) \\
&= e^T(t)(cMDM \otimes PBB^T P)e(t) \\
&\quad - e^T(t)(cMLM \otimes PBB^T P)e(t) \\
&\ge -e^T(t)(cMLM \otimes PBB^T P)e(t) \\
&= \sum_{i=1}^{N}\sum_{j=1}^{N} w_{ij} m_{ij}^2 e_i^T(t) PBB^T P[e_i(t) - e_j(t)] \\
&\ge -\frac{3c}{2N^2}\sum_{i=1}^{N}\sum_{j=1}^{N} w_{ij} e_i^T(t) PBB^T P e_i(t) \\
&\quad - \frac{c}{2N^2}\sum_{i=1}^{N}\sum_{j=1}^{N} w_{ij} e_j^T(t) PBB^T P e_j(t) \\
&= -\frac{2c}{N^2}\sum_{i=1}^{N} d_i \|K e_i(t)\|^2
\end{aligned} \tag{3.3}
$$

$$
\begin{aligned}
& x^T(t_k - \tau)(cMWM \otimes PBB^T P)x(t_k - \tau) \\
&\quad \le -x^T(t_k - \tau)(cMLM \otimes PBB^T P)x(t_k - \tau) \\
&= -\sum_{i=1}^{N}\sum_{j=1}^{N} w_{ij} m_{ij}^2 x_i^T(t_k^i - \tau) PBB^T P[x_i(t_{k_i(t)}^i - \tau(t)) - x_j^T(t_{k_j(t)}^j - \tau(t)) \\
&\le -\frac{c}{2N^2}\sum_{i=1}^{N}\sum_{j=1}^{N} w_{ij} \left\| K[x_i(t_{k_i(t)}^i - \tau_i(t)) - x_j(t_{k_j(t)}^j - \tau_j(t))] \right\|^2
\end{aligned} \tag{3.4}
$$

So we have:

$$\begin{aligned}\dot{V}(t) &\le \zeta^T(t)[(I_N \otimes (PA + A^T P) - 2cL \otimes PBB^T P]\zeta(t) \\ &+ \frac{c}{N^2}\sum_{i=1}^{N}[2d_i\|Ke_i(t)\|^2 - \frac{1}{2}\sum_{j=1}^{N} w_{ij}\left\| K[x_i(t^i_{k_i(t)} - \tau_i(t)) - x_j(t^j_{k_j(t)} - \tau_j(t))]\right\|^2] \\ &\le \zeta^T(t)[(I_N \otimes (PA + A^T P) - 2cL \otimes PBB^T P]\zeta(t) + \frac{c}{N}\mu e^{-\upsilon t}\end{aligned} \tag{3.5}$$

We can get the following inequation from the lemma 1:

$$\zeta^T(L \otimes PBB^T P)\zeta \ge \lambda_2(L)\zeta^T(L \otimes PBB^T P)\zeta \tag{3.6}$$

If it is designed:

$$2c \ge \frac{1}{\lambda_2(L)} \tag{3.7}$$

then:

$$\begin{aligned}\dot{V}(t) &\le \zeta^T(t)[(I_N \otimes (PA + A^T P \\ &- PBB^T P)]\zeta(t) + cN\mu e^{-\upsilon t} \\ &\le -\zeta^T(t)(I_N \otimes Q)\zeta(t) + cN\mu e^{-\upsilon t} \\ &\le -\lambda_{\max}(Q)\zeta^T(t)\zeta(t) + cN\mu e^{-\upsilon t}\end{aligned} \tag{3.8}$$

From the above relationship, we have

$$0 \le V(t) \le cN\mu \int_0^t e^{-\upsilon\tau} d\tau \tag{3.9}$$

We can know that ζ and $\dot{\zeta}$ are both bounded as $V(t)$ is bounded.
Then we get:

$$V(\infty) - V(0) \le -\int_0^\infty \lambda_{\min}(Q)\zeta^T(t)\zeta(t)dt + \frac{c\mu}{N\upsilon}$$

$$\int_0^\infty \zeta^T(t)\zeta(t)dt \le \frac{1}{\lambda_{\min}(Q)}[(V(0) - V(\infty) + \frac{c\mu}{N\upsilon}]$$

As ζ is bounded, we can get:

$$\lim_{t\to\infty} \zeta(t) = 0 \tag{3.10}$$

which means that the consensus problem of the multi-agent system is solved.

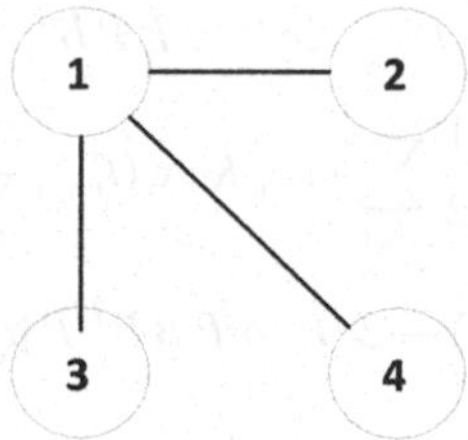

Fig. 1 Communication graph

4 Example

The demonstration of the effectiveness of above theory will be displayed in this section by numerical simulation. Consider a multi-agent system with four agents, each agent can be described by a general linear dynamic as (2.1) with $A = \begin{bmatrix} -2 & 2 \\ -1 & 1 \end{bmatrix}$, $B = \begin{bmatrix} 1 \\ 0 \end{bmatrix}$.

The communication graph between the different agents can be described as follow (Fig. 1).

We can have the Laplace matrix $L = \begin{bmatrix} 3 & -1 & -1 & -1 \\ -1 & 1 & 0 & 0 \\ -1 & 0 & 1 & 0 \\ -1 & 0 & 0 & 1 \end{bmatrix}$ when $w_{ij} = 1$.

Design the parameter as follow:

$$Q = \begin{bmatrix} 1 & 0 \\ 0 & 1 \end{bmatrix}, P = \begin{bmatrix} 1.1974 & -2.6116 \\ -2.6116 & 8.1333 \end{bmatrix}, K = \begin{bmatrix} -1.1974 & 2.6116 \end{bmatrix}, \mu = 2, v = 0.5$$

Then we can get the consensus error of the system as Fig. 2, from which we can know that consensus problem is solved.

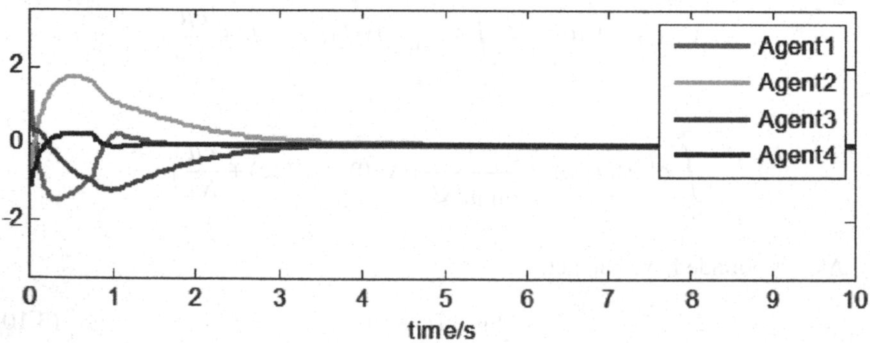

Fig. 2 The consensus error

5 Conclusion

In this paper, we solved the consensus problem of general linear multi-agent systems with time-varing delays by event-triggered control protocol. For designed control protocol, we only need the state of agent itself and its neighbors, which means the choose of sampling time needs neither global information nor continuous communications among the neighboring agents. As we can see from the proof in detail, the proposed control protocol can solve the consensus problem of multi-agent system.

The considering of the time-varing communication delay makes the consensus of multi-agent systems more useful in practical as it is closer to reality, which will get more accurate simulations before practical application.

References

1. M.H. DeGroot, Reaching a consensus. J. Am. Stat. Assoc. **69**(345), 118–121 (1974)
2. H.J. Gao, T.W. Chen, T.Y. Chai, Passivity and passification for net worked control systems. SIAM J. Control Optimation **46**(4), 1299–1322 (2007)
3. K.Q. Gu, V.L. Kharitonov, J. Chen, *Stability of time-delay systems* (Spring-Verlag, Berlin, 2003)
4. W. Michiels, S.L. Niculescu, Characterization of delay-independant stability and delay interference phenomena. SIAM J. Control Optimation **45**(6), 2138–2155 (2007)
5. S. Xu, J. Lam, Y. Zou, Improved conditions on delay-dependent robust stability and stabilization of uncertain discrete time-delay systems. Asian J. Control **7**(3), 344–348 (2005)
6. S. Xu, J. Lam, Y. Zou, New results on delay-dependent robust H ∞ control for systems with time-varying delays. Automatica **42**(2), 343–348 (2006)
7. H.S. Zhang, D. Zhang, H.L. Xie, An innovation approach to H ∞ prediction with application to systems with delayed measurement. Automatica **40**(7), 1253–1261 (2004)
8. Z.D. Wang, D.W.C. Ho, Filtering on nonlinear time-delay stochastic systems. Automatica **39**(1), 101–109 (2003)
9. D. Yang, W. Ren, X. Liu, et al.: Decentralized event-triggered consensus for linear multi-agent systems under general directed graphs. Automatica **69**(C), 242–249 (2016)
10. P. Tabuada, Event-triggered real-time scheduling of stabilizing control tasks. IEEE Trans. Autom. Control **52**(9), 1680–1685 (2007)
11. D. Xie, S. Xu, B. Zhang et al., Consensus for multi-agent systems with distributed adaptive control and an event-triggered communication strategy. IET Control Theory Appl. **10**(13), 1547–1555 (2016)
12. D.V. Dimarogonas, E. Frazzoli, K.H. Johansson, Distributed event-triggered control for multi-agent systems. IEEE Trans. Autom. Control **57**(5), 1291–1297 (2012)
13. G.S. Seyboth, D.V. Dimarogonas, K.H. Johansson, Event-based broadcasting for multi-agent average consensus. Automatica **49**(1), 245–252 (2013)
14. E. Garcia, Y. Cao, H. Yu et al., Decentralised event-triggered cooperative control with limited communication. Int. J. Control **86**(9), 1479–1488 (2013)
15. X. Meng, T. Chen, Event based agreement protocols for multi-agent networks ☆. Automatica **49**(7), 2125–2132 (2013)
16. W. Zhu, Z.P. Jiang, G. Feng, Event-based consensus of multi-agent systems with general linear models ☆. Automatica **50**(2), 552–558 (2014)
17. G. Guo, L. Ding, Q.L. Han, A distributed event-triggered transmission strategy for sampled-data consensus of multi-agent systems ☆. Automatica **50**(5), 1489–1496 (2014)

18. D. Liuzza, Dimarogonas D.V., M.D. Bernardo, et al., Distributed model based event-triggered control for synchronization of multi-agent systems, in *Distributed Model Based Event-Triggered Control for Synchronization of Multi-Agent Systems,* pp. 329–334 (2014)
19. Z. Zhang, F. Hao, L. Zhang et al., Consensus of linear multi-agent systems via event-triggered control. Int. J. Control **87**(6), 1243–1251 (2014)
20. H. Zhang, G. Feng, H. Yan et al., Observer-based output feedback event-triggered control for consensus of multi-agent systems. IEEE Trans. Industr. Electron. **61**(9), 4885–4894 (2014)
21. Z.K. Li, Z.S. Duan, *Cooperative control of multi-agent systems: a consensus region approach* (CRC Press, BocaRaton, FL, 2014)
22. P.A. Ioannou, J. Sun, *Robust Adaptive Control* (Springer, London, 2015)

Hierarchical Modelling and Simulation of a Novel Integrated Electro-Mechanical Hydrostatic Actuator Based on Bond Graph

Xudong Yan, Liming Yu, Jiang'ao Zhao, Jian Fu and Yongling Fu

Abstract Thrust vector control (TVC) actuation system is an important part for swing angle control of the nozzle of the launch vehicle. This communication studied a new type of powered electrically actuator: electro-mechanical hydrostatic actuator (EMHA) for driving the nozzle. The key components of the system were firstly introduced and then the mathematical models were presented. The hierarchical method was used for modelling, the models were built from the functional and behavioral—based on the bond graph. Finally, the simulation analysis of the two models is carried out in AMESim virtual prototype simulation environment. This paper shows different levels of models can be used when consider the different engineering needs such as controller design and the energy loss analysis of the system.

Keywords Electro mechanical hydrostatic actuator (EMHA) · Bond graph
Hierarchical modeling · AMESim

1 Introduction

As the development of energy-saving and environmental protection in the aerospace, the concept of "more-electric" or "all-electric" has become increasingly attractive. In

---

X. Yan
School of Automation Science and Electrical Engineering,
Beihang University, Beijing 100191, China
e-mail: xdyanbuaa@163.com

L. Yu
Flying College, Beihang University, Beijing 100191, China
e-mail: yuliming@buaa.edu.cn

J. Zhao · J. Fu (✉) · Y. Fu
School of Mechanical Engineering and Automation,
Beihang University, 100191 Beijing, China
e-mail: fujianbuaa@126.com

Y. Jia et al. (eds.), *Proceedings of 2018 Chinese Intelligent Systems Conference*, Lecture Notes in Electrical Engineering 529,
https://doi.org/10.1007/978-981-13-2291-4_59

many aerospace researches, power-by-wire (PbW) has gained a lot of applications. It is an inevitable trend that PbW replaces traditional hydraulic technology [1].

The traditional servo hydraulic actuator (SHA) is powered by a central hydraulic source. It has the advantages of large load and high-power density, but it also has many disadvantages, such as complex oil circuit, low efficiency, fussy maintenance, oil pollution, and large noise [2]. To replace SHA, there are currently two types of PbW actuators, electro hydrostatic actuators (EHA) and electro mechanical actuators (EMA). EHA consists of a high-speed motor, a hydraulic pump, a hydraulic cylinder and some hydraulic accessories. Normally, the motor drives the hydraulic pump and the hydraulic cylinder drives the load. Compared with SHA, EHA eliminates the servo valve that can easily cause pollution and jam. What's more, it removes the traditional oil tank and external pipes, greatly improving the sealing and maintenance, and has a good application prospect [3]. Considering the maturity of EHA, to meet the reliability requirements, the electric backup hydraulic actuator (EBHA) is produced, which combines SHA and EHA. It generally works in the state of SHA, while EHA serves as a backup and toggles state through the mode switching valve. EMA completely removes hydraulic circuits and converts the electrical power to mechanical power by a motor, and then transmits mechanical power to drive the load through an optional gear box and nut-screw mechanism. EMA eliminates the leakage in hydraulic systems but has limits for large load and high power.

Actuation systems are applied in many places in aircrafts, such as landing gear and flight surfaces. Replacing heavy hydraulic pipes with lighter cables in the actuation systems can reduce energy loss. Because the actuator can supply power on demand, the efficiency of the actuator is improved. The EHA is used as a backup for primary flight control in the aircraft and has been used on Airbus A350 and A380. In the Boeing B787, EMA is successfully applied to secondary flight control. It has not been widely used due to mechanical jamming and free play etc. However, EMA has advantages in economic performance and is the development trend of more electric aircraft in the future.

In launch vehicles, TVC system is the actuator of the control system and usually includes a servo actuator and an engine. To achieve high precision control, high control accuracy and fast dynamic response are important to TVC systems [4]. Compared with aircraft actuators, TVC's mission duration is very short. Like aircraft, traditional SHA was employed at beginning. With the development of PbW, SHA is gradually replaced with EMA and EHA. In Atlas V of the United States and Vega of Europe, EMA has been applied. EHA has the advantage of high power density, so it also has a broad application prospect in TVC [5].

EHA can be divided into three types according to the variables of pump or motor: EHA with fixed displacement pump and variable speed motor (EHA-FPVM), EHA with variable displacement pump and fixed speed motor (EHA-VPFM), and EHA with variable displacement pump and variable speed motor (EHA-VPVM) [6]. Nowadays, EHA-FPVM is more mature and efficient, however poor rigidity and slow response restrict its application. EHA-VPFM alters the displacement of pump by adjusting the swash plate angle, so the dynamic response is faster. EHA-VPVM combines the advantages of the former two, but its control algorithms are complicate.

In this paper, a new type of integrated electro mechanical hydrostatic actuator is studied for hierarchical modeling and simulation analysis. It is one type of EHA-VPFM and can meet the high dynamic response requirements of TVC systems.

2 System Description

TVC is used to control the torque, which regulates the flight attitude and direction of launch vehicle, by changing the swing angle of the engine nozzle [7]. EMHA is used to drive the nozzle, as shown in Fig. 1.

A. System structure

EMHA is characterized by high efficiency, light weight, high reliability, and good maintainability. Compared to throttle governing, its power loss is smaller, and the generated heat is reduced. So, it is suitable for high power systems. Its structure is shown in Fig. 2.

The EMHA is mainly composed of a power unit, an electro mechanical servo pump, safety valves, check valves, relief valves, a hydraulic cylinder, a bootstrap reservoir, and sensors etc. The output of the servo pump is directly connected to the actuator to drive the piston rod. Servo motor drives the mechanical power transformation (MPT), then changing the swash plate angle of the variable pump, to drive the hydraulic cylinder to push the load.

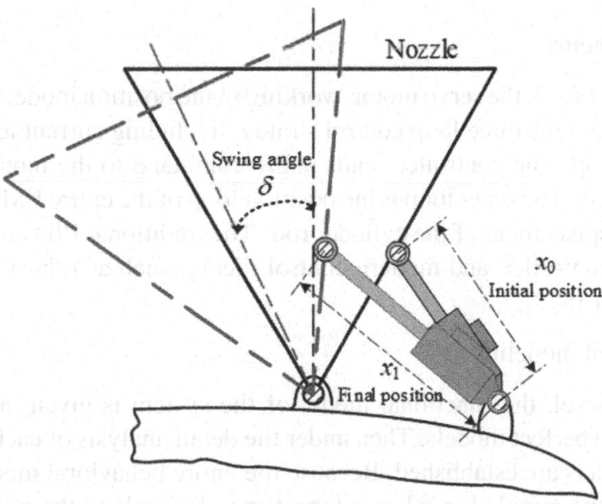

Fig. 1 The structure of TVC system

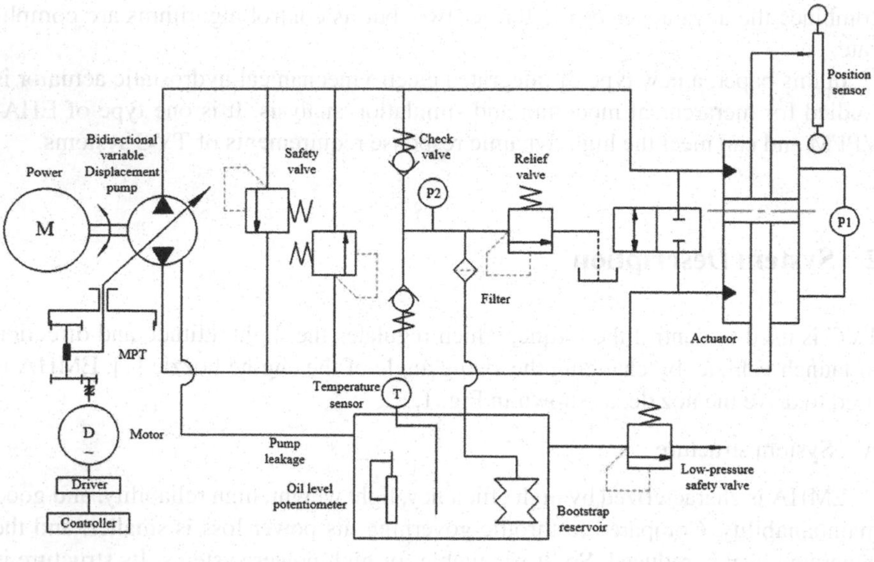

Fig. 2 Principle structure of EMHA

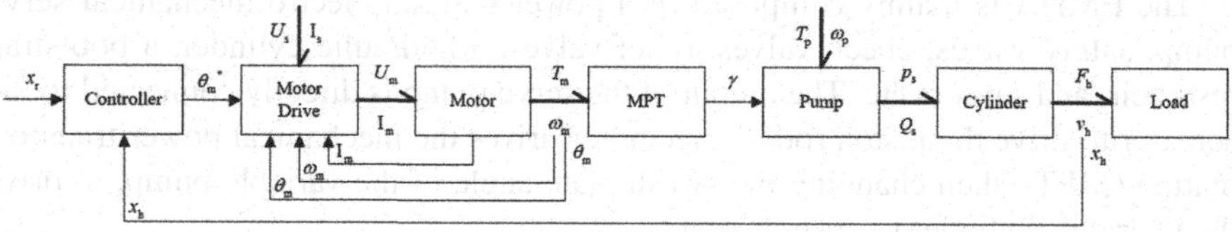

Fig. 3 Control scheme of EMHA

B. Control scheme

As shown in Fig. 3, the servo motor, working in the position mode, is controlled by a driver which adopts three loop control strategy including current loop, speed loop and position loop. The controller sends angle command to the motor to adjust the swash plate angle. The outer loop is the position loop of the entire EMHA system and controls the displacement of the cylinder rod. The traditional PID control algorithm is used in the controller, and modern control theory, such as robust control, can be considered later [8, 9].

C. Hierarchical modeling

In the first level, the functional model of the system is given, in which all the components are perfect models. Then under the detail analysis of each element, their behavioral models are established. Because the entire behavioral model considering many factors, the simulation takes a long time. To analyze the performance of a certain component, only the behavioral model of concerned component is replaced to compose the critical component system model, so that details of the concerned

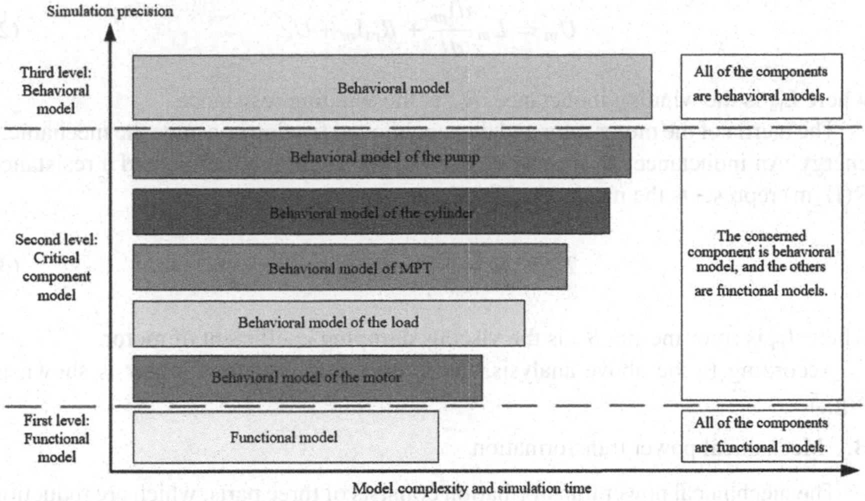

Fig. 4 Hierarchical modeling of the system

part can be obtained in less time. Figure 4 shows the hierarchical modeling of the system.

3 Mathematical Model of Major Components

The EMHA is a complex system including multi-energy domains, such as mechanical, electrical and hydraulic, and conversion between different energy form is various. Bond graph (BG), based on the law of conservation of energy, can express various energy forms in a unified form [10]. Therefore, EMHA will be firstly modeled by bond graph and then actually built in AMESim.

A. Electric motor

A permanent magnet synchronous motor (PMSM) is selected in the study. Ideally, it converts electrical energy into mechanical energy, which can be represented by a gyrator GY(k_t).

$$\begin{cases} T_{\mathrm{m}} = k_{\mathrm{t}} \mathrm{I}_{m} \\ \omega_{\mathrm{m}} = U_{\mathrm{m}} / k_{\mathrm{t}} \end{cases} \tag{1}$$

where T_{m} is motor torque, ω_{m} is motor angular velocity, k_{t} is electromagnetic torque constant, U_{m} is motor armature voltage, and I_{m} is motor current.

The resistance and inductance of stator windings are respectively represented by a resistance R(R_m) and an inductance I(L_m).

$$U_{\mathrm{m}} = L_m \frac{d\mathrm{I}_m}{dt} + R_{\mathrm{m}}\mathrm{I}_m + U_{\mathrm{s}} \tag{2}$$

where L_{m} is the winding inductance, R_{m} is the winding resistance.

The inertia of the motor rotor and the mechanical friction consume the mechanical energy. An inductance I(J_m) can be used to represent the inertia, and a resistance R(B_m) represents the mechanical friction.

$$T_{\mathrm{m}} = J_{\mathrm{m}} \frac{d\omega_{\mathrm{m}}}{dt} + B_{\mathrm{m}}\omega_{\mathrm{m}} + T_{\mathrm{r}} \tag{3}$$

where J_{m} is rotor inertia, B_{m} is the viscous damping coefficient of motor.

According to the above analysis, the bond graph model of motor is shown in Fig. 5.

B. Mechanical power transformation

The mechanical power transformation consists of three parts, which are reduction gear, nut-screw and link mechanism.

The mathematical equation of reduction gear is given as:

$$\begin{cases} T_{\mathrm{c}} = k_{\mathrm{i}} T_{\mathrm{r}} \\ \omega_{\mathrm{c}} = \omega_{\mathrm{r}} / k_{\mathrm{i}} \end{cases} \tag{4}$$

where k_{i} is the gear ratio.

The torque and speed transmitted from reduction gear are converted into force and speed by nut-screw. The mechanical conversion ratio is determined by the screw pitch of lead l.

$$\begin{cases} F_{\mathrm{s}} = (2\pi/l) \cdot T_{\mathrm{c}} \\ v_{\mathrm{s}} = \omega_{\mathrm{c}} (2\pi/l)^{-1} \end{cases} \tag{5}$$

The screw rod drives the swash plate through the link mechanism to change its angle. The relationship between swash plate angle and the displacement of the screw can be simplified into linear relationship.

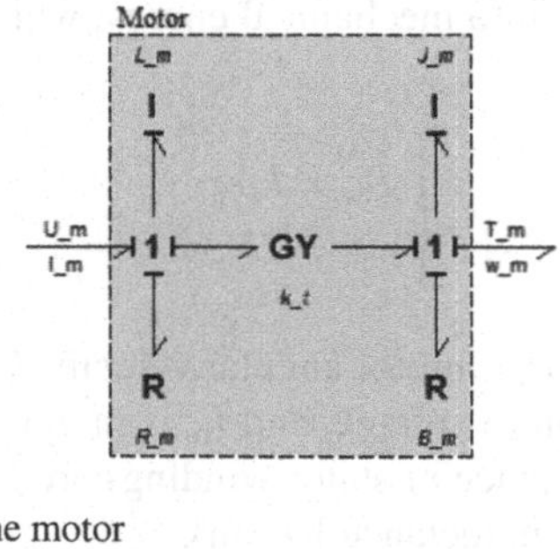

Fig. 5 Casual BG model of the motor

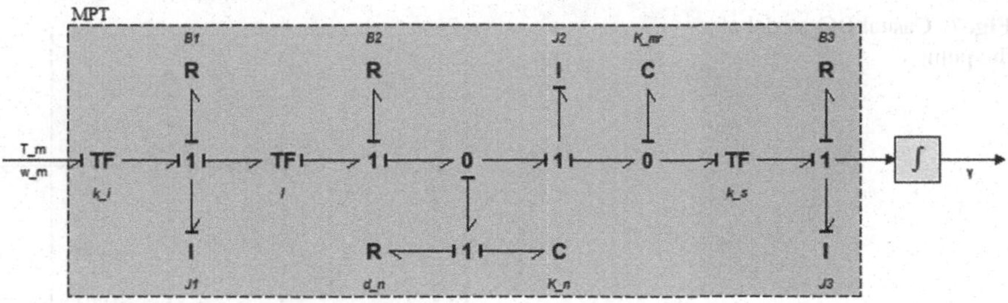

Fig. 6 Casual BG model of MPT

$$\gamma = k_s x_s \tag{6}$$

where k_s is the coefficient of the linear relationship.

Therefore, in an ideal situation, each part of MPT can be represented by a transformer TF (k_i, l, k_s).

The inertia and mechanical friction of the transmission mechanism are added by inductances I and resistances R. In addition, the rigidity in the nut screw can be represented by a capacitance C (K_n), and considering its flexibility, a resistance R (d_n) needs to be added. And they are connected by a 1-node, then jointed to a 0-junction after friction. The stiffness between the screw and the link mechanism is represented by a capacitance C (K_mr). Finally, the swash plate angle is integrated by its speed. The bond graph model of MPT is shown in Fig. 6.

C. Hydraulic pump

A hydraulic pump converts mechanical energy into hydraulic energy.

$$\begin{cases} Q_s = D\omega_p \\ p_s = T_p / D \end{cases} \tag{7}$$

where D is the displacement of pump.

The hydraulic pump satisfies the characteristic equation of transformer TF. Since the displacement of the pump is adjustable, it is expressed by a modulation transformer MTF (D).

When external power source drives the pump, the inertia and friction of the rotating part of the pump need to be considered, which are represented by a resistance R(B_p) and an inductance I(J_p). As there are two channels, feeding and return, in the hydraulic system, two MTF elements are needed to represent the hydraulic pump in the bond graph model. In actual operation, the hydraulic pump also has internal leakage and external leakage. Internal leakage which changes the pressure difference between two chambers can be regarded as a resistance R (R_pil) connected to a 1-junction. External leakage which leads to flow loss is represented as a resistance R (R_pel) linked to a 0-junction. Based on the above analysis, the bond graph model of the hydraulic pump is shown as Fig. 7.

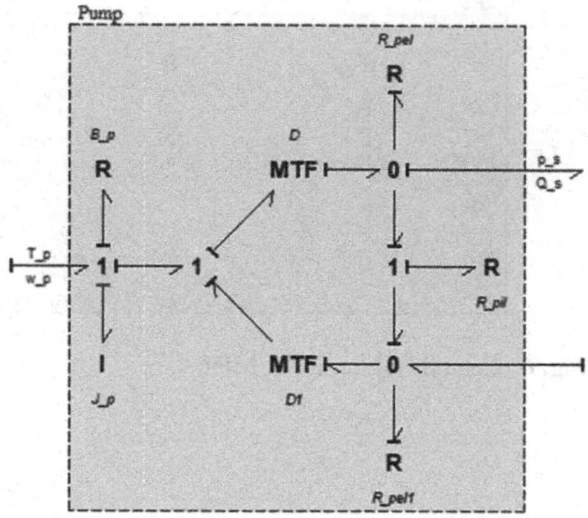

Fig. 7 Casual BG model of the pump

D. Hydraulic cylinder

The symmetrical hydraulic cylinder acting as a transformer TF can be modeled as follow:

$$\begin{cases} F_a = A_h p_a \\ v_h = Q_a / A_h \end{cases} \tag{8}$$

$$\begin{cases} p_b = F_b / A_h \\ Q_b = v_h A_h \end{cases} \tag{9}$$

$$F_h = F_a - F_b \tag{10}$$

where A_h is the area of piston.

Leakage in hydraulic cylinders is mainly classified into external leakage and internal leakage. The external leakage can be represented by a resistance R (R_ael) and the internal leakage can be expressed by a resistance R (R_ail). Considering fluid compressibility in the hydraulic chamber, a capacitance C (b_e) needs to be appended. An inductance I(m_t) is added to represent the mass of the cylinder moving part. Considering the damp of the piston, a resistance R(B_t) can be connected to the 1-junction. The bond graph model of the hydraulic cylinder is shown in Fig. 8.

E. Load

The load of the system is a large inertia load that rotates around a fixed axis. Its dynamic equation is:

$$T_L = J_L \frac{d\omega_L}{dt} + B_L \omega_L \tag{11}$$

where J_L is rotary inertia of the load, and B_m is the viscous damping coefficient of load.

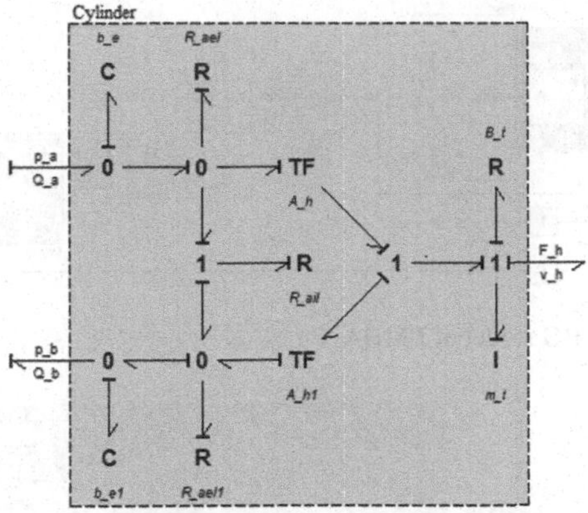

Fig. 8 Casual BG model of the cylinder

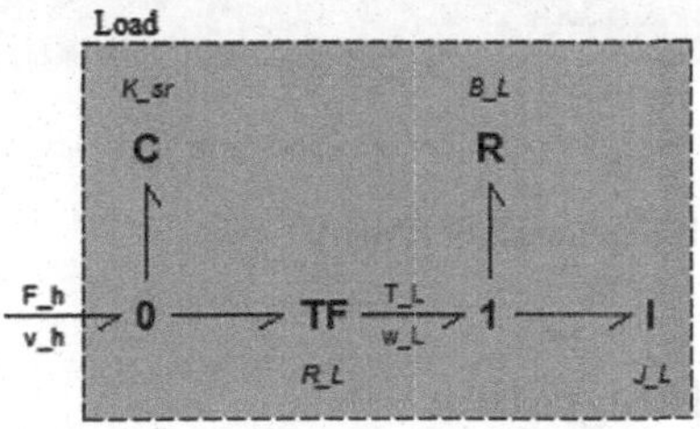

Fig. 9 Casual BG model of the load

Inertia and damping are represented by an inductance I (J_L) and a resistance R (B_L).

The motion of the hydraulic cylinder is translation, while the load's is rotation. The conversion relationship is square to the arm of force and can be represented by a transformer TF (R_L). Considering that the load rotates at a small angle, to facilitate the processing, the rotating load can be equivalent to a translational load. So, there is no need for a transformer, and the equivalent equation is

$$\begin{cases} M_{\mathrm{L}} = J_{\mathrm{L}}/R_{\mathrm{L}}^2 \\ B'_{\mathrm{L}} = B_{\mathrm{L}}/R_{\mathrm{L}}^2 \end{cases} \quad (12)$$

where M_{L} is the equivalent mass of the load, and B'_{L} is the equivalent friction of the load.

Because of the large load inertia, the joint between the hydraulic cylinder and the load needs to consider the coupling stiffness, which can be regarded as a spring which is represented as a capacitance C (K_sr) in the bond graph as shown in Fig. 9.

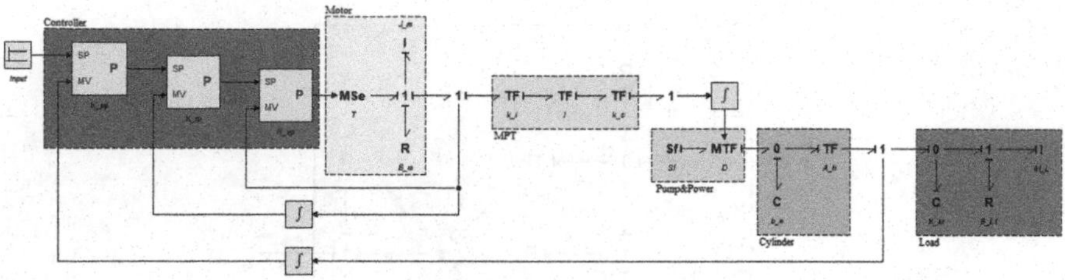

Fig. 10 The functional BG model of EMHA

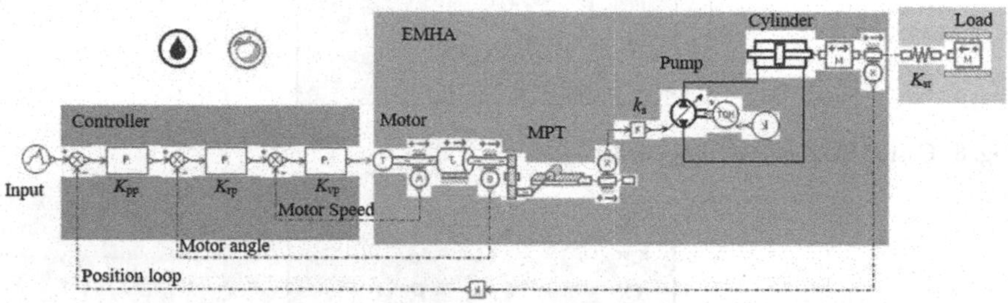

Fig. 11 The functional AMESim model of EMHA

4 System Model Integration

Up to now, the major components of the EMHA have been analyzed and modeled under ideal condition and complex condition. The modeling and simulation of the entire system are completed below.

A. Functional model

In the functional model, only the perfect model of each component is considered. The torque reference value of the driver can be directly applied to the motor shaft as a torque source, and the load is in the form of the equivalent translational motion. The components are spliced together, and the controller is designed according to the control scheme. The complete bond graph model is shown in Fig. 10. A functional model is also built in AMESim to get simulation results, as shown in Fig. 11.

B. Behavioral model

The model of each component in the functional model is replaced with a corresponding behavioral model. The hydraulic resistance of pipes, the accumulator and the torque of the swash plate acting on the motor are also taken into consideration. Figures 12 and 13 are EMHA behavioral models in bond graph and AMESim.

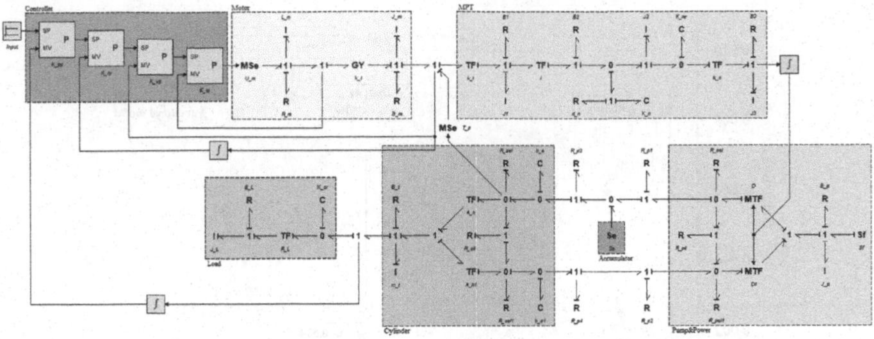

Fig. 12 The behavioral BG model of EMHA

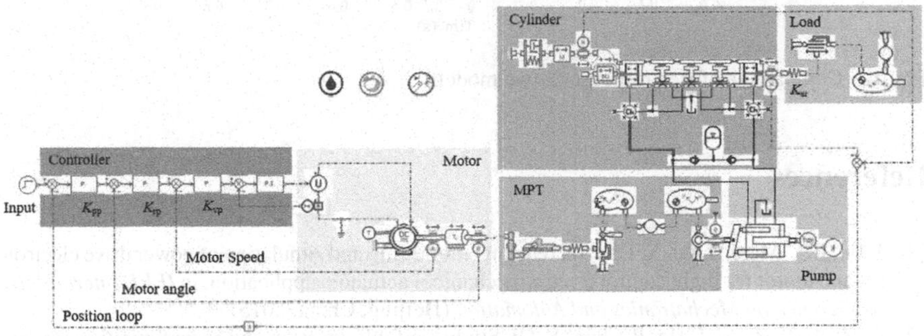

Fig. 13 The behavioral AMESim model of EMHA

C. Simulation analysis

Figure 14 shows the comparison of the step response of two models. When the position command is 10 cm, the step response curves are basically identical in the case of the same control parameters of two models.

5 Conclusion

A hierarchical modeling method of EMHA has been proposed in this paper. This method establishes the functional model and behavior model of the EMHA system. Considering different engineering needs, the functional model is a perfect model, which can be used to design the controller, while the behavior model considers various factors of the system, and the simulation accuracy is higher. The behavior model can be used for energy and efficiency analysis of the system, and the critical component models can be used to study the impact of key components on the system performance.

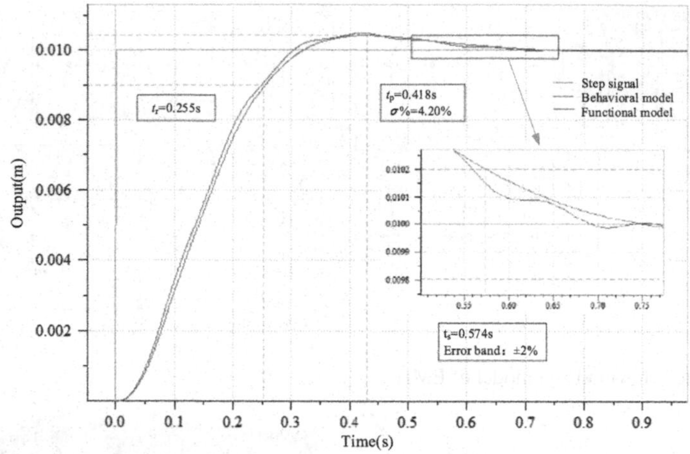

Fig. 14 Comparison of step response of two models

References

1. J. Fu, J.C., Maré, Y. Fu, X. Han, Incremental modelling and simulation of power drive electronics and motor for flight control electro-mechanical actuators application, in *IEEE International Conference on Mechatronics and Automation* (Beijing, China, 2015)
2. C. Zhao, S. Zhao, J. He, P. Chen, Y. Qi, Study on fault isolation and reconfiguration for dual redundancy electro-hydrostatic actuators. Manned Space **21**(3), 205–211 (2015)
3. S. Zeng, S. Zhao, X. Zhang, Analysis of the development of servo system of manned launch vehicle in China. Manned Spaceflight **19**(4), 3–10 (2013)
4. S. Zhao, J. He, Y. Zhang, The study on the dynamic capability of an electro-hydrostatic actuator to drive a large inertia load, in *IEEE/CSAA International Conference on Aircraft Utility Systems* (Beijing, China, 2016)
5. J. Fu, J.C. Maré, Y. Fu, Modelling and simulation of flight control electromechanical actuators with special focus on model architecting, multidisciplinary effects and power flows. J. Chin. J. Aeronautics **30**(1), 47–65 (2017)
6. Z. Wang et al., *Modern electric hydraulic servo control* (Beihang University Press, Beijing, 2005), pp. 145–149
7. C. Binbin, C. Guobiao, T. Hui, Z. Hao, Study on thrust vector control and nozzle efficiency for hybrid rocket motor. J. Astronautics, **38**(10), (2017)
8. Y. Jia, Robust control with decoupling performance for steering and traction of 4WS vehicles under velocity-varying motion, in IEEE Transactions on Control Systems Technology, **8**(3), 554–569 (2000)
9. Y. Jia, Alternative proofs for improved LMI representations for the analysis and the design of continuous-time systems with polytopic type uncertainty: a predictive approach, in *IEEE Transactions on Automatic Control*, **48**(8), 1413–1416 (2003)
10. L. Yu, Y. Zhang H. Liu. Performance analysis and optimization design of dissimilar redundant hybrid electro-hydraulic actuation system, in *IEEE/CSAA International Conference on Aircraft Utility Systems* (Beijing, China, 2016)

Matlab-Based Myocardial Ischemia Detection System Design via Deterministic Learning

Hongji Lai, Muqing Deng, Min Tang and Cong Wang

Abstract In this paper, based on a recently presented myocardial ischemia detection method via deterministic learning theory, a flexible computer-aided-diagnosis system for ischemic heart disease is proposed. Surface 12-lead ECG signals are collected and cardiac dynamics are extracted via deterministic learning. This kind of cardiac dynamics is shown to be sensitive to the variance during myocardial ischemia, and is used for myocardial ischemia detection. Benefit from the powerful matrix computing capabilities and data visualization capabilities of Matlab platform, we develop an effective myocardial ischemia detection system, facilitate the application of deterministic learning and Matlab programming language in myocardial ischemia detection. The effectiveness of the proposed system is verified at Chinese National Center of Cardiovascular Diseases, which is helpful for building up a real-time software tool towards assisting the physician in cardiology departments.

Keywords Myocardial ischemia · Matlab · Deterministic learning

H. Lai · C. Wang (✉)
College of Automation, South China University of Technology,
Guangzhou 510640, China
e-mail: wangcong@scut.edu.cn

M. Deng
Institute of Information and Control,
Hangzhou Dianzi University, Hangzhou 310018, China

M. Tang
State Key Laboratory of Cardiovascular Disease, Fuwai Hospital,
National Center for Cardiovascular Diseases, Beijing 100000, China

Y. Jia et al. (eds.), *Proceedings of 2018 Chinese Intelligent Systems Conference*, Lecture Notes in Electrical Engineering 529,
https://doi.org/10.1007/978-981-13-2291-4_60

1 Introduction

Myocardial ischemia is the most common type of heart disease and cause of heart attacks, which is even deadly. Therefore, early detection of myocardial ischemia and minimizing ischemic time are of great significance [1–3].

Deterministic learning theory is proposed for identification of nonlinear dynamical systems undergoing periodic or recurrent motions [4, 5]. A partial persistence of excitation (PE) condition can be satisfied by using the localized radial basis function (RBF) neural network. The PE condition is tempting because it can guarantee the exponential stability of identification error along the periodic trajectory. As a result, accurate approximation of system dynamics can be achieved. Recently, deterministic learning has been applied in cardiac dynamics extraction and myocardial ischemia detection [6–8]. The cardiac dynamics is extracted via deterministic learning and plotted into three-dimensional space. By analyzing morphology of the three-dimensional cardiac dynamics information, significant correlations can be found between cardiac dynamics and ischemia.

Matlab is a popular programming language of technical computing, which is powerful in algorithm development, data analysis, visualization, numeric computation and graphical user interface. With the advantages such as flexible programming capability and powerful reliability, it is easy to get started and shorten the development period [9].

In this paper, based on the myocardial ischemia detection method via deterministic learning theory and powerful data processing capabilities of Matlab platform, an effective myocardial ischemia detection system is proposed. It provides a new software tool towards assisting the physician in cardiology departments. The extracted cardiac dynamics reflects the essential dynamic characteristics of electrocardiography (ECG) [10], which is more sensitive than the original ECG signals. It is user-friendly, simple and noninvasive. The effectiveness of the proposed system is verified at Chinese National Center of Cardiovascular Diseases. Experimental results show that a total accuracy of 84.58%, with the sensitivity of 87.92% and specificity of 81.25% is achieved in the proposed system.

The rest of the paper is organized as follows: Sect. 2 introduces the generation of cardiodynamicsgram (CDG) as well as its application in myocardial ischemia detection. Section 3 introduces the design of myocardial ischemia detection system based on the Matlab platform. In Sect. 4, experiments are designed to evaluate the effectiveness of the proposed system. Finally, conclusion of this paper is given in Sect. 5.

2 Myocardial Ischemia Detection via Deterministic Learning

The cardiac dynamics can be extracted via deterministic learning by modeling 12-lead ECG, CDG is generated by plotting the cardiac dynamics into three-dimensional space. Myocardial ischemia can be diagnosed by analyzing the CDG morphology.

2.1 Acquisition of Cardiodynamicsgram via Deterministic Learning

The nonlinear cardiac dynamical system can be described as (1):

$$\dot{x} = F\left(x; p\right), x\left(t_0\right) = x_0, \tag{1}$$

where $x(t) = [x_1, \ldots, x_{12}]^T \in R^{12}$ represents the 12-lead ECG signals, p is a vector which represents system parameters, $F\left(x; p\right) = \left[f_1\left(x; p\right), \ldots, f_{12}\left(x; p\right)\right]^T$ is a vector which represents smooth but unknown nonlinear function.

Since transformation between twelve-dimensional ECG and three-dimensional vectorcardiography(VCG) will not lead to loss of information [11], using $v = [v1, v2, v3]^T$ to represent VCG, the input dimension can be reduced according to conversion rules (2):

$$\begin{cases} v_1 = 0.38x_1 - 0.07x_2 - 0.13x_7 + 0.15x_8 \\ \quad - 0.01x_9 + 0.14x_{10} + 0.06x_{11} + 0.54x_{12} \\ v_2 = -0.07x_1 + 0.93x_2 + 0.06x_7 - 0.02x_8 \\ \quad - 0.05x_9 + 0.06x_{10} - 0.17x_{11} + 0.13x_{12} \\ v_3 = 0.11x_1 - 0.23x_2 - 0.43x_7 - 0.06x_8 \\ \quad - 0.14x_9 - 0.20x_{10} - 0.11x_{11} + 0.31x_{12} \end{cases} \tag{2}$$

The cardiac dynamics system after conversion can be represented as:

$$\dot{v} = \phi(v; p') \tag{3}$$

where p' is a vector which represents system parameters.

We employ the following RBF network to model the unknown system dynamics $\phi\left(v; p'\right) = \left[\phi_1\left(v; p'\right), \phi_2\left(v; p'\right), \phi_3\left(v; p'\right)\right]^T$ implicit in φ_ς:

$$\dot{\hat{v}}_i = -a_i\left(\hat{v}_i - v_i\right) + \hat{W}_i^T S_i\left(v\right), \tag{4}$$

where φ_ζ is a periodic trajectory, $\hat{v}_i$ is an estimation of v_i, v_i is the state of system (3), a_i is a positive constant designed by ourselves, RBF network $\hat{W}_i^T S_i(v)$ is used for approximating the unknown $\phi_i(v; p)$ in (3) with $\hat{W}_i^T = [w_{i1}, \ldots, w_{iN}]$ and $S_i(v) = [s_{i1}(\|v - \zeta_1\|), \ldots, s_{iN}(\|v - \zeta_N\|)]^T$. We adopt Gaussian function $s_{ij}(\cdot)\ (i = 1, \ldots, N)$ and distinct points in state ζ_j in $S_i(v)$.

The state estimation error is defined as $\tilde{v}_i = \hat{v}_i - v_i$, its derivative satisfies

$$\dot{\tilde{v}}_i = -a_i \tilde{v}_i + \hat{W}_i^T S_i(v) - \phi_i(v; p'), \quad (5)$$

we denote $\tilde{W}_i = \hat{W}_i - W_i^*$, W_i^* is the ideal constant weight vector. And the ideal approximation error is denoted as $\varepsilon_i = \phi_i(v; p') - \hat{W}_i^{T*} S_i(v)$, so (5) can be represented as:

$$\dot{\tilde{v}}_i = -a_i \tilde{v}_i + \tilde{W}_i^T S_i(v) - \varepsilon_i. \quad (6)$$

We update the weight estimate according to the following rule:

$$\dot{\hat{W}}_i = \dot{\tilde{W}}_i = -\Gamma_i S_i(v) \tilde{v}_i - \sigma_i \Gamma_i \hat{W}_i, \quad (7)$$

where $\Gamma_i = \Gamma_i^T > 0$, and σ_i is small and positive.

We set the initial value $\hat{W}_i(0) = 0$ to achieve the goals of locally accurate modeling of the unknown $\phi_i(v; p')$ along almost every recurrent trajectory φ_ζ:

$$\phi_i(\varphi_\zeta; p') = \hat{W}_i^T S_i(\varphi_\zeta) + \varepsilon_{\zeta i} = \overline{W}_i^T S_i(\varphi_\zeta) + \varepsilon_{\zeta i1}, \quad (8)$$

where $\overline{W}_\zeta$ is the arithmetic mean of $\hat{W}_\zeta(t)$, $t \in [t_a, t_b]$, $0 < t_a < t_b$ represent a continuous time starts at moment t_a and ends at moment t_b. Approximation error $\varepsilon_{\zeta i1} = O(\varepsilon_{\zeta i}) = O(\varepsilon_i)$ will converge to 0, which indicates that dynamics $\phi_i(\varphi_\zeta; p')$ underlying almost every recurrent trajectory φ_ζ can be accurately modeled via deterministic learning.

Thus, CDG can be generated by plotting $\phi(v; p')$ into three-dimensional space coordinate system along the trajectory φ_ζ.

2.2 Application of CDG in Myocardial Ischemia Detection

Early detection of myocardial ischemia detection can be achieved by analyzing CDG morphology. Regular shapes always mean healthy, while irregular shapes always represent unhealthy.

For quantitative analysis, we extract Lyapunov indicator [12] *idlya* and Fourier indicator [13] *idfft* to reflect the CDG morphological characteristics in time and frequency domain, respectively. The Lyapunov indicator and Fourier indicators are defined as follows:

For a time sequence $X(n), n = 1, 2, \ldots, N$, we denote d_{i1} as the Euclid distance between the ith data point and its nearest data point, d_{i2} as the Euclid distance between the ith data point and its nearest data point after Δ step, where Δ is a positive parameter adjustable. The Lyapunov indicator can be calculated as (9).

$$idlya = \frac{1}{N}\sum_{n=1}^{N} ln\left(\frac{d_{n2}}{d_{n1}}\right) \tag{9}$$

We get a Fourier transform result F of $X(n), n = 1, 2, \ldots, N$, and use a function f_λ as the fitting of F with the eigenvalue equal to λ, The Fourier indicator can be calculated as (10). This pair of indicators can be classified by support vector machines (SVM) [14].

$$idfft = argmin_\lambda |F - f_\lambda|. \tag{10}$$

3 Design of Myocardial Ischemia Detection System Based on Matlab

3.1 Implementation Platform

Matlab software is a commercial mathematics software provided by MathWorks, USA. The syntax of Matlab is simple, it is easy to master as long as the user understand some basic knowledge of advanced programming languages. Matlab functions are powerful, users can pay more attention on the creation, while the complicated problems are left to the internal functions to solve. What's more, the powerful graphics and symbol processing capabilities are also the advantages of Matlab. Matlab graphical user interface(GUI) development environment is actually a set of Matlab tools, this set of tools greatly simplifies the GUI design and generation process, eliminating the need for developers to bother with cumbersome code and simplifying the process, thus the development cycle can be shorten greatly [15, 16].

3.2 Modular Design for Myocardial Ischemia Detection System

The proposed myocardial ischemia detection system consists of three modules: ECG data collection and preprocessing, VCG transformation and T wave extraction, cardiodynamics calculation and diagnosis. Figure 1 shows the flow chart of the system.

(1) Standard 12-lead ECG collection and preprocessing: we collect 12-lead ECG by using commercially available electrocardiograph, the time for each measurement

is 20 seconds. Then median filter is used for baseline drift correction, wavelet transform is used to filter the electromyography interference, the related panel is shown in Fig. 2.

(2) VCG transformation and T wave extraction: transformed 3-lead VCG is converted from the 12-lead ECG after preprocessing, and the T wave is extracted from VCG using bump operator and dynamic search window method [17], T wave extraction can also be adjust manually if the automatic interception effect is not ideal. The related panel is shown in Fig. 3.

(3) Cardiodynamics calculation and diagnosis: cardiodynamics can be obtained by modeling the T wave via deterministic learning. Then CDG is plotted in three-dimensional space, and indicators are calculated to feature the morphology of CDG. Auxiliary diagnosis is also given. The related panel is shown in Fig. 4.

The file organization structure is shown in Fig. 5. There are three types of files under the detection system directory. The data storage files store the data files collected and processed in the analysis process. Code files in the Matlab GUI development environment include fig file and m file. The fig file is a Matlab interface graphic file that contains controls and their attributes on the interface, while the definition of the callback function associated with the control is contained in the m file. Other files store pictures displayed on the interface which are used for illustrations or beautifications.

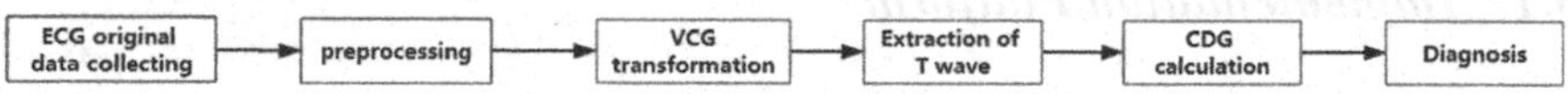

Fig. 1 Flow chart of the system

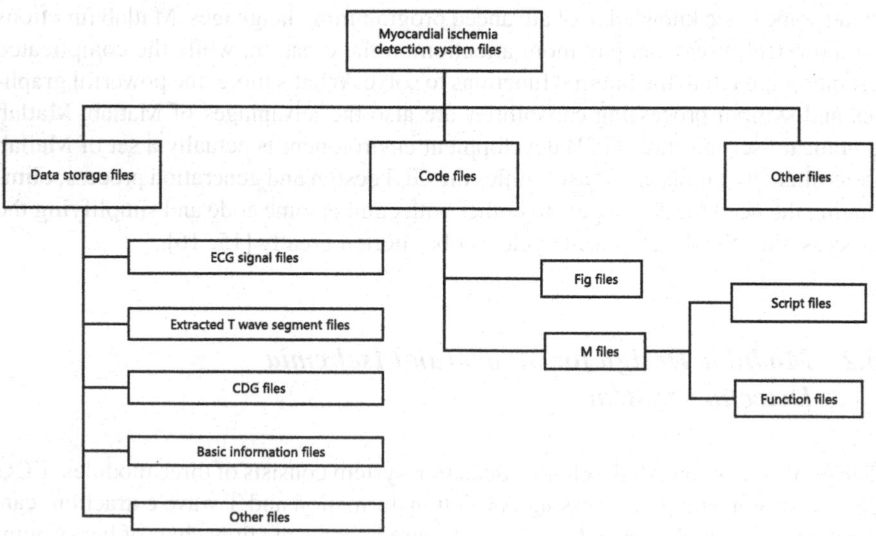

Fig. 2 Standard 12-lead ECG obtain and preprocessing

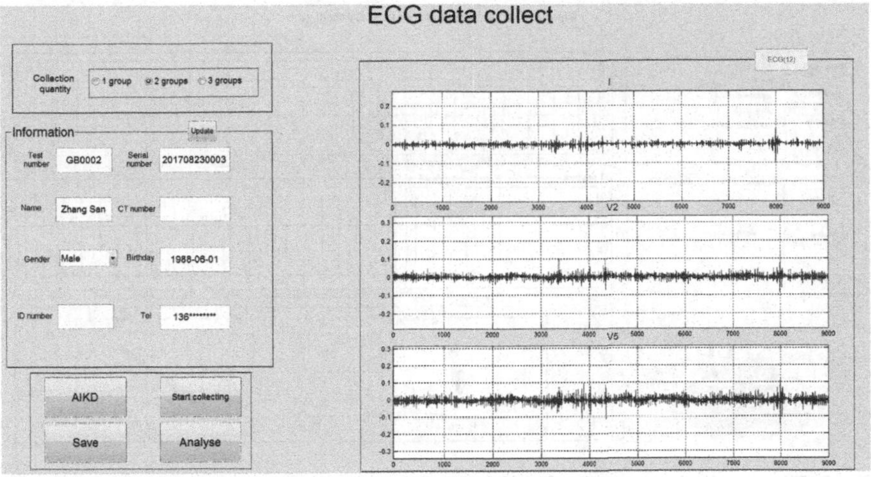

Fig. 3 Transformed VCG calculation and T wave extraction

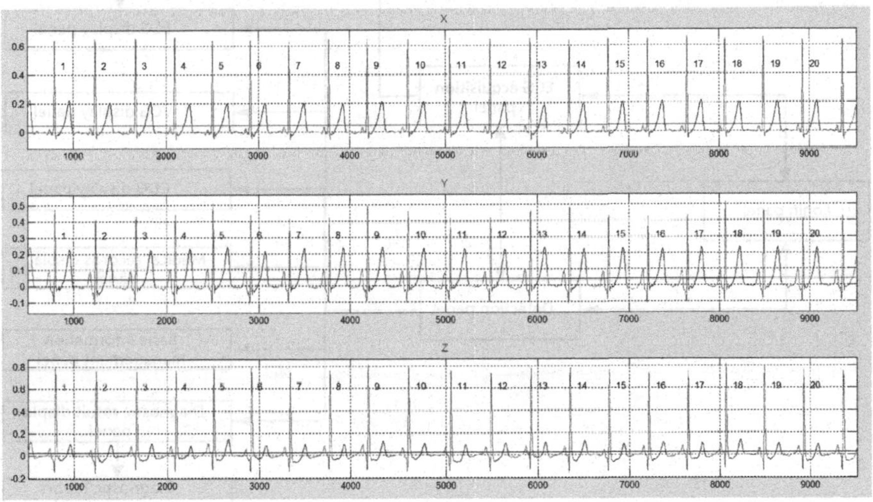

Fig. 4 Cardiodynamics calculation and diagnosis

Furthermore, the relationship between different panels are given in Fig. 6. Arrows indicate the link between different panels.

4 Experiments

The myocardial ischemia detection system is tested in FuWai Hospital, Beijing from December, 2015 to April, 2017. The study has been approved by the local ethics committee and informed consent has been obtained before the study. A total of 320

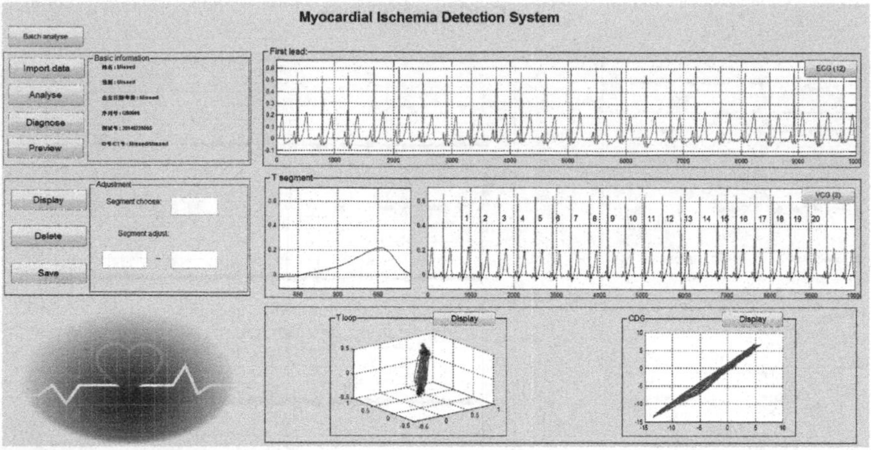

Fig. 5 File organization structure of Matlab-based detection system

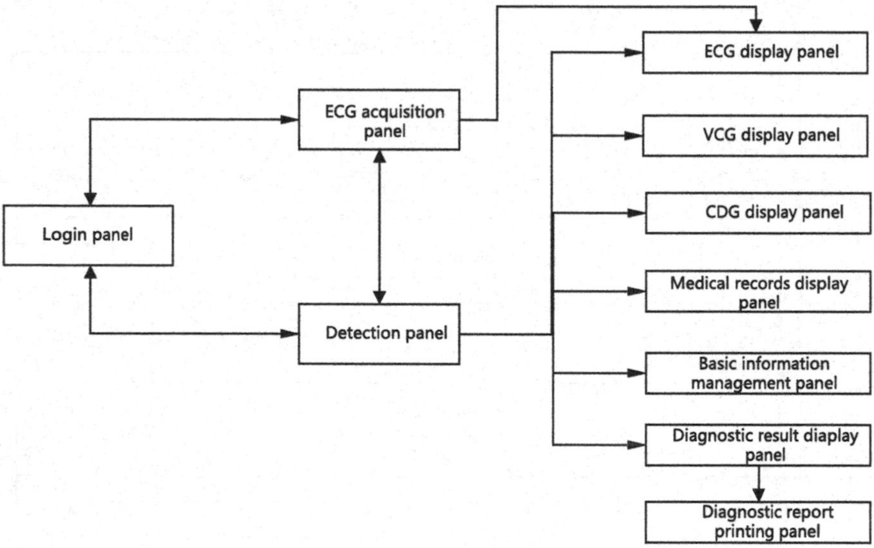

Fig. 6 Relationship between different panels of Matlab-based detection system

inpatients are recruited and their coronary angiography results are collected as the gold standard of diagnosis. The whole population is divided into four groups, each group contains 80 records. Group 1 are records recruited from patients suffered from stenosis of >75% in only 1 coronary artery of three main coronary arteries: left anterior decending artery (LAD), left circumflex artery (LCA) and right coronary artery (RCA). Group 2 are records recruited from patients suffered from stenosis of >75% in two main coronary arteries. Group 3 contains records recruited from

Table 1 Accuracy of CDG under the condition of patients with stenosis in only one coronary artery

CDG	Coronary angiography		Total
	Positive (+)	Negative (−)	
Positive (+)	68(85%)	17(21.25%)	85
Negative (−)	12(15%)	63(78.75%)	75
Total	80	80	160

Table 2 Accuracy of CDG under the condition of patients with stenosis in two coronary arteries

CDG	Coronary angiography		Total
	Positive (+)	Negative (−)	
Positive (+)	70(87.5%)	17(21.25%)	87
Negative (−)	10(12.5%)	63(78.75%)	73
Total	80	80	160

patients suffered from stenosis of >75% in all three main coronary arteries. Group 4 contains records recruited from people without coronary artery stenosis. It is kindly noticed that the standard 12-lead ECG records of each group are measured before the coronary angiography examination. We use sensitivity, specificity and accuracy to evaluate the effectiveness of CDG. Sensitivity is defined as the ratio of true positives to the sum of true positives and false negatives, specificity is defined as the ratio of true negatives to the sum of true negatives and false positives, accuracy is defined as the ratio of the sum of true positives and true negatives to the total amount of population. Sensitivity and specificity are always used to measure the quality of diagnostic methods in clinical, generally speaking, the effectiveness of a diagnostic method is proportional to its sensitivity and specificity.

Firstly, the accuracy of the myocardial ischemia detection system is verified by using group 1 and group 4. Results show that CDG can obtain a total accuracy of 81.875%, sensitivity of 85%, specificity of 78.75% (Table 1). Secondly, the accuracy of the myocardial ischemia detection system is verified by using group 2 and group 4. Results show that CDG can obtain a total accuracy of 83.125%, sensitivity of 87.5%, specificity of 78.75% (Table 2). Finally, the accuracy of the myocardial ischemia detection system is verified by using group 3 and group 4. Results show that CDG can obtain a total accuracy of 88.75%, sensitivity of 91.25%, specificity of 86.25% (Table 3).

In fact, CDG reflects the cardiac electrical conduction, that is, CDG is negative for normal conduction, and otherwise it will be positive. However, besides myocardial ischemia there are other factors can change the stability of cardiac electrical conduction, such as myocardial injury arrhythmia, Purkinje fiber injury, which may be the reason for false positive cases. In addition, ECG noise is another reason for false positive cases. For some patients with local old myocardial infarction or ischemia diseases persisted for a long time, the instability of cardiac electrical conduction will

Table 3 Accuracy of CDG under the condition of patients with stenosis in three coronary arteries

CDG	Coronary angiography		Total
	Positive (+)	Negative (−)	
Positive (+)	73(91.25%)	11(13.75%)	84
Negative (−)	7(8.75%)	69(86.25%)	76
Total	80	80	160

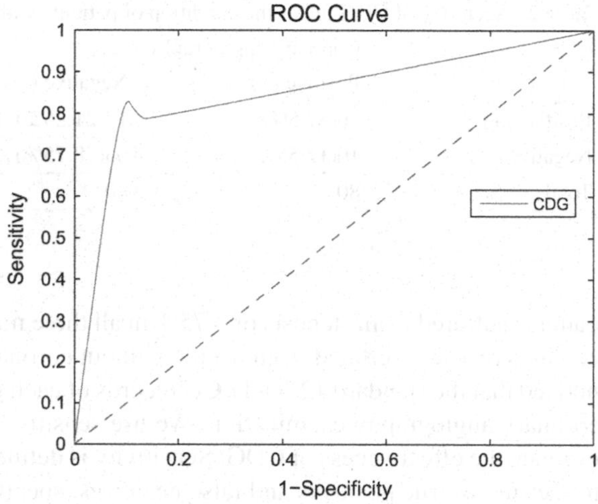

Fig. 7 ROC curves of CDG and VCG

be turned into stability because of the myocardial remodeling, in these cases, the ischemia is exist while the CDG is negative, that is the main reason for the false negative cases of CDG. But overall, CDG still has a very high accuracy with other advantages such as convenient and noninvasive.

It can be observed that CDG can obtain an average accuracy of 84.58%, and with the number of coronary arteries having stenosis increasing, the accuracy of CDG increases obviously. Receiver operating characteristic(ROC) curves of CDG are give in Fig. 7 to show the effectiveness. As CDG is generated by 12-lead ECG signals and can be interpreted directly by objective indicators, it is easy to operate like ECG, but can obtain much more satisfactory results.

5 Conclusions

In this paper, an effective myocardial ischemia detection system is proposed. The system is constructed based on deterministic learning algorithm and Matlab platform. Combined with the powerful calculation capabilities and convenient GUI controls of Matlab platform, the proposed system provides a noninvasive, convenient, cost-effective and accurate method for myocardial ischemia detection. Easy manipulation of the entire diagnostic analysis process is achieved. The experiments indicate that the system is effective, the requirements of myocardial ischemia detection is fulfilled. With the high accuracy of CDG, the system proposed in this paper is helpful in myocardial ischemia detection.

References

1. H. Teragawa, Y. Fukuda, K. Matsuda, H. Hirao, Y. Higashi, T. Yamagata, T. Oshima, H. Matsuura, K. Chayama, Myocardial bridging increases the risk of coronary spasm. Clin. Cardiology **26**(8), 377–383 (2003)
2. G. Teofilovskiparapid, R. Jankovic, V. Kanjuh, R. Virmani, N. Danchin, N. Prates, D.V. Simic, B. Parapid, Myocardial bridges, neither rare nor isolated - autopsy study. Annals of Anatomy - Anatomischer Anzeiger **210**, 25–31 (2016)
3. D. Çiçek, N. Kalay, H. Müderrisoğlu, Incidence, clinical characteristics, and 4-year follow-up of patients with isolated myocardial bridge: a retrospective, single-center, epidemiologic, coronary arteriographic follow-up study in southern turkey. Cardiovasc Revasc Med. **12**(1), 25–28 (2011)
4. C. Wang, D.J. Hill, Deterministic learning and rapid dynamical pattern recognition. IEEE Trans. Neural Networks **18**(3), 617–630 (2007)
5. C. Wang, T. Chen, Rapid detection of small oscillation faults via deterministic learning. IEEE Trans. Neural Networks **28**(3–5), 1284–1296 (2011)
6. C. Wang, X. Dong, O.U. Shanxing, W. Wang, H.U. Junmin, F. Yang, A new method for early detection of myocardial ischemia:cardiodynamicsgram(cdg). Sci. China (Information Sciences) **59**(1), 1–11 (2016)
7. M. Deng, M. Tang, C. Wang, L. Shan, L. Zhang, J. Zhang, W. Wu, L. Xia, Cardiodynamicsgram as a new diagnostic tool in coronary artery disease patients with nondiagnostic electrocardiograms. Am. J. Cardiology **119**(5), 698–704 (2017)
8. M. Deng, W. Cong, T. Min, T. Zheng, Extracting cardiac dynamics within ecg signal for human identification and cardiovascular diseases classification. Neural Networks **100**, 70 (2018)
9. J.H. Qiu, Y.H. Wang, L.I. Zhen-Quan, A new way to develop interface based on matlab/gui. Hebei J. Ind. Sci. Technol. (2008)
10. J.A. Purcell, L. Haynes, Using the ecg to detect mi. Am. J. Nursing **84**(5), 627–642 (1984)
11. J.A. Kors, G.V. Herpen, A.C. Sittig, J.H.V. Bemmel, Reconstruction of the frank vectorcardiogram from standard electrocardiographic leads: diagnostic comparison of different methods. Eur. Heart J. **11**(12), 1083 (1990)
12. C. Froeschlé, R. Gonczi, E. Lega, The fast lyapunov indicator: a simple tool to detect weak chaos. application to the structure of the main asteroidal belt. Planetary Space Sci. **45**(7), 881–886 (1997)
13. R.N. Bracewell, R.N. Bracewell, The Fourier transform and its Applications. (McGraw-Hill New York, 1986), vol. 31999
14. A. Rakotomamonjy, Variable selection using svm-based criteria. J. Machine Learning Res. **3**, 1357–1370 (2003)
15. G. Xian-cai, Realization of bp networks and their applications on matlab. J. Zhanjiang Teachers College **3**, 79–83 (2004)
16. B. Tibor, V. Fedak, F. Durovský, Modeling and simulation of the bldc motor in matlab gui, in *IEEE International Symposium on Industrial Electronics (ISIE)*, IEEE **2011**, pp. 1403–1407 (2011)
17. Y. Liu, M. Deng, C. Wang, Extraction of ecg st-t segment and its system implementation. Comput. Eng. Appl. (2017)

A Method Based on Pseudo Inverse of Image Jacobian Matrix on Uncalibrated Visual Control for Robot

Zhen Yang, Fang Wang, Jiaguo Lv and Xishang Dong

Abstract It is necessary for controlling the robot manipulator to acquire an image Jacobian matrix in the field of uncalibrated visual servo for image-based. In fact, it is difficult to do. In this article, a new controller of uncalibrated visual servo is proposed, which uses pseudo inverse of image Jacobian matrix of robot manipulators. On the basis of the mathematical model of robot, asymptotic convergence is proved by the Lyapunov theory, which means the image errors to zero. Simulations is carried out, which verify performances of the presented scheme.

Keywords Uncalibrated · Visual servo · Robot manipulators · Jacobian matrix

1 Introduction

For robots, vision is just as important as eyes for man. In the field of intelligent industrial robot, such as welding, painting, assembling, handling work etc., vision is mainly used to control and measure the pose and position for end-effectors of robot manipulator. In the field of intelligent mobile robot, such as visual localization, target tracking, visual obstacle avoidance and so on, all need visual information.

Visual control has been one of the most active topics in the field of robotics since 1990s. According to different control models, there are three basic schemes, such as position-based control, image-based control and hybrid visual servo [1–6]. Advantages of position-based control are: the path planning of manipulator is easier, the controller implementing simply etc. But this scheme needs to calibrate the camera. Whenever parameters of camera such as position, focal length have been changed, it is necessary to calibrate again [7]. It is not necessary to calibrate cameras for image-based control, and it can realize controlling precisely. But this scheme needs to estimate Jacobian matrix (interaction matrix) which is difficult. It will be failure

Z. Yang (✉) · F. Wang · J. Lv · X. Dong
School of Information Science and Engineering,
Zaozhuang University, Zaozhuang 277100, China
e-mail: yang_zh99@163.com

Y. Jia et al. (eds.), *Proceedings of 2018 Chinese Intelligent Systems Conference*, Lecture Notes in Electrical Engineering 529,
https://doi.org/10.1007/978-981-13-2291-4_61

to servo by visioning because control variables in singular points of Jacobian matrix can not been obtained [8–10]. Hybrid visual servo can combine advantages of both, which employ a part of degrees of freedom from scheme of position-based control, and others from scheme of image-based control. But this method needs to calculate and decompose the Homography matrix from current image features to ideal one. So it is much more complex and wastes a large amount of calculation [4, 5, 11].

In order to avoid the tedious camera calibration, recently, many scholars in the field of robot focus on uncalibrated visual control. Researchers have proposed a variety of schemes on uncalibrated visual control based on estimation of image Jacobian matrix. All of those are not need to estimate parameters of camera which are integrated into image Jacobian matrix with robot's parameters. These systems based on image information can control robot's movements directly without 3-D reconstruction [11, 12]. Bishop [13], Wang [14, 15] achieved controlling the end-effectors of robot in speed using the information of image. Smith [16] and Miura [17] controlled robots to move in small step by making use of characteristic differences between the current and desired image, then made them to the same.

Presently, visual servo based on kinematics is determined by the speed of joints or end-effectors in robot. Through the control variable from the inner loop, drive the robot to move. Namely, many systems of visual control are devices for positioning completely. Therefore, for requirements of high-speed and high-precision for controlling robot, only considering the control strategy based on kinematics is impossible to achieve good performances.

In this paper, on analysis of above schemes, a method based on uncalibrated visual servo for robot manipulator is proposed. The contribution of this paper can be summarized as follows: (1) a method on uncalibrated visual control based on pseudo inverse of image Jacobian matrix is proposed without robot manipulator's joint speed; (2) stability of this controller is analyzed and asymptotic convergence is proved; (3) for fixed targets and moving targets, respectively, it shows the scheme performs well by simulation.

2 Visual Servo Based on Dynamics of Robot

2.1 Dynamics of Robot

The dynamic equation of a manipulator without friction and other interference has the form:

$$M(\theta)\ddot{\theta} + C(\theta, \dot{\theta})\dot{\theta} + g(\theta) = \tau \tag{1}$$

Here $\theta \in R^{n\times 1}$ is joint vector; $\tau \in R^{n\times 1}$ is torque vector; $M(\theta) \in R^{n\times n}$ is positive symmetric Inertia Matrix; $C(\theta, \dot{\theta})$ is Coriolis and Centrifugal force; $g(\theta)$ is Gravity Moment. According to (1), a well-known characteristic follows:

$$\dot{\theta}^T[\frac{1}{2}\dot{M}(\theta) - C(\theta, \dot{\theta})]\dot{\theta} = 0 \tag{2}$$

2.2 Problem Statement

In image plane of camera, setting the desired image target is $f^*(t)$, the actual image feature is $f(t)$, errors $e(t) = f(t) - f^*(t)$ come from features between $f^*(t)$ and $f(t)$, torques of robot manipulator is τ. The aim of visual servo is: ensuring system's asymptotical stability, make $f(t)$ to be close to $f^*(t)$, namely:

$$\lim_{t\to\infty} f(t) = f^*(t) \Leftrightarrow e(t) = 0 \tag{3}$$

2.3 Visual Control

From previous references, there exits:

$$\dot{e} = J\dot{\theta} \tag{4}$$

where J is the image Jacobian matrix, and $\dot{\theta}$ is joint's speed of robot.

3 Designing the Visual Controller

Generally, it is necessary to acquire the information of image feature and measure the robot's joint speed. However, joint speed can be neglected by the pseudo-inverse estimation of image Jacobian matrix. So a new controller based on pseudo-inverse of image Jacobian matrix can been proposed:

$$\begin{cases} \tau = g(\theta) - J^T K_p e - K_d\dot{\theta} \\ \dot{\theta} = J^+\dot{e}(t) \end{cases} \tag{5}$$

It can be seen from (5), the scheme that may achieve the visual controller based on dynamic of robot under the condition of image feature' errors, their variation and Jacobian matrix, and its pseudo-inverse.

Theorem *For dynamic system of a manipulator without friction and other interference (1), if choosing the scheme (5), Closed-loop system based visual servo will be stable, and image errors for fixed target will converge to zero. Here, K_p and K_d are $n \times n$ positive definite matrix.*

Proof Selecting the following Lyapunov-funtion:

$$V(\dot{\theta}, e) = \frac{1}{2}(\dot{\theta}^T M(\theta)\dot{\theta} + e^T K_p e) > 0 \tag{6}$$

where $\dot{\theta}$ and e are not zero vector. In (6), the first item is kinetic energy, and the second item is the potential of image feature. The derivative of Eq. (6), combine (1) and (4):

$$\begin{aligned}\dot{V}(\dot{\theta}, e) &= \dot{\theta}^T \cdot M(\theta) \cdot \ddot{\theta} + \frac{1}{2}\dot{\theta}^T \cdot \dot{M}(\theta) \cdot \dot{\theta} + e^T K_p \dot{e} \\ &= \dot{\theta}^T[\tau - g(\theta) - C(\theta, \dot{\theta})\dot{\theta} + \frac{1}{2}\dot{M}(\theta) \cdot \dot{\theta}] + \dot{e}^T K_p e \\ &= \dot{\theta}^T[\tau - g(\theta) + J^T K_p e]\end{aligned}$$

Substitute $\tau = g(\theta) - J^T K_p e - K_d \dot{\theta}$:

$$\dot{V}(\dot{\theta}, e) = -\dot{\theta}^T K_d \dot{\theta} < 0, \forall \dot{\theta} \neq 0 \tag{7}$$

Substitute $\dot{\theta} = J^+ \dot{e}(t)$ into (7):

$$\dot{V}(\dot{\theta}, e) = -[J^+ \dot{e}(t)]^T K_d [J^+ \dot{e}(t)] \leq 0, \forall [J^+ \dot{e}(t)] \neq 0 \tag{8}$$

where J^+ is nonsingular. Scheme (5) can ensure the system stability. For the fixed target, in equilibrium: $\dot{\theta} = \ddot{\theta} = \dot{e} = 0$.

According to the scheme (5), dynamics of robot (1) follows:

$$J^T K_p e = 0 \tag{9}$$

When the Jacobian matrix J is nonsingular, from (9):

$$e(t) = f(t) - f^*(t) = 0 \tag{10}$$

It shows that control scheme (5) can make image errors for fixed target converge to zero.

4 Simulation

In simulation, a planar manipulator robot with two links is used to illustrate performances of above schemes. Parameters of robot is followed: $m_1 = 2\,\text{kg}$, $m_2 = 1\,\text{kg}$ (mass of two links); $a_1 = a_2 = 1\,\text{m}$ (length of two links); $z = 2\,\text{m}$, $\lambda = 0.08$ (focal length of camera).

From control scheme (5):

$$\tau_1 = \frac{\lambda}{z}k_p[(\sin(\theta_1) + \sin(\theta_1 + \theta_2))e_1 - (\cos(\theta_1) + \cos(\theta_1 + \theta_2))e_2] + g_1(\theta) - k_d\frac{z}{\lambda \cdot \sin(\theta_2)}[\cos(\theta_1 + \theta_2)\dot{e}_1 + \sin(\theta_1 + \theta_2)\dot{e}_2] \quad (11)$$

$$\tau_2 = \frac{\lambda}{z}k_p[\sin(\theta_1 + \theta_2)e_1 - \cos(\theta_1 + \theta_2)e_2] + g_2(\theta) + k_d \cdot \frac{z}{\lambda \cdot \sin(\theta_2)}[(\cos(\theta_1) + \cos(\theta_1 + \theta_2))\dot{e}_1 + (\sin(\theta_1) + \sin(\theta_1 + \theta_2))\dot{e}_2] \quad (12)$$

On the platform Matlab/Simulink, the schemes (5) is tested. Here, $K_p = 3I, K_d = 0.2I$; image resolution in X and Y direction is 8000 pixels/m.

Fig. 1 Tracking 160 pixels

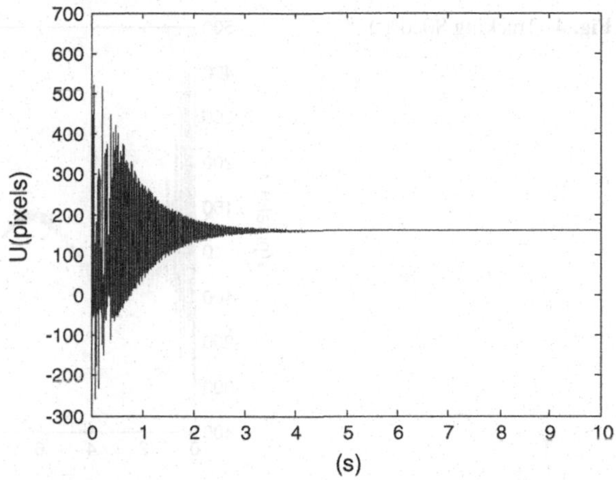

Fig. 2 Tracking −64 pixels

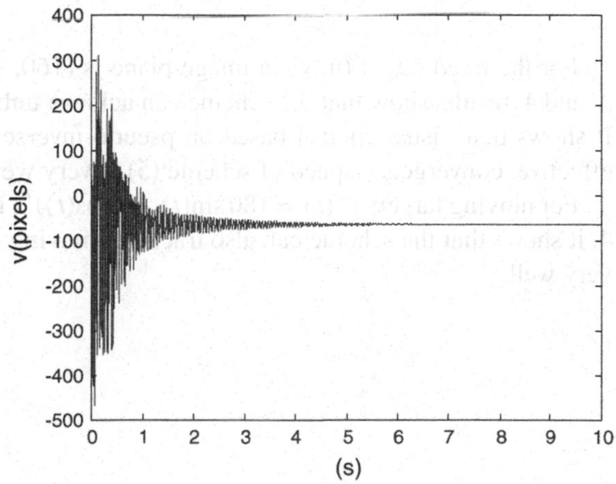

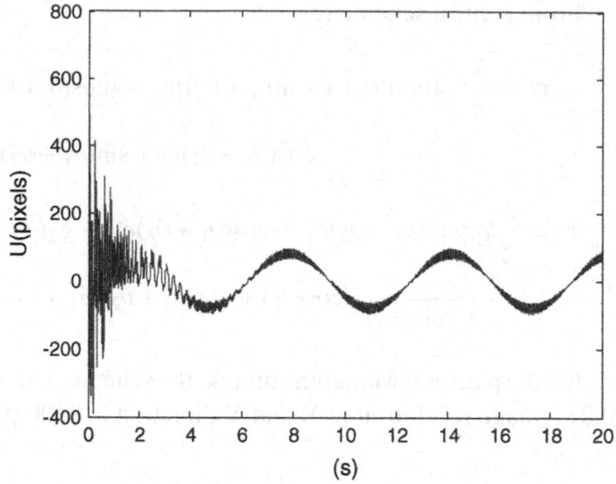

Fig. 3 Tracking 80sin(t)

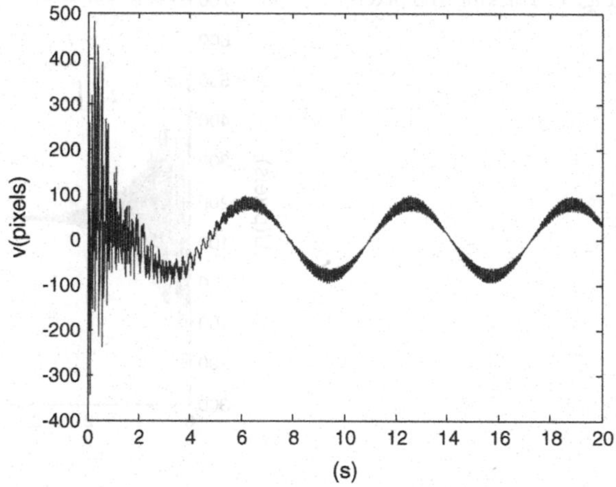

Fig. 4 Tracking 80cos(t)

For the fixed target (u, v) in image plane is (160, −64) pixels. From Figs. 1, 2, 3, and 4, results show that the scheme can achieve unbiased tracking in fixed target. It shows that visual control based on pseudo inverse of image Jacobian matrix is effective, convergence speed of scheme (5) is very well.

For moving target: $f^{*}(t) = (80\sin(t), 80\cos(t))^T$. From Figs. 1, and 2, and 3 and 4, it shows that the scheme can also track the moving target, but performance is not very well.

5 Conclusion

It is unable to meet requirements of engineering for uncalibrated visual servo with high-speed and high-precision if dynamics of robot is ignored. In this paper, the uncalibrated visual control is discussed. A method of visual control based on pseudo inverse of image Jacobian matrix is presented and its stability is proved. By simulation, performance of the scheme is very well. It shows that it can locate the fixed target and track the moving target successfully. However, for moving targets, the scheme performs not very well, which we strive to in the next step.

Acknowledgements This work was supported by Natural Science Foundations of Shandong province (ZR2017LF010), and the Project between industry and education of Ministry of education for Zaozhuang University-Ugrow Co.

References

1. G.D. Hager, S. Hutchinson, P.I. Corke, A tutorial on visual servo control. IEEE Trans. Robot. Autom. **5**, 651–670 (1996)
2. F. Chaumette, S. Hutchinson, Visual servo control, part I: basic approaches. IEEE Robot. Auto Mag. **4**, 82–90 (2006)
3. C. Copot, C. Lazar, A. Burlacu, Predictive control of nonlinear visual servoing systems using image moments. IET Control Theory Appl. **10**, 1486–1496 (2012)
4. F. Chaumette, E. Malis, 2D 1/2 visual servoing: a possible solution to improve image-based and position-based visual servings, in *IEEE International Conference on Robotic and Automation*, San Francisco, pp. 630–635 (2000)
5. E. Malis, F. Chaumette, S. Boudet, Positioning a coarse-calibrated camera with respect to an unknown object by 2D 1/2 visual serving, in *IEEE International Conference on Robotics and Automation*, Leuven, pp. 1352–1359 (1998)
6. C. Cosmin, A. Burlacu, C.M. Ionescu, C. Lazar, R. De Keyser, A fractional order control strategy for visual servoing systems. Mechatronics, 848–855 (2013)
7. P.I. Corke, S.A. Hutchinson, A new partitioned approach to image-based visual servo control, in *Proceeding of the 31st International Symposium on Robotics*, Montreal, pp. 507–515 (2000)
8. C.C. Cheaha, C. Liub, J.J.E. Slotine c, Adaptive Jacobian vision based control for robots with uncertain depth information. Automatica, 1228–1233 (2010)
9. J.A. Piepmeier, G.V. McMurray, H. Lipkin, Uncalibrated dynamic visual servoing. IEEE Trans. Robot. Autom. **1**, 143–147 (2004)
10. X. Gratal, J. Romero, J. Bohg, D. Kragic. Visual servoing on unknown objects. Mechatronics, 423–435 (2012)
11. J.A. Piepmeier, H. Lipkin, Uncalibrated eye-in-hand visual servoing. Int. J. Robot. Res. **10–11**, 805–819 (2003)
12. J.B. Su, W.B. Qiu, Robotic calibrated-free hand-eye coordination based on auto disturbances rejection controller. Acta Autom. Sin. **2**, 161–167 (2003)
13. B.E. Bishop, M.W. Spong, Toward 3D uncalibrated monocular visual servo, in *Proceedings of IEEE International Conference on Robotics and Automations*, pp. 2664–2669 (1998)
14. H. Wang, Y.-H. Liu, W. Chen, Z. Wang, A new approach to dynamic eye-in-hand visual tracking using nonlinear observer. IEEE/ASME Trans. Mechatron. **2**, 387–394 (2011)
15. H. Wang, Y.-H. Liu, D. Zhou, Dynamic visual tracking for manipulators using an uncalibrated fixed camera. IEEE Trans. Rob. **3**, 610–617 (2007)

16. C.E. Smith, N.P. Papanikolopoulos, Grasping of static and moving objects using a vision-based control approach. J. Intell. Robot. Syst. Theory Appl. **3**, 237–270 (1997)
17. K. Miura, J. Gangloff, M.D. Mathelin, et al., Visual servoing without Jacobian using modified simplex optimization, in *Proceedings of the SICE Annual Conference*, pp. 1313–1318 (2004)

Adaptive Switching Control for Robotic Manipulators with Unknown Disturbances

Fang Wang, Zhen Yang, Jiaguo Lv, Chenggan Shan and Huaizhi Ma

Abstract In this paper, an adaptive switching controller is designed for the problem of trajectory tracking in the field of robotic manipulator. The proposed controller which consists of a PD scheme and an adaptive switching law is under the condition that the robotic manipulator's supremum of bounded disturbance is not known. According to the Lyapunov stability theorem, it shows that the presented controller can not only ensure the robot to track the desired trajectory and stabilize the robotic system, but also enhance the ability to varying loads. Finally, Simulations is studied. It shows that the presented controller has an advantage in avoiding the overlarge input torque.

Keywords Bounded disturbance · Robot · Adaptive control · Switch control

1 Introduction

In the field of robotic control, it is important for the system to have the excellent performance for tracking trajectory because of its highly nonlinear, multi-variable, coupling strongly and time-varying systems. A lot of schemes have been proposed in past decades [1, 2]. Generally, in the process of operating robotic manipulators, there are many uncertainties and disturbances, such as the non-linear friction, varying payloads and etc. So it is difficult to find out an accurate mathematical model for a robotic manipulator. Usually, those uncertainties cause the robot system unstable and deteriorate the system performance.

Because the linear PD control has been widely used in the field of industrial robots [3, 4], it is one of the most simple and effective control methods. However, it shows from the field of application that a very large initial output requirement for the driving mechanism is a defect to further the application of the linear PD control. Actually,

F. Wang (✉) · Z. Yang · J. Lv · C. Shan · H. Ma
School of Information Science and Engineering,
Zaozhuang University, Zaozhuang 277100, China
e-mail: yang_zh99@163.com

Y. Jia et al. (eds.), *Proceedings of 2018 Chinese Intelligent Systems Conference*, Lecture Notes in Electrical Engineering 529,
https://doi.org/10.1007/978-981-13-2291-4_62

driving mechanism generally cannot provide the larger initial torque for the linear PD control. Moreover, the maximum torque from robot manipulator is limited, which restricts to further improve the system performance by adjusting the coefficient of PD control. As a result, many schemes for nonlinear PD control are brought up; but most of them, there are only parameters for PD, which means that the coefficient of proportional and differential is still larger, and the output of torque is still overlarge.

The adaptive control is very popular to cope with the parameter uncertainty of robotic system [5]. But for the traditional adaptive control, the uncertain parameter must be constant. Practically, robots often must pick up or lay down some objects and the load for manipulator is not constant. Therefore, parameter jumping exits in this system. So it is difficult for the traditional adaptive control to solve the above problem. In fact, a system with a jumping parameter can be considered to be a switching system. Because of displaying switching features, robotic manipulator can be considered as switching systems which are used to model many physical systems. There are a few of works combining the adaptive control with the switched system in order to deal with the above problem [6–9].

In this paper, we present an adaptive switched controllers with PD parameters for the robotic manipulator, which is designed for the robotic manipulator's supremum of bounded disturbance being unknown. Via Lyapunov stability theorem, the proposed controller can ensure the robot to track the desired trajectory and stabilize the robotic system. In the end, the simulation is carried out to demonstrate the feasibility and validity of the scheme.

This paper is organized as follows: in Sect. 2, the switched robot model and some properties are given. In Sect. 3, the adaptive switching controller with PD parameters is designed. In Sect. 4, simulation are shown. In the end, conclusions are given.

2 Problem Statement and Preliminaries

Considering an n-link robotic manipulator:

$$D(q)\ddot{q} + C(q, \dot{q})\dot{q} + G(q) = \tau + r \tag{1}$$

Here $q \in R^n$, $\dot{q} \in R^n$, $\ddot{q} \in R^n$ are joint angles, velocity and acceleration, respectively. $\tau \in R^n$ is the torque input vector, $r \in R^n$ is the disturbance input and errors, $D(q) \in R^{n\times n}$ is the inertial matrix which is symmetric positive definite, $C(q, \dot{q}) \in R^n$ is the vector of centripetal and Coriolis torques, and $G(q) \in R^n$ stands for the vector gravitational forces.

Following, some properties and assumptions of the robot (1) is listed, they will be useful in stability analysis.

Property 1 *Dynamic of robot (1) can be shown as:*

$$D(q)\ddot{q} + C(q, \dot{q})\dot{q} + G(q) = W(q, \dot{q}, \ddot{q})\rho \tag{2}$$

Here $W(q, \dot{q}, \ddot{q}) \in R^{n\times m}$ is the regression matrix on the vector of joints, and ρ is the load vector of robotic manipulator which is unknown constant.

Assumption 1 $q_d(t), \dot{q}_d(t), \ddot{q}_d(t) \in R^n$ are the desired vector of joint position, joint speed, joint acceleration, which are bounded.

Assumption 2 The disturbance input and errors r is satisfied with:

$$\|r\| \leq t_1 + t_2\|e\| + t_3\|\dot{e}\| \tag{3}$$

Here t_1, t_2, t_3 are positive constants, and $e = q - q_d$ is the tracking error.

Considering the varying load, the following switching model of robotic manipulator for subsystems is used:

$$D_\sigma(q)\ddot{q} + C_\sigma(q, \dot{q})\dot{q} + G_\sigma(q) = \tau + r = W(q, \dot{q}, \ddot{q})\rho_\sigma \tag{4}$$

Here $\sigma(t) : [0, +\infty) \to \Lambda = \{1, 2, \ldots N\}$, which is the switching signal according to the varying load. Designing an adaptive switching controller which can ensure the tracking error e converges to zero is our next step.

3 Adaptive Switching Controller Design

This section introduces the adaptive switching controller applied to robotic manipulator. For the model (4), the proposed controller follows:

$$\begin{aligned} \tau &= -K_d\dot{e} - K_p e + W(q, \dot{q}, \ddot{q})\hat{\theta}_\sigma + u \\ u &= [u_1, u_2 \ldots u_n]^T, u_i = -(t_1 + t_2\|e\| + t_3\|\dot{e}\|)\mathrm{sgn}(e_i) \end{aligned} \tag{5}$$

where K_p and K_d are the proportional gain matrix and derivative gain matrix respectively, which are positive definite. $\hat{\rho}_i$ is the estimation of ρ_i. Only when ith subsystem is active, $\hat{\rho}_i$ will work on it. The presented adaptive law is:

$$\begin{aligned} \dot{\hat{\rho}}_i^T &= -e^T W(q, \dot{q}, \ddot{q})\Gamma_i^{-1}, i = \sigma \\ \dot{\hat{\rho}}_i^T &= 0, i \neq \sigma \end{aligned} \tag{6}$$

where $\tilde{\rho}_\sigma = \hat{\rho}_\sigma - \theta_\sigma$.

Theorem *For robotic system (4), the following control law (7), (8), (9), (10):*

$$\tau = -K_d\dot{e} - K_p e + W(q, \dot{q}, \ddot{q})\hat{\theta}_\sigma + u \tag{7}$$

$$u = -\frac{(\hat{d}f)^2}{df\|e\| + \varepsilon^2}e \tag{8}$$

$$\dot{\tilde{\rho}}_i^T = -e^T W(q, \dot{q}, \ddot{q}) \Gamma_i^{-1}, i = \sigma \quad (9)$$

$$\dot{\tilde{\rho}}_i^T = 0, i \neq \sigma (\tilde{\rho}_\sigma = \hat{\rho}_\sigma - \theta_\sigma)$$

$$\dot{\hat{d}} = \gamma_1 f \|e\|, \hat{d}(0) = 0$$

$$\dot{\varepsilon} = -\gamma_2 \varepsilon, \varepsilon(0) = 0 \quad (10)$$

can guarantee the system (4) to obtain global asymptotic stability. Where K_p and K_d are the proportional gain matrix and derivative gain matrix respectively, which are positive definite; $d = t_1 + t_2 + t_3$, $\tilde{d} = d - \hat{d}$, $f = max(1, \|e\|, \|\dot{e}\|)$, and $\hat{d}$ is the estimated value of d; γ_1 and γ_2 are all the positive constant value.

Proof The Lyapunov function candidate is selected:

$$V(e, H) = \frac{1}{2}[e^T K_d e + tr(H^T \Gamma H)] + \frac{1}{2}(\gamma_1^{-1} \tilde{d}^2 + \gamma_2^{-1} \varepsilon^2)$$

where $H^T = [\tilde{\rho}_1 \ldots \tilde{\rho}_N]$ and $\Gamma = diag(\Gamma_1 \ldots \Gamma_N)$ with $\Gamma_i > 0, i = 1, 2 \ldots N$
Combining (4) with (5):

$$- K_d \dot{e} - K_p e + W(q, \dot{q}, \ddot{q}) \hat{\rho}_\sigma + u + r = W(q, \dot{q}, \ddot{q}) \rho_\sigma$$

$$K_d \dot{e} = -K_p e + W(q, \dot{q}, \ddot{q}) \tilde{\rho}_\sigma + u + r \quad (11)$$

For $V(e, H)$, taking the derivative, using (9) (11), we have:

$$\begin{aligned}
\dot{V}(e, H) &= e^T K_d \dot{e} + \dot{\tilde{\rho}}_\sigma^T \Gamma_\sigma \tilde{\rho}_\sigma + \gamma_1^{-1} \tilde{d} \dot{\tilde{d}} + \gamma_2^{-1} \varepsilon \dot{\varepsilon} \\
&= e^T (-K_p e + W(q, \dot{q}, \ddot{q}) \tilde{\rho}_\sigma + u + r) + \dot{\tilde{\rho}}_\sigma^T \Gamma_\sigma \tilde{\rho}_\sigma + \gamma_1^{-1} \tilde{d} \dot{\tilde{d}} + \gamma_2^{-1} \varepsilon \dot{\varepsilon} \\
&= -e^T K_p e + e^T u + e^T r + (e^T W(q, \dot{q}, \ddot{q}) + \dot{\tilde{\rho}}_\sigma^T \Gamma) \tilde{\rho}_\sigma + \gamma_1^{-1} \tilde{d} \dot{\tilde{d}} + \gamma_2^{-1} \varepsilon \dot{\varepsilon} \\
&= -e^T K_p e + e^T u + e^T r + \gamma_1^{-1} \tilde{d} \dot{\tilde{d}} + \gamma_2^{-1} \varepsilon \dot{\varepsilon}
\end{aligned}$$

According to (8), we have:

$$\dot{V}(e, H) = -e^T K_p e - e^T \frac{(\hat{d} f)^2}{\hat{d} f \|e\| + \varepsilon^2} e + e^T r + \gamma_1^{-1} \tilde{d} \dot{\tilde{d}} + \gamma_2^{-1} \varepsilon \dot{\varepsilon}$$

Because: $e^T r \leq \|e\| \cdot \|r\|$, $\|r\| \leq t_1 + t_2 \|e\| + t_3 \|\dot{e}\| \leq df$, $e^T e = \|e\|^2$,and $\dot{\tilde{d}} = -\dot{\hat{d}} = -\gamma_1 f \|e\|$
There is:

$$\begin{aligned}
\dot{V}(e, H) &\leq -e^T K_p e - \frac{(\hat{d} f)^2}{\hat{d} f \|e\| + \varepsilon^2} \|e\|^2 + df \|e\| + \gamma_1^{-1} \tilde{d} \dot{\tilde{d}} + \gamma_2^{-1} \varepsilon \dot{\varepsilon} \\
&= -e^T K_p e - \frac{(\hat{d} f)^2}{\hat{d} f \|e\| + \varepsilon^2} \|e\|^2 + df \|e\| - \tilde{d} f \|e\| - \varepsilon^2
\end{aligned}$$

$$
\begin{aligned}
&= -e^T K_p e - \frac{(\hat{d} f)^2}{\hat{d} f \|e\| + \varepsilon^2} \|e\|^2 + \hat{d} f \|e\| - \varepsilon^2 \\
&= -e^T K_p e + \frac{-(\hat{d} f)^2 \|e\|^2 + (\hat{d} f)^2 \|e\|^2 - \varepsilon^4}{\hat{d} f \|e\| + \varepsilon^2} \\
&= -e^T K_p e - \frac{\varepsilon^4}{\hat{d} f \|e\| + \varepsilon^2}
\end{aligned}
$$

Because of $\hat{d} > 0$ (it can be seen from the definition of $\hat{d}$),

$$
\text{So} : \dot{V}(e, H) \le -e^T K_p e < 0. \quad \forall e \ne 0 \tag{12}
$$

We have $\lim_{t \to \infty} e(t) = 0$. The proof has been completed.

4 Simulation

In order to illustrate the performance of the proposed adaptive switching strategy, a two-DOF planar manipulator is used for simulations whose load is persistently varying. Referring to [10], the mathematic model of robotic dynamics is:

$$
\begin{bmatrix} D_{11}(q_2) & D_{21}(q_2) \\ D_{12}(q_2) & D_{22}(q_2) \end{bmatrix} \begin{bmatrix} \ddot{q}_1 \\ \ddot{q}_2 \end{bmatrix} + \begin{bmatrix} -C_{12}(q_2)\dot{q}_2 & -C_{12}q_2(\dot{q}_1 + \dot{q}_2) \\ C_{12}(q_2)\dot{q}_1 & 0 \end{bmatrix} \begin{bmatrix} \dot{q}_1 \\ \dot{q}_2 \end{bmatrix} + \begin{bmatrix} G_1(q_1, q_2)g \\ G_2(q_1, q_2)g \end{bmatrix} + \begin{bmatrix} \omega_1 \\ \omega_2 \end{bmatrix} = \begin{bmatrix} \tau_1 \\ \tau_2 \end{bmatrix}
$$

Here:

$$
\begin{aligned}
D_{11}(q_2) &= (m_1 + m_2) r_1^2 + m_2 r_2^2 + 2 m_2 r_1 r_2 \cos q_2 \\
D_{12}(q_2) &= m_2 r_2^2 + m_2 r_1 r_2 \cos q_2 \\
D_{22}(q_2) &= m_2 r_2^2 \\
C_{12}(q_2) &= m_2 r_1 r_2 \sin q_2 \\
G_1(q_1, q_2) &= (m_1 + m_2) r_1 \cos q_2 + m_2 r_2 \cos(q_1 + q_2) \\
G_2(q_1, q_2) &= m_2 r_2 \cos(q_1 + q_2)
\end{aligned}
$$

The link consists of two rigid beams with actuators mounted on the joints. The load can be considered as part of the second link. Parameters of dynamic model (1) follow:

The length and mass of robot: $r_1 = 1, r_2 = 0.8; m_1 = 0.5, m_2 = 0.5(\sigma = 1)$, $m_2 = 1(\sigma = 2)$.

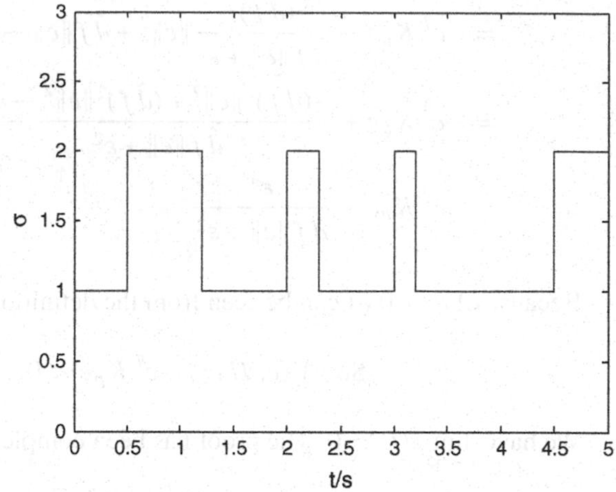

Fig. 1 The switching signal

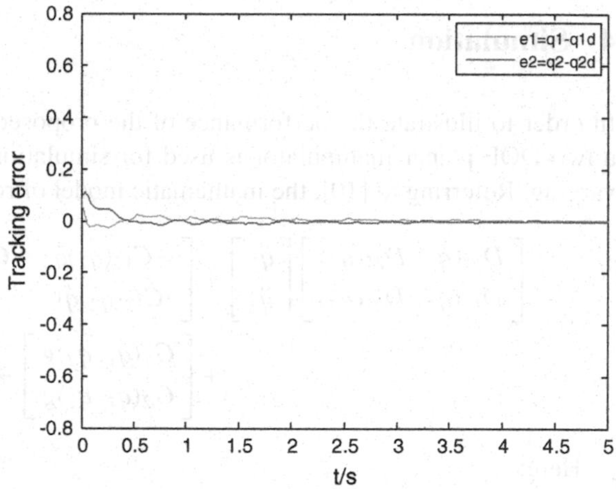

Fig. 2 Tracking error

The given reference trajectory and initial state of system:

$$\begin{cases} q_{1d} = \sin(2\pi t) \\ q_{2d} = \sin(2\pi t) \end{cases}, \begin{bmatrix} q_1 \\ q_2 \\ q_3 \\ q_4 \end{bmatrix} = \begin{bmatrix} 0.1 \\ 0 \\ 0.1 \\ 0 \end{bmatrix}$$

Control parameters:

$$K_p = diag(50, 50), K_d = diag(180, 180), \Gamma = diag(5, 5)$$

Choosing $\gamma_1 = 20$, $\gamma_2 = 20$.

Fig. 3 Torque for link 1

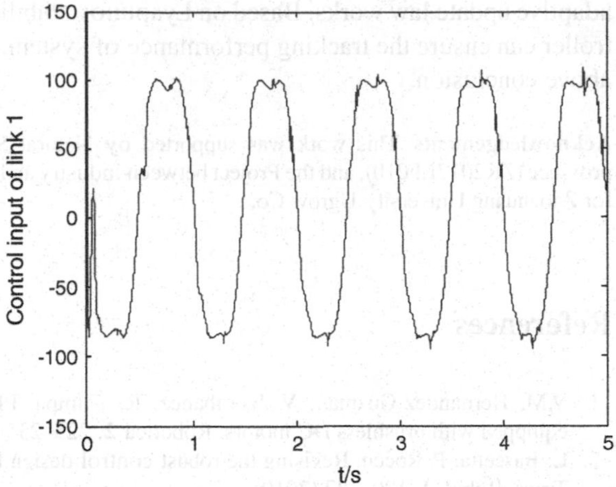

Fig. 4 Torque for link 2

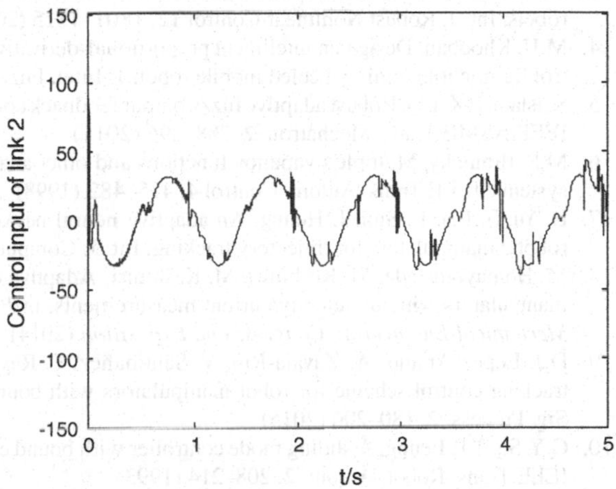

The switch signal is arbitrary as the Fig. 1, and simulations are shown in Figs. 2, 3, and 4. The tracking error of two links which converge to zero is shown in Fig. 2. Which illustrates that the proposed controller in this paper can ensure the robot system to follow the desired output signal, and the tracking performance is very well.

5 Conclusion

In this paper, an adaptive switching controller has been studied for robotic manipulator with varying loads. When the corresponding subsystem is activated, the proposed adaptive update law works. Based on Lyapunov stability theorem, the proposed controller can ensure the tracking performance of system. Simulations have verified the above conclusion.

Acknowledgements This work was supported by Natural Science Foundations of Shandong province (ZR2017LF010), and the Project between industry and education of Ministry of education for Zaozhuang University-Ugrow Co.

References

1. V.M. Hernandez-Guzman, V. Santibanez, R. Campa, PID control of robot manipulators equipped with brushless DC motors. Robotica **2**, 225–233 (2009)
2. L. Bascetta, P. Rocco, Revising the robust-control design for rigid robot manipulators. IEEE Trans. Robot. **1**, 180–187 (2010)
3. E. Slawiñski, M. Vicente, PD-like controllers for delayed bilateral teleoperation of manipulators robots. Int. J. Robust Nonlinear Control **12**, 1801–1815 (2015)
4. M.H. Khooban, Design an intelligent proportional-derivative (PD) feedback linearization control for nonholonomic-wheeled mobile robot. J. Intel. Fuzzy Syst. **4**, 1833–1843 (2014)
5. S. Islam, P.X. Liu, Robust adaptive fuzzy output feedback control system for robot manipulators. IEEE/ASME Trans. Mechatron. **2**, 288–296 (2011)
6. M.S. Branicky, Multiple Lyapunov functions and other analysis tools for switched and hybrid systems. IEEE Trans. Autom. Control **4**, 475–482 (1998)
7. L. Yu, S. Fei, L. Sun, J. Huang, An adaptive neural network switching control approach of robotic manipulators for trajectory tracking. Int. J. Comput. Math. **5**, 983–995 (2014)
8. M. Homayounzade, M. Keshmiri, M. Keshmiri, Adaptive control of electrically-driven robot manipulators without velocity/current measurements, in *Proceedings of ASME International Mechanical Engineering Congress and Exposition* (2014)
9. D.J. López-Araujo, A. Zavala-Río, V. Santibáñez, F. Reyes, A generalized global adaptive tracking control scheme for robot manipulators with bounded inputs. Int. J. Adapt. Control Sig. Process. **2**, 180–200 (2015)
10. C.Y. Su, T.P. Leung, A sliding mode controller with bound estimation for robotic manipulators. IEEE Trans. Robot. Autom. **2**, 208–214 (1993)

Research of CDG to Identify Individuals via Deterministic Learning Theory

Guohui Yan, Tong Geng, Muqing Deng and Cong Wang

Abstract An approach of human identification based on cardiodynamicsgram (CDG) is proposed in this paper. Algorithm design for the electrocardiogram (ECG) is carried out to achieve human identification, which includes collecting ECG data from the PTB database, filtering the ECG, synthesizing the VCG by using the filtered 12-lead ECG, and intercepting the ST-T segment from the VCGs. Then the CDGs are obtained by using radial basis function (RBF) neural networks (NNs) to model the ST-T segment through deterministic learning (DL). The obtained knowledge is stored in constant RBF networks. Finally stable features-Spatial heterogeneity index and temporal heterogeneity index are extracted to characterize the uniqueness of an individual, and SVM is used to train the classification model in MATLAB. The test data is fed to the obtained model to evaluate our proposed method. The results show that the proposed method can achieve more than 85% correct classification rate.

Keywords CDG · ECG · Human identification · Deterministic learning · SVM

1 Introduction

As a newly merged biometrics, ECG has attracted considerable attention in the area of pattern recognition. It's greatest feature is that ECG is hard to forge, highly deceptive, and a natural tool for living confirmation. It's merits overcome the shortcomings of the existing biometric identity identification technologies. At the same time, ECG identification is an effective supplement to the current biometric identification system and can be integrated with existing biometrics to further increase information security.

G. Yan · T. Geng · C. Wang (✉)
School of Automation Science and Engineering,
South China University of Technology, Guangzhou 510640, China
e-mail: wangcong@scut.edu.cn

M. Deng
Institute of Information and Control,
Hangzhou Dianzi University, Hangzhou 310018, China

Y. Jia et al. (eds.), *Proceedings of 2018 Chinese Intelligent Systems Conference*, Lecture Notes in Electrical Engineering 529,
https://doi.org/10.1007/978-981-13-2291-4_63

Biel et al. [1] collected 12-lead information from healthy individuals, extracted 30 features about width, interval, and amplitude from each lead, and used Principal Component Analysis (PCA) to get 10 irrelevant features. Soft Independent Modeling Class Analysis (SIMCA) was designed to identify individuals and the accuracy was 80.1%. Kyoso et al. [2] extracted the P and QRS wave width, PQ and QT interval of healthy people as features. The discriminant function was used to calculate the mean, difference, and covariance of each feature, then the Mahalanobis distance was calculated and the minimum distance was taken as evidence of identity matching. In [3], the authors divided the cardiac cycle by positioning the R-wave peak point in the ECG waveform, using the DB (Daubechies) wavelet basis for each cardiac cycle waveform to perform 8-layer decomposition, resulting in D3, D4 and A4 wavelet coefficients are used as wavelet features. Independent component analysis (ICA) algorithm is used to extract morphological information for single lead signals and classify them using support vector machine classifier. For these methods mentioned above, there mainly exists two problems: (1) in the experimental analysis, data samples of healthy individuals are constantly used, and no experiments are performed on data containing lesions, so the practicality is limited; (2) the above experiments only analyzed a few leads of the 12 lead ECG, and didn't fully extract the information contained in the ECG signal; (3) autocorrelation coefficient or the calculation of the distance measure is based on the data matrix calculation when the experimental sample increases, the calculation amount increases dramatically.

In this paper, a new approach of human identification based on CDG via DL is proposed. Since the ECG signal is a non-stationary, non-linear, low-frequency, weak signal, it is susceptible to interference from the instrument or external noise during the acquisition process. Therefore, the ECG data is first subjected to median filtering and wavelet filtering. Synthesizing the VCG data by using the filtered 12-lead ECG data, and the ST-T segment in the VCG data is extracted. The CDGs are obtained by using RBF NNs to model the ST-T segment through DL. The obtained knowledge is stored in constant RBF networks. Finally, we extract two stable features, including CDG-spatial heterogeneity index and temporal heterogeneity index. CDG-spatial heterogeneity index is based on the Lyapunov exponent [4] and describes the spatial dispersion of the CDG; Temporal heterogeneity index is based on the Fourier coefficient [5] and represents the time-period of the CDG, and SVM is used to train the classification model in MATLAB. Experiments show that the proposed algorithm can achieve encouraging recognition accuracy.

2 Preliminaries and Feasibility of the Biometric

2.1 Deterministic Learning Theory

Deterministic Learning Theory is a method to solve the problems of knowledge acquisition, expression and utilization in the unknown dynamic environment. This method uses a constant-valued radial basis function (RBF) neural network obtained in the learning phase to construct a dynamic estimator for each training pattern to represent these already trained dynamic patterns. These constant-valued RBF neural networks can quickly recall the dynamic knowledge of the system acquired during the training phase, thereby locally and accurately approaching the system dynamics of the training model, and realizing the rapid and effective reuse of knowledge.

The basic form of RBF neural network is shown in Eq. (1):

$$f_{nn}(Z) = \sum_{i=1}^{N} \omega_i s_i(Z) = W^T S(Z) \tag{1}$$

where $Z \in \Omega_z \subset R^q$ is the neural input vector, $N > 1$ is the number of neural network nodes, $W = [\omega_1, \omega_2, ..., \omega_N] \in R^N$ is the weight vector, and $S(Z) = [s_1(||Z - \zeta_1||), ...s_N(||Z - \zeta_N||)]^T$ is the Radial basis vector, with $s_i(.)$ is the Radial basis function (RBF), $\zeta_i (i = 1, ...N)$ is the center vector of the i-th node of the network in the state space. Here we choose commonly Gaussian function here as radial basis function.Mathematical expressions are shown in Eq. (2).

$$s_i(||Z - \zeta_i||) = \exp[-||Z - \zeta_i||^2/\eta_i^2], i = 1, 2, ..., N \tag{2}$$

where $\zeta_i = [\zeta_{i1}, \zeta_{i2}, ..., \zeta_{iq}]^T$ is the center of the receptive field, $\eta_i > 0$ is the width of the receptive field.

It has been proved in [4] that when the RBF nodes in the neural network are large enough and the distribution of the center point ζ_i and the width of the RBF nodes are appropriate. The RBF neural network can approximate any continuous function $f(Z)$ defined in a compact set $\Omega_Z \subset R^q$ with arbitrary precision as:

$$f(Z) = W^{*T} S(Z) + \varepsilon(Z), \forall Z \in \Omega_Z \tag{3}$$

where W^* is the ideal constant weight vector,$\varepsilon(Z)$ is the accuracy of approximation which satisfies $|\varepsilon(Z)| < \varepsilon^*$ with $\varepsilon^* > 0$ for all $Z \in \Omega_Z$.

For a localized RBF NNs, each RBF only affects the local network output [5, 6]. Therefore, for any bounded trajectory $Z(t)$ defined on a compact set, $f(Z)$ can be approximated by a local neuron along the region near the trajectory and we abbreviate $\varepsilon(Z)$ as ε:

$$f(Z) = W_\zeta^{*T} S_\zeta(Z) + \varepsilon_\zeta \tag{4}$$

where ε_ζ is the approximation error,and with $||\varepsilon_\zeta| - |\varepsilon||$ being small enough. $S_\zeta(Z) = |s_{j1}(Z), ..., s_{j\zeta}(Z)|^T \in R^{N\zeta}$ is the subvector of $S(Z)$, and $W_\zeta^* = |\omega_{j1}^*, ..., \omega_{j\zeta}^*|^T \in R^{N\zeta}$ with $N_\zeta < N$.

It has been proved in [5, 7] that the regressor subvector $S_\zeta(Z)$, along any recurrent trajectory $Z_\zeta(t)$, satisfies the PE condition.

2.2 CDG as a Biometric

In principle, any physiological or behavioral feature of a person can be used as a biometric for human identification as long as the following conditions are met:

- universality: everyone has;
- uniqueness: any two people are not the same;
- stability: this feature is unchanged at least for a period of time;
- measurability: features must be quantifiable and readily available.

However, the biometrics satisfying the above conditions may not be feasible for an actual system, because the actual system must also consider:

- performance: namely, the accuracy, speed, robustness of the recognition, and the required accuracy;
- acceptability: people's acceptance of a particular biometric identification in daily life;
- anti-deception: the ability to distinguish fraud biometric.

Therefore, an actual biometric identification system should meet the requirements: Achieve acceptable accuracy and speed of recognition under reasonable resource requirements; Be harmless and acceptable to people; It's robust enough to all kinds of deceptive methods. The ECG signal belongs to the endogenous physiological signal of human body, which is periodic. The ECG curve of each cycle includes one P wave, one QRS wave group, one T wave, and one U wave, these basic bands contain the characteristic parameters such as the peak of R wave, the RR interval, the S-T segment and the P-R interval, which reflect the contraction and relaxation of the heart at different stages. The basic band and characteristic parameters are the basic components of the ECG waveforms, which together complete the ECG. Studies have shown that the ECG varies from one individual to another due to factors such as the location, size, configuration, age, sex, weight, and thoracic structure of the individual's heart [8]. It also contains information related to individual identification and meets the four important characteristics that biometrics require for identification.

- Universality: the heart of each living body produces ECG signals at all times.
- Uniqueness: the difference between individual's ECG is mainly influenced by body shape, age, weight, emotion, sex, heart position, heart size, heart geometry, heart physiological characteristics, chest structure and so on, and the ECG signals from different people are also unique.
- Stability: the cardiac structure and size of adults are basically stereotyped. Studies have shown that changes in time, heart rate, and mood do not fundamentally alter ECG waveforms. Several experiments have shown stability.
- Measurability: the electrocardiograph generally exists in hospital, and with the development of science and technology, the collection of 12 lead ECG is more convenient and rapid, especially the appearance of the portable ECG acquisition equipment, which facilitates the collection of ECG.

Table 1 Comparison of performance of different biometric technologies [9]

Biometrics	Universality	Uniqueness	Stability	Feasibility	Acceptability	Anti-deception
Gait	M	L	L	L	H	M
DNA	H	H	H	H	L	L
Fingerprint	M	H	H	H	M	M
Face	H	L	M	L	H	M
Iris	H	H	H	H	L	L
Voice	M	L	L	L	H	H
Hand shape	M	M	M	M	M	M
Signature	L	L	L	L	H	H
Palm print	M	H	H	H	L	M

Anil K. Jain, a distinguished professor of biometrics and a professor at Michigan State University in the United States, has done a lot of research for this and a lot of experiments to compare the performance of these current biometric technologies and got convincing results [9], as shown in Table 1 (H, M, and L denote High, Medium, and Low respectively).

There are various defects in these biometric identification methods. The emergence of many counterfeit biometric systems (such as false iris, fake fingerprints, false faces, voice recordings, etc.) makes the ulterior person can sneak into the system of recognition through the acquisition process or the transmission process of the sensor, then leads to the confusion of the system and eventually makes a wrong decision. However, because ECG signals cover a large amount of physiological information, which meets all the traits of biometrics. It is a chance to be a new and powerful tool for identity recognition. Most importantly, Alan Kaplan, a research engineer at the Lawrence Livermore National Laboratory, has done a lot of experiments and has publicly published his findings, who once said: "The ECG is very hard to fake."

3 Identification Based on CDG

3.1 Preprocessing of ECG

Because ECG signal is a kind of non-stationary, non-linear, low-frequency, and weak signal, it's extremely easy to be interfered by the instrument or external factors during the acquisition process. The common disturbance noise includes baseline drift, power frequency and myoelectric interference, which poses great difficulties for the subsequent modeling and identification, so they need to be filtered and calibrated.

Median filtering, a non-linear smoothing technique, is used for baseline calibration, the principle of which is simple: median filtering for a series of numerical sequences is to replace the values of every point in the sequence to the median of all values in the neighborhood of that point, and the neighborhood of filter point is called filter window,whose length setting is the most critical step. The system adopts two median filtering to obtain the drift position information of the original ECG baseline, and then removes the baseline drift by getting the difference between the original signal and the drift position, so as to achieve the purpose of baseline calibration. The window length of the first median filter is set to 100, but at this time, the amplitude of R-wave and T-wave are greatly reduced. Therefore, based on the results of the first filtering, the window length of the second filter is set to 300, forcing amplitude of the R-wave, T-wave to be pulled back to the baseline drift. Now baseline correction can be achieved very well.

Multi-scale analysis in wavelet transform is designed to remove power frequency and myoelectric interference, which can perform localized analysis of time (space) frequency [10]. It's also called multi-resolution analysis, proposed by the French scholar Mallat. The Mallat algorithm is a fast algorithm for wavelet decomposition and reconstruction, which mentioned the theorem: if f(k) is the discrete sampling data of the continuous signal f(t), the orthogonal wavelet transform decomposition formula of the signal f(t) is as follows.

$$\begin{cases} c_{j,k} = \sum_m c_{j-1,m} h(m-2k), \\ d_{j,k} = \sum_m c_{j-1,m} g(m-2k), \end{cases} \tag{5}$$

where $c_{j,k}$ is scale factor, $d_{j,k}$ is wavelet factor, h and g are the high-pass and low-pass filters respectively, j is the scale of the scale space. The process of wavelet reconstruction is the inverse operation of wavelet decomposition, the formula of which is as follows.

$$c_{j-1,m} = \sum_m c_{j,k} h(m-2k) + \sum_k d_{j,k} g(m-2k) \tag{6}$$

In this paper, Coif 4 in the Coifiet wavelet is used as the wavelet basis. Referring to the pass band of the Coif4 wavelet, an 8-level Coif4 decomposition is performed on the ECG signal. Since the energy of the power frequency and myoelectric interference is mainly distributed on the low-level 1-3, after the decomposition, the soft-threshold filter is applied to the low-scale wavelet coefficients, and then the ECG signal is reconstructed by using the wavelet inverse transform to remove the signal frequency and myoelectric interference. The filtering effect is shown in Fig. 1.

3.2 Dynamic Modeling

This section mainly implements the dynamic modeling of ST-T segment. The first step is to synthesize VCG from filtered ECG data. The conversion formula is as follows [11, 12]:

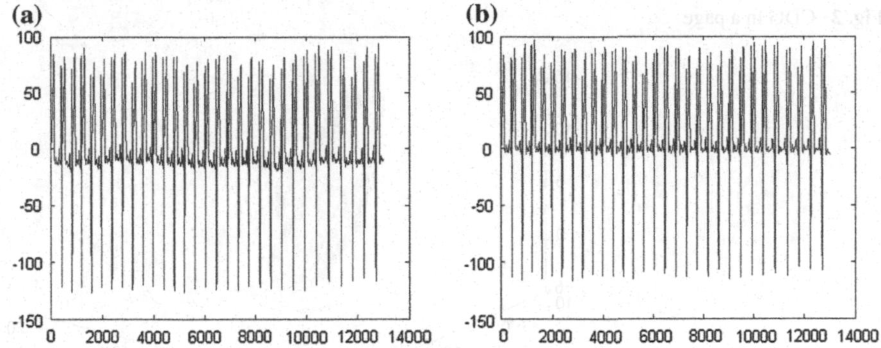

Fig. 1 **a** Raw ECG signal. **b** ECG with median filtering and wavelet transform

$$\begin{cases} V_x = 0.38I - 0.07II - 0.13V_1 + 0.05V_2 - 0.01V_3 + 0.14V_4 + 0.06V_5 + 0.54V_6 \\ V_y = -0.07I + 0.93II + 0.06V_1 - 0.12V_2 - 0.05V_3 + 0.06V_4 - 0.17V_5 + 0.13V_6 \\ V_z = 0.11I - 0.23II - 0.43V_1 - 0.06V_2 - 0.14V_3 - 0.20V_4 - 0.11V_5 + 0.31V_6 \end{cases} \tag{7}$$

It can be seen from the above equation that three-dimensional data of the VCG is the weighted sum of the 12-lead ECG data, and the ST-T segment of the VCG data is intercepted based on DL and recorded as $[v_x, v_y, v_z]$, Wang et al. further promoted it and proposed data-based modeling [13, 14], this modeling method only requires the signal of the ECG vector to achieve accurate identification of the ECG intrinsic dynamics information. According to the method, the ECG dynamics can be modeled in the following general dynamical system with using three RBF NNs $\hat{W}_i^T S_i(\phi_\zeta)(i \in X, Y, Z)$:

$$\hat{v}_i(k+1) = a_i(\hat{v}_i(k) - v_i(k)) + \hat{W}_i^T(k+1)S(v(k)) \tag{8}$$

where a_i is a constant with $0 < ||a_i|| < 1$, $S(v(k))$ is Gaussian RBF with $S(v(k)) = [s_1(||v(k) - \xi_1||), ..., s_N(||v(k) - \xi_n||)]$, $N > 1$ is the number of nodes in the RBF NNs and $\xi_j (j = 1, 2, ..., N)$ is the central point of the neuron.

According to the RBF NNs weight updating rule of the discrete system, the following formula can be obtained:

$$\hat{W}_i(k+1) = \hat{W}_i(k) + \Gamma S_k v_i(k) - \sigma \hat{W}_i(k) \tag{9}$$

The above formula is the regulation rate, and it can be proved that learning and training are conducted under the condition of $\hat{W}_i(0) = 0$, $\hat{W}_i$ will converge to a small neighborhood of its optimal value W_i^*. The dynamic expression of the final VCG is as follows:

$$f_i(\phi v) = \hat{W}_i^T S_i(\phi v) + \varepsilon_{i1} = \bar{W}_i^T S_i(\phi v) + \varepsilon_{i2} \tag{10}$$

Fig. 2 CDG in a page

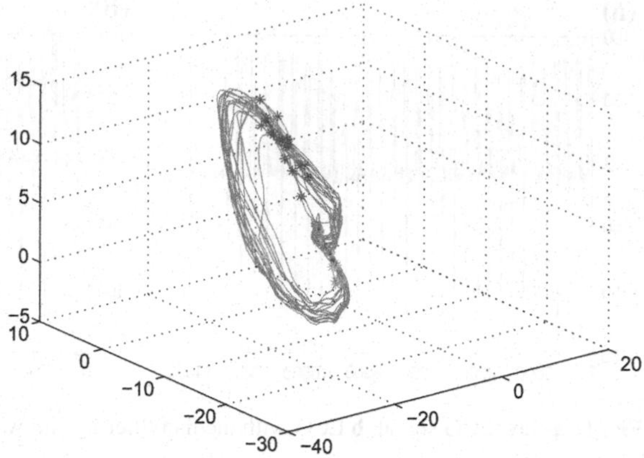

where $i \in X, Y, Z$, and $S(\nu)$ is a matrix of $N \times M$ with N being the number of RBF neurons and M being the length of the vector ν_i, $\hat{W}_i$ is a matrix of $1 \times N$, $\bar{W}_i = mean_{t \in [t_a, t_b]} \hat{W}_i(t)$ with mean being arithmetic average, and $t_a > t_b > 0$ is a short period of time after learning and training.

The above is the process of dynamic system identification of ECG via the DL, the identification result will be stored in the constant weights of the RBF NNs. The CDG can be obtained by performing a 3d-visualization of the identification result, as Fig. 2 show. It can be regarded as a mapping of the VCG in the three-dimensional space. The CDG shape of the same person is very similar, but that of different people is quite different.

3.3 *Feature Extraction*

The human identification approach based on CDG designed in this paper is the same as the general pattern recognition system. Feature extraction is an extremely important part of the identification method. The foremost aim to perform this module is that converting the CDG ring to some type of parametric representation for further study and analysis. In order to implement an efficient human identification system, it is a great challenge to extract features that can truly represent the characteristics of a subject. This paper describes the degree of CDG's dispersion from two aspects: time dispersion and spatial dispersion. The first feature extracted is the spatial heterogeneity index (SHI) which is based on the Lyapunov exponent [13] and describes the spatial dispersion of the electrocardiogram:

$$SHI = \frac{1}{N} \sum_{n=1}^{N} \ln(d_{n2}/d_{n1}) \qquad (11)$$

where N denotes the number of data points in the CDG, d_{n1} denotes the distance between the nth point and its closest point, and d_{n2} denotes the distance between the nth point and its nearest point after 10 steps. The second one is the temporal heterogeneity index (THI) which is based on the Fourier coefficients [14] and describes the time periodicity of the electrocardiogram.

$$THI = \arg\min_{\lambda_i} |F.\exp(-0.1\lambda_i)| \tag{12}$$

where F denotes the Fourier transform of the CDG.

4 Classification

The class here refers to the individual. Since our classification program is applied to the extracted features, it can also be called feature matching, the advanced application of feature matching technology in electrocardiogram recognition mainly includes dynamic time warping (DTW), Hidden Markov Modeling (HMM) neural network (NN) [15], vector quantization. And the classification method adopted in the paper is the support vector machine (SVM) [16] for the reason that it is easy to implement and high accuracy, high-precision and high generalization performance, since SVM originally was used for binary classification, it could not handle multi-class classification directly. For multi-class classification problem, the problem was generally decomposed into several two-class classification problems. We use the One-versus-One Strategy (OVO), where one set of binary classifiers is constructed by corresponding data from two classes. In the test, we use the vote. The "Marx win" strategy produces output and using a suitable core can get better results. Figure 3 demonstrates the block diagram of the proposed system.

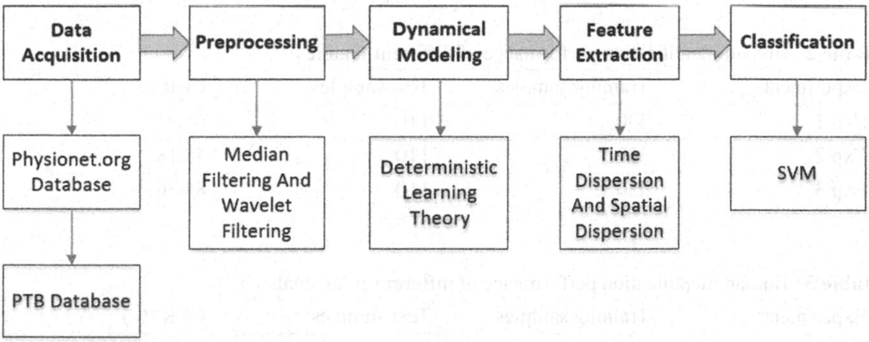

Fig. 3 Overview of the CDG identification approach

5 Experiment

The study has been approved by the local ethics committee and informed consent has been obtained before the study. To evaluate the performance of the proposed method, Correct Classification Rate (CCR), False Acceptance Rate (FAR) and False Rejection Rate are used as the indicator of classification performance. In this paper, we use the data of 10 individuals in the PTB database to verify the method, including 5 males and 5 females, all of whom are between 30 and 50 years old, and among them there are 3 healthy individuals and 7 individuals with myocardial ischemia, each person has 44 sets of data, with a total of 440 sets of data, which are randomly divided into 75% training sets and 25% test sets, but in order to prevent skewed classification, the partition is done individually by each class, so they can be proportionately presented in the training data and test data. This experiment is based on one of the following platforms, with the following parameters: CPU:Intel Core i-4790, RAM:16 GB, 64 bit Windows8.1 enterprise version operating system. In the implementation of the specific experimental code, this paper uses the MATLAB language to encode. In order to ensure the accuracy of the experiment, this paper calls the SVM packet in the library to train the data set characteristic matrix of the upper layer, and finally obtains the decision model.

In the course of the experiment, the two sets of features extracted from the CDG are divided into three experiments. The experiment by using the first set of features is recorded as the Exp1, and the experiment is recorded as Exp2 with the second features, the two sets of features are fused and then the experiment is recorded as Exp3, and the training samples and test samples of the three experiments are the same, which are 330 training samples and 110 test samples. The CCR of the proposed method is presented in Table 2. Then we conduct experiments on health testers and those with myocardial ischemia (MI) respectively, the CCR is shown in Table 3.

In addition, FAR and FRR are also used to evaluate the performance of the human identification method proposed in this paper, the formulas of FAR and FRR is as

Table 2 Human identification performance of different features

Experiments	Training samples	Test samples	CCR (%)
Exp 1	330	110	65.45
Exp 2	330	110	78.18
Exp 3	330	110	86.36

Table 3 Human identification performance of different individuals

Experiments	Training samples	Test samples	CCR (%)
healthy individuals	165	55	94.55
individuals with MI	165	55	81.82
total	330	110	88.18

Table 4 The results of FAR and FRR

Indicator	DB1 (%)	DB2 (%)
FAR	0.67	4.67
FRR	0.76	3.50

follows, and 300 sets of data different from the above individual are used to calculate FAR. 200 sets of data are extracted from the 440 sets of data in the original database to calculate FRR. In order to test the impact of MI on human identification, FAR and FRR were calculated in two databases. The 132 sets of data from healthy individuals are included in the DB1, the 132 sets of data from healthy individuals and 68 sets of data from individuals with MI are included in the DB2. The results of FAR and FRR are shown in Table 4.

$$\begin{cases} FAR = NFA \div NIRA \times 100\% \\ FRR = NFR \div NGRA \times 100\% \end{cases} \tag{13}$$

where NIRA is the total number of tests between classes, NGRA is the total number of tests within the class, and NFA and NFR are the number of false rejections and the number of incorrect acceptances respectively.

It can be seen from Table 2 that the CCR of Exp2 is higher than that of the Exp1, and the CCR of the Exp3 is the highest. From the Table 3, we can conclude that the CCR of healthy individuals is higher than that of patients with MI, and Comparing Tables 2 and 3, we can see that for the same number of testing samples, the more healthy individuals there are, the higher the recognition accuracy is. The time required for each set of experiments we conducted was within 2 s. Therefore, compared with some current experiments, this paper adopts the patients to perform the experiment, which makes the experiment more universal, and the CCR is higher than the method used in [1], the recognition speed is within 2 s, it's very fast.

6 Conclusions and Future Work

Biometrics recognition is the key to solving many problems, including information security, access control, authorization, digital and online usage. In addition to using biometric technology such as biometric fingerprint, iris and face recognition, there is a promising way to identify multiple factors. Current research mainly focuses on the use of ECG data for human identification. In this paper, we introduced a human identification method base on CDG via DL, we can conclude that the accuracy was improved significantly by conducting experiments with the two features, and the CCR was increasing with the increase of the healthy individuals in the experiment. However, the data and features we used are too little, we will collect more ECG data and extract more distinguishing characteristics to improve the accuracy in future work.

References

1. L. Biel, O. Pettersson, L. Philipson, P. Wide, Ecg analysis: a new approach in human identification. IEEE Trans. Instrum. Measur. **50**(3), 808–812 (2001)
2. M. Kyoso, A. Uchiyama, Development of an ecg identification system, in *Proceedings of the 23rd Annual International Conference of the IEEE Engineering in Medicine and Biology Society*, vol. 4. IEEE, pp. 3721–3723 (2001)
3. C. Ye, M.T. Coimbra, B.V. Kumar, Investigation of human identification using two-lead electrocardiogram (ecg) signals, in *2010 Fourth IEEE International Conference on Biometrics: Theory Applications and Systems (BTAS)*. IEEE, pp. 1–8 (2010)
4. T. Liu, C. Wang, D.J. Hill, Deterministic learning and rapid dynamical pattern recognition of discrete-time systems, in *IEEE International Symposium on Intelligent Control, ISIC*. IEEE, pp. 1091–1096 (2008)
5. C. Wang, D.J. Hill, Learning from neural control. IEEE Trans. Neural Netw. **17**(1), 130–146 (2006)
6. C. Wang, T.-R. Chen, T.-F. Liu, Deterministic learning and data-based modeling and control. Acta Automatica Sin. **35**(6), 693–706 (2009)
7. C. Wang, D.J. Hill, *Deterministic Learning Theory for Identification, Recognition, and Control*, vol. 32 (CRC Press, 2009)
8. J. Irvine, B. Wiederhold, L. Gavshon, S. Israel, S. McGehee, R. Meyer, M. Wiederhold, Heart rate variability: a new biometric for human identification, in *Proceedings of the International Conference on Artificial Intelligence (IC-AI'01)*, pp. 1106–1111 (2001)
9. A.K. Jain, A. Ross, S. Prabhakar, An introduction to biometric recognition. IEEE Trans. Circ. Syst. Video Technol. **14**(1), 4–20 (2004)
10. P. Das, Wavelets and some applications, in *Differential Equations and Dynamical Systems*, p. 1, (2005)
11. J. Kors, G. Van Herpen, A. Sittig, J. Van Bemmel, Reconstruction of the frank vectorcardiogram from standard electrocardiographic leads: diagnostic comparison of different methods. Eur. Heart J. **11**(12), 1083–1092 (1990)
12. D. Cortez, N. Sharma, C. Devers, E. Devers, T.T. Schlegel, Visual transform applications for estimating the spatial qrs-t angle from the conventional 12-lead ecg: Kors is still most frank. J. Electrocardiol. **47**(1), 12–19 (2014)
13. A. Wolf, J.B. Swift, H.L. Swinney, J.A. Vastano, Determining Lyapunov exponents from a time series. Phys. D: Nonlinear Phenom. **16**(3), 285–317 (1985)
14. L.B. Almeida, The fractional Fourier transform and time-frequency representations. IEEE Trans. Sig. Process. **42**(11), 3084–3091 (1994)
15. Y. Wan, J. Yao, et al., A neural network to identify human subjects with electrocardiogram signals, in *Proceedings of the World Congress on Engineering and Computer Science*, pp. 1–4 (2008)
16. V. Vapnik, *The Nature of Statistical Learning Theory* (Springer science & business media, 2013)

Estimating the Perturbation Origin in Networked Dynamical Systems with Sparse Observation

Chaoyi Shi, Qi Zhang and Tianguang Chu

Abstract We consider the problem of estimating the location of the source and the start time of the perturbation diffusion in networked dynamical systems, under the condition that only a subset of nodes can be observed. A maximum likelihood (ML) estimator is formulated, which taking advantage of the linear correlation between the time at which a node receives the perturbation and its delay time from the source. Experiments verify the effectiveness of the proposed algorithm in scale-free (BA) and small-world (WS) networks.

Keywords Perturbation diffusion · Sparse observation · Maximized likelihood (ML) estimator · Correlation coefficient

1 Introduction

Recent years have witnessed major advances in study of diffusion processes over networks in different fields, such as the epidemiology, bioinformatics, communication systems, and sociology. In particular, many studies focus on the forward problem of understanding the diffusion process on various networks [1, 2]. Here, we intend to consider the inverse problem of inferring the origin of perturbation diffusion in complex networks when the observation data gathered at partial nodes are available in a network consisting of large amount of individuals or nodes. Evidently, finding location of the source or origin of diffusion is crucial in interfering or controlling the spread of disturbances or detrimental influences in a network and has potential application in many interesting problems [3, 4].

C. Shi · T. Chu (✉)
College of Engineering, Peking University, Beijing 100871, China
e-mail: chutg@pku.edu.cn

Q. Zhang
School of Information Technology & Management,
University of International Business & Economics, Beijing 100029, China

Y. Jia et al. (eds.), *Proceedings of 2018 Chinese Intelligent Systems Conference*, Lecture Notes in Electrical Engineering 529,
https://doi.org/10.1007/978-981-13-2291-4_64

In [5, 6], the location of diffusion source is determined based on the measurement of centricity, including the degree, betweenness, closeness, eigenvector and rumor centrality, under the condition of complete observation of the state of all nodes in a network. A probabilistic algorithm was also developed based on the dynamic message passing equations in [7]. However, the complete observation conditions are severe in applications especially for the case of large scale networks where the costs for observing all the nodes are very expensive. To overcome this difficulty, several methods were proposed for source identification based on incomplete observations where only a subset of the nodes are observed. For example, Ref. [8] proposed an effective distance-based method to identify the source in weighted networks based on wavefront observations, and [9] introduced the Jordan centrality as a metric to be maximized for the infection graph. In addition, a two-stage algorithm was proposed in [10–12], which first identifies the candidate sources based on reverse dissemination method, and then locates the source using a maximum likelihood (ML) estimator.

In this paper, we consider the source identification problem of perturbation diffusion in a dynamic system with transmission delay, where only a small subset of nodes can be monitored by sensors recording the infection time of the nodes. Our aim is to locate the source and estimate the start time of perturbation diffusion. Similar problems have been studied by several heuristic methods in literature, including the Gaussian method [13], time-reversal backward spreading algorithm [14], and importance sampling [15], etc. However, the complexity of identification algorithms mentioned above is usually very high under partial observations because almost all nodes in the networks are potential sources. Here, we will present a source identification method by calculating the correlation coefficient of the infection time of a node and its delay time from the source, and perform numerical simulations on different networks to verify the effectiveness of the method.

The rest of this paper is organized as follows. Section 2 gives preliminaries and the problem formulation. Section 3 presents the method for solving the ML estimator, and Sect. 4 offers numerical simulation results. Finally, Sect. 5 concludes the paper.

2 Problem Formulation

In this section, we first introduce preliminary knowledge, including the dynamical system, perturbation diffusion and sparse observations, and then formulate the source identification problem as a maximum likelihood estimator.

2.1 Networked Dynamical System

We consider a system of n agents (nodes) linked via a digraph $G = (V, E, A)$, where $V = \{v_1, \ldots, v_n\}$ is a set of nodes, $E \subset V \times V$ is a set of edges, and an adjacency matrix $A = [a_{ij}] \in R^{n \times n}$ contains elements of 0 or 1. The node indexes belong to a finite index set $I = \{1, \ldots, n\}$. The set of neighbors of node v_i is denoted by $N_i = \{v_j \in V : (v_i, v_j) \in E\}$. Moreover, we assume $a_{ii} = 0$ for all $i \in I$.

The networked dynamical system is modeled as follows: at time t, each node is characterized by a time dependent state $x_i(t)$, $i = 1, \ldots, n$, a networked system with communication delay is driven by

$$\dot{x}_i(t) = \sum_{v_j \in N_i} a_{ij}(x_j(t - T_{ij}) - x_i(t)), \quad i \in I, \tag{1}$$

where T_{ij} represents the transmission delay along the link between node v_i and node v_j, and the right-hand side of Eq. (1) captures the interactions of v_i with its neighbors N_i.

2.2 Perturbation Diffusion

A source node s initiates the perturbation diffusion at time t_s. We track the propagation of the signal by introducing a permanent perturbation dx_s on the steady-state activity of a source node s, observing the response of all remaining nodes and recording the time at which the state of node deviates from the steady state. Let t_i is the time at which the node v_i receives the perturbation signal, $t_i \in [0, T] \cup \{+\infty\}$, where T is the upper bound of observation window, and $+\infty$ denotes that the node has not received the perturbation signal during the observation window. Figure 1 shows an example of the perturbation diffusion in a network. The network topology with the source node labeled by S and the transmission delays along links are depicted in Fig. 1a. At time 5, we induce a permanent perturbation on node S, the rest of the nodes $\{A, B, C, D, E, F\}$ receive the perturbation at time $\{10, 10, 9, 7, 8, +\infty\}$, as shown in Fig. 1b.

2.3 Observation

In real-world scenario, the location of the source and the start time of perturbation diffusion are unlikely to directly obtain, since the states of some nodes are hidden and observing the states of all the nodes are expensive. Here only the partial nodes were observed by sensors. Let $O = \{o_1, o_2, o_3, \ldots, o_K\}$ denote the set of K observers, the localization and perturbation time of any observer $o_j \in O$ are known. The observation set is composed of tuples of the locations and time measurements, i.e., $\mathcal{O} = \{(o_j, t_{o_j})\}_{j=1}^{k}$, where observer $o_j \in O$ receives the perturbation at time t_{o_j}. Our objective is to construct an estimator to identify the source node s and the start time t_s of perturbation diffusion based on the partial observations $\mathcal{O}$ and the network topology G.

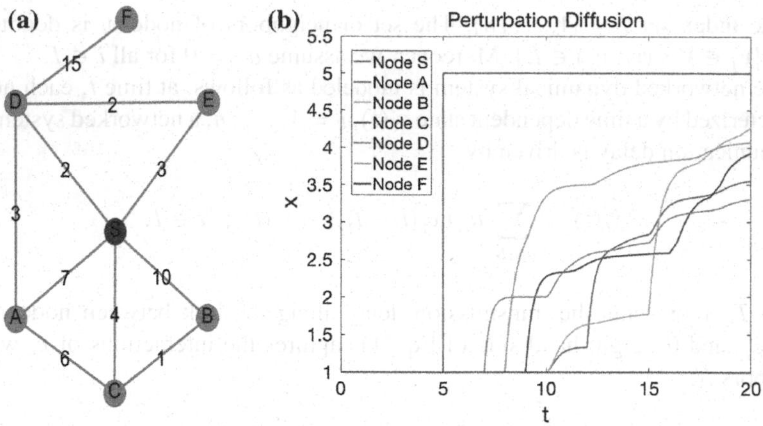

Fig. 1 An example of the perturbation diffusion in a network: **a** the network topology with the source node labeled by S and transmission delays along links; **b** the source node S begin to spread a permanent perturbation at time 5, the rest of the nodes $\{A, B, C, D, E, F\}$ receive the perturbation at time $\{10, 10, 9, 7, 8, +\infty\}$

2.4 Maximum Likelihood Estimator

Each node $v_i \in V$ has the same prior probability to be a source. A maximized likelihood (ML) estimator $\hat{s}$ of the random variable s^* is designed to locate the source and estimate the start time such that the joint probability of the observations $\mathcal{O}$ is maximized. By definition, the ML estimator is rewritten as

$$\hat{s} = arg \max_{v_i \in V} \max_{t_s \in (-\infty, \min_{o_j \in O} t_{o_j}]} P(\mathcal{O}|s^* = v_i, t_s), \quad (2)$$

where $P(\mathcal{O}|s^* = v_i, t_s)$ is the joint probability of the observations $\mathcal{O}$ assuming v_i is the source, t_s is the start time of perturbation diffusion, and $t_s \le \min_{o_j \in O} t_{o_j}$.

The computation of ML estimator (2) is difficult and complex. In the following, we will design algorithms to efficiently optimize the estimator (2).

3 Method

A diffusion process is initiated by a single node s at an unknown time point t_s. Let $d(s, o_j)$ be the shortest transmission delay between node $s \in V$ and $o_j \in O$, the observer o_j receives a perturbation at time $t_{o_j} = t_s + d(s, o_j)$, which shows that the time t_{o_j} is linear with shortest transmission delay $d(s, o_j)$. Thus, the ML estimator is defined as maximizing the correlation coefficient between the shortest transmission delays and the times at which observers receive the perturbation, the specific method is as follows and the estimation process is shown in Algorithm 1.

(i) Obtain the vector $\mathbf{T} = \{t_{o_1}, t_{o_2}, \ldots, t_{o_k}\}^T$ recording the times when observers receive the perturbation.
(ii) Define the vector of transmission delays from node v_i to all other observers

$$\mathbf{D}(v_i) = \{d(v_i, o_1), d(v_i, o_2), \ldots, d(v_i, o_K)\}^T,$$

where v_i is a node that initiates the diffusion of perturbation, and $d(v_i, o_j)$ represents the shortest transmission delay between the node v_i and the observer o_j.
(iii) Compute the correlation coefficient $Cov(\mathbf{D}(v_i), \mathbf{T})$ of the arrival time $\mathbf{T}$ and the shortest transmission delays $\mathbf{D}(v_i)$ for any node v_i.
(iv) Locate the source node by the estimator $\hat{s} = arg\max_{v_i \in V} Cov(\mathbf{D}(v_i), \mathbf{T})$.
(v) Compute the start time $t_s = \frac{\sum_{o_i \in O}(t_{o_i} - d(\hat{s}, o_i))}{K}$.

Algorithm 1 Estimating the Perturbation Origin

Input: $G, \mathcal{O}, \mathbf{T}$
Output: $\hat{s}, t_s$
  $\hat{s} \leftarrow \{\}, \rho \leftarrow -\infty$
for $v_i \in V$ **do**
  for each observer $o_j \in O$ **do**
    Calculate the shortest transmission delay $d(v_i, o_j)$
  end for
  Calculate the correlation coefficient between $\mathbf{D}(v_i)$ and $\mathbf{T}$ as follows:
  $$\hat{\rho} = Cov(\mathbf{D}(v_i), \mathbf{T})$$
  if $\hat{\rho} < \rho$ **then**
    $\hat{s} \leftarrow v_i, \rho \leftarrow \hat{\rho}$
  end if
end for
Compute the start time as follows:
$$t_s = \frac{\sum_{o_j \in O}(t_{o_j} - d(\hat{s}, o_j))}{k}$$
Return the estimated diffusion source $\hat{s}$ and the estimated start time t_s.

4 Simulation

We simulate two types of networks including scale-free (BA) and small-world (WS) networks. Each edge $(v_i, v_j) \in E$ is assigned with a delay $d(v_i, v_j)$, which is uniformly distributed in the range of $[1, 5]$. Firstly, a node is randomly selected to propagate the perturbation by using Eq. (1). The observations $\mathcal{O}$ are obtained with sample ratio γ, which indicate that each node can be observed with probability γ. Suppose the observations used in the simulations are nonempty. Then, we can locate

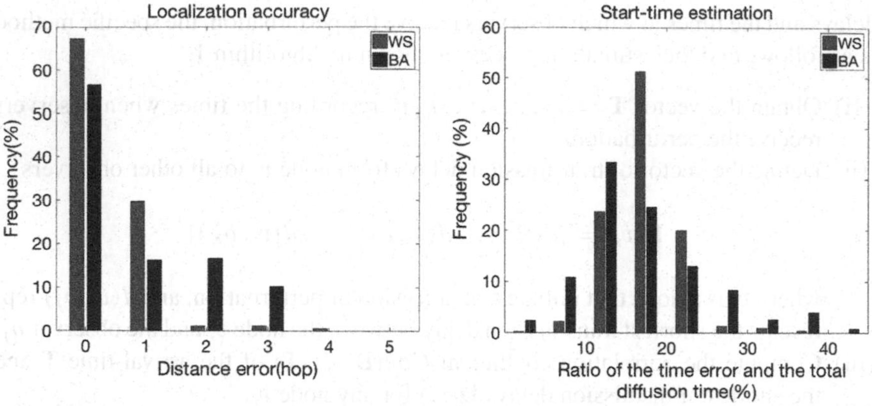

Fig. 2 A histogram of the location estimation (left) and start-time estimation (right) on BA and WS networks

the source and estimate the start time by using algorithm 1. Finally, the performance metric is selected, our approach achieves high accuracy in different networks.

The scale-free and small-world networks with the number of nodes $N = 100$ and the average node degree $\langle k \rangle = 4$. We repeated the simulation 500 times to get the location and time estimation accuracy. Figure 2(left) shows that the estimator locates the source with a distance error at most 3 hops while observing only 10% of nodes. The diffusion start time t_s is unknown, we measure the time estimation using time ratio of the time error and the total diffusion time, where the time error is the number of time units between the estimated start time and the actual t_s. Figure 2(right) shows the estimated start time is only 20% away from the initial time with a probability of 76 and 71%.

5 Conclusion

We presented a ML estimator for identification of perturbation source in networked dynamical systems with partially-observed nodes. The maximum likelihood of each candidate source node is calculated by using the linear correlation between the perturbation times of observers and their transmission delays from the source. Simulation results indicate that our algorithm is efficient with high detection probability for source identification and start time estimation in different network topologies.

Acknowledgements This work was supported by NSFC (No. 61673027), National Basic Research Program of China (973 Program, No. 2012CB821200).

References

1. V. Colizza, R. Pastor-Satorras, A. Vespignani, Reaction-diffusion processes and metapopulation models in heterogeneous networks. Nat. Phys. **3**(4), 276 (2007)
2. M. Tizzoni, P. Bajardi, A. Decuyper, G.K.K. King, C.M. Schneider, V. Blondel, Z. Smoreda, M.C. González, V. Colizza, On the use of human mobility proxies for modeling epidemics. PLoS Comput. Biol. **10**(7), e1003716 (2014)
3. Y. Jia, Robust control with decoupling performance for steering and traction of 4ws vehicles under velocity-varying motion. IEEE Trans. Control Syst. Technol. **8**(3), 554–569 (2000)
4. Y. Jia, Alternative proofs for improved lmi representations for the analysis and the design of continuous-time systems with polytopic type uncertainty: a predictive approach. IEEE Trans. Autom. Control **48**(8), 1413–1416 (2003)
5. C.H. Comin, L. da Fontoura Costa, Identifying the starting point of a spreading process in complex networks. Phys. Rev. E **84**(5), 056105 (2011)
6. D. Shah, T. Zaman, Detecting sources of computer viruses in networks: theory and experiment. ACM SIGMETRICS Perform. Eval. Rev. **38**, 203–214 (2010)
7. D. Shah, T. Zaman, Rumors in a network: who's the culprit? IEEE Trans. Inf. Theory **57**(8), 5163–5181 (2011)
8. D. Brockmann, D. Helbing, The hidden geometry of complex, network-driven contagion phenomena. Science **342**(6164), 1337–1342 (2013)
9. K. Zhu, L. Ying, A robust information source estimator with sparse observations. Comput. Soc. Netw. **1**(3), 1–21 (2014)
10. A. Louni, K. Subbalakshmi, A two-stage algorithm to estimate the source of information diffusion in social media networks, in *INFOCOM Workshops*, pp. 329–333 (2014)
11. J. Jiang, W. Sheng, S. Yu, Y. Xiang, W. Zhou, Rumor source identification in social networks with time-varying topology. IEEE Trans. Dependable Sec. Comput. **15**, 166–179 (2016)
12. C. Shi, Q. Zhang, T. Chu, Source identification of network diffusion processes with partial observations, in *Chinese Control Conference*, pp. 11296–11300 (2017)
13. P.C. Pinto, P. Thiran, M. Vetterli, Locating the source of diffusion in large-scale networks. Phys. Rev. Lett. **109**(6), 068702 (2012)
14. Z. Shen, S. Cao, W.X. Wang, Z. Di, H.E. Stanley, Locating the source of diffusion in complex networks by time-reversal backward spreading. Phys. Rev. E **93**(3), 032301 (2016)
15. M. Farajtabar, M. Gomez-Rodriguez, N. Du, M. Zamani, H. Zha, L. Song, Back to the past: Source identification in diffusion networks from partially observed cascades, in *Artificial Intelligence and Statistics* (2015)

The Modeling and Implementation of Non-rigid Motion of Carcasses

Shutian Fang, Zhehao Xu, Hanqi He and Qiyun Zheng

Abstract The unmanned aerial vehicle (UAV) air show is most popular today, but it has drawbacks when displaying the image. Aiming to investigate the image scatter sample, based on bionic movement, this article uses the dragon as an experimental object to find the multi-link device rotation system such as 3d rotation transformations. Ultimately, considering the skeleton's characteristics, the methods used are to construct both complex and non-rigid images and both three-dimensional motion and plane trajectory models. The result of the MATLAB simulation is vivid, and thus, the method discussed in this article provides a general solution for unmanned aerial vehicle (UAV) air shows.

Keywords UAV aerial show · Non-rigid motion · Carcass traits
Three-dimensional motion model · Plane trajectory model

1 Introduction

Since Intel Corporation has achieved the Intel® Shooting Star ™ [1] for flight shows and created a stunning nighttime aerial landscape, the flight control system of the unmanned aerial vehicle (UAV) group is widely utilized in the arts, entertainment, and at festivals. The success of the air show is mostly accredited to the precise control of the graphic shave, the various ways of presenting artistic visual effects and the functional features of environmental protection.

S. Fang (✉) · Z. Xu · H. He · Q. Zheng
YK PAO School, Shanghai, China
e-mail: s17515@stu.ykpaoschool.cn

Z. Xu
e-mail: s13036@stu.ykpaoschool.cn

H. He
e-mail: s13015@stu.ykpaoschool.cn

Q. Zheng
e-mail: s15583@stu.ykpaoschool.cn

Y. Jia et al. (eds.), *Proceedings of 2018 Chinese Intelligent Systems Conference*, Lecture Notes in Electrical Engineering 529,
https://doi.org/10.1007/978-981-13-2291-4_65

The core technology is designing and operating UAVs flight according to the coordination of every point and motion functions in the image if it were put in a coordinate system. By doing so, most of the motion models, the analysis of image characters, can be materialized.

In real life, some common UAV air show design images are two-dimensional, such as the alphabet, numbers, etc., and it is rare to have irregular figures displayed. Furthermore, irregular and three-dimensional schemes simply have not been reported yet.

The foundation of this model is to investigate the skeletal characteristics and the complex motion of non-rigid images in a horizontal trajectory of carcass traits. Taking the bionic movement of the dragon, as an example, based on the foundation, this article discusses how to build models of non-rigid and three-dimensional objects moving in a trajectory. And then the models are processed in MATLAB. One achievement is that the model can be applied to any image of an unmanned aerial vehicle (UAV) air show.

Moreover, the critical technology of the scheme includes image preprocessing, fixed-point locating, skeleton constructing, skin adhering, flat trajectory and 3D rotation trajectory analyzation, etc.

2 Image Preprocessing

In this article, the purpose of image preprocessing is to remove the noise and edge glitches from the original input image, thus obtaining a clear image edge and better scatter sampling. We can therefore determine the position and quantity of the UAV motion. The result by processing the output image through the steps is shown in Fig. 1 and the steps are listed below.

Step 1: Transform the input images into a grayscale image [3], and acquire a binary image by using the threshold transformation method [4], and then invert the binary image;

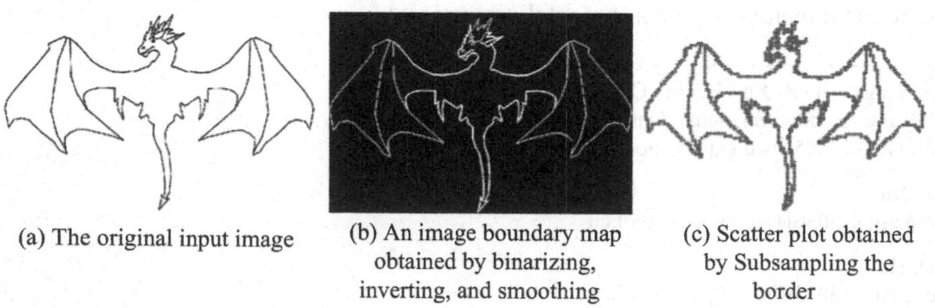

(a) The original input image

(b) An image boundary map obtained by binarizing, inverting, and smoothing

(c) Scatter plot obtained by Subsampling the border

Fig. 1 Effect of image preprocessing

Step 2: Use the linear spatial filter function [5, 6] to filter the image and remove the noise points. Use the smoothing function [7] to remove the glitches at the edges of the image;
Step 3: Use the Gonzalez's boundary function [8, 9] to extract the contour in a binary image;
Step 4: Use the Gonzales bsubsamp function [10] to subsample the outline boundaries and obtain the scatter distribution on the image boundaries.

3 The Modeling of Non-rigid Motion in Complex Image Plane

The dragon is an irregular object. The movement of various joints, which is a typical non-rigid motion, makes the moving pattern more complicated to analyze.

In order to solve this problem, we first simplify the linkage of the dragon's bones and joints to a multi-link system [11]. Therefore, the next key issue is to select the key points in the skeleton and to re-establish the multi-link system. Before constructing the movement model, we need to determine the plane coordinate system of this object.

To properly manage the dragon's movement in the air, we chose the graphic center of the dragon's body as a fixed point. Then we established the 2-dimensional coordinate system with this fixed point as the origin and each point in the outline is calculated and shown as in Fig. 2a.

4 The Moving Trajectory of Non-rigid Motion

We divide the planar motion model into two implements. First, it sets up a multi-linked rotation model for the skeleton trajectory. Based on this, the complete two-

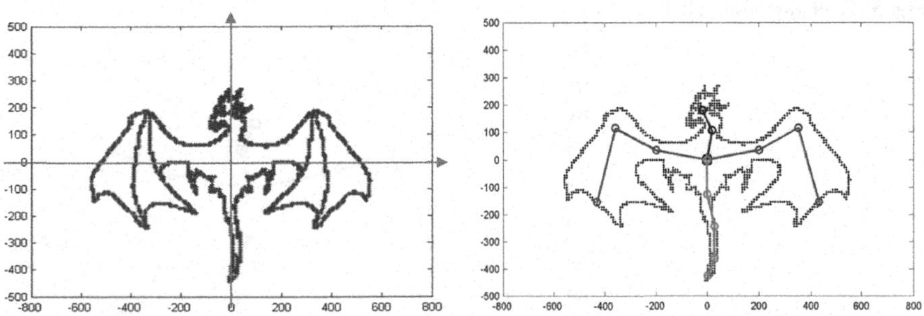

(a) Dragon's outline coordinates (b) The key point of the dragon and its skeleton

Fig. 2 Outline points and skeleton of the dragon

dimensional trajectory of complex images is constructed by implementing adhesion for the skin. The other steps, to achieve the trajectories of skeleton motion, are listed below.

Step 1: Pin the 13 key points in the coordinate system and set up coordinates graph for dragon skeleton: points that are related to head movement, trunk movement, wing movement and tail movement. Then, the points are collected in a chart. The coordinate chart is shown in Fig. 2b, Table 1. The figure, comprising key points and coordinates, is shown in Fig. 3.

Step 2: Separate every motion and render the motion track. Taking the right wing, for instance, the detailed process is shown here:

As shown in Fig. 4, the dragon's right wing has 4 key points A_0, A_1, A_2, A_3 which forms a three-joint link system, and takes A_0 as the fixed point. First, A_1, A_2, A_3 rotates around A_0 and this is defined as movement Z_1; then, A_2, A_3 makes a rotation around A_1 which is defined as movement Z_2; finally, A_3 makes a rotation around A_2 which is defined as the movement. A new set of coordinates A_1, A_2, A_3 is marked as A_1', A_2', A_3'. The mathematical relationship between them can be described as:

$$\begin{cases} A_1' - A_0 = Z_1(A_1 - A_0) \\ A_2' - A_1' = Z_2 Z_1(A_2 - A_1) \\ A_3' - A_2' = Z_3 Z_2 Z_1(A_3 - A_2) \end{cases} \tag{1}$$

In this case, $Z_i (i = 1, 2, 3)$ is the rotation-variation matrix [12], called:

Table 1 The sequence of key points involved in the movement of each component

Sports branch	Key point sequence
Head movement	0-1-2
Body movement	0-1, 0-3
Left wing movement	0-7-8-9
Right wing movement	0-10-11-12
Tail movement	0-3-4-5-6

Fig. 3 Keel key point label

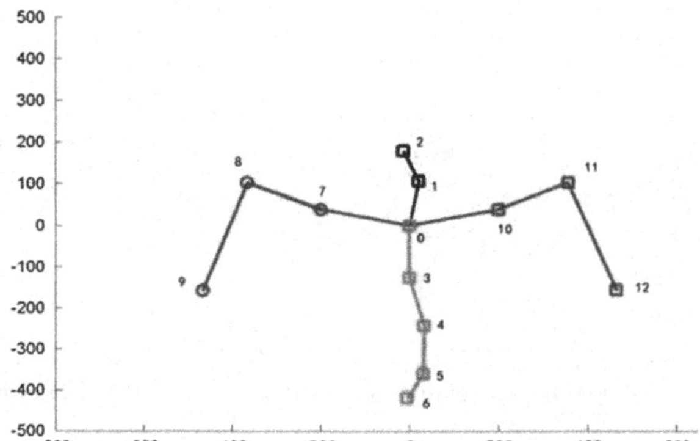

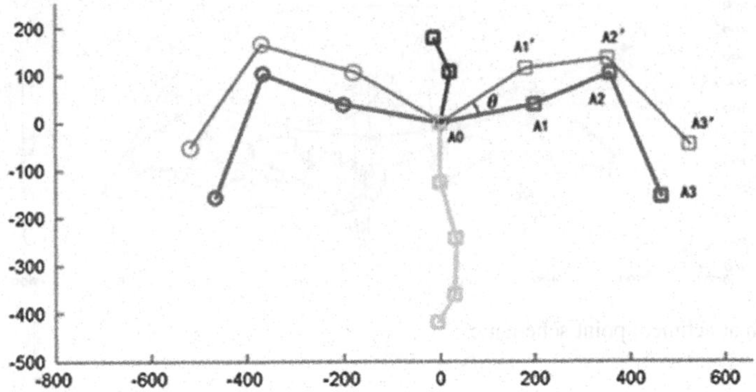

Fig. 4 The rotative diagram of right-wing skeleton

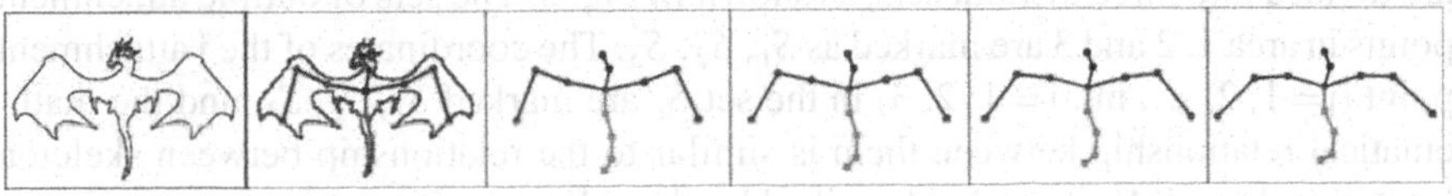

Fig. 5 Dragon skeleton kinematic trajectory

$$Z_i = \begin{bmatrix} \cos\theta_i & -\sin\theta_i \\ \sin\theta_i & \cos\theta_i \end{bmatrix} \tag{2}$$

In this case, θ_i is an angle of counterclockwise rotation. The motion track of the key points can be derived by (2)

$$\begin{cases} A_1' = Z_1(A_1 - A_0) + A_0 \\ A_2' = Z_2Z_1(A_2 - A_1) + Z_1(A_1 - A_0) + A_0 \\ A_3' = Z_3Z_2Z_1(A_3 - A_2) + Z_2Z_1(A_2 - A_1) + Z_1(A_1 - A_0) + A_0 \end{cases} \tag{3}$$

Step 3: Following a similar procedure as step 2, each motion of the head, body, and wings are deduced. The motion of the synthesized skeleton is then obtained, as shown in Fig. 5.

5 Skin Attachment Model

We define the drones which synchronize the movement of the skeleton as the outline attachment point of the skeleton. Since the movement of the skeleton and its outline attachment points are synchronized, setting attachment points for each skeleton will enable the movement of the outline. Taking the right wing, for example, we show the motion of its outline below.

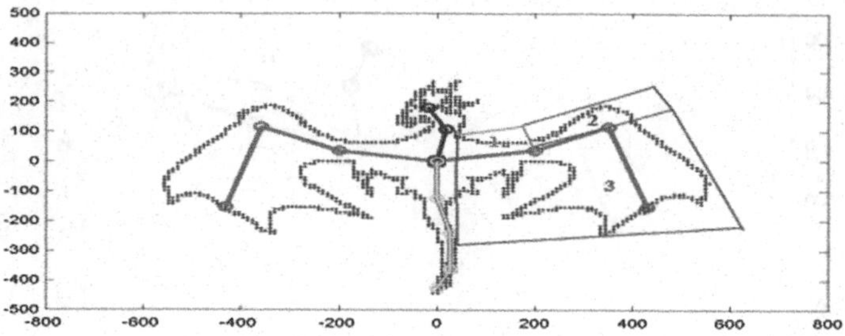

Fig. 6 Skin attachment point schematic

Thanks to the coverage of each skeleton, the right wing's skin attachment points are divided into three disjoint sets, as shown in Fig. 6. The sets of outline attachment points in area 1, 2 and 3 are marked as S_1, S_2, S_3. The coordinates of the j attachment point (j = 1, 2, ..., mi, i = 1, 2, 3) in the set S_i are marked S_1, S_2, S_3, and the mathematical relationship between them is similar to the relationship between skeleton key point A_i and A_i', it can be described by the following formula:

$$\tilde{x}_{ij} = \prod_{\tau=1}^{i} Z_\tau (x_{ij} - A_{i-1}) + \sum_{k=1}^{i-1} \prod_{\tau=1}^{k} Z_\tau (A_k - A_{k-1}) + x_0 \tag{4}$$

By implementing this method to the body part and the left wing, it can obtain one motion trajectory image of this dragon. And since then, we have accomplished the investigation and modeling of non-rigid motion of an irregular shape in the two-dimensional plane.

6 The 3D Motion Model

After the investigation of the 2-dimensional motional model, we further extended the movement of this western dragon into three-dimensional space. Having a thorough consideration that the dragon image is relatively complicated, we split the dragon into the body, left wing, right wing, and tail. We initialized the position point of the dragon as the yoz plane, as shown in Fig. 7, and assume that the body of the dragon is placed in the yoz coordinate system. The wings' movement can be defined as rotational movements with Z as their axis. This three-dimensional movement is therefore animating the air show.

In the 3D motion model, above, we convert the 2-dimensional coordinates of the outline points into 3-dimensional coordinates. Because the original plane of the dragon is yoz, 2D to 3D conversion can be achieved by the formula of:

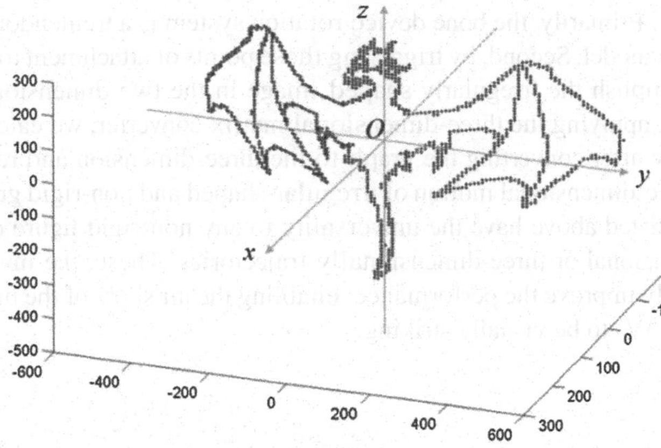

Fig. 7 Dragon in three-dimensional space diagram

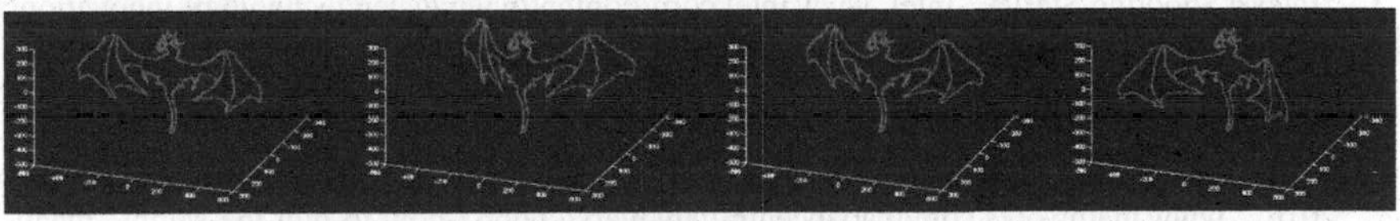

Fig. 8 Dragon's trajectory dynamic diagram in three-dimensional space motion

$$R^2 \rightarrow R^3$$
$$(x, y) \rightarrow (0, x, y) \tag{5}$$

Assuming that a random point $P(x, y, z)$ on the wings is rotated by θ degrees around z-axes and the coordinate is $\tilde{P}(\tilde{x}, \tilde{y}, \tilde{z})$, then the relationship between them can be expressed as

$$\begin{pmatrix} \tilde{x} \\ \tilde{y} \\ \tilde{z} \end{pmatrix} = \begin{pmatrix} \cos\theta & -\sin\theta & 0 \\ \sin\theta & \cos\theta & 0 \\ 0 & 0 & 1 \end{pmatrix} \begin{pmatrix} x \\ y \\ z \end{pmatrix} \tag{6}$$

According to (6), we can calculate the track of the dragon's wings during the movement. The simulation in MATLAB is shown in Fig. 8.

7 Conclusions

To achieve the model for an irregular and non-rigid figure, we first scatter the image contours obtained through the pretreatment of samples. After that, we establish a plane coordinate graph and select several key points that are closely related to the graph movement. Primarily, the bone device rotation system is a tremendous step to the frame motion model. Second, by triggering those points of attachment to the skin, we further accomplish the irregularly shaped image in the two-dimensional plane trajectory. Third, applying the three-dimensional matrix converter, we calculate the motion trajectory after converting the graph to the three-dimension and realize the model of the three-dimensional motion of irregular-shaped and non-rigid graph. The models that are listed above have the universality to any non-rigid figure complex, either two-dimensional or three-dimensionally trajectories. These, the investigated models, massively improve the performance, enabling the air show of the unmanned aerial vehicle (UAV) to be visually striking.

References

1. "Intel® Shooting Star™." Intel, www.intel.com/content/www/us/en/technology-innovation/aerial-technology-light-show.html. Accessed 18 Nov 2017
2. A.R. Dill, M.D. Levine, in *Non-Rigid Body Motion*. http://graphicsinterface.org/wp-content/uploads/gi1985-55.pdf ed., Montreal. Quebec, McGill University
3. Agustin, *How to Plot the Sample Mean in a Scatter Plot?* MathWorks, 4 Aug 2016, www.mathworks.com/matlabcentral/answers/266667-how-to-plot-the-sample-mean-in-a-scatter-plot? Accessed 18 Nov 2017
4. P. van Den Driessche, J. Watmough, Reproduction numbers and sub-threshold endemic equilibria for compartmental models of disease transmission, in *Reproduction Numbers and Subthreshold Endemic Equilibria for Compartmental Models of Disease Transmission, Digital*, pp. 1–21
5. E. Esser, UCI interdisciplinary computational and applied mathematics program, in *Linear Spatial Filtering*, ed. by S. Ahn, PDF, UCI. Excerpt originally published in ICAMP, pp. 1–4 (2014)
6. T. Fletcher, (ed.), in *Spatial Filtering* (Image Processing Basics). www.coe.utah.edu, Utah, 31 Jan 2012, www.coe.utah.edu/~cs4640/slides/Lecture5.pdf. Accessed 18 Nov 2017
7. T. Finnigan, in *Blending Function*. Exploring Blending Function, PDF ed., pp. 39–44
8. "Function B = Boundaries(BW, Conn, Dir)." Fourier.eng.hmc, 9 Mar. 2015, www.fourier.eng.hmc.edu/e161/dipum/boundaries.m. Accessed 18 Nov 2017
9. A. Bultheel, E. Hendriksen, Gonzales's Boundaries Function. Orthogonal Rational Functions, 4th ed. (Cambridge, Cambridge University, 2007), pp. 155–172
10. "Function B = Bsubsamp(A, Conn, Dir)." Fourier.eng.hmc, 9 Mar. 2015, www.fourier.ng.hmc.edu/e161/dipum/boundaries.m. Accessed 18 Nov 2017
11. J.E. Lagnese, et al., in *Splitting of Eigenvalues*, 34 vols. (Boston, Birkhäuser Boston, 2012)
12. J. Ramsay, B.W. Silverman, in *Visualizing the Results*. PDF ed. (New York, Springer New York, 2006)

Dynamic Modeling of a Variable Structure Two-Wheeled Robot During the Mode Switching Process Between Segway Mode and Bicycle Mode

Lei Guo, Hongquan Wu and Yuan Song

Abstract The variable structure two-wheeled robot (VSTWR) is a nonlinear system. The dynamic model between the VSTWR during the mode switching process of Segway mode and Bicycle mode is proposed in the paper. According to kinematic and energy analysis, dynamical models based on Appell equations and Chaplygin equations are built respectively. The computer numerical simulation based on Matlab is achieved. And the validity of the two dynamic models are testified by the simulation.

Keywords Mode switching · Appell equations · Chaplygin equations
Matlab simulation

1 Introduction

The VSTWR is an important research object in robotics. Compared with other robots, it requires higher demands for system modeling and attitude controlling. It can switching between Segway mode and Bicycle Mode. These two different modes are complementary in performance and application scenarios. For example, a Segway is suitable be applied in flat and open road conditions for low-speed cruising. Conversely, bicycle is more suitable for a winding and narrow road condition for high-speed flexible driving. Therefore, the research on VSTWR has strong theoretical significance and practical value.

Due to the importance of dynamical models for simulation and control, several methods have been developed to carry out dynamic analysis of Bicycle robot and Segway robot, in order to achieve good performances in terms of self-balancing. Those methods are range from Lagrange equations, to Appell equations or Chaplygin equations. In [1], a kind of dynamic model for Bicycle robot based on Lagrange

L. Guo · H. Wu (✉) · Y. Song
School of Automation, Beijing University of Posts and Telecommunications, Beijing 100876, China
e-mail: 451697284@qq.com

Y. Jia et al. (eds.), *Proceedings of 2018 Chinese Intelligent Systems Conference*, Lecture Notes in Electrical Engineering 529,
https://doi.org/10.1007/978-981-13-2291-4_66

equations was built. In [2], Gibbs-Appell equations was applied to build the dynamic model of Segway robot. As in [3], a dynamic model based on Chaplygin equations for unicycle robot was put forward. In terms of the control strategy of bicycle robot or Segway robot, in [4], a self-balanced bicycle that uses a flywheel to change the center of mass of the bicycle was designed. Shafiei, MH used neural networks to establish the relationship between the roll angle and the rotational angle of the bicycle, and a tracking controller was finished in [5]. Because the camber angle effect in the slip ratio of the bicycle was considered, the stability was improved in [6]. Zhao [7] proposed a self-balanced control algorithm based on PID neural network. In [8], an improved PCA error detection strategy based on the automatic adjustment of the robot's center of mass was proposed. Chih-Hui [9] proposed a position and angle decoupled intelligent back-stepping control system for two-wheel robot.

In summary, the research results on bicycle robots and Segway robots are fruitful. However, there is a large gap in the research on VSTWR. This paper focuses on the dynamic model of switching process of Segway mode and bicycle mode. And the mode switching process of VSTWR is firstly modeled by Appell method and Chaplygin method, and then the rationality of the two kind of models is compared and verified by Matlab.

2 Kinematic Analysis

The structural diagram of the VSTWR is shown in Fig. 1. As can be seen from the figure, the VSTWR is composed of five components: the frame, the left fork, the left wheel, the right fork, and the right wheel.

The following coordinate systems can be established according to the structural features of VSTWR:

(1) $o - e_1^{(0)}, e_2^{(0)}, e_3^{(0)}$ is the coordinate system fixed on the ground, and $e_3^{(0)}$ is perpendicular to the ground;

Fig. 1 VSTWR mode switching structure diagram

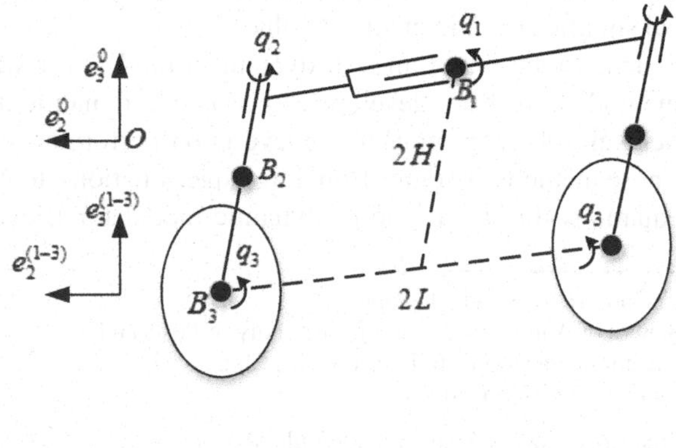

Table 1 Nomenclature of VSTWR

Parameters	Definition
$2H$	The distance between the frame center and the center of mass of the two wheels
r	Left and right wheels' radius
m_1, m_2, m_3	Masses of frame, fork and wheel
$q_1, \dot{q}_1$	Pitch angle, rate of the frame
$q_2, \dot{q}_2$	Rotational angle, rate of the left fork
$q_3, \dot{q}_3$	Rotational angle, rate of the left wheel
J_{1x}, J_{1y}, J_{1z}	The frame's moment of inertia around $e_1^{(1)}, e_2^{(1)}, e_3^{(1)}$
J_{2x}, J_{2y}, J_{2z}	The left fork's moment of inertia around $e_1^{(2)}, e_2^{(2)}, e_3^{(2)}$
J_{3x}, J_{3y}, J_{3z}	The left wheel's moment of inertia around $e_1^{(3)}, e_2^{(3)}, e_3^{(3)}$
L	The horizontal distance from B_1 to B_2
τ_2, τ_3	Torques applied on the forks, wheels
g	Gravity acceleration (9.8 kg m/s^2)

(2) $B_1 - e_1^{(1)}, e_2^{(1)}, e_3^{(1)}$ represents the coordinate system attached to the frame, and its origin is also located at the geometric center of the frame;
(3) $B_2 - e_1^{(2)}, e_2^{(2)}, e_3^{(2)}$ represents the coordinate system fixed on the center of mass of the left fork, and it is initially parallel to the inertial coordinate system;
(4) $B_3 - e_1^{(3)}, e_2^{(3)}, e_3^{(3)}$ represents the coordinate system attached to the center of mass of the left wheel, and it is initially parallel to the inertial coordinate system.

In order to facilitate kinematics analysis, the following reasonable assumptions can be made:

(1) Five components in the VSTWR are rigid bodies; (2) Two wheels do pure rolling on the ground; (3) VSTWR moves on horizontal ground.

The variables that will be used for dynamics analysis are shown in Table 1.

In order to build the dynamic model of the VSTWR, the kinematics of each component of the VSTWR must be analyzed firstly, including the translational speed and rotation speed of each component.

The angle-velocity vector of the frame can be presented as:

$$\omega_1 = (0\,\dot{q}_1\,0)^{\mathrm{T}} \tag{1}$$

The rotation transformation matrix of coordinate system $\{2\}$ to coordinate system $\{1\}$ is denoted as ${}_1^2R$, and the following transformation matrix has a similar definition. The angle-velocity vector of the left fork can be presented as:

$$\omega_2 = {}_1^2\boldsymbol{R}\omega_1 + (0\,0\,\dot{q}_2)^{\mathrm{T}}$$
$$= (\sin q_2\dot{q}_1\ \cos q_2\dot{q}_1\ \dot{q}_2)^{\mathrm{T}} \quad (2)$$

Similarly, the angle-velocity vector of the left wheel can be presented as:

$$\omega_3 = {}_2^3\boldsymbol{R}\omega_2 + (0\quad \dot{q}_3\quad 0)^{\mathrm{T}} \quad (3)$$

Because the wheels' movement on the ground is pure rolling, the velocity of the wheels' grounding point equals 0, thus the velocity of the left wheel's center of mass can be expressed as:

$$v_3 = -\omega_3 \times (0\quad 0\quad -r)^{\mathrm{T}} \quad (4)$$

Similarly, the translation speed of the frame and the left fork can be presented as:

$$v_1 = {}_2^1\boldsymbol{R}_3^2\boldsymbol{R}v_3 + \omega_1 \times (0\quad L\quad 2H)^{\mathrm{T}} \quad (5)$$
$$v_2 = {}_3^2\boldsymbol{R}v_3 + \omega_2 \times (0\quad 0\quad H)^{\mathrm{T}} \quad (6)$$

Due to the pure rolling of the wheel on the ground, non-holonomic constraints can be obtained by combining (4):

$$\dot{q}_4 = r\ \cos q_2\dot{q}_1 + r\dot{q}_3 \quad (7)$$
$$\dot{q}_5 = -r\ \sin q_2\dot{q}_1 \quad (8)$$

3 Appell Equations of the System

The standard Appell equations is shown as (9):

$$\sum_{j=1}^{m}(Q_j - \frac{\partial G}{\partial \ddot{q}_j})\delta q_j = 0, (j = 1, 2, \ldots, m) \quad (9)$$

where Q_j denotes the generalized force of the system, G denotes the Gibbs function which also known as "acceleration energy", q_j denotes the generalized variables. The Appell equations can also be represented as (10):

$$\frac{\partial G}{\partial \ddot{q}_j} - \frac{\partial P}{\partial q_j} = \tau_j, (j = 1, 2, \ldots, m) \quad (10)$$

in which, p denotes the potential energy of the system, and τ_j is the generalized force corresponding to q_j.

To compute the Appell equations, the Gibbs function must be computed firstly. Defining the diagonal matrix $\boldsymbol{M}_i(m_i, m_i, m_i, J_{ix}, J_{iy}, J_{iz})\ \ (i = 1, 2, 3)$ as inertial matrix, setting the acceleration vector of the frame, fork, and wheel as aa_1, aa_2, aa_3, according to (1)–(7) we can get:

$$aa_1 = \begin{pmatrix} \dot{v}_1 \\ \dot{\omega}_1 \end{pmatrix},\ aa_2 = \begin{pmatrix} \dot{v}_2 \\ \dot{\omega}_2 \end{pmatrix},\ aa_3 = \begin{pmatrix} \dot{v}_3 \\ \dot{\omega}_3 \end{pmatrix}$$

Because the entire VSTWR is completely symmetrical, the left and right wheels are in exactly the same state during the switching process between the bicycle and Segway modes. Similarly, the left and right forks are in exactly the same state of motion. Therefore, we can separately obtain the acceleration energy of the frame, two forks and two wheels as G_1, G_2, G_3:

$$G_1 = \frac{1}{2} aa_1^{\mathrm{T}} M_1 aa_1,\quad G_2 = aa_2^{\mathrm{T}} M_2 aa_2,\quad G_3 = aa_3^{\mathrm{T}} M_3 aa_3$$

the total acceleration energy of the system can be expressed as:

$$G = G_1 + G_2 + G_3 \tag{11}$$

Taking the horizontal plane of the two wheel axes as the zero potential energy reference plane, the total potential energy of the system can be expressed as:

$$P = 2(m_1 + m_2)gH\cos q_1 \tag{12}$$

Substituting (11) and (12) into Eq. (10), we can get the dynamic model of the system as follows:

$$\begin{bmatrix} D_{11} & D_{12} & D_{13} \\ D_{21} & D_{22} & D_{23} \\ D_{31} & D_{32} & D_{33} \end{bmatrix} \begin{bmatrix} \ddot{q}_1 \\ \ddot{q}_2 \\ \ddot{q}_3 \end{bmatrix} + \begin{bmatrix} F_1(q,\dot{q}) \\ F_2(q,\dot{q}) \\ F_3(q,\dot{q}) \end{bmatrix} = \begin{bmatrix} 0 \\ \tau_2 \\ \tau_3 \end{bmatrix} \tag{13}$$

where, D_{ij} is a function of system mass, moment of inertia, and generalized coordinates; F_i is a term related to centrifugal force, Coriolis force, gravity; τ_2 and τ_3 are the driving torque of the fork and the wheel respectively.

4 Chaplygin Equations of the System

For the dynamical system of this paper, the specific form of the Chaplygin equation can be expressed as follows:

$$\frac{d}{dt}\frac{\partial \tilde{T}}{\partial \dot{q}_\sigma} - \frac{\partial \tilde{T}}{\partial q_\sigma} - \frac{\partial P}{\partial q_\sigma} + \sum_{\beta=1}^{\gamma} \frac{\partial T}{\partial \dot{q}_{3+\beta}} \sum_{\upsilon=1}^{3} \left(\frac{\partial B_{3+\beta,\upsilon}}{\partial q_\sigma} - \frac{\partial B_{3+\beta,\sigma}}{\partial q_\upsilon} \right) \dot{q}_\upsilon = \tilde{Q}_\sigma,\quad (\sigma = 1, 2, 3) \tag{14}$$

in which, γ is the number of non-holonomic constraints of the system, $\partial B_{3+\beta,v}$ is the independent coefficient of generalized velocity $\dot{q}_v$ in the non-holonomic constraint. $\dot{q}_\sigma$ is the independent generalized velocity. $\tilde{Q}_\sigma$ is the generalized force corresponding to the independent generalized coordinate.

The kinetic energy of the frame T_1, the kinetic energy of the forks T_2, and the kinetic energy of the wheels T_3 can be presented as:

$$T_1 = \frac{1}{2}\boldsymbol{vv}_1^{\mathrm{T}}.\boldsymbol{M}_1.\boldsymbol{vv}_1, \quad T_2 = \boldsymbol{vv}_2^{\mathrm{T}}.\boldsymbol{M}_2.\boldsymbol{vv}_2, \quad T_3 = \boldsymbol{vv}_3^{\mathrm{T}}.\boldsymbol{M}_3.\boldsymbol{vv}_3$$

where, $\boldsymbol{vv}_1 = \begin{pmatrix} \boldsymbol{v}_1 \\ \omega_1 \end{pmatrix}$, $\boldsymbol{vv}_2 = \begin{pmatrix} \boldsymbol{v}_2 \\ \omega_2 \end{pmatrix}$, $\boldsymbol{vv}_3 = \begin{pmatrix} \boldsymbol{v}_3 \\ \omega_3 \end{pmatrix}$. The total kinetic energy of the system is T, therefore:

$$T(q_\sigma, \dot{q}_\sigma, \dot{q}_4, \dot{q}_5) = T_1 + T_2 + T_3 \tag{15}$$

Substituting the non-holonomic constraints (7) and (8) into (15), we can get kinetic energy $\tilde{T}(q_\sigma, \dot{q}_\sigma)$ without $(\dot{q}_4, \dot{q}_5)$. Substituting $\tilde{T}$, P and T into Eq. (14), the kinetic model can also be presented as formula (13):

$$\begin{bmatrix} D_{11} & D_{12} & D_{13} \\ D_{21} & D_{22} & D_{23} \\ D_{31} & D_{32} & D_{33} \end{bmatrix} \begin{bmatrix} \ddot{q}_1 \\ \ddot{q}_2 \\ \ddot{q}_3 \end{bmatrix} + \begin{bmatrix} F_1(\boldsymbol{q}, \dot{\boldsymbol{q}}) \\ F_2(\boldsymbol{q}, \dot{\boldsymbol{q}}) \\ F_3(\boldsymbol{q}, \dot{\boldsymbol{q}}) \end{bmatrix} = \begin{bmatrix} 0 \\ \tau_2 \\ \tau_3 \end{bmatrix} \tag{16}$$

5 Matlab Numerical Simulation

The physical parameters of the VSTWR are shown in Table 2.

And let $x_1 = q_1$, $x_2 = \dot{q}_1$, $x_3 = q_2$, $x_4 = \dot{q}_2$, $x_5 = q_3$ and $x_6 = \dot{q}_3$ in (13) and (16), then the affine nonlinear system of the robot can be presented as (17) and (18).

Table 2 The actual value of parameters

Parameters	Value
r, H	0.17, 0.122 m
m_1, m_2, m_3	17.05, 2.66, 3.8 kg
J_{1x}, J_{1y}, J_{1z}	1.18, 0.06, 1.21 kg m^2
J_{2x}, J_{2y}, J_{2z}	0.014, 0.017, 0.0046 kg m^2
J_{3x}, J_{3y}, J_{3z}	0.03, 0.05, 0.02 kg m^2

$$\begin{cases} \dot{x}_1 = x_2 \\ \dot{x}_2 = \frac{0.56\tau_3 \cos x_3}{-1.14+\cos^2 x_3} + \frac{-14.3x_1+0.16\sin(2x_3)x_2x_4-0.36\sin x_3x_4x_6}{-1.14+\cos^2 x_3} \\ \dot{x}_3 = x_4 \\ \dot{x}_4 = 20.3\tau_2 \\ \dot{x}_5 = x_6 \\ \dot{x}_6 = \frac{-1.18\tau_3}{-1.14+\cos^2 x_3} + \frac{26.4\cos x_3x_1-0.69\sin x_3+0.34\sin(2x_3)x_4x_6}{-1.14+\cos^2 x_3} \end{cases} \tag{17}$$

$$\begin{cases} \dot{x}_1 = x_2 \\ \dot{x}_2 = \frac{0.74\tau_3 \cos x_3}{-1.5+1.3\cos^2 x_3} + \frac{-18.96x_1+0.64\sin(2x_3)x_2x_4-0.72\sin x_3x_4x_6}{-1.5+1.3\cos^2 x_3} \\ \dot{x}_3 = x_4 \\ \dot{x}_4 = 20.3\tau_2 - 0.47\sin(2x_3)x_2^2 - 36.3\sin x_3x_2x_6 \\ \dot{x}_5 = x_6 \\ \dot{x}_6 = \frac{-1.56\tau_3}{-1.5+1.3\cos^2 x_3} + \frac{35\cos x_3x_1-3.38\sin x_3x_2x_4+0.66\sin(2x_3)x_4x_6}{-1.5+1.3\cos^2 x_3} \end{cases} \tag{18}$$

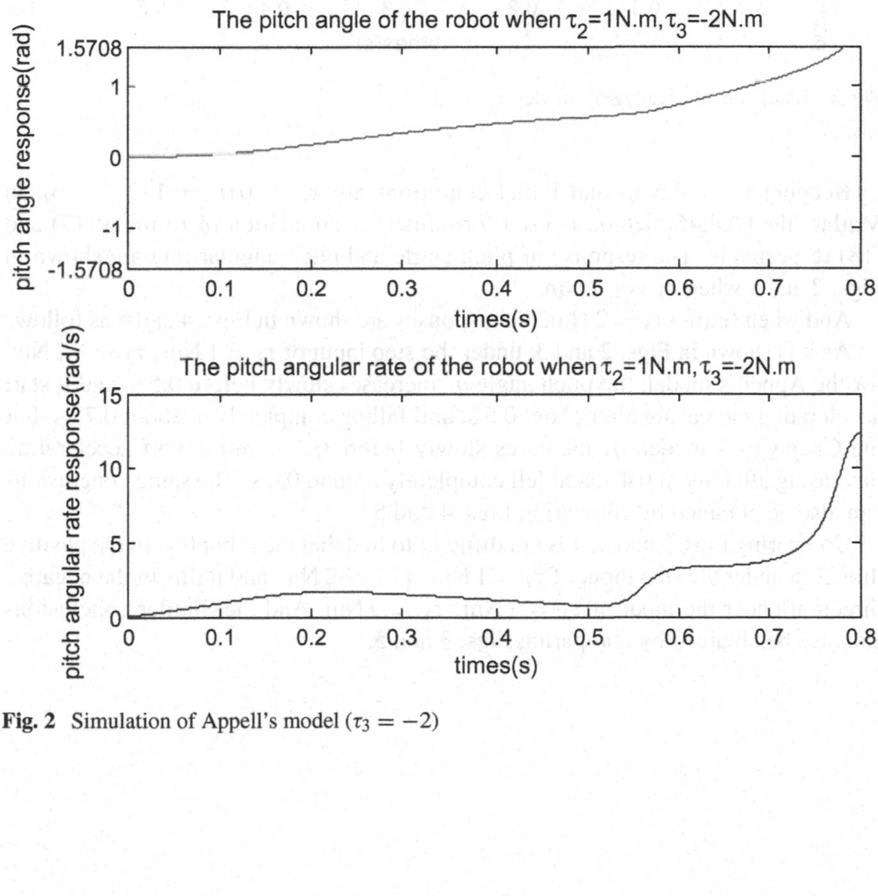

Fig. 2 Simulation of Appell's model ($\tau_3 = -2$)

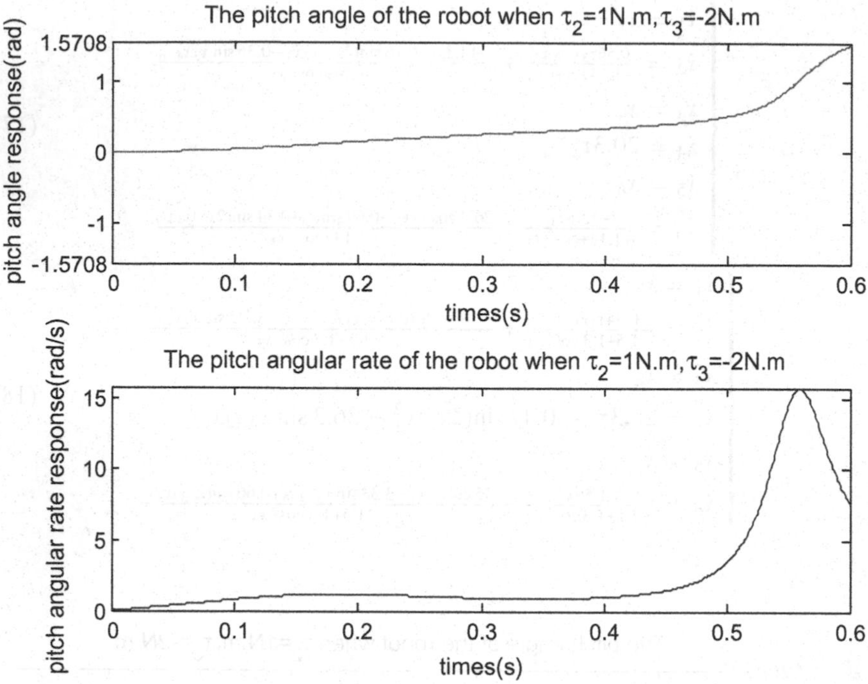

Fig. 3 Simulation of Chaplygin's model ($\tau_3 = -2$)

Keeping $\tau_2 = 1\,\text{Nm}$, and initial conditions are $x_i = 0\,(i = 1, 2, \ldots, 6)$. In Matlab, the ODE45 method is used for numerical simulation of formula (17) and (18) respectively. The response of pitch angle and pitch angular rate are shown in Figs. 2 and 3 when $\tau_3 = -2\,\text{Nm}$.

And when setting $\tau_3 = 2\,\text{Nm}$, the responses are shown in Figs. 4 and 5 as follow.

As it is shown in Figs. 2 and 3, under the step input of $\tau_2 = 1\,\text{Nm}$, $\tau_3 = -2\,\text{Nm}$, for the Appell's model, the pitch angle q_1 increases slowly before 0.5 s, and it start accelerating increasing after about 0.5 s, and falling completely at about 0.78 s. For the Chaplygin's model, q_1 increases slowly before 0.4 s, and it start accelerating increasing after about 0.4 s, and fell completely around 0.6 s. The same conclusions can also be obtained by comparing Figs. 4 and 5.

Comparing Figs. 2 and 4, it is not difficult to find that the robot tilts in the positive direction under the step input of $\tau_2 = 1\,\text{Nm}$, $\tau_3 = -2\,\text{Nm}$, and it tilts in the negative direction under the input of $\tau_2 = 1\,\text{Nm}$, $\tau_3 = 2\,\text{Nm}$. And the Similar conclusions can also be obtained by comparing Figs. 3 and 5.

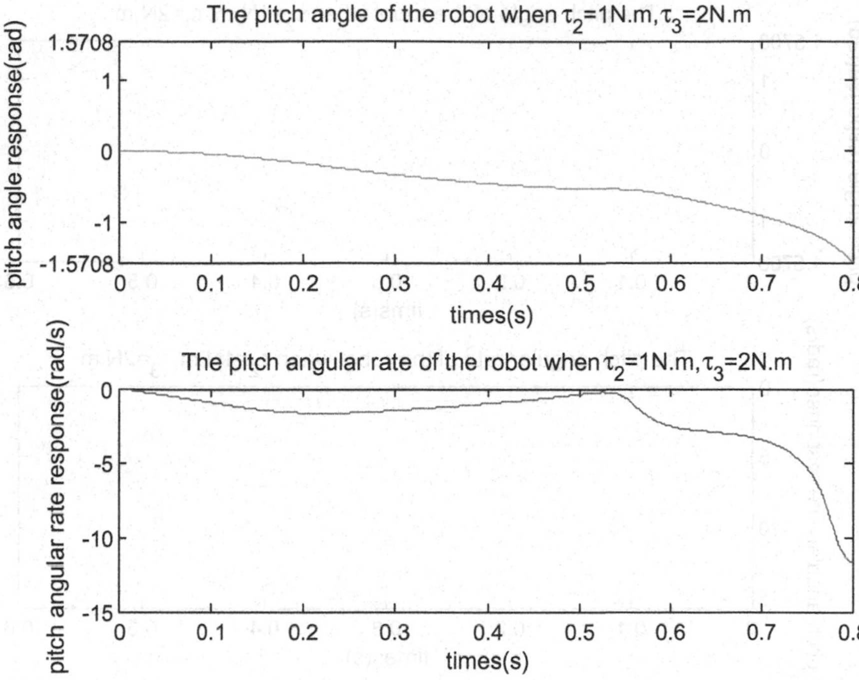

Fig. 4 Simulation of Appell's model ($\tau_3 = 2$)

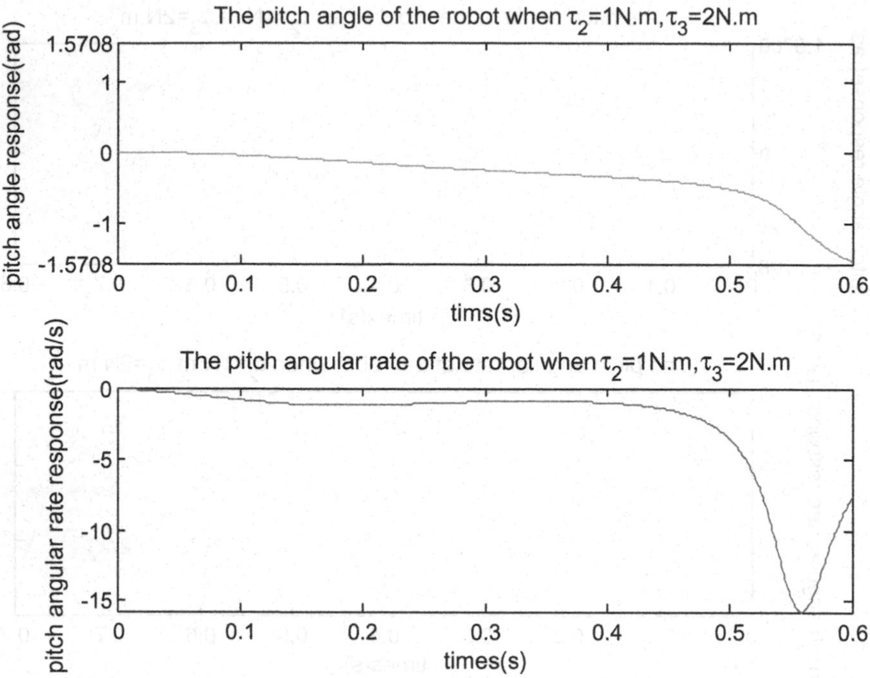

Fig. 5 Simulation of Chaplygin's model ($\tau_3 = 2$)

6 Conclusion

The kinematics of the mode switching process (between Segway mode and Bicycle mode) of the VSTWR is analyzed in the paper. Then, the dynamic models based on Appell equations and Chaplygin equations are achieved separately, and the mechanical coupling relationship among the frame, the fork and the wheel of the VSTWR is effectively revealed. Finally, the numerical simulations of the two dynamic models were performed based on Matlab. And the rationality and difference of the two dynamic models are analyzed.

Acknowledgements This work is supported by Beijing Municipal Natural Science Foundation #L172031, #3162021.

References

1. Y. Huang, Y. Huang, Track-stand motion of a front-wheel drive bicycle robot under 45° front-bar turning angle. J. Mech. Eng. **48**(7), 16 (2012)
2. S. Yuan, G. Lei, X. Bin, Dynamic modeling and sliding mode controller design of a two-wheeled self-balancing robot, in *IEEE International Conference on Mechatronics and Automation*. IEEE, pp. 2437–2442 (2016)
3. W. Zhuang, H. Jiang, C. Liu, et al., Dynamic model and balanced lateral rolling motion control of a unicycle robot, in *IEEE International Conference on Information and Automation*. IEEE, pp. 164–168 (2015)
4. L. Keo, K. Yoshino, M. Kawaguchi, et al., Experimental results for stabilizing of a bicycle with a flywheel balancer, in *IEEE International Conference on Robotics and Automation*. IEEE, pp. 6150–6155 (2011)
5. M.H. Shafiei, M. Emami, T. Binazadeh, Design of a tracking controller for an unmanned bicycle using neural networks. J. Comput. Intell. Electron. Syst. **3**(3), 193–199 (2014)
6. T. Abumi, T. Murakami, Posture stabilization of two-wheel drive electric motorcycle by slip ratio control considering camber angle, in *IEEE International Conference on Mechatronics*. IEEE, pp. 353–358 (2015)
7. Y. Zhao, J.H. Wang, Q. Wang, Algorithm and achieve of self-balancing two-wheeled control system based on PID neural network, in *International Conference on Electronics and Information Engineering. International Society for Optics and Photonics*, 97942Y (2015)
8. Y. Liu, X. Gao, Y. Mu, et al., Fault detection of two wheel inverted pendulum robot with center of gravity self-adjusting mechanism, in *IEEE International Conference on Robotics and Biomimetics*. IEEE, pp. 230–235 (2017)
9. C.H. Chih-Hui, Y.-F. Peng, Position and angle control for a two-wheel robot. Int. J. Control Autom. Syst. **15**(1), 1–12 (2017)

Design of a Kind of Trajectory Optimization Algorithm for a Manipulator Based on Genetic Algorithm

Qiang Fu, Yuan Song, Shimin Wei and Lei Guo

Abstract According to a existing 3-DOF manipulator, its kinematic constraints are analyzed. A manipulator trajectory optimization algorithm is proposed based on genetic algorithm. Considered with several obstacles, both the running time and the energy consumption are taken as the optimization targets with the speed and acceleration constraints. The fourth-fifth-order polynomial is used as the interpolation curve to fit the joint trajectory, which is to ensure the continuity of joint operation. Based on genetic algorithm, the optimal trajectory under the relevant constraints is obtained. The algorithm can solve the problem of trajectory optimization in the environment with several obstacles very well. The experiment results show the efficiency of the algorithm.

Keywords 3-DOF manipulator · Trajectory optimization · Genetic algorithm
Time optimization · Energy consumption optimization

1 Introduction

The research of trajectory planning of the manipulator is focused on how to make the manipulator move smoothly from the initial position to the target position. In practical applications, if we only focus on the smoothness then it will not meet the requirements usually. Some other factors such as efficiency, energy consumption, and stability are also need to be considered [1]. A good trajectory in a specific application scene not only will satisfy the motion requirements, but also will ensure the smoothness of the movements. The trajectory can reduce the mechanical wear. And the motion time and the energy loss decreased too.

In the study of the trajectory optimization of manipulators, a multi-objective genetic algorithm was proposed in [2]. The algorithm has a certain reference value for the scholars to design the derivative algorithms. A specific implementation pro-

---

Q. Fu · Y. Song (✉) · S. Wei · L. Guo
School of Automation, Beijing University of Posts and Telecommunications, Beijing, China
e-mail: songyuan@bupt.edu.cn

Y. Jia et al. (eds.), *Proceedings of 2018 Chinese Intelligent Systems Conference*, Lecture Notes in Electrical Engineering 529,
https://doi.org/10.1007/978-981-13-2291-4_67

cess of optimization algorithm based on genetic algorithm was introduced in [3]. However, they did not consider the presence of obstacles in the workspace. Combined the traditional annealing algorithm with the genetic algorithm, an improved simulated annealing genetic algorithm was proposed in [4]. Compared with the traditional genetic algorithm, the algorithm can converge faster. However, its optimized parameter is only the running time. In [5], Gan Yahui put forward an optimal trajectory planning method for multi-robot systems. However, the object of the analysis is only a plane 2-DOF manipulator. To achieve multi-objective optimization, an idea to mix the optimal time and impact into a single fitness function was proposed in [6]. And the algorithm lacked the initial restriction on the running time. It is possible that the results obtained are not accurate enough. A point-to-point trajectory planning method was designed for a three-link redundant manipulator in [7]. It does not consider whether there is an obstacle between the two points. And the manipulator is a planar structure. Besides genetic algorithms, there are many other heuristic search methods that play a role in trajectory optimization. In [8], the particle swarm optimization algorithm is used. The goal of the optimization is to reduce the running time. However, this method only considers time optimal. It is applicable to the case of intermediate interpolation point are determined only. Xia Hongwei proposed an algorithm based on chaotic particle swarm optimization for the trajectory planning of space manipulator in [9]. But the method considered the pedestal posture only so that the application range is narrow. An idea to use seven-degree B-spline curves for joint interpolation is presented in [10]. The experiment results show that the optimization method can effectively reduce the trajectory tracking error. But compared with other interpolation methods, the calculation of the seven B-spline interpolation is too complex to be realized in the application.

This paper takes a 3-DOF manipulator as object to study the trajectory optimization algorithm. The running time and the energy consumption are taken as the optimization target with the velocity and acceleration constraints. Based on Genetic Algorithm, a trajectory optimization algorithm is proposed. To make the joints move with a smooth and continue way, a fourth-fifth-order polynomial is chosen as interpolation curve. Compared with the other papers, the research object is a 3-DOF manipulator instead of the planar robot. The algorithm takes the obstacles into the consideration also. By designing the penalty function, the collision can avoid efficiently. The algorithm is very suitable to find the best trajectory in a complicated environment. The performance of the algorithm is verified by the simulation based on MATLAB.

2 Kinematic Analysis of the First Three Joints of the UR10

For the UR10 manipulator, the D-H method is mainly applied to establish the kinematic model. The coordinate system for links is established which is shown in Fig. 1. For the UR10 manipulator, the first three joints are mainly used to determine the position, the latter three joints are mainly used to determine the orientation. There-

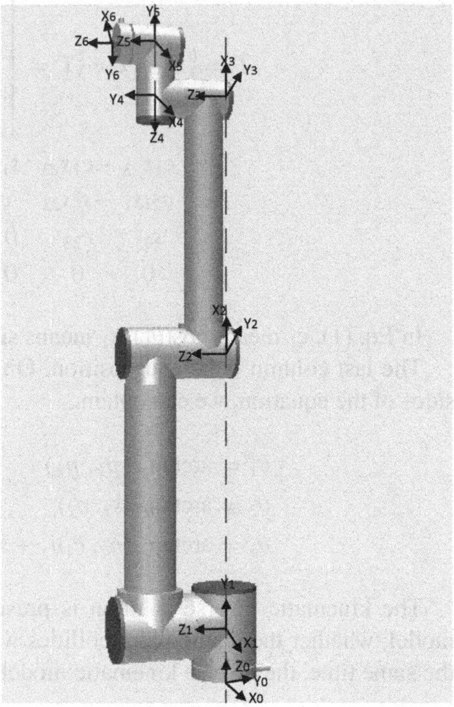

Fig. 1 UR10 manipulator coordinate system

fore, only the first three joints are studied in this article. The D-H parameters are shown in Table 1.

According to the transformation matrix formula for links, the kinematic positive solution of the 3-DOF can be obtained as follows.

Table 1 D-H parameters of the first three joints of the UR10 manipulator

Link	a_i(mm)	α_i(°)	d_i(mm)	θ_i(°)
1	0	90	127.3	$\theta_1(0)$
2	612.0	0	0	$\theta_2(-90)$
3	572.3	0	0	$\theta_3(0)$

$$
\boldsymbol{T} = {}^0_1\boldsymbol{T} * {}^1_2\boldsymbol{T} * {}^2_3\boldsymbol{T} = \begin{bmatrix} n_x & o_x & a_x & p_x \\ n_y & o_y & a_y & p_y \\ n_z & o_z & a_z & p_z \\ 0 & 0 & 0 & 1 \end{bmatrix}
$$

$$
= \begin{bmatrix} c_1c_{23} & -c_1s_{23} & s_1 & a_2c_1c_2 + a_3c_1c_{23} \\ c_{23}s_1 & -s_1s_{23} & -c_1 & a_2c_2s_1 + a_3c_{23}s_1 \\ s_{23} & c_{23} & 0 & a_2s_2 + a_3s_{23} + d_1 \\ 0 & 0 & 0 & 1 \end{bmatrix} \tag{1}
$$

In Eq. (1), c_1 means $\cos(\theta_1)$, s_1 means $\sin(\theta_1)$, c_{23} means $\cos(\theta_2 + \theta_3)$, and etc. The last column is its end position. On that basis, multiplied by ${}^0_1\boldsymbol{T}^{-1}$ on both sides of the equation, we can obtain

$$
\begin{aligned}
\theta_1 &= \arctan 2(p_y, p_x) \\
\theta_2 &= \arctan 2(s_2, c_2) \\
\theta_3 &= \arctan 2(n_z, c_1n_x + s_1n_y) - \arctan 2(s_2, c_2)
\end{aligned}
$$

The kinematic inverse solution is presented. Based on the positive kinematic model, whether the manipulator collides with an obstacle or not can be judged. At the same time, the inverse kinematic model is ready for trajectory planning.

3 Method of Trajectory Planning

3.1 *Fourth-Fifth-Order Polynomial Interpolation Curve*

The trajectory planning of the manipulator can be divided into two kinds, trajectory planning in the joints space and in the Cartesian space. In this paper, a fourth-fifth-order polynomial curve is chosen to fit the joints movements. Between the start point and interpolation point, a fourth-order polynomial curve is chosen, and between interpolation point and the target point, we choose a fifth-order polynomial curve to fit. Assuming that the freedom of the manipulator is n, the number of intermediate points is m. The fourth-order polynomial can be expressed as follow,

$$
\begin{aligned}
\theta_{i,j}(t) = a_{i0} + a_{i1}t_i + a_{i2}t_i^2 + a_{i3}t_i^3 \\
+ a_{i4}t_i^4 \quad (i = 0, \ldots, m-1, \quad j = 1, \ldots n)
\end{aligned} \tag{2}
$$

In Eq. 2, t_i is the time moving from point i to point $i + 1$. Parameters $a_{i0} \ldots a_{i4}$ can be determined based on the initial conditions. The fifth-order polynomial can be expressed similarly.

3.2 Determination of Optimization Parameters

The initial conditions satisfy the following equations,

$$\begin{aligned}
\theta_i &= a_{i0}, \theta_i' = a_{i1}, \theta_i'' = 2a_{i2} \\
\theta_{i+1} &= a_{i0} + a_{i1}T_i + a_{i2}T_i^2 + a_{i3}T_i^3 + a_{i4}T_i^4 \\
\theta_{i+1}' &= a_{i1} + 2a_{i2}T_i + 3a_{i3}T_i^2 + 4a_{i4}T_i^3
\end{aligned}$$

The value of $a_{i0} \ldots a_{i4}$ can be calculated as follows,

$$\begin{aligned}
a_{i0} &= \theta_i, a_{i1} = \theta_i', a_{i2} = \frac{\theta_i''}{2} \\
a_{i3} &= (-4\theta_i + 4\theta_{i+1} - 3\theta_i' T_i - \theta_{i+1}' T_i - \theta_i'' T_i^2)/T_i^3 \\
a_{i4} &= (3\theta_i - 3\theta_{i+1} + 2\theta_i' T_i + \theta_{i+1}' T_i + \theta_i'' T_i^2/2)/T_i^4
\end{aligned}$$

The acceleration of point $i + 1$ is also determined.

$$\theta_{i+1}'' = 2a_{i2} + 6a_{i3}T_i + 12a_{i4}T_i^2$$

In this study, the degree of freedom n is 3, and the intermediate point m is 1. The joint angle, angular velocity, and angular acceleration of the starting point and the target point are all given. According to the equation above, the following parameters are required to be determined, including the angles of the joints in the interpolation point, the sum of the movement time of the two moving sections, and the angular velocity values of the joints at the interpolation point.

$$\begin{bmatrix} \theta_{m0} \ \theta_{m1} \ \theta_{m2} \ t_0 \ t_1 \ \theta_{m0}' \ \theta_{m1}' \ \theta_{m2}' \end{bmatrix}$$

4 Design of Trajectory Optimization Algorithm

4.1 Introduction of Genetic Algorithm

Genetic algorithm is a computational model which simulates natural selection and genetic mechanisms. It is an effective method to find the optimal solution in the global working space. The algorithm generally includes the following steps, determining the range of variables, coding the parameters, setting initial population, designing fitness function, performing genetic operations such as selection, crossover, mutation, obtaining new populations, and continuously iterating until the fitness value reaches the requirements or the number of iterations reaches the set value. In this paper, the genetic algorithm is used to find the optimal trajectory of the manipulator and the optimization parameter is,

$$\left[\theta_{m0}\ \theta_{m1}\ \theta_{m2}\ t_0\ t_1\ \theta'_{m0}\ \theta'_{m1}\ \theta'_{m2}\right]$$

4.2 Process of Algorithm Design

Step 1 Determine the range of related variables. In order to simplify the calculation, the base coordinate system of the manipulator and the world coordinate system of the entire system are coincident. There is an obstacle in the world coordinate system. The outer contour of the obstacle is a cube with a center length of (0, –0.4, 0) and a side length of 0.2 m. The initial position of the robot is

$$\left[\theta_{s0}\ \theta_{s1}\ \theta_{s2}\right] = \left[0\ 0\ -0.5pi\right]$$

The target position is

$$\left[\theta_{g0}\ \theta_{g1}\ \theta_{g2}\right] = \left[1/2pi\ -1/6pi\ 1/3pi\right]$$

The range of θ_{m0} , θ_{m1}, θ_{m2} is defined as $\left[0\ pi\right]$, $\left[-pi\ 0\right]$, $\left[0\ pi\right]$, parameters t_0 and t_1 are defined as $\left[0\ 8\right]$,θ'_{m0}, θ'_{m1} and θ'_{m2} are all defined as $\left[0\ 1/4pi\right]$rad/s.

Step 2 Determine coding method. Since there are 8 optimization parameters, if we used binary coding, the search space would expand wildly. Therefore, the real number coding method is chosen. Each variable is represented as a floating-point number within the value range. The eight numbers represent an individual.

Step 3 Generate the initial population. The population size was chosen as 80, that is, each generation was composed by 80 individuals. Those individuals were randomly generated by the random number function in MATLAB.

Step 4 Define the fitness function. The fitness function is a criterion for evaluating the performance of each individual. The fitness function is designed as,

$$F_f = \omega_1 T + \omega_2 f_\theta$$

Among them, ω_1, ω_2 are constant coefficient which is used to adjust the time and angle to the same level. Set $\omega_1 = 2$, $\omega_2 = 1$. T is the sum of the time of two movements. f_θ is the sum of each joint angle value of the movement. The movement cost of each joint is different. Define the cost ratio of Joint 1, Joint 2, and Joint 3 as 2:1.5:1. According to the established fitness function, the optimal individual in each generation of population would be found. The individual with the smallest fitness value is chosen as the best individual of each generation.

Step 5 Generate new population. Through the selection, crossover, and mutation operation, a new population will be generated. The selection operation is based on the individual fitness value. The roulette method is chosen to generate new populations. After that, crossover operation will be used to exchange some genes between two chromosomes. Every crossover operation will generate two new individuals. The crossover operation can improve the global search ability of the algorithm. Mutation operation will change the genes of individuals in a very small possibility. It is also an auxiliary method for generating new individuals.

Step 6 Set the number of iterations. Generally, we set the number between 100 and 500, depending on the complexity of the problem. The number of iterations defined in this paper is 200.

According to the above steps, an algorithm for trajectory optimization is designed with the M language of MATLAB.

5 Simulation Based on MATLAB Software

When the start point and target point are determined, every joint moves straightly to the target angle is the shortest trajectory. In this case, the fitness value is 14.45. However, the trajectory has a high probability to collide with obstacles. To avoid the collision, we can plan the movement of each joints. Two interpolation points are determined, and at each point, the angles are set as follow.

$$\left[\theta_{10}\ \theta_{11}\ \theta_{12}\right] = \left[0\ -0.5pi\ \ -0.5pi\right]$$

$$\left[\theta_{20}\ \theta_{21}\ \theta_{22}\right] = \left[0.5pi\ \ -0.5pi\ \ -0.5pi\right]$$

So the trajectory can be divided into three movements. Every movement's running time is set as 2.5 s. The fitness value is 24.69 which means we need to pay a high consumption for the movements. The trajectory optimization algorithm we proposed can well solve the problem.

In order to verify the effectiveness of the algorithm, simulation experiments were conducted in MATLAB. When calculating the fitness value, collision between the manipulator and the obstacle need to be checked. If there is a collision, the penalty function would get work to avoid the collision. Through the genetic algorithm, the values of the eight optimization parameters are determined as follow.

$$\left[1.1784\ -0.6901\ 1.4802\ 2.2831\ 2.5038\ 0.4576\ 0.0057\ 0.7609\right]$$

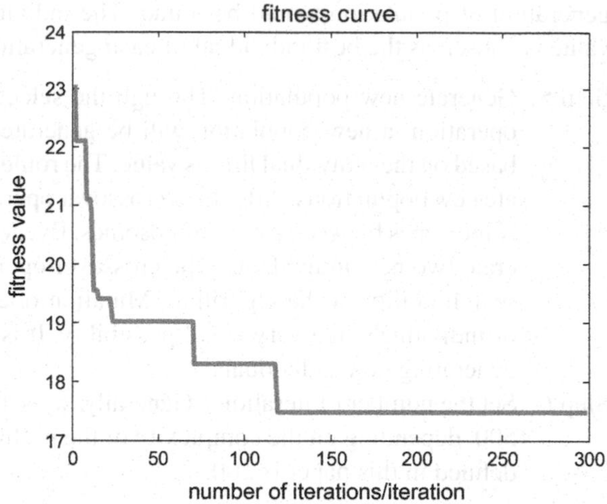

Fig. 2 Best fitness value in each generation

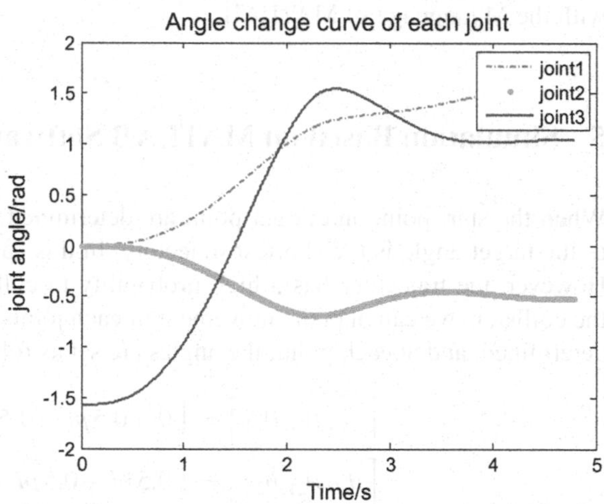

Fig. 3 The angle of each joint

Figure 2 shows the best fitness value obtained in the each iteration. At the beginning of the iterations, the fitness value decreased rapidly. After the iterations running to the 120th generation, the fitness value gets stable. The best fitness value obtained is 17.484. The effectiveness of the algorithm is confirmed.

It can be found out that a best trajectory is determined by the algorithm from Figs. 3, 4 and 5. Each joint's angle, angular velocity, and angular acceleration are all smooth and continuous. Because of the existence of the penalty function, the collision between manipulator and obstacle would not happen. If the collision is checked, a new individual will be generated to replace it. Compared with the existing result, the obtained trajectory can reduce the operation consumption obviously, and avoid colliding with obstacles.

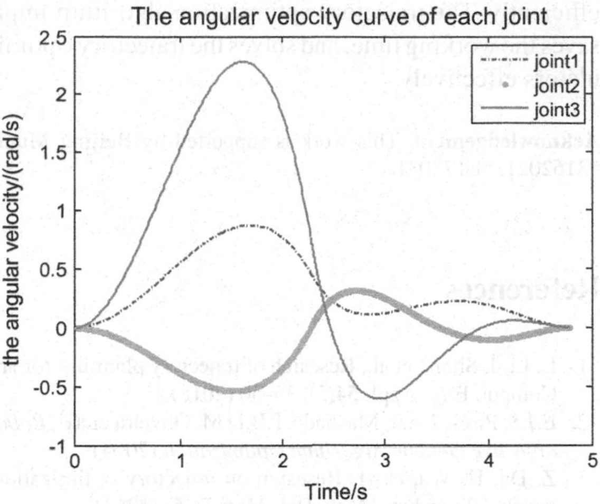

Fig. 4 The angular velocity of each joint

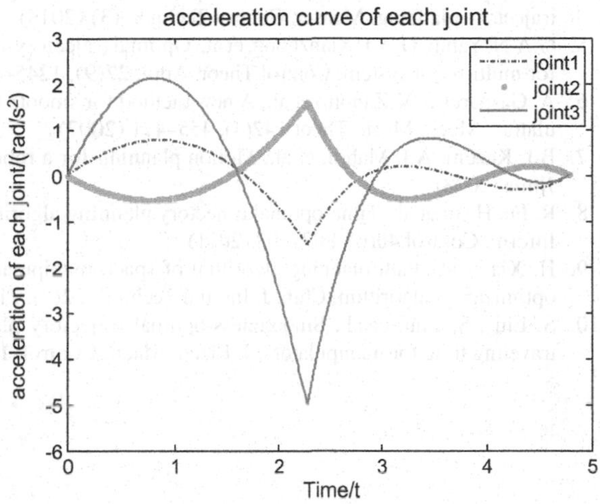

Fig. 5 The angular acceleration of each joint

6 Conclusion

In order to solve the trajectory optimization problem of manipulators, a trajectory optimization algorithm based on genetic algorithm is proposed. A 3-DOF manipulator is taken as the platform. The positive and inverse kinematic models are established. The using of the fourth-fifth-order polynomial interpolation curve guarantees the smoothness and continuity of the joints' movements. Finally, the computer simulation is achieved based on MATLAB. According to the simulation results, an optimal trajectory with the shortest running time and the lowest energy consumption is found. The smoothness of joints angle, angular velocity and angular acceleration are proved. The smoothness can reduce the mechanical wear to the manipulator efficiently. The trajectory optimization algorithm improves the operating efficiency, saves the working time, and solves the trajectory optimization problems of the manipulators effectively.

Acknowledgement This work is supported by Beijing Municipal Natural Science Foundation #3162021, #L172031.

References

1. L. Li, J. Shang et al., Research of trajectory planning for articulated industrial robot: a review. Comput. Eng. Appl. **54**(5), 36–50 (2018)
2. E.J.S. Pires, J.A.T. Machado, P.B.D.M. Oliveira et al., *Robot Trajectory Planning Using Multi-objective Genetic Algorithm Optimization* (2004)
3. Z. Du, H. Wu, et al. Research on trajectory optimization of mobile manipulator based on genetic algorithm. Mach. Des. Manuf. (5) (2013)
4. Y. Zong, J. Cui, et al., Application of simulated annealing genetic algorithm in manipulator's trajectory planning. Measur. Control Technol. (3) (2018)
5. G.A.N. Yahui, D.A.I. Xianzhong, et al., Optimal trajectory-planning based on genetic algorithm for multi-robot system. Control Theor. Appl. **27**(9), 1245–1252 (2010)
6. A. Gasparetto, V. Zanotto et al., A new method for smooth trajectory planning of robot manipulators. Mech. Mach. Theory **42**(4), 455–471 (2007)
7. B.I. Kazem, A.I. Mahdi, et al., Motion planning for a robot arm by using genetic algorithm. Jjmie (2008)
8. R. Fu, H. Ju et al., Time-optimal trajectory planning algorithm for manipulator based on PSO. Inform. Control **40**(6), 802–808 (2011)
9. H. Xia et al., Path planning algorithm of space manipulator based on chaos particle swarm optimization algorithm. Chin. J. Inertial Technol. **22**(2), 211–216 (2014)
10. S. Liu , S. Zhu, et al., Smoothness-optimal trajectory planning method with constraint on traveling time for manipulators. J. Electr. Mach. Control, **13**(6), 897–902 (2009)

Feature Points Designing and Images Merging in the Final Approaching Phase of Rendezvous and Docking

Wenjing Pei and Yingmin Jia

Abstract In this paper, three feature points are designed on the fore-end cabin of the target spacecraft model. Some feature points are regarded as control points. A global affine transformation based on control points is used to image merging. The feature which extracted on the front-end cabin and on two solar panels is merged with the same image. That is to say, three feature points and one characteristic circle on the fore-end cabin and 48 feature points on two solar panels are merged into the same image. In the experiment, a global affine transformation is applied for images merging based on three control points and five control points respectively. The experimental results achieved images translation, rotation, scaling and shearing. They are merged into the one image in the end. It lays the foundation for the measurement of the position and attitude of the target spacecraft.

Keywords Feature points · Images merging · Control points
Global affine transformation

1 Introduction

Space rendezvous and docking is the general term of space rendezvous and space docking, which is the orbit movement of two spacecraft. Generally speaking, the orbiting spacecraft is the target spacecraft and the rendezvous and docking with the target spacecraft is tracking spacecraft. In space, the process of two or more spacecraft through the coordination of orbital parameters at the same time to reach the same location is called rendezvous. Docking is based on rendezvous to connect the two aircraft into a whole by using the special docking mechanism.

W. Pei · Y. Jia (✉)
Seventh Research Division and Center for Information and Control,
School of Automation Science and Electrical Engineering,
Beihang University (BUAA), Beijing 100191, China
e-mail: ymjia@buaa.edu.cn

Y. Jia et al. (eds.), *Proceedings of 2018 Chinese Intelligent Systems Conference*, Lecture Notes in Electrical Engineering 529,
https://doi.org/10.1007/978-981-13-2291-4_68

One of the most fundamental conditions is to ensure that two aircraft can't collide in any case for spacecraft rendezvous and docking. Due to the high precision of rendezvous and docking, the difficulty of controlling the multi-degree of freedom and the requirements of safety factors, the key technology of the rendezvous and docking is studied by the space scientific research institutions in many countries. In the final approaching phase, the distance between the tracking spacecraft and the target spacecraft will get closer and closer. The accuracy of position and attitude measurement will directly affect the success or failure of the docking process.

At present, feature points are designed on the target aircraft and the CCD sensor camera is installed on the tracking aircraft to obtain the images of the target aircraft. The image real-time processing system is used to analyze and calculate the relative position between these two spacecraft. The vision measurement system is limited to the distance between 100 and 3 m. If the distance is too far, the images of the target spacecraft are so small the camera plane and the images of feature points are so concentrated that the vision system can't correctly extract the useful information. If the distance is too close, the area of the target aircraft fore-end cabin and two solar are so large that all of feature points will not be on the camera and the system can't extract enough information from the feature points.

Images merging are required to calibrate numbers of control points [1–3]. The images can be shifted in the x-directions and y-directions in relation to the reference image based on measured points [4]. Points are generated from the images based on physical sensor model with accurately determined sensor alignment parameters [5]. Although lot of methods has been proposed for images merging, there are few studies in the final approaching phase of rendezvous and docking. An image registration algorithm is proposed based on the corner detection and affine transformation model in [6]. An adaptive technique against each reference pattern used global or local affine transformation in a hierarchical manner to normalize an input pattern [7]. In [8], a promising method of gray scale character recognition is proposed, which offers both noise tolerance and affine invariance and shows the high matching ability.

In this paper, three feature points are designed and one characteristic circle is extracted on the fore-end cabin of the target spacecraft model. A global affine transformation based on control points is used to image merging. The extracted front-end cabin is merged with the extracted solar panels. The front-end cabin contained three feature points and 48 feature points on two solar panels, which are merged into the same image.

The rest of this paper is organized as follows. Section 2 presents three feature points designing on the fore-end cabin. In Sect. 3, images registration based on control points is used to the reference and to be register images. Then images are merged using a global affine transformation based on control points. In Sect. 4, the results of simulation experiments are presented. Section 5 shows conclusion of this paper and proposes the future work of this research.

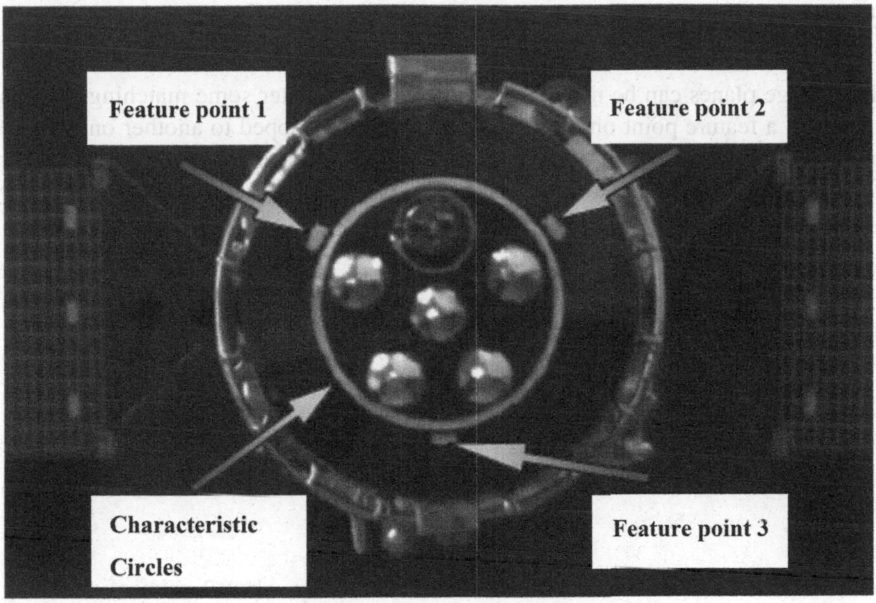

Fig. 1 Feature points and the characteristic circle on the fore-end cabin

2 Feature Point Designing on the Fore-End Cabin

In order to extract the various features from two images easily and quickly, it is necessary to design the feature points of the target spacecraft. The area of the front-end of the aircraft is small, and there are many parts in the front end structure. Therefore, it is not suitable to set too many feature points. We need know the coordinates of at least three points on the plane, which determine a spatial plane, according to the knowledge of spatial analytic geometry. Based on this principle, three characteristic points and a characteristic circle are designed on the front-end cabin. Three feature points are arranged evenly around the feature circle at the angle of 120°. The characteristic circle is designed to locate the three feature points precisely, because it is much easier to detect circles directly than three feature points in the front-end cabin. Figure 1 shows three feature points and the characteristic circle on the fore-end cabin of the target spacecraft model. The plane formed with three characteristic points is perpendicular to the central axis of the cabin, which is called the characteristic plane of the front-end cabin.

3 Global Affine Transformation

Two image planes can be used as two affine spaces. After some matching point is obtained, a feature point on the image space can be mapped to another one. There are two affine spaces A and B. Affine transformation is applied for point and vector in affine space, which includes translation (tx, ty), rotation θ, scale (sx, sy) and shear (shx, shy).

$$\begin{bmatrix} u \\ v \end{bmatrix} = \begin{bmatrix} 1 & 0 \\ shy & 1 \end{bmatrix} \begin{bmatrix} 1 & shy \\ 0 & 1 \end{bmatrix} \begin{bmatrix} sx & 0 \\ 0 & sy \end{bmatrix} + \begin{bmatrix} \cos\theta & \sin\theta \\ \sin\theta & \cos\theta \end{bmatrix} \begin{bmatrix} x \\ y \end{bmatrix} + \begin{bmatrix} tx \\ ty \end{bmatrix} \quad (1)$$

Homogeneous coordinate is as follows.

$$\begin{bmatrix} u \\ v \\ 1 \end{bmatrix} = \begin{bmatrix} a & b & c \\ d & e & f \\ 0 & 0 & 1 \end{bmatrix} \quad (2)$$

There are 6 unknowns in the format 2. In this way, when 3 pairs of matching points are known, transformation relations between affine spaces can be obtained.

Affine transformation can maintain the same relationship:

① Points colinearity and curves intersection are maintaining relations;
② Parallel lines are still parallel;
③ The midpoints is still midpoints;
④ The proportion of line segments on the same line is constant.

However, affine transformation can't keep the same segment length and the same angle.

4 Image Merge Using a Global Affine Transformation Based on Control Points

A global affine transformation based on control points required that some reliable matching points as control points are confirmed. In this paper, feature points have been extracted in the previous studies. In this way, some of feature points are used as control points. Afterwards, construct a global affine transformation method. The feature points on the one image can be mapped to the matched images. Finally, these two images are merged into the same image.

In two dimensional coordinates, the point coordinates are (x, y) and new point coordinates are (x', y') after the linear transformation. The details are as follows:

$$\begin{cases} x' = a_1 x + b_1 y + c_1 \\ y' = a_2 x + b_2 y + c_2 \end{cases} \quad (3)$$

where a_1, b_1, c_1, a_2, b_2 and c_2 are 6 unknowns in the format 3. (x', y') is an affine transformation of (x, y). Matrix representation is

$$\begin{bmatrix} x' \\ y' \end{bmatrix} = \begin{bmatrix} a_1 & b_1 \\ a_2 & b_2 \end{bmatrix} \begin{bmatrix} x \\ y \end{bmatrix} + \begin{bmatrix} c_1 \\ c_2 \end{bmatrix} \tag{4}$$

In the format 3, $\begin{bmatrix} a_1 & b_1 \\ a_2 & b_2 \end{bmatrix}$ is the transformation matrix, which represents rotation, scale and shear of images. Besides, $\begin{bmatrix} c_1 \\ c_2 \end{bmatrix}$ is the translation vector, which represents images translation.

In order to solve these 6 parameters, at least 3 control points are required:

$$\begin{bmatrix} a_1 & a_2 \\ b_1 & b_2 \\ c_1 & c_2 \end{bmatrix} = \begin{bmatrix} x_1 & y_1 & 1 \\ x_2 & y_2 & 1 \\ x_3 & y_3 & 1 \end{bmatrix}^{-1} \begin{bmatrix} x_1' & y_1' \\ x_2' & y_2' \\ x_3' & y_3' \end{bmatrix} \tag{5}$$

$(x_i, y_i), (x_i', y_i')$ are 3 pairs of control points. When the number of matching points is more than 3, the least squares method is used to solve.

5 Experimental Results

In this experiment, images are merged based on three control points and five control points respectively. Feature points are regard as control points, which have been extracted in [9] and are considered as the original images to complete images merging.

5.1 Image Merging Based on Three Control Points

There are three feature points and the characteristic circle on the fore-end cabin of the target spacecraft. In the experiment, three feature points are considered as three control points. A global affine transformation is used to image merging. Coordinate affine transformation matrix is as follows:

$$H = \begin{bmatrix} 0.2817 & -1.4539 & 300.6483 \\ 1.3358 & 0.3348 & -254.2036 \\ 0 & 0 & 1.0000 \end{bmatrix}$$

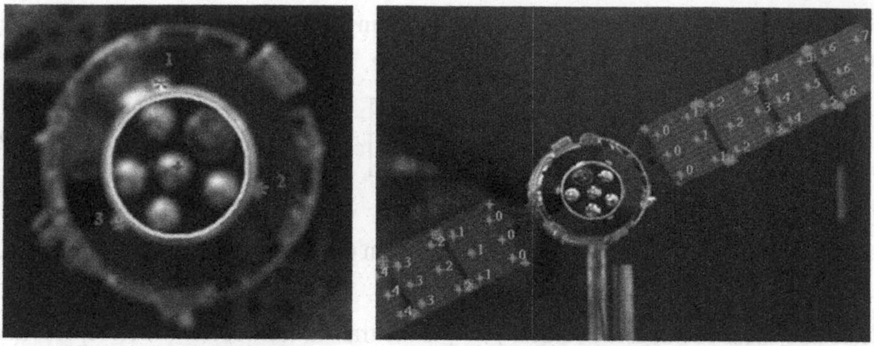

(a) on the fore-end cabin (b) on two solar panels

Fig. 2 The original images

(a) **(b)**

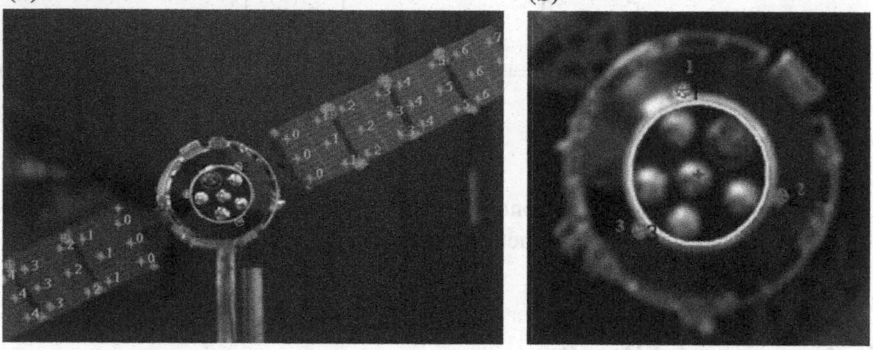

Fig. 3 Selecting control points

Fig. 4 The result of images merging

The original images are as follows. Figure 2a shows three feature points and the characteristic circle on the fore-end cabin. Figure 2b shows 48 feature points on two solar panels. Three control points are selected clockwise from two original images respectively in Fig. 3a, b. In particular, green and red small circles represent three control points. Figure 4 shows the results of images merging, which indicates that features on the fore-end cabin and two solar panels of the target aircraft are merged the same images. It can be applied for extracting motion information based on these feature points.

5.2 Image Merging Based on Five Control Points

When the number of matching points is more than 3, the least squares method is used to solve. In the experiment, five feature points are considered as five control points. Coordinate affine transformation pseudo inverse matrix is as follows:

$$Hinv = \begin{bmatrix} 0.3321 & -1.7882 & 637.7555 \\ 1.6506 & 0.4383 & -154.8800 \\ 0 & 0 & 1.0000 \end{bmatrix}$$

The original images are in Figs. 5 and 2b. Five control points are selected clockwise from two original images respectively in Fig. 6a, b. In particular, green and red small circles represent five control points, three feature points on the fore-end cabin and two feature points on the left and right solar. Figure 7 shows the experimental results of images merging, which achieve images translation, rotation, scale and shear.

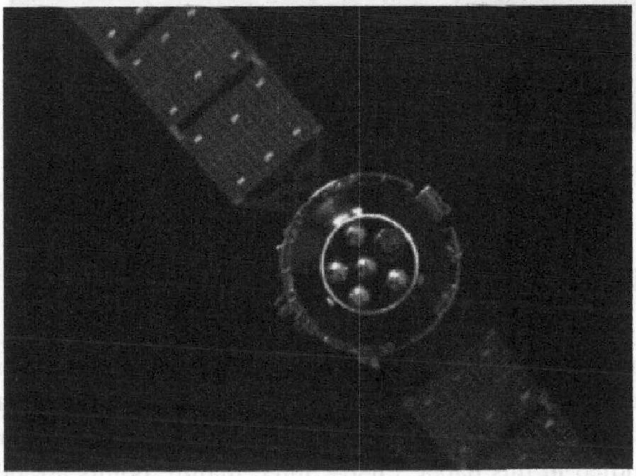

Fig. 5 The original images

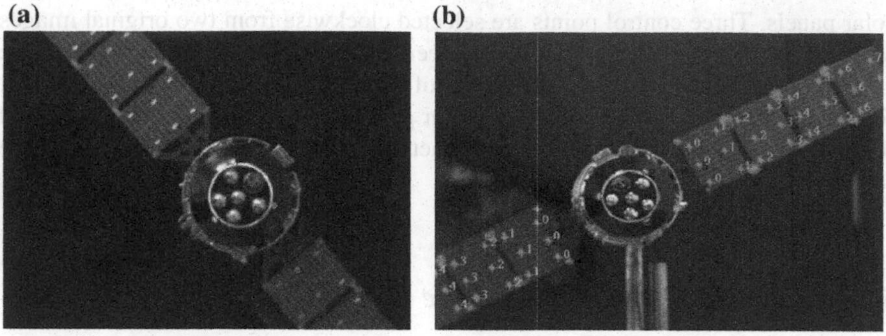

Fig. 6 Selecting control points

Fig. 7 The result of images merging

6 Conclusion

In order to accomplish the rendezvous and docking successfully, the visual measurement system played a key role. The motion information extraction of target spacecraft is achieved based on feature points. Therefore, feature point designing and images merging are studied in this paper. Firstly, three feature points are designed on the fore-end cabin of the target spacecraft model, which are arranged evenly around the feature circle at the angle of 120°. Then some of extracted feature points are regard as control points and a global affine transformation is utilized to image merging. The feature extracted on the front-end cabin and on the two solar panels are merged. Finally, the image of on the fore-end cabin contained three feature points and the one characteristic circle and 48 feature points on the two solar panels are merged into the same image. In the experiment, a global affine transformation is applied for images merging based on three control points and five control points, which achieve images translation, rotation, scale and shear well. In the future work, we will extract motion information of the target aircraft base on these features and design ground simulation vision system.

Acknowledgements This work was supported by the NSFC (61327807, 6152 1091, 61520106010, 61134005), and the National Basic Research Program of China (973 Program: 2012CB821200, 2012CB821201).

References

1. W. Kornus, M. Lehner, M. Schroeder, Geometric inflight-calibration by block adjustment using MOMS-2P imagery of three intersecting stereo-strips, in *Proceedings of ISPRS Workshop "Sensors and Mapping from Space* (1999)
2. P. Tao, L. Lu, Y. Zhang et al., On-orbit geometric calibration of the panchromatic/multispectral camera of the ZY-1 02C Satellite based on public geographic data. Photogram. Eng. Remote Sens. **80**(6), 505–517 (2014)
3. T. Toutin, E. Blondel, K. Rother et al., In-flight calibration of SPOT5 and formosat2. Rev. Fr. De Photogrammétrie Et De Télédétection **184**, 107–114 (2006)
4. K. Jacobsen, Calibration of imaging satellite sensors. Int. Arch. Photogramm. Remote Sens. (2012)
5. D.C. Seo, C. Lee, J. Oh, Merge of sub-images from two PAN CCD lines of Kompsat-3 AEISS. Ksce J. Civil Eng. **20**(2), 863–872 (2016)
6. L. Hui, Image registration based on corner detection and affine transformation. Int. Congr. Image Sig. Process. IEEE, 2184–2188 (2010)
7. T. Wakahara, K. Odaka, Adaptive normalization of handwritten characters using global/local affine transformation. IEEE Trans. Pattern Anal. Mach. Intell. **20**(12), 1332–1341 (1998)
8. T. Wakahara, Y. Kimura, A. Tomono, Affine-invariant recognition of gray-scale characters using global affine transformation correlation. IEEE Trans. Pattern Anal. Mach. Intell. **23**(4), 384–395 (2001)
9. W. Pei, Y. Jia, Feature points designing and matching for the target spacecraft in the final approaching phase of rendezvous and docking, in *Proceeding of the International Conference on Artificial Life and Robotics*, Oita, Japan (2018)

Computing Derivatives in Fuzzy Logic Systems with Consequents as Fuzzy Weighted Averages of Antecedents

Yuqing Liu, Qiye Zhang and Chunwei Wen

Abstract Zhang, Liu and Tian have proposed a fuzzy logic system (called FWA) with consequents as fuzzy weighted averages of antecedents [5]. However, it is generally complicated to compute the derivatives that are needed to implement steepest-descent parameter tuning algorithms for such systems. Therefore, in this paper, we provide mathematical formulas for computing the derivatives of trapezoidal membership functions used in [5].

Keywords Fuzzy logic system · Trapezoidal fuzzy number · Fuzzy weighted average · Error back-propagation · Steepest descent algorithm

1 Introduction

Mamdai fuzzy logic systems (FLSs) [1] and Takagi-Sugeno-Kang (TSK) systems [2, 3] have been developed rapidly and utilized to various fields, such as system identification, pattern recognition, signal processing, communication network and financial investment, etc, [4]. Both of them are characterized by IF-THEN rules, but neither of them establishes the connection between the consequents and antecedents. They take consequents arbitrarily, or only describe the relationship between consequents and inputs. From this reason, Zhang, Liu and Tian have proposed a fuzzy logic

Y. Liu · Q. Zhang (✉) · C. Wen
The School of Mathematics and Systems Science and LMIB of the Ministry of Education, Beihang University, Beijing 100191, China
e-mail: zhangqiye@buaa.edu.cn

Y. Liu
e-mail: 13998127561@163.com

C. Wen
e-mail: zzzabc159@qq.com

Y. Jia et al. (eds.), *Proceedings of 2018 Chinese Intelligent Systems Conference*, Lecture Notes in Electrical Engineering 529,
https://doi.org/10.1007/978-981-13-2291-4_69

system (called FWA) with consequents as fuzzy weighted averages of antecedents in [5]. The proposed rules establish some relationship between consequents and antecedents in advance, and the proposed FWA FLS can reduce the training time, improve training efficiency, and optimize parameters faster.

Generally speaking, the adjustment of the parameters of antecedent and consequent fuzzy sets in such a system is very difficult. However, for trapezoidal fuzzy numbers, the computation of derivatives is feasible. Therefore, in this paper, we provides mathematical formulas for computing the derivatives of trapezoidal membership functions that are needed to implement steepest-descent algorithms for such systems. The computation method is also used in [5] for their parameters adjustment.

The rest of this paper is organized as follows. Section 2 introduces the FWA FLS. Section 3 provides the detailed computation of the derivatives of trapezoidal membership functions.

2 The FWA Fuzzy Logic System

As mentioned in [5], the structure of a FWA FLS is composed of four modules: fuzzier, IF-THEN rules, inference and defuzzifier, as shown in Fig. 1. They are described in detail next.

Fuzzifier In [5], for computation complexity, singleton fuzzifier is used, which is nothing but a fuzzy singleton, i.e., $A_{\mathbf{x}'}$ is a *fuzzy singleton* with support $\mathbf{x}' = (x'_1, x'_2, \ldots, x'_p)$ if

$$\mu_{X_k}(x_k) = \begin{cases} 1, & x_k = x'_k, \\ 0, & x_k \neq x'_k, \end{cases} \qquad k = 1, \ldots, p$$

Rules Like a Mamdani FLS, the FWA FLS has p inputs $x_j \in [\alpha_j, \beta_j]$, $(j = 1, 2, \ldots, p)$, and one output $y \in Y$. For each $j = 1, 2, \ldots, p$, the jth input space is divided into N_j fuzzy sets $F_j^{l_j}$ $(l_j = 1, 2, \ldots, N_j)$ on $[\alpha_j, \beta_j]$. Thus, there are $M = N_1 N_2 \ldots N_p$ rules in the system:

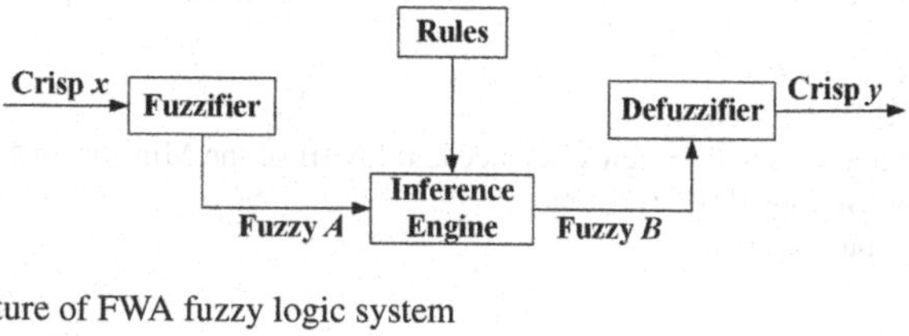

Fig. 1 The structure of FWA fuzzy logic system

$$\mathbf{R}^{l_1 l_2 \ldots l_p}: \text{IF } x_1 \text{ is } F_1^{l_1} \text{ and } \ldots \text{ and } x_p \text{ is } F_p^{l_p}, \text{THEN}$$
$$y^{l_1 l_2 \ldots l_p} \text{ is } Y^{l_1 l_2 \ldots l_p} = \frac{\sum_{j=1}^{p} W_j^{l_j} F_j^{l_j}}{\sum_{j=1}^{p} W_j^{l_j}} + B^{l_1 l_2 \ldots l_p} \tag{1}$$

where $j = 1, 2, \ldots, p$, $l_j = 1, 2, \ldots, N_j$; $W_j^{l_j}$ is a fuzzy set, viewed as the "fuzzy weight", and $B^{l_1 l_2 \ldots l_p}$ is also a fuzzy set, viewed as the "fuzzy bias". To simplify notation, we set $l_1 l_2 \ldots l_p$ to be l. Then the lth ($l = 1, \ldots, M$) rule can be described as

$$\mathbf{R}^l: \text{IF } x_1 \text{ is } F_1^{l_1} \text{ and } \ldots \text{ and } x_p \text{ is } F_p^{l_p}, \text{THEN } y^l \text{ is } Y^l = \frac{\sum_{j=1}^{p} W_j^{l_j} F_j^{l_j}}{\sum_{j=1}^{p} W_j^{l_j}} + B^l \tag{2}$$

Inference and defuzzifier To get a compromise between accuracy and computational complexity, COS defuzzifier is used in the FWA FLS. The final output of the FWA FLS is calculated by the following numerical weighted average:

$$y_{FWA}(\mathbf{x}) = \frac{\sum_{l_1=1}^{N_1} \sum_{l_2=1}^{N_2} \cdots \sum_{l_p=1}^{N_p} f^{l_1 l_2 \ldots l_p}(\mathbf{x}) y^{l_1 l_2 \ldots l_p}}{\sum_{l_1=1}^{N_1} \sum_{l_2=1}^{N_2} \cdots \sum_{l_p=1}^{N_p} f^{l_1 l_2 \ldots l_p}(\mathbf{x})} \triangleq \frac{\sum_{l=1}^{M} y^l f^l(\mathbf{x})}{\sum_{l=1}^{M} f^l(\mathbf{x})} \tag{3}$$

where $f^l(\mathbf{x}) \triangleq f^{l_1 l_2 \ldots l_p}(\mathbf{x}) = \prod_{j=1}^{p} \mu_{F_j^{l_j}}(x_j)$ is the firing level (or firing strength) of the $l \triangleq l_1 l_2 \ldots l_p$ th rule; $y^l \triangleq y^{l_1 l_2 \ldots l_p}$ is the GMIR (graded mean integration representation [6]) of $Y^l \triangleq Y^{l_1 l_2 \ldots l_p}$ (substitute for the centroid of Y^l), and $M = N_1 N_2 \ldots N_p$.

The output of the FWA FLS $y_{FWA}(\mathbf{x})$ can be also written by using FBFs as follows:

$$y_{FWA}(\mathbf{x}) = \sum_{l=1}^{M} y^l \phi^l(\mathbf{x}) \tag{4}$$

where the FBFs $\phi^l(\mathbf{x})$ are calculated by $\phi^l(\mathbf{x}) = \frac{f^l(\mathbf{x})}{\sum_{l=1}^{M} f^l(\mathbf{x})}$, $l = 1, \ldots, M$.

Again, for computation simplicity, in [5], trapezoidal fuzzy numbers are taken for both antecedent and consequent fuzzy sets. Let $Y_{FWA}^l \triangleq Y_{FWA}^{l_1 l_2 \ldots l_p}$ denote the fuzzy weighted average term $\frac{\sum_{j=1}^{p} W_j^{l_j} F_j^{l_j}}{\sum_{j=1}^{p} W_j^{l_j}}$ in Y^l, which was calculated through the method introduced in [7]. By that one first computes the α-cuts of the involved fuzzy sets and then calculates the corresponding interval weighted averages using KM algorithm [8] or EKM algorithm [9]. The resulted fuzzy set is also a trapezoidal fuzzy number. According to the results in [10], the GMIR y^l of the consequent $Y^l, l = 1, \ldots, M$ can be computed by

$$y^l = P(Y_{FWA}^l) + P(B^l) = \frac{1}{6}\left(a_{Y_{FWA}^l} + 2b_{Y_{FWA}^l} + 2c_{Y_{FWA}^l} + d_{Y_{FWA}^l} + a_{B^l} + 2b_{B^l} + 2c_{B^l} + d_{B^l}\right) \tag{5}$$

where $P(Y^l_{FWA})$ and $P(B^l)$ denote the GMIR of Y^l_{FWA} and B^l, respectively. Thus, the output of the FWA FLS $y_{FWA}(\mathbf{x})$ can be rewritten as follows:

$$y_{FWA}(\mathbf{x}) = \frac{1}{6}\sum_{l=1}^{M}\phi^l(\mathbf{x})\left(a_{Y^l_{FWA}} + 2b_{Y^l_{FWA}} + 2c_{Y^l_{FWA}} + d_{Y^l_{FWA}} + a_{B^l} + 2b_{B^l} + 2c_{B^l} + d_{B^l}\right) \tag{6}$$

In the following, we will derive the analytical expressions of the parameters $a_{Y^l_{FWA}}$, $b_{Y^l_{FWA}}$, $c_{Y^l_{FWA}}$, $d_{Y^l_{FWA}}$ for the trapezoidal fuzzy set Y^l_{FWA}.

Specifically, for $\alpha \in [0, 1]$, let $\underline{y}^l_{FWA}(\alpha)$ and $\overline{y}^l_{FWA}(\alpha)$ denote the left and right end points of Y^l_{FWA}'s α-cut $Y^l_{FWA}(\alpha)$, respectively; $\underline{W}^{l_j}_j(\alpha)$, $\overline{W}^{l_j}_j(\alpha)$, $(j = 1, \ldots, p)$ represent the left and right ends of $W^{l_j}_j$'s α-cut $W^{l_j}_j(\alpha)$, respectively; $F^{l_j}_{jL}(\alpha), F^{l_j}_{jR}(\alpha)$, $(j = 1, \ldots, p)$ denote the left and right ends of $F^{l_j}_j$'s α-cut $F^{l_j}_j(\alpha)$, respectively. According to KM algorithm, $\underline{y}^l_{FWA}(\alpha)$ and $\overline{y}^l_{FWA}(\alpha)$ can be expressed as follows:

$$\underline{y}^l_{FWA}(\alpha) = \frac{\sum_{i=1}^{L} F^{l_i}_{iL}(\alpha)\overline{W}^{l_i}_i(\alpha) + \sum_{j=L+1}^{p} F^{l_j}_{jL}(\alpha)\underline{W}^{l_j}_j(\alpha)}{\sum_{i=1}^{L} \overline{W}^{l_i}_i(\alpha) + \sum_{j=L+1}^{p} \underline{W}^{l_j}_j(\alpha)} \tag{7}$$

$$\overline{y}^l_{FWA}(\alpha) = \frac{\sum_{i=1}^{R} F^{l_i}_{iR}(\alpha)\underline{W}^{l_i}_i(\alpha) + \sum_{j=R+1}^{p} F^{l_j}_{jR}(\alpha)\overline{W}^{l_j}_j(\alpha)}{\sum_{i=1}^{R} \underline{W}^{l_i}_i(\alpha) + \sum_{j=R+1}^{p} \overline{W}^{l_j}_j(\alpha)} \tag{8}$$

where L, R are the switch points in the process of $\underline{y}^l_{FWA}(\alpha)$ and $\overline{y}^l_{FWA}(\alpha)$ by using KM algorithm, respectively. It is worthy to know that the formulas in Eqs. (7) and (8) cannot be used because the $\underline{W}^{l_i}_i(\alpha)$, $F^{l_j}_{jL}(\alpha)$, $\overline{W}^{l_j}_j(\alpha)$ and $F^{l_j}_{jR}(\alpha)$ have been reordered during step-1 of the two iterative procedures used to compute $\underline{y}^l_{FWA}(\alpha)$ and $\overline{y}^l_{FWA}(\alpha)$ [11]. In order to compute derivatives of $\underline{y}^l_{FWA}(\alpha)$ and $\overline{y}^l_{FWA}(\alpha)$ with respect to MF parameters, we will adopt the procedures suggested in [11].

First, we deduce the analytic expressions of $\underline{y}^l_{FWA}(\alpha)$ and $\overline{y}^l_{FWA}(\alpha)$. For this, we introduce the following notations (see Table 1):

Refer to the method in [11], we can obtain the analytic formulas of $\underline{y}^l_{FWA}(\alpha)$ and $\overline{y}^l_{FWA}(\alpha)$ as follows:

$$\begin{aligned}\underline{y}^l_{FWA}(\alpha) &= \frac{\overline{\mathbf{W}}^{\mathrm{T}}(\alpha)\mathbf{M}^l_{L1}(\alpha)\mathbf{F}^l_L(\alpha) + \underline{\mathbf{W}}^{\mathrm{T}}(\alpha)\mathbf{M}^l_{L2}(\alpha)\mathbf{F}^l_L(\alpha)}{\overline{\mathbf{W}}^{\mathrm{T}}(\alpha)\mathbf{r}^{\mathrm{T}}_L\mathbf{Q}_L(\alpha) + \underline{\mathbf{W}}^{\mathrm{T}}(\alpha)\mathbf{s}^{\mathrm{T}}_L\mathbf{Q}_L(\alpha)} = \frac{\overline{\mathbf{W}}^{\mathrm{T}}(\alpha)\mathbf{m}^l_L(\alpha) + \underline{\mathbf{W}}^{\mathrm{T}}(\alpha)\mathbf{k}^l_L(\alpha)}{\overline{\mathbf{W}}^{\mathrm{T}}(\alpha)\mathbf{p}^l_L(\alpha) + \underline{\mathbf{W}}^{\mathrm{T}}(\alpha)\mathbf{q}^l_L(\alpha)} \\ &= \frac{\sum_{i=1}^{p} m^l_{L,i}(\alpha)\overline{W}^{l_i}_i(\alpha) + \sum_{j=1}^{p} k^l_{L,j}(\alpha)\underline{W}^{l_j}_j(\alpha)}{\sum_{i=1}^{p} p^l_{L,i}(\alpha)\overline{W}^{l_i}_i(\alpha) + \sum_{j=1}^{p} q^l_{L,j}(\alpha)\underline{W}^{l_j}_j(\alpha)}\end{aligned} \tag{9}$$

Table 1 Notations and formulas involved in calculating $\underline{y}^l_{FWA}(\alpha)$ and $\overline{y}^l_{FWA}(\alpha)$

Calculation of $\underline{y}^l_{FWA}(\alpha)$	Calculation of $\overline{y}^l_{FWA}(\alpha)$
$\underline{\mathbf{g}}^l = (\underline{g}^l_1, \underline{g}^l_2, \ldots, \underline{g}^l_p) \equiv \mathbf{Q}^l_L(\alpha)\underline{\mathbf{W}}(\alpha)$	$\underline{\mathbf{g}}^l = (\underline{g}^l_1, \underline{g}^l_2, \ldots, \underline{g}^l_p) \equiv \mathbf{Q}^l_R(\alpha)\underline{\mathbf{W}}(\alpha)$
$\overline{\mathbf{g}}^l = (\overline{g}^l_1, \overline{g}^l_2, \ldots, \overline{g}^l_p) \equiv \mathbf{Q}^l_L(\alpha)\overline{\mathbf{W}}(\alpha)$	$\overline{\mathbf{g}}^l = (\overline{g}^l_1, \overline{g}^l_2, \ldots, \overline{g}^l_p) \equiv \mathbf{Q}^l_R(\alpha)\overline{\mathbf{W}}(\alpha)$
$\mathbf{E}_1 \equiv (\mathbf{e}_1\vert\mathbf{e}_2\vert\ldots\vert\mathbf{e}_{K_l}\vert\mathbf{0}\vert\ldots\vert\mathbf{0})^{\mathrm{T}}\quad K_l \times p$	$\mathbf{E}_1 \equiv (\mathbf{e}_1\vert\mathbf{e}_2\vert\ldots\vert\mathbf{e}_{K_r}\vert\mathbf{0}\vert\ldots\vert\mathbf{0})^{\mathrm{T}}\quad K_r \times p$
$\mathbf{e}_i = K_l \times 1$ elementary vector	$\mathbf{e}_i = K_r \times 1$ elementary vector
$\mathbf{E}_2 \equiv (\mathbf{0}\vert\ldots\vert\mathbf{0}\vert{''}_1\vert{''}_2\vert\ldots\vert{''}_{\mathbf{p-K_l}})^{\mathrm{T}}$ $(p-K_l)\times 1$	$\mathbf{E}_2 \equiv (\mathbf{0}\vert\ldots\vert\mathbf{0}\vert{''}_1\vert{''}_2\vert\ldots\vert{''}_{\mathbf{p-K_r}})^{\mathrm{T}}$ $(p-K_r)\times 1$
$\varepsilon_i = (p-K_l)\times 1$ elementary vector	$\varepsilon_i = (p-K_r)\times 1$ elementary vector
$\mathbf{M}^l_{L1}(\alpha) \equiv \mathbf{Q}^{l\mathrm{T}}_L(\alpha)\mathbf{E}^{\mathrm{T}}_1\mathbf{E}_1\mathbf{Q}^l_L(\alpha)\quad p\times p$	$\mathbf{M}^l_{R1}(\alpha) \equiv \mathbf{Q}^{l\mathrm{T}}_R(\alpha)\mathbf{E}^{\mathrm{T}}_1\mathbf{E}_1\mathbf{Q}^l_R(\alpha)\quad p\times p$
$\mathbf{M}^l_{L2}(\alpha) \equiv \mathbf{Q}^{l\mathrm{T}}_L(\alpha)\mathbf{E}^{\mathrm{T}}_2\mathbf{E}_2\mathbf{Q}^l_L(\alpha)\quad p\times p$	$\mathbf{M}^l_{R2}(\alpha) \equiv \mathbf{Q}^{l\mathrm{T}}_R(\alpha)\mathbf{E}^{\mathrm{T}}_2\mathbf{E}_2\mathbf{Q}^l_R(\alpha)\quad p\times p$
$\mathbf{r}_L \equiv (\underbrace{1,1,\ldots,1}_{K_l},0,\ldots,0)^{\mathrm{T}}\quad p\times 1$	$\mathbf{s}_R \equiv (0,0,\ldots,\overbrace{1,\ldots,1}^{K_r})^{\mathrm{T}}\quad p\times 1$
$\mathbf{F}^l_L(\alpha) = (F^{l_1}_{1L}(\alpha), F^{l_2}_{2L}(\alpha), \ldots, F^{l_p}_{pL}(\alpha))^{\mathrm{T}}$	$\mathbf{F}^l_R(\alpha) = (F^{l_1}_{1R}(\alpha), F^{l_2}_{2R}(\alpha), \ldots, F^{l_p}_{pR}(\alpha))^{\mathrm{T}}$
$\underline{\mathbf{W}}(\alpha) = (\underline{W}^{l_1}_1(\alpha), \underline{W}^{l_2}_2(\alpha), \ldots, \underline{W}^{l_p}_p(\alpha))^{\mathrm{T}}$	$\overline{\mathbf{W}}(\alpha) = (\overline{W}^{l_1}_1(\alpha), \overline{W}^{l_2}_2(\alpha), \ldots, \overline{W}^{l_p}_p(\alpha))^{\mathrm{T}}$
$\mathbf{m}^l_L(\alpha) \equiv \mathbf{M}^l_{L1}(\alpha)\mathbf{F}^l_L(\alpha)\quad p\times 1$	$\mathbf{m}^l_R(\alpha) \equiv \mathbf{M}^l_{R1}(\alpha)\mathbf{F}^l_R(\alpha)\quad p\times 1$
$\mathbf{k}^l_L(\alpha) \equiv \mathbf{M}^l_{L2}(\alpha)\mathbf{F}^l_L(\alpha)\quad p\times 1$	$\mathbf{k}^l_R(\alpha) \equiv \mathbf{M}^l_{R2}(\alpha)\mathbf{F}^l_R(\alpha)\quad p\times 1$
$\mathbf{p}^l_L(\alpha) \equiv \mathbf{r}^{\mathrm{T}}_L\mathbf{Q}_L(\alpha)\quad 1\times p$	$\mathbf{p}^l_R(\alpha) \equiv \mathbf{r}^{\mathrm{T}}_R\mathbf{Q}_R(\alpha)\quad 1\times p$
$\mathbf{q}^l_L(\alpha) \equiv \mathbf{s}^{\mathrm{T}}_L\mathbf{Q}_L(\alpha)\quad 1\times p$	$\mathbf{q}^l_R(\alpha) \equiv \mathbf{s}^{\mathrm{T}}_R\mathbf{Q}_R(\alpha)\quad 1\times p$

$$
\begin{aligned}
\overline{y}^l_{FWA}(\alpha) &= \frac{\underline{\mathbf{W}}^{\mathrm{T}}(\alpha)\mathbf{M}^l_{R1}(\alpha)\mathbf{F}^l_R(\alpha) + \overline{\mathbf{W}}^{\mathrm{T}}(\alpha)\mathbf{M}^l_{R2}(\alpha)\mathbf{F}^l_R(\alpha)}{\underline{\mathbf{W}}^{\mathrm{T}}(\alpha)\mathbf{r}^{\mathrm{T}}_R\mathbf{Q}_R(\alpha) + \overline{\mathbf{W}}^{\mathrm{T}}(\alpha)\mathbf{s}^{\mathrm{T}}_R\mathbf{Q}_R(\alpha)} = \frac{\underline{\mathbf{W}}^{\mathrm{T}}(\alpha)\mathbf{m}^l_R(\alpha) + \overline{\mathbf{W}}^{\mathrm{T}}(\alpha)\mathbf{k}^l_R(\alpha)}{\underline{\mathbf{W}}^{\mathrm{T}}(\alpha)\mathbf{p}^l_R(\alpha) + \overline{\mathbf{W}}^{\mathrm{T}}(\alpha)\mathbf{q}^l_R(\alpha)} \\
&= \frac{\sum_{i=1}^{p} m^l_{R,i}(\alpha)\underline{W}^{l_i}_i(\alpha) + \sum_{j=1}^{p} k^l_{R,j}(\alpha)\overline{W}^{l_j}_j(\alpha)}{\sum_{i=1}^{p} p^l_{R,i}(\alpha)\underline{W}^{l_i}_i(\alpha) + \sum_{j=1}^{p} q^l_{R,j}(\alpha)\overline{W}^{l_j}_j(\alpha)}
\end{aligned}
\tag{10}
$$

By taking α in Eqs. (9) and (10) to be 0 and 1 in turn, we can obtain the analytic expressions of $a_{Y^l_{FWA}}, b_{Y^l_{FWA}}, c_{Y^l_{FWA}}, d_{Y^l_{FWA}}$ immediately:

$$
a_{Y^l_{FWA}} = \underline{y}^l_{FWA}(0) = \frac{\overline{\mathbf{W}}^{\mathrm{T}}(0)\mathbf{m}^l_L(0) + \underline{\mathbf{W}}^{\mathrm{T}}(0)\mathbf{k}^l_L(0)}{\overline{\mathbf{W}}^{\mathrm{T}}(0)\mathbf{p}^l_L(0) + \underline{\mathbf{W}}^{\mathrm{T}}(0)\mathbf{q}^l_L(0)} = \frac{\sum_{i=1}^{p} m^l_{L,i}(0)\overline{W}^{l_i}_i(0) + \sum_{j=1}^{p} k^l_{L,j}(0)\underline{W}^{l_j}_j(0)}{\sum_{i=1}^{p} p^l_{L,i}(0)\overline{W}^{l_i}_i(0) + \sum_{j=1}^{p} q^l_{L,j}(0)\underline{W}^{l_j}_j(0)}
\tag{11}
$$

$$
b_{Y^l_{FWA}} = \underline{y}^l_{FWA}(1) = \frac{\overline{\mathbf{W}}^{\mathrm{T}}(1)\mathbf{m}^l_L(1) + \underline{\mathbf{W}}^{\mathrm{T}}(1)\mathbf{k}^l_L(1)}{\overline{\mathbf{W}}^{\mathrm{T}}(1)\mathbf{p}^l_L(1) + \underline{\mathbf{W}}^{\mathrm{T}}(1)\mathbf{q}^l_L(1)} = \frac{\sum_{i=1}^{p} m^l_{L,i}(1)\overline{W}^{l_i}_i(1) + \sum_{j=1}^{p} k^l_{L,j}(1)\underline{W}^{l_j}_j(1)}{\sum_{i=1}^{p} p^l_{L,i}(1)\overline{W}^{l_i}_i(1) + \sum_{j=1}^{p} q^l_{L,j}(1)\underline{W}^{l_j}_j(1)}
\tag{12}
$$

$$c_{y^l_{FWA}} = \overline{y}^l_{FWA}(1) = \frac{\underline{\mathbf{W}}^{\mathrm{T}}(1)\mathbf{m}^l_R(1) + \overline{\mathbf{W}}^{\mathrm{T}}(1)\mathbf{k}^l_R(1)}{\underline{\mathbf{W}}^{\mathrm{T}}(1)\mathbf{p}^l_R(1) + \overline{\mathbf{W}}^{\mathrm{T}}(1)\mathbf{q}^l_R(1)} = \frac{\sum_{i=1}^{p} m^l_{R,i}(1)\underline{W}^{l_i}_i(1) + \sum_{j=1}^{p} k^l_{R,j}(1)\overline{W}^{l_j}_j(1)}{\sum_{i=1}^{p} p^l_{R,i}(1)\underline{W}^{l_i}_i(1) + \sum_{j=1}^{p} q^l_{R,j}(1)\overline{W}^{l_j}_j(1)} \tag{13}$$

$$d_{y^l_{FWA}} = \overline{y}^l_{FWA}(0) = \frac{\underline{\mathbf{W}}^{\mathrm{T}}(0)\mathbf{m}^l_R(0) + \overline{\mathbf{W}}^{\mathrm{T}}(0)\mathbf{k}^l_R(0)}{\underline{\mathbf{W}}^{\mathrm{T}}(0)\mathbf{p}^l_R(0) + \overline{\mathbf{W}}^{\mathrm{T}}(0)\mathbf{q}^l_R(0)} = \frac{\sum_{i=1}^{p} m^l_{R,i}(0)\underline{W}^{l_i}_i(0) + \sum_{j=1}^{p} k^l_{R,j}(0)\overline{W}^{l_j}_j(0)}{\sum_{i=1}^{p} p^l_{R,i}(0)\underline{W}^{l_i}_i(0) + \sum_{j=1}^{p} q^l_{R,j}(0)\overline{W}^{l_j}_j(0)} \tag{14}$$

3 Parameters Learning Based on the Steepest Descent Method

In this section, we derive the parameter updating formulas that used by the steepest descent method in the FWA FLS. Let θ denote a generic symbol that represents all parameters to be optimized for the generic objective function E by the following steepest descent algorithm:

$$\theta_{new} = \theta_{old} - \alpha \frac{\partial E}{\partial \theta}|_{\theta_{old}} \tag{15}$$

where $\alpha > 0$ is the learning parameter, and $\partial E/\partial \alpha$ is computed as

$$\frac{\partial E}{\partial \theta} = \frac{\partial E}{\partial y(t|\theta)} \frac{\partial y(t|\theta)}{\partial \theta} \tag{16}$$

The calculation of $\frac{\partial E}{\partial \theta}$ depends on the specific FLS and requires a very careful use of the chain rule.

In the FWA FLS, there are $4Mp$ antecedent parameters and $4Mp + 4M$ consequent parameters to be tuned, they are $a_{F^{l_j}_j}, b_{F^{l_j}_j}, c_{F^{l_j}_j}, d_{F^{l_j}_j}, a_{W^{l_j}_j}, b_{W^{l_j}_j}, c_{W^{l_j}_j}, d_{W^{l_j}_j}$, $(j = 1, 2, \ldots, p, l_1 = 1, \ldots, N_1, l_2 = 1, \ldots, N_2, \ldots, l_p = 1, \ldots, N_p)$, and $a_{B^l}, b_{B^l}, c_{B^l}, d_{B^l}$, $(l = 1, \ldots, M)$, $M = N_1 N_2 \ldots N_p$. We will optimize all aforementioned parameters based on the following error function:

$$E^{(t)}_{FWA} = \frac{1}{2}\left(y_{FWA}(\mathbf{x}^{(t)}) - y^{(t)}\right)^2 \tag{17}$$

where $(\mathbf{x}^{(t)}, y^{(t)})$, $(t = 1, \ldots, T)$ are T training data pairs, $y_{FWA}(\mathbf{x}^{(t)})$ is the crisp output of the FWA FLS. For notational simplicity, we shorten $y_{FWA}(\mathbf{x}^{(t)})$ to y_{FWA}.

3.1 The updating formulas of the consequent parameters

There are $4Mp + 4M$ consequent parameters to be tuned, they are $a_{W_j^{l_j}}$, $b_{W_j^{l_j}}$, $c_{W_j^{l_j}}$, $d_{W_j^{l_j}}$, $(j = 1, 2, \ldots, p, l_1 = 1, \ldots, N_1, l_2 = 1, \ldots, N_2, \ldots, l_p = 1, \ldots, N_p)$, and a_{B^l}, $b_{B^l}, c_{B^l}, d_{B^l}, (l = 1, \ldots, M), M = N_1 N_2 \ldots N_p$. To determine the parameters by using steepest descent method, based on δ-rule we have

$$a_{W_j^{l_j}}(q+1) = a_{W_j^{l_j}}(q) - \eta \cdot \left.\frac{\partial E_{FWA}^{(t)}}{\partial a_{W_j^{l_j}}}\right|_q, \quad a_{B^l}(q+1) = a_{B^l}(q) - \eta \cdot \left.\frac{\partial E_{FWA}^{(t)}}{\partial a_{B^l}}\right|_q \tag{18}$$

We can get $b_{W_j^{l_j}}, c_{W_j^{l_j}}, d_{W_j^{l_j}}$ and $b_{B^l}, c_{B^l}, d_{B^l}$ in a similar way. From Eq. (17) and compound function derivation rule, we can replace $\frac{\partial E_{FWA}^{(t)}}{\partial a_{W_j^{l_j}}}$ and $\frac{\partial E_{FWA}^{(t)}}{\partial a_{B^l}}$ with the following formulas

$$\frac{\partial E_{FWA}^{(t)}}{\partial a_{W_j^{l_j}}} = \left(y_{FWA} - y^{(t)}\right) \cdot \frac{\partial y_{FWA}}{\partial a_{W_j^{l_j}}}, \quad \frac{\partial E_{FWA}^{(t)}}{\partial a_{B^l}} = \left(y_{FWA} - y^{(t)}\right) \cdot \frac{\partial y_{FWA}}{\partial a_{B^l}} \tag{19}$$

Similarly, we can do the same thing with $\frac{\partial E_{FWA}^{(t)}}{\partial b_{W_j^{l_j}}}, \frac{\partial E_{FWA}^{(t)}}{\partial c_{W_j^{l_j}}}, \frac{\partial E_{FWA}^{(t)}}{\partial d_{W_j^{l_j}}}$ and $\frac{\partial E_{FWA}^{(t)}}{\partial b_{B^l}}, \frac{\partial E_{FWA}^{(t)}}{\partial c_{B^l}}, \frac{\partial E_{FWA}^{(t)}}{\partial d_{B^l}}$.

From (6), it is easily obtained that

$$\frac{\partial y_{FWA}}{\partial a_{B^l}} = \frac{\partial y_{FWA}}{\partial d_{B^l}} = \frac{1}{6}\phi^l, \quad \frac{\partial y_{FWA}}{\partial b_{B^l}} = \frac{\partial y_{FWA}}{\partial c_{B^l}} = \frac{1}{3}\phi^l \tag{20}$$

Then according to Eqs. (11)–(14) and (6), we can calculate the derivatives of y_{FWA} to $a_{W_j^{l_j}}, b_{W_j^{l_j}}, c_{W_j^{l_j}}, d_{W_j^{l_j}}$ as follows:

$$\begin{aligned}
\frac{\partial y_{FWA}}{\partial a_{W_j^{l_j}}} &= \frac{1}{6}\phi^l \left(\frac{\partial a_{Y_{FWA}^l}}{\partial \underline{W}_j^{l_j}(0)} + \frac{\partial d_{Y_{FWA}^l}}{\partial \underline{W}_j^{l_j}(0)} \right), \\
\frac{\partial y_{FWA}}{\partial b_{W_j^{l_j}}} &= \frac{1}{3}\phi^l \left(\frac{\partial b_{Y_{FWA}^l}}{\partial \underline{W}_j^{l_j}(1)} + \frac{\partial c_{Y_{FWA}^l}}{\partial \underline{W}_j^{l_j}(1)} \right), \\
\frac{\partial y_{FWA}}{\partial c_{W_j^{l_j}}} &= \frac{1}{3}\phi^l \left(\frac{\partial b_{Y_{FWA}^l}}{\partial \overline{W}_j^{l_j}(1)} + \frac{\partial c_{Y_{FWA}^l}}{\partial \overline{W}_j^{l_j}(1)} \right), \\
\frac{\partial y_{FWA}}{\partial d_{W_j^{l_j}}} &= \frac{1}{6}\phi^l \left(\frac{\partial a_{Y_{FWA}^l}}{\partial \overline{W}_j^{l_j}(0)} + \frac{\partial d_{Y_{FWA}^l}}{\partial \overline{W}_j^{l_j}(0)} \right)
\end{aligned} \tag{21}$$

Again, by Eqs. (11)–(14), we can calculate the following derivatives:

$$\frac{\partial a_{Y^l_{FWA}}}{\partial \underline{W}^{l_j}_j(0)} = \frac{k^l_{L,i}(0) - q^l_{L,i}(0)a_{Y^l_{FWA}}}{\sum_{j=1}^{p} p^l_{L,j}(0)\overline{W}^{l_j}_j(0) + \sum_{i=1}^{p} q^l_{L,i}(0)\underline{W}^{l_i}_i(0)} \tag{22}$$

$$\frac{\partial a_{Y^l_{FWA}}}{\partial \overline{W}^{l_j}_j(0)} = \frac{m^l_{L,i}(0) - p^l_{L,i}(0)a_{Y^l_{FWA}}}{\sum_{j=1}^{p} p^l_{L,j}(0)\overline{W}^{l_j}_j(0) + \sum_{i=1}^{p} q^l_{L,i}(0)\underline{W}^{l_i}_i(0)} \tag{23}$$

we can get $\frac{\partial b_{Y^l_{FWA}}}{\partial \underline{W}^{l_j}_j(1)}$, $\frac{\partial b_{Y^l_{FWA}}}{\partial \overline{W}^{l_j}_j(1)}$, $\frac{\partial c_{Y^l_{FWA}}}{\partial \underline{W}^{l_j}_j(1)}$, $\frac{\partial c_{Y^l_{FWA}}}{\partial \overline{W}^{l_j}_j(1)}$, $\frac{\partial d_{Y^l_{FWA}}}{\partial \underline{W}^{l_j}_j(0)}$ and $\frac{\partial d_{Y^l_{FWA}}}{\partial \overline{W}^{l_j}_j(0)}$ in the same way.

Finally, we substitute Eqs. (22)–(23) for Eqs. (21), and substitute the results of Eqs. (20) and (21) for (19), then put the results into (18), we get the final updating formulas for consequent parameters as follows:

$$a_{W^{l_j}_j}(q+1) = a_{W^{l_j}_j}(q) - \eta \cdot (y_{FWA} - y^{(t)}) \cdot \frac{1}{6}\phi^l\left[\frac{k^l_{L,i}(0) - q^l_{L,i}(0)a_{Y^l_{FWA}}}{\sum_{j=1}^{p} p^l_{L,j}(0)\overline{W}^{l_j}_j(0) + \sum_{i=1}^{p} q^l_{L,i}(0)\underline{W}^{l_i}_i(0)} + \frac{m^l_{R,i}(0) - p^l_{R,i}(0)d_{Y^l_{FWA}}}{\sum_{i=1}^{p} p^l_{R,i}(0)\underline{W}^{l_i}_i(0) + \sum_{j=1}^{p} q^l_{R,j}(0)\overline{W}^{l_j}_j(0)}\right]\Bigg|_q \tag{24}$$

$$b_{W^{l_j}_j}(q+1) = b_{W^{l_j}_j}(q) - \eta \cdot (y_{FWA} - y^{(t)}) \cdot \frac{1}{3}\phi^l\left[\frac{k^l_{L,i}(1) - q^l_{L,i}(1)b_{Y^l_{FWA}}}{\sum_{j=1}^{p} p^l_{L,j}(1)\overline{W}^{l_j}_j(1) + \sum_{i=1}^{p} q^l_{L,i}(1)\underline{W}^{l_i}_i(1)} + \frac{m^l_{R,i}(1) - p^l_{R,i}(1)c_{Y^l_{FWA}}}{\sum_{i=1}^{p} p^l_{R,i}(1)\underline{W}^{l_i}_i(1) + \sum_{j=1}^{p} q^l_{R,j}(1)\overline{W}^{l_j}_j(1)}\right]\Bigg|_q \tag{25}$$

$$c_{W^{l_j}_j}(q+1) = c_{W^{l_j}_j}(q) - \eta \cdot (y_{FWA} - y^{(t)}) \cdot \frac{1}{3}\phi^l\left[\frac{m^l_{L,i}(1) - p^l_{L,i}(1)b_{Y^l_{FWA}}}{\sum_{j=1}^{p} p^l_{L,j}(1)\overline{W}^{l_j}_j(1) + \sum_{i=1}^{p} q^l_{L,i}(1)\underline{W}^{l_i}_i(1)} + \frac{k^l_{R,i}(1) - q^l_{R,i}(1)c_{Y^l_{FWA}}}{\sum_{i=1}^{p} p^l_{R,i}(1)\underline{W}^{l_i}_i(1) + \sum_{j=1}^{p} q^l_{R,j}(1)\overline{W}^{l_j}_j(1)}\right]\Bigg|_q \tag{26}$$

$$d_{W^{l_j}_j}(q+1) = d_{W^{l_j}_j}(q) - \eta \cdot (y_{FWA} - y^{(t)}) \cdot \frac{1}{6}\phi^l\left[\frac{m^l_{L,i}(0) - p^l_{L,i}(0)a_{Y^l_{FWA}}}{\sum_{j=1}^{p} p^l_{L,j}(0)\overline{W}^{l_j}_j(0) + \sum_{i=1}^{p} q^l_{L,i}(0)\underline{W}^{l_i}_i(0)} + \frac{k^l_{R,i}(0) - q^l_{R,i}(0)d_{Y^l_{FWA}}}{\sum_{i=1}^{p} p^l_{R,i}(0)\underline{W}^{l_i}_i(0) + \sum_{j=1}^{p} q^l_{R,j}(0)\overline{W}^{l_j}_j(0)}\right]\Bigg|_q \tag{27}$$

$$
\begin{aligned}
a_{B^l}(q+1) &= a_{B^l}(q) - \eta \cdot (y_{FWA} - y^{(t)}) \cdot \frac{1}{6}\phi^l\Big|_q, \\
b_{B^l}(q+1) &= b_{B^l}(q) - \eta \cdot (y_{FWA} - y^{(t)}) \cdot \frac{1}{3}\phi^l\Big|_q, \\
c_{B^l}(q+1) &= c_{B^l}(q) - \eta \cdot (y_{FWA} - y^{(t)}) \cdot \frac{1}{3}\phi^l\Big|_q, \\
d_{B^l}(q+1) &= d_{B^l}(q) - \eta \cdot (y_{FWA} - y^{(t)}) \cdot \frac{1}{6}\phi^l\Big|_q
\end{aligned}
\tag{28}
$$

3.2 The updating formulas of antecedent parameters

There are $4Mp$ antecedent parameters to be tuned, they are $a_{F_j^{l_j}}$, ($j = 1, 2, \ldots, p$, $l_1 = 1, \ldots, N_1$, $l_2 = 1, \ldots, N_2, \ldots, l_p = 1, \ldots, N_p$), and $M = N_1 N_2 \ldots N_p$. Again, based on δ-rule, we have

$$
a_{F_j^{l_j}}(q+1) = a_{F_j^{l_j}}(q) - \eta \cdot \frac{\partial E_{FWA}^{(t)}}{\partial a_{F_j^{l_j}}}\Bigg|_q \tag{29}
$$

and $b_{F_j^{l_j}}, c_{F_j^{l_j}}, d_{F_j^{l_j}}$ can be expressed in the same way. According to error function (17) and chain rule, $\frac{\partial E_{FWA}^{(t)}}{\partial a_{F_j^{l_j}}}$ can be computed as follows:

$$
\frac{\partial E_{FWA}^{(t)}}{\partial a_{F_j^{l_j}}} = \left(y_{FWA} - y^{(t)}\right) \cdot \frac{\partial y_{FWA}}{\partial a_{F_j^{l_j}}} \tag{30}
$$

and $\frac{\partial E_{FWA}^{(t)}}{\partial b_{F_j^{l_j}}}, \frac{\partial E_{FWA}^{(t)}}{\partial c_{F_j^{l_j}}}, \frac{\partial E_{FWA}^{(t)}}{\partial d_{F_j^{l_j}}}$ are similar to $\frac{\partial E_{FWA}^{(t)}}{\partial a_{F_j^{l_j}}}$.

Again, by chain rule, we have

$$
\begin{aligned}
\frac{\partial y_{FWA}}{\partial a_{F_j^{l_j}}} &= \frac{\partial y_{FWA}}{\partial y^l} \cdot \frac{\partial y^l}{\partial a_{F_j^{l_j}}} + \frac{\partial y_{FWA}}{\partial f^l} \cdot \frac{\partial f^l}{\partial a_{F_j^{l_j}}} \\
&= \phi^l \cdot \frac{\partial y^l}{\partial a_{F_j^{l_j}}} + \frac{y^l - y_{FWA}}{\sum_{l=1}^{M} f^l} \cdot \frac{\partial f^l}{\partial a_{F_j^{l_j}}} = \phi^l \cdot \left[\frac{\partial y^l}{\partial a_{F_j^{l_j}}} + \frac{y^l - y_{FWA}}{\mu_{F_j^{l_j}}(x_j)} \cdot \frac{\partial \mu_{F_j^{l_j}}(x_j)}{\partial a_{F_j^{l_j}}}\right]
\end{aligned}
\tag{31}
$$

where

$$\frac{\partial \mu_{F_j^{l_j}}(x_j)}{\partial a_{F_j^{l_j}}} = \begin{cases} \frac{x_j - b}{(b-a)^2}, & a_{F_j^{l_j}} \le x_j \le b_{F_j^{l_j}} \\ 0, & \text{others} \end{cases}$$

Similarly, we can get the derivatives of y_{FWA} to $b_{F_j^{l_j}}, c_{F_j^{l_j}}, d_{F_j^{l_j}}$. And according to Eqs. (11)–(14) and (5), the derivatives of y^l to $a_{F_j^{l_j}}, b_{F_j^{l_j}}, c_{F_j^{l_j}}, d_{F_j^{l_j}}$ can be written as

$$\frac{\partial y^l}{\partial a_{F_j^{l_j}}} = \frac{\partial y^l}{\partial a_{Y^l}} \cdot \frac{\partial a_{Y^l}}{\partial a_{F_j^{l_j}}} = \frac{1}{6} \cdot \frac{\partial a_{Y^l_{FWA}}}{\partial a_{F_j^{l_j}}} \tag{32}$$

Suppose that $\mathbf{M}^l_{L1}(\alpha) = (a_{ij}(\alpha))$, $\mathbf{M}^l_{L2}(\alpha) = (b_{ij}(\alpha))$, $\mathbf{M}^l_{R1}(\alpha) = (c_{ij}(\alpha))$, $\mathbf{M}^l_{R2}(\alpha) = (d_{ij}(\alpha))$. Then we have

$$\begin{aligned} \frac{\partial a_{Y^l_{FWA}}}{\partial a_{F_j^{l_j}}} &= \sum_{j=1}^{p} \frac{\partial a_{Y^l}}{\partial m^l_{L,j}(0)} \cdot \frac{\partial m^l_{L,j}(0)}{\partial a_{F_j^{l_j}}} + \sum_{j=1}^{p} \frac{\partial a_{Y^l}}{\partial k^l_{L,j}(0)} \cdot \frac{\partial k^l_{L,j}(0)}{\partial a_{F_j^{l_j}}} \\ &= \frac{\sum_{j=1}^{p}(a_{ij}(0)\overline{W}_j^{l_j}(0) + b_{ij}(0)\underline{W}_j^{l_j}(0))}{\sum_{i=1}^{n} p^l_{L,i}(0)\overline{W}_i^{l_i}(0) + \sum_{j=1}^{n} q^l_{L,j}(0)\underline{W}_j^{l_j}(0)} \end{aligned} \tag{33}$$

$$\begin{aligned} \frac{\partial b_{Y^l_{FWA}}}{\partial b_{F_j^{l_j}}} &= \sum_{j=1}^{p} \frac{\partial b_{Y^l}}{\partial m^l_{L,j}(1)} \cdot \frac{\partial m^l_{L,j}(1)}{\partial b_{F_j^{l_j}}} + \sum_{j=1}^{p} \frac{\partial b_{Y^l}}{\partial k^l_{L,j}(1)} \cdot \frac{\partial k^l_{L,j}(1)}{\partial b_{F_j^{l_j}}} \\ &= \frac{\sum_{j=1}^{p}(a_{ij}(1)\overline{W}_j^{l_j}(1) + b_{ij}(1)\underline{W}_j^{l_j}(1))}{\sum_{i=1}^{n} p^l_{L,i}(1)\overline{W}_i^{l_i}(1) + \sum_{j=1}^{n} q^l_{L,j}(1)\underline{W}_j^{l_j}(1)} \end{aligned} \tag{34}$$

$$\begin{aligned} \frac{\partial c_{Y^l_{FWA}}}{\partial c_{F_j^{l_j}}} &= \sum_{j=1}^{p} \frac{\partial c_{Y^l}}{\partial m^l_{R,j}(1)} \cdot \frac{\partial m^l_{R,j}(1)}{\partial c_{F_j^{l_j}}} + \sum_{j=1}^{p} \frac{\partial c_{Y^l}}{\partial k^l_{R,j}(1)} \cdot \frac{\partial k^l_{R,j}(1)}{\partial c_{F_j^{l_j}}} \\ &= \frac{\sum_{j=1}^{p}(c_{ij}(1)\underline{W}_j^{l_j}(1) + d_{ij}(1)\overline{W}_j^{l_j}(1))}{\sum_{i=1}^{n} p^l_{R,i}(1)\underline{W}_i^{l_i}(1) + \sum_{j=1}^{n} q^l_{R,j}(1)\overline{W}_j^{l_j}(1)} \end{aligned} \tag{35}$$

$$\begin{aligned} \frac{\partial d_{Y^l_{FWA}}}{\partial d_{F_j^{l_j}}} &= \sum_{j=1}^{p} \frac{\partial d_{Y^l}}{\partial m^l_{R,j}(0)} \cdot \frac{\partial m^l_{R,j}(0)}{\partial d_{F_j^{l_j}}} + \sum_{j=1}^{p} \frac{\partial d_{Y^l}}{\partial k^l_{R,j}(0)} \cdot \frac{\partial k^l_{R,j}(0)}{\partial d_{F_j^{l_j}}} \\ &= \frac{\sum_{j=1}^{p}(c_{ij}(0)\underline{W}_j^{l_j}(0) + d_{ij}(0)\overline{W}_j^{l_j}(0))}{\sum_{i=1}^{n} p^l_{R,i}(0)\underline{W}_i^{l_i}(0) + \sum_{j=1}^{n} q^l_{R,j}(0)\overline{W}_j^{l_j}(0)} \end{aligned} \tag{36}$$

According to Eqs. (31)–(36), we can get the final updating formulas for antecedent parameters as follows:

$$a_{F_j^{l_j}}(q+1) = a_{F_j^{l_j}}(q) - \eta \cdot (y_{FWA} - y^{(t)}) \cdot \phi^l \cdot \begin{cases} \left[\frac{1}{6}\frac{\partial d_{Y_{FWA}^l}}{\partial a_{F_j^{l_j}}} + \frac{y^l - y_{FWA}}{\mu_{F_j^{l_j}}(x_j)} \cdot \frac{x_j - b}{(b-a)^2}\right]\Bigg|_q, & a_{F_j^{l_j}} \le x_j \le b_{F_j^{l_j}} \\ \frac{1}{6}P(W_j^{l_j})\Big|_q, & \text{others} \end{cases} \tag{37}$$

$$b_{F_j^{l_j}}(q+1) = b_{F_j^{l_j}}(q) - \eta \cdot (y_{FWA} - y^{(t)}) \cdot \phi^l \cdot \begin{cases} \left[\frac{1}{3}\frac{\partial d_{Y_{FWA}^l}}{\partial b_{F_j^{l_j}}} + \frac{y^l - y_{FWA}}{\mu_{F_j^{l_j}}(x_j)} \cdot \frac{a - x_j}{(b-a)^2}\right]\Bigg|_q, & a_{F_j^{l_j}} \le x_j \le b_{F_j^{l_j}} \\ \frac{1}{3}P(W_j^{l_j})\Big|_q, & \text{others} \end{cases} \tag{38}$$

$$c_{F_j^{l_j}}(q+1) = c_{F_j^{l_j}}(q) - \eta \cdot (y_{FWA} - y^{(t)}) \cdot \phi^l \cdot \begin{cases} \left[\frac{1}{3}\frac{\partial d_{Y_{FWA}^l}}{\partial c_{F_j^{l_j}}} + \frac{y^l - y_{FWA}}{\mu_{F_j^{l_j}}(x_j)} \cdot \frac{d - x_j}{(d-c)^2}\right]\Bigg|_q, & c_{F_j^{l_j}} \le x_j \le d_{F_j^{l_j}} \\ \frac{1}{3}P(W_j^{l_j})\Big|_q, & \text{others} \end{cases} \tag{39}$$

$$d_{F_j^{l_j}}(q+1) = d_{F_j^{l_j}}(q) - \eta \cdot (y_{FWA} - y^{(t)}) \cdot \phi^l \cdot \begin{cases} \left[\frac{1}{6}\frac{\partial d_{Y_{FWA}^l}}{\partial d_{F_j^{l_j}}} + \frac{y^l - y_{FWA}}{\mu_{F_j^{l_j}}(x_j)} \cdot \frac{x_j - c}{(d-c)^2}\right]\Bigg|_q, & c_{F_j^{l_j}} \le x_j \le d_{F_j^{l_j}} \\ \frac{1}{6}P(W_j^{l_j})\Big|_q, & \text{others} \end{cases} \tag{40}$$

At last, we summarize the parameters updating steps as follows:

Step 1: Initialization, set the input number p, rule number M, epoch number *epoch*, Monte-Carlo number n, and training and testing data pair numbers;

Step 2: Given the training input-output data pairs $(\mathbf{x}^{(t)}, y^{(t)})$, $(t = 1, \ldots, T_{\text{train}})$ and calculate the FWA FLS output $y_{FWA}(\mathbf{x}^{(t)})$ using (3) or (4);

Step 3: For each training pair, update the weight parameters $a_{W_j^{l_j}}$, $b_{W_j^{l_j}}$, $c_{W_j^{l_j}}$, $d_{W_j^{l_j}}$ and a_{B^l}, b_{B^l}, c_{B^l}, d_{B^l} using Eqs. (24)–(27) and (28), and antecedent parameters $a_{F_j^{l_j}}$, $b_{F_j^{l_j}}$, $c_{F_j^{l_j}}$, $d_{F_j^{l_j}}$ using Eqs. (37)–(40);

Step 4: For each epoch e, execute Step 3, until $e = epoch$, return the finally updated parameters;

Step 5: Given the testing input-output data pairs $(\mathbf{x}^{(t)}, y^{(t)})$, $(t = 1, \ldots, T_{\text{test}})$, and calculate the FWA FLS output $y_{FWA}(\mathbf{x}^{(t)})$ using (3) or (4) with finally updated parameters.

4 Conclusions

In this paper, we provide the detailed computation of derivatives for trapezoidal membership functions used in [5] when using the steepest descent method to optimize the parameters. It can also be used to triangular MFs.

Acknowledgements This work is supported by National Natural Science Foundation of China (61403011).

References

1. E.H. Mamdani, Applications of fuzzy algorithms for simple dynamic plant. Proc. IEE **121**(12), 1585–1588 (1974)
2. T. Takagi, M. Sugeno, Fuzzy identification of systems and its applications to modeling and control. IEEE Trans. Syst. Man Cybern. **15**(1), 116–132 (1985)
3. M. Sugeno, G.T. Kang, Structure identification of fuzzy model. Fuzzy Sets Syst. **28**, 15–33 (1988)
4. L.X. Wang, *A Course in Fuzzy Systems and Control* (Prentice-Hall, 1996)
5. Q-Y. Zhang, Y.-Q. Liu, X. Tian, A novel fuzzy logic system with consequents as fuzzy weighted averages of antecedents, in *Submitted to the 14th Chinese Intelligent Systems Conference*, Wenzhou, China, 13–14 Oct 2018
6. S.H. Chen, G.C. Li, Representation, ranking, and distance of fuzzy number with exponential membership function using graded mean integration method. Tamsui Oxford J. Math. Sci. **16**(2), 123–131 (2000)
7. J. M. Mendel, D.R. Wu, *Perceptual Computing: Aiding People in Making Subjective Judgments* (Wiley and Sons, 2010)
8. F.L. Liu, J.M. Mendel, Aggregation using the fuzzy weighted average as computed by the KarnikCMendel algorithms. IEEE Trans. Fuzzy Syst. **16**(1), 1–12 (2008)
9. D.R. Wu, J.M. Mendel, Enhanced Karnik-Mendel algorithms. IEEE Trans. Fuzzy Syst. **17**(4), 923–934 (2009)
10. C.C. Chou, in *The Representation of Multiplication Operation on Fuzzy numbersand Application to Solving Fuzzy Multiple Criteria Decision Making Problems*, ed. byQ. Yang, G. Webb, PRICAI 2006, LNCS(LNAI), vol. 4099 (Springer Heidelberg, 2006), pp. 161-169
11. J.M. Mendel, Computing derivatives in interval type-2 fuzzy logic systems. Trans. Fuzzy Syst. **12**(1), 84–98 (2004)

Robust Stability Analysis for Uncertain Polynomial Fuzzy Systems with Time-Varying Delay via Delay-Partitioning Approach

Jiafeng Yu, Qinsheng Li and Chunsong Han

Abstract In this paper, a new robust stability criteria by delay-partitioning approach is presented for uncertain polynomial fuzzy system with time-varying delay. The parameter-dependent Lyapunov-Krasovskii functional is employed for the stability analysis, which is constructed in the formwork of state vector augmentation. All the conditions in the proposed approach can be represented as sum-of-squares (SOS) problems.

Keywords Polynomial fuzzy systems · Sum of squares · Delay dependent Delay-partitioning approach

1 Introduction

In recent years, polynomial fuzzy models have been proposed by using polynomial system matrices for modeling of nonlinear plants in [1, 2]. The polynomial fuzzy models can be applied to nonlinear systems with polynomial terms. Therefore, polynomial fuzzy models are very attractive.

J. Yu (✉) · Q. Li
Jiangsu Maritime Institute, Nanjing 211170, China
e-mail: yyujie99@163.com

Q. Li
Shanghai University, Shanghai 200072, China

C. Han
Qiqihar University, Qiqihar 161006, China
e-mail: hanchunsong@126.com

C. Han
Bohai University, Jinzhou 121001, China

Y. Jia et al. (eds.), *Proceedings of 2018 Chinese Intelligent Systems Conference*, Lecture Notes in Electrical Engineering 529,
https://doi.org/10.1007/978-981-13-2291-4_70

Sum-of-squares (SOS) approach have been presented for polynomial fuzzy control system design in [3–6]. Compared with linear matrix inequality (LMI) approach to T-S fuzzy model control, SOS method provided more extensive results. However, to the best of our knowledge, there exists no literature on SOS-based robust stability analysis for uncertain polynomial fuzzy systems with time-varying delay via delay-partitioning approach in [7, 8].

This paper presents an SOS-based delay-dependent robust Stability criteria for uncertain time-delay polynomial fuzzy systems via delay-partitioning method. The parameter-dependent Lyapunov-Krasovskii functional is employed for the stability analysis. All the conditions for the proposed approach can be converted to SOS problems. The SOS problems can be solved by SOSTOOLS [9].

The outline of the paper is as follows. In Sect. 2, the time-delay polynomial fuzzy system is introduced. The robust stability criteria for uncertain polynomial fuzzy system with time-varying delay is presented by the SOS optimization in Sect. 3. Finally, concluding remarks are given.

To reduce the redaction, we will omit the notation with respect to time t in the part formulas. For example, we will use x instead of $x(t)$. $\mathrm{Sy}(A)$ denotes $A + A^T$.

2 Time-Delay Polynomial Fuzzy System

Consider a continuous time-delay nonlinear system that is represented by the following polynomial fuzzy model with time-varying delay:

Model Rule i:

$$\text{If } z_1 \text{ is } M_{i1} \text{ and } \cdots \text{ and } z_p \text{ is } M_{ip}$$
$$\text{Then} \begin{cases} \dot{x}(t) = A_i(x)x + A_{di}(x)x(t-h(t)) + B_i(x)u \\ x(t) = \phi(t),\ t \in [-\bar{h}, 0] \end{cases} \tag{1}$$

where $i = 1, 2, \ldots, r$. z_j $(j = 1, 2, \ldots, p)$ is the premise variable. M_{ij} denotes the member function. r denotes the number of model rules. $x \in \mathbb{R}^n$ denotes the state vector, and $u \in \mathbb{R}^m$ is the control input. $A_i(x) \in \mathbb{R}^{n\times n}$, $A_{di}(x) \in \mathbb{R}^{n\times n}$ and $B_i(x) \in \mathbb{R}^{n\times m}$ are polynomial matrices in x. Time-varying delay $h(t)$ is a continuous function subject to

$$0 \le h(t) \le \bar{h} < \infty,\ \forall t \ge 0, \tag{2}$$

$$\dot{h}(t) < \mu, \tag{3}$$

where $\bar{h}$ and μ are nonnegative constants assumed to exist. Initial state $\phi(t)$ denotes a continuous vector function for $t \in [-\bar{h}, 0]$.

Therefore, $A_i(x)x + A_{di}(x)x(t-h(t)) + B_i(x)u$ is a polynomial vector. Thus, the time delay polynomial fuzzy system (1) has a polynomial model consequence in each consequence part.

The defuzzification process of the time-delay polynomial fuzzy model (1) can be written as

$$\dot{x} = \sum_{i=1}^{r} h_i(z)\{A_i(x)x + A_{di}(x)x(t-h(t)) + B_i(x)u\}. \tag{4}$$

where

$$h_i(z) = \frac{\prod_{j=1}^{p} M_{ij}(z_j)}{\sum_{i=1}^{r}\prod_{j=1}^{p} M_{ij}(z_j)}, \tag{5}$$

By the properties of membership functions, the following relations satisfy

$$h_i(z) \geq 0, \sum_{i=1}^{r} h_i(z) = 1.$$

Thus, the overall polynomial fuzzy model is obtained by fuzzy blending of the polynomial system model.

Lemma 1 [10] *For a given scalar $\lambda > 0$, and arbitrary constant matrix $S = S^T > 0$, vector function $\omega : [0, \lambda] \rightarrow \mathbb{R}^n$ and is integrable, then*

$$\left(\int_0^{\lambda} \omega(\alpha)d\alpha\right)^T S \left(\int_0^{\lambda} \omega(\alpha)d\alpha\right) \leq \lambda \left(\int_0^{\lambda} \omega^T(\alpha)S\omega(\alpha)d\alpha\right). \tag{6}$$

Lemma 2 (Finsler's lemma) [11] *Let $\zeta \in \mathbb{R}^n$, $\Phi = \Phi^T \in \mathbb{R}^{n\times n}$, and $B \in \mathbb{R}^{m\times n}$ such that rank$(B) < n$. Then the following statements are equivalent*

(1) $\zeta^T \Phi \zeta < 0, \forall B\zeta = 0, \zeta \neq 0,$

(2) $B^{\perp^T} \Phi B^{\perp} < 0,$

(3) $\exists Y \in \mathbb{R}^{n\times m}, \Phi + Sy(YB) < 0,$

in which $B^{\perp} \in \mathbb{R}^{n\times(n-rank(B))}$ denotes the right orthogonal complement of B.

Lemma 3 (Peng-Park's integral inequality) [12] *For any matrix $\begin{bmatrix} Y & Z \\ * & Y \end{bmatrix} \geq 0$, positive scalars $\bar{h}$ and $h(t)$ satisfying $0 < h(t) < \bar{h}$, vector function $\dot{\omega} : [-\bar{h}, 0] \rightarrow \mathbb{R}^n$ such that the integrations are defined, then*

$$-\bar{h}\int_{t-\bar{h}}^{t} \dot{\omega}(\alpha)^T Y \dot{\omega}(\alpha)d\alpha \leq \eta^T(t)\Omega\eta(t), \tag{7}$$

where $\eta(t) = \begin{bmatrix} x(t) \\ x(t-h(t)) \\ x(t-\bar{h}) \end{bmatrix}, \Omega = \begin{bmatrix} -Y & Y-Z & Z \\ * & -2Y+Sy(Z) & -Z+Y \\ * & * & -Y \end{bmatrix}.$

Lemma 4 (Seuret-Wirtinger's integral inequality) [13] *For any matrix* $Q > 0$, *continuously differentiable function* $x : [\tau_1, \tau_2] \to \mathbb{R}^n$, *the following inequality holds*

$$\int_{\tau_1}^{\tau_2} \dot{x}^T(s)Q\dot{x}(s)ds \geq \frac{1}{\tau_2 - \tau_1}\nu^T(t)\Phi\nu(t), \tag{8}$$

where $\nu(t) = \begin{bmatrix} x(\tau_2) \\ x(\tau_1) \\ \frac{1}{\tau_2-\tau_1}\int_{\tau_1}^{\tau_2} x(s)ds \end{bmatrix}$, $\Phi = \begin{bmatrix} 4Q & 2Q & -6Q \\ * & 4Q & -6Q \\ * & * & 12Q \end{bmatrix}$.

Lemma 5 [14] *Let* $Z = Z^T$, M, N *and* $F(t)$ *are appropriately dimensional matrices,* $F^T(t)F(t) \leq I$, *then the following inequality*

$$Z + Sy\{MF(t)N\} < 0 \tag{9}$$

is true, if and only if the following inequality holds for any scalar $\varepsilon > 0$,

$$Z + \varepsilon^{-1}MM^T + \varepsilon N^T N < 0. \tag{10}$$

3 Robust Stability Analysis of Uncertain Polynomial Fuzzy System with Time-Varying Delay

In this section, our goal is to develop robust stability criteria for uncertain polynomial fuzzy system with time-varying delay by using delay-partitioning approach.

Consider the following time-delay polynomial fuzzy system with $u(t) = 0$:

$$\begin{cases} \dot{x}(t) = \sum_{i=1}^{r} h_i(z)\{(A_i(x) + \Delta A_i)x + (A_{di}(x) + \Delta A_{di})x(t - h(t))\} \\ x(t) = \phi(t), t \in [-\bar{h}, 0] \end{cases} \tag{11}$$

where $[\Delta A_i, \Delta A_{di}] = H_i F_i(t) \left[E_i, E_{di}\right]$, H_i, E_i, and E_{di} are known constant real matrices with appropriate dimensions, $F_i(t)^T F_i(t) \leq I$.

For any integer $m > 1$, define $\delta = \frac{\bar{h}}{m}$, then $[0, \bar{h}]$ can be divided into m segments, i.e.,

$$[0, \bar{h}] = \cup_{j=1}^{m}[(j-1)\delta, j\delta]. \tag{12}$$

For notation simplification, motivated by [7, 8, 15], let

$$\begin{cases} e_s = [\underbrace{0, \ldots, 0}_{s-1}, I, \underbrace{0, \ldots, 0}_{m+4-s}]^T, s = 1, 2, \ldots, m+4, \\ \zeta(t) = [x^T(t - h(t)), \zeta_1^T(t), x^T(t - m\delta), \frac{1}{\delta}\int_{t-\delta}^{t} x^T(s)ds, \dot{x}^T(t)]^T, \end{cases} \tag{13}$$

in which $\zeta_1(t) = [x^T(t), x^T(t-\delta), \ldots, x^T(t-(m-1)\delta)]^T$.

Remark 1 $A_i^k(x)$ denotes the kth row of polynomial matrix $A_i(x)$. $K = \{k_1, k_2, \ldots, k_m\}$ denote the row indices of $B_i(x)$ and $A_{di}(x)$ whose corresponding row are zero, and define $\tilde{x} = (x_{k_1}, x_{k_2}, \ldots, x_{k_m})$.

We state the following robust stability criteria for the polynomial fuzzy system (11).

Theorem 1 *For a given integer $m > 1$, scalars μ, $\bar{h} \geq 0$, and $\delta = \frac{\bar{h}}{m}$, then the polynomial fuzzy system (11) with a time-varying delay $h(t)$ satisfying (2) and (3) is asymptotically stable if there exist scalars $\varepsilon_{ik} > 0$ $(i = 1, 2, \ldots, r; k = 1, 2, \ldots, m)$, symmetric positive polynomial matrix $P(\tilde{x})$, symmetric positive matrices X, Q_j, Z_0, Z_j, R_l $(l = 1, 2, \ldots, m-1,\ j = 1, 2, \ldots, m)$, and any matrices Y and S_{ik} with appropriate dimensions, such that the following* ***SOS*** *conditions hold*

$$\sigma_1^T(P(\tilde{x}) - \epsilon_1(x)I_1)\sigma_1 \text{ is SOS} \tag{14}$$

$$-\sigma_2^T\left(\begin{bmatrix} \Theta & YH_i & \varepsilon_{ik}(e_2E_i^T + e_1E_{di}^T) \\ * & -\varepsilon_{ik}I_4 & 0 \\ * & * & -\varepsilon_{ik}I_5 \end{bmatrix} + \epsilon_2(x)I_2\right)\sigma_2 \text{ is SOS} \tag{15}$$

$$\sigma_3^T\left(\begin{bmatrix} Z_k & S_{ik} \\ * & Z_k \end{bmatrix} - \epsilon_3(x)I_3\right)\sigma_3 \text{ is SOS} \tag{16}$$

where σ_1, σ_2, and σ_3 are arbitrary vectors that are independent of x. $I_1, I_2, \ldots, I_5$ are identity matrix with appropriate dimensions. $\epsilon_1(x)$, $\epsilon_2(x)$, and $\epsilon_3(x)$ are nonnegative polynomials for $x \neq 0$.

$$\Theta = \Phi(i,k) + Sy(Y\Gamma_i)$$

$$\Phi(i,k) = \left\{\sum_{j=0}^{3}\Phi_j + \Phi_4(k) + \Phi_5(i,k) + e_{m+4}(\delta^2\sum_{j=0}^{m}Z_j)e_{m+4}^T\right\},$$

$$\Gamma_i = A_i(x(t))e_2^T + Ad_i(x(t))e_1^T - e_{m+4}^T,$$

$$\Phi_0 = \begin{bmatrix} e_2^T \\ e_3^T \\ e_{m+3}^T \end{bmatrix}^T \begin{bmatrix} -4Z_0 & -2Z_0 & 6Z_0 \\ * & -4Z_0 & 6Z_0 \\ * & * & -12Z_0 \end{bmatrix} \begin{bmatrix} e_2^T \\ e_3^T \\ e_{m+3}^T \end{bmatrix},$$

$$\Phi_1 = \left\{\begin{array}{c} \begin{bmatrix} e_2^T \\ \delta e_{m+3}^T \end{bmatrix}^T \begin{bmatrix} \sum\limits_{k\in K} \frac{\partial P(\tilde{x})}{\partial x_k}\{A_i^k(x)x\} \end{bmatrix} \cdot \begin{bmatrix} e_2^T \\ \delta e_{m+3}^T \end{bmatrix} \\ +\mathbf{Sy}\left(\begin{bmatrix} e_2^T \\ \delta e_{m+3}^T \end{bmatrix}^T P(\tilde{x}) \begin{bmatrix} e_{m+4}^T \\ e_2^T - e_3^T \end{bmatrix}\right) \end{array}\right\},$$

$$\Phi_2 = \left\{\begin{bmatrix} e_2^T \\ e_3^T \\ \vdots \\ e_{m+1}^T \end{bmatrix}^T X \begin{bmatrix} e_2^T \\ e_3^T \\ \vdots \\ e_{m+1}^T \end{bmatrix} - \begin{bmatrix} e_3^T \\ e_4^T \\ \vdots \\ e_{m+2}^T \end{bmatrix}^T X \begin{bmatrix} e_3^T \\ e_4^T \\ \vdots \\ e_{m+2}^T \end{bmatrix}\right\},$$

$$\Phi_3 = \sum_{j=1}^{m-1} \left(\begin{bmatrix} e_{j+1}^T \\ e_{j+2}^T \end{bmatrix}^T R_j \begin{bmatrix} e_{j+1}^T \\ e_{j+2}^T \end{bmatrix} - \begin{bmatrix} e_{j+2}^T \\ e_{j+3}^T \end{bmatrix}^T R_j \begin{bmatrix} e_{j+2}^T \\ e_{j+3}^T \end{bmatrix} \right),$$

$$\Phi_4(k) = \left\{ \sum_{j=1}^{k-1} \left[e_{j+1} Q_j e_{j+1}^T - e_{j+2} Q_j e_{j+2}^T \right] + e_{k+1} Q_k e_{k+1}^T - (1-\mu) e_1 Q_k e_1^T \right\},$$

$$\Phi_5(i,k) = \left\{ \begin{array}{c} \sum\limits_{j=1, j \neq k}^{m} \begin{bmatrix} e_{j+1}^T \\ e_{j+2}^T \end{bmatrix}^T \begin{bmatrix} -Z_j & Z_j \\ * & -Z_j \end{bmatrix} \cdot \begin{bmatrix} e_{j+1}^T \\ e_{j+2}^T \end{bmatrix} \\ + \begin{bmatrix} e_{k+1}^T \\ e_1^T \\ e_{k+2}^T \end{bmatrix}^T \cdot \begin{bmatrix} -Z_k & Z_k - S_{ik} & S_{ik} \\ * & Sy(S_{ik}) - 2Z_k & Z_k - S_{ik} \\ * & * & -Z_k \end{bmatrix} \cdot \begin{bmatrix} e_{k+1}^T \\ e_1^T \\ e_{k+2}^T \end{bmatrix} \end{array} \right\}.$$

Proof Consider the following augmented parameter-dependent Lyapunov-Krasovskii functional for stability analysis of the polynomial fuzzy system (11):

$$V(x(t)_{|\{h(t) \in [(k-1)\delta,\, k\delta]\}} = V_1(x) + V_2(x) + V_3(x) + V_4(x) + V_5(x) \tag{17}$$

where

$$\begin{aligned}
V_1(x) &= \eta_0^T(t) P(\tilde{x}(t)) \eta_0(t), \\
V_2(x) &= \int_{t-\delta}^{t} \zeta_1^T(s) X \zeta_1(s) ds, \\
V_3(x) &= \sum_{j=1}^{m-1} \int_{t-\delta}^{t} \eta_j^T(s) R_j \eta_j(s) ds, \\
V_4(x) &= \sum_{j=1}^{k-1} \int_{t-j\delta}^{t-(j-1)\delta} x^T(s) Q_j x(s) ds + \int_{t-h(t)}^{t-(k-1)\delta} x^T(s) Q_k x(s) ds, \\
V_5(x) &= \sum_{j=1}^{m} \delta \int_{-j\delta}^{-(j-1)\delta} \int_{t-\theta}^{t} \dot{x}^T(s) Z_j \dot{x}(s) ds d\theta + \delta \int_{-\delta}^{0} \int_{t+\theta}^{t} \dot{x}^T(s) Z_0 \dot{x}(s) ds d\theta,
\end{aligned}$$

in which $P(\tilde{x}(t))$ denotes symmetric positive definition polynomial matrix.

$\eta_0(t) = \left[x^T(t), \int_{t-\delta}^{t} x^T(s) ds \right]^T,$

$\eta_j(t) = \left[x^T(s-(j-1)\delta), x^T(s-j\delta) \right]^T.$

The time derivative of $V(x(t))$ along the solution of the polynomial fuzzy system (11) is obtained as

$$\dot{V}(x(t)) = \dot{V}_1(x) + \dot{V}_2(x) + \dot{V}_3(x) + \dot{V}_4(x) + \dot{V}_5(x) \tag{18}$$

where

$$\dot{V}_1(x) = \eta_0^T(t)\dot{P}(\tilde{x})\eta_0(t) + 2\eta_0^T(t)P(\tilde{x})\dot{\eta}_0(t) \\ = \zeta^T \Phi_1 \zeta \tag{19}$$

Since $k \in K$, $A_{di}^k(x) = 0$, then

$$\dot{x}_k = \sum_{i=1}^{r} h_i(z)\{A_i^k(x(t))x(t)\}. \tag{20}$$

On the other hand, $\frac{\partial P(\tilde{x})}{\partial x_i} = 0$ for $i \notin K$, we obtain

$$\dot{P}(\tilde{x}(t)) = \sum_{i=1}^{r} h_i(z) \sum_{k \in K} \frac{\partial P(\tilde{x})}{\partial x_k}\{A_i^k(x(t))x(t)\}. \tag{21}$$

Therefore,

$$\dot{V}_2(x) = \zeta_1^T X \zeta_1 - \zeta_1^T(t-\delta)X\zeta_1(t-\delta) = \zeta^T \Phi_2 \zeta, \tag{22}$$

$$\dot{V}_3(x) = \sum_{j=1}^{m-1}\left[\eta_j^T R_j \eta_j - \eta_j^T(t-\delta)R_j\eta_j(t-\delta)\right] = \zeta^T \Phi_3 \zeta, \tag{23}$$

$$\dot{V}_4(x) \le \left\{ \begin{array}{c} \sum_{j=1}^{k-1}\begin{bmatrix} x^T(t-(j-1)\delta)\cdot Q_j \cdot x(t-(j-1)\delta) \\ -x^T(t-j\delta)\cdot Q_j \cdot x(t-j\delta) \end{bmatrix} \\ +\begin{bmatrix} x^T(t-(k-1)\delta)\cdot Q_k.x(t-(k-1)\delta) \\ -(1-\mu)\cdot x^T(t-h(t))\cdot Q_k \cdot x(t-h(t)) \end{bmatrix} \end{array} \right\} \tag{24}$$

$$= \zeta^T \Phi_4(k)\zeta$$

$$\dot{V}_5(x) = \left\{ \begin{array}{c} \sum_{j=0}^{m}\delta^2\cdot \dot{x}^T Z_j \dot{x} - \sum_{j=1}^{m}\delta\int_{t-j\delta}^{t-(j-1)\delta}\dot{x}^T(\alpha)Z_j\dot{x}(\alpha)d\alpha \\ -\delta\int_{t-\delta}^{t}\dot{x}^T(\alpha)Z_0\dot{x}(\alpha)d\alpha \end{array} \right\}. \tag{25}$$

Dealing with the last item in (25) employs Lemmas 3 and 4, respectively. It can be deduced for $\begin{bmatrix} Z_k & \hat{S}_K \\ * & Z_k \end{bmatrix} \ge 0$, in which $\hat{S}_K = \sum_{i=1} h_i S_{ik}$.
From Lemma 3, we have

$$-\sum_{j=1}^{m}\delta\int_{t-j\delta}^{t-(j-1)\delta}\dot{x}^T(\alpha)Z_j\dot{x}(\alpha)d\alpha$$

$$\le \sum_{j=1, j\ne k}^{m} v_1^T\begin{bmatrix} -Z_j & Z_j \\ * & -Z_j \end{bmatrix} v_1 + w_1^T\begin{bmatrix} -Z_k & Z_k-\hat{S}_K & \hat{S}_K \\ * & \mathrm{Sy}(\hat{S}_K)-2Z_k & Z_k-\hat{S}_K \\ * & * & -Z_k \end{bmatrix} w_1$$

$$= \sum_{i=1}^{r} h_i \zeta^T \Phi_5(i,k)\zeta \tag{26}$$

where
$v_1 = [x^T(t-(j-1)\delta), x^T(t-j\delta)]^T,$
$w_1 = [x^T(t-(k-1)\delta), x^T(t-h(t)), x^T(t-k\delta)]^T.$

From Lemma 4 it follows that

$$-\delta \int_{t-\delta}^{t} \dot{x}^T(\alpha) Z_0 \dot{x}(\alpha) d\alpha \leq v_2^T \begin{bmatrix} -4Z_0 & -2Z_0 & 6Z_0 \\ * & -4Z_0 & 6Z_0 \\ * & * & 12Z_0 \end{bmatrix} v_2$$
$$= \zeta^T \Phi_0 \zeta \tag{27}$$

where $v_2 = [x^T, x^T(t-\delta), \frac{1}{\delta}\int_{t-\delta}^{t} x^T(s)ds]^T$.

Therefore, we have

$$\dot{V}(x) \leq \sum_{i=1}^{r} h_i \zeta^T \Phi(i,k) \zeta. \tag{28}$$

Applying the augmented vector ζ, the polynomial fuzzy system (11) can be represented by

$$\sum_{i=1}^{r} h_i \Pi_i \zeta = 0,$$

where $\Pi_i = \Gamma_i + H_i F_i(t) E_i e_2^T + H_i F_i(t) E_{di} e_1^T$, $\Gamma_i = A_i(x(t)e_2^T + Ad_i(x(t)e_1^T - e_{m+4}^T$.
Therefore, the robust stability conditions for the uncertain polynomial fuzzy system (11) can be written as

$$\sum_{i=1}^{r} h_i \zeta^T \Phi(i,k) \zeta < 0,$$
$$\text{s.t.} \sum_{i=1}^{r} h_i \Pi_i \zeta = 0. \tag{29}$$

From Lemma 2, the conditions in (29) are equivalent to

$$\sum_{i=1}^{r} h_i \zeta^T [\Phi(i,k) + \mathrm{Sy}(Y\Pi_i)] \zeta < 0. \tag{30}$$

Therefore, by (30), we obtain

$$\sum_{i=1}^{r} h_i(z) \zeta^T \left[\Theta + \mathrm{Sy}\left[YH_i F_i(t)(E_i e_2^T + E_{di} e_1^T)\right]\right] \zeta < 0. \tag{31}$$

By Lemma 5 and Schur complement, (31) can be represented as

$$\sum_{i=1}^{r} h_i(z)\zeta^T \begin{bmatrix} \Theta & YH_i & \varepsilon_{ik}(E_i e_2^T + E_{di} e_1^T)^T \\ * & -\varepsilon_{ik} I & 0 \\ * & * & -\varepsilon_{ik} I \end{bmatrix} \zeta < 0, \quad (32)$$

By (28)–(32) and (15), $\dot{V}(x) < 0$. We have

$$\dot{V}(x(t)) < -\gamma \left\| x(t) \right\|^2, \gamma > 0.$$

Therefore, the polynomial fuzzy system (11) is globally asymptotically stable. This completes the proof.

4 Conclusions

This paper focuses on robust stability criteria for uncertain time-delay polynomial fuzzy systems by delay-partitioning approach. The stability analysis for the polynomial fuzzy systems with time-varying delay employ parameter-dependent Lyapunov-Krasovskii functional. All the analysis conditions in the presented approach can be expressed as SOS problems.

References

1. K. Tanaka, H. Yoshida, H. Ohtake, H.O. Wang, A sum of squares approach to stability analysis of polynomial fuzzy systems, in *Proceedings of America Control Conerence*, pp. 4071–4076 (2007)
2. K. Tanaka, H. Yoshida, H. Ohtake, H.O. Wang, A sum of squares approach to modeling and control of nonlinear dynamical systems with polynomial fuzzy systems. IEEE Trans. Fuzzy Syst. **17**(4), 911–922 (2009)
3. K. Tanaka, H. Ohtake, T. Seo, M. Tanaka, H.O. Wang, Polynomial fuzzy obsever designs: a sum-of-squares approach. IEEE Trans. Syst. Man Cybern. Part B Cybern. Publ. IEEE Syst. Man Cybern. Soc. **42**(5), 1330–1342 (2012)
4. F. Zhang, L. Li, W. Wang, Stability and stabilization for a class of polynomial discrete fuzzy system with time delay by sum of squares optimization, in *Proceedings of the 8th International Conference on Fuzzy Systems and Knowledge Discovery*, pp. 713–717 (2012)
5. H.K. Lam, S. Member, Stability analysis of polynomial-fuzzy-model-based control systems with mismatched premise. IEEE Trans. Fuzzy Syst. **22**(1), 223–229 (2014)
6. K. Tanaka, H.O. Wang, *Fuzzy Control Systems Design and Analysis: A Linear Matrix Inequality Approach* (Wiley, NJ, 2001)
7. C. Peng, M.R. Fei, An improved result on the stability of uncertain T-S fuzzy systems with interval time-varying delay. Fuzzy Sets Syst. **212**, 97–109 (2013)
8. J. Yang, W.P. Luo, Y.H. Wang, C.S. Duan, Improved stability criteria for T-S fuzzy systems with Time-varying delay-partitioning approach. Int. J. Control Autom. Syst. **13**(6), 1521–1529 (2015)
9. S. Prajna, A. Papachristodoulou, P.A. Parrilo, Introducing SOSTOOLS: a general purpose sum of squares programming solver, in *Proceedings of the 41st Conference on Decision and Control*, pp. 741–746 (2002)

10. K.Q. Gu, V.L. Kharitonov, J. Chen, Stability of time-delay systems. Control Eng. **26**(4), 951–953 (2003)
11. M.C. de Oliveira, R.E. Skelton, *Stability Tests for Constrained Linear Systems* (Springer, Berlin, 2001)
12. P.G. Park, J.W. Ko, C.K. Jeong, Reciprocally convex approach to stability of systems with time-varying delays. Automatica **47**(1), 235–238 (2011)
13. A. Seuret, F. Gouaisbaut, Wirtinger-based integral inequality: application to time-delay systems. Automatica **49**(9), 2860–2866 (2013)
14. I.R. Petersen, C.V. Hollot, A Riccati equation approach to the stabilization of uncertain linear systems. Automatica **22**(4), 397–411 (1986)
15. H.B. Zeng, J.H. Park, J.W. Xi, S.P. Xiao, Improved delay-dependent stability criteria for T-S fuzzy systems with time-varying delay. Appl. Math. Comput. **235**, 492–501 (2014)

Full-State Stabilization of Hovercraft Based on Discrete Constant Control

Yan Lixia and Ma Baoli

Abstract The full-state stabilization problem of hovercraft is studied in this work. Under the mild assumption that no damping terms contained in the system dynamics, we propose a new discrete constant control scheme guaranteeing the hovercraft moving towards reference states with a prescribed distance in each control cycle and moving to an arbitrarily small neighborhood of references in finite control updates. Several simulation examples including stabilization and way-point assignments cases are carried out to verify the proposed controller.

Keywords Hovercraft · Discrete control · Stabilization · Way-point assignment

1 Introduction

The control of underactuated surface vessel is difficult and challenging, and has been one of the hottest research topics in the control community since years ago [1–3]. In particular, there has appeared plentiful results of vessel control focusing on the problem of stabilization [4–6, 12–14], trajectory tracking [7–16] and path-following [12, 17, 18]. However, all these mentioned results are constructed basing on the assumption that the sway dynamics of vessels contain linear/nonlinear damping terms, in absence of which the sway velocity will grow unbounded and the closed-loop system will become unstable. Imagine a vessel sailing in the open water, the water resistance acts as large damping, making the vessel hardly to sway. While for

Y. Lixia · M. Baoli (✉)
The Seventh Research Division, School of Automation Science
and Electrical Engineering, Beihang University, Beijing 100191, China
e-mail: mabaoli@buaa.edu.cn

Y. Lixia
e-mail: yanlixia@buaa.edu.cn

Y. Jia et al. (eds.), *Proceedings of 2018 Chinese Intelligent Systems Conference*, Lecture Notes in Electrical Engineering 529,
https://doi.org/10.1007/978-981-13-2291-4_71

a hovercraft moving above the surface of earth or water, the sway motion is only hindered by very small air friction, which can be neglected. And hence, the existing vessel controllers are not applicable to hovercraft, it is urgently needed to design new control algorithms.

Given a model of hovercraft without damping dynamics, its stabilization problem is firstly investigated in [4], in which the velocities and position are stabilized by a smooth continuous controller and a non-smooth controller respectively, however, the full-state stabilization is not discussed. Even till now, the full-state stabilization problem of hovercraft remains open and unsolved.

In this work, we plan to present some initial results regrading the full-state stabilization of hovercraft basing on the discrete constant control. Our control scheme involves several control cycles and each control cycle contains identical time length. Before starting every control cycle, we acquire control constants by two steps: (1) Integrate the system dynamics by viewing the control inputs as to-be-determined constants, and get the analytic expression of system states by a group of nonlinear equations; (2) Adopt gradient descent method to solve the nonlinear equations and get the control constants. The simulation results show that the proposed control scheme well regulate hovercraft to a neighborhood of reference states and this neighboring region can be arbitrarily small, realizing full-state stabilization of hovercraft.

The organization of this contribution is as follows, Sect. 2 addresses problem formulation and controller design, the simulation results are introduced in Sect. 3, conclusion is presented in Sect. 4.

2 Main Results

2.1 Problem Formulation

Consider a hovercraft with its kinetic model in the form of [4]

$$\begin{aligned} \dot{x} &= u\cos\psi - v\sin\psi \\ \dot{y} &= u\sin\psi + v\cos\psi \\ \dot{\psi} &= r \\ \dot{v} &= -ur \end{aligned} \tag{1}$$

where (x, y) denotes the position in Cartesian coordinate, ψ is orientation, (u, v, r) are surge, sway and yaw velocity respectively. Variables (u, r) are considered as control inputs.

Remark 1 The kinetic model (1), absence of model parameters, helps to catch essential nonlinear characteristics of the hovercraft. And one is able to generalize the method presented here when model parameters are involved.

Let $\xi = [x, y, \psi, v]^T$ and define the reference states by

$$\xi_r = [x_r, y_r, \psi_r, 0]^T \tag{2}$$

where the fourth entry of ξ_r denotes the reference sway velocity that equals zero identically.

Select the control cycle by $T_c = 2h$, $h \in \mathbb{R}$. To facilitate the control, we split T_c into two subcycles averagely. More precisely, the full time frame of a control cycle equals $[t_k + h) \bigcup [t_k + h, t_k + T_c]$, where $k \in \mathbb{Z}_{\geq 0}$, $t_k = kT_c$. Let the to-be-determined control constants be $U_k = [u_{1k}, r_{1k}, u_{2k}, r_{2k}]^T$ and the components of U_k appear in each subcycle as the following manner:

$$\begin{cases} u = u_{1k}, r = r_{1k}, \forall t \in [t_k, t_k + h) \\ u = u_{2k}, r = r_{2k}, \forall t \in [t_k + h, t_k + T_c] \end{cases} \tag{3}$$

Considering the full-state stabilization of (1), we aim to find U_k successively such that there exists $\bar{k} \in \mathbb{Z}_+$ and arbitrarily small $\bar{\varepsilon} \in \mathbb{R}_+$ guaranteeing

$$\| \xi_r - \xi(t) \| \leq \bar{\varepsilon}, \forall t \geq T_f = \bar{k} T_c \tag{4}$$

Regarding the control goal (4) that ξ moves to the arbitrarily small neighborhood of ξ_r after finite control updates, the state variations for each control cycle need to be interpreted primarily. Suppose that $\| \xi_r - \xi_0 \| \neq 0$ and we find an arbitrarily small constant $\varepsilon \in \mathbb{R}_+$ such that $\varepsilon < \| \xi_r - \xi_0 \|$, the desirable state variation for the $k-$th control cycle $t \in [t_k, t_k + T_c]$ is selected as

$$\Delta_k = \xi_{k+1} - \xi_k = \varepsilon \frac{\xi_r - \xi_k}{\| \xi_r - \xi_k \|} \tag{5}$$

where $\Delta_k \in \mathbb{R}^4$, $\xi_{k+1} = \xi(t_k + T_c)$, $\xi_k = \xi(t_k)$. As $\Delta_k = \xi_{k+1} - \xi_k = \xi_r - \xi_k - (\xi_r - \xi_{k+1})$, we have the following conditions

$$\begin{aligned} \xi_r - \xi_k - \varepsilon \frac{\xi_r - \xi_k}{\|\xi_r - \xi_k\|} = (\xi_r - \xi_{k+1}) \Rightarrow \left\| \xi_r - \xi_k - \varepsilon \frac{\xi_r - \xi_k}{\|\xi_r - \xi_k\|} \right\| = \|(\xi_r - \xi_{k+1})\| \\ \Rightarrow \|(\xi_r - \xi_{k+1})\| = \| \|\xi_r - \xi_k\| - \varepsilon \| \end{aligned} \tag{6}$$

It is easy to verify that ξ would move to the ball $B_r = \{\xi \in \mathbb{R}^4 \mid \| \xi_r - \xi \| \leq \varepsilon\}$ in finite control updates provided that we find U_k guaranteeing (5) for all $k \in \mathbb{Z}_{\geq 0}$. Furthermore, a tough estimation of $\bar{k}$ can be given by $\bar{k} = \dfrac{\|\xi_r - \xi_0\|}{\varepsilon}$.

For simplicity, we use subscripts k to denote the state value at $t = t_k$ and $k + 1$ denote that at $t = t_k + T_c$, for example, $x_k = x(t_k)$, $x_{k+1} = x(t_k + T_c)$.

2.2 Controller Development

Considering (3) and integrating (1) from t_k to $t_k + T_c$, one obtains the analytic state variations with entries given by

$$\begin{aligned}
x_{k+1} - x_k &= 2\frac{u_{1k}}{r_{1k}}\left[\sin\left(\psi_k + r_{1k}h\right) - \sin\psi_k\right] + \frac{v_k}{r_{1k}}\left[\cos\left(\psi_k + r_{1k}h\right) - \cos\psi_k\right] \\
&- u_{1k}h\cos\left(\psi_k + r_{1k}h\right) + 2\frac{u_{2k}}{r_{2k}}\left[\sin\left(\psi_k + r_{1k}h + r_{2k}h\right) - \sin\left(\psi_k + r_{1k}h\right)\right] \\
&+ \frac{v_k - u_{1k}r_{1k}h}{r_{2k}}\left[\cos\left(\psi_k + r_{1k}h + r_{2k}h\right) - \cos\left(\psi_k + r_{1k}h\right)\right] \\
&- u_{2k}h\cos\left(\psi_k + r_{1k}h + r_{2k}h\right) \\
y_{k+1} - y_k &= -2\frac{u_{1k}}{r_{1k}}\left[\cos\left(\psi_k + r_{1k}h\right) - \cos\psi_k\right] + \frac{v_k}{r_{1k}}\left[\sin\left(\psi_k + r_{1k}h\right) - \sin\psi_k\right] \\
&- u_{1k}h\sin\left(\psi_k + r_{1k}h\right) - 2\frac{u_{2k}}{r_{2k}}\left[\cos\left(\psi_k + r_{1k}h + r_{2k}h\right) - \cos\left(\psi_k + r_{1k}h\right)\right] \\
&+ \frac{v_k - u_{1k}r_{1k}h}{r_{2k}}\left[\sin\left(\psi_k + r_{1k}h + r_{2k}h\right) - \sin\left(\psi_k + r_{1k}h\right)\right] \\
&- u_{2k}h\sin\left(\psi_k + r_{1k}h + r_{2k}h\right) \\
\psi_{k+1} - \psi_k &= \left(r_{1k} + r_{2k}\right)h \\
v_{k+1} - v_k &= -\left(u_{1k}r_{1k} + u_{2k}r_{2k}\right)h
\end{aligned} \tag{7}$$

Since nonlinear terms coupled together in (7) obstruct solving U_k, we use Taylor series to expand the cosine and sine functions contained in (7) at ψ_k and rearrange the elements therein, which results

$$\begin{aligned}
x_{k+1} - x_k &= \left(u_{1k} + u_{2k}\right)h\cos\psi_k + r_{1k}\left(u_{1k} - u_{2k}\right)h^2\sin\psi_k \\
&+ v_k\left(-\frac{3}{2}r_{1k}h^2\cos\psi_k - \frac{r_{2k}h^2}{2}\cos\psi_k\right) - 2v_kh\sin\psi_k + o\left(h^3\right) \\
y_{k+1} - y_k &= \left(u_{1k} + u_{2k}\right)h\sin\psi_k - r_{1k}\left(u_{1k} - u_{2k}\right)h^2\cos\psi_k \\
&+ v_k\left(-\frac{3r_{1k}h^2}{2}\sin\psi_k - \frac{r_{2k}h^2}{2}\sin\psi_k\right) + 2v_kh\cos\psi_k + o\left(h^3\right) \\
\psi_{k+1} - \psi_k &= \left(r_{1k} + r_{2k}\right)h \\
v_{k+1} - v_k &= -\left(u_{1k}r_{1k} + u_{2k}r_{2k}\right)h
\end{aligned} \tag{8}$$

where $o(h^3)$ denotes the high-order terms of h^3 and the following equations are used iterately to reach (8).

$$\begin{aligned}
\sin\left(\psi_k + r_{1k}h\right) &= \sin\psi_k + r_{1k}h\cos\psi_k - \frac{r_{1k}^2h^2}{2}\sin\psi_k + o\left(h^3\right) \\
\cos\left(\psi_k + r_{1k}h\right) &= \cos\psi_k - r_{1k}h\sin\psi_k - \frac{r_{1k}^2h^2}{2}\cos\psi_k + o\left(h^3\right) \\
\sin\left(\psi_k + r_{1k}h + r_{2k}h\right) &= \sin\left(\psi_k + r_{1k}h\right) + r_{2k}h\cos\left(\psi_k + r_{1k}h\right) \\
&- \frac{r_{2k}^2h^2}{2}\sin\left(\psi_k + r_{1k}h\right) + o\left(h^3\right) \\
\cos\left(\psi_k + r_{1k}h + r_{2k}h\right) &= \cos\left(\psi_k + r_{1k}h\right) - r_{2k}h\sin\left(\psi_k + r_{1k}h\right) \\
&- \frac{r_{2k}^2h^2}{2}\cos\left(\psi_k + r_{1k}h\right) + o\left(h^3\right)
\end{aligned} \tag{9}$$

By combine (5) and (8), we obtain the following nonlinear equations

$$\bar{F}(U_k) = [F_1, F_2, F_3, F_4]^T = \xi_{k+1} - \xi_k - \Delta_k = 0_{4\times 1} \tag{10}$$

To determine U_k, we are able to adopt the gradient descent method to solve nonlinear equations (10). Let U_k^p denote the $p-$th iteration value of U_k and U_k^{p+1} is generated by

$$U_k^{p+1} = U_k^p - \left(\nabla \bar{F}_k^p\right)^T \bar{F}_k^p \tag{11}$$

where the Jacobian matrix $\nabla \bar{F}$ is given by

$$\nabla \bar{F} = \frac{\delta \bar{F}}{\delta U_k} = \begin{bmatrix} \frac{\delta F_1}{\delta U_k} \\ \vdots \\ \frac{\delta F_4}{\delta U_k} \end{bmatrix}, \frac{\delta F_j}{\delta U_k} = \left[\frac{\delta F_j}{\delta u_{1k}}, \frac{\delta F_j}{\delta r_{1k}}, \frac{\delta F_j}{\delta u_{2k}}, \frac{\delta F_j}{\delta r_{2k}} \right] \tag{12}$$

and

$$\begin{cases} \begin{cases} \frac{\delta F_1}{\delta u_{1k}} = h \cos \psi_k + r_{1k} h^2 \sin \psi_k \\ \frac{\delta F_1}{\delta r_{1k}} = -\frac{3}{2} v_k h^2 \cos \psi_k - u_{2k} h^2 \sin \psi_k + u_{1k} h^2 \sin \psi_k \\ \frac{\delta F_1}{\delta u_{2k}} = h \cos \psi_k - r_{1k} h^2 \sin \psi_k \\ \frac{\delta F_1}{\delta r_{2k}} = -\frac{1}{2} v_k h^2 \cos \psi_k \\ \frac{\delta F_2}{\delta u_{1k}} = h \sin \psi_k - r_{1k} h^2 \cos \psi_k \\ \frac{\delta F_2}{\delta r_{1k}} = -\frac{3}{2} v_k h^2 \sin \psi_k + u_{2k} h^2 \cos \psi_k - u_{1k} h^2 \cos \psi_k \\ \frac{\delta F_2}{\delta u_{2k}} = h \sin \psi_k + r_{1k} h^2 \cos \psi_k \\ \frac{\delta F_2}{\delta r_{2k}} = -\frac{1}{2} v_k h^2 \sin \psi_k \end{cases} \\ \frac{\delta F_3}{\delta u_{1k}} = 0, \frac{\delta F_3}{\delta r_{1k}} = h, \frac{\delta F_3}{\delta u_{2k}} = 0, \frac{\delta F_3}{\delta r_{2k}} = h \\ \frac{\delta F_4}{\delta u_{1k}} = -r_{1k} h, \frac{\delta F_4}{\delta r_{1k}} = -u_{1k} h, \frac{\delta F_4}{\delta u_{2k}} = -r_{2k} h, \frac{\delta F_4}{\delta r_{2k}} = -u_{2k} h \end{cases} \tag{13}$$

Choose a reasonable constant $\sigma \in \mathbb{R}_+$ and set the following iteration end condition

$$\left\| U_k^{p+1} - U_k^p \right\| \leq \sigma \tag{14}$$

Till now, our control scheme is fully presented.

Remark 2 To solve nonlinear equations (10), we use the transverse of $\nabla \bar{F}$, rather than its inverse [19], due to that the zero-point searching for (10) is equivalent to minimum-value finding for scalar $M(U_k) = 0.5\bar{F}^T \bar{F}$. Adopting the gradient descent method to search the minimum-value point of $M(U_k)$, we obtain iteration approach (11).

Remark 3 We do not present the conditions under which the solution of (10) exists when using gradient descent method and we are going to introduce some more convincing and solid proofs in the future.

Remark 4 The selection of pair (h, ε) effects the accurate level that (8) approximates (7), more precisely, a smaller εh contributes to more accurate approximation. As for the real applications, this selection problem has to be discussed together with system stability, accuracy level and hardware costs as well as other factors, which will be revealed in our later work too.

Remark 5 To implement the proposed controller by means of way-point assignment when ξ_r is time-varying, one should renew the reference signal ξ_r in the k-th control cycle by

$$\xi_{r,k} = \xi_r(t_k + T_c) \tag{15}$$

For slow-varying ξ_r (small $|\dot{\xi}_r|$), choose ε such that $\varepsilon > \|\dot{\xi}_r\|$. When ξ_r varies fast (large $\dot{\xi}_r$), some modifications are needed, which is also one part of our future works. Also note that the reference sequence $\{\xi_{r,1}, \xi_{r,2}, ..., \xi_{r,k}, ...\}$ would well coincide with time-varying reference trajectory ξ_r if T_c is small enough.

Remark 6 The proposed controller in this work is capable of handling full-state stabilization of hovercraft, seems better than results such as [7] in which the orientation error is ignored and the reference results that rely on the damping terms of $\dot{v}$.

3 Numerical Simulation

In this section, we present several simulation results of the proposed control scheme regarding one stabilization case and two way-point assignment cases. The initial condition and necessary variables of three cases are given below.
Case 1. Stabilization:

$$\begin{aligned} x_0 &= 5, y_0 = -5, \psi_0 = -\frac{\pi}{6}, v_0 = 1 \\ x_r &= 0, y_r = 0, \psi_r = \frac{\pi}{8}, v_r = 0 \\ T_c &= 0.5\text{s}, h = 0.3\text{s}, \varepsilon = 0.06, \sigma = 0.005, T = 100\,\text{s} \end{aligned} \tag{16}$$

Case 2. Way-point assignment of straight line tracking:

$$\begin{aligned} x_0 &= 5, y_0 = -5, \psi_0 = -\pi, v_0 = 1 \\ \dot{x}_r &= 0.02, \dot{y}_r = 0.02, \dot{\psi}_r = 0, \dot{v}_r = 0 \\ x_r(0) &= 0, y_r(0) = 0, \psi_r(0) = \frac{\pi}{4}, v_r(0) = 0 \\ T_c &= 0.4\text{s}, h = 0.2\text{s}, \varepsilon = 0.02, \sigma = 0.005, T = 1000\,\text{s} \end{aligned} \tag{17}$$

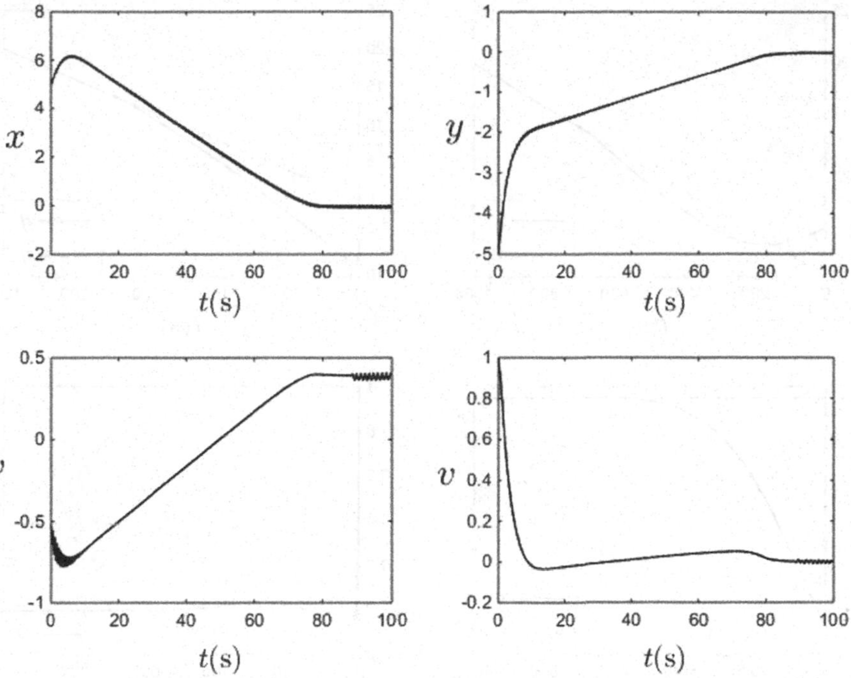

Fig. 1 The simulation results of stabilization

Case 3. Way-point assignment of circular trajectory tracking:

$$\begin{aligned} x_0 &= 5, y_0 = -5, \psi_0 = -\pi, v_0 = 1 \\ \dot{x}_r &= 0.02\cos\psi_r, \dot{y}_r = 0.02\sin\psi_r, \dot{\psi}_r = 0.003, \dot{v}_r = 0 \\ x_r(0) &= 0, y_r(0) = -6, \psi_r(0) = 0, v_r(0) = 0 \\ T_c &= 0.5\text{s}, h = 0.25\text{s}, \varepsilon = 0.02, \sigma = 0.005, T = 2000\,\text{s} \end{aligned} \tag{18}$$

where T in above equations denotes simulation time length. For the all three considered cases, we start iteration (11) by setting $U_1^1 = [-0.2, -0.2, 0.2, 0.2]^T$ and $U_k^1 = U_{k-1}, \forall k > 1$. The simulation results are shown in Figs. 1, 2 and 3.

The simulation results displayed in Figs. 1, 2 and 3 show that all states move to the reference in both stabilization and way-point assignments cases, which well verify the effectiveness of the proposed controller.

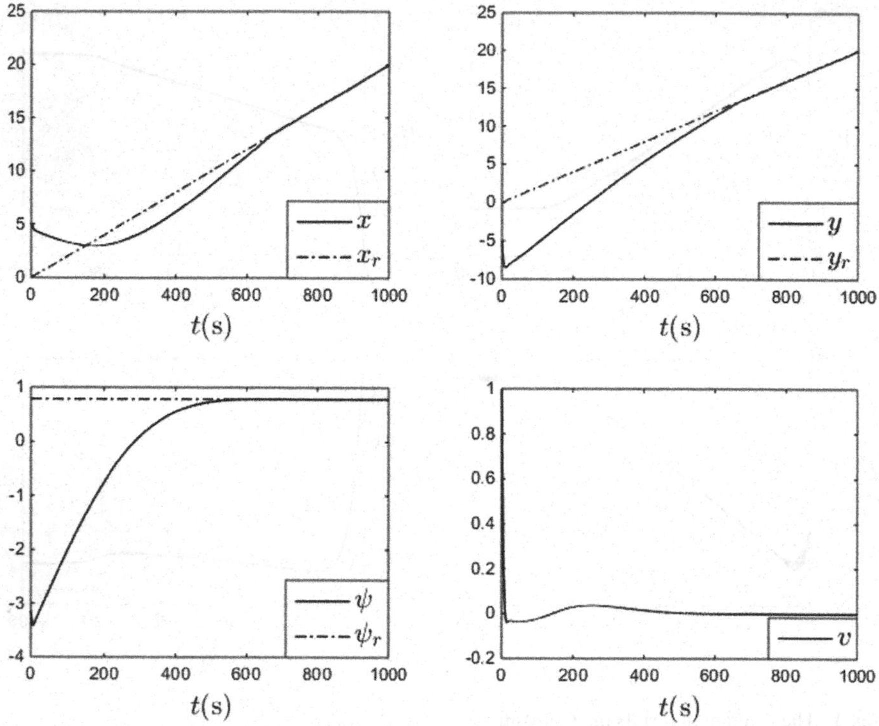

Fig. 2 Way-point assignment of straight line tracking

4 Conclusion

The control of hovercraft concerned in this work successfully solve the stabilization problem. Thanks to the special form of system dynamics, we are able to obtain analytic solution of states. To steer the states to a prescribed length towards reference states, we use iteration method to obtain real controls for each control cycle. The high performance of our proposed controllers are verified by several simulations. As for the future work, we are going to investigate the unreveal details in this contribution, such as the existence problem of U_k in Eq. (10) and the selection criterions of (h, ε).

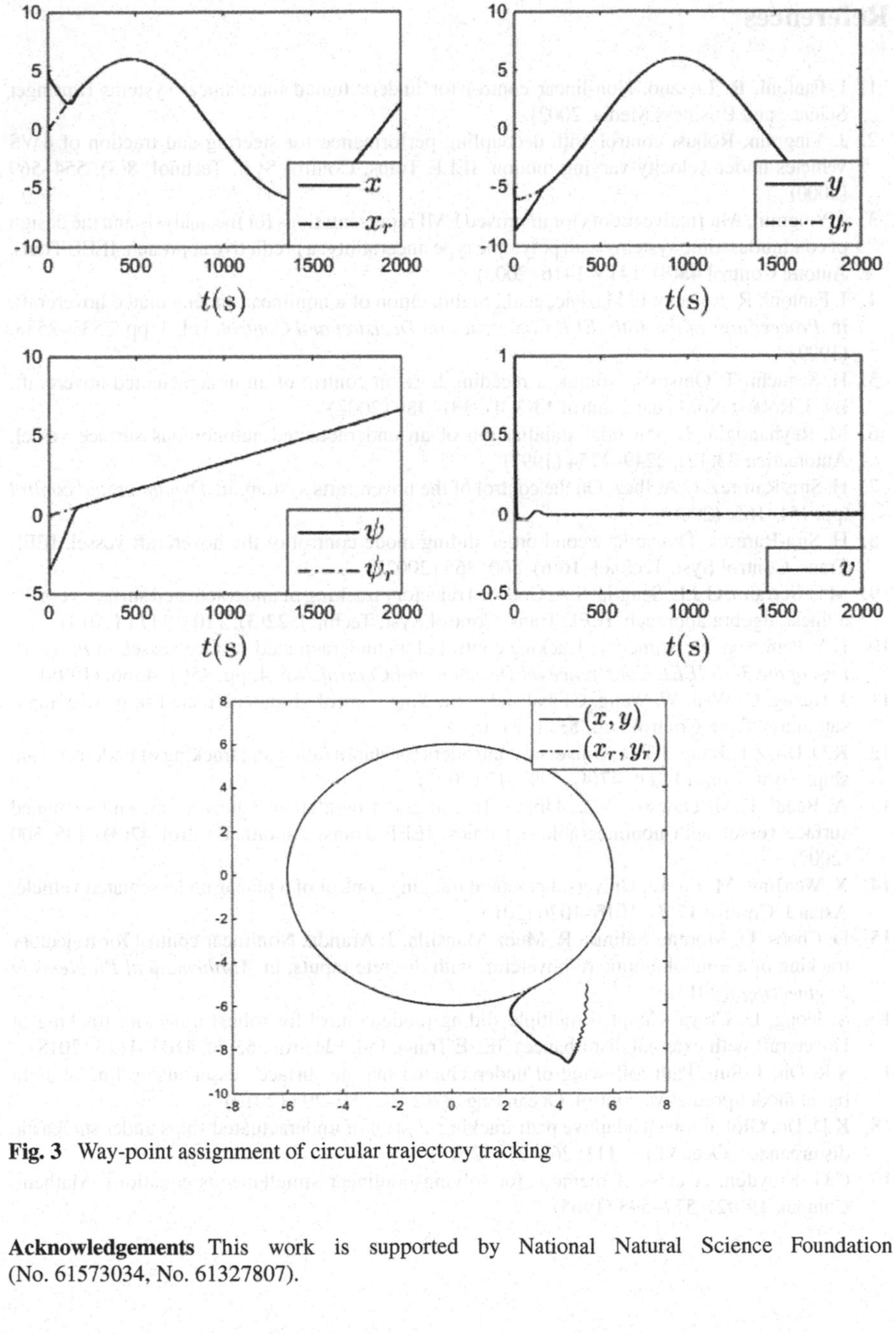

Fig. 3 Way-point assignment of circular trajectory tracking

Acknowledgements This work is supported by National Natural Science Foundation (No. 61573034, No. 61327807).

References

1. I. Fantoni, R. Lozano, Non-linear control for underactuated mechanical systems (Springer Science and Business Media, 2002)
2. J. Yingmin, Robust control with decoupling performance for steering and traction of 4WS vehicles under velocity-varying motion. IEEE Trans. Control Syst. Technol. **8**(3), 554–569 (2000)
3. J. Yingmin, Alternative proofs for improved LMI representations for the analysis and the design of continuous-time systems with polytopic type uncertainty: a predictive approach. IEEE Trans. Autom. Control **48**(8), 1413–1416 (2003)
4. I. Fantoni, R. Lozano, F. Mazenc, et al., Stabilization of a nonlinear underactuated hovercraft, in *Proceedings of the 38th IEEE Conference on Decision and Control*, vol. 3, pp. 2533–2538, (1999)
5. H. Seguchi, T. Ohtsuka, Nonlinear receding horizon control of an underactuated hovercraft. Int. J. Robust Nonlinear Control **13**(3–4), 381–389 (2003)
6. M. Reyhanoglu, Exponential stabilization of an underactuated autonomous surface vessel. Automatica **33**(12), 2249–2254 (1997)
7. H. Sira-Ramrez, C.A. Ibez, On the control of the hovercrafts system, in *Dynamics and control* ,pp. 151–163, (2000)
8. H. Sira-Ramrez, Dynamic second-order sliding mode control of the hovercraft vessel. IEEE Trans. Control Syst. Technol. **10**(6), 860–865 (2002)
9. M.E. Serrano, G.J.E. Scaglia, S.A. Godoy, Trajectory tracking of underactuated surface vessels: a linear algebra approach. IEEE Trans. Control Syst. Technol. **22**(3), 1103–1111 (2014)
10. K.Y. Pettersen, H. Nijmeijer, Tracking control of an underactuated surface vessel, in *Proceedings of the 37th IEEE Conference on Decision and Control*, vol. 4, pp. 4561–4566, (1998)
11. J. Huang, C. Wen, W. Wang, Global stable tracking control of underactuated ships with input saturation. Syst. Control Lett. **85**, 1–7 (2015)
12. K.D. Do, Z.P. Jiang, J. Pan, Universal controllers for stabilization and tracking of underactuated ships. Syst. Control Lett. **47**(4), 299–317 (2002)
13. A. Behal, D.M. Dawson, W.E. Dixon, Tracking and regulation control of an underactuated surface vessel with nonintegrable dynamics. IEEE Trans. Autom. Control **47**(3), 495–500 (2002)
14. X. WenJing, M. BaoLi, Universal practical tracking control of a planar underactuated vehicle. Asian J. Control **17**(3), 1016–1026 (2015)
15. D. Chaos, D. Moreno-Salinas, R. Muoz-Mansilla, J. Aranda, Nonlinear control for trajectory tracking of a nonholonomic rc-hovercraft with discrete inputs, in *Mathematical Problems in Engineering*, (2013)
16. S. Jeong, D. Chwa, Coupled multiple sliding-mode control for robust trajectory tracking of Hovercraft with external disturbances. IEEE Trans. Ind. Electron. **65**(5), 4103–4113 (2018)
17. S.R. Oh, J. Sun, Path following of underactuated marine surface vessels using line-of-sight based model predictive control. Ocean Eng. **37**(2–3), 289–295 (2010)
18. K.D. Do, Global robust adaptive path-tracking control of underactuated ships under stochastic disturbances. Ocean Eng. **111**, 267–278 (2016)
19. C.G. Broyden, A class of methods for solving nonlinear simultaneous equations. Mathem. Comput. **19**(92), 577–593 (1965)

Anti-disturbance Tracking Controller Design for PMSM via T-S Disturbance Modeling

Mintai Wang, Chengbo Niu, Bei Liu and Yang Yi

Abstract This brief proposes a rotor angular velocity anti-disturbance control way for typical permanent magnet synchronous motor (PMSM) models by using fuzzy disturbance modeling. Following T-S fuzzy description for unknown exogenous disturbances, a nonlinear observer (DO) is used to estimate unknown load disturbances existed in PMSM models. Furthermore, based on the estimation of disturbances, a composite control input is discussed to ensure the PMSM stability. Meanwhile, the dynamical trajectory of angular velocity can track to the desired value. It is noted that corresponding theorem proof can be achieved by applying Lyapunov analysis method with optimization theory.

Keywords PMSM systems · Tracking control · Anti-disturbance control
Convex optimization · T-S fuzzy models

1 Introduction

In addition to the basic features of general synchronous motors, PMSM also have the characteristics of high efficiency, small size, relatively simple structure and easy control [1]. In recent years, it is reported that the PMSM system has applied many advanced control approaches, such as predictive control [1], adaptive learning algorithm [2, 3], observer design [4], iterative algorithm [5], fuzzy-NN theories [2, 6], variable structure control [6, 7].

In traditional control methods, PID control, robust control and DOBC control are widely used to suppress disturbances. However, these methods can only be used to compensate regular disturbances such as harmonic or constant disturbances, which cannot solve the estimation and rejection for those unknown irregular disturbances. On the other hand, for some gray-box or black-box systems (see [8]), T-S models are

M. Wang · C. Niu · B. Liu · Y. Yi (✉)
College of Information Engineering,
Yangzhou University, Yangzhou 225127, China
e-mail: yiyangcontrol@163.com

Y. Jia et al. (eds.), *Proceedings of 2018 Chinese Intelligent Systems Conference*, Lecture Notes in Electrical Engineering 529,
https://doi.org/10.1007/978-981-13-2291-4_72

very effective. By using T-S modeling, many complex physical models or dynamical models can be modeled or analyzed, such as network models [9], descriptor models [10] and postpone systems [11]. It is in [11] that the fault detection and tolerant control problem as well as the fault-estimation were proposed for typical T-S fuzzy modeling.

Based on the above motivation, in this brief, the main aim is to design a new T-S modeling method for those unknown perturbations, and further discuss the design of anti-disturbance controller design for PMSM system. By using transformation of coordinates, the unknown loads and the controlled input in PMSM are adjusted in the same gallery. By combining PI state controller with the estimate of disturbances, the anti-disturbance is proposed such that the PMSM model stability.

2 Model Description of Surface-Mounted PMSM

Under the d-q rotary coordinate, the mathematical description of PMSM is expressed as:

$$\begin{aligned}
\frac{d\omega}{dt} &= \frac{K_t}{J} i_q - \frac{B}{J}\omega - \frac{T_L}{J} \\
\frac{di_d}{dt} &= \frac{1}{L_d} u_d + p\omega i_q - \frac{R}{L_d} i_d \\
\frac{di_q}{dt} &= \frac{1}{L_q} u_q - \frac{R}{L_q} i_q - p\omega i_d - \frac{p\psi_f}{L_q}\omega \\
T_e &= 1.5 p\psi_f i_q
\end{aligned} \tag{1}$$

where u_q is the q axis voltage, i_q is the q axis current, ω is the angular velocity, J is the rotor moment of inertia. L_q is q axial inductance component. B is the viscous friction coefficient. R is stator resistance. T_L is the load torque, ψ_f is the rotor flux linkage. $K_t = 1.5p\psi_f$ is representative torque constant. These parameters are considered known when the PMSM is identified.

It is noted that the coupling term in the PMSM makes the system difficult to control. To solving the coupling between angular velocity and current, this article will discuss the basis of vector control ($i_d = 0$). Therefore, the model is rebuilt as:

$$\begin{pmatrix} \dot{i}_q \\ \dot{\omega} \end{pmatrix} = \begin{pmatrix} -\frac{R}{L_q} & -\frac{p\psi_f}{L_q} \\ \frac{K_t}{J} & -\frac{B}{J} \end{pmatrix} \begin{pmatrix} i_q \\ \omega \end{pmatrix} + \begin{pmatrix} \frac{u_q}{L_q} \\ 0 \end{pmatrix} + \begin{pmatrix} 0 \\ -\frac{T_L}{J} \end{pmatrix} \tag{2}$$

To facilitate the design of DO, make:

$$x_1 = \omega, x_2 = \dot{\omega}, x_3 = \xi \tag{3}$$

From (2) and (3), we can get:

$$\dot{x}_1 = x_2 \quad \dot{x}_3 = -x_1 + \omega^*$$
$$\dot{x}_2 = \left(-\frac{RB}{JL_q} - \frac{K_t p\psi_f}{JL_q}\right)x_1 + \left(-\frac{R}{L_q} - \frac{B}{J}\right)x_2$$
$$+ \frac{K_t}{JL_q}u - \frac{R}{JL_q}T_L - \frac{1}{J}\dot{T}_L \tag{4}$$

where $\xi(t)$ is subsidiary variables with the presentation $\dot{\xi}(t) = \omega^* - \omega$, ω^* is designed target angle speed. Defining $u(t) = u_q(t)$ is control input, new state variables is $x(t) = \left[\omega^T, \dot{\omega}^T, \xi^T\right]^T$, external load disturbance is $d_0(t) = T_L$, $\dot{d}_0(t) = \dot{T}_L$ is first derivative of load disturbance and $d_1(t) = \omega^*$ is unknown bounded disturbance item.

3 T-S Disturbance Modeling

The model with new state variables is described as follows:

$$\dot{x}(t) = A_0 x(t) + B_0 u(t) + H_0 d_0(t) + B_2 \dot{d}_0(t) + B_1 d_1(t) \tag{5}$$

where

$$A_0 = \begin{bmatrix} 0 & 1 & 0 \\ -\frac{RB+K_t p\psi_f}{JL_q} & \left(-\frac{R}{L_q} - \frac{B}{J}\right) & 0 \\ -1 & 0 & 0 \end{bmatrix} \quad H_0 = \begin{bmatrix} 0 \\ -\frac{R}{JL_q} \\ 0 \end{bmatrix}$$
$$B_0 = \begin{bmatrix} 0 \\ \frac{K_t}{JL_q} \\ 0 \end{bmatrix} \quad B_1 = \begin{bmatrix} 0 \\ 0 \\ 1 \end{bmatrix} \quad B_2 = \begin{bmatrix} 0 \\ -\frac{1}{J} \\ 0 \end{bmatrix}$$

Fuzzy If–Then rules whose collection can be seen as a good representation of modeling of the nonlinear dynamics expressed the T–S fuzzy model. Assume that the following T-S fuzzy model with rules can describe $d(t)$ in controlled input channel.

Plant Rule j: If θ_1 is μ_{1j}, θ_2 is μ_{2j}, ..., θ_n is μ_{nj}, the model is expressed as:

$$\begin{cases} \dot{w}(t) = W_j w(t) \\ d(t) = V_j w(t) \end{cases} \tag{6}$$

wherein the known coefficient matrices are W_j and V_j. The number of If–Then rules is R and the number of premise variables is N. The precondition variables and fuzzy sets characterized by membership functions are θ_i and μ_{ij}.

Through fuzzy mixing, the total T-S model can be deduced as:

$$\begin{cases} \dot{w}(t) = \sum_{j=1}^{r} h_j(\theta) W_j w(t) \\ d(t) = \sum_{j=1}^{r} h_j(\theta) V_j w(t) \end{cases} \tag{7}$$

where $\theta = [\theta_1, \theta_2, \cdots, \theta_n], \sigma_j(\theta) = \prod_{i=1}^{n} \mu_{ij}(\theta_j), h_j(\theta) = \frac{\sigma_j(\theta)}{\sum_{j=1}^{r} \sigma_j(\theta)} . \mu_{ij}(.)$ are the level of membership functions. This article assumes that:

$$\sigma_j(\theta) \geq 0, j = 1, \ldots, r, \sum_{j=1}^{r} \sigma_j(\theta) > 0$$

for any θ. Hence, $h_j(\theta)$ satisfies

$$h_j(\theta) \geq 0, j = 1, \ldots, r, \sum_{j=1}^{r} h_j(\theta) = 1 \tag{8}$$

for any θ.

4 Design of DO

In order to ensure the anti-disturbance performance of PMSM, it is necessary to construct a effective nonlinear DO. Then the designed DO is expressed as:

$$\begin{aligned} \hat{d}_0(t) &= \sum_{j=1}^{r} h_j(\theta) V_j \hat{w}(t) + \delta \\ \hat{w}(t) &= v(t) - Lx(t) \\ \dot{v}(t) &= \sum_{j=1}^{r} h_j(\theta) \big(W_j + L H_0 V_j + L B_2 V_j W_j \big)(v(t) - Lx(t)) \\ &\quad + L(A_0 x(t) + B_0 u(t)) + L H_0 \delta \end{aligned} \tag{9}$$

where the estimated values of $d_0(t)$ and $\dot{d}_0(t)$ are $\hat{d}_0(t)$ and $\dot{\hat{d}}_0(t)$ respectively. The estimated values of $w(t)$ is $\hat{w}(t)$. The observer gain L will be identified later. The auxiliary vector for the DO is $v(t)$. $e_w(t) = w(t) - \hat{w}(t)$ is used to represent the estimation error. By comparing (5) and (7) with (9), we can get the following expression:

$$\dot{e}_w(t) = \sum_{j=1}^{r} h_j(\theta)\big(W_j + LH_0V_j + LB_2V_jW_j\big)e_w(t) + LB_1d_1(t) \tag{10}$$

The main purpose is to compute the observer gain, so that the system (10) meet the desired stability and robustness. It is noted that the first derivative of $d_0(t)$ can also be estimated by the designed DO (9). The specific form is $\dot{\hat{d}}_0(t) = VW\hat{w}(t)$.

5 Design of Compound Controller and Related Theorem

To solve the angular velocity tracking of PMSM, a state feedback controller with generalized PI structure is designed as:

$$u_0(t) = K_{p1}\omega(t) + K_{p2}\dot{\omega}(t) + K_I\xi(t) \tag{11}$$

where K_{P1}, K_{P2}, K_I are PI control gains. $\dot{\xi}(t) = \omega^* - \omega$. The compound controller based on Eq. (11) is designed as:

$$u(t) = u_0(t) + \frac{R}{K_t}\hat{d}_0(t) + \frac{L_q}{K_t}\dot{\hat{d}}_0(t) \tag{12}$$

Substituting (12) into PMSM augmented system (5), the following closed-loop system can be obtained as:

$$\dot{x}(t) = (A_0 + B_0K)x(t) + \sum_{j=1}^{r} h_j(\theta)\big(H_0V_j + B_2V_jW_j\big)e_w(t) + B_1d_1(t) \tag{13}$$

In addition, an integrated disturbance observation error model (10) and a closed-loop system model (11) are available:

$$\dot{\rho}(t) = \bar{A}\rho(t) + H_1d_1(t) \tag{14}$$

where $\rho(t) = \left[x^T(t), e_w^T(t)\right]^T$, and

$$\bar{A} = \begin{bmatrix} A_0 + B_0K & \sum_{j=1}^{r} h_j(\theta)\big(H_0V_j + B_2V_jW_j\big) \\ 0 & W + \sum_{j=1}^{r} h_j(\theta)\big(LH_0V_j + LB_2V_jW_j\big) \end{bmatrix} \quad H_1 = \begin{bmatrix} B_1 \\ LB_1 \end{bmatrix}$$

To sum up, a composite system including the closed-loop PMSM model (13) and the disturbance observation error model (10) is available. The following theorem gives the solution to compute the controller gain and the observer gain by using convex optimization theory.

Theorem 1 *Consider the close-loop composite system (14) containing T-S disturbance model (7), for known parameters* $\mu_i (i = 1, 2, 3, 4)$, $\mu_3 > 0$, $\mu_4 > 0$ *and* $\lambda_i (i = 1, 2)$, *if there exist matrices* $Q_1 = P_1^{-1} > 0$, $P_2 > 0$, $T > 0$ *and* R_1, R_2 *with the appropriate dimension meet the following inequalities*

$$\begin{bmatrix} \Xi_1 & \lambda_1^{-1} B_1 & Q_1 \\ * & -I & 0 \\ * & * & -\mu_3^{-1} I \end{bmatrix} < 0 \tag{15}$$

with $\Xi_1 = sym(A_0 Q_1 + B_0 R_1) + \mu_2^2 Q_1$

$$\begin{bmatrix} \Xi_2 & R_2 B_1 \\ B_1^T R_2^T & -\lambda_2^2 I \end{bmatrix} < 0 \tag{16}$$

$$\Xi_2 = sym\left[\sum_{j=1}^{r} h_j(\theta) P_2 W_j + R_2 \sum_{j=1}^{r} h_j(\theta)\big(H_0 V_j + B_2 V_j W_j\big)\right] + \mu_1^2 T + \mu_4 I$$

are solvable, then the enlarged closed-loop system (14) is asymptotically stable under the action of the composite control law (12), tracking error satisfied $\lim_{t\to\infty} \omega(t) = \omega^*$. *And* $K = R_1 Q_1^{-1}$ *is the PI state feedback control gain and the disturbance observer gain is* $L = P_2^{-1} R_2$.

6 Simulation Results

The specification of the PMSM system is shown in Table 1.

Case i: Harmonic disturbance with decay properties

Usually, linear exogenous system describes the common disturbance type in engineering by harmonic disturbance or constant disturbance. However, those irregular

Table 1 Specification of the PMSM system

Stator resistance R	0.54 Ω
Stator inductance L	0.0096 H
Pole pairs p	4
Viscous coefficient B	0.0001 Nm/rad/s
Rotor inertia J	0.016 kg m^2
Flux of linkage ϕ_f	0.61 wb

harmonic disturbances cannot be modelled by mentioned linear system in [5, 9], for example, disturbance with fading characteristics. The following T-S model is provided to depict the harmonic disturbance of decay properties.

Disturbance model:

Rule 1: If w_1 is A_1^1, then $\dot{w}(t) = [-1\,1; -5\,0]w(t)$, $d(t) = [-4\,0]w(t)$

Rule 2: If w_1 is A_1^2, then $\dot{w}(t) = [0\,10; -5\,0]w(t)$, $d(t) = [4\,0]w(t)$

In this paper, in order to better model the dynamics of unknown disturbances, the gaussian-type member functions are chosen as follows:

$$A_1^1 = \frac{\exp(\frac{-(w_1-1.5)^2}{2\sigma_1^2})}{\exp(\frac{-(w_1-1.5)^2}{2\sigma_1^2}) + \exp(\frac{-(w_1-1)^2}{2\sigma_2^2})}$$

$$A_1^2 = \frac{\exp(\frac{-(w_1-1)^2}{2\sigma_2^2})}{\exp(\frac{-(w_1-1.5)^2}{2\sigma_1^2}) + \exp(\frac{-(w_1-1)^2}{2\sigma_2^2})}$$

where $\sigma_1^2 = 0.5$, $\sigma_2^2 = 1$. The observer gain L and the control gains K_P, K_I are found to be

$$K_P = \begin{bmatrix} 2.4368 \; 0.0019 \end{bmatrix}; K_I = 0.0067; L = \begin{bmatrix} 0 & -0.0428 & 0 \\ 0 & 0.4276 & 0 \end{bmatrix}$$

Case ii: Irregular harmonic disturbances

In this section, the T-S model is imported to identity the irregular harmonic disturbances. The system matrices are defined as

$$W_1 = \begin{bmatrix} -1 & 2 \\ -5 & 0 \end{bmatrix}, V_1 = \begin{bmatrix} 0 \; 4 \end{bmatrix}, W_2 = \begin{bmatrix} 0 & -8 \\ 3.8 & 0 \end{bmatrix}, V_2 = \begin{bmatrix} 0 \; 4 \end{bmatrix}$$

The following member functions are considered

$$A_1^1 = \frac{\exp(\frac{-(w_1-1.2)^2}{2\sigma_1^2})}{\exp(\frac{-(w_1-1.2)^2}{2\sigma_1^2}) + \exp(\frac{-(w_1-1)^2}{2\sigma_2^2})}$$

$$A_1^2 = \frac{\exp(\frac{-(w_1-1)^2}{2\sigma_2^2})}{\exp(\frac{-(w_1-1.2)^2}{2\sigma_1^2}) + \exp(\frac{-(w_1-1)^2}{2\sigma_2^2})}$$

where $\sigma_1^2 = 0.5$, $\sigma_2^2 = 1$. Based on Theorem 1, the observer gain L and the control gains K_P, K_I can be computed as

$$K_P = \begin{bmatrix} 2.4367 \; 0.0019 \end{bmatrix}; K_I = 0.0067; L = \begin{bmatrix} 0 & -0.2615 & 0 \\ 0 & 0.0096 & 0 \end{bmatrix};$$

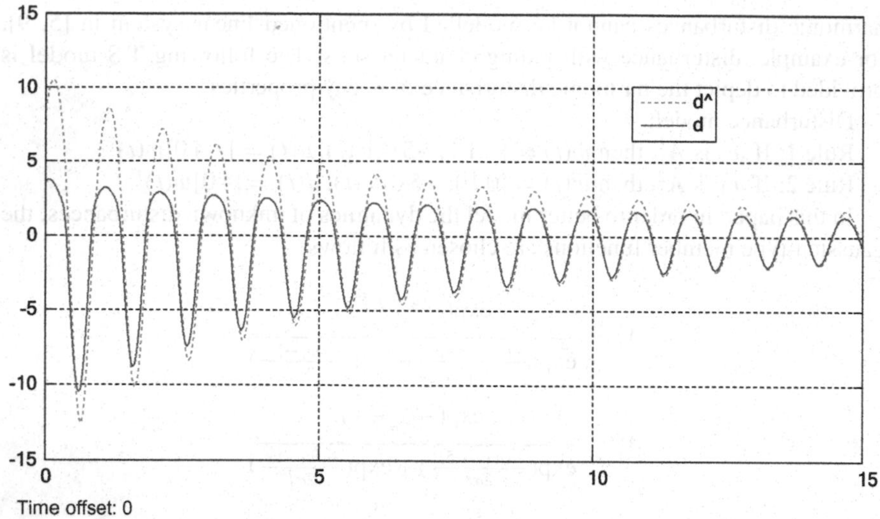

Fig. 1 Disturbance and disturbance estimation in model 1

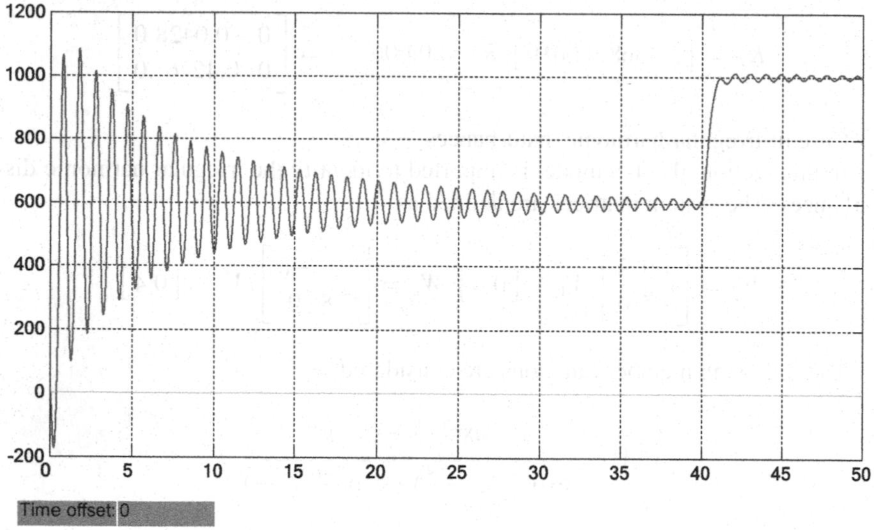

Fig. 2 Responses of angular velocity without DO in model 1

The above figures show the dynamical responses of PMSM and the disturbance estimation performance in different exogenous disturbances. When harmonic disturbances with attenuation characteristics is considered, Fig. 1 is the dynamics of disturbance and its estimation. Figures 2 and 3 show the responses of angular velocity with DO and without DO respectively, which reflects the importance of DO. When considering irregular harmonic disturbances, similar simulation results can

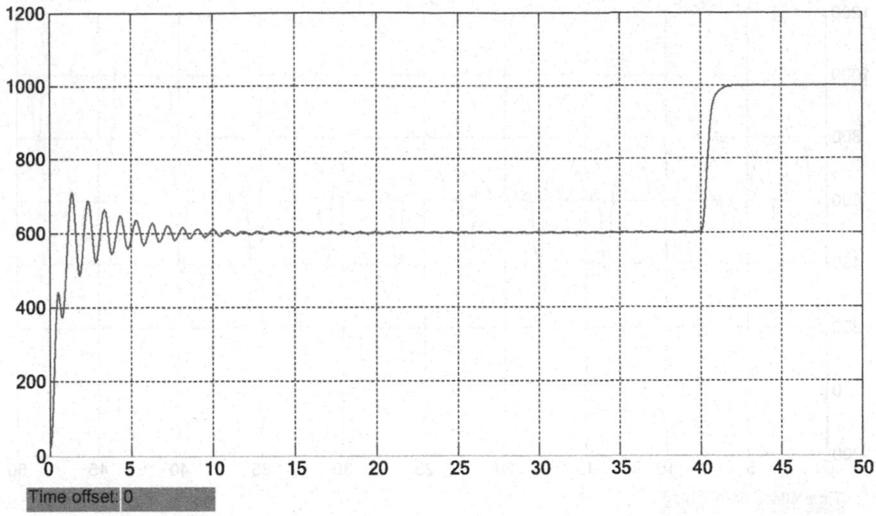

Fig. 3 Responses of angular velocity DO in model 1

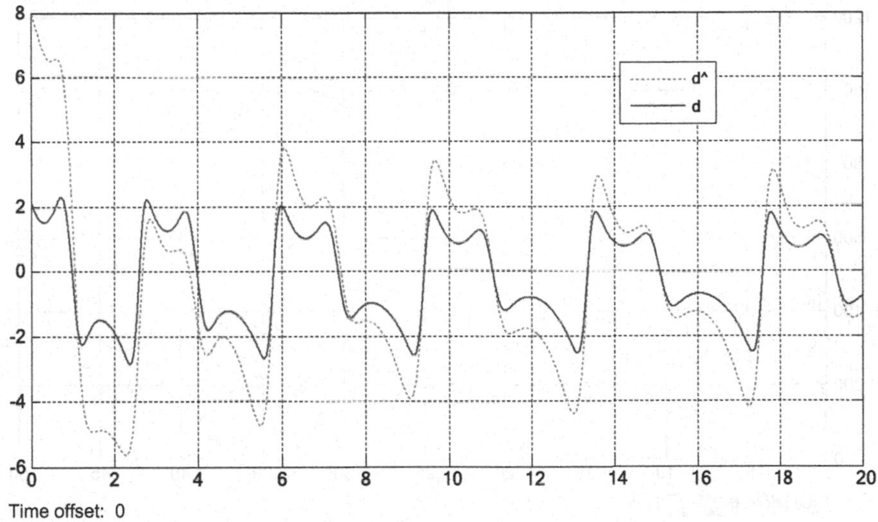

Fig. 4 Disturbance and disturbance estimation in with model 2

be found in Figs. 4, 5 and 6. The satisfactory stability, tacking performance and anti-disturbance capabilities can be demonstrated from Figs. 1, 2, 3, 4, 5 and 6.

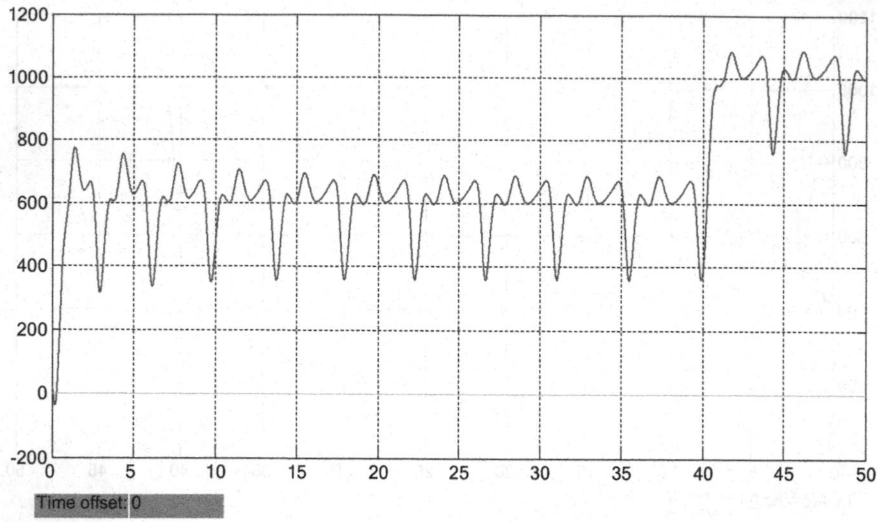

Fig. 5 Responses of angular velocity without DO in model 2

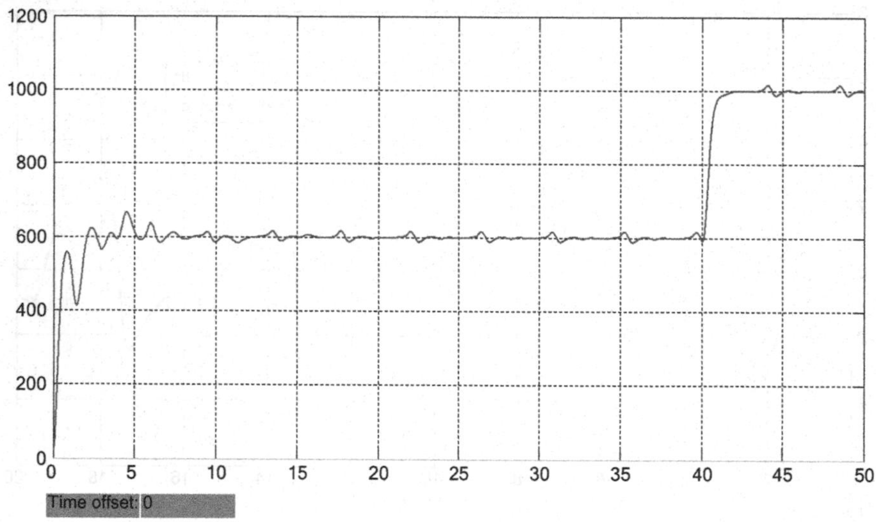

Fig. 6 Responses of angular velocity with DO in model 2

7 Conclusion

The tracking problem of rotor angular velocity of PMSM system is studied by combining T-S fuzzy disturbance modelling with DO design method. By using convex optimization techniques and Lyapunov analysis method, the disturbance observer gain and controller gain can be computed such that multi-objective requirements including stabilization, the angular velocity tracking and robustness of the PMSM system can be ensured at the same time. The corresponding simulation results show that this method is effective and advanced.

Acknowledgements This work was supported in part by the National Nature Science Foundation of China under Grants (61473249, 61503329), the Nature Science Foundation of Jiangsu Province under Grant (BK2017515) and the QingLan Project of Jiangsu Province.

References

1. H. Liu, S. Li, Speed control for PMSM servo system using predictive functional control and extended state observer. IEEE Trans. Ind. Electron. **59**(2), 1171–1183 (2012)
2. S. Li, H. Gu, Fuzzy adaptive internal model control schemes for PMSM speed-regulation system. IEEE Trans. Ind. Inform. **8**(4), 767–779 (2012)
3. S. Li, Z. Liu, Adaptive speed control for permanent magnet synchronous motor system with variations of load inertia. IEEE Trans. Ind. Electron. **56**(8), 3050–3059 (2009)
4. Y.A.R.I. Mohamed, Design and implementation of a robust current control scheme for a PMSM vector drive with a simple adaptive disturbance observer. IEEE Trans. Ind. Electron. **54**(4), 1981–1988 (2007)
5. Y. Yi, X.K. Sun, X.L. Li, X.X. Fan, Disturbance observer based composite speed controller design for PMSM system with mismatched disturbances. Trans. Inst. Meas. Cont. **38**(6), 742–750 (2016)
6. F.F.M. El-Sousy, Robust wavelet-neural-network sliding mode control system for permanent magnet synchronous motor drive. IET Electr. Power Appl. **5**(1), 113–132 (2011)
7. X.G. Zhang, L.Z. Sun, K. Zhao, L. Sun, Nonlinear speed control for PMSM system using sliding-mode control and disturbance compensation techniques. IEEE Trans. Power Electron. **28**(3), 1358–1365 (2013)
8. G. Feng, A survey on analysis and design of model-based fuzzy control systems. IEEE Trans. Fuzzy Syst. **14**(5), 676–697 (2006)
9. Y. Yi, W.X. Zheng, C.Y. Sun, L. Guo, DOB fuzzy controller design for non-Gaussian stochastic distribution systems using two-step fuzzy identification. IEEE Trans. Fuzzy Syst. **24**(2), 401–418 (2016)
10. S.J. Huang, G.H. Yang, Fault tolerant controller design for T-S fuzzy systems with time-varying delay and actuator faults: a K-step fault estimation approach. IEEE Trans. Fuzzy Syst. **22**(6), 1526–1540 (2014)
11. X.M. Zhang, Z.J. Zhang, G.P. Lu, Fault detection for state-delay fuzzy systems subject to random communication delay. Int. J. Innov. Comput. Inf. Control **8**(4), 2439–2451 (2012)

The Fuzzy Control of Electro-hydraulic Servo System Based on DE Algorithm

Meng Dong, Xiting Luan, Baoyuan Wu and Junlong Liang

Abstract In order to improve the control performance of the electro-hydraulic servo system, this paper firstly presents an intelligent fuzzy controller relying on experience. Then, the differential evolution (DE) algorithm is used to optimize the membership function and control rules to overcome the shortcoming that the fuzzy controller based on the human experience does not reach the optimal performance. The simulation results show that, for the optimized fuzzy controller, the control intensity applied to the object in the early period is more increased, the system has almost no overshoot, the response time is shorter, and the tracking performance is improved.

Keywords Fuzzy control · Differential evolution algorithm
Membership function · Control rules

1 Introduction

The electro-hydraulic servo control system has the characteristics of high precision, fast response, and high output power of the hydraulic system. At the same time, it has the advantages of flexibly processing electrical signal and easily forming parameter feedback. It is particularly suitable for the use of computer-implemented complex control strategies, and widely used in industrial production. The main components of electro-hydraulic servo system are electro-hydraulic servo valve and asymmetric hydraulic cylinder, called valve-controlled asymmetric cylinder system. Ye redefined the load flow and load pressure, used an idea of weighted average to unify the linear

M. Dong (✉) · B. Wu
Science and Technology on Liquid Rocket Engine Laboratory, Xi'an Aerospace Propulsion Institute, Shaanxi, China
e-mail: mengdong73@163.com

J. Liang
Xi'an Aerospace Propulsion Institute, Shaanxi, China

X. Luan · B. Wu
Academy of Aerospace Propulsion Technology, Shaanxi, China

Y. Jia et al. (eds.), *Proceedings of 2018 Chinese Intelligent Systems Conference*, Lecture Notes in Electrical Engineering 529,
https://doi.org/10.1007/978-981-13-2291-4_73

models in both directions, and finally designed the internal model variable structure controller [1]. Zheng established a nonlinear model of a valve-controlled asymmetric cylinder, applies localized linearization, and designed a sliding mode controller [2].

Fuzzy control is widely used because it has the advantages of no need of system-accurate model and strong robustness. However, the design of fuzzy control membership functions and control rules heavily relies on designer's experience, so it is not guaranteed that the controller‘s performance is optimal. If the optimization algorithm is integrated into the fuzzy control, the performance of the system can be greatly improved [3–5]. Differential evolution (DE) algorithm is a stochastic heuristic search algorithm, which retains global search strategy based on the population. Also it adopts real coding, differential-based simple mutation operation and one-to-one competitive survival strategy, which reduces the complexity of genetic operation. Due to its less pending parameters, higher speed of convergence, and difficulty in falling into local solutions, the DE algorithm is widely used in various fields [6]. Applying the DE algorithm to the optimization of membership functions and control rules is of great significance to design a fuzzy controller with a global optimal performance index.

2 Model of System

2.1 *Model of Electro-hydraulic Servo Valve*

For the dual-nozzle baffle force feedback electro-hydraulic servo valve, the current input as control signal makes the torque motor produce torque. The torque causes the armature-bourdon tube assembly to deflect. At the same time, the displacement of nozzle-flapper valve is passed to form the displacement of the slide valve. The spring rod as feedback acts on the armature-bourdon tube assembly in the form of force to configure a negative feedback system [7]. When the requirements of dynamic response of system are high, the natural frequency of the electro-hydraulic servo valve and the hydraulic power element is equivalent, so the model of electro-hydraulic servo valve is regarded as the second-order oscillation elements just like Eq. (1).

$$\Phi_{sv}(s) = \frac{Q}{I} = \frac{K_{sv}}{\frac{s^2}{\omega_{mf}^2} + \frac{2\xi_{mf}}{\omega_{mf}}s + 1} \tag{1}$$

2.2 *Model of Asymmetric Cylinder*

When the piston of valve-controlled asymmetric cylinder moves forward and backward, the model structure is the same, but the model coefficient values are different.

Take the piston moving forward of asymmetrical cylinder as an example to establish a system model.

(1) Linearized flow equation

$$q_L = K_q x_v - K_c p_L \tag{2}$$

where, q_L is load flow; p_L is load pressure; x_v is displacement of slide valve; K_q is flow gain; K_c is flow-pressure coefficient.

(2) Flow continuity equation of asymmetric cylinder

According to the law of conservation of mass, the inlet of the hydraulic chamber 1 and the piston of hydraulic chamber 2 are respectively bounded, and the flow before the boundary is equal to the flow after the boundary, as follows.

$$q_L = C_L p_L + C_s p_s + \frac{V_t}{4\beta_e}\frac{\mathrm{d}p_L}{\mathrm{d}t} + A_1\frac{\mathrm{d}x_p}{\mathrm{d}t} \tag{3}$$

where, p_s is pressure of bump outlet; C_L, C_s are equivalent leakage coefficients; V_t is equivalent volume; β_e is equivalent bulk modulus; x_p is the displacement of piston.

(3) The balanced force equation between the cylinder and load

$$A_1 p_L = M_t\frac{\mathrm{d}^2 x_p}{\mathrm{d}t^2} + B_c\frac{\mathrm{d}x_p}{\mathrm{d}t} + Kx_p + F_L \tag{4}$$

where, M_t is the total mass of piston and load; B_c is viscous damping coefficient of piston and load; K is the stiffness of load; F_L is external load force on the piston.

(4) Model of asymmetric cylinder

Through the three Eqs. (2), (3) and (4) are Laplace transformed, the open loop transfer function of the asymmetric cylinder is as follow

$$X_p = \frac{\frac{K_q}{A_e}X_v - \frac{1}{A_e^2}\left(K_{cL} + \frac{V_t}{4\beta_e}s\right)F_L}{\frac{V_t M_t}{4\beta_e A_e^2}s^3 + \left(\frac{K_{cL}M_t}{A_e^2} + \frac{V_t B_c}{4\beta_e A_e^2}\right)s^2 + \left(\frac{K_{cL}B_c}{A_e^2} + \frac{V_t K}{4\beta_e A_e^2} + 1\right)s + K_{cL}K}$$

$$\underline{\underline{K=0}}\ \frac{\frac{K_q}{A_e}X_v - \frac{1}{A_e^2}\left(K_{cL} + \frac{V_t}{4\beta_e}s\right)F_L}{\left(\frac{s^2}{\omega_h^2} + \frac{2\xi_h}{\omega_h}s + 1\right)s} \tag{5}$$

where, when the piston moves forward and backward, the model structure is the same, but the values of model coefficients are different. When moving forward, A_e equals A_1, otherwise, A_e equals A_2; $K_{cL} = K_c + C_L$ is total coefficient of flow pressure; ω_h is natural frequency of hydraulic asymmetric cylinder; ξ_h is damping ratio of hydraulic asymmetric cylinder.

2.3 Model of Electro-hydraulic Servo Control System

Assume that the system is unloaded, that is $F_L = 0$. The controller, servo amplifier, electro-hydraulic servo valve, and asymmetric cylinder model are integrated to form an electro-hydraulic servo control system as shown in Fig. 1.

3 Design of Fuzzy Controller

The two-dimensional fuzzy controller is selected. And the mode of fuzzy inference is mamdani. The structure is shown in the Fig. 2. The true error is converted into real control output by three stages: fuzzification, make fuzzy inference and defuzzification.

(1) Fuzzification of inputs and outputs
The fuzzy set of e, ec, u: {NB, NM, NS, ZO, PS, PM, PB};
The basic domain of e, ec: [−1, 1];
The basic domain of u: [−30, 30];
The fuzzy domain of e, ec, u: {−3, −2, −1, 0, 1, 2, 3};
Quantification factor: $K_e = 3$, $K_{ec} = 3$; Scale factor: $K_u = 10$.

(2) Selecting membership function
Z-shaped membership function is adopted at both ends of basic domain, otherwise Triangular-shaped membership function, as the Fig. 3 shows. The figure

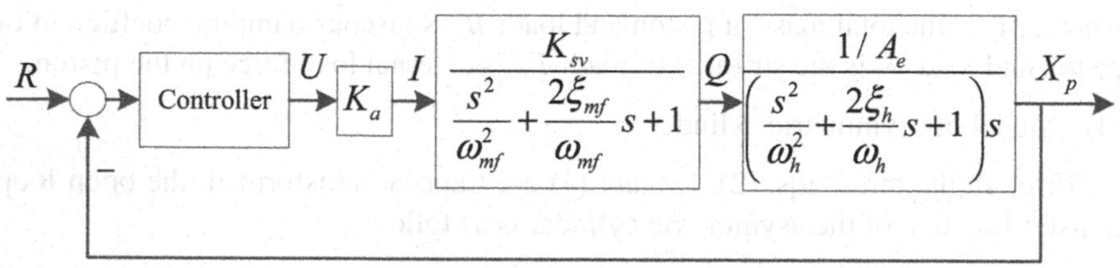

Fig. 1 The structure diagram of closed-loop transfer function for electrohydraulic servo system

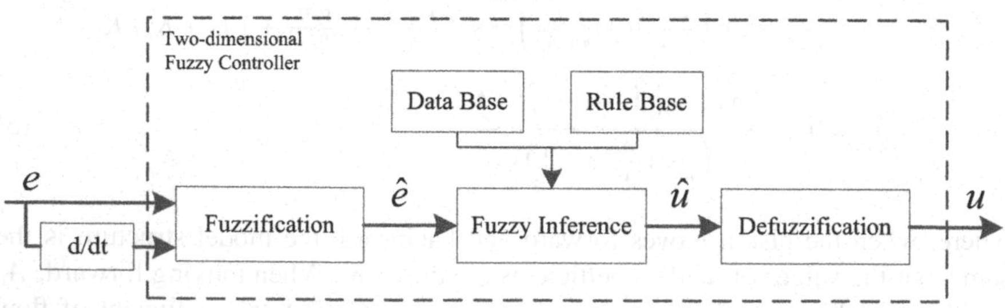

Fig. 2 Fuzzy control system block diagram

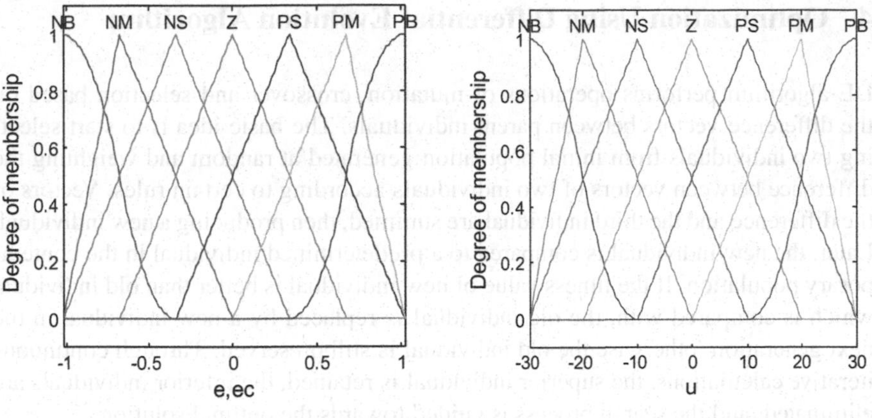

Fig. 3 Membership functions for input and output

Table 1 Fuzzy control rules

ec \ e	NB	NM	NS	ZO	PS	PM	PB
NB	PB	PB	PM	PM	PS	PS	ZO
NM	PB	PM	PM	PS	PS	PS	NS
NS	PM	PM	PS	PS	ZO	ZO	NS
ZO	PM	PS	PS	ZO	NS	NS	NM
PS	PS	PS	ZO	NS	NS	NM	NM
PM	PS	ZO	NS	NS	NM	NM	NB
PB	ZO	NS	NS	NM	NM	NB	NB

presents the basic domain of true error and control output. Fuzzy domain forms after the basic domains are scaled by quantification factor and scale factor.

(3) Creating a table of fuzzy control rules (Table 1)

(4) Defuzzification

The method of defuzzification is selected as the strategy of center of gravity in order to develop a control signal that is non-fuzzy, as shown in Eq. (6).

$$u = \frac{\sum_{i=1}^{n} u_i \mu(i)}{\sum_{i=1}^{n} \mu(i)} \tag{6}$$

where, $\mu(i)$ is the degree of membership function.

4 Optimization Using Differential Evolution Algorithm

DE algorithm performs operations of mutation, crossover and selection based on the difference vectors between parent individuals. The basic idea is to start selecting two individuals from initial population generated at random and weighting the difference between vectors of two individuals according to certain rules. Vectors of the difference and the third individual are summed, then producing a new individual. Later, the new individual is compared to a predetermined individual in the contemporary population. If the fitness value of new individual is better than old individual which is compared with, the old individual is replaced by a new individual in the next generation, otherwise the old individual is still preserved. Through continuous iterative calculations, the superior individual is retained, the inferior individuals are eliminated, and the search process is guided towards the optimal solution.

The fuzzy control structure diagram based on DE algorithm is shown in the Fig. 4. Some scholars have proved that the shape of the membership function curve has little effect on the control performance, so this paper will optimize the domain distribution of membership functions and control rules. Based on the objective function composed of error and control output, DE algorithm constantly optimizes the domain distribution of the membership function and control rules. Each domain of membership functions is divided into seven grades, and each domain distribution can be simply determined by values of the three points due to the symmetry. Nine parameters of the membership functions e, ec, and u, and 49 control rules parameters, together form 58 parameters to be optimized. Line it up as an individual and perform DE algorithm.

DE algorithm is an evolutionary algorithm based on real coding. The overall structure is similar to other evolutionary algorithms and consists of three basic operations: variation, crossover and selection. The steps of designing fuzzy controller optimization based on DE algorithm are as follows:

(1) Determining the parameter range

When the DE algorithm optimizes the fuzzy controller, firstly design a set of domain points of membership function and control rules according to the designers' experience. And then the ranges of the domain parameters are generated near its domain points. The set of domain points of fuzzy controller relying on experience is {1,2,3}, so the ranges of parameters to be optimized are set as {[0,1], [1, 3], [3, 4]}. Finally, the control rule parameters is an integer from 1 to 7.

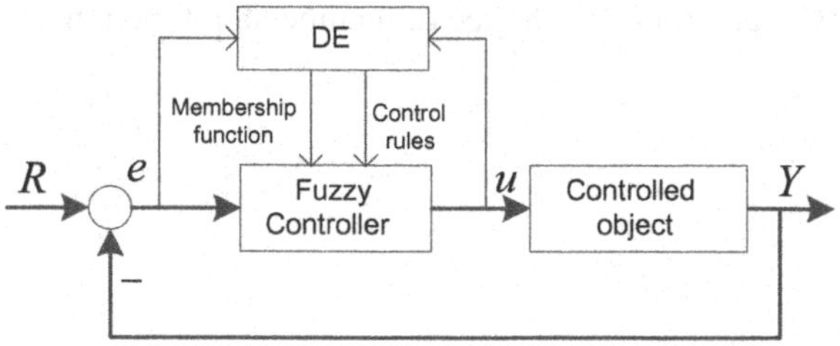

Fig. 4 The fuzzy control structure diagram based on the DE algorithm

(2) Selecting initial population

M individuals satisfying the constraint conditions are randomly generated, as shown in Eq. (7). And the empirical membership function domain designed and control rules are taken as one of the individuals. A larger M means the stronger diversity of population and the greater probability of obtaining the optimal solution, but the calculation time is longer, generally taking 20–50.

$$x_i(0) = x_{i\min} + \left(x_{i\max} - x_{i\min}\right) \times rand(0, 1) \tag{7}$$

(3) Selecting fitness function

On the one hand, considering the rapidity and accuracy of the control system, the absolute error moment integral (ITAE) is used as the minimum objective function in order to reduce the overshoot and steady state error. On the other hand, a control quantity is added to the objective function in order to prevent the control volume being too large and affecting the stability of the system. Fitness function is the reciprocal of the objective function, and similarly, the larger fitness function means that objective function is smaller. Finally, the determined objective function is

$$J = \int_0^\infty (w_1 t|e(t)| + w_2|u(t)|)\mathrm{d}t \tag{8}$$

(4) Mutating

Three individuals selected at random from the population are used to mutate.

$$h_i(t+1) = x_{p1}(t) + F \times \left(x_{p2}(t) - x_{p3}(t)\right) \tag{9}$$

If there is no local optimization problem, the above formula can be written as

$$h_i(t+1) = x_{bi}(t) + F \times \left(x_{p2}(t) - x_{p3}(t)\right) \tag{10}$$

where, $x_{bi}(t)$ is the best individual in the current generation; $x_{p2}(t) - x_{p3}(t)$ is Differentiated vector; p_1, p_2, p_3 are random integers; F is scale factor, controlling population diversity and convergence, and the range of value is [0,2].

(5) Crossovering

Crossover is as follows in order to increase the diversity of the group.

$$v_i(t+1) = \begin{cases} h_i(t+1), & rand_i \le P_c \\ x_i(t), & rand_i > P_c \end{cases} \tag{11}$$

where, P_c is crossover probability, and the range of value is [0.6, 0.9]. The larger the P_c and the smaller F speed up the convergence of the population. However, as the crossover factor P_c increases, the sensitivity of convergence to the variability factor F increases gradually.

(6) Selecting

Select next generation members by comparison of evaluation functions.

$$x_i(t+1) = \begin{cases} v_i(t+1), & f(v_i(t+1)) \geq f(x_i(t)) \\ x_i(t), & f(v_i(t+1)) < f(x_i(t)) \end{cases} \tag{12}$$

5 Simulation Results

By calculating, the values of the parameters of the electro-hydraulic servo system are shown in Table 2, and the parameters of optimization algorithm are shown in Table 3. Where, M is individual population of per generation, G is hereditary algebra in Table 3.

The results of simulation are as follows. using MATLAB programming to discretize and simulate the system. In order to clearly present the control output of the fuzzy control system, the map is partially enlarged. The domain factor optimized of membership function is {0.900, 1.100, 3.361, 0.571, 2.900, 3.793, 0.432, 2.340, 3.461}. The optimization parameters of control rule are {4 5 6 6 7 6 2, 2 5 7 6 7 1 6, 7 4 6 7 5 4 7, 7 1 7 4 4 3 2, 7 7 3 1 3 3 1, 6 6 3 4 4 1 3, 7 3 5 2 1 4 1}. The domain of output membership functions is denser than control system by experience, indicating that the control intensity of early time applied to the object increases for the optimized fuzzy control system, as shown in Figs. 6 and 8, thus making the system response time shorter (Figs. 5 and 7).

It can be seen from the Fig. 5 that the uncontrolled electro-hydraulic servo system has a long response time and cannot achieve the purpose of fast tracking commands. The fuzzy controller based on experience can greatly speed up the response time of the servo system to the command. Besides, it can be seen from the Fig. 7 that the response time of the fuzzy controller optimized by the DE algorithm is shorter than fuzzy controller, and the system can more accurately and quickly track the

Table 2 Parameters of electro-hydraulic servo system

Parameter	Value	Parameter	Value	Parameter	Value
K_{sv}	0.02	ω_{mf}	300	ξ_{mf}	0.6
A_e	1.2×10^{-4}	ω_h	200	ξ_h	0.2

Table 3 Parameters of optimization algorithm

Parameter	M	G	P_c	F	w_1	w_2
Value	30	50	0.9	0.6	0.9	0.1

sinusoidal signal as shown in Fig. 9. Generally, the control performance of fuzzy control system optimized by DE algorithm is improved greatly compared with the no optimized control system.

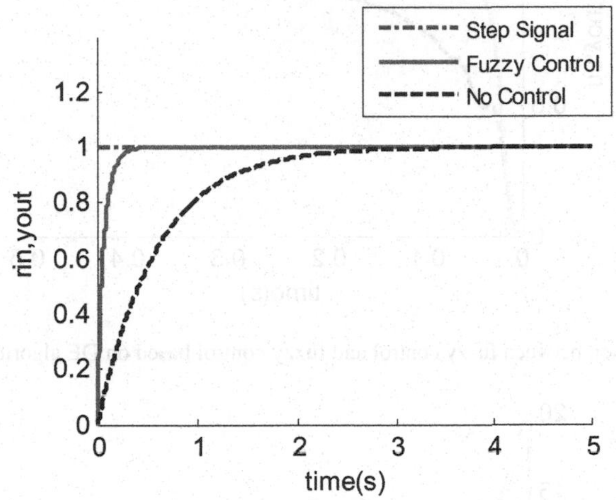

Fig. 5 Comparison between fuzzy control and non-control

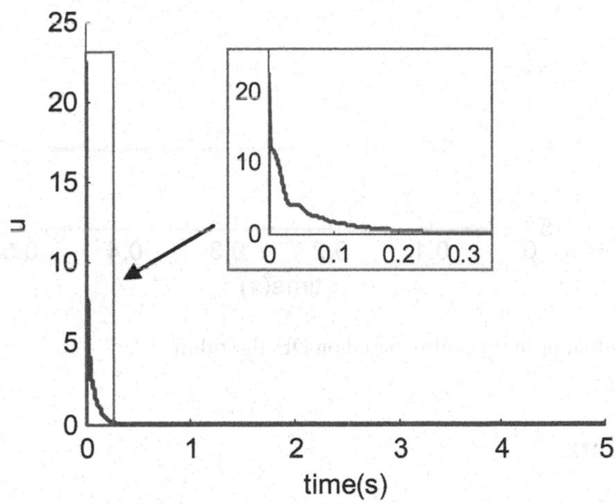

Fig. 6 Control output of fuzzy control

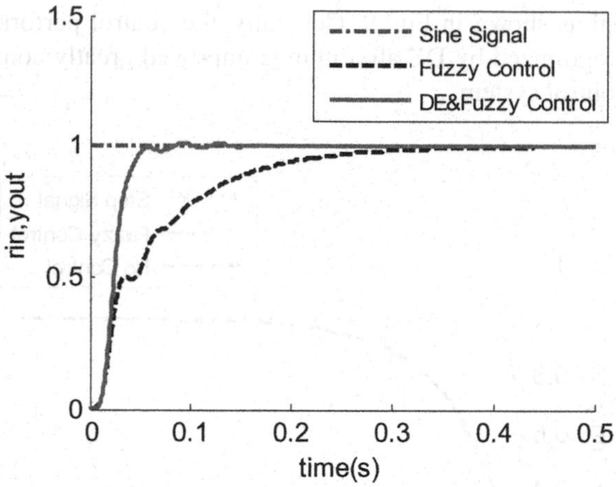

Fig. 7 Comparison between fuzzy control and fuzzy control based on DE algorithm

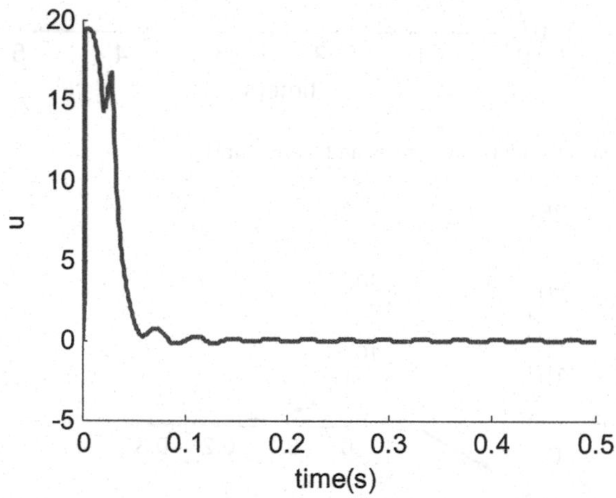

Fig. 8 Control output of fuzzy control based on DE algorithm

6 Conclusions

(1) Compared to the original non-controller closed-loop system, fuzzy control designed by experience in the paper can reduce the response time, and make the system almost no overshoot.
(2) Compared to the fuzzy control system by experience, for control system optimized by DE algorithm, the control intensity applied to the object in the early period is more increased. Besides, the system optimized has shorter response

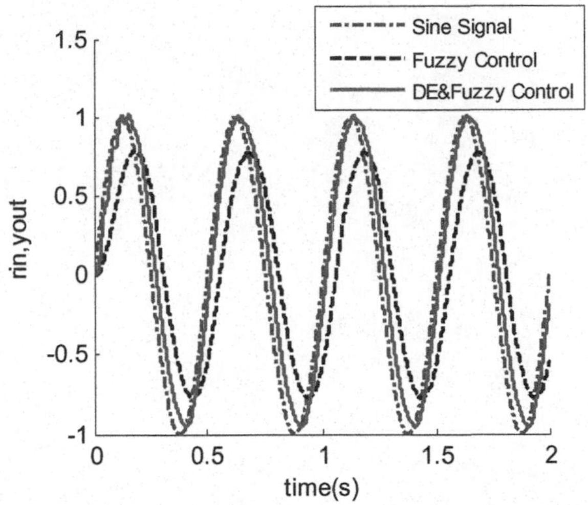

Fig. 9 Tracking performance of sine signal

time, remains almost no overshoot for the step signal, and presents better tracking performance for the sine signal. In a word, the performance of the fuzzy control system based on DE algorithm can get optimal or suboptimal in electro-hydraulic servo system.

References

1. X. Ye, Modeling and control method of valve-controlled asymmetric cylinder system. Hefei University of Technology, (2015)
2. K. Zheng, G. Yang, J. Fang et al., Feedback linearization sliding mode control for valve controlled asymmetric cylinder system. Mach. Tools Hydraul. **45**(5), 151–154 (2017)
3. M. Magzoub, N. Saad, R. Ibrahim, et al., A genetic algorithm optimization of hybrid fuzzy-fuzzy rules in induction motor control, in *International Conference on Intelligent and Advanced Systems*. IEEE, pp. 1–6 (2017)
4. A. Nabi, N.A. Singh, GA optimization of fuzzy logic controller for voltage sag improvement in power systems, in *International Conference on Intelligent Systems and Control*. IEEE, pp. 1–5 (2016)
5. J. Liu, *Intelligent Control*, 3rd edn. (Publishing House of Electronics Industry, 2014)
6. S. Wang, L. Ding, W. Zhang et al., Progress in differential evolution algorithms. J. Wuhan Univ. (Sci. Edn.) **60**(4), 283–292 (2014)
7. Z. Song. *Simulation of MATLAB/Simulink and Hydraulic Control System*. (National Defense Industry Press, 2012)

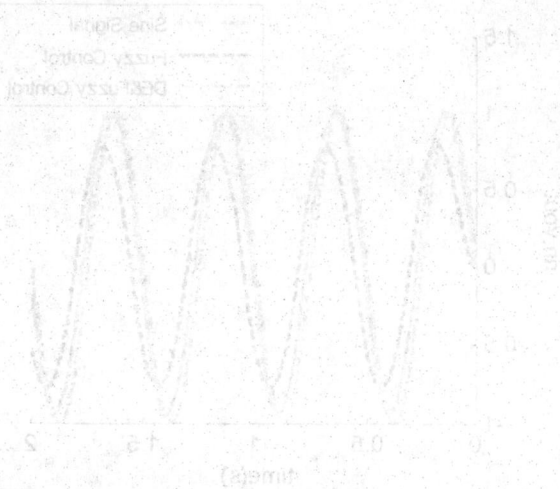

Plant/Controller Integrated Design for Dual-Motor Servo Systems with Backlash

Zimei Sun, Xuemei Ren and Minlin Wang

Abstract This paper proposes an integrated design of the plant, finite-time controller and bias torque for the dual-motor servo systems with backlash. To achieve the finite-time error convergence of the tracking error, a recursive fast terminal sliding mode controller (FTSMC) is proposed, which contains two sliding mode control laws for the situations that both of motors driving the load and only one motor driving the load. Since the backlash will result in the load uncontrollable problem, the time-varying bias torque is designed to eliminate the backlash nonlinearity. Finally, an integrated design uses the particle swarm optimization algorithm to optimize the parameters of all the controller, bias torque and backlash. Simulation results are conducted to validate the desired load tracking performance of the proposed integrated design.

Keywords Dual-motor servo systems · SMC · Time-varying bias torque Optimization · Plant/controller integrated design

1 Introduction

Backlash is a kind of nonlinear which is inevitable in the process of gear transmission, and it is also an important factor influencing the system dynamic performance and steady-state precision. The traditional method is to design a controller for compensating the nonlinear of backlash in the closed-loop feedback system [1]. In [2], a switching control strategy was implemented, which used a proportional-integral controller for the mode of contact and a proportional controller for the mode of backlash. For a class of multivariable nonlinear system with sandwiched backlash, Tao [3] presented a combined optimal and nonlinear feedback linearization decoupling control design. Although the above researches can eliminate the backlash nonlinearity, a high performance controller is rarely concerned to improve the steady-state responses of the tracking error.

Z. Sun · X. Ren (✉) · M. Wang
School of Automation, Beijing Institute of Technology, Beijing 100081, China
e-mail: xmren@bit.edu.cn

Y. Jia et al. (eds.), *Proceedings of 2018 Chinese Intelligent Systems Conference*, Lecture Notes in Electrical Engineering 529,
https://doi.org/10.1007/978-981-13-2291-4_74

Sliding mode control (SMC) is a well-known powerful control scheme which has been successfully and widely applied for linear and nonlinear systems [4]. Compared with the conventional SMC with linear sliding surface, terminal sliding mode control (TSMC) offers some superior properties such as faster, finite time convergence, and higher control precision [5]. By combining the nonsingular terminal sliding mode (NTSM) with the high-order sliding mode (HOSM) method, Feng et al. [6] proposed a hybrid terminal sliding-mode observer for the rotor position and speed estimation in PMSM systems. And many other controls are designed for the dual-motor servo system, for example adaptive control [7, 8].

In the practical system, the structure and control parameters are coupled with each other. Wang [9] developed the optimization model of the coupling between the structure subsystem and the control subsystem. Mayzus and Grigoriadis [10] adopted cyclic iteration in the integrated design of structure/control. In this paper, an integrated design scheme is developed to improve the performance of the dual-motor servo systems.

2 Dual-Motor Servo Systems Model

The dynamic equation of dual-motor systems can be expressed as

$$\begin{cases} J_1\ddot{\theta}_1 + b_1\dot{\theta}_1 = u_1 - T_1 \\ J_2\ddot{\theta}_2 + b_2\dot{\theta}_2 = u_2 - T_2 \\ J_m\ddot{\theta}_m + b_m\dot{\theta}_m = \sum_{i=1}^{2} T_i \end{cases} \tag{1}$$

where b_i, $i = 1, 2$ are the viscous friction coefficients of the motor shafts, b_m is the viscous friction coefficient of the load shaft; θ_i, $i = 1, 2$ are the motor angular positions, θ_m is the load angular position; J_i, $i = 1, 2$, are the motor inertia, J_m is the load inertia; u_i, $i = 1, 2$, are the control inputs.

Affected by the backlash, the motor-load interaction torques T_i can be expressed as

$$T_i = \begin{cases} k(\Delta\theta_i - \alpha) + c\Delta\theta_i & \Delta\theta_i \geq \alpha \\ 0 & |\Delta\theta_i| < \alpha \\ k(\Delta\theta_i + \alpha) + c\Delta\theta_i & \Delta\theta_i \leq -\alpha \end{cases} \tag{2}$$

where $\Delta\theta_i = \theta_i - \theta_m$, $k > 0$ is the stiffness coefficient, $c > 0$ is the damping coefficient of the contact force between the load and motor, and $2\alpha > 0$ is the backlash width parameter which is assumed to be known.

Remark 1 Through applying the bias torque in the whole control process, the load O_0 can stay in the contact phase, which means that the load O_0 can be controlled by the motor O_1 or the motor O_2 or both of them.

According to Assumption 1, both of $T_i (i = 1, 2)$ are not equal to zero simultaneously. In order to reflect the influence of backlash nonlinearity on the system, the $\chi_i \in R$ is defined as

$$\chi_i = \begin{cases} 1 \ |\Delta\theta_i| \geq \alpha \\ 0 \ |\Delta\theta_i| \ < \ \alpha \end{cases}, \tag{3}$$

which indicates the motor O_i drives the load O_0 alone and $\chi_1 \cup \chi_2 \neq 0$. In order to facilitate the control design, we choose x_1, x_2, x_{3i}, x_{4i} as the state variables

$$\begin{cases} x_1 = \theta_m(t), x_2 = \dot{\theta}_m(t) \\ x_{3i} = \Delta\theta_i(t) - \mathrm{sgn}(\Delta\theta_i(t))\alpha \\ x_{4i} = \Delta\theta_i(t) \end{cases} \tag{4}$$

and $f(t) = \sum_{i=1}^{2} T_i = \sum_{i=1}^{2} \chi_i [kx_{3i} + cx_{4i}]$.

Then the state space equation is expressed as

$$\begin{cases} \dot{x}_1 = x_2 \\ \dot{x}_2 = \frac{b_m}{J_m} x_2 + \frac{1}{J_m} f(t) \\ \dot{x}_{3i} = x_{4i} \\ \dot{x}_{4i} = a_{3i}x_{3i} + a_{4i}x_{4i} + a_i x_2 + \frac{1}{J_i} u_i - \frac{1}{J_m} \chi_j (kx_{3j} + cx_{4j}) \\ y = x_1 \end{cases} \tag{5}$$

where

$$a_{3i} = -\frac{k}{J_i} - \frac{k}{J_m}, a_{4i} = -\frac{b_i + c}{J_i} - \frac{c}{J_m}, a_i = \frac{b_m}{J_m} - \frac{b_i}{J_i} \text{ and } j = 2 - i (i = 1, 2).$$

3 Finite-Time Tracking Controller Design

In this section, we present the recursive FTSMC design for system (5) in the contact phase, the load O_0 contacts with either the motor O_1 or the motor O_2 or both of them. The control task in this phase is to control the system output y to track the desired output y_d asymptotically and stably.

To achieve the output asymptotically stable tracking with guaranteed performance, we study the steady-state performance of tracking error $e(t) = y(t) - y_d(t)$ and further deduce the error equation as

$$\dot{e}(t) = \dot{y}(t) - \dot{y}_d(t) = x_2(t) - \dot{y}_d(t) \tag{6}$$

Moreover, the second derivative term can be derived as

$$\begin{aligned}\ddot{e}(t) &= \ddot{y}(t) - \ddot{y}_d(t) = \dot{x}_2 - \ddot{y}_d(t)\\ &= -\frac{b_m}{J_m}x_2 + \frac{1}{J_m}f(t) - \ddot{y}_d(t)\end{aligned} \tag{7}$$

Similar to above analysis, the third derivative of $e(t)$ can be given by

$$\begin{aligned}\dddot{e}(t) &= \ddot{x}_2 - \dddot{y}_d(t)\\ &= -\frac{b_m}{J_m}\dot{x}_2 + \frac{1}{J_m}\dot{f}(t) - \dddot{y}_d(t)\\ &= \frac{b_m^2}{J_m^2}x_2 - \frac{b_m}{J_m^2}f(t) + \frac{1}{J_m}\sum_{i=1}^{2}\chi_i[kx_{4i} + a_i x_2\\ &\quad + c(a_{3i}x_{3i} + a_{4i}x_{4i} - \frac{\chi_j}{J_m}(kx_{3j} + cx_{4j}) + \frac{u_i}{J_i})] - \dddot{y}_d(t)\end{aligned} \tag{8}$$

Then the fast terminal sliding surface is defined as follows:

$$\begin{cases} s_0 = e\\ s_1 = \dot{s}_0 + \alpha_0 s_0 + \beta_0 s_0^{q_0/p_0}\\ s_2 = \dot{s}_1 + \alpha_1 s_1 + \beta_1 s_1^{q_1/p_1}\end{cases} \tag{9}$$

where $p_i > q_i$ are positive odd integers, and $\alpha_i > 0, \beta_i > 0, (i = 0, 1)$. Apparently, if $s_2 = 0$ is reached in finite time, s_1 and s_0 will also reach zeros in finite time. When the load O_0 is driven by only one motor (O_1 or O_2), the system (5) in the contact phase will be asymptotically stable if the FTSMC control law $u_{ai}(t)$ is adopted as the control effort and designed as

$$u_{ai}(t) = \frac{J_i}{c}(\sigma - \omega) \tag{10}$$

where

$$\sigma = J_m[-\phi s_2 - \gamma s_2^{q/p} - \alpha_0\ddot{s}_0 - \left(\beta_0 s_0^{q_0/p_0}\right)'' - \alpha_1\dot{s}_1 - \left(\beta_1 s_1^{q_1/p_1}\right)'],$$

$$\omega = c(a_{3i}x_{3i} + a_{4i}x_{4i}) + a_i x_2 - J_m\dddot{y}_d + kx_{4i} + \frac{b_m^2}{J_m}x_2 - \frac{b_m}{J_m}f(t),$$

$\phi, \gamma > 0$ are weight constants, $p > q$ are positive odd integers.

Moreover, while the load O_0 is driven by both motors (O_1 or O_2), the control equation is derived as:

$$\frac{u_{c1}(t)}{J_1} + \frac{u_{c2}(t)}{J_2} = \frac{1}{c}(\sigma - \varphi) \tag{11}$$

where $\varphi = \sum_{i=1}^{2}\left[\omega - \frac{c}{J_m}(kx_{3i} + cx_{4i})\right]$.

Theorem 1 *For the dual-motor driving system (5), if we design the controller as (10) and (11), the tracking error will converge to zero in finite time.*

Proof The following Lyapunov function is selected to prove the synchronization and the tracking error $e(t)$ convergence to zero.

$$V_1 = \frac{1}{2}s_2^2 + \frac{1}{2}s_E^2 \tag{12}$$

From the fast terminal sliding mode (9), $\dot{V}_1$ can be written as

$$\begin{aligned}\dot{V}_1 &= s_2\dot{s}_2 + s_E\dot{s}_E \\ &= s_2(\dot{s}_1 + \alpha_1 s_1 + \beta_1 s_1^{q_1/p_1})' + s_E\dot{e}_1(t) \\ &= s_2[\dddot{e} + \alpha_0\ddot{s}_0 + (\beta_0 s_0^{q_1/p_1})'' \\ &\quad + \alpha_1\dot{s}_1 + (\beta_1 s_1^{q_1/p_1})'] + s_E\dot{e}_1(t)\end{aligned} \tag{13}$$

As the load contacts with both motors, $\chi_1, \chi_2 = 1$ and the third derivation of e can be rewritten as:

$$\begin{aligned}\dddot{e}(t) &= \ddot{x}_2 - \dddot{y}_d(t) \\ &= -\frac{b_m}{J_m}\dot{x}_2 + \frac{1}{J_m}\dot{f}(t) - \dddot{y}_d(t) \\ &= \frac{b_m^2}{J_m^2}x_2 - \frac{b_m}{J_m^2}f(t) + \frac{1}{J_m}\sum_{i=1}^{2}[kx_{4i} + a_i x_2 \\ &\quad + c(a_{3i}x_{3i} + a_{4i}x_{4i} - \frac{1}{J_m}(kx_{3j} + cx_{4j}) + \frac{u_i}{J_i})] - \dddot{y}_d(t)\end{aligned} \tag{14}$$

Substituting the Eq. (14) and the control (16) into (13) yields

$$\begin{aligned}\dot{V}_1 &= s_2[\frac{\varphi}{J_m} + \frac{c}{J_m}(\frac{u_1}{J_1} + \frac{u_2}{J_2}) + \alpha_0\ddot{s}_0 + \left(\beta_0 s_0^{q_0/p_0}\right)'' \\ &\quad + \alpha_1\dot{s}_1 + \left(\beta_1 s_1^{q_1/p_1}\right)'] + s_E(\frac{u_1}{J_1} - \frac{u_2}{J_2} + \delta) \\ &= -\phi s_2^2 - \gamma s_2^{(q+p)/p} - \varepsilon s_E \mathrm{sgn}(s_E) - ks_E^2\end{aligned} \tag{15}$$

As $\phi, \gamma, \varepsilon, k > 0$, it is easy to obtain $\dot{V}_1 \le 0$.

Then the stability for only one motor contacting with the load is discussed in the following.

$$
\begin{aligned}
\dddot{e}(t) &= \dddot{x}_2 - \dddot{y}_d(t) \\
&= -\frac{b_m}{J_m}\ddot{x}_2 + \frac{1}{J_m}\dot{f}(t) - \dddot{y}_d(t) \\
&= \frac{b_m^2}{J_m^2}x_2 - \frac{b_m}{J_m^2}f(t) + \frac{1}{J_m}[kx_{4i} + a_i x_2 + c(a_{3i}x_{3i} + a_{4i}x_{4i} + \frac{u_i}{J_i})] - \dddot{y}_d(t)
\end{aligned}
\tag{16}
$$

Then, choose another Lyapunov function to prove the tracking performance.

$$V_2 = \frac{1}{2}s_2^2 \tag{17}$$

Hence, its derivative term is derived as:

$$
\begin{aligned}
\dot{V}_2 &= s_2\dot{s}_2 = s_2(\dot{s}_1 + \alpha_1 s_1 + \beta_1 s_1^{q_1/p_1})' \\
&= s_2[\dddot{e} + \alpha_0\ddot{s}_0 + (\beta_0 s_0^{q_1/p_1})'' + \alpha_1\dot{s}_1 + (\beta_1 s_1^{q_1/p_1})']
\end{aligned}
\tag{18}
$$

Substituting the Eq. (16) and the control (10) into (18), it is easy to find that

$$
\begin{aligned}
\dot{V}_2 &= s_2[\frac{w}{J_m} + \frac{c}{J_m}(\frac{u_i}{J_j}) + \alpha_0\ddot{s}_0 + \left(\beta_0 s_0^{q_0/p_0}\right)'' + \alpha_1\dot{s}_1 + \left(\beta_1 s_1^{q_1/p_1}\right)'] \\
&= -\phi s_2^2 - \gamma s_2 s_2^{q/p}
\end{aligned}
\tag{19}
$$

Obviously, the first term $-\phi s_2^2$ is not positive. Moreover, s_2 and $s_2^{q/p}$ have the same symbol, hence the second term is also not positive. Then $\dot{V}_2 \le 0$.

Remark 2 From Eq. (9), it is noted that because $p_0 - q_0 < 0$ and $p_1 - q_1 < 0$, $|s_0|^{(p_0-q_0)/q_0}\mathrm{sign}(s_0)$ and $|s_1|^{(p_1-q_1)/q_1}\mathrm{sign}(s_1)$ will lead to the singularity problem at $s_0 = s_1 = 0$ and $\dot{s}_0 \ne \dot{s}_1 \ne 0$, that is, $\lim_{s_0\to 0}|s_0|^{(p_0-q_0)/q_0}\mathrm{sign}(s_0) \to \infty$ and $\lim_{s_1\to 0}|s_1|^{(p_1-q_1)/q_1}\mathrm{sign}(s_1) \to \infty$.

To overcome the singularity problem, the following function *sag* is introduced

$$
sag^{(p_0-q_0)/q_0}(s_0) = \begin{cases} |s_0|^{(p_0-q_0)/q_0}\mathrm{sign}(s_0), & s_0 \ne 0 \;\&\; \dot{s}_0 \ne 0 \\ \Theta^{(p_0-q_0)/q_0}\dot{s}_0, & s_0 = 0 \;\&\; \dot{s}_0 \ne 0 \\ 0, & \dot{s}_0 = 0 \end{cases}
\tag{20}
$$

where $\Theta > 0$ is a small positive constant. The same substitution is also applied for the sliding manifold $|s_1|^{(p_1-q_1)/q_1}\mathrm{sign}(s_1)$.

4 Time-Varying Bias Torque Design

The constant bias torque usually apply a large bias torque to stabilize the system as quickly as possible, but it would cause large energy consumption. This paper proposed time-varying biasing moment based on hyperbolic tangent function.

$$u_{vi} = (-1)^{i+1} w \tanh\left(e^{-k_w(|\hat{z}_i|-\alpha)}\right) \tag{21}$$

where w is a positive constant, $|\hat{z}_i|$ is the angle difference among the motors and the load. It can utilize the hyperbolic tangent function to make the anti-backlash process smoother and reduce system oscillations. For another, the bias torque is large when entering the backlash, which makes the system stable as soon as possible. And the bias torque will keep at a minimum value when the backlash is exited so the energy consumption is less.

5 Simulation Results

In this section, simulation example is performed to illustrate the tracking performance of the FTSMC with bias torque for the dual-motor servo systems with backlash. The reference signal is chosen as $y_d = 2\sin(0.4\pi t)$. The sampling time is 0.01 s, and the initial values of the system are selected as $x_1 = 0, x_2 = 0, x_{31} = 0, x_{32} = 0, x_{41} = 0, x_{42} = 0$. The physical parameters of system (1) are selected as $b_1 = b_2 = 0.015, b_m = 0.02, k = 1, c = 0.15, J_1 = J_2 = 0.0026\,\mathrm{kg\,m^2}$ and $J_m = 0.0113\,\mathrm{kg\,m^2}$.

In the following simulation, we have compared the proposed FTSMC without bias torque and the proposed FTSMC with time-varying bias torque.

All the control parameters and the backlash width parameter α are searched as $\phi = 78.46, \gamma = 10.00, \alpha_0 = 7.86, \alpha_1 = 14.10$ and $\alpha = 0.37$ by the PSO to make the performance index minimized.

The comparative simulation results are shown in Figs. 1, 2, 3 and 4. Figures 1 and 2 depict the tracking output and the tracking error of the FTSMC without bias torque. It is noted that there are serious vibrations appearing in the tracking error profile due to the backlash nonlinearity. Figures 3 and 4 show the tracking output and the tracking error of the FTSMC with the time-varying bias torque. Compared with the FTSMC without bias torque, the FTSMC with the time-varying bias torque imposes a much smaller steady-state error and avoids the vibration phenomenon, which demonstrates that the effectiveness of the proposed control scheme.

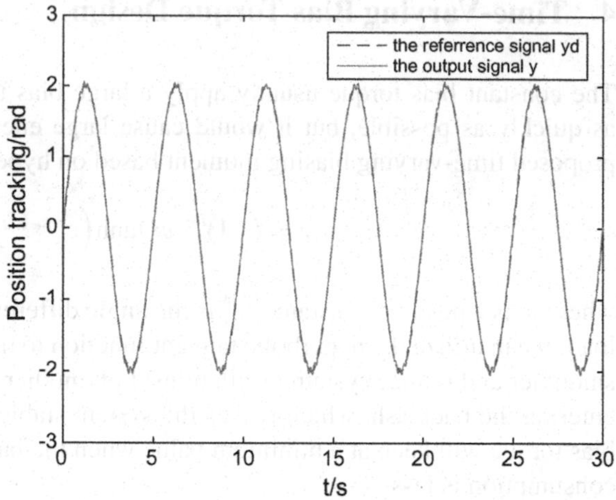

Fig. 1 The tracking output of the FTSMC without bias torque

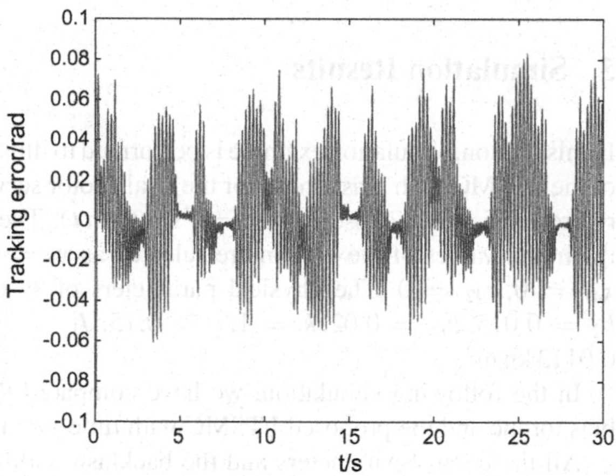

Fig. 2 The tracking error of the FTSMC without bias torque

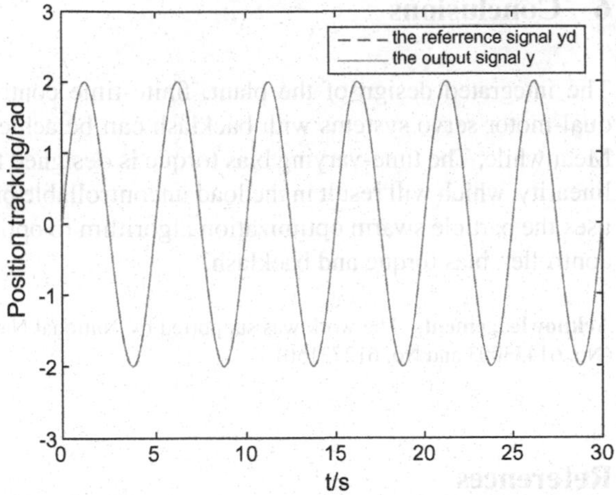

Fig. 3 The tracking output of the FTSMC with bias torque

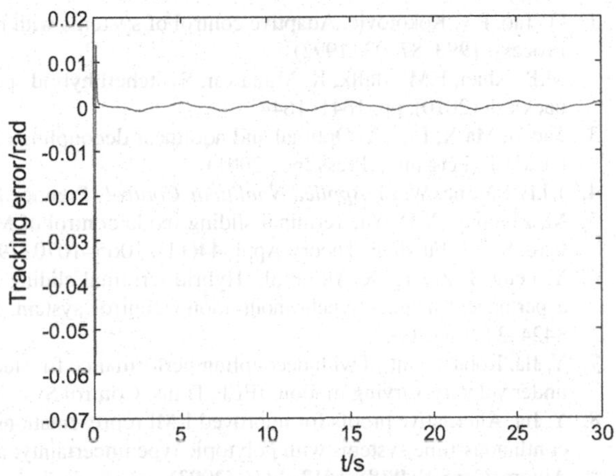

Fig. 4 The tracking error of the FTSMC with bias torque

6 Conclusions

The integrated design of the plant, finite-time controller and bias torque for the dual-motor servo systems with backlash can be achieved by the method proposed. Meanwhile, The time-varying bias torque is designed to eliminate the backlash nonlinearity, which will result in the load uncontrollable problem. The integrated design uses the particle swarm optimization algorithm to optimize the parameters of all the controller, bias torque and backlash.

Acknowledgements The work was supported by National Natural Science Foundation of China (No. 61433003 and No. 61273150).

References

1. G. Tao, P.V. Kokotović, Adaptive control of systems with backlash. Adapt. Syst. Control Sig. Process. **1993**, 87–93 (1992)
2. M.B. Khan, F.M. Malik, K. Munawar, Switched hybrid speed control of elastic systems with backlash (2010), pp. 1641–1644
3. Tao G, Ma X, Ling Y. Optimal and nonlinear decoupling control of systems with sandwiched backlash (Pergamon Press Inc., 2001)
4. J.J.E. Slotine, W. Li, *Applied Nonlinear Control* (Prentice Hall, Englewood Cliffs, NJ, 1991)
5. M. Zhihong, X.H. Yu, Terminal sliding mode control of MIMO linear systems. IEEE Trans. Circ. Sys. I: Fundam. Theory Appl. **44**(11), 1065–1070 (1997)
6. Y. Feng, J. Zheng, X. Yu, et al., Hybrid terminal sliding-mode observer design method for a permanent-magnet synchronous motor control system. IEEE Trans. Ind. Electron. **56**(9), 3424–3431.4 (2009)
7. Y. Jia, Robust control with decoupling performance for steering and traction of 4WS vehicles under velocity-varying motion. IEEE Trans. Control Syst. Technol. **8**(3), 554–569 (2000)
8. Y. Jia, Alternative proofs for improved LMI representations for the analysis and the design of continuous-time systems with polytopic type uncertainty: a predictive approach. IEEE Trans. Autom. Control **48**(8), 1413–1416 (2003)
9. H. Wang, Radar servo system structure and control integrated design (Xi'an Electronic and Science University, 2008)
10. A. Mayzus, K. Grigoriadis, Integrated structural and control design for structural systems via LMIs, in *Proceedings of the 1999 IEEE International Conference on Control Applications, 1999*, vol. 1 (IEEE, 1999), pp. 75–79

Robust Fixed-Time Tracking Control of Wheeled Mobile Robots

Liming Chen and Yingmin Jia

Abstract This paper considers the problem of trajectory tracking control of a two-wheeled mobile robot in fixed time. By using the differential flatness property, the system model is linearized with several input transformations and an input prolongation. Then a feedback control law is designed to ensure the convergence of tracking errors in fixed time. The dynamic model with disturbances and unmodeled dynamics is also derived, and it is controlled in fixed time by designing a novel integral sliding mode surface. Theoretical results are finally verified by numerical simulations.

Keywords wheeled mobile robot · fixed-time control · integral sliding mode · differential flatness

1 Introduction

Over the last few decades, motion control of nonholonomic wheeled mobile robot has received much attention. Researches have been carried out in two aspects, one is stabilization and the other is tracking control. Usually, stabilization can be seen as a special case of tracking, but this is not true for the nonholonomic mobile robot. For the problem of stabilization, according to Brocket condition [1], stability can not be achieved by any smooth or continuous time-invariant state feedback control law. So methods in literature include discontinuous control laws [2], time-varying control

L. Chen · Y. Jia (✉)
The Seventh Research Division and the Center for Information and Control,
School of Automation Science and Electrical Engineering,
Beihang University (BUAA), Beijing 100191, China
e-mail: ymjia@buaa.edu.cn

L. Chen
e-mail: clmtest@126.com

Y. Jia et al. (eds.), *Proceedings of 2018 Chinese Intelligent Systems Conference*, Lecture Notes in Electrical Engineering 529,
https://doi.org/10.1007/978-981-13-2291-4_75

laws [3] and hybrid control laws [4]. The problem of tracking which is considered in this paper seems to be more useful in practice, and the methods include backstepping [5], adaptive control [6], linearization [7], neural network-based control [8], fuzzy control [9], etc.

Most of the previous researches only consider the asymptotic tracking or exponentially asymptotic tracking. Compared to this, finite-time control and fixed-time control [10] have faster convergence and better robustness. In [11], finite-time tracking control of nonholonomic mobile robots with extended chained form was realized using recursive terminal sliding mode control. In [12], finite-time tracking control was considered in a cascaded systems form. And robust finite-time tracking control of nonholonomic mobile robots without measurements of velocity was investigated in [13]. Besides, fixed-time tracking control of mobile robots was studied in [14].

The existing approaches of finite-time or fixed-time control of nonholonomic mobile robots require that the angular velocity can not be zero, so even a straight line can not be tracked. In this paper, this assumption is not required. Our method is based on differential flatness of the model. Disturbances and unmodeled dynamics are also considered and partially eliminated by designing an integral siding mode surface.

The rest of this paper is arranged as follows. In Sect. 2, the dynamic model of mobile robots is analyzed using differential flatness, considering disturbances and unmodeled dynamics, and some definition and lemmas about fixed-time control are introduced. The proposed controllers are given in Sect. 3. Simulations are shown in Sect. 4. Finally, conclusion is drawn in Sect. 5.

2 Preliminaries

In this section, models of wheeled mobile robots are analyzed using differential flatness, and some definitions and lemmas about fixed-time control are given.

2.1 Model of Two-Wheeled Mobile Robots

Figure 1 shows a mobile robot with two driving wheels. With no slip, the two wheels, which are driven by two motors, can only move forward or backward and they give the robot a forward velocity v at the midpoint (x, y) between the two wheels and an angular velocity ω. Denoting the orientation angle of the robot as θ, the kinematic model is given by

$$\dot{\boldsymbol{q}} = S(\boldsymbol{q})\boldsymbol{v} \tag{1}$$

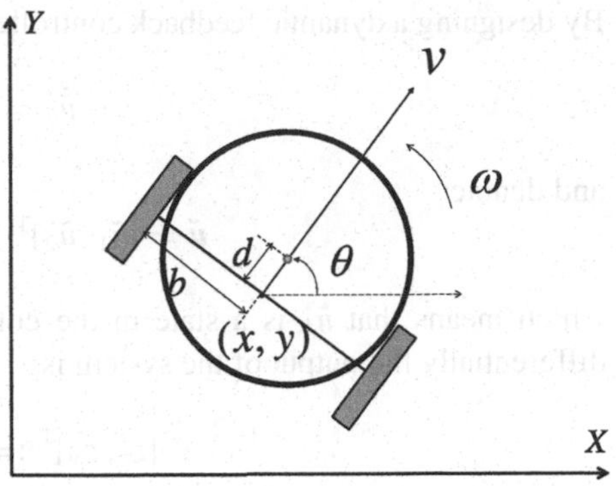

Fig. 1 The configuration of a two-wheeled mobile robot

where $\boldsymbol{q} = [x, y, \theta]^{\mathrm{T}}$, $\boldsymbol{v} = [v, \omega]^{\mathrm{T}}$ and

$$S(\boldsymbol{q}) = \begin{bmatrix} \cos\theta & 0 \\ \sin\theta & 0 \\ 0 & 1 \end{bmatrix}. \tag{2}$$

As shown in [15], the dynamic model can be derived as

$$\dot{\boldsymbol{v}} = A(\boldsymbol{v}) + B\boldsymbol{\tau} \tag{3}$$

where

$$A(\boldsymbol{v}) = \begin{bmatrix} d\omega^2 \\ -\frac{dm\omega v}{d^2m+I} \end{bmatrix},\ B = \begin{bmatrix} \frac{1}{mr} & \frac{1}{mr} \\ \frac{b}{r(d^2m+I)} & -\frac{b}{r(d^2m+I)} \end{bmatrix}, \tag{4}$$

m is the mass of the robot, I the rotational inertia of the robot about its center of mass, d the distance between (x, y) and the center of mass, r the wheel radius, b the half distance between the two wheels, and the inputs $\boldsymbol{\tau} = [\tau_r, \tau_l]^{\mathrm{T}}$ are the motor torques on the right and left wheels.

The problem of fixed-time trajectory tracking is considered in this paper, which means that given a reference trajectory $(x_d(t), y_d(t))$ which is smooth enough, control law of $\boldsymbol{\tau}$ is designed, such that the tracking errors between (x, y) and $(x_d(t), y_d(t))$ converge to zero in fixed time.

Differential flatness property of the system model has been shown in [15]. First an input transformation is introduced as

$$A(\boldsymbol{v}) + B\boldsymbol{\tau} := \bar{\boldsymbol{u}} = [\bar{u}_1, \bar{u}_2]^{\mathrm{T}}, \tag{5}$$

so if $\bar{\boldsymbol{u}}$ has been designed, then input $\boldsymbol{\tau}$ can be calculated as

$$\boldsymbol{\tau} = B^{-1}(\bar{\boldsymbol{u}} - A(\boldsymbol{v})). \tag{6}$$

By designing a dynamic feedback controller, an input prolongation of $\bar{u}_1$ is given as

$$\dot{\bar{u}}_1 = \tilde{u}_1 \tag{7}$$

and denote

$$\tilde{\boldsymbol{u}} = [\tilde{u}_1, \tilde{u}_2]^{\mathrm{T}} := [\dot{\bar{u}}_1, \bar{u}_2]^{\mathrm{T}} \tag{8}$$

which means that $\bar{u}_1$ is a state in the controller, and we need to design $\tilde{\boldsymbol{u}}$. The differentially flat output of the system is

$$[z_x, z_y]^{\mathrm{T}} := [x, y]^{\mathrm{T}} \tag{9}$$

and using (1), (3), (5), (8), we obtain

$$[\dddot{z}_x, \dddot{z}_y]^{\mathrm{T}} = C + D\tilde{\boldsymbol{u}} \tag{10}$$

where

$$C = \begin{bmatrix} -2\bar{u}_1\omega\sin\theta - v\omega^2\cos\theta \\ 2\bar{u}_1\omega\cos\theta - v\omega^2\sin\theta \end{bmatrix}, D = \begin{bmatrix} \cos\theta & -v\sin\theta \\ \sin\theta & v\cos\theta \end{bmatrix}. \tag{11}$$

Another input transformation is introduced as

$$C + D\tilde{\boldsymbol{u}} := \boldsymbol{u} = [u_1, u_2]^{\mathrm{T}}, \tag{12}$$

which yields

$$\tilde{\boldsymbol{u}} = D^{-1}(\boldsymbol{u} - C), \tag{13}$$

and (10) becomes

$$[\dddot{z}_x, \dddot{z}_y]^{\mathrm{T}} = \boldsymbol{u}. \tag{14}$$

so finally, after prolongation, the dynamics (1), (3) and (7) is equivalent to (14), and their states can be expressed by each other:

$$[x, y, \theta, v, \omega, \bar{u}_1]^{\mathrm{T}} \longleftrightarrow [z_x, \dot{z}_x, \ddot{z}_x, z_y, \dot{z}_y, \ddot{z}_y]^{\mathrm{T}}. \tag{15}$$

Denote the tracking error as

$$\begin{aligned} \boldsymbol{e} &= [e_{x1}, e_{x2}, e_{x3}, e_{y1}, e_{y2}, e_{y3}]^{\mathrm{T}} \\ &= [z_x - x_d, \dot{z}_x - \dot{x}_d, \ddot{z}_x - \ddot{x}_d, z_y - y_d, \dot{z}_y - \dot{y}_d, \ddot{z}_y - \ddot{y}_d]^{\mathrm{T}}, \end{aligned} \tag{16}$$

then the tracking error dynamics is derived from (14), as

$$\dot{\boldsymbol{e}} = \left[\begin{array}{ccc|ccc} 0 & 1 & 0 & & & \\ 0 & 0 & 1 & & \boldsymbol{0} & \\ 0 & 0 & 0 & & & \\ \hline & & & 0 & 1 & 0 \\ & \boldsymbol{0} & & 0 & 0 & 1 \\ & & & 0 & 0 & 0 \end{array}\right] \boldsymbol{e} + \left[\begin{array}{cc} 0 & 0 \\ 0 & 0 \\ 1 & 0 \\ \hline 0 & 0 \\ 0 & 0 \\ 0 & 1 \end{array}\right] \left(\boldsymbol{u} - \begin{bmatrix} \dddot{x}_d \\ \dddot{y}_d \end{bmatrix}\right). \tag{17}$$

The state $\boldsymbol{e}$ can be calculated from $[x, y, \theta, v, \omega, \bar{u}_1]^{\mathrm{T}}$ using the relationship (15).

2.2 Model with Disturbances and Unmodeled Dynamics

Consider that if there exist disturbances in the input channel $\boldsymbol{\tau}$, the dynamics (1), (3) and (7) can be rewritten as

$$\begin{aligned} \dot{x} &= v\cos\theta \\ \dot{y} &= v\sin\theta \\ \dot{\theta} &= \omega \\ \dot{v} &= \bar{u}_1 + \delta_{1l} + \delta_{1h} \\ \dot{\bar{u}}_1 &= \tilde{u}_1 \\ \dot{\omega} &= \tilde{u}_2 + \delta_2. \end{aligned} \tag{18}$$

The disturbance in $\dot{v}$ is divided into two parts for analysis purposes. By setting

$$\begin{aligned} x_1 &= x \\ x_2 &= v\cos\theta \\ x_3 &= (\bar{u}_1 + \delta_{1l})\cos\theta - \omega v\sin\theta \\ y_1 &= y \\ y_2 &= v\sin\theta \\ y_3 &= (\bar{u}_1 + \delta_{1l})\sin\theta + \omega v\cos\theta, \end{aligned} \tag{19}$$

one obtains that

$$\begin{aligned} \dot{x}_1 &= x_2 \\ \dot{x}_2 &= x_3 + \delta_{1h}\cos\theta \\ \dot{x}_3 &= u_1 + \delta_3 - 2\delta_{1l}\omega\sin\theta + \dot{\delta}_{1l}\cos\theta - \delta_2 v\sin\theta - \delta_{1h}\omega\sin\theta \\ \dot{y}_1 &= y_2 \\ \dot{y}_2 &= y_3 + \delta_{1h}\sin\theta \\ \dot{y}_3 &= u_2 + \delta_4 + 2\delta_{1l}\omega\cos\theta + \dot{\delta}_{1l}\sin\theta + \delta_2 v\cos\theta + \delta_{1h}\omega\cos\theta \end{aligned} \tag{20}$$

where we have added δ_3 and δ_4 considering the effects of unmodeled dynamics especially uncertain parameters. We can see that uncertain parameters such as d, m, I, b and r can only affect u_1 and u_2, and that is why we put δ_3 and δ_4 there. Denote

$$\begin{aligned}\delta_5 &:= \delta_3 - 2\delta_{1l}\omega\sin\theta + \dot{\delta}_{1l}\cos\theta - \delta_2 v\sin\theta \\ \delta_6 &:= \delta_4 + 2\delta_{1l}\omega\cos\theta + \dot{\delta}_{1l}\sin\theta + \delta_2 v\cos\theta\end{aligned} \tag{21}$$

and the tracking error

$$\begin{aligned}\boldsymbol{e} &= [e_{x1}, e_{x2}, e_{x3}, e_{y1}, e_{y2}, e_{y3}]^{\mathrm{T}} \\ &= [x_1 - x_d, x_2 - \dot{x}_d, x_3 - \ddot{x}_d, y_1 - y_d, y_2 - \dot{y}_d, y_3 - \ddot{y}_d]^{\mathrm{T}},\end{aligned} \tag{22}$$

one obtains the tracking error dynamics

$$\dot{\boldsymbol{e}} = \left[\begin{array}{ccc|ccc} 0 & 1 & 0 & & & \\ 0 & 0 & 1 & & \mathbf{0} & \\ 0 & 0 & 0 & & & \\ \hline & & & 0 & 1 & 0 \\ & \mathbf{0} & & 0 & 0 & 1 \\ & & & 0 & 0 & 0 \end{array}\right] \boldsymbol{e} + \left[\begin{array}{cc} 0 & 0 \\ 0 & 0 \\ 1 & 0 \\ \hline 0 & 0 \\ 0 & 0 \\ 0 & 1 \end{array}\right] \left(\boldsymbol{u} - \begin{bmatrix} \dddot{x}_d \\ \dddot{y}_d \end{bmatrix} + \begin{bmatrix} \delta_5 \\ \delta_6 \end{bmatrix}\right) + \begin{bmatrix} 0 \\ \cos\theta \\ -\omega\sin\theta \\ 0 \\ \sin\theta \\ \omega\cos\theta \end{bmatrix} \delta_{1h}. \tag{23}$$

For our controller, some assumptions are needed:

Assumption 1 There exist positive constants ϵ_1, ϵ_2, ϵ_3 and ϵ_4, such that δ_5 and δ_6 are bounded by

$$\begin{aligned}|\delta_5| &\le \epsilon_1 + \epsilon_2(|v| + |\omega|) \\ |\delta_6| &\le \epsilon_3 + \epsilon_4(|v| + |\omega|).\end{aligned} \tag{24}$$

Assumption 2 x_1, x_2, x_3, y_1, y_2, y_3 can all be measured, and $\delta_{1h} = 0$; if x_1, x_2, x_3, y_1, y_2, y_3 can only be calculated by (19) using states $(x, y, \theta, v, \omega)$ of the original system, then $\delta_{1l} = 0$ should be furthermore assumed.

From (23) we can see that δ_{1h} exists in the 2nd and 5th equations, so it can not be fully eliminated by our controller. Since the closed system is a low pass filter, δ_{1h} will slightly affect the tracking performance.

2.3 Fixed-Time Control

Consider a nonlinear system

$$\dot{\boldsymbol{x}} = \boldsymbol{f}(t, \boldsymbol{x}), \quad \boldsymbol{x}(0) = \boldsymbol{x}_0 \tag{25}$$

where states $\boldsymbol{x} \in \mathbb{R}^n$, vector field $\boldsymbol{f}: \mathbb{R}_+ \times \mathbb{R}^n \to \mathbb{R}^n$ may be discontinuous with respect to the states. Then the solutions $\boldsymbol{x}(t, \boldsymbol{x}_0)$ of the system are understood in the sense of Filippov.

Definition 1 ([16]) The origin of system (25) is said to have finite-time attractivity if there exists a locally bounded function $T: \mathbb{R}^n \setminus \{0\} \to \mathbb{R}_+$, such that $\lim_{t \to T(\boldsymbol{x}_0)} \boldsymbol{x}(t, \boldsymbol{x}_0) = 0$ and $\boldsymbol{x}(t, \boldsymbol{x}_0) = 0$, $\forall t > T(\boldsymbol{x}_0)$, for all $\boldsymbol{x}_0 \in \mathbb{R}^n \setminus \{0\}$. T is called the settling-time function. The origin of system (25) is said to be globally finite-time stable if it is Lyapunov stable and has finite-time attractivity. The origin of system (25) is said to be globally fixed-time stable if it is globally finite-time stable and the settling-time function is bounded, that is $\exists T_{max} \in \mathbb{R}_+$: $T(\boldsymbol{x}_0) \le T_{max}$, $\forall \boldsymbol{x}_0 \in \mathbb{R}^n$.

Lemma 1 ([17]) *Consider an nth-order integrator system*

$$\dot{\boldsymbol{x}} = \left[\begin{array}{c|ccc} 0 & 1 & \cdots & 0 \\ \vdots & \vdots & \ddots & \vdots \\ 0 & 0 & \cdots & 1 \\ \hline 0 & 0 & \cdots & 0 \end{array}\right] \boldsymbol{x} + \left[\begin{array}{c} 0 \\ \vdots \\ 0 \\ \hline 1 \end{array}\right] u, \quad \boldsymbol{x}(0) = \boldsymbol{x}_0 \tag{26}$$

where $\boldsymbol{x} = [x_1, x_2, \ldots, x_n]^{\mathrm{T}} \in \mathbb{R}^n$ is the state, and $u \in \mathbb{R}$ is the input. The origin of system (26) is fixed-time stable if

$$u = -\sum_{i=1}^{n} k_i |x_i|^{\alpha_i} \operatorname{sgn}(x_i) - \sum_{i=1}^{n} \kappa_i |x_i|^{\beta_i} \operatorname{sgn}(x_i). \tag{27}$$

Here, exponents $\alpha_i \in (0, 1)$, $\beta_i \in (1, +\infty)$, $i = 1, \ldots, n$, and satisfy the recurrent relations

$$\alpha_{i-1} = \frac{\alpha_i \alpha_{i+1}}{2\alpha_{i+1} - \alpha_i}, \quad \beta_{i-1} = \frac{\beta_i \beta_{i+1}}{2\beta_{i+1} - \beta_i}, \quad i = 2, \ldots, n \tag{28}$$

where $\alpha_{n+1} = \beta_{n+1} = 1$, and $\alpha_n = \alpha \in (1 - \varepsilon, 1)$, $\beta_n = \beta \in (1, 1 + \varepsilon)$ for a sufficiently small $\varepsilon > 0$. Control gains k_i and κ_i, $i = 1, \ldots, n$, are assigned such that polynomials $s^n + k_n s^{n-1} + \cdots + k_1$ and $s^n + \kappa_n s^{n-1} + \cdots + \kappa_1$ are Hurwitz. In addition, the settling-time function

$$T \le T_{max} = \frac{\alpha \lambda_{max}(P_1)}{(1 - \alpha)\lambda_{min}(Q_1)} \cdot \lambda_{max}^{\frac{1-\alpha}{\alpha}}(P_1) + \frac{\beta \lambda_{max}(P_2)}{(\beta - 1)\lambda_{min}(Q_2)} \cdot \frac{1}{\Upsilon^{\frac{\beta-1}{\beta}}} \tag{29}$$

where $0 < \Upsilon \le \lambda_{min}(P_2)$, symmetric positive definite matrices P_i and Q_i satisfy the Lyapunov equations

$$P_i A_i + A_i^{\mathrm{T}} P_i = -Q_i, \quad i = 1, 2 \tag{30}$$

and A_i are in controllable canonical forms as

$$A_1 = \left[\begin{array}{c|ccc} 0 & 1 & \cdots & 0 \\ \vdots & \vdots & \ddots & \vdots \\ 0 & 0 & \cdots & 1 \\ \hline -k_1 & -k_2 & \cdots & -k_n \end{array}\right], \ A_2 = \left[\begin{array}{c|ccc} 0 & 1 & \cdots & 0 \\ \vdots & \vdots & \ddots & \vdots \\ 0 & 0 & \cdots & 1 \\ \hline -\kappa_1 & -\kappa_2 & \cdots & -\kappa_n \end{array}\right]. \quad (31)$$

Lemma 2 ([18]) *If there exists a continuous positive definite and radially unbounded function $V(\boldsymbol{x})$ such that*

$$\dot{V}(\boldsymbol{x}) \leq -(aV^p(\boldsymbol{x}) + bV^q(\boldsymbol{x}))^k \quad (32)$$

for some $a, b, p, q, k > 0$ satisfying $pk < 1$ and $qk > 1$, then the origin of system (25) is fixed-time stable and the settling-time function

$$T \leq T_{max} = \frac{1}{a^k(1-pk)} + \frac{1}{b^k(qk-1)}. \quad (33)$$

Lemma 3 ([19]) *For $x_i \geq 0$, $i = 1, \ldots, N$, then*

$$\sum_{i=1}^{N} x_i^p \geq \begin{cases} (\sum_{i=1}^{N} x_i)^p & 0 < p \leq 1 \\ N^{1-p}(\sum_{i=1}^{N} x_i)^p & 1 < p < \infty \end{cases} \quad (34)$$

3 Controller Design

In this section, fixed-time controllers are designed. After the analysis in Sect. 2, what we need to do is to design $\boldsymbol{u}$ in (17) and (23).

3.1 Controller for Ideal Model

By using Lemma 1, a control law can be designed as

$$\boldsymbol{u}_0 = -\begin{bmatrix} \sum_{i=1}^{3}(k_i|e_{xi}|^{\alpha_i}\operatorname{sgn}(e_{xi}) + \kappa_i|e_{xi}|^{\beta_i}\operatorname{sgn}(e_{xi})) \\ \sum_{i=1}^{3}(k_i|e_{yi}|^{\alpha_i}\operatorname{sgn}(e_{yi}) + \kappa_i|e_{yi}|^{\beta_i}\operatorname{sgn}(e_{yi})) \end{bmatrix} \quad (35)$$

where exponents $\alpha_1, \alpha_2, \alpha_3 \in (0, 1)$, $\beta_1, \beta_2, \beta_3 \in (1, +\infty)$, satisfying

$$\alpha_1 = \frac{\alpha_2\alpha_3}{2\alpha_3 - \alpha_2}, \quad \alpha_2 = \frac{\alpha_3}{2-\alpha_3}, \quad \beta_1 = \frac{\beta_2\beta_3}{2\beta_3 - \beta_2}, \quad \beta_2 = \frac{\beta_3}{2-\beta_3}, \quad (36)$$

control gains $k_1, k_2, k_3, \kappa_1, \kappa_2, \kappa_3$ are assigned such that $s^3 + k_3 s^2 + k_2 s + k_1$ and $s^3 + \kappa_3 s^2 + \kappa_2 s + \kappa_1$ are Hurwitz. Then $\boldsymbol{u}$ can be designed as

$$\boldsymbol{u} = \begin{bmatrix} \dddot{x}_d \\ \dddot{y}_d \end{bmatrix} + \boldsymbol{u}_0. \tag{37}$$

Theorem 1 *Consider the ideal system* (1), (3), *then fixed-time trajectory tracking can be achieved with the dynamic feedback controller* (6), (8), (13), (35) *and* (37).

Proof From the analysis in Sect. 2, it has been shown that the trajectory tracking of system (1), (3) with control laws (6), (8), (13) is equivalent to (17), then substitute (35), (37) into (17), using Lemma 1, the fixed-time trajectory tracking is achieved.

Remark 1 Control law (13) requires that D is invertible, and this can not be satisfied when $v = 0$. So an open loop control should be applied to activate the system when $v = 0$, and the reference trajectory should avoid the situation that the forward velocity is zero.

3.2 Controller Considering Disturbances and Unmodeled Dynamics

Under the Assumptions 1 and 2, what we need to do is to eliminate δ_5 and δ_6 in (23). For this purpose, an integral sliding mode surface is designed as

$$\boldsymbol{s} = \begin{bmatrix} s_1 \\ s_2 \end{bmatrix} = \begin{bmatrix} e_{x3} \\ e_{y3} \end{bmatrix} - \int_0^t \boldsymbol{u}_0(\tau)\mathrm{d}\tau \tag{38}$$

and a compensation law is given by

$$\boldsymbol{\Delta u} = -\bar{a}\begin{bmatrix} |s_1|^p \mathrm{sgn}(s_1) \\ |s_2|^p \mathrm{sgn}(s_2) \end{bmatrix} - \bar{b}\begin{bmatrix} |s_1|^q \mathrm{sgn}(s_1) \\ |s_2|^q \mathrm{sgn}(s_2) \end{bmatrix} - \begin{bmatrix} (\epsilon_1 + \epsilon_2(|v| + |\omega|))\mathrm{sgn}(s_1) \\ (\epsilon_3 + \epsilon_4(|v| + |\omega|))\mathrm{sgn}(s_2) \end{bmatrix} \tag{39}$$

where $\bar{a} > 0, \bar{b} > 0, 0 < p < 1, q > 1$. Then $\boldsymbol{u}$ is designed as

$$\boldsymbol{u} = \begin{bmatrix} \dddot{x}_d \\ \dddot{y}_d \end{bmatrix} + \boldsymbol{u}_0 + \boldsymbol{\Delta u}. \tag{40}$$

Theorem 2 *Consider the system (1), (3), with disturbances $\delta_{1l}, \delta_{1h}, \delta_2$ in input channels (like* (18)*), and unmodeled dynamics δ_3, δ_4 (like* (20)*), then fixed-time trajectory tracking can be achieved with dynamic feedback controller* (6), (8), (13), (35), (38), (39) *and* (40), *under Assumptions* 1 *and* 2.

Proof The proof includes two steps. First, it will be shown that the sliding mode surface (38) can be reached in a fixed time. Then on the sliding mode surface $\boldsymbol{s} = 0$, the tracking error $\boldsymbol{e}$ will converge to zero in a fixed time.

Consider the Lyapunov function

$$V = \frac{1}{2}\boldsymbol{s}^{\mathrm{T}}\boldsymbol{s}. \tag{41}$$

By using (23), (38), (40) and the assumptions, taking the derivative of V yields

$$\begin{aligned}
\dot{V} &= \boldsymbol{s}^{\mathrm{T}}\dot{\boldsymbol{s}} \\
&= \boldsymbol{s}^{\mathrm{T}}([\dot{e}_{x3}, \dot{e}_{y3}]^{\mathrm{T}} - \boldsymbol{u}_0) \\
&= \boldsymbol{s}^{\mathrm{T}}(\boldsymbol{\Delta u} + [\delta_5, \delta_6]^{\mathrm{T}}) \\
&= -\bar{a}(|s_1|^{p+1} + |s_2|^{p+1}) - \bar{b}(|s_1|^{q+1} + |s_2|^{q+1}) + \delta_5 s_1 + \delta_6 s_2 \\
&\quad - (\epsilon_1 + \epsilon_2(|v| + |\omega|))|s_1| - (\epsilon_3 + \epsilon_4(|v| + |\omega|))|s_2| \\
&\le -\bar{a}(|s_1|^{p+1} + |s_2|^{p+1}) - \bar{b}(|s_1|^{q+1} + |s_2|^{q+1}) \\
&= -\bar{a}((|s_1|^2)^{\frac{p+1}{2}} + (|s_2|^2)^{\frac{p+1}{2}}) - \bar{b}((|s_1|^2)^{\frac{q+1}{2}} + (|s_2|^2)^{\frac{q+1}{2}}).
\end{aligned} \tag{42}$$

It follows from Lemma 3 that

$$\begin{aligned}
\dot{V} &\le -\bar{a}(s_1^2 + s_2^2)^{\frac{p+1}{2}} - 2^{1-\frac{q+1}{2}} \times \bar{b}(s_1^2 + s_2^2)^{\frac{q+1}{2}} \\
&= -2^{\frac{p+1}{2}}\bar{a}V^{\frac{p+1}{2}} - 2\bar{b}V^{\frac{q+1}{2}}.
\end{aligned} \tag{43}$$

Then it follows from Lemma 2 that sliding mode surface $\boldsymbol{s} = 0$ can be reached in a fixed time.

When $\boldsymbol{s} = 0$ is reached, $\dot{\boldsymbol{s}} = 0$ is also satisfied, which means that

$$\begin{bmatrix} \dot{e}_{x3} \\ \dot{e}_{y3} \end{bmatrix} = \boldsymbol{u}_0. \tag{44}$$

By using Lemma 1, the tracking error will converge to zero in a fixed time. This completes the proof.

Remark 2 In control law (39), the sign functions $\mathrm{sgn}(s)$ contained in the last term make the controller discontinuous around the sliding mode surface, which results in the chattering phenomenon. So in applications, the sign function can be approximated by a continuous function

$$\mathrm{sat}(s) = \begin{cases} \mathrm{sgn}(s) & |s| \ge \Phi \\ s/\Phi & |s| < \Phi \end{cases} \tag{45}$$

where Φ is a small positive constant. And this will sacrifice some performance of the controller.

4 Simulation Results

In this section, simulation results are presented to illustrate the efficiency and effectiveness of the proposed controller.

First, we choose an ideal model with the nominal parameters in Table 1. Then a fixed-time tracking controller is designed according to Theorem 1, using the same nominal parameters, and other parameters in $\boldsymbol{u}_0$ are set as that in Tabel 2.

Table 1 Parameters of model

	d	b	r	I	m
Nominal	0.30	0.75	0.15	15.625	30
Real	0.35	0.80	0.17	12.625	31

Table 2 Parameters of controllers

$\boldsymbol{u}_0$	k_1	k_2	k_3	κ_1	κ_2	κ_3	α_1	α_2	α_3	β_1	β_2	β_3
	8	12	6	8	12	6	$\frac{4}{7}$	$\frac{2}{3}$	$\frac{4}{5}$	2	$\frac{3}{2}$	$\frac{6}{5}$
$\boldsymbol{\Delta u}$	$\bar{a}$	$\bar{b}$	p	q	ϵ_1	ϵ_2	ϵ_3	ϵ_4	Φ			
	2	2	$\frac{1}{2}$	$\frac{3}{2}$	1	1	1	1	$\frac{1}{5}$			

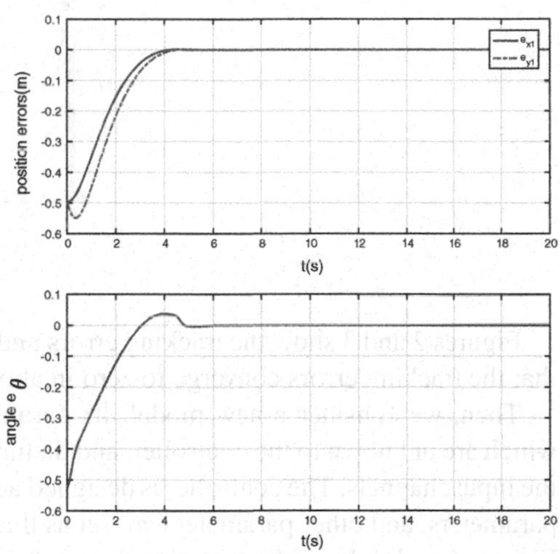

Fig. 2 Tracking errors of x, y and θ

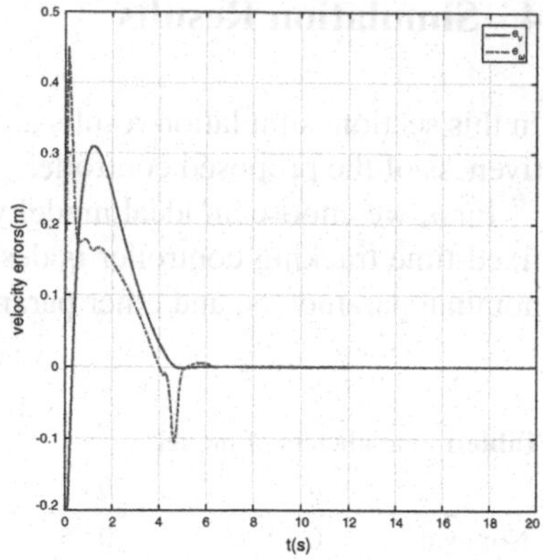

Fig. 3 Tracking errors of v and ω

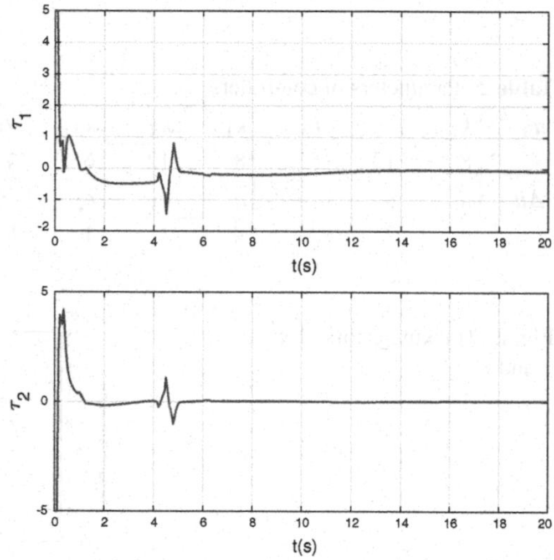

Fig. 4 Control input τ

Figures 2 and 3 show the tracking errors and Fig. 4 is control input. It can be seen that the tracking errors converge to zero in about 6 seconds.

Then, we consider a new model. Its parameters are the real values in Table 1, which are unknown to the controller, and disturbances of magnitude 0.2 are added to the input channels. The controller is designed according to Theorem 2, using nominal parameters, and other parameters are set as that in Table 2. The results are shown in Figs. 5, 6 and 7. It can be seen that the tracking errors converge nearby zero. Small deviations exist because of δ_{1h} and the approximation of the sign function.

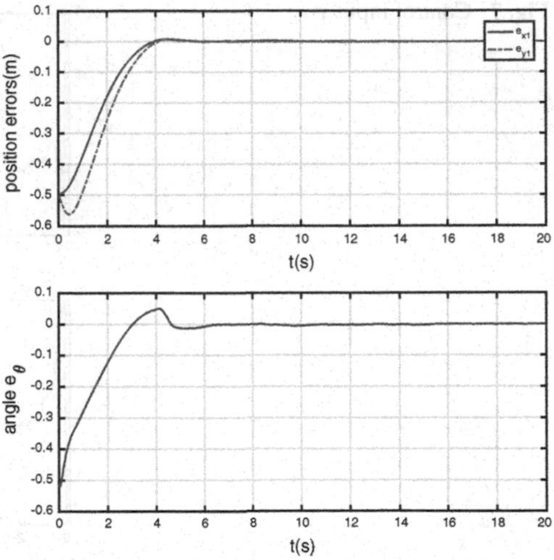

Fig. 5 Tracking errors of x, y and θ

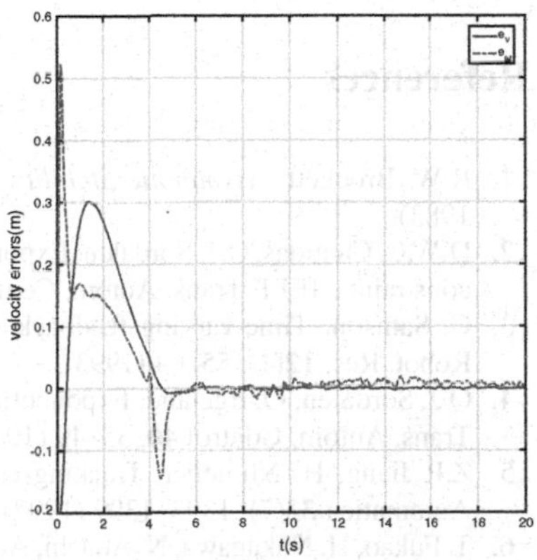

Fig. 6 Tracking errors of v and ω

5 Conclusions

This paper is devoted to the fixed-time trajectory tracking control of a two-wheeled mobile robot. The controller is designed based on differential flatness. The model with disturbances and uncertain parameters is also analyzed. Some of their influences are eliminated by designing an integral sliding mode surface.

Acknowledgements This work was supported by the NSFC (61327807, 61521091, 61520106010, 61134005) and the National Basic Research Program of China (973 Program: 2012CB821200, 2012CB821201).

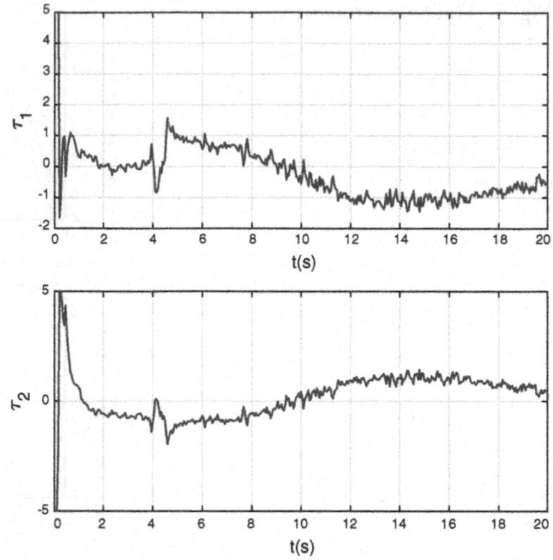

Fig. 7 Control input $\boldsymbol{\tau}$

References

1. R.W. Brockett, *Asymptotic Stability and Feedback Stabilization* (Birkhauser, Boston, MA, 1983)
2. D.W.C. Canudas, O.J. Sordalen, Exponential stabilization of mobile robots with nonholonomic constraints. IEEE Trans. Autom. Control **37**(11), 1791–1797 (1992)
3. C. Samson, Time-varying feedback stabilization of car-like wheeled mobile robots. Int. J. Robot. Res. **12**(1), 55–64 (1993)
4. O.J. Sordalen, O. Egeland, Exponential stabilization of nonholonomic chained systems. IEEE Trans. Autom. Control **40**, 35–49 (1995)
5. Z.P. Jiang, H. Nijmeijer, Tracking control of mobile robots: a case study in backstepping. Automatica **33**(7), 1393–1399 (1997)
6. T. Fukao, H. Nakagawa, N. Adachi, Adaptive tracking control of a nonholonomic mobile robot. IEEE Trans. Robot. Autom. **16**(5), 609–615 (2000)
7. D.H. Kim, J.H. Oh, Tracking control of a two-wheeled mobile robot using input-output linearization. Control Eng. Pract. **7**(3), 369–373 (1999)
8. L. Boquete, R.L. Garcia, R. Barea et al., Neural control of the movements of a wheelchair. J. Intell. Robot. Syst. **25**(3), 213–226 (1999)
9. T. Das, I.N. Kar, Design and implementation of an adaptive fuzzy logic-based controller for wheeled mobile robots. IEEE Trans. Control Syst. Technol. **14**(3), 501–510 (2006)
10. S.P. Bhat, D.S. Bernstein, Finite-time stability of continuous autonomous systems. SIAM J. Control Optim. **38**(3), 751–766 (2000)
11. Y. Wu, B. Wang, G.D. Zong, Finite-time tracking controller design for nonholonomic systems wiht extended chained form. IEEE Trans. Circuits Syst. II: Express. Briefs **52**(11), 798–802 (2005)
12. Y. Zhang, G. Liu, B. Luo, Finite-time cascaded tracking control approach for mobile robots. Inf. Sci. **284**, 31–43 (2014)
13. S. Shi, X. Yu, S. Khoo, Robust finite-time tracking control of nonholonomic mobile robots without velocity measurements. Int. J. Control **89**(2), 411–423 (2016)
14. W. Huang, Y. Yang and C. Hua. Fixed-time tracking control approach design for nonholonomic mobile robot, in *Proceedings of the 35th Chinese Control Conference, July 27–29, 2016, Chengdu, China*, pp. 3423–3428

15. L. Chen, Y. Jia, Variable-poled tracking control of a two-wheeled mobile robot using differential flatness. J. Robot. Netw. Artif. Life **1**(1), 12–16 (2014)
16. A. Polyakov, D. Efimov, W. Perruquetti, Robust stabilization of MIMO systems in finite/fixed time. Int. J. Robust Nonlinear Control **26**, 69–90 (2016)
17. M. Basin, Y. Shtessel, F. Aldukali, Continuous finite- and fixed-time high-order regulators. J. Franklin Inst. **353**, 5001–5012 (2016)
18. A. Polyakov, Nonlinear feedback design for fixed-time stabilization of linear control systems. IEEE Trans. Autom. Control **57**(8), 2106–2110 (2012)
19. Z. Zuo, Nonsigular fixed-time consensus tracking for second-order multi-agent networks. Automatica **54**, 305–309 (2015)

Optimal Robust Guaranteed Cost Backstepping Control for Multi-motor Driving System

Minlin Wang, Xuemei Ren and Tianyi Zeng

Abstract This paper proposes an optimal robust guaranteed cost backstepping control for the multi-motor driving system with parameter uncertainties. First, the state representation of the multi-motor driving system with parameter uncertainties is established, and a feedforward controller is designed based on the backstepping control technique. Then, based on the feedforward control, an optimal robust guaranteed cost feedback controller is designed to achieve the asymptotically stability of the error system. The proposed controller not only can increase the system robustness to the parameter uncertainties, but also can make the cost function limited by a certain upper bound. Lyapunov theory proves the stability of the control system. Finally, simulation results based on a four-motor driving system demonstrate the effectiveness of the proposed control scheme.

Keywords Multi-motor driving system · Parameter uncertainties
Optimal robust guaranteed cost · Backstepping control

1 Introduction

With the rapid development of servo system, single-motor driving system has been widely used in military and industry. However, for some large inertia and high power loads, such as the radar antenna system [1] and the artillery control system [2], the single-motor servo system cannot meet its demand on driving force and output power. Therefore, it is necessary and important for such systems to adopt multiple motors to drive the load. The control target of the multi-motor driving system is to make the load output track a reference command. However, the conventional controller (PID control) cannot ensure that the system has better tracking performance because of the nonlinearities and parameter uncertainties in the system. At present, a variety of advanced control algorithms have been applied on the multi-motor driving system

M. Wang · X. Ren (✉) · T. Zeng
School of Automation, Beijing Institute of Technology, Beijing 100081, China
e-mail: xmren@bit.edu.cn

Y. Jia et al. (eds.), *Proceedings of 2018 Chinese Intelligent Systems Conference*, Lecture Notes in Electrical Engineering 529,
https://doi.org/10.1007/978-981-13-2291-4_76

to realize the precise tracking control, such as optimal control, adaptive control and robust control.

Robust control technology is an effective tool to deal with the system parameter uncertainties, which has usually combined with other control techniques to ensure the system stability. By using the extended state observer to estimate the unknown system states, Ren et al. [3] proposed an adaptive robust controller to ensure the tracking performance of the dual-motor driving system. Zhao et al. [4] combined the robust technique with the sliding mode controller to overcome the on the backlash and friction nonlinear influence on the multi-motor driving system. However, a good controller can not only guarantee the system stability, but also make the system cost function minimized. The robust guaranteed cost control proposed by Chang and Peng [5] is that controller. It can make the system cost function reach a certain upper bound under the condition of the system uncertainties. Liu et al. [6] designed an optimal guaranteed cost robust controller based on neural network for time-varying nonlinear systems with parameter uncertainties. Yang and Jian-Ming [7] derived the optimal guaranteed cost control law for large-scale interconnected systems with time-delay by using the linear matrix inequalities.

In this paper, an optimal robust guaranteed cost controller is proposed for multi-motor driving systems with parameter uncertainties. The controller is composed of a feedforward controller and a feedback controller. The feedforward controller is designed based on the backstepping control technique, and the feedback controller is developed via the optimal robust guaranteed cost control. The designed controller can not only guarantee the stability of the system, but also make the cost function limited by a given upper bound. The simulation results verify the effectiveness of the proposed method.

2 The Dynamic Model of Multi-motor Driving Systems

The test rig of the four-motor driving system is given in Fig. 1.

In general, the multi-motor driving system [8] can be described by the following dynamics.

$$\begin{cases} J_i\ddot{\theta}_i + b_i\dot{\theta}_i = u_i - T_i \\ J_m\ddot{\theta}_m + b_m\dot{\theta}_m = \sum\limits_{i=1}^{n} T_i \end{cases} \tag{1}$$

where θ_i, θ_m denotes the motor angular position and load angular position respectively; $\dot{\theta}_i$, $\dot{\theta}_m$ represents the motor angular velocity and load angular velocity respectively; J_i, b_i represents the motor inertia and viscous friction coefficient; J_m, b_m represents the load inertia and viscous friction coefficient; u_i is the control input, T_i is the transmission torque between the load and motor.

Affected by the backlash, T_i is expressed as

Fig. 1 Schematic diagram of four-motor driving system

$$T_i = kf(z_i(t)) = \begin{cases} k(z_i(t) - \alpha), & z_i(t) \geq \alpha \\ 0, & |z_i(t)| < \alpha \\ k(z_i(t) + \alpha), & z_i(t) \leq -\alpha \end{cases} \tag{2}$$

where $k > 0$ is the stiffness coefficient, $z_i(t) = \theta_i - \theta_m$ is the position difference between the load and motor, $2\alpha > 0$ is the backlash width. Then, $f(z_i(t))$ can be written as

$$f(z_i(t)) = z_i(t) + d_\alpha(z_i(t)) \tag{3}$$

where

$$d_\alpha(z_i(t)) = \begin{cases} -\alpha & z_i(t) \geq \alpha \\ -z_i(t) & |z_i(t)| < \alpha \\ \alpha & z_i(t) \leq -\alpha \end{cases}. \tag{4}$$

Choose the state variables as $x_1 = \theta_m, x_2 = \dot{\theta}_m, x_{3i} = \theta_i, x_{4i} = \dot{\theta}_i$, the state representation of the multi-motor driving system is given as:

$$\begin{cases} \dot{x}_1 = x_2 \\ \dot{x}_2 = -\frac{b_m}{J_m}x_2 + \frac{k}{J_m}\sum\limits_{i=1}^{n}(x_{3i} - x_1) + \frac{k}{J_m}\sum\limits_{i=1}^{n}d_\alpha(z_i(t)) \\ \dot{x}_{3i} = x_{4i} \\ \dot{x}_{4i} = \frac{1}{J_i}u_i - \frac{b_i}{J_i}x_{4i} - \frac{k}{J_i}(x_{3i} - x_1) - \frac{k}{J_i}d_\alpha(z_i(t)) \\ y = x_1 \end{cases} \quad . \tag{5}$$

We assume that $J = J_1 = \cdots = J_n$ and $b = b_1 = \cdots = b_n$. By defining the sum of motor states as $x_3 = \sum_{i=1}^{n} x_{3i}$, $x_4 = \sum_{i=1}^{n} x_{4i}$ and $u = \sum_{i=1}^{n} u_i$, the system (5) is transformed into

$$\begin{cases} \dot{x}_1 = x_2 \\ \dot{x}_2 = -(a_1 + \Delta a_1)x_2 + (a_2 + \Delta a_2)(x_3 - nx_1 + \sum\limits_{i=1}^{n} d_\alpha(z_i(t))) \\ \dot{x}_3 = x_4 \\ \dot{x}_4 = (a_0 + \Delta a_0)u - (a_3 + \Delta a_3)x_4 - (a_4 + \Delta a_4)(x_3 - nx_1 + d_\alpha(z_i(t))) \\ y = x_1 \end{cases} \tag{6}$$

where $a_0 = \frac{1}{J}$, $a_1 = \frac{b_m}{J_m}$, $a_2 = \frac{k}{J_m}$, $a_3 = \frac{b}{J}$, $a_4 = \frac{k}{J}$, Δa_0, Δa_1, Δa_2, Δa_3 and Δa_4 are the parameter uncertainties.

Assumption 1 All the parameters $a_j (j = 0, 1, 2, 3, 4)$ and parameter uncertainties Δa_j are bounded, i.e., $a_{j\min} < a_j < a_{j\max}$, $\Delta a_{j\min} < \Delta a_j < \Delta a_{j\max}$.

The control objective of this paper is to make the system output y track a reference command y_d and guarantee that all the signals of the close-loop system are uniformly ultimately bounded.

3 The Controller Design

The proposed controller consists of a feedforward controller U^a and a feedback controller U^*. The feedforward controller is derived by a backstepping control method, and the feedback controller is designed based on a robust guaranteed cost controller that is, the final controller is $U = U^a + U^*$.

Backstepping control is used to design feedforward controller. In order to facilitate controller design, the following error system is designed:

$$\begin{cases} e_1 = y - y_d \\ e_i = x_i - \eta_{i-1} \\ i = 2, 3, 4 \end{cases} \tag{7}$$

where η_{i-1} is the virtual control law expressed as $\eta_{i-1} = \eta_{i-1}^a + \eta_{i-1}^*$, η_{i-1}^a is the virtual feedforward control law, η_{i-1}^* is the virtual backforward control law.

Step 1: According to (6) and (7), the derivative of tracking error can be obtained.

$$\dot{e}_1 = x_2 - \dot{y}_d = e_2 + \eta_1^a + \eta_1^* - \dot{y}_d \tag{8}$$

Consider the following Lyapunov function $V_1 = \frac{1}{2}e_1^2$ whose derivative is

$$\dot{V}_1 = e_1(e_2 + \eta_1^a + \eta_1^* - \dot{y}_d). \tag{9}$$

The feedforward virtual control can be designed as:

$$\eta_1^a = -c_1 e_1 + \dot{y}_d. \tag{10}$$

Substituting the control (10) into (9), we have

$$\dot{V}_1 = -c_1 e_1^2 + e_1 e_2 + e_1 \eta_1^*. \tag{11}$$

Step 2: Consider the second Lyapunov function $V_2 = V_1 + \frac{1}{2}e_2^2$ whose derivative is

$$\begin{aligned}\dot{V}_2 = {} & -c_1 e_1^2 + e_1 e_2 + e_1 \eta_1^* + e_2[-(a_1 + \Delta a_1)x_2 + (a_2 + \Delta a_2)(e_3 + \eta_2^a + \eta_2^* \\ & - n x_1 + \sum_{i=1}^{n} d_\alpha(z_i(t))) - \dot{\eta}_1].\end{aligned} \tag{12}$$

The second feedforward virtual control signals can be designed as

$$\eta_2^a = \frac{1}{a_2}(-c_2 e_2 + a_1 x_2 + \dot{\eta}_1) + n x_1. \tag{13}$$

Substituting (13) into (12), one obtains

$$\dot{V}_2 = -c_1 e_1^2 + e_1 e_2 + e_1 \eta_1^* - c_2 e_2^2 + a_2 e_2 e_3 + a_2 e_2 \eta_2^* + e_2 \Delta f_1 \tag{14}$$

where

$$\begin{aligned}\Delta f_1 = {} & -\Delta a_1 x_2 + \frac{\Delta a_2}{a_2}(-c_2 e_2 + a_1 x_2 + \dot{\eta}_1) \\ & + (a_2 + \Delta a_2)\sum_{i=1}^{n} d_\alpha(z_i(t)) + \Delta a_2(e_3 + \eta_2^*).\end{aligned}$$

Step 3: Consider the third Lyapunov function $V_3 = V_2 + \frac{1}{2}e_3^2$ whose derivative is

$$\dot{V}_3 = -c_1 e_1^2 + e_1 e_2 + e_1 \eta_1^* - c_2 e_2^2 + a_2 e_2 e_3 + a_2 e_2 \eta_2^* + e_2 \Delta f_1 + e_3(e_4 + \eta_3^a + \eta_3^* - \dot{\eta}_2^*), \quad (15)$$

The third feedforward virtual control signals can be designed as

$$\eta_3^a = -c_3 e_3 + \dot{\eta}_2. \quad (16)$$

Substituting (16) into (15), one obtains

$$\dot{V}_3 = -c_1 e_1^2 + e_1 e_2 + e_1 \eta_1^* - c_2 e_2^2 + a_2 e_2 e_3 + a_2 e_2 \eta_2^* + e_2 \Delta f_1 - c_3 e_3^2 + e_3 e_4 + e_3 \eta_3^*. \quad (17)$$

Step 4: Consider the fourth Lyapunov function $V_4 = V_3 + \frac{1}{2} e_4^2$ whose derivative is

$$\dot{V}_4 = -c_1 e_1^2 + e_1 e_2 + e_1 \eta_1^* - c_2 e_2^2 + a_2 e_2 e_3 + a_2 e_2 \eta_2^* + e_2 \Delta f_1 - c_3 e_3^2 + e_3 e_4 + e_3 \eta_3^* + e_4[(a_0 + \Delta a_0)u - (a_3 + \Delta a_3)x_4 - (a_4 + \Delta a_4)(x_3 - n x_1 + d_\alpha(z_i(t))) - \dot{\eta}_3]. \quad (18)$$

The finial control law is given as

$$u^a = \frac{1}{a_0}[-c_4 e_4 + a_3 x_4 + a_4(x_3 - n x_1) + \dot{\eta}_3]. \quad (19)$$

Substituting (19) into (18), one obtains

$$\dot{V}_4 = -c_1 e_1^2 + e_1 e_2 + e_1 \eta_1^* - c_2 e_2^2 + a_2 e_2 e_3 + a_2 e_2 \eta_2^* + e_2 \Delta f_1 - c_3 e_3^2 + e_3 e_4 + e_3 \eta_3^* - c_4 e_4^2 + a_0 e_4 u^* + e_4 \Delta f_2 \quad (20)$$

where

$$\Delta f_2 = \Delta a_0 u^* + \frac{\Delta a_0}{a_0}[-c_4 e_4 + a_3 x_4 + a_4(x_3 - n x_1) + \dot{\eta}_3] - \Delta a_3 x_4 - \Delta a_4(x_3 - n x_1) + (a_4 + \Delta a_4) d_\alpha(z_i(t)).$$

Define the error system as $e = [\, e_1 \; e_2 \; e_3 \; e_4 \,]^T$, (20) can be rewritten as:

$$\dot{V}_4 = -\sum_{i=1}^{4} c_i e_i^2 + e(Ae + BU^* + \Delta f) \quad (21)$$

where $A = \begin{bmatrix} 0 & 1 & 0 & 0 \\ 0 & 0 & a_2 & 0 \\ 0 & 0 & 0 & 1 \\ 0 & 0 & 0 & 0 \end{bmatrix}^T$, $B = \begin{bmatrix} 1 & 0 & 0 & 0 \\ 0 & a_2 & 0 & 0 \\ 0 & 0 & 1 & 0 \\ 0 & 0 & 0 & a_0 \end{bmatrix}^T$, $U^* = \begin{bmatrix} \eta_1^* \\ \eta_2^* \\ \eta_3^* \\ u^* \end{bmatrix}$ and $\Delta f = \begin{bmatrix} 0 \\ \Delta f_1 \\ 0 \\ \Delta f_2 \end{bmatrix}$.

According to (21), the following subsystem with parameter uncertainties is constructed.

$$\dot{e} = Ae + BU^* + \Delta f(e, U^*) \tag{22}$$

where Δf represents parameter uncertainty related to the error e and U^*, that is, $\Delta f(e, U^*)$.

According to the Assumption 1, since the system parameters and uncertainties are all bounded, the uncertainties $\Delta f(e, U^*)$ can be expressed by the following equation

$$\Delta f(e, U^*) = Gd(e, U^*) \tag{23}$$

where G is a positive definite matrix, $d(e, U^*)$ is a function satisfying

$$d^T(e, U^*)d(e, U^*) \leq e^T L_1 e + U^{*T} L_2 U^* \tag{24}$$

where L_1, L_2 are positive definite matrices.

Lemma 1 *For the parameter uncertainties $\Delta f(e, U^*)$, we can find a positive function as*

$$\Gamma(e, U^*) = e^T L_1 e + U^{*T} L_2 U^* + \frac{1}{4}\lambda^T G G^T \lambda \tag{25}$$

to make the following equation achieved

$$\lambda^T \Delta f(e, U^*) \leq \Gamma(e, U^*) \tag{26}$$

where λ is the Lagrange multiplier.

Proof According to (23)–(26), one has

$$\begin{aligned}
\Gamma(e, U^*) - \lambda^T \Delta f(e, U^*) &= e^T L_1 e + U^{*T} L_2 U^* + \frac{1}{4}\lambda^T G G^T \lambda - \lambda^T \Delta f(e, U^*) \\
&\geq d^T(e, U^*)d(e, U^*) + \frac{1}{4}\lambda^T G G^T \lambda - \lambda^T G d(e, U^*) \\
&= (d(e, U^*) - \frac{1}{2}G^T\lambda)^T (d(e, U^*) - \frac{1}{2}G^T\lambda) \geq 0
\end{aligned} \tag{27}$$

That completes the proof.

For the system (22) with parameter uncertainties, the control target is to design a robust guarantee cost control

$$U^* = Ke \tag{28}$$

such that the following performance index is minimized

$$J = \int_{t_0}^{\infty} (e^T Qe + U^{*T} RU^*)dt \tag{29}$$

where Q, R are the positive definite matrices.

Definition 1 For the system (22) with parameter uncertainty, if there is a controller (28) and a positive number J^* such that the uncertainty (24) is satisfied, the closed loop system (22) is regular, impulsive, stable and satisfies the $J \leq J^*$, J^* is the system performance index and the control (28) is called a robust guaranteed cost control.

Therefore, the nominal system for uncertain systems (22) is

$$\dot{e} = Ae + BU^*. \tag{30}$$

The Hamilton function is derived as

$$H = e^T Qe + U^{*T} RU^* + \Gamma(e, U^*) + \lambda^T(Ae + BU^*) = 0 \tag{31}$$

and it satisfies

$$\sup J(t_0, e(t_0), U^*) \leq J^*(t_0, e(t_0)) \tag{32}$$

where

$$J^*(t_0, e(t_0)) = \int_{t_0}^{\infty} (e^T Qe + U^{*T} RU^* + \Gamma(e, U^*))dt \tag{33}$$

is the performance index for the nominal system (30).

For the Hamilton function (31), the following equations can be obtained

$$\begin{bmatrix} \dot{e} \\ \dot{\lambda} \end{bmatrix} = \begin{bmatrix} A & -\frac{1}{2}B(R + L_2)B^T \\ -2(Q + L_1) & -A^T \end{bmatrix} \begin{bmatrix} e \\ \lambda \end{bmatrix} \tag{34}$$

According to the optimization principle, the solution for (34) is

$$U^* = -\frac{1}{2}(R + L_2)^{-1}B^T\lambda = -\frac{1}{2}(R + L_2)^{-1}B^T Pe = -Ke \tag{35}$$

where P is a positive definite matrix solved by the following Riccati equation

$$A^T P + PA - \frac{1}{2}PB(R + L_2)B^T P + 2(Q + L_1) = 0. \tag{36}$$

4 Simulation Results

The control algorithm is verified by the simulation results. The four-motor driving system is employed as the test rig whose parameters are set as $J_m = 0.113\,\mathrm{kg\,m^2}, b_m = 0.2\,\mathrm{Nm\,s/rad}, J = 0.026\,\mathrm{kg\,m^2}, b = 0.15, \mathrm{Nm\,s/rad}, k = 1\,\mathrm{Nm}$. The backstepping controller is designed as (10), (13), (16) and (19), and its parameters are chosen as $c_1 = 4, c_2 = 10, c_3 = 20, c_4 = 15$. The robust guaranteed cost controller is designed as (35) and its parameters are $Q = diag\{0.3, 0.1, 3, 1\}, R = diag\ \{1, 1, 1, 1\}, L_1 = diag\{0.05, 0.2, 1, 0.6\}, L_2 = diag\{0, 0.5, 0, 2\}$.

The simulation time interval is 0.01 s, the initial state is set to 0, and the sinusoidal wave $y_d = 2\sin(t)$ is selected as the reference signal. Figures 2 and 3 show the tracking performance under the backstepping control and the proposed robust guaranteed cost control. From Figs. 2 and 3, it can be seen that the robust guaranteed cost control can improve the transient and steady-state tracking performance compared with the conventional backstepping controller. Therefore, for the multi-motor driving system with parameter uncertainties, the proposed robust guaranteed cost controller can effectively suppress the system parameter uncertainties and achieve a satisfactory performance of the load tracking.

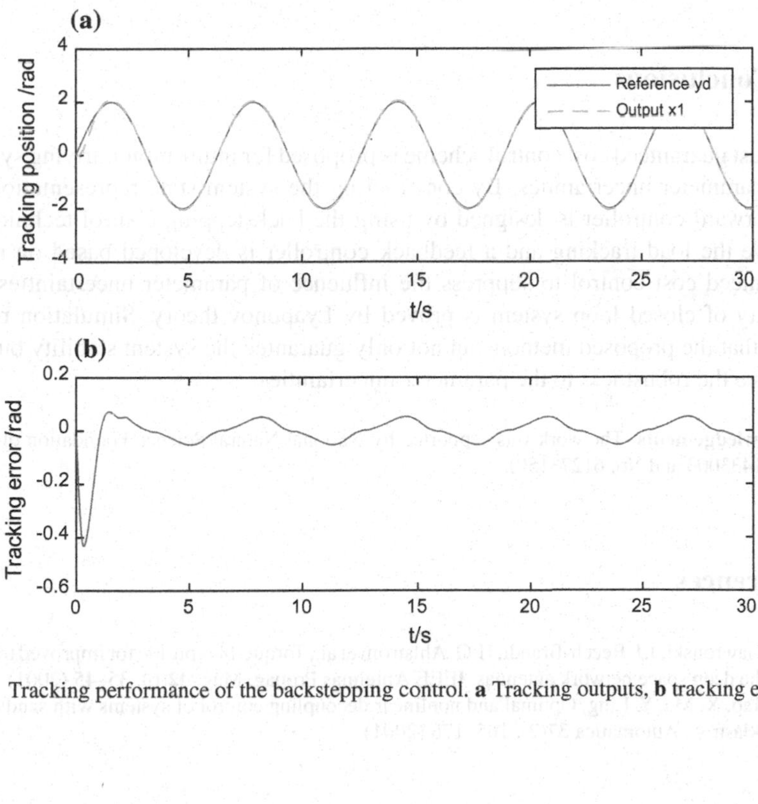

Fig. 2 Tracking performance of the backstepping control. **a** Tracking outputs, **b** tracking error

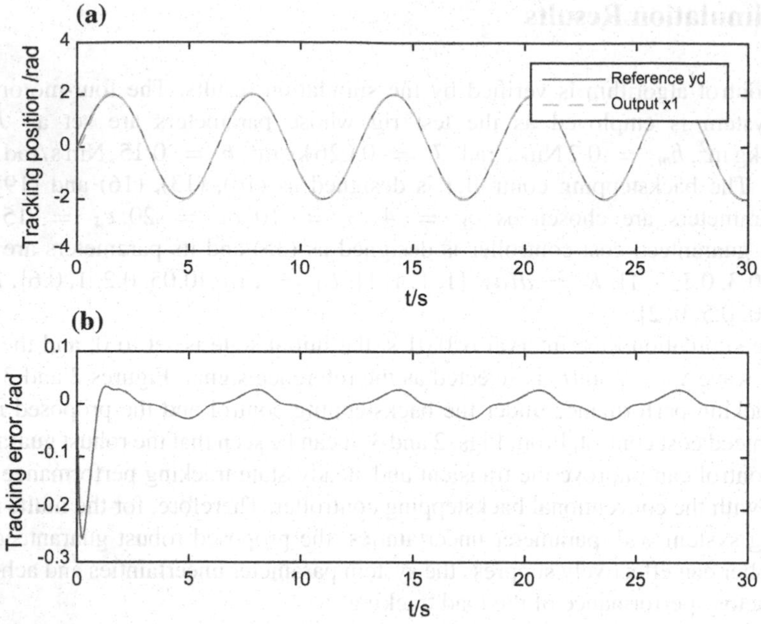

Fig. 3 Tracking performance of the proposed control. **a** Tracking outputs, **b** tracking error

5 Conclusions

A robust guaranteed cost control scheme is proposed for multi-motor driving systems with parameter uncertainties. By constructing the system state representation, the feedforward controller is designed by using the backstepping control technique to achieve the load tracking and a feedback controller is developed based on robust guaranteed cost control to suppress the influence of parameter uncertainties. The stability of closed loop system is proved by Lyapunov theory. Simulation results show that the proposed method can not only guarantee the system stability but also enhance the robustness to the parameter uncertainties.

Acknowledgements The work was supported by National Natural Science Foundation of China (No. 61433003 and No. 61273150).

References

1. W. Gawronski, J.J. Beech-Brandt, H.G. Ahlstrom et al., Torque-bias profile for improved tracking of the deep space network antennas. IEEE Antennas Propag. Mag. **42**(6), 35–45 (2001)
2. G. Tao, X. Ma, Y. Ling, Optimal and nonlinear decoupling control of systems with sandwiched backlash ☆. Automatica **37**(2), 165–176 (2001)

3. X. Ren, D. Li, G. Sun et al., Eso-based adaptive robust control of dual motor driving servo system. Asian J. Control **18**(6), 2358–2365 (2016)
4. W. Zhao, X. Ren, X. Gao, Synchronization and tracking control for multi-motor driving servo systems with backlash and friction. Int. J. Robust Nonlinear Control **26**(13), 2745–2766 (2016)
5. S. Chang, T. Peng, Adaptive guaranteed cost control of systems with uncertain parameters. IEEE Trans. Autom. Control **17**(4), 474–483 (2003)
6. D. Liu, D. Wang, F.Y. Wang et al., Neural-network-based online HJB solution for optimal robust guaranteed cost control of continuous-time uncertain nonlinear systems. IEEE Trans. Cybern. **44**(12), 2834 (2014)
7. X.H. Yang, X.U. Jian-Ming, Decentralized guaranteed cost control for uncertain large-scale interconnected systems with multiple delays. J. Central S. Univ. (2009)
8. H.B. Zhao, X.H. Zhou, Backstepping adaptive control of dual-motor driving servo system. Comput. Eng. Appl. (2012)

Fault Detection for Discrete-Time Systems Over Signal to Noise Ratio Constrained Channels

Fumin Guo and Xuemei Ren

Abstract This paper studies fault detection for the discrete-time systems with signal to noise ratio constrained channels. By using the descriptor system method, a fault detection strategy is proposed. First, in order to guarantee the residual generation system is admissible, an augmented robust fault detection filter is constructed, and the fault is estimated; then, a constant threshold is designed to detect the fault. Finally, a simulation example is given to show the applicability of the proposed approach.

Keywords fault detection · signal to noise ratio · descriptor system approach

1 Introduction

Due to the significance for the safety and reliability of the complex modern systems, fault detection (FD) has been becoming an important research topic [1–4]. After several decades of the development, some FD methods have been proposed, such as observer-based methods [5], filter-based methods [6], data-driven approaches [7]. For a nonliear switched system, a robust mean-square exponential stable fault detection filter was designed in [8], and the solvability condition of filter was also given. The authors in [9] were concerned with the fault detection problem of discrete-time linear periodic systems, and a residual generator was constructed based on the unknown input observer. In addition, for the nonlinear systems with norm bounded unknown

F. Guo
National Computer Network Emergency Response Technical Team/Coordination Center of China, Beijing 100029, China

X. Ren (✉)
School of Automation, Beijing Institute of Technology, Beijing 100081, China
e-mail: xmren@bit.edu.cn

Y. Jia et al. (eds.), *Proceedings of 2018 Chinese Intelligent Systems Conference*, Lecture Notes in Electrical Engineering 529,
https://doi.org/10.1007/978-981-13-2291-4_77

input, a robust fault detection was discussed in [10], and the evaluation function and threshold determination as well as false alarm rate verification were both designed.

Signal to noise ratio (SNR) constraint as a nonnegligible constraint usually exists in communication systems, so it is necessary to research the SNR constrant. The authors in [11] investigated the stability of discrete-time systems over the SNR constrained channels, and static feedback controllers were designed to stabilize the unstable plant.

This paper studies fault detection for the discrete-time systems with signal to noise ratio constrained channels. By using the descriptor system method, a fault detection strategy is proposed. First, in order to guarantee the residual generation system is admissible, an augmented robust fault detection filter is constructed, and then a constant threshold is designed to detect the fault. The rest of this paper is organized as follows. Section 2 discusses problem formulation. The design of fault detection strategy is given in Sect. 3. Simulation results are provided in Sect. 4, and a conclusion is shown in Sect. 5.

2 Problem Formulation

Consider the discrete-time systems over the channels subject to SNR constrants as shown in Fig. 1:

$$\begin{cases} x(k+1) = Ax(k) + Bu(k) + E_d d(k) + E_f f(k) \\ y(k) = Cx(k) + F_d d(k) \end{cases} \tag{1}$$

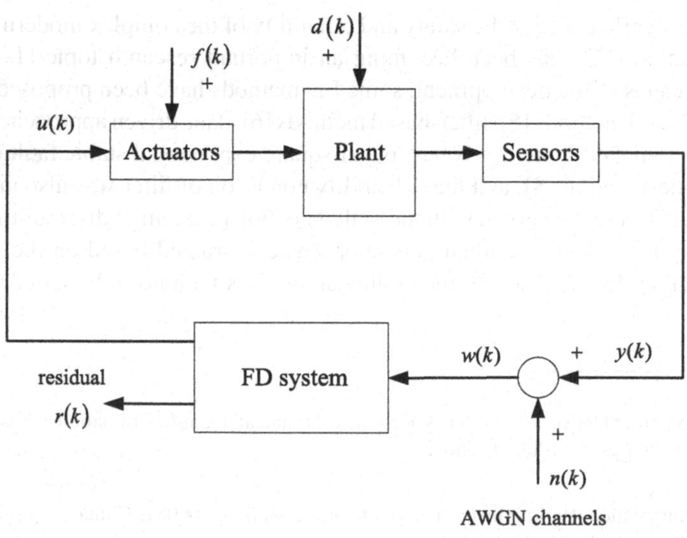

Fig. 1 Structure of FD for the feedback control systems with SNR constrained channels

where $x(k) \in R^n$ is system state; $u(k) \in R^p$ is system input; $y(k) \in R^m$ is the system output; $d(k) \in R^{n_d}$ and $f(k) \in R^{n_f}$ are the bounded unknown disturbance and the fault, respectively. A, B, C, E_d, E_f and F_d are known constant matrices with the appropriate dimensions.

Assumption 1 The channel is described by $w(k) = y(k) + n(k)$, where $w(k) = [w_1(k)\ w_2(k)\ \ldots\ w_m(k)]^T$ is the channel output, the additive white gaussian channel noise $n(k) = [n_1(k)\ n_2(k)\ \ldots\ n_m(k)]^T$ is the zero mean with the positive diagonal covariance matrix $P_{nc} = diag\left(\sigma_{n_1}^2, \sigma_{n_2}^2, \ldots, \sigma_{n_m}^2\right)$, and each channel is subject to a SNR constraint:

$$S_j = \frac{P_{y_j}}{\sigma_{n_j}^2}, j = 1, 2, \ldots, m \tag{2}$$

where S_j is SNR of the jth channel, and P_{y_j} is the channel input power of the jth channel.

This paper is devoted to studying FD of system (1) subject to SNR constrained channels imposed by (2), and a fault detection strategy including an augmented robust FD filter and a constant threshold will be constructed.

3 Fault Detection

This section will discuss the FD including an augmented robust FD filter and a threshold. In the following text, we assume (A, C) is detectable.

3.1 Fault Detection Filter

Let $\overline{x}(k) = [x(k); d(k)]$, then a descriptor system for system (1) can be given:

$$\begin{cases} \overline{E}\overline{x}(k+1) = \overline{A}\overline{x}(k) + \overline{B}u(k) + \overline{E}_d d(k) + \overline{E}_f f(k) \\ \overline{y}(k) = y(k) = \overline{C}\overline{x}(k) \end{cases} \tag{3}$$

where

$$\overline{E} = \begin{bmatrix} I & 0 \\ 0 & 0 \end{bmatrix}, \overline{A} = \begin{bmatrix} A & 0 \\ 0 & -I \end{bmatrix}, \overline{B} = \begin{bmatrix} B \\ 0 \end{bmatrix}, \overline{E}_d = \begin{bmatrix} E_d \\ I \end{bmatrix}, \overline{E}_f = \begin{bmatrix} E_f \\ 0 \end{bmatrix}, \overline{C} = [C\ \ F_d].$$

Then, a FD filter can be constructed to generate the so-called residual signal:

$$\begin{cases} \widehat{\overline{x}}(k+1) = A_F\widehat{\overline{x}}(k) + B_F w(k) \\ r(k) = C_F\widehat{\overline{x}}(k) + D_F w(k) \\ w(k) = y(k) + n(k) \end{cases} \tag{4}$$

where $\widehat{\overline{x}}(k) = \left[\widehat{x}(k); \widehat{d}(k)\right] \in R^{n+n_d}$ is the state; $r(k)$ is residual signal; A_F, B_F, C_F and D_F are free design matrices.

A weighted fault $\widehat{f_w}(z) = W_f(z) f(z)$ can be considered as the estimation of $f(k)$, where $W_f(z)$ being a given stable weighting matrix. Then, a minimal realization of $\widehat{f_w}(z)$ is:

$$\begin{cases} x_w(k+1) = A_w x_w(k) + B_w f(k) \\ \widehat{f_w}(k) = C_w x_w(k) + D_w f(k) \end{cases} \tag{5}$$

where $x_w \in R^{n_w}$ is the state; $\widehat{f_w}(k) \in R^{n_f}$ is the weighted fault; $f(k) \in R^{n_f}$ is the original fault; A_w, B_w, C_w and D_w are the known matrices.

Denote $r_e(k) = r(k) - \widehat{f_w}(k)$, then the augmented system is:

$$\begin{cases} \overline{E}_e \overline{e}(k+1) = \overline{A}_e \overline{e}(k) + \overline{B}_e v(k) \\ r_e(k) = \overline{C}_e \overline{e}(k) + \overline{D}_e v(k) \end{cases} \tag{6}$$

where

$$\overline{e}(k) = \begin{bmatrix} \overline{x}(k) \\ \widehat{\overline{x}}(k) \\ x_w(k) \end{bmatrix}, v(k) = \begin{bmatrix} u(k) \\ d(k) \\ n(k) \\ f(k) \end{bmatrix}, \overline{E}_e = \begin{bmatrix} \overline{E} & 0 & 0 \\ 0 & I & 0 \\ 0 & 0 & I \end{bmatrix}, \overline{A}_e = \begin{bmatrix} \overline{A} & 0 & 0 \\ B_F \overline{C} & A_F & 0 \\ 0 & 0 & A_w \end{bmatrix},$$

$$\overline{B}_e = \begin{bmatrix} \overline{B} & \overline{E}_d & 0 & \overline{E}_f \\ 0 & 0 & B_F & 0 \\ 0 & 0 & 0 & B_w \end{bmatrix}, \overline{C}_e = \begin{bmatrix} D_F \overline{C} & C_F & -C_w \end{bmatrix}, \overline{D}_e = [0 \;\; 0 \;\; D_F \;\; -D_w].$$

Then, Theorem 1 will show a sufficient condition for the admissibility of system (6).

Theorem 1 *For system* (6) *and given* $\gamma > 0$, *if there exist positive definite matrices* P *and* Q *such that the following matrix inequalities hold:*

$$\overline{E}_e^T P \overline{E}_e \geq 0 \tag{7}$$

$$\begin{bmatrix} -Q^{-1} & 0 & 0 & I & 0 \\ * & -I & 0 & \overline{C}_e & \overline{D}_e \\ * & * & -P^{-1} & \overline{A}_e & \overline{B}_e \\ * & * & * & -\overline{P} & 0 \\ * & * & * & * & -\gamma^2 I \end{bmatrix} < 0 \tag{8}$$

where $\overline{P} = \overline{E}_e^T P \overline{E}_e + Q$, *thus system* (6) *is admissible and satisfies the following* H_∞ *performance:*

$$\mathbb{E}\left\{\sqrt{\sum_{k=0}^{\infty}\|r_e(k)\|^2}\right\} < \gamma\mathbb{E}\left\{\sqrt{\sum_{k=0}^{\infty}\|v(k)\|^2}\right\}$$

where $\mathbb{E}$ is the mathematical expectation.

The proof is omited.

Because of the coupling form in inequality (8), *it is difficult to design the robust fault detection filter. Then, we will introduce an auxiliary slack nonsingular matrix Φ.*

For inequality (8), *pre-multiplying diag $\left\{\Phi^T, I, \Phi^T, I, I\right\}$ and post-multiplying diag $\{\Phi, I, \Phi, I, I\}$,*

$$\begin{bmatrix} -\Phi^T Q^{-1}\Phi & 0 & 0 & \Phi^T & 0 \\ * & -I & 0 & \overline{C}_e & \overline{D}_e \\ * & * & -\Phi^T P^{-1}\Phi & \Phi^T\overline{A}_e & \Phi^T\overline{B}_e \\ * & * & * & -\overline{E}_e^T P\overline{E}_e - Q & 0 \\ * & * & * & * & -\gamma^2 I \end{bmatrix} < 0 \tag{9}$$

Because P is positive definite and Φ is nonsingular, we obtain

$$(P-\Phi)^T P^{-1}(P-\Phi) \geq 0$$

i.e.,

$$-\Phi^T P^{-1}\Phi \leq P - \left(\Phi^T + \Phi\right) \tag{10}$$

In the same way

$$-\Phi^T Q^{-1}\Phi \leq Q - \left(\Phi^T + \Phi\right) \tag{11}$$

Thus, substitute inequalities (10) *and* (12) *into inequality* (9), *we can obtain the following decoupling matrix inequality*

$$\begin{bmatrix} Q - \left(\Phi + \Phi^T\right) & 0 & 0 & \Phi^T & 0 \\ * & -I & 0 & \overline{C}_e & \overline{D}_e \\ * & * & P - \left(\Phi + \Phi^T\right) & \Phi^T\overline{A}_e & \Phi^T\overline{B}_e \\ * & * & * & -\overline{E}_e^T P\overline{E}_e - Q & 0 \\ * & * & * & * & -\gamma^2 I \end{bmatrix} < 0 \tag{12}$$

Then, Theorem 2 *will give the design of fault detection filter* (4).

Theorem 2 *For system (6) and given $\gamma > 0$, if there exist positive definite symmetric matrices P_{11}, P_{22}, P_{33}, Q_{11}, Q_{22}, Q_{33}, Δ_{11}, Δ_{22}, Δ_{33} and the matrices P_{12}, P_{13}, P_{23}, Q_{12}, Q_{13}, Q_{23}, Δ_{12}, Δ_{13}, Δ_{23}, X, Y, G, Z, $\overline{A}_f$, $\overline{B}_f$, $\overline{C}_f$, $\overline{D}_f$ such that*

$$\begin{bmatrix} \Psi_{11} & 0 & 0 & \Psi_{14} & 0 \\ * & -I & 0 & \Psi_{24} & \Psi_{25} \\ * & * & \Psi_{33} & \Psi_{34} & \Psi_{35} \\ * & * & * & \Psi_{44} & 0 \\ * & * & * & * & -\gamma^2 I \end{bmatrix} < 0 \tag{13}$$

holds, where

$$\Psi_{11} = \begin{bmatrix} Q_{11} - \overline{X} & Q_{12} - \overline{M} & Q_{13} \\ * & Q_{22} - \overline{Y} & Q_{23} \\ * & * & Q_{33} - \overline{Z} \end{bmatrix}, \Psi_{14} = \begin{bmatrix} X & X & 0 \\ Y+G & Y & 0 \\ 0 & 0 & Z \end{bmatrix}$$

$$\Psi_{24} = \begin{bmatrix} \overline{D}_f\overline{C} + \overline{C}_f & \overline{D}_f\overline{C} & -C_w \end{bmatrix}, \Psi_{25} = \begin{bmatrix} 0 & 0 & \overline{D}_f & -D_w \end{bmatrix}$$

$$\Psi_{33} = \begin{bmatrix} P_{11} - \overline{X} & P_{12} - \overline{M} & P_{13} \\ * & P_{22} - \overline{Y} & P_{23} \\ * & * & P_{33} - \overline{Z} \end{bmatrix},$$

$$\Psi_{34} = \begin{bmatrix} X\overline{A} & X\overline{A} & 0 \\ Y\overline{A} + \overline{B}_f\overline{C} + \overline{A}_f & Y\overline{A} + \overline{B}_f\overline{C} & 0 \\ 0 & 0 & ZA_w \end{bmatrix},$$

$$\Psi_{35} = \begin{bmatrix} X\overline{B} & X\overline{E}_d & 0 & X\overline{E}_f \\ Y\overline{B} & Y\overline{E}_d & B_f & Y\overline{E}_f \\ 0 & 0 & 0 & ZB_w \end{bmatrix}, \Psi_{44} = \begin{bmatrix} -\Delta_{11} - Q_{11} & -\Delta_{12} - Q_{12} & -\Delta_{13} - Q_{13} \\ * & -\Delta_{22} - Q_{22} & -\Delta_{23} - Q_{23} \\ * & * & -\Delta_{33} - Q_{33} \end{bmatrix}$$

and $\overline{X} = X + X^T$, $\overline{Y} = Y + Y^T$, $\overline{Z} = Z + Z^T$, $\overline{M} = X + Y^T + G^T$, then the fault detection filter can be gained as: $A_F = \overline{A}_f G^{-1}$, $B_F = \overline{B}_f$, $C_F = \overline{C}_f G^{-1}$, $D_F = \overline{D}_f$.

The proof is omited.

3.2 Threshold Computing

The threshold J_{th} ($J_{th} > 0$) can be designed based on the following fault detection logic

$$\begin{cases} \left\| r_{ef}(k) \right\|_{2,\rho} < J_{th}, & fault\text{-}free \\ \left\| r_{ef}(k) \right\|_{2,\rho} \geq J_{th}, & alarm\ \ for\ \ fault \end{cases} \tag{14}$$

where the evaluation function $\left\| r_{ef}(k) \right\|_{2,\rho}$ is

$$\left\| r_{ef}(k) \right\|_{2,\rho} = \frac{1}{\rho} \sum_{i=k-\rho+1}^{k} r^T(i)\, r(i) \tag{15}$$

According to systems (1) and (4), we have

$$\| r(k) \|_{2,\rho} = \left\| r_{u,d,n}(k) + r_f(k) \right\|_{2,\rho} \tag{16}$$

where

$$r_{u,d,n}(k) = r(k)\,|_{f=0}$$

$$r_f(k) = r(k)\,|_{u=0,d=0,n=0}$$

Therefore, the threshold J_{th} can be designed as

$$\begin{aligned} J_{th} &= \sup_{u,d,n} \left\| r_{u,d,n}(k) \right\|_{2,\rho} \\ &= \sup_{u,d,n} \left(\frac{1}{\rho} \sum_{i=k-\rho+1}^{k} r^T_{u,d,n}(i)\, r_{u,d,n}(i) \right) \end{aligned} \tag{17}$$

4 Simulation Results

Consider the following two dimensional system

$$\begin{aligned} \dot{x}(t) &= \begin{bmatrix} -3 & 3 \\ 3 & -6 \end{bmatrix} x(t) + \begin{bmatrix} 3 & 0 \\ 0 & 3 \end{bmatrix} u(t) + \begin{bmatrix} 0.7 & 0.3 \\ 0.3 & 0.7 \end{bmatrix} d(t) + \begin{bmatrix} 0 \\ 1 \end{bmatrix} f(t) \\ y(t) &= \begin{bmatrix} 1 & 0 \\ 0 & 1 \end{bmatrix} x(t) + \begin{bmatrix} 0.2 & 0.1 \\ 0.5 & 0.2 \end{bmatrix} d(t) \end{aligned} \tag{18}$$

Assume the sampling period $h = 0.1s$, $SNR = 35dB$, $\rho = 20$, $d(k)$ is a random signal uniformly distributed between [0, 0.025], and $W_f(z) = 0.5z/(z - 0.5)$, i.e., $A_w = 0.5$, $B_w = 0.25$, $C_w = 1$, and $D_w = 0.5$.

The fault is assumed as

$$f(k) = \begin{cases} 0 & 0 \le k < 20 \\ 2\left(1 - e^{-0.6(k-20)}\right) & 20 \le k < 100 \end{cases}$$

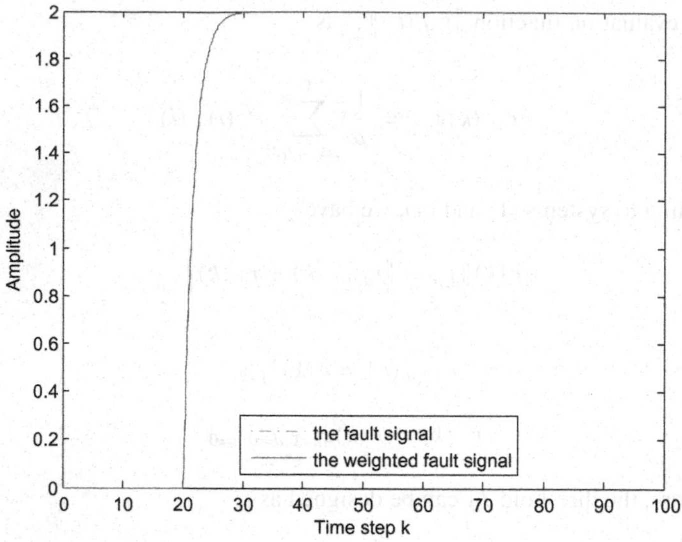

Fig. 2 The fault signal and weighted fault signal

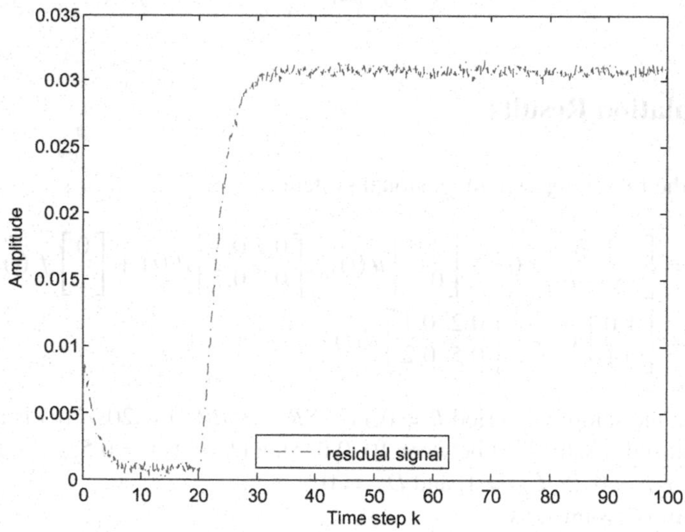

Fig. 3 Residual signal

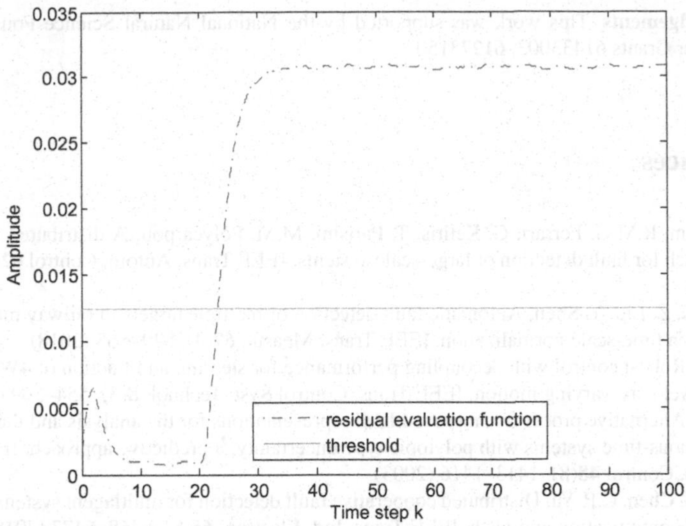

Fig. 4 Residual evaluation function and threshold

The FD filter gain matrices are

$$A_F = \begin{bmatrix} 0.5275 & 0.1201 & -0.0338 & -0.0124 \\ 0.1151 & 0.4238 & -0.1091 & -0.0089 \\ -0.0141 & 0.0525 & -0.8738 & 0.0398 \\ -0.0093 & 0.0004 & 0.0405 & -0.9517 \end{bmatrix}, B_F = \begin{bmatrix} -0.3261 & -0.0498 \\ -0.0509 & -0.2530 \\ 0.0594 & 0.1158 \\ 0.0303 & 0.0496 \end{bmatrix},$$

$$C_F = [-0.0076 \quad -0.0197 \quad 0.0037 \quad -0.0005], D_F = [0.0017 \quad 0.0058].$$

Figures 2, 3 and 4 show the simulation results.

As shown in Fig. 2, the fault can be estimated accurately. Then, from Fig. 4, we can see that the proposed method can detect the fault when it occurs.

5 Conclusion

The fault detection problem for the discrete-time systems with SNR channels was studied in this paper. By establishing a descriptor system, a fault detection strategy was given. An augmented robust fault detection filter was proposed to guarantee the residual generation system was admissible. Then, a constant threshold was designed to detect the fault. Finally, simulation results showed the applicability of the proposed approach.

Acknowledgements This work was supported by the National Natural Science Foundation of China under Grants 61433003, 61273150.

References

1. F. Boem, R.M.G. Ferrari, C. Keliris, T. Parisini, M.M. Polycarpou, A distributed networked approach for fault detection of large-scale systems. IEEE Trans. Autom. Control **62**(1), 18–33 (2017)
2. S.F. Lu, Z. Liu, Y. Shen, Automatic fault detection of multiple targets in railway maintenance based on time-scale normalization. IEEE Trans. Measur. **67**(4), 849–865 (2018)
3. Y. Jia, Robust control with decoupling performance for steering and traction of 4WS vehicles under velocity-varying motion. IEEE Trans. Control Syst. Technol. **8**(3), 554–569 (2000)
4. Y. Jia, Alternative proofs for improved LMI representations for the analysis and the design of continuous-time systems with polytopic type uncertainty: a predictive approach. IEEE Trans. Autom. Control **48**(8), 1413–1416 (2003)
5. Y. Li, J. Chen, C.P. Yu, Distributed cooperative fault detection for multiagent systems: a mixed H_∞/H_2 optimization approach. IEEE Trans. Ind. Electron. **65**(8), 6468–6477 (2018)
6. T.B. Wu, C.H. Yang, W.H. Gui, Event-based fault detection filtering for complex networked jump systems. IEEE Trans. Mechatron. **23**(2), 497–505 (2018)
7. Q.C. Jiang, S.X. Ding, Y. Wang, X.F. Yan, Data-driven distributed local fault detection for large-scale processes based on the GA-regularized canonical correlation analysis. IEEE Trans. Ind. Electron. **64**(10), 8148–8157 (2017)
8. X.J. Su, P. Shi, L.G. Wu, Y.D. Song, Fault detection filtering for nonlinear switched stochastic systems. IEEE Trans. Autom. Control **61**(5), 1310–1315 (2016)
9. S. Longhi, A. Monteriu, Fault detection and isolation of linear discrete-time periodic systems using the geometric approach. IEEE Trans. Autom. Control **62**(3), 1518–1523 (2017)
10. M.Y. Zhong, L.G. Zhang, S.X. Ding, D.H. Zhou, A probabilistic approach to robust fault detection for a class of nonlinear systems. IEEE Trans. Ind. Electron. **64**(5), 3930–3939 (2017)
11. J.H. Braslavsky, R.H. Middleton, J.S. Freudenberg, Feedback stabilization over signal-to-noise ratio constrained channels. IEEE Trans. Autom. Control **52**(8), 1391–1403 (2007)

Tri-self-taught Learning of Artificial Neural Networks

Feng Liu and Shuling Dai

Abstract We present a conceptually simple and general framework for self-taught learning, and this method can modify weights of neural networks when making prediction on samples according to the previous knowledge. The method called Tri-STNN. Based on Tri-training, it adds a Judge Network to make a higher accuracy of prediction. Tri-STNN is simple to be trained and the training datasets for Judge Network is easy to obtain. Moreover, Tri-STNN is capable of lots of tasks. We test Tri-STNN on the datasets of CIFAR-10, and the result shows that Tri-STNN can constantly keep self-taught learning and improve the generalization ability for new samples.

Keywords Artificial neural networks · Self-taught learning · Lifelong- learning
Judge network

1 Introduction

Under normal circumstances, artificial neural networks are used to do prediction on new samples, such as classification task. Samples in training sets are used to train neural network, and samples in testing sets are used to evaluate the performance of neural networks. But usually the prediction accuracy on testing sets is less than that on training sets [1]. We did an experiment where the training sets of Cats versus Dogs which is released by Kaggle were applied to train a classified neural network. After 10,000 iterations of training, the network was used to classify on training sets and testing sets separately. The prediction accuracy of training sets is 78.67% and the prediction accuracy of testing sets is 71.21%. The different detailed features of the images in these two data sets lead to the difference of prediction accuracy, though these images in training sets and testing sets have the same distribution.

F. Liu (✉) · S. Dai
School of Automation Science and Electrical Engineering,
Beihang University (BUAA), Beijing 100191, China
e-mail: feng.liu@realflytech.com

Y. Jia et al. (eds.), *Proceedings of 2018 Chinese Intelligent Systems Conference*, Lecture Notes in Electrical Engineering 529,
https://doi.org/10.1007/978-981-13-2291-4_78

As the duty of a well-trained artificial neural network is to do prediction on unlabeled datasets, an effective method that can increase the accuracy of prediction is to retrain the neural network by unlabeled datasets [2, 3]. But, how to train a neural network with samples without labels? In order to solve this problem, Zhou and Li proposed Tri-Training algorithm [4], which give a label to an unlabeled sample through the collaborative working of three agents. Tri-STNN has similar structures to Tri-Training, and the difference between them is that Tri-STNN gets the final result by using an artificial neural network called Judge Network, instead of voting method in Tri-Training. In this paper, Judge Network is described in detail in Sect. 3.

Lifelong-learning is an important field of machine learning, and it is a continuous learning processes [5]. In Lifelong-learning, the learning machines have carried out a task sequence, including N tasks of T_1, T_2,..., T_N. When facing the tasks of T_{N+1} and the corresponding dataset of D_{N+1}, the learning machines can utilize the previous experiences to learn these tasks [1, 6]. In knowledge base, the knowledge that has been accumulated during finishing tasks N will be updated according to the result of Task T_{N+1} [7, 8]. Tri-STNN belongs to the field of Lifelong-learning because it can improve the generalization ability of new samples during predicting on new samples. However, Lifelong-learning in deep neural networks suffers from a phenomenon called catastrophic forgetting [9]. Specifically, the weights of networks will be changed during learning new knowledge so as to change the understanding of previous knowledge. In this paper, when retraining, parts of labeled samples are taken randomly to mix with new samples, and then the new batch is used to retrain agents. Catastrophic forgetting would be avoid through the method of replaying samples when retraining.

Overall, Tri-STNN can improve the generalization of new samples according to previous experiences. Finally, we show the experiment where Tri-STNN worked on Cifar-10 datasets in classification task.

2 Related Work

Tri-training In many conditions, it is easy to get samples without labels and hard to get labeled ones, because they require human effort [10]. Therefore, semi-supervised learning that exploits unlabeled samples in addition to labeled ones has become a hot topic. Tri-training [4] was put forward by Zhou and Li, and it is a cooperative training algorithm. The striking feature of this algorithm is to adopt 3 classifiers not only to deal with the prediction problems of unlabeled samples but also to improve the generalization ability by ensemble learning.

Tri-training adopts bootstrap sampling to labeled samples to obtain 3 labeled datasets and then generate a classifier with each dataset. Differences must follow after initial training, even though it is not demand the three classifiers to be different types. During the test section, if two classifiers get the same prediction on a certain unlabeled sample, this sample is supposed to be with higher confidence of mark, and it will be set into retraining sets of the rest Classifier [4]. In this way, Tri-training can take advantages of ensemble learning and improve every classifier.

3 Tri-STNN

Similar to Tri-Training, Tri-STNN adopts three artificial neural networks with same functions to make prediction. The three agents have the same structure, as shown in Fig. 1. They are given different weights, because the training datasets are disturbed randomly before training, and random gradient descent algorithm is used in the initial training. So, different prediction may be given from three agents about a certain sample. The difference of prediction between three agents follows up with the request of Co-Training.

Judge Network Tri-Training adopt voting method to determine the classification results, but there are some limitations in the simple decision procedure. In an experiment, one classifier with 2 convolution layers, 2 full connected layers and a Softmax layer was be taken to predict the Cats versus Dogs datasets. As results shown in Table 1, the item 5 display that there is wrong decision by the method of max-voting and it is right by adopting the method of max-probability, while in item 3, there is right result by the method of max-voting and it is wrong by the method of max-

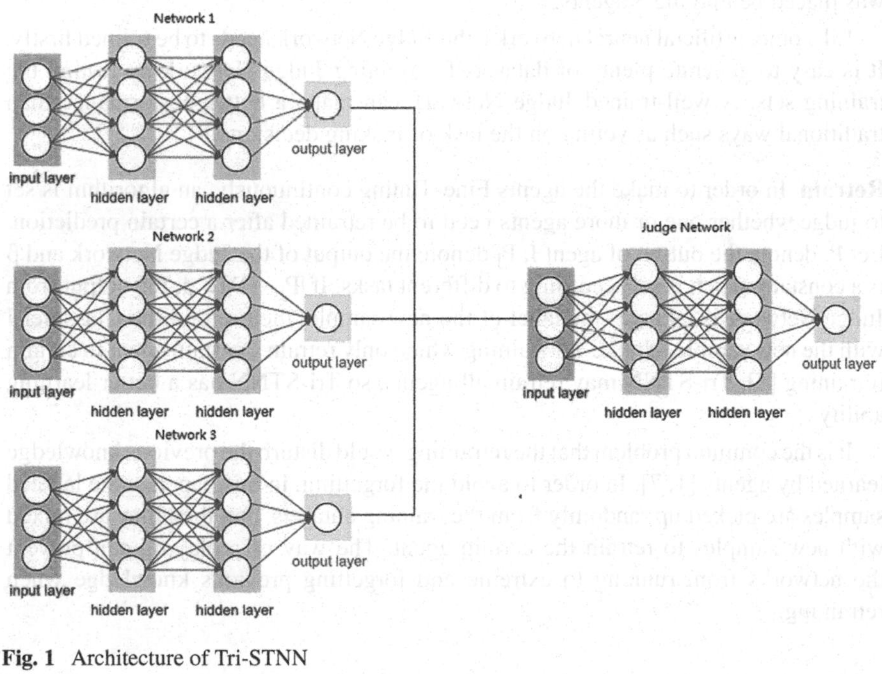

Fig. 1 Architecture of Tri-STNN

Table 1 Results of prediction on cats versus dogs datasets

Item	Real label	Network 1	Network 2	Network 3	Max-voting	Max-probability
1	Dog	90.11% Dog	99.71% Dog	87.56% Dog	Dog	Dog
2	Cat	91.67% Cat	99.25% Cat	99.97% Cat	Cat	Cat
3	Cat	83.52% Cat	76.05% Cat	90.85% Dog	Cat	Dog
4	Dog	78.13% Dog	95.72% Dog	88.51% Dog	Dog	Dog
5	Dog	98.66% Dog	86.23% Cat	70.15% Cat	Cat	Dog

probability. Which method would give a most accurate result? At this time, it seems very difficult to artificially set an optimum method, because, usually, the relationship between variates is polybasic, higher order and nonlinear. As Fig. 2 shows, the kind of helically curve relationship can be learned well by the relation network while the artificial metric is completely out of work, so it's hard to generalize an effective method based on artificial logic [11, 12], in many cases. Therefore, Tri-STNN use a single neural network called Judge Network to learn the complicated relationship between outputs of 3 agents and the true labels. As Fig. 1 shows, the Judge Network was placed behind the 3 agents.

Like other artificial neural networks, the Judge Network needs to be trained firstly. It is easy to generate plenty of datasets for training Judge Network by testing the training sets. A well-trained Judge Network can make a better performance than traditional ways such as voting on the task of making decisions.

Retrain In order to make the agents Fine-Tuning continuously, an algorithm is set to judge whether one or more agents need to be retrained after a certain prediction. Let P_i denote the output of agent I, P_j denote the output of the Judge Network and β is a constant which is set according to different tasks. If $|P_j - P_i| > \beta$, the output from Judge Network is taken as the label of the new sample, then retrain the Classifier I with the new sample. Unlike Tri-training which only retrain a certain agent in certain retraining [4], Tri-STNN may retrain all agents, so Tri-STNN has a better learning ability.

It is the common problem that the retraining would disturb the previous knowledge learned by agents [1, 7]. In order to avoid the forgetting, in this paper, some labeled samples are picked up randomly from the training datasets, and then they are mixed with new samples to retrain the certain agent. The way of replaying can prevent the networks from running to extreme and forgetting previous knowledge when retraining.

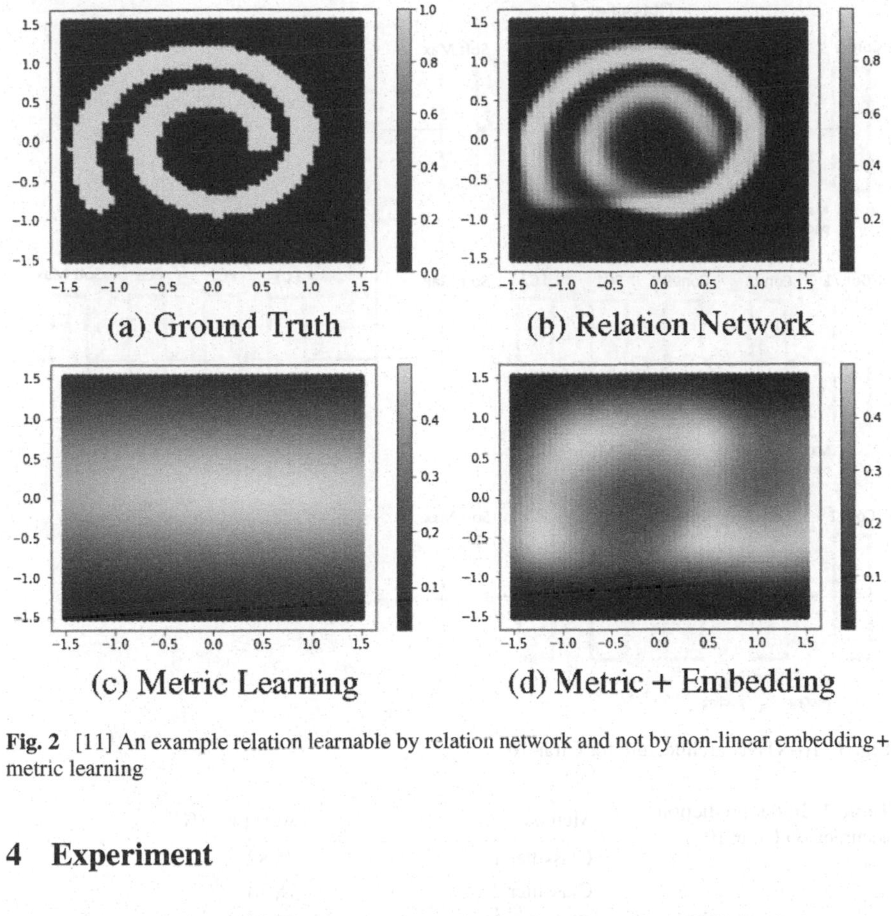

Fig. 2 [11] An example relation learnable by relation network and not by non-linear embedding + metric learning

4 Experiment

We evaluate our approaches on a related task: Classification on CIFAR-10 data sets. Three classifiers and a Judge Network were designed based on Tri-STNN, and the experiment was implemented based on TensorFlow.

Network Architecture Similar to most of image classifiers, convolutional neural networks are used as classifiers [13], in this experiment. As Fig. 3 shows, we utilized 3 convolutional blocks, and each convolutional block contains a 16-filter 3×3 convolution, a batch normalization and a ReLU nonlinearity layer respectively. The first two blocks also contain a 3×3 max-pooling layer while the latter do not. Two full connection layers and a Softmax layer were placed behind the convolutional blocks. The judge network is composed of 3 full connection layers and a Softmax layer.

Cifar-10 datasets There are 60,000 color images with the size 32×32 in this datasets which are divided into 10 kinds. 50,000 of these are used to train classifiers, and the other 10,000 images are used to test classifiers.

Training parameters When classifiers are trained, the learning rate was set to 0.005, and the max-step was set to 20,000. When the judge network is trained, the learning

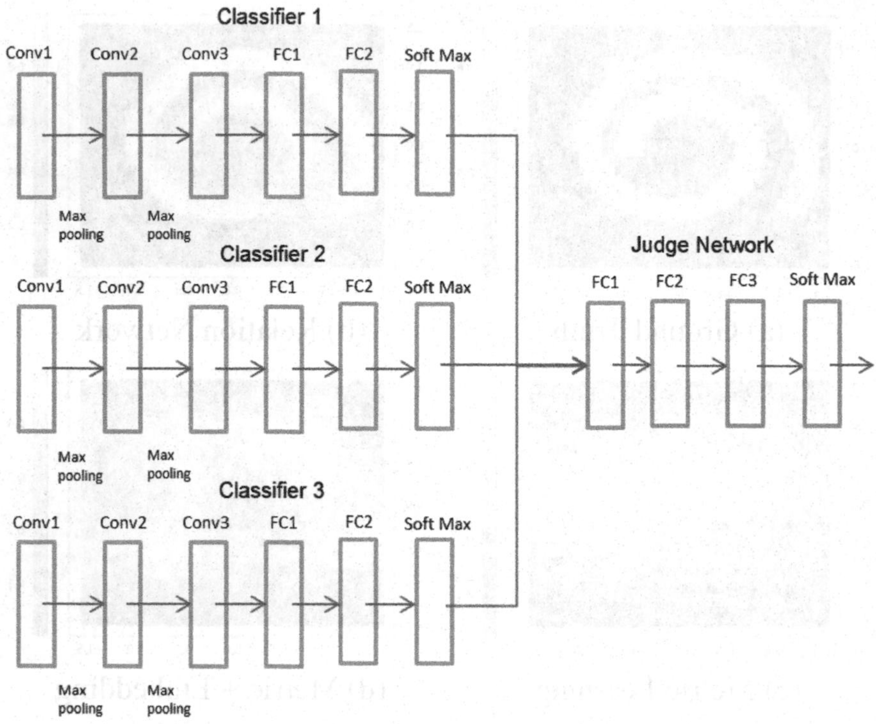

Fig. 3 Tri-STNN architecture for Cifar-10

Table 2 Initial prediction accuracy on Cifar-10

Method	Accuracy (%)
Classifier 1	78.82
Classifier 2	78.33
Classifier 3	77.96
Max-voting	82.35
Judge net	86.51

rate is set to 0.0001, and the max-step is set to 10,000. At the section of retraining, β is set to 0.35, 100 steps are allowed during each retraining.

Result Table 2 shows the prediction results about testing sets. Obviously, the prediction accuracy of Judge Network is higher than the method of voting and each of the 3 classifiers.

As Table 3 shows, the prediction accuracy keep rising continually as the number of iterations increases. The result shows that the generalization ability to the test datasets of Tri-STNN become better and better, at former stage. And yet the prediction accuracy is hard to increase with the quantity of test over 28,000. The 30,000 iteration and the 32,000 iteration show the similar accuracy, because it is limited by the depth of artificial neural networks and the quantity of previous knowledge.

Table 3 Prediction accuracy following iterations on Cifar-10

Iterations	Classifier 1 (%)	Classifier 2 (%)	Classifier 3 (%)	Max-voting (%)	Judge net (%)
2000	78.87	78.53	77.97	82.42	86.77
4000	78.95	78.74	80.19	82.52	86.81
6000	80.05	79.22	80.19	82.55	87.35
8000	80.26	80.12	80.27	82.60	87.12
10,000	80.51	80.37	80.46	83.91	87.99
12,000	80.56	80.56	80.51	84.03	88.27
28,000	83.75	83.72	83.73	85.09	90.15
30,000	83.72	83.73	83.75	85.09	90.13
32,000	83.73	83.73	83.74	85.08	90.15

5 Conclusion

In this paper, Tri-STNN was described in detail. Tri-STNN is constituted by three agents and a Judge Network. Every agent can make a prediction for a certain sample independently. The Judge Network is capable of deciding which prediction is right from the outputs of three agents. From the experiment on Cifar-10, it is easy to draw the conclusion that Judge Network has a better performance than traditional methods in multi-agents system. The method of taking three agents working together to promote each other is very effective for Lifelong-learning and self-taught learning.

Acknowledgements We would like to acknowledge Kevin Xu for release his TensorFlow Tutorial code, that help us finish the experiment quickly.

References

1. G.I. Parisi, R. Kemker, J.L, Part, et al., Continual lifelong learning with neural networks: a review (2018)
2. H. Edwards, A. Storkey, Towards a neural statistician (2016)
3. K. Nigam, R. Ghani, Analyzing the effectiveness and applicability of co-training, in *International Conference on Information and Knowledge Management* (ACM, 2000), pp. 86–93
4. Z.H. Zhou, M. Li, Tri-Training: exploiting unlabeled data using three classifiers. IEEE Trans. Knowl. Data Eng. **17**(11), 1529–1541 (2005)
5. C. Cortes, X. Gonzalvo, V. Kuznetsov, et al., AdaNet: adaptive structural learning of artificial neural networks (2016)
6. Y. Bengio, Deep learning of representations for unsupervised and transfer learning, in *Workshop on Unsupervised and Transfer Learning*, vol. 7 (2012), pp. 1–20
7. I.J. Goodfellow, M. Mirza, D. Xiao et al., An empirical investigation of catastrophic forgetting in gradient-based neural networks. Comput. Sci. **84**(12), 1387–1391 (2013)
8. Y. Jia, Robust control with decoupling performance for steering and traction of 4WS vehicles under velocity-varying motion. IEEE Trans. Control Syst. Technol. **8**(3), 554–569 (2000)
9. H. Shin, J.K. Lee, J. Kim, et al., Continual learning with deep generative replay (2017)

10. J. Turian, L. Ratinov, Y. Bengio, Word representations: a simple and general method for semi-supervised learning, in *ACL 2010, Proceedings of the, Meeting of the Association for Computational Linguistics, July 11–16, 2010* (DBLP, Uppsala, Sweden, 2010), pp. 384–394
11. F. Sung, Y. Yang, L. Zhang, et al., Learning to compare: relation network for few-shot learning (2017)
12. Y. Jia, Alternative proofs for improved LMI representations for the analysis and the design of continuous-time systems with polytopic type uncertainty: a predictive approach. IEEE Trans. Autom. Control **48**(8), 1413–1416 (2003)
13. A. Krizhevsky, I. Sutskever, G.E. Hinton, ImageNet classification with deep convolutional neural networks, in *International Conference on Neural Information Processing Systems* (Curran Associates Inc. 2012), pp. 1097–1105
14. K.P. Bennett, A. Demiriz, R. Maclin, Exploiting unlabeled data in ensemble methods, in *ACM International Conference on Knowledge Discovery and Data Mining* (2007), pp. 289–296
15. A. Soltoggio, K.O. Stanley, S. Risi, Born to learn: the inspiration, progress, and future of evolved plastic artificial neural networks (2017)
16. Z. Akata, F. Perronnin, Z. Harchaoui et al., Label-embedding for image classification. IEEE Trans. Pattern Anal. Mach. Intell. **38**(7), 1425–1438 (2016)
17. S.W. Lee, J.-H. Kim, J. Jun, et al., Overcoming catastrophic forgetting by incremental moment matching (2017)

A TT&C Resources Schedule Method Based on Markov Decision Process

Yi Wu, Maoyun Guo, Yi Chai, Haoxing Liang and Yiyao An

Abstract For the problem of how to improve the efficiency in the scheduling of Tracking, Telemetry and Command (TT&C) resources, the paper proposes a model of TT&C resources scheduling based on the Markov decision process. In this model, the total work time of TT&C equipment for the spacecraft is considered as the decision criterion and the adjustment range of TT&C equipment are introduced. When a finite-stage backward recursive iterative algorithm is applied to solve the model, the optimal TT&C equipment scheduling strategy, which leads to effectively completion of the TT&C tasks, is obtained. And the TT&C equipment is adjusted according to the optimal strategy. Finally, simulation cases illustrate the proposed method.

Keywords TT&C resources · Markov decision process · Scheduling

1 Introduction

As the key of mission completion and flight safety of the spacecraft, the scheduling of TT&C resources has been concerned by scholars and engineering technicians. With the vigorous development of space exploration activities, the contradiction between the increasing demand for TT&C tasks and the limitation of available TT&C resources has become increasingly prominent. It is impractical to meet the demand of TT&C tasks relying solely on increasing the quantity of TT&C equipment. Therefore, research on scheduling methods for TT&C resources is needed to provide technical support for completion of TT&C tasks.

Y. Wu · M. Guo (✉) · Y. Chai · H. Liang · Y. An
School of Automation, Chongqing University, Chongqing 400044, China
e-mail: gmy@cqu.edu.cn

Y. Wu · M. Guo · Y. Chai · H. Liang · Y. An
The Key Laboratory of Complex System Safety and Autonomous Control, Ministry of Education, Chongqing University, 400044 Chongqing, China

Y. Jia et al. (eds.), *Proceedings of 2018 Chinese Intelligent Systems Conference*, Lecture Notes in Electrical Engineering 529,
https://doi.org/10.1007/978-981-13-2291-4_79

The biggest difference between the scheduling of TT&C resources and other resources is that it has a strict constraint of time window and is subject to constraints such as terrain. At present, the algorithms for the TT&C resources scheduling are mainly categorized as heuristic algorithms and intelligent search algorithms. LIU Yang provided the local optimal solution of the mission planning problem for the ground station in the satellite system using greedy algorithm, the goal of which is the maximum utilization of the resources [1]. Zhang presented a new algorithm of multi-satellite TT&C resource scheduling using ant colony optimization to minimize the working burden [2]. Hongsheng proposed a regression model based on SVM to describe the relationship of satellite station resource allocation [3]. Stephen Mack put forward an optimization strategy for planning satellite earth station based on genetic algorithm [4].

The paper takes various factors which causes the uncertainty into consideration, including the spacecraft's flight environment, different orbits and so on. A model of TT&C resources scheduling based on Markov decision process is proposed, and the decision criterion is total work time of all TT&C equipment for all spacecraft. Then the optimal adjustment strategy, related to task, is obtained as a result of solving the model by a finite-stage backward recursive iterative algorithm.

2 Introduction of Markov Decision Process and Problem Description

The Markov decision model is a sequential decision-making issue based on uncertainty [5]. For the Markov decision process, the decision maker selects an available decision according to the current system state at each decision-making moment, and then the system state is going to transfer to another with a certain transition probability and obtain a reward. In the Markov process, the reward received and the next state of the system are only related to the current state and decision while are independent of decision and state of the previous stage [6].

The model mentioned in the paper makes full use of the uncertainty of the TT&C state transition of the system, and obtains a scheduling strategy for TT&C resources. At a certain moment, only the TT&C equipment which has the longest visible time is selected for the same spacecraft. In consequence, the system can obtain the longest effective time of TT&C process.

As described in Fig. 1, TT&C equipment located in different places has different coverage range for the same spacecraft.

The TT&C resources need to be scheduled to meet the task requirements of different spacecraft. Lots of factors, such as the influence of terrain, TT&C equipment's azimuthal angles, the change of spacecraft's flight trajectory, external environmental factors, and whether the TT&C equipment and spacecraft works in a normal state or not, will affect the TT&C results. These factors should be taken into consideration in the TT&C resources scheduling. Then the paper presents a TT&C resources scheduling model based on the Markov decision.

3 Markov Decision Model for TT&C Resources Scheduling

For the TT&C system mentioned in this paper, it is assumed that one TT&C equipment can see many spacecraft. The TT&C equipment's azimuthal angle will affect the visual range of the TT&C equipment to the spacecraft, resulting in whether TT&C tasks can be completed or not. So it is feasible to complete the TT&C tasks by adjusting the azimuthal angle of the TT&C equipment. The research process is as following.

3.1 State Space

Assuming that the TT&C system currently needs to complete multiple TT&C tasks of n spacecraft, and m TT&C equipment can participate in the task. For the state of the TT&C system, set the state space as follows [7]:

$$S = \{s_{b_1 b_2 \cdots b_n} = (\theta_{11}, \theta_{12}, \theta_{13}, \ldots \theta_{1\mathrm{m}}, \theta_{21}, \theta_{22}, \theta_{23}, \ldots \theta_{2\mathrm{m}}, \ldots \theta_{\mathrm{n}1}, \theta_{\mathrm{n}2}, \theta_{\mathrm{n}3}, \ldots \theta_{\mathrm{nm}})\} \tag{1}$$

In formula (1), θ_{ij} denotes the state of the TT&C equipment j for aircraft i

i indicates the number of the aircraft
j indicates the number of the TT&C equipment.
In order to compress the state space, set the following rules:

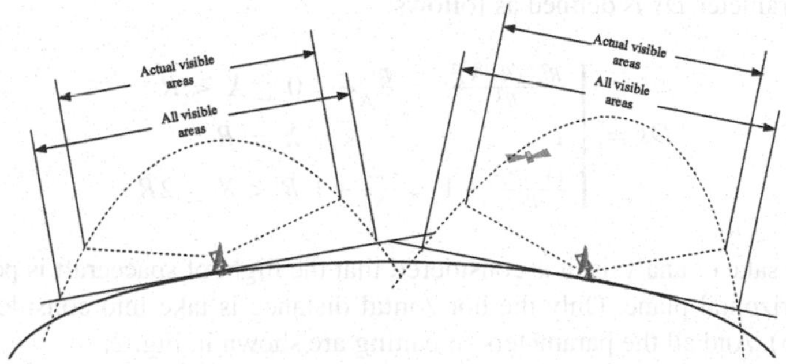

Fig. 1 Different visibility of different TT&C equipment

$$\theta_{ij} == \begin{cases} 0 \; the\; jth\; equipment\; does\; not\; obtain\; the\; flight\; data\; of\; spacecraft\; i \\ 1 \; the\; jth\; equipment\; obtains\; the\; flight\; data\; of\; spacecraft\; i \end{cases} \tag{2}$$

For the purpose of distinguishing different states in a more concise way, we make the following rules for $b_1, b_2 \ldots b_n$ in formula (1):

$$b_i = \begin{cases} j \; \theta_{ij} = 1 \; in \; formula(2), \forall i \in [1, n], j \in [1, m] \\ 0 \; \theta_{ij} = 0 \; in \; formula(2), \forall i \in [1, n], j \in [1, m] \end{cases} \tag{3}$$

In order to improve the utilization rate of TT&C resources and balance the burden of each TT&C equipment, only one TT&C equipment is occupied in this paper, which means every spacecraft of the system should satisfy:

$$\forall i \in [1, n], \sum_{j=1}^{m} \theta_{ij} \in [0, 1] \tag{4}$$

In the case of the same spacecraft simultaneously possibly visible to different TT&C equipment, in order to figure out which TT&C equipment is used for a specific spacecraft at a certain moment in the definition of state, following rules are established to make judgement.

It is necessary to process and extract information, including the orbits of the spacecraft and the position of TT&C equipment as well as spacecraft, before using parameters Ds (showed as formula 5) to determine the state of the TT&C system. Finally, the TT&C equipment which has the largest parameter Ds for certain spacecraft is selected, and the corresponding TT&C state for this spacecraft is set to 1, while other TT&C states for this spacecraft which come from other TT&C equipment are set as 0. If $\theta_{ij} = 1$, the TT&C equipment j is used to conduct an effective TT&C task of the spacecraft i. That is, the flight data of TT&C equipment j for the spacecraft i at the corresponding moment is loaded in actual application.

The parameter Ds is defined as follows:

$$Ds = \begin{cases} \frac{R^2-(R-X)^2}{R^2} = \frac{R^2-r^2}{R^2} & 0 \le X < R \\ 1 & X = R \\ \frac{(X-R)^2}{R^2} + 1 = \frac{r^2}{R^2} + 1 & R < X \le 2R \end{cases} \tag{5}$$

For the sake of analysis, it is considered that the flight of spacecraft is performed in one horizontal plane. Only the horizontal distance is take into consideration in formula (5). And all the parameters' meaning are shown in Fig. 2.

R is the maximum radius of the coverage range of the equipment for a specific spacecraft.

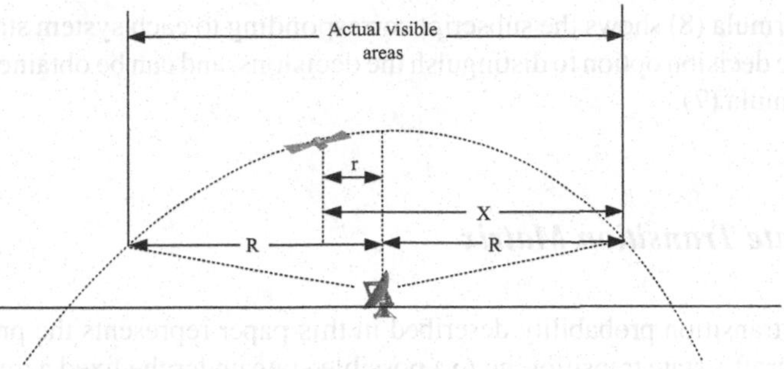

Fig. 2 The determination of the related parameter of the system state

r is the distance between the spacecraft and the TT&C equipment in the current horizontal plane of the spacecraft.

X is the horizontal distance between the spacecraft's current position and the farthest coverage boundary of the TT&C equipment to be passed.

3.2 The Set of Feasible Decision Option

The combination of azimuthal angles adjustment in one task at a moment is taken as an element in the set of feasible decision option. The limitation of the azimuthal angle adjustment range of the TT&C equipment can ensure the minimum adjustment in the TT&C resource scheduling effectively. It is assumed that the adjustment strategy for each TT&C equipment is divided into two cases: increase or decrease in the range of $(0°, \theta°)$. As a consequence, a reasonable model can be established on the basis of reasonable set of θs in which adjustment angles of the TT&C equipment can be constrained within a certain range. The set of feasible decision option is defined as follows:

$$A = \{a_d = (d_1, d_2, \ldots, d_n)\} \tag{6}$$

Among them

$$d_j = \begin{cases} 0\ the\ azimuth\ of\ jth\ equipment\ decreases(0°, \theta°) \\ 1\ the\ azimuth\ of\ jth\ equipment\ increases(0°, \theta°) \end{cases} \tag{7}$$

$$d = \sum_{j=1}^{n} 2^{n-j} d_j + 1 \tag{8}$$

The formula (8) shows the subscript corresponding to each system state in the set of feasible decision option to distinguish the decisions, and can be obtained according to the formula (7).

3.3 State Transition Matrix

The state transition probability described in this paper represents the probability of the spacecraft's state transitioning to a possible state under the fixed azimuthal angle of a TT&C equipment, it plays an important role in the scheduling of the TT&C resources. The state transition probability of the TT&C system can be obtained from relevant historical data of the relationship between trajectory of the spacecraft and the azimuthal angle of TT&C equipment. The specific generation of state transition probabilities should take the current state, the state transforms to with the taken decision, and the decisions taken to move to the following state into account.

3.4 Reward

The scheduling model of TT&C resource in this paper schedules TT&C equipment aiming to achieve the longest total time of TT&C process. Therefore, the total time of the effective TT&C process of each spacecraft from all the TT&C equipment is selected as the decision criteria. The longest time of a specific spacecraft under a TT&C equipment is expressed by t_{lij}. The actual time t_{ij} of a certain TT&C equipment for a specific spacecraft at the current azimuthal angle can be given precise description using parameters in Sect. 1.3.1 as following:

$$t_{ij} = \frac{Ds \bullet t_{lij}}{2} = \begin{cases} 0 & X = 0 \\ \frac{R^2 - r^2}{2R^2} \bullet t_{lij} & 0 < X < R \\ \frac{1}{2} \bullet t_{lij} & X = R \\ (\frac{r^2}{2R^2} + \frac{1}{2}) \bullet t_{lij} & R < X < 2R \\ t_{lij} & X = 2R \end{cases} \quad (9)$$

For the accurate modeling, the paper use θ_{ij} in formula (1) as a weight for the above-mentioned length of time interval t_{ij}. Under the current azimuthal angles of all the TT&C equipment, the system's reward, the total time of the TT&C process can be described as:

$$r_k = \sum_{i=1}^{n} \sum_{j=1}^{m} \theta_{ij} t_{ij} \quad (10)$$

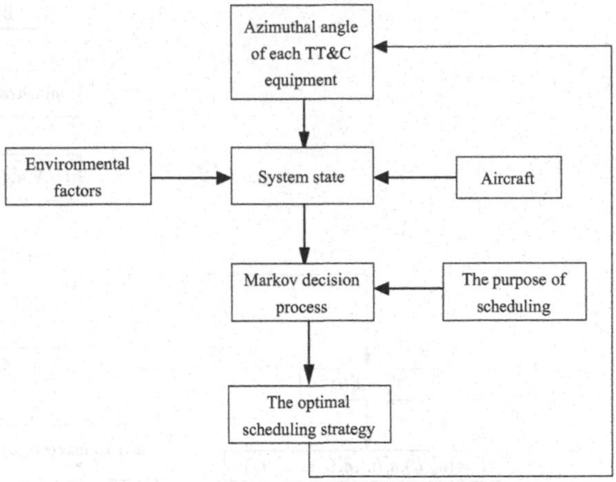

Fig. 3 Diagram of scheduling

In formula (10), r_k means the reward of system at the kth decision-making moment.

3.5 *Decision-Making Moment and Decision Cycle*

The choice of decision-making moment and decision cycle depends on the actual conditions such as the number of TT&C equipment, the average flight time of the spacecraft and so on. If the decision cycle is too long, it is probable to make the system model unable to obtain the optimal solution which means that the optimal scheduling strategy for TT&C resource is not able to generate. If the decision cycle is too short, the optimal solution can be obtained, but it will increase the calculation of the model with a lot of unnecessary calculations. The number of decision cycles N in the decision process should satisfy $N = A \times m$, A is any positive integer and m can be found in Sect. 1.3.

3.6 *The Process of TT&C Resources Scheduling*

According to the model in the paper, essentially, the selection of a decision is a selection of a group of azimuthal angles of the TT&C equipment, under which the defined system state will change with a certain transition probability. Simultaneously, a reward can be obtained and it can be considered as a 'feedback' that participate in the selection of the next TT&C decision at the next decision-making moment.

To make the Markov decision model established for the scheduling of TT&C resources more clear, the following scheduling procedures of TT&C resources and the algorithm for solving the optimal scheduling strategy are determined (Fig. 3).

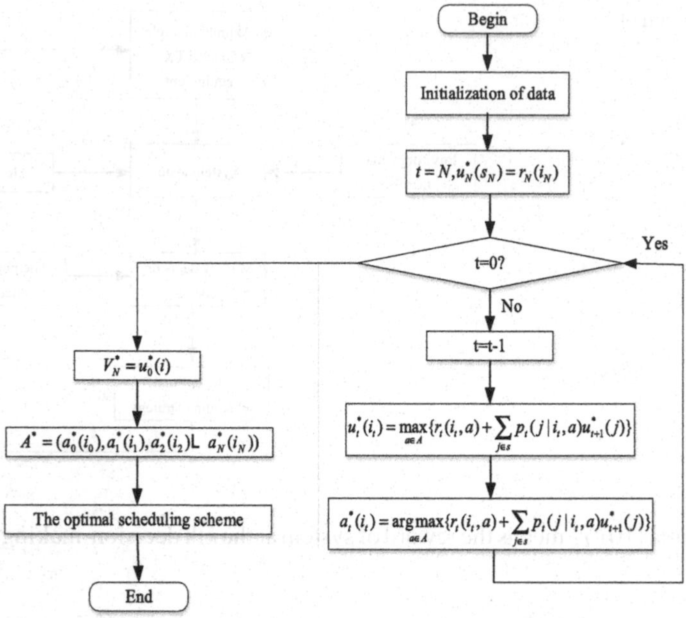

Fig. 4 Diagram of Markov decision process

Figure 4 shows the decision process of the proposed model in the paper, that is how to use the finite-stage backward recursive iterative algorithm to solve the Markov process's finite-stage model.

4 Simulation and Analysis

The method and algorithm mentioned in this paper were implemented in MATLAB.

In the simulation, it is supposed that there are two spacecraft, five TT&C equipment in the TT&C system, the departure time of the two spacecraft is not the same as start time of the system simulation, and the start time of different spacecraft may have a certain time interval, so the uncertainty of the system can be highlighted.

4.1 Initial Data

① The state space is $S = \{s_{b_1 b_2} = (\theta_{11}, \theta_{12}, \theta_{13}, \theta_{14}, \theta_{15}, \theta_{21}, \theta_{22}, \theta_{23}, \theta_{24}, \theta_{25})\}$, all the possible states of the TT&C system are listed as following, with a simple representation described in part Sect. 1.3:

Table 1 The initial data

t_{ij}/min	No measurement	$j = 1$	$j = 2$	$j = 3$	$j = 4$	$j = 5$
$i = 1$	0	7	6	5	7	8
$i = 2$	0	6	5	9	6	7

$$\begin{gathered} s_{00}, s_{01}, s_{02}, s_{03}, s_{04}, s_{05}, s_{10}, s_{11}, s_{12}, \\ s_{13}, s_{14}, s_{15}, s_{20}, s_{21}, s_{22}, s_{23}, s_{24}, s_{25}, \\ s_{30}, s_{31}, s_{32}, s_{33}, s_{34}, s_{35}, s_{40}, s_{41}, s_{42}, \\ s_{43}, s_{44}, s_{45}, s_{50}, s_{51}, s_{52}, s_{53}, s_{54}, s_{55} \end{gathered} \tag{11}$$

② The set of feasible decision option is $A = \{a_d = (d_1, d_2, d_3, d_4, d_5)\}$, all the possible decisions which the system can take according to the mentioned regulations are listed as following, with a simple representation described in part Sect. 1.3

$$\begin{gathered} a_1, a_2, a_3, a_4, a_5, a_6, a_7, a_8, a_9, \\ a_{10}, a_{11}, a_{12}, a_{13}, a_{14}, a_{15}, a_{16}, a_{17}, \\ a_{18}, a_{19}, a_{20}, a_{21}, a_{22}, a_{23}, a_{24}, a_{25}, \\ a_{26}, a_{27}, a_{28}, a_{29}, a_{30}, a_{31}, a_{32} \end{gathered} \tag{12}$$

The tentative value of θ is set as 30°.

③ In the simulation, the transition probability is set manually. It satisfies some basic transfer rules while satisfy the basic condition, $\sum_{j \in S} p(j|i, a) = 1$, of the state transition matrix.

④ Due to the blindness of manual set, *Ds* is set as a fixed value at any moment in the simulation so that $t_{lij} = t_{ij}$, finally the reward is described as:

$$r_k = \sum_{i=1}^{n} \sum_{j=1}^{m} \theta_{ij} t_{ij} = \sum_{i=1}^{n} \sum_{j=1}^{m} \theta_{ij} t_{lij} \tag{13}$$

⑤ During the flight of the spacecraft, 10 decision cycles are set to make decisions in the scheduling of the TT&C resources.

4.2 Experimental Results and Analysis

Some initial simulation data are shown in Table 1.

For the case that the state of the system at the time of beginning is characterized as s_{45}, the optimal decisions of the system at 10 decision-making moments can be obtained as $(a_{32}, a_2, a_2, a_2, a_2, a_2, a_{26}, a_2, a_{26}, a_{26})$. In other words, the system takes decision a_{32} at the first decision-making moment, takes decision a_2 at the second,

Table 2 Comparisons of simulation results

Decisions		Take the optimal adjustment strategy	Take decision a_{26} at every needed moment	Take decision a_{27} at every needed moment	Take decision a_{29} at every needed moment
Total reward/min	Initial state is s_{45}	142.1035	137.6135	138.5621	138.4859
	Initial state is s_{41}	141.6884	135.6971	136.8584	137.3936

third, fourth, fifth, sixth, and eighth decision-making moment, and takes decision a_{26} at the seventh, ninth and tenth decision-making moment. According to the definition in Sect. 1.3, it can be concluded that when the system takes the decision a_{26}, the TT&C equipment numbered 1, 2, 5 adopts the adjustment of increasing the azimuthal angle of $(0°, 30°)$, and the TT&C equipment numbered 3 and 4 adopts the adjustment of decreasing the azimuthal angle of $(0°, 30°)$, the rest details can be obtained in the same way.

For the case that the state of the system at the time of beginning is characterized as s_{41}, the optimal decisions of the system at 10 decision-making moments can be obtained as $(a_{25}, a_9, a_{25}, a_9, a_{25}, a_{15}, a_9, a_{15}, a_9, a_9)$. The detailed explanation of this optimal strategy can be referred using the method of the above case.

In order to demonstrate the optimality of the decisions based on the established model and related algorithms as well as the longest total time of TT&C process, the following comparisons are made.

From the Table 2, it can be seen that, for the current data, the optimal scheduling strategy obtained by solving the scheduling model described in the paper can be used to make the total time of the TT&C process for multiple spacecraft the longest among different strategies. It can be obtained from:

$$142.1035 > 138.5621 > 138.4859 > 137.6135$$
$$141.6884 > 137.3936 > 136.8584 > 135.6971$$

Based on the above data, it is concluded that the optimal scheduling strategy of the TT&C system can be obtained by solving the scheduling model of TT&C resources based on the Markov decision process, succeeding at the goal of the longest total effective time of TT&C process. During modeling, the total scheduling adjustment of the TT&C system is in the minimum level due to the constraints of the system state as well as the adjustment angles of the TT&C equipment which are less than 30°, at the same time, the system's resource utilization rate rises while the burden on individual TT&C resources decreases.

5 Conclusion

This paper proposes a method of establishing a model based on Markov decision process for the complex issue of TT&C resource scheduling. In order to obtain the optimal scheduling strategy of the TT&C resources, it solves the model with a finite-stage backward recursive iterative algorithm. The paper provides a new idea of solving the problem of TT&C resources scheduling. However, due to the complexity of the scheduling problem itself, the optimization of the state space and the determination of the optimal decision cycle still need to be further explored and studied.

References

1. L. Yang, C. Y.-w, T. Yue-jin, Mission planning method of the satellite ground station based on the greedy algorithm. Syst. Eng. Electron. **25**(10), 1239–1241 (2003)
2. Z. Zhang, N. Zhang, Z. Feng, Multi-satellite control resource scheduling based on ant colony optimization (Pergamon Press Inc., 2014)
3. F. Hongsheng, C. Yang, W. Xiaoyue, SVM regression model for satellite ground station resources allocation. J. Spacecraft TT&C Technol. **30**(2), 15–19 (2011)
4. S. Mack, Optimization of a satellite earth station layout using genetic algorithms. Int. J. Electron. Comput. Commun. Technol. **2**(4) (2012)
5. W. Wei, L. Mao, W. Li, The dynamic optimal method of emergency resources deployment planning based on markov decision process. Acta Scientiarum Naturalium University Nankaiensis **3**, 18–23 (2010)
6. K. Liu, *Applied Markov Decision Processes* (Tsinghua University Press, 2004)
7. A. Liu, K. Liu, G. Liu, Modeling scheduling problem by markov decision process, in *The National Youth Conference on Information and Management Sciences* (2011)
8. Y. Qingqing, S. Huairong, S. Qiongling, Research overview on modeling and solution of aerospace TT&C scheduling problem. J. Syst. Simul. **27**(1), 1–12 (2015)
9. H. Guan, Y.-Z. Ning, The development course and an outlook on chinese satellite TT&C. Space Int. 2018(1)

On Finite-Time Stability of Switched Homogeneous Systems

Bin Zhang

Abstract The finite-time stability is investigated for switched homogeneous systems. It is assumed that each subsystem possesses a homogeneous Lyapunov-like function. The derivative of the function is with hybrid homogenous degrees. Two substantially different situations are considered and different sufficient conditions are provided, respectively.

Keywords Finite-time stability · switched systems · homogeneous degree

1 Introduction

Switched nonlinear systems are widely considered in engineering practice to represent a system with parameter jump and device conversion [1, 2]. A switched system is essentially a hybrid system that consists of a family of subsystems and a switching law. The stability of the switching system is determined both by the individual stability of each mode and the logic of the switching law. The research achievements on the stability problem of switched systems are fruitful in the recent years, refer to the excellent works [3–5] and references therein.

Two general approaches to the stability problem of switched systems are common Lyapunov function (CLF) technique and multiple Lyapunov functions (MLFs) technique. The CLF technique has been effectively used in many situations [6, 7]. A switched system with a CLF remains stable for any switching laws. Therefore, the CLF technique is naturally used when there is no a priori hypothesis of the switching law. However, the constructive problem of a CLF for general switched systems has not been solved.

B. Zhang (✉)
The School of Automation, Beijing University of Posts and Telecommunications, Beijing 100876, China
e-mail: zb362301@126.com

Y. Jia et al. (eds.), *Proceedings of 2018 Chinese Intelligent Systems Conference*, Lecture Notes in Electrical Engineering 529,
https://doi.org/10.1007/978-981-13-2291-4_80

MLFs technique relaxes the constraint conditions of CLF. In [8], it is shown that if the Lyapunov function of each mode is decreasing and the energy is decreasing at switching times, then the switched system is asymptotically stable. In [9], the MLFs condition is relaxed by introducing the concept of weak Lyapunov functions (WLFs). An extension of the invariance principle is provided relative to dwell time switched solutions. In [10], union/intersection WLFs techniques are presented, where more accurate convergence region is obtained. In these works, maximal ratio coefficient is required among the Lyapunov functions. More specifically, for any subsystems i and j, it is assumed that $V_i \leq \phi V_j$ with $\phi \geq 1$. However, it is not easy to get the estimation of ϕ. Especially, the existence of ϕ is not clear in many situations.

It is worth noting that homogenous theory can give simplified conditions for stability analysis of switched nonlinear systems, where the value of ϕ is obtained accurately. In [11], stability problem of switched homogeneous systems is addressed using semi-tensor product of matrices and LMI conditions are achieved. In [12], homogeneous Lyapunov function is constructed and stability analysis via both CLF and MLFs are given. Some other results on this topic can be found in [13–15]. In comparison with the existing results where single homogenous degree is considered, in this paper, we consider switched homogenous systems with hybrid homogenous degrees. That is to say, we consider homogenous switched systems with Lyapunov function $\dot{V}_i(x) \leq -pV_i^{\theta}(x) \pm qV_i(x)$, $0 < \theta < 1$, rather than $\dot{V}_i(x) \leq -pV_i^{\theta}(x)$. Recently, nonlinear systems with hybrid homogenous degrees have attracted a considerable attention [16, 17]. However, such systems under switched conditions have not been investigated. This problem is treated in this paper. We extend the homogenous results to the case with hybrid homogenous degrees and sufficient conditions are obtained for finite-time stability.

2 Preliminaries

Consider the following switched nonlinear system

$$\dot{x} = f_{\sigma}(x), \quad t \geq t_0, \quad t_0 \in \mathbb{R} \tag{1}$$

where $x \in \mathbb{R}^n$ is the state vector, $\sigma(t) : [0, \infty) \rightarrow \mathcal{P} = \{1, 2, \ldots, p\}$ denotes the piecewise constant switching signal, which is continuous from the right, i.e., $\sigma(t+) = \lim_{s \downarrow t} \sigma(s)$. For $\forall i \in \mathcal{P}, f_i$ is a smooth function with $f_i(0) = 0$. Let $\{t_1, t_2, \ldots\}$ be the switching sequence. Then, it follows that $\sigma(t_i)$ is active in $[t_i, t_{i+1})$. Throughout, we adopt the following assumption.

Assumption 1 For the switching sequence $\{t_1, t_2, \ldots\}$, there exists positive constant τ such that $t_i - t_{i-1} = \tau$, $\forall i \in \{1, 2, \ldots\}$.

Definition 1 Let $x = (x_1, \ldots, x_n)^T \in \mathbb{R}^n$. A function $V : \mathbb{R}^n \rightarrow \mathbb{R}$ is called homogeneous of degree $\mu \in \mathbb{R}$ with respect to $r = (r_1, \ldots, r_n)^T \in \mathbb{R}_+^n$ if

$$V(\lambda^{r_1}x_1, \ldots, \lambda^{r_n}x_n) = \lambda^{\mu} V(x_1, \ldots, x_n), \quad \forall x \in \mathbb{R}^n, \quad \forall \lambda \in \mathbb{R}_+ \tag{2}$$

Definition 2 Let $x = (x_1, \ldots, x_n)^T \in \mathbb{R}^n$. A vector field $f(x) = (f_1(x), \ldots, f_n(x))^T : \mathbb{R}^n \to \mathbb{R}^n$ is called homogeneous with respect to $r = (r_1, \ldots, r_n)^T \in \mathbb{R}^n_+$ if for each $i \in \{1, \ldots, n\}$

$$f_i\left(\lambda^{r_1}x_1, \ldots, \lambda^{r_n}x_n\right) = \lambda^{\kappa + r_i} f_i(x_1, \ldots, x_n), \quad \forall x \in \mathbb{R}^n, \quad \forall \lambda \in \mathbb{R}_+ \tag{3}$$

holds for some constant $\kappa \geq -\min_{1 \leq i \leq n} r_i$. The constant κ is called the degree of homogeneity. A time-invariant system $\dot{x} = f(x)$ is called homogeneous if its vector filed $f(x)$ is homogeneous.

Assumption 2 For each $i \in \mathcal{P}$ of switched system (1), there exists $\mathcal{C}^1$ homogeneous function $V_i(x) : \mathbb{R}^n \to \mathbb{R}$ of degree $\mu \in \mathbb{R}$ with respect to $r = (r_1, \ldots, r_n)^T \in \mathbb{R}^n_+$ such that

$$\alpha \Lambda^{\mu}(x) \leq V_i(x) \leq \beta \Lambda^{\mu}(x) \tag{4}$$

$$V_i(x) \leq \phi V_j(x), \quad \forall i, j \in \mathcal{P} \tag{5}$$

where $\Lambda(x) = \sum\limits_{i=1}^{n} |x_i|^{\frac{1}{r_i}}$, $\alpha = \min\limits_{i \in \mathcal{P}} \min\limits_{\Lambda(x)=1} V_i(x)$, $\beta = \max\limits_{i \in \mathcal{P}} \max\limits_{\Lambda(x)=1} V_i(x)$, and $\phi = \max\limits_{i,j \in \mathcal{P}} \max\limits_{\Lambda(x)=1} \frac{V_i(x)}{V_j(x)}$.

Remark 1 Condition (4) is a direct corollary of function $V_i(x)$, in view of the homogeneous property. In (2), we choose $\lambda = \frac{1}{\Lambda(x)}$, it follows that

$$V_i(x_1, \ldots, x_n) = \frac{1}{\lambda^{\mu}} V_i(\lambda^{r_1}x_1, \ldots, \lambda^{r_n}x_n) = \Lambda^{\mu}(x) V_i(\xi_1, \ldots, \xi_n) \tag{6}$$

where $\xi_i = \lambda^{r_i}x_i$. Let $\xi = (\xi_1, \ldots, \xi_n)^T$. We can get $\Lambda(\xi) = \sum\limits_{i=1}^{n} |\xi_i|^{\frac{1}{r_i}} = \lambda \sum\limits_{i=1}^{n} |x_i|^{\frac{1}{r_i}} = \lambda \Lambda(x) = 1$. Therefore, we conclude that $\alpha \leq V_i(\xi_1, \ldots, \xi_n) \leq \beta$, which implies (4) holds. Similarly, we have

$$\frac{V_i(x)}{V_j(x)} = \frac{\Lambda^{\mu}(x) V_i(\xi_1, \ldots, \xi_n)}{\Lambda^{\mu}(x) V_j(\xi_1, \ldots, \xi_n)} \leq \max_{i,j \in \mathcal{P}} \max_{\Lambda(x)=1} \frac{V_i(x)}{V_j(x)} \tag{7}$$

which implies (5) holds.

Remark 2 In the existing literature, switched systems with $\dot{V}_i \leq -pV_i^{\theta}$ for some $p, \theta \in \mathbb{R}$ are widely considered. Different kinds of sufficient conditions for switched stability have been presented. In this paper, we consider switched systems with hybrid homogeneous degrees, and as far as we know, there are no results on this kind of switched systems. Specifically, the following two different situations are considered.

Assumption 3 For the homogeneous function $V_i(x)$, $i \in \mathcal{P}$, defined in Assumption 2, we have

$$\frac{\partial V_i}{\partial x} f_i(x) \leq -pV_i^{\theta}(x) + qV_i(x) \tag{8}$$

where $p > 0$, $q > 0$, and $0 < \theta < 1$.

Assumption 4 For the homogeneous function $V_i(x)$, $i \in \mathcal{P}$, defined in Assumption 2, we have

$$\frac{\partial V_i}{\partial x} f_i(x) \leq -pV_i^{\theta}(x) - qV_i(x) \tag{9}$$

where $p > 0$, $q > 0$, and $0 < \theta < 1$.

3 Main Results

3.1 Stability with (8)

Lemma 1 *Let* $t^{''} > t^{'} \geq t_0$. *If function* $y(t) : \mathbb{R} \to \mathbb{R}$ *satisfies*

$$\dot{y}(t) = -py^{\theta}(t) + qy(t) \tag{10}$$

for $p > 0$, $q > 0$, *and* $0 < \theta < 1$. *Then, we can get that*

$$y(t^{''}) = e^{q(t^{''}-t^{'})}\left(y^{1-\theta}(t^{'}) - \frac{p}{q} + \frac{p}{q}e^{-(1-\theta)q(t^{''}-t^{'})}\right)^{\frac{1}{1-\theta}}. \tag{11}$$

Proof Multiplying (10) by e^{-qt}, we can get that

$$e^{-qt}\dot{y}(y) - qe^{-qt}y(t) = -pe^{-qt}y^{\theta}(t) \tag{12}$$

which implies

$$\frac{d(e^{-qt}y(t))}{dt} = -pe^{-qt}y^{\theta}(t). \tag{13}$$

Let $z(t) = e^{-qt}y(t)$. Then, we have that

$$\frac{dz}{dt} = -pe^{-(1-\theta)qt}z^{\theta}. \tag{14}$$

Integrating (14) on the time interval $(t^{'}, t^{''})$, we can get that

$$z^{1-\theta}(t^{''}) = z^{1-\theta}(t^{'}) + \frac{p}{q}\left(e^{-(1-\theta)qt^{''}} - e^{-(1-\theta)qt^{'}}\right) \tag{15}$$

which implies

$$y^{1-\theta}(t'') = e^{(1-\theta)q(t''-t')}y^{1-\theta}(t') + \frac{p}{q}\left(1 - e^{(1-\theta)q(t''-t')}\right). \tag{16}$$

It follows that

$$y(t'') = e^{q(t''-t')}\left(y^{1-\theta}(t') - \frac{p}{q} + \frac{p}{q}e^{-(1-\theta)q(t''-t')}\right)^{\frac{1}{1-\theta}} \tag{17}$$

which completes the proof.

Theorem 1 *Consider switched system* (1). *Assume that Assumptions* 1–3 *hold. Then, the origin of* (1) *is finite-time stable with initial value* $x(t_0)$ *if the time interval* τ *satisfies*

$$\frac{1}{e^{q\tau}}\left(\frac{p}{q}\frac{\phi(e^{(1-\theta)q\tau}-1)}{\phi e^{(1-\theta)q\tau}-1}\right)^{\frac{1}{1-\theta}} \geq V_{\sigma(t_0)}(x(t_0)). \tag{18}$$

Moreover, the setting time is $T = t_{\Pi}$, *where* $\Pi = \left[\frac{\ln(-\frac{\Upsilon}{\Psi})}{\ln(\phi e^{(1-\theta)q\tau})}\right]$.

Proof Using the homogeneous function $V_i(x), i \in \mathcal{P}$, we construct the multiple Lyapunov function $V_{\sigma(t)}(x)$ corresponding to switching law $\sigma(t)$. We can see $V_{\sigma(t)}(x)$ is piecewise smooth.

For any $t \geq t_0$, there exists time interval $[t_{j-1}, t_j), j \geq 1$, such that $t_{j-1} \leq t < t_j$. Integrating differential inequalities (8) successively on the intervals $[t_0, t_1)$, $[t_1, t_2)$, …, $[t_{j-1}, t]$, we can get

$$\begin{aligned}
V_{\sigma(t_{j-1})}^{1-\theta}(x(t)) &\leq e^{(1-\theta)q(t-t_{j-1})}V_{\sigma(t_{j-1})}^{1-\theta}(x(t_{j-1})) + \frac{p}{q}\left(1 - e^{(1-\theta)q(t-t_{j-1})}\right)\\
V_{\sigma(t_{j-2})}^{1-\theta}(x(t_{j-1})) &\leq e^{(1-\theta)q(t_{j-1}-t_{j-2})}V_{\sigma(t_{j-2})}^{1-\theta}(x(t_{j-2})) + \frac{p}{q}\left(1 - e^{(1-\theta)q(t_{j-1}-t_{j-2})}\right)\\
&\cdots\cdots\\
V_{\sigma(t_0)}^{1-\theta}(x(t_1)) &\leq e^{(1-\theta)q(t_1-t_0)}V_{\sigma(t_0)}^{1-\theta}(x(t_0)) + \frac{p}{q}\left(1 - e^{(1-\theta)q(t_1-t_0)}\right).
\end{aligned} \tag{19}$$

Taking into account condition (5), we obtain

$$V_{\sigma(t_i)}^{1-\alpha}(x(t_i)) \leq \phi V_{\sigma(t_{i-1})}^{1-\alpha}(x(t_i)) \tag{20}$$

for any $i = 1, \ldots, j-1$. Substituting (20) into (19) yields

$$
\begin{aligned}
V^{1-\theta}_{\sigma(t_{j-1})}(x(t)) &\le \phi e^{(1-\theta)q(t-t_{j-1})} V^{1-\theta}_{\sigma(t_{j-2})}(x(t_{j-1})) + \frac{p}{q}\left(1 - e^{(1-\theta)q(t-t_{j-1})}\right) \\
V^{1-\theta}_{\sigma(t_{j-2})}(x(t_{j-1})) &\le \phi e^{(1-\theta)q(t_{j-1}-t_{j-2})} V^{1-\theta}_{\sigma(t_{j-3})}(x(t_{j-2})) + \frac{p}{q}\left(1 - e^{(1-\theta)q(t_{j-1}-t_{j-2})}\right) \\
&\cdots\cdots \\
V^{1-\theta}_{\sigma(t_0)}(x(t_1)) &\le e^{(1-\theta)q(t_1-t_0)} V^{1-\theta}_{\sigma(t_0)}(x(t_0)) + \frac{p}{q}\left(1 - e^{(1-\theta)q(t_1-t_0)}\right).
\end{aligned}
\tag{21}
$$

It follows that

$$
\begin{aligned}
V^{1-\theta}_{\sigma(t_{j-1})}(x(t)) &\le \phi^{j-1} e^{(1-\theta)q(t-t_0)} V^{1-\theta}_{\sigma(t_0)}(x(t_0)) \\
&\quad + \frac{p}{q}\left(1 - e^{(1-\theta)q(t-t_{j-1})}\right) \\
&\quad + \frac{p}{q}\phi\left(1 - e^{(1-\theta)q(t_{j-1}-t_{j-2})}\right) e^{(1-\theta)q(t-t_{j-1})} \\
&\quad + \frac{p}{q}\phi^2\left(1 - e^{(1-\theta)q(t_{j-2}-t_{j-3})}\right) e^{(1-\theta)q(t-t_{j-2})} \\
&\quad \cdots\cdots \\
&\quad + \frac{p}{q}\phi^{j-2}\left(1 - e^{(1-\theta)q(t_2-t_1)}\right) e^{(1-\theta)q(t-t_2)} \\
&\quad + \frac{p}{q}\phi^{j-1}\left(1 - e^{(1-\theta)q(t_1-t_0)}\right) e^{(1-\theta)q(t-t_1)} \\
&\le \phi^{j-1} e^{(1-\theta)qj\tau} V^{1-\theta}_{\sigma(t_0)}(x(t_0)) \\
&\quad + \frac{p}{q}\frac{\phi(e^{(1-\theta)q\tau} - 1)}{\phi e^{(1-\theta)q\tau} - 1}(1 - \phi^{j-1} e^{(1-\theta)q(j-1)\tau}).
\end{aligned}
\tag{22}
$$

Therefore, we obtain that

$$
0 \le V^{1-\theta}_{\sigma(t_{j-1})}(x(t)) \le \Psi \phi^{j-1} e^{(1-\theta)q(j-1)\tau} + \Upsilon \tag{23}
$$

where $\Upsilon = \frac{p}{q}\frac{\phi(e^{(1-\theta)q\tau}-1)}{\phi e^{(1-\theta)q\tau}-1}$ and $\Psi = e^{(1-\theta)q\tau} V^{1-\theta}_{\sigma(t_0)}(x(t_0)) - \Upsilon$. It follows that

$$
0 \le \Lambda(x(t)) \le \frac{1}{\alpha^{\frac{1}{\mu}}}\left(\Psi \phi^{j-1} e^{(1-\theta)q(j-1)\tau} + \Upsilon\right)^{\frac{1}{(1-\theta)\mu}}. \tag{24}
$$

From (24) we can see that $\Psi < 0$, which implies $\Lambda(x(t)) = 0$ for $j > 1 + \left[\frac{\ln(-\frac{\Upsilon}{\Psi})}{\ln(\phi e^{(1-\theta)q\tau})}\right]$. Therefore, we conclude that $x(t) = 0$ for $t > t_\Pi$.

3.2 *Stability with* (9)

Lemma 2 *Let* $t'' > t' \geq t_0$. *If function* $y(t) : \mathbb{R} \to \mathbb{R}$ *satisfies*

$$\dot{y}(t) = -py^{\theta}(t) - qy(t) \tag{25}$$

for $p > 0$, $q > 0$, *and* $0 < \theta < 1$. *Then, we can get that*

$$y(t'') = e^{-q(t''-t')}\left(y^{1-\theta}(t') + \frac{p}{q} - \frac{p}{q}e^{(1-\theta)q(t''-t')}\right)^{\frac{1}{1-\theta}}. \tag{26}$$

Proof Multiplying (25) by e^{qt}, we can get that

$$e^{qt}\dot{y}(y) + qe^{qt}y(t) = -pe^{qt}y^{\theta}(t) \tag{27}$$

which implies

$$\frac{d(e^{qt}y(t))}{dt} = -pe^{qt}y^{\theta}(t). \tag{28}$$

Let $z(t) = e^{qt}y(t)$. Then, we have that

$$\frac{dz}{dt} = -pe^{(1-\theta)qt}z^{\theta}. \tag{29}$$

Integrating (29) on the time interval (t', t''), we can get that

$$z^{1-\theta}(t'') = z^{1-\theta}(t') - \frac{p}{q}\left(e^{(1-\theta)qt''} - e^{(1-\theta)qt'}\right) \tag{30}$$

which implies

$$y^{1-\theta}(t'') = e^{-(1-\theta)q(t''-t')}y^{1-\theta}(t') - \frac{p}{q}\left(1 - e^{-(1-\theta)q(t''-t')}\right). \tag{31}$$

It follows that

$$y(t'') = e^{-q(t''-t')}\left(y^{1-\theta}(t') + \frac{p}{q} - \frac{p}{q}e^{(1-\theta)q(t''-t')}\right)^{\frac{1}{1-\theta}} \tag{32}$$

which completes the proof.

Theorem 2 *Consider switched system* (1). *Assume that Assumptions* 1, 2 *and* 4 *hold. Then, we can have that:*

(1) If $\phi e^{-(1-\theta)q\tau} - 1 < 0$*, then the origin of* (1) *is globally finite-time stable for any initial values.*

(2) If $\phi e^{-(1-\theta)q\tau} - 1 > 0$*, then the origin of* (1) *is locally finite-time stable for initial value* $x(t_0)$ *satisfying* $V^{1-\theta}_{\sigma(t_0)}(x(t_0)) \le -\frac{p}{q}\frac{\phi e^{-(1-\theta)q\tau}(e^{-(1-\theta)q\tau}-1)}{\phi e^{-(1-\theta)q\tau}-1}$.

Moreover, the setting times for cases (1) and (2) are both $t_{\Pi'}$*, where* $\Pi' = \left[\frac{\ln\left(\frac{\Upsilon'}{\Psi'}\right)}{\ln\left(\frac{\phi}{e^{(1-\theta)q\tau}}\right)}\right]$.

Proof Using the homogeneous function $V_i(x), i \in \mathcal{P}$, we construct the multiple Lyapunov function $V_{\sigma(t)}(x)$ corresponding to switching law $\sigma(t)$. We can see $V_{\sigma(t)}(x)$ is piecewise smooth.

For any $t \ge t_0$, there exists time interval $[t_{j-1}, t_j), j \ge 1$, such that $t_{j-1} \le t < t_j$. Integrating differential inequalities (9) successively on the intervals $[t_0, t_1)$, $[t_1, t_2), \ldots,$ $[t_{j-1}, t]$, we can get

$$
\begin{aligned}
V^{1-\theta}_{\sigma(t_{j-1})}(x(t)) &\le e^{-(1-\theta)q(t-t_{j-1})} V^{1-\theta}_{\sigma(t_{j-1})}(x(t_{j-1})) - \frac{p}{q}\left(1 - e^{-(1-\theta)q(t-t_{j-1})}\right)\\
V^{1-\theta}_{\sigma(t_{j-2})}(x(t_{j-1})) &\le e^{-(1-\theta)q(t_{j-1}-t_{j-2})} V^{1-\theta}_{\sigma(t_{j-2})}(x(t_{j-2})) - \frac{p}{q}\left(1 - e^{-(1-\theta)q(t_{j-1}-t_{j-2})}\right)\\
&\cdots\cdots\\
V^{1-\theta}_{\sigma(t_0)}(x(t_1)) &\le e^{-(1-\theta)q(t_1-t_0)} V^{1-\theta}_{\sigma(t_0)}(x(t_0)) - \frac{p}{q}\left(1 - e^{-(1-\theta)q(t_1-t_0)}\right).
\end{aligned}
\tag{33}
$$

Taking into account condition (5), we obtain

$$
V^{1-\alpha}_{\sigma(t_i)}(x(t_i)) \le \phi V^{1-\alpha}_{\sigma(t_{i-1})}(x(t_i)) \tag{34}
$$

for any $i = 1, \ldots, j-1$. Substituting (34) into (33) yields

$$
\begin{aligned}
V^{1-\theta}_{\sigma(t_{j-1})}(x(t)) &\le \phi e^{-(1-\theta)q(t-t_{j-1})} V^{1-\theta}_{\sigma(t_{j-2})}(x(t_{j-1})) - \frac{p}{q}\left(1 - e^{-(1-\theta)q(t-t_{j-1})}\right)\\
V^{1-\theta}_{\sigma(t_{j-2})}(x(t_{j-1})) &\le \phi e^{-(1-\theta)q(t_{j-1}-t_{j-2})} V^{1-\theta}_{\sigma(t_{j-3})}(x(t_{j-2})) - \frac{p}{q}\left(1 - e^{-(1-\theta)q(t_{j-1}-t_{j-2})}\right)\\
&\cdots\cdots\\
V^{1-\theta}_{\sigma(t_0)}(x(t_1)) &\le e^{-(1-\theta)q(t_1-t_0)} V^{1-\theta}_{\sigma(t_0)}(x(t_0)) - \frac{p}{q}\left(1 - e^{-(1-\theta)q(t_1-t_0)}\right).
\end{aligned}
\tag{35}
$$

It follows that

$$\begin{aligned} V_{\sigma(t_{j-1})}^{1-\theta}(x(t)) &\le \phi^{j-1} e^{-(1-\theta)q(t-t_0)} V_{\sigma(t_0)}^{1-\theta}(x(t_0)) \\ &- \frac{p}{q}\left(1 - e^{-(1-\theta)q(t-t_{j-1})}\right) \\ &- \frac{p}{q}\phi\left(1 - e^{-(1-\theta)q(t_{j-1}-t_{j-2})}\right) e^{-(1-\theta)q(t-t_{j-1})} \\ &- \frac{p}{q}\phi^2\left(1 - e^{-(1-\theta)q(t_{j-2}-t_{j-3})}\right) e^{-(1-\theta)q(t-t_{j-2})} \\ &\cdots\cdots \\ &- \frac{p}{q}\phi^{j-2}\left(1 - e^{-(1-\theta)q(t_2-t_1)}\right) e^{-(1-\theta)q(t-t_2)} \\ &- \frac{p}{q}\phi^{j-1}\left(1 - e^{-(1-\theta)q(t_1-t_0)}\right) e^{-(1-\theta)q(t-t_1)} \\ &\le \phi^{j-1} e^{-(1-\theta)q(j-1)\tau} V_{\sigma(t_0)}^{1-\theta}(x(t_0)) \\ &- \frac{p}{q}\frac{\phi e^{-(1-\theta)q\tau}(e^{-(1-\theta)q\tau}-1)}{\phi e^{-(1-\theta)q\tau}-1}(1-\phi^{j-1}e^{-(1-\theta)q(j-1)\tau}). \end{aligned} \tag{36}$$

Therefore, we obtain that

$$0 \le V_{\sigma(t_{j-1})}^{1-\theta}(x(t)) \le \Psi' \phi^{j-1} e^{-(1-\theta)q(j-1)\tau} - \Upsilon' \tag{37}$$

where $\Upsilon' = \frac{p}{q}\frac{\phi e^{-(1-\theta)q\tau}(e^{-(1-\theta)q\tau}-1)}{\phi e^{-(1-\theta)q\tau}-1}$ and $\Psi' = V_{\sigma(t_0)}^{1-\theta}(x(t_0)) + \Upsilon'$. It follows that

$$0 \le \Lambda(x(t)) \le \frac{1}{\alpha^{\frac{1}{\mu}}}\left(\Psi' \phi^{j-1} e^{-(1-\theta)q(j-1)\tau} - \Upsilon'\right)^{\frac{1}{(1-\theta)\mu}}. \tag{38}$$

Then, we consider the following two cases:

Case 1: $\phi e^{-(1-\theta)q\tau} - 1 < 0$. We can see that the switched system (1) is globally finite-time stable with any initial value, and the setting time is $t_{\Pi'}$.

Case 2: $\phi e^{-(1-\theta)q\tau} - 1 > 0$. We can see that if $\Psi' < 0$, then switched system (1) is locally finite-time stable, and the setting time also is $t_{\Pi'}$.

4 Conclusion

In this paper, the finite-time stability problem of switched homogenous systems have been studied using the dwell-time scheme. Two substantially different situations with hybrid homogenous degrees have been introduced. Sufficient conditions and estimations of the setting time have been given under both situations.

References

1. R. Cardim, M.M. Teixeira, E. Assuncao, M.R. Covacic, Variable-structure control design of switched systems with an application to a DC-DC power converter. IEEE Trans. Ind. Electron. **56**(9), 3505–3513 (2009)
2. R. Olfati-Saber, R.M. Murray, Consensus problems in networks of agents with switching topolory and time-delays. IEEE Trans. Autom. Control **49**(9), 1520–1533 (2004)
3. D. Liberzon, A.S. Morse, Basic problems in stability and design of switched systems. IEEE Control Syst. Mag. **19**(5), 59–70 (1999)
4. D. Liberzon, *Switching in Systems and Control, Boston* (Birkhauser, MA, 2003)
5. Z. Sun, S.S. Ge, *Switched Linear Systems: Control and Design* (Spronger, London, UK, 2005)
6. K.S. Narendra, J. Balakrishnan, A common Lyapunov function for stable LTI systems with commuting A-matrices. IEEE Trans. Autom. Control **39**(12), 2469–2471 (1994)
7. L. Vu, D. Liberzon, Common Lyapunov functions for families of commuting nonlinear systems. Syst. Control Lett. **54**(5), 405–416 (2005)
8. M.S. Branicky, Multiple Lyapunov functions and other analysis tools for switched and hybrid systems. IEEE Trans. Autom. Control **43**(4), 475–482 (1998)
9. A. Bacciotti, L. Mazzi, An invariance principle for nonlinear switched systems. Syst. Control Lett. **54**(11), 1109–1119 (2005)
10. B. Zhang, Y. Jia, On weak-invariance principles for nonlinear switched systems. IEEE Trans. Autom. Control **59**(6), 1600–1605 (2014)
11. L. Zhang, S. Liu, H. Lan, On stability of switched homogeneous nonlinear systems. J. Math. Anal. Appl. **334**(1), 414–430 (2007)
12. A.Y. Aleksandrov, A.A. Kosov, A.V. Platonov, On the asymptotic stability of switched homogeneous systems. Syst. Control Lett. **61**(1), 127–133 (2012)
13. R. Kuiava, B.K. Matos, G.R. Razente, Finite-time stability of a class of continuous-time non-homogeneous switched systems. Nonlinear Anal.: Hybrid Syst. **26**, 101–114 (2017)
14. D. Holcman, M. Margaliot, Stability analysis of second-order switched homogeneous systems. SIAM J. Control Optim. **41**(5), 1609–1625 (2002)
15. Y. Orlov, Finite time stability of homogeneous switched systems, in *Proceedings of the 42nd IEEE Conference on Decision and Control* (Maui Hawaii USA, December 9–12, 2003), pp. 4271–4276
16. Y. Shen, X. Xia, Semi-global finite-time observers for nonlinear systems. Automatica **44**(12), 3152–3156 (2008)
17. Y. Shen, Y. Huang, J. Gu, Global finite-time observers for Lipschitz nonlinear systems. IEEE Trans. Autom. Control **56**(2), 418–424 (2011)

Data Acquisition Technology of Time Tagging for Pulsar Navigation

Hanwen Zhang, Qiang Chen, Xiaomin Bei and Hengbin Zhang

Abstract The spacecraft position in space can be calculated by time-of-arrival of X-ray photon signals. In paper a simulation test system is designed to verify the X-ray pulsar navigation principle on the ground. In the system integrated data acquisition device is used for the weak light pulse signal detection with high precisely timing. The device is composed of three circuit boards including time reference, data acquisition and signal processing which can carry out data acquisition of great magnitude and speed. Compared with the other simulation systems there are several aspects to be improved such as data acquisition speed, time tagging precision, data processing ability continuously, noise filtering customized, and data wireless communication. The technology is very significant for on-orbit flight application of X-ray pulsar navigation satellites in future. It can be also applied to other data acquisition systems with high speed, such as radar signal processing system, atomic nuclear physics signal processing system and weak light signal processing system.

Keywords Data acquisition · Time tagging · Pulsar navigation

H. Zhang
Measurement and Support Center,
Beijing Institute of Control Engineering, Beijing 100190, China
e-mail: hanwen96888@163.com

Q. Chen (✉) · X. Bei · H. Zhang
Qian Xuesen Laboratory of Space Technology, China Academy of Space Technology, Beijing 100094, China
e-mail: cq_01@sina.com

X. Bei
e-mail: beixiaomin@qxslab.cn

H. Zhang
e-mail: zhanghengbin@qxslab.cn

Y. Jia et al. (eds.), *Proceedings of 2018 Chinese Intelligent Systems Conference*, Lecture Notes in Electrical Engineering 529,
https://doi.org/10.1007/978-981-13-2291-4_81

1 Introduction

The spacecraft position in space can be calculated by time-of-arrival (TOA) of X-ray photon signals [1–4]. In order to verify this navigation principle on the ground, there were several simulation systems constructed. A small X-ray tube ground test environment based on a desktop platform was carried out in Godard Center of NASA [5, 6]. Liu proposed a semi-physical pulsar navigation simulation system [7, 8]. Due to the attenuation of X-ray in the atmosphere of the Earth and radiation hazards to the human body, Sun, Su and Wang proposed a single photon detection simulation system with visible light [9–11]. Peng invented an X-ray detector for pulsar navigation which converted X-ray signal into electrical signal and tagged the arrival time [12]. Sun designed a frequency simulator outputting signals like the pulsars which a GPS timing module was used for synchronizing by the pulse 1PPS.

But in the above systems, there are existing several aspects to be improved. Firstly, the data acquisition speed is limited that the high-speed and multi-channel signal acquisition at the same time are not realized, and the ADC acquisition speed is no more than GHz. Secondly, the signal processing speed is limited that there is no ping pang buffer with large capacity data, and the digital circuit bottleneck such as CPU and RAM chip is below GHz (bus frequency). Thirdly, the time tagging accuracy is limited because there are no accurate time synchronization and atomic clock maintenance. Finally, there are no wireless network to transmit high-speed data.

For the verification of spacecraft orbital position determination, a pulsar navigation ground test system is designed in this paper. Specially, a data acquisition device with wireless transmitting, large capacity data storage and processing, high accuracy time tagging is used for acquire the photons TOA. The hardware is divided into three parts as clock reference circuit, data acquisition circuit, and signal processing circuit.

2 Pulsar Navigation Ground Test System

The pulsar navigation ground test system is composed of four sections: two-dimensions control turntable, spacecraft orbit simulation device, optical system (light sources and optical detectors) and integrated data acquisition device, as shown in Figs. 1 and 2. In the system light sources are used to simulate the pulsars, single-axis turntable is used to simulate the satellite orbit, 2D control turntable is used to simulate the satellite attitude and the detectors pointing to the pulsar.

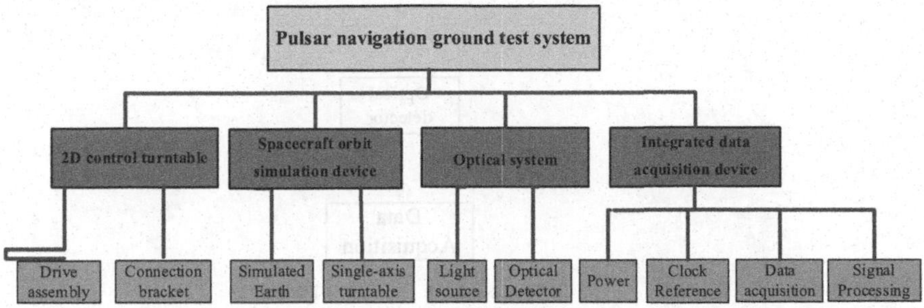

Fig. 1 Structure illustration of the pulsar navigation ground test system

Fig. 2 Physics field of the pulsar navigation ground test system

3 Hardware Design of Data Acquisition Device

In order to meet the requirements of pulsar navigation applications, a data acquisition device for weak light signals with time tagging precisely is designed. The device is used for data acquisition of photoelectric pulse signals, synchronization with GPS/BD time, and maintenance of atomic clock time. Noise filtering is performed on the board. large-capacity data for high-speed data acquisition is ping pang stored, and transmitted to the user terminal through wireless WIFI, as shown in Fig. 3.

3.1 Clock Reference Circuit

The functions of clock reference circuit are precise time synchronization and atomic clock time maintenance. Clock reference circuit includes atomic clock, standard source, phase detector, VCO (voltage controlled oscillator), DDS (direct digital synthesizer), and atomic clock control module, as shown in Fig. 4.

Clock reference circuit output a frequency signal with three-level stability. Firstly in the absence of atomic clock module the crystal oscillator on the board is used

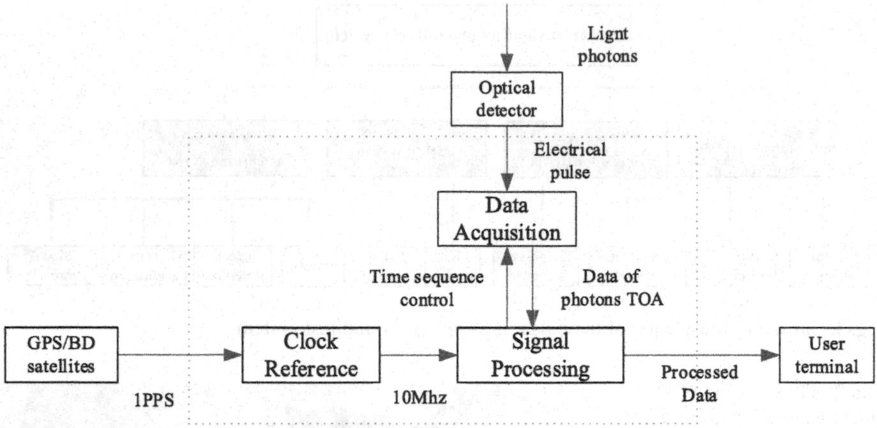

Fig. 3 Hardware structure of data acquisition device

Fig. 4 Structure illustration of clock reference circuit

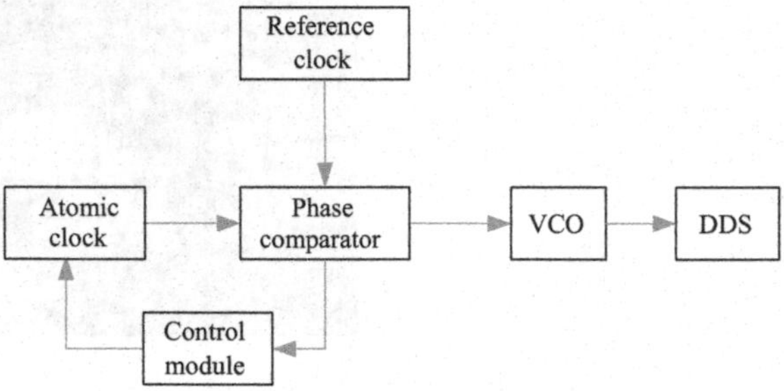

Fig. 5 Reference clock of 1PPS and 10 MHz

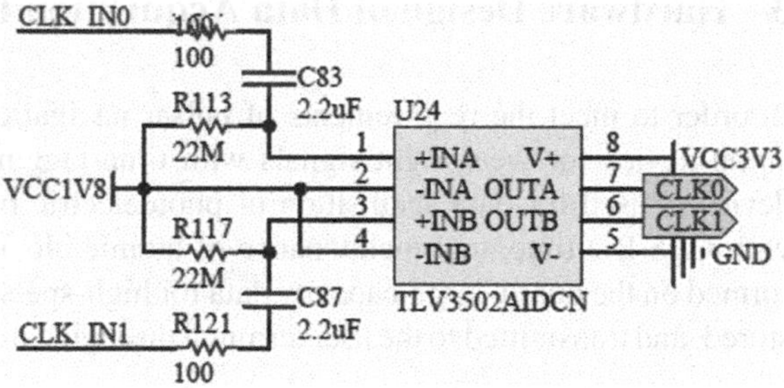

as a 10 M frequency reference, and the day stability of output signal is about 10^{-6}. Secondly while the atomic clock module is working and there is no GPS or BD signal, the day stability is about 10^{-11}. Thirdly while atomic clock is tamed by the GPS or BD signal, the day stability is about 10^{-13}. The control module for atomic clock performs PD control law according to the phase difference signal and its rate of change (Figs. 5, 6 and 7).

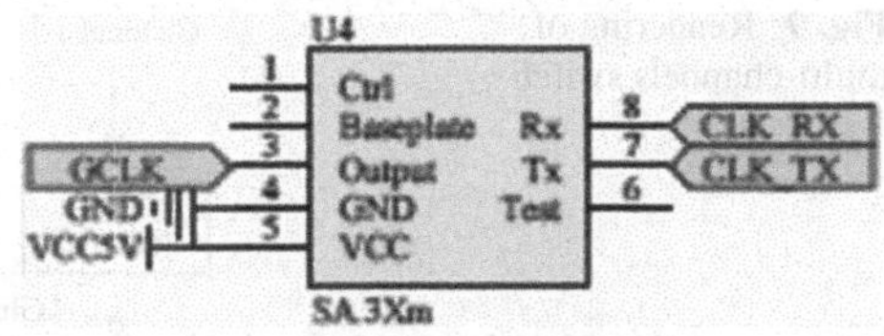

Fig. 6 SA.3Xm rubidium atomic clock

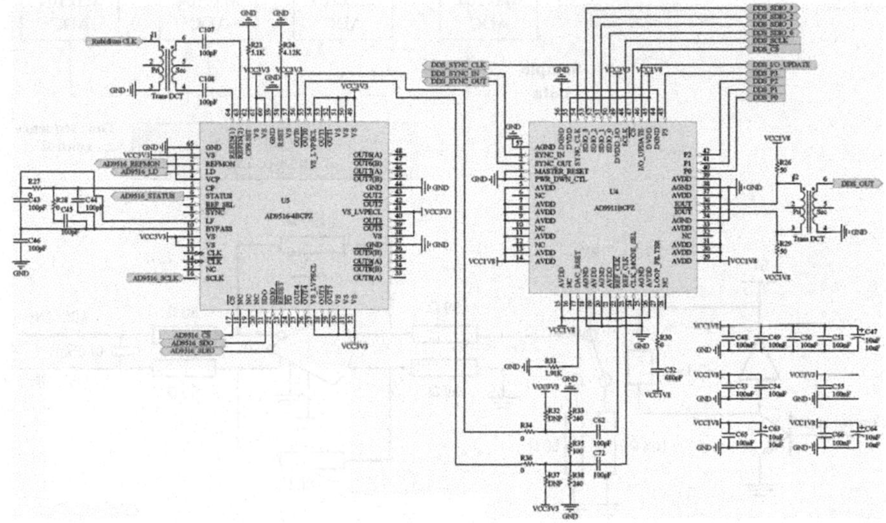

Fig. 7 10 MHz standard sine wave reference clock output

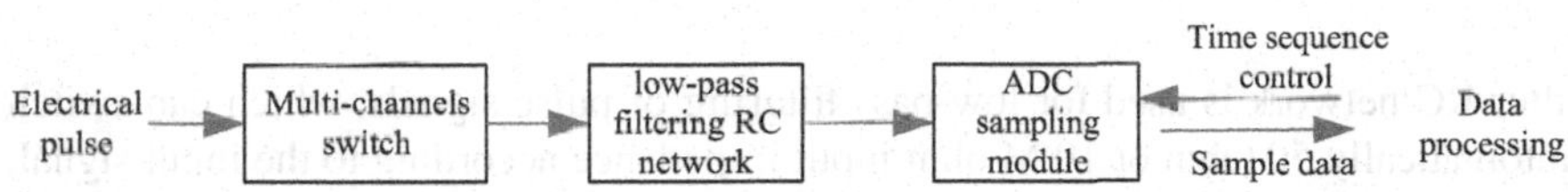

Fig. 8 Structure illustration of data acquisition circuit

3.2 *Data Acquisition Circuit*

The functions of data acquisition circuit are to perform low-pass filtering on multiple optical pulse signals, and real-time data acquisition of the filtered multiple optical pulse signals at a sampling rate of 1 GHz according to the timing control logic of the FPGA. Data acquisition circuit includes ADC sampling module, low-pass filtering RC network and multi-channels switch, as shown in Fig. 8.

Four high-speed ADC chips are used to perform 4 channels data acquisition for the photoelectric pulse at 1 GHz sampling rate. The measurement accuracy of the pulse time is 1 ns while every channel is independent. The accuracy can be reached to 0.25 ns while 4 channels are connected in series (Jumper settings on the circuit board). The ADC chip are connected with FPGA through 8bits data bus. Low-pass

Fig. 9 Rendering of multi-channels switch

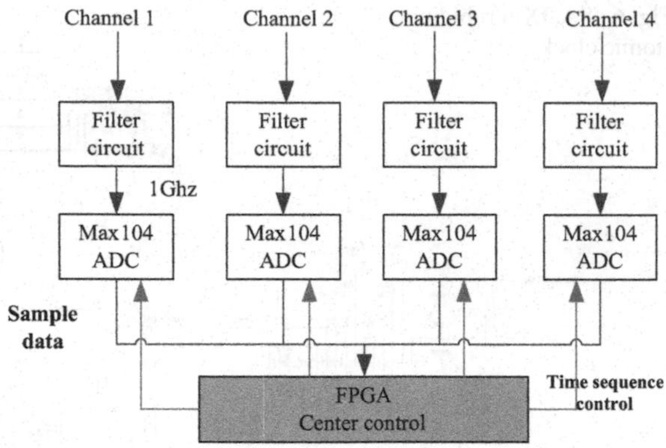

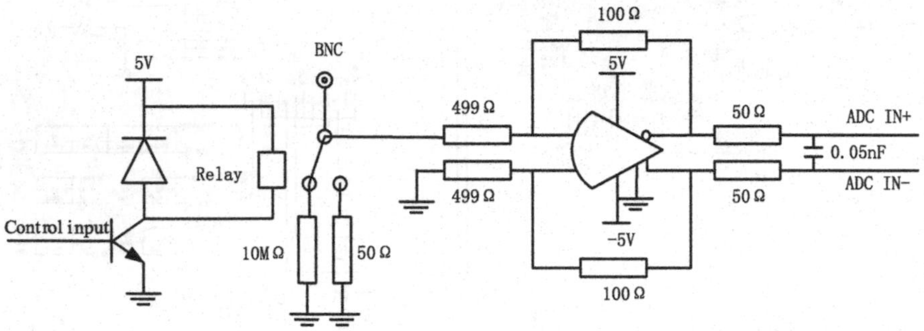

Fig. 10 schematic of low-pass filter network

filter RC network is used for low-pass filtering of pulse signals, which can switch automatically 50 Ohm or 10 M ohm input impedance according to the input signal, as shown in Figs. 9 and 10.

3.3 Signal Processing Circuit

The functions of signal processing circuit is to perform data processing on the pulse signals, including time tagging, noise filtering, ping pang storage, and wireless transmission. Signal processing circuit includes an FPGA module and a wireless transmission network, as shown in Fig. 11.

The filtering module performs FIR or wavelet transform filtering on the time tagging data. Ping pang storage module performs time sequence control of on-chip RAM reading and writing. If the first on-chip RAM is full, then the second on-chip RAM is written. Meanwhile the first on-chip RAM data is read out and written to the external RAM, and vice versa. Xilinx's V6 series FPGA chip is used to generate a micro-blaze CPU on-chip. The CPU and each on-chip module use the AXI4 bus

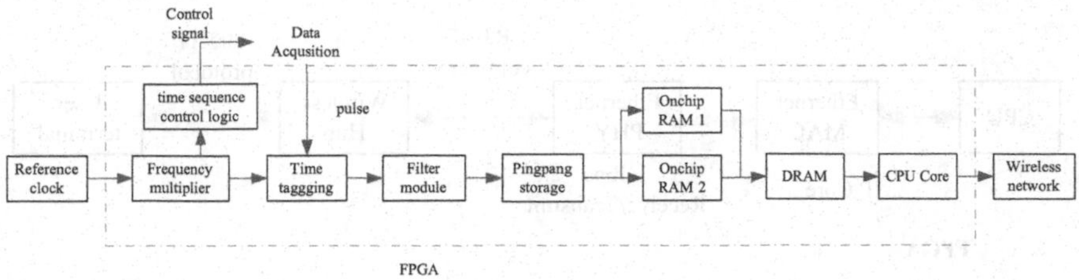

Fig. 11 Structure illustration of signal processing circuit

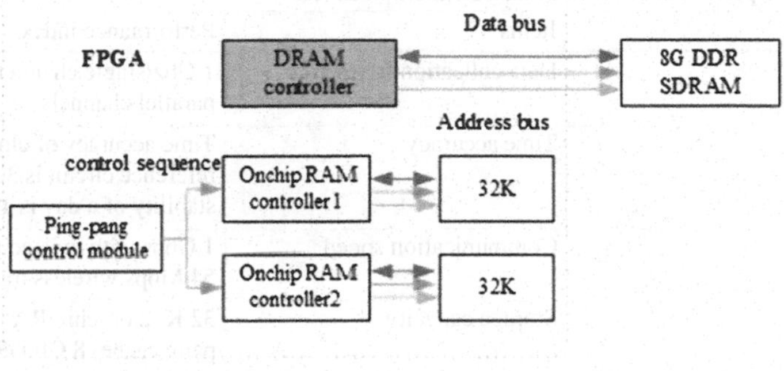

Fig. 12 Inter time sequence control logic of FPGA

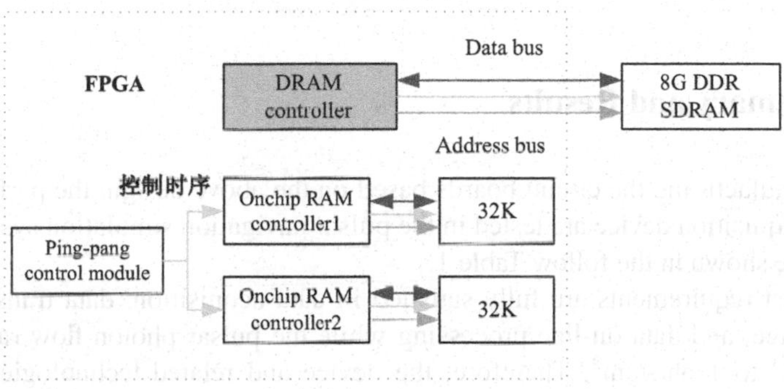

Fig. 13 Structure of ping pang storage with high speed

for data and command transmission. The Ethernet protocol can be carried out with standard sockets (Figs. 12, 13 and 14).

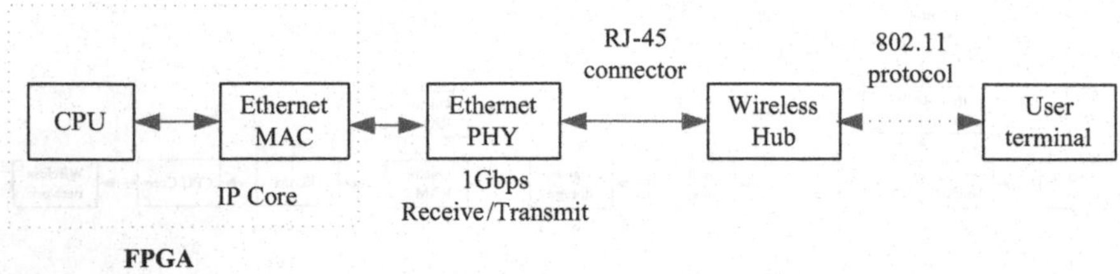

Fig. 14 Structure of Ethernet network

Table 1 The performance index of data acquisition device

No.	Items	Performance index
1	Data collection frequency	1 Ghz/single channel, 4 parallel channels
2	Time accuracy	Time accuracy of clock reference circuit is 30 ns signal stability of a day is 10^{-13}
3	Communication speed	1 Gbps with twisted pair cable 54 Mbps wireless transmission
4	Storage capacity	32 K*2 on-chip RAM (ping pang cache) 8 Gb DRAM
5	System weight	≤2 kg
6	Power consumption	≤10 W

4 Summary and Results

After manufacturing the circuit boards based on the above design, the performance of data acquisition device are tested in the pulsar navigation simulation system. The results are shown in the follow Table 1.

The test requirements are fully satisfied in data acquisition, data transmission, data storage, and data on-line processing while the pulsar photon flow rate varies from 10^{-5} to 1 ph/s/cm^2. Therefore, the device and related technologies can be also applied to other data acquisition systems with high speed, such as radar signal processing system, atomic nuclear physics signal processing system and weak light signal processing system.

Acknowledgements Supported by the National Key Research and Development Program of China (Grant No. 2017YFB0503300, 2017YFB0503304).

References

1. P. Shuai, L. Ming, S. Chen, Z. Huang, *Principles and methods of X-ray pulsar navigation system* (China Aerospace Press, Beijing, 2009)
2. S. Sheikh, The use of variable celestial X-ray sources for spacecraft navigation University of Maryland, 2005
3. A.A. Emadzadeh, J.L. Speyer, Navigation in Space by X-ray Pulsars, 1st edn. (Springer, 2011)
4. J.E. Hanson, *Principles of X-ray Navigation* (Stanford University, Stanford, 1996)
5. Y. Jia, Robust control with decoupling performance for steering and traction of 4WS vehicles under velocity-varying motion]. IEEE Trans. Control Syst. Technol. **3**, 554–569 (2000)
6. K.C. Gendreau, The neutron star interior composition explorer (NICER): design and development, in *Society of photo-optical instrumentation engineers (SPIE) Conference Series* (2016) (9905)
7. L. Li, Z. Wei, T. Guojian, S. Shouming, Semi-physical simulation system for navigation based on X-ray pulsar. Changsha: J. Natl. Univ. Defense Technol. **5**, 10–14 (2012)
8. Y. Jia, Alternative proofs for improved LMI representations for the analysis and the design of continuous-time systems with polytopic type uncertainty: a predictive approach. IEEE Trans. Autom. Control **8**, 1413–1416 (2003)
9. S. Haifeng, Simulation of high stability X-ray pulsar signal. Beijing: Acta Phys. **10**, 510–520 (2013)
10. S. Zhe, *X-ray pulsar navigation signal processing method and simulation experiment system* (Xidian University, Xi'an, 2011)
11. W. Jing, *Design and implementation of high-precision X-ray pulsar ground simulation system* (Xidian University, Xi'an, 2013)
12. P. Jilong, L. Baoquan, Z. X, W. Fei, Low noise signal process system for solar X-ray imaging camera. Beijing: Nucl. Electron. Det. Technol. **4**, 839–843 (2009)

Nested Optimization Based Co-design Method for Motor Driving System

Tianyi Zeng, Xuemei Ren and Zimei Sun

Abstract Due to the complex structure of the motor driving system, the control performance can be influenced significantly by the plant design. To improve the control performance, a plant/controller co-design method is developed in this paper. A combined optimization index considering both the plant design and the controller design is developed for the motor driving system. By solving the proposed co-design problem, the largest load's moment of inertia can be achieved with the same control performance. A nested optimization strategy is adopted to simplified the co-design problem to achieve the system optimality reliably and effectively. Simulation results illustrate the effectiveness of the proposed plant/controller co-design method.

Keywords motor driving system · co-design method · nested optimization

1 Introduction

With the development of the technology, the modern control system design is not only a feasible solution, but also an optimal solution. The conventional system design is always a sequential procedure: the plant design followed be the controller design [1, 2]. However, due to the coupling between the plant and the controller, the overall system optimality cannot be guaranteed. Because of the complex structure of the motor driving system, the control performance will be significantly influenced by the plant design. To achieve the overall system optimality, the plant/controller co-design method should be investigated.

The plant/controller co-design method has found its application from space structure design [3] to elevator design [4]. To improve the control performance of the system, combined optimization indexes considering both the plant and the controller were developed. By solving the combined optimization problem, the effect of the plant design is taken into account in the controller design. In [5], the wind tur-

T. Zeng · X. Ren (✉) · Z. Sun
School of Automation, Beijing Institute of Technology, Beijing 100081, China
e-mail: xmren@bit.edu.cn

Y. Jia et al. (eds.), *Proceedings of 2018 Chinese Intelligent Systems Conference*, Lecture Notes in Electrical Engineering 529,
https://doi.org/10.1007/978-981-13-2291-4_82

bine structure design and the linear parameter-varying (LPV) controller design are combined to achieve a better close-loop control performance. To achieve the high accuracy, high speed and high stiffness, a co-design procedure of mechanical structure and the control method is adopted for the four-bar linkage system [6]. In this paper, we develop a co-design index for the motor driving system to drive the largest load's moment of inertia.

To achieve the system optimum effectively, a nested optimization strategy is introduced in the co-design scheme. Due to the coupling term between the plant and the controller, optimize the plant and the controller sequentially cannot guarantee the truly system optimality [7]. In addition, the co-design optimization problem is normally a nonconvex optimization problem, even if performance indexes of the plant and the controller are convex respectively, due to the coupling between the two optimization problems [8]. Hence, solving the co-design optimization problem directly may cause the computation burden. The nested optimization strategy, also named as the bi-level optimization strategy, can make the combined optimization simplified with two embedded optimization loops [9]. A two-level nonlinear programming method is used for the facility selection and the production planning [10]. By introducing the nested optimization strategy, the computation complexity of the co-design problem can be reduced so that the overall system optimality can ge guaranteed.

In this paper, a co-design scheme is developed for the motor driving system to seek the largest load's moment of inertia with the same control performance. A LQR based tracking controller is developed and taken into consideration into the plant/controller co-design scheme. The nested optimization strategy is adopted so that the overall system optimality can be achieved reliably and effectively.

The rest of this paper is organized as follows: The model of the motor driving system is developed in Sect. 2; The LQR based tracking controller is designed and the co-design scheme is presented in Sect. 3; To illustrate the effectiveness of the proposed co-design scheme, simulation results are demonstrated in Sect. 3; Finally, this paper is concluded in Sect. 4.

2 Problem Formulation

In this section, control-oriented models for both single motor and dual-motor system are presented. The nonlinearity caused by the gear clearance is represented by a deadzone model. Finally, the model is transformed to a second-order model with reasonable approximation for controller design.

The schematic diagram of the motor driving system is shown in Fig. 1.

Here, O_1 is the driving motor and O_0 is the load to be driven. For single motor driving system, it can be treated as a two-mass system linked by a minor gear and a bull gear with a nonlinearity represented by a deadzone nonlinearity between them. The block diagram of single motor driving system is shown in Fig 2.

The load and the motor are linked by gears with clearance. The model can be shown as follows:

Fig. 1 Schematic diagram of motor driving system

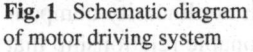

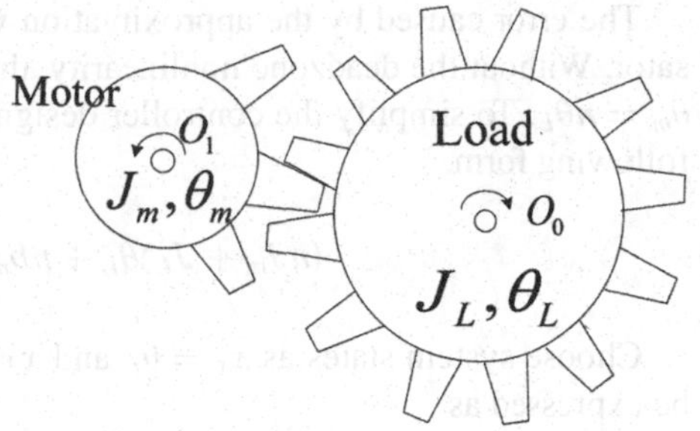

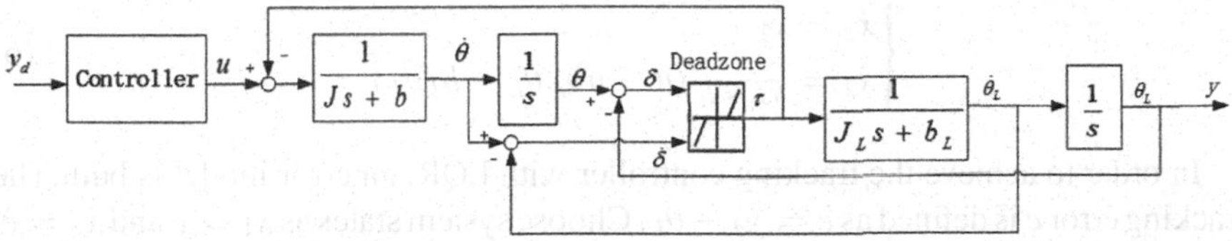

Fig. 2 Block diagram of motor driving system

$$\begin{cases} J\ddot{\theta}_m + b_m\dot{\theta}_m = u - \tau \\ J_L\ddot{\theta}_L + b_L\dot{\theta}_L = \tau \end{cases} \tag{1}$$

Here θ_m and θ_L are angular positions of motor and load respectively; J and J_L are moments of inertia of the motor and load respectively; b_m is the viscous friction coefficient of motor; u is the control input; $f_L(\dot{\theta}_L)$ is the friction torque acting on the load; τ is the transmission torque of the gear box which can be expressed as

$$\tau = kf(\delta) + cf(\dot{\delta}) \tag{2}$$

Here k and c are the torsional coefficient and the damping coefficient; δ is the angle position error between the motor and the load, i.e. $\delta = \theta_m - \theta_L$; $f(\delta)$ can be expressed by a deadzone model as

$$f(\delta) = \begin{cases} \delta - \alpha, & \delta \geq \alpha \\ 0, & |\delta| < \alpha \\ \delta + \alpha, & \delta \leq -\alpha \end{cases} \tag{3}$$

where α is the gear clearance width parameter. The purpose of this paper is to control the system output $y = \theta_L$ to track the expected output y_d. If we ignore the gear clearance nonlinearity, (2) can be approximated by

$$\tau' = k\delta + c\dot{\delta} \tag{4}$$

The error caused by the approximation will be dealt with nonlinearity compensator. Without the deadzone nonlinearity, there is a proportionable relationship that $\theta_m = n\theta_L$. To simplify the controller design, the system (1) can be replaced by the following form.

$$(nJ_m + J_L)\ddot{\theta}_L + nb_m\dot{\theta}_L + b_L\dot{\theta}_L = u \tag{5}$$

Choose system states as $x_1 = \theta_L$ and $x_2 = \dot{\theta}_L$, and the state space equation can be expressed as

$$\begin{cases} \dot{x}_1 = x_2 \\ \dot{x}_2 = \frac{1}{nJ_m+J_L}(u - nb_m\dot{\theta}_L - b_L\dot{\theta}_L) \end{cases} \tag{6}$$

In order to achieve the tracking controller with LQR, an error model is built.The tracking error e is defined as $e = y_d - \theta_L$. Choose system states as $x_1 = e$ and $x_2 = \dot{e}$. According to (5) the state space equation of error system can be expressed as

$$\begin{cases} \dot{x}_1 = x_2 \\ \dot{x}_2 = \ddot{e} \\ \quad = \ddot{y}_d - \ddot{\theta}_L \\ \quad = \ddot{y}_d - \frac{1}{nJ_m+J_L}(u - nb_m\dot{\theta}_L - b_L\dot{\theta}_L) \\ \quad = \ddot{y}_d - \frac{1}{nJ_m+J_L}(u - nb_m(\dot{y}_d - x_2)_L - b_L(\dot{y}_d - x_2)) \\ \quad = \ddot{y}_d - (\frac{1}{nJ_m+J_L})u + (\frac{1}{nJ_m+J_L})(nb_m + b_L)\dot{y}_d \\ \quad\quad -(\frac{1}{nJ_m+J_L})(nb_m + b_L)x_2 \end{cases} \tag{7}$$

$\dot{x} = Ax + Bu + b_d d$ where $A = \begin{bmatrix} 0 & 1 \\ 0 & \frac{-nb_m+b_L}{nJ_m+J_L} \end{bmatrix}$; $B = \begin{bmatrix} 0 \\ \frac{-1}{nJ_m+J_L} \end{bmatrix}$; $B_L = \begin{bmatrix} 0 \\ 1 \end{bmatrix}$ and $d = \begin{bmatrix} 0 \; \ddot{y}_d + \frac{nb_m+b_L}{nJ_m+J_L}\dot{y}_d \end{bmatrix}$

3 Plant/controller Co-design Scheme

In this section, a combined optimization problem considering both the plant and the controller is developed for the motor driving system. The controller is developed based on LQR and the load's moment of inertia is chosen as the plant parameters. Finally, the nested optimization strategy is adopted in the co-design scheme.

Since $\ddot{y}_d + (\frac{1}{nJ_m+J_L})(nb_m + b_L)\dot{y}_d$ can be known directly, the controller can be designed as $u = u_d + u_{LQR}$. Here $u_d = (\frac{1}{nJ_m+J_L})(\ddot{y}_d + (\frac{1}{nJ_m+J_L})(nb_m + b_L)\dot{y}_d)$ and u_{LQR} can be achieved by using LQR strategy. The co-design method is adopted to the design of u_{LQR} and J_L as controller and plant parameters.

3.1 Co-design Method

Consider the following controller/plant combined optimization problem:

$$
\begin{aligned}
F &= \min_{J_L,S}\left\{w_p\frac{1}{J_L} + w_c\int_0^{\infty}(\mathbf{x}^{\mathrm{T}}\mathbf{Q}\mathbf{x} + \mathbf{u}^{\mathrm{T}}\mathbf{R}\mathbf{u})\right\} \\
& s.t.\, 0.0035 \le J_L \le 0.0516\,\mathrm{kg\,m^2} \\
& \dot{x} = A(J_L)x + B(J_L)u_{LQR}, x(t_0) = x_0
\end{aligned}
\tag{8}
$$

where w_p and w_c are weight parameters; Q and R are assumed positive definite and positive semidefinite. All the plant states are assumed to be directly measured. The physical meaning of the proposed combined optimization problem is driving the largest load's moment of inertia with the same control performance.

A novel co-design method to solve the problem in (8) is given as follow:

Outer Loop:

$$
J_L^* = \arg\min_{J_L,S}\left\{n_p\frac{1}{J_L} + n_c f_c^*(J_L)\right\} \tag{9}
$$

s.t. $0.0035 \le J_L \le 0.0516\mathrm{kg\,m^2}$, $f_c^*(J_L) =$ output of the inner loop.

Inner Loop:

$$
\begin{aligned}
& f_c^*(J_L) = \arg\min_{x,u}\int_0^{\infty}(\mathbf{x}^{\mathrm{T}}\mathbf{Q}\mathbf{x} + \mathbf{u}^{\mathrm{T}}\mathbf{R}\mathbf{u}) \\
& \text{s.t.}\, \dot{x} = A(J_L)x + B(J_L)u, x(t_0) = x_0.
\end{aligned}
\tag{10}
$$

Here the outer loop is for plant optimization; the inner loop is for controller optimization and a standard LQR problem with the solution:

$$
\begin{aligned}
& f_c^*(J_L) = x_0^{\mathrm{T}} S x_0, u_{LQR} = R^{-1}B^{\mathrm{T}}Sx, \\
& A^{\mathrm{T}}S + SA - SBR^{-1}B^{\mathrm{T}}S + Q = 0
\end{aligned}
\tag{11}
$$

Lemma 1 *Define* arg F^* *as the solution to* (8), *i.e. the optimal solution considering both plant and controller. In addition, let* Ψ^* *be the solution of the optimization method proposed in* (9) *and* (10). *Then* $\Psi^* = \arg F^*$.

Remark 1 By introducing the nested optimization strategy, the weight parameters, i.e. n_c and n_p can be tuned based on different scenarios and we do not have to make the necessary condition developed in [9] be satisfied. Hence, a flexible weight parameters selection can be achieved with the co-design scheme based on nested optimization strategy.

3.2 Simulation Results

In this section, simulation results with parameters designed by the co-design method is demonstrated. Contrastive simulation results are also presented to show the effectiveness of the proposed plant/controller co-design method.

Choose the weight parameters as $w_p = 1$, $w_c = 1$ and the reference signal as $y_d = \sin(t)$. The system parameters are shown in Table 1.

The optimal solution of the load can be achieved as $J_L^* = 0.0723\,\mathrm{kg\,m^2}$ by solving (9) and (10). The initial position of the load is set as 0.5 rad. The tracking performance, velocity and controller output are shown in Figs. 3, 4 and 5.

With simulation results above, we can see that the load can track the reference signal driven by the developed controller under the proposed co-design scheme.

Theoretically, the tracking performance should be influenced by the initial moment of the load significantly. The tracking performance with different initial moments of the load is shown in Fig. 6. It can be seen that with different plant parameters, the same tracking performance can be achieved. $J_L = 0.0113\,\mathrm{kg\,m^2}$ is the frequently-

Table 1 System parameters

Parameter	Value	Unit
J_m	0.0026	$\mathrm{kg\,m^2}$
b_m	0.015	Nm s/rad
b_l	0.02	Nm s/rad
k	560	Nm/rad
α	0.2	rad

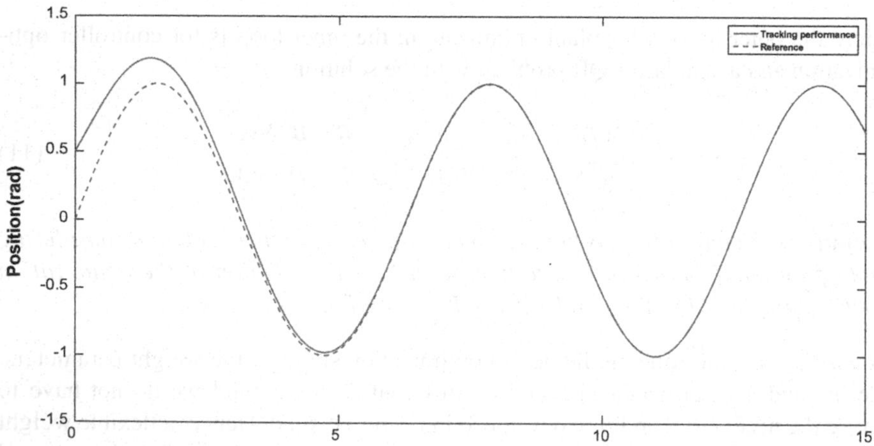

Fig. 3 Tracking performance with LQR

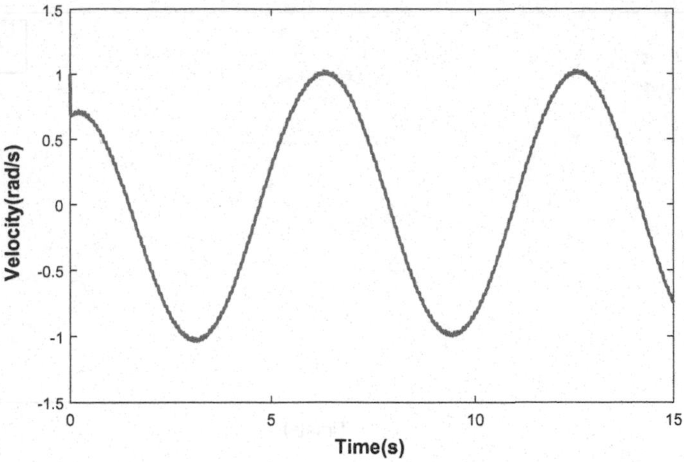

Fig. 4 Velocity of the load

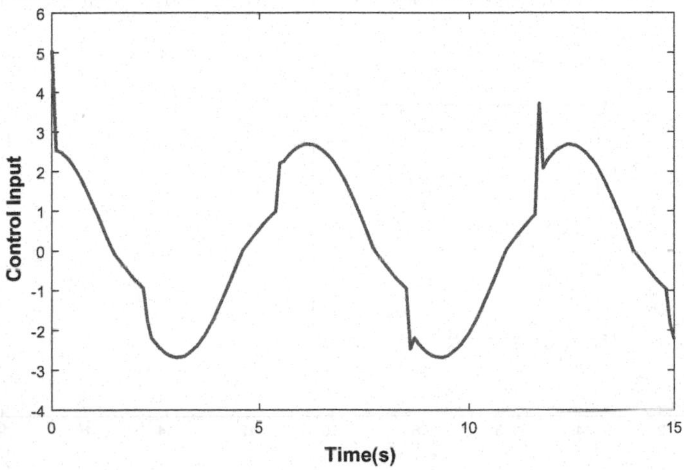

Fig. 5 Control input

used plant parameter. With co-design method, the extra initial moment of the load can be added with the same tracking performance.

It can be seen in Fig. 6 that the similar tracking performance can be achieved with different inertia moments of the load. Hence, extra inertia moments of the load can be driven with the same control performance which can prove the effectiveness of the proposed plant/controller co-design method. On the other hand, if the load's moment of inertia is tuned larger than than $J_L^* = 0.0723\,\text{kg}\,\text{m}^2$, the control performance is shown in Fig. 7.

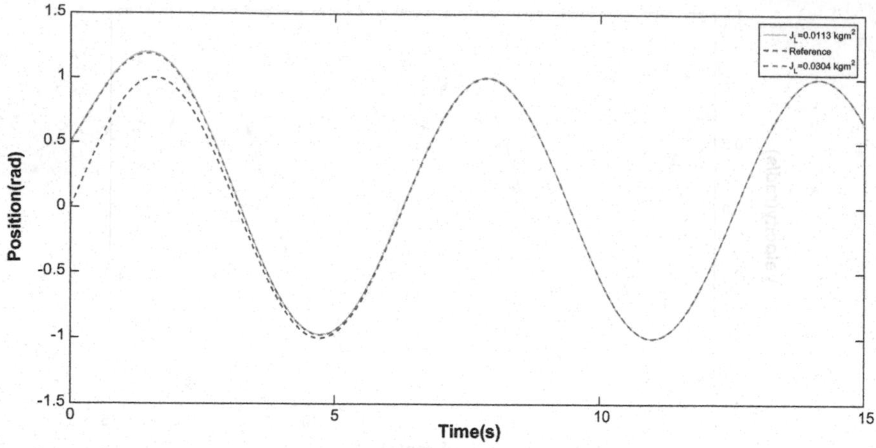

Fig. 6 Tracking performance with different initial moments of the load

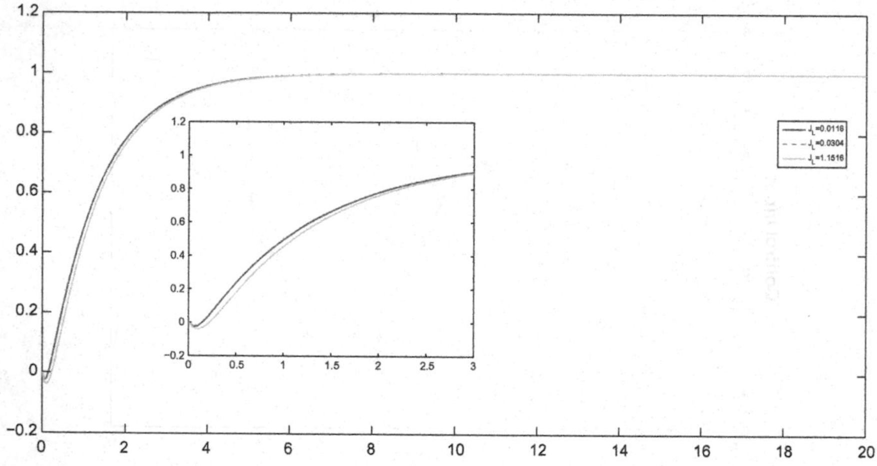

Fig. 7 Tracking performance with co-design method

In Fig.7, we can see that the control performance is influence with the load's moment of inertia larger than than $J_L^* = 0.0723\,\mathrm{kg\,m^2}$ which is achieved by the proposed co-design method.

4 Conclusions

To improve the control performance of the motor driving system, a plant/controller co-design scheme is developed in this paper. A combined co-design optimization problem considering both the plant and the controller is designed to drive the largest load's moment of inertia with the same control performance. The nested optimization strategy is introduced so that the co-design problem can be simplified and the co-design solution can be achieved reliably and effectively. Simulation results are demonstrated to illustrate the effectiveness of the proposed plant/controller co-design method.

Acknowledgements This work is sponsored by National Natural Science Foundation of China (No. 61433003, No. 61273150 and No. 61321002).

References

1. Y. Jia, Robust control with decoupling performance for steering and traction of 4ws vehicles under velocity-varying motion. IEEE Trans. Control Syst. Technol. **8**(3), 554–569 (2000)
2. Y. Jia, Alternative proofs for improved lmi representations for the analysis and the design of continuous-time systems with polytopic type uncertainty: a predictive approach. IEEE Trans. Autom. Control **48**(8), 1413–1416 (2003)
3. J.A. Perez, D. Alazard, T. Loquen, C. Pittet, Mechanical/control integrated design of a flexible planar rotatory spacecraft, in *Advances in Aerospace Guidance, Navigation and Control* (2018), pp. 651–667
4. H.K. Fathy, S.A. Bortoff, G.S. Copeland, P.Y. Papalambros, A.G. Ulsoy, Nested optimization of an elevator and its gain-scheduled lqg controller, in ASME. International Mechanical Engineering Congress and Exposition **2002**, 119–126 (2002)
5. F.A. Shirazi, K.M. Grigoriadis, D. Viassolo, Wind turbine integrated structural and lpv control design for improved closed-loop performance. Int. J. Control **85**(8), 1178–1196 (2012)
6. W.J. Zhang, Q. Li, L.S. Guo, Integrated design of mechanical structure and control algorithm for a programmable four-bar linkage. IEEE/ASME Trans. Mechatron. **4**(4), 354–362 (2002)
7. H.K. Fathy, J.A. Reyer, P.Y. Papalambros, A.G. Ulsov, On the coupling between the plant and controller optimization problems. American Control Conference **3**, 1864–1869 (2001)
8. H.K. Fathy, P.Y. Papalambros, A.G. Ulsoy, Integrated plant, observer, and controller optimization with application to combined passive/active automotive suspensions, in *Proceedings of the 2003 ASME*, vol. 4 (2003) pp. 3375–3380
9. J.F. Bard, Practical Bilevel Optimization: Algorithms and Applications (1998), pp. 144–146
10. D. Cao, M. Chen, Capacitated plant selection in a decentralized manufacturing environment: a bilevel optimization approach. Eur. J. Oper. Res. **169**(1), 97–110 (2006)

Adaptive Neural Control for Nonlinear Systems in Non-strict-feedback Form

Chao Yang and Yingmin Jia

Abstract This article studies the ANC problem for nonlinear systems. Unlike the classical backstepping strategy, the control issue of nonlinear system in non-strict feedback(NSF) form is more challenging. In the design process, neural networks and high-gain observers are applied to tackle with the issues of unknown nonlinearity and unmeasured states, respectively. Adaptive backstepping technique and a high-order sliding mode (HOSM) differentiator are combined to present a novel ANC algorithm. In the stability analysis, signals in the considered systems turn to be SGGB with appropriately designed parameters. Finally, a numerical example is practiced. The results of the numerical simulation further illustrate the usefulness of the new algorithm.

Keywords adaptive neural control · backstepping · nonstrict-feedback systems

1 Introduction

In recent years, adaptive backstepping technology has received extensive attention and been applied in many fields such as aerospace, marine. By means of this technology, many meaningful results were published [1–6]. Kokotovic [1] proposed an symmetric adaptive control strategy for feedback linearizable nonlinear systems. A modified Lyapunov function is introduced in [2]. And for the considered linear system, a novel adaptive design scheme is developed. Observer-based adaptive backstepping technique is also studied. With this method, [3] solved the control problem of uncertain nonlinear systems. In the field of switched control, adaptive backstep-

C. Yang · Y. Jia (✉)
The Seventh Research Division and the Center for Information and Control,
School of Automation Science and Electrical Engineering,
Beihang University (BUAA), Beijing 100191, China
e-mail: ymjia@buaa.edu.cn

C. Yang
e-mail: yangchao_2011@sina.cn

Y. Jia et al. (eds.), *Proceedings of 2018 Chinese Intelligent Systems Conference*, Lecture Notes in Electrical Engineering 529,
https://doi.org/10.1007/978-981-13-2291-4_83

ping method is viewed as a powerful tool. A tracking control scheme applying this method is provided for switched nonlinear systems in [4]. Similar ways are used in the issue of stochastic control. For instance, the tracking control problem of stochastic system is settled in [5]. In addition, backstepping based strategy is employed in time-delay systems in [6].

On the other hand, research on ANC and other intelligent strategy for systems including nonlinearity has made significant achievements. However, the above research results are all obtained in a nonlinear system with a strict feedback structure. For non-linear systems that are not strictly feedback structures, these methods will no longer apply. Moreover, for the classical backstepping control technology, there is a so-called computational expansion problem. Since the virtual control design information in the current step needs the information of the previous step of the virtual control derivative, the calculation is complicated, especially for high-order systems. In order to relax the restrictions on the system structure, there are many meaningful explorations [7–10]. In view of the above discussion, this work aims to settle the ANC problem of nonlinear systems in specific form like the NSF form.

This article has the following structure. Section 2 provides description of the problem and assumptions of the investigated subject. In Sect. 3, an high-gain observer and adaptive neural controller are designed and the theoretical analysis is exhibited. Furthermore, through conducting a numerical example, Sect. 4 validates the designed method. Section 5 draws the conclusion of the work.

2 Problem Description and Preliminaries

Consider nonlinear systems in NSF:

$$\begin{cases} \dot{x}_1 = f_1(x) + x_2, \\ \dot{x}_i = f_i(x) + x_{i+1}, 2 \le i \le n-1, \\ \dot{x}_n = f_n(x) + u(t), \\ y = x_1 \end{cases} \tag{1}$$

where $x = [x_1, x_2, \ldots, x_n]^T \in R^n$ stands for the state vector, $u \in R$ is the system input, and $y \in R$ is the output, respectively, $\bar{x}_i = [x_1, x_2, \ldots, x_i]^T \in R^i$; the unknown function $f_i(.)$ is smooth.

The target of the work is to provide ANC algorithm for the considered system (1) with high-gain observer and HOSM differentiator technique, and make sure that all the signals in the system are SUUB. Before design of the ANC scheme, some necessary presumptions and lemmas are given:

Lemma 1 *For* $\forall \nu \in R$, $\forall \mu \in R$, *the following inequality is true:*

$$\nu\mu \le \frac{\kappa^a |\nu|^a}{a} + \frac{|\mu|^b}{b\kappa^b}$$

where $\kappa > 0, a > 1, b > 1,$ *and* $\frac{1}{a} + \frac{1}{b} = 1.$

Assumption 1 Assume that there exist constants L_i satisfies the inequality:

$$|f_i(x) - f_i(x')| \leq L_i \|x - x'\|$$

where $\| \cdot \|$ means the Euclidean norm.

Using the RBFNNs, continuous nonlinear functions can be approximated by arbitrary precision. The RBFNNs are described in the following form:

$$\Xi = \theta^T \psi(X) \tag{2}$$

where X stands for the network input, Ξ means the network output. θ is designed to represent the weight vector of the artificial neural network. $\psi(X)$ is a function vector contains of Gaussian functions as its elements, which means

$$\psi_i(X) = \exp\left[-\frac{(X - c_i)^T (X - c_i)}{w_i^2} \right], \tag{3}$$

where c_i and w_i are design parameter vector and constant, respectively. The former vector stands for the center and the posterior means value of the width. Therefore, smooth function may be approached through ANN. In detail, that is

$$f(X) = \theta^{*T} \psi(X) + \delta^*, \tag{4}$$

where the ideal weight vector of the NN is represented by θ^*. Moreover, the vector here is defined

$$\theta^* := arg \min_{\theta \in \bar{R}^l} \left\{ \sup_{X \in \bar{\Omega}} |f(X) - \theta^T \psi(X)| \right\}$$

Also, the approximation error δ^* should satisfy $|\delta^*| \leq \delta_M$.

3 State Observer and ANC Algorithm Design

In the beginning of this section, a high-gain state observer is constructed in the following form.

$$\begin{aligned} \dot{\hat{x}}_i &= \hat{x}_{i+1} + \hat{f}_i(\hat{x}|\theta_i) + l_i(y - \hat{x}_1), 1 \leq i \leq n-1, \\ &\vdots \\ \dot{\hat{x}}_n &= u + \hat{f}_n(\hat{x}|\theta_n) + l_n(y - \hat{x}_1) \\ y &= x_1 \end{aligned} \tag{5}$$

Then it is feasible to calculate the observer error. It is easy to obtain

$$e = x - \hat{x} \tag{6}$$

Using (1), (5) and the above equation, we have

$$\dot{e} = Ae + F(x) - \hat{F}(\hat{x}|\theta) \tag{7}$$

where $F(x) = [f_1(x), \ldots, f_n(x)]^T$, $\hat{F}(\hat{x}|\theta) = [\hat{f}_1(\hat{x}|\theta_1) \ldots, \hat{f}_n(\hat{x}|\theta_n)]$, $\hat{f}_i(\hat{x}|\theta_i) = \theta_i^T \psi_i(\hat{x})$

$$A = \begin{bmatrix} -l_1 & \\ \vdots & I_{n-1} \\ -l_n & 0 \cdots 0 \end{bmatrix} \tag{8}$$

For proper parameter l_i, A can be designed as a stable matrix which satisfies

$$A^T P + PA \le -Q \tag{9}$$

Here, with a design matrix $Q = Q^T > 0$, the matrix $P = P^T > 0$ can be found out.

A candidate Lyapunov function is developed as

$$V_e = e^T Pe \tag{10}$$

Then the derivative of V_e is derived as

$$\begin{aligned} V_e &= \dot{e}^T Pe + e^T P\dot{e} \\ &\le -e^T Qe + 2e^T P(\tilde{\theta}^T \psi(\hat{x}) + \delta + F - F(\hat{x})) \end{aligned} \tag{11}$$

By means of Lemma 1, one can obtain

$$e^T P(\tilde{\theta}^T \psi(\hat{x})) \le 2\|Pe\|^2 + \frac{1}{2}N^2 \sum_{i=1}^{n} \tilde{\theta}_i^T \tilde{\theta}_i \tag{12}$$

$$2e^T P\tilde{F} \le \|Pe\|^2 + \sum_{i=1}^{n} L_i^2 \|e\|^2 \tag{13}$$

Submitting (12) and (13) into (11) it yields

$$V_e \le -\lambda_0 \|e\|^2 + \frac{1}{2}\|\bar{\epsilon}\|^2 + \frac{N^2}{2} \sum_{i=1}^{n} \tilde{\theta}_i^T \tilde{\theta}_i \tag{14}$$

where$\lambda_0 = \lambda_{min} Q - 6\|P\|^2 - 0.5 \sum_{i=1}^{n} l_i^2$

Define the error surface of system (1)

$$\begin{cases} z_1 = x_1, \\ z_2 = x_2 - \alpha_1, \\ \vdots \\ z_n = x_n - \alpha_{n-1}, \end{cases} \tag{15}$$

where α_i is the the virtual control of the subsystem. The virtual control and adaptive parameters are designed as

$$\alpha_1 = -c_1 z_1 - 2z_1 - \theta_1^T \psi_1(\hat{x}_1) \tag{16}$$

$$\dot{\theta}_1 = z_1 \psi_1(\hat{x}_1) - \gamma_1 \theta_1 \tag{17}$$

$$\alpha_i = -c_i z_i - 2.5 z_i - \theta_i^T \psi_1(\hat{x}_i) \tag{18}$$

$$\dot{\theta}_i = \sigma_i z_i \psi_i(\hat{\bar{x}}_i) - \gamma_i \theta_i, 2 \le i \le n-1 \tag{19}$$

Control input u and $\dot{\theta}_n$ are described as

$$u = -c_n z_n - 2.5 z_n - \theta_n^T \psi_n(\hat{x}) \tag{20}$$

$$\dot{\theta}_n = \sigma_n z_n \psi_n(\hat{x}) - \gamma_n \theta_n \tag{21}$$

Then the main achievement of this work is obtained:

Theorem 1 *For the considered nonlinear system* (1) *with previous assumption and lemma, the virtual control and adaptive law designed in equalities* (16)–(21), *all signals in the system are SUUB and the system output can be remained in arbitrary small scale.*

Proof

Step 1

Calculate $\dot{z}_1$, we have

$$\dot{z}_1 = f_1(x_1) + x_2 \tag{22}$$

Define a candidate Lyapunov function as

$$V_1 = V_e + \frac{1}{2} z_1^2 + \frac{1}{2\sigma_1} \tilde{\theta}_1^T \tilde{\theta}_1 \tag{23}$$

Calculation of $\dot{V}_1$ is

$$\begin{aligned} \dot{V}_1 &= \dot{V}_e + z_1 \dot{z}_1 - \frac{1}{2\sigma_1} \tilde{\theta}_1^T \dot{\theta}_1 \\ &\le -\lambda_1 \|e\|^2 - c_1 z_1^2 + \frac{1}{\sigma_1} \tilde{\theta}_1^T \theta_1 + z_1 z_2 + \frac{N^2}{2} \sum_{i=1}^{n} \tilde{\theta}_i^T \theta_i + M_1 \end{aligned} \tag{24}$$

where $\lambda_1 = \lambda_0 - 0.25\|\theta_1\|^2 - 1$, $M_1 = 0.5\|\bar{\epsilon}_1\|^2 + \delta_1^2 + 0.5N^2\|\theta^*\|^2$ One problem of the conventional backstepping method is the so-called explosion of complexity. To tackle with this issue, we introduce the Levants HOSM differentiator as follows

$$\begin{aligned} \dot{\xi}_1 &= \tau_1 \\ \tau_1 &= -r_1\|\xi_1 - \alpha\|^{0.5} sign(\xi_1 - \alpha) + \xi_2 \\ \dot{\xi}_2 &= -r_2\, sign(\xi_2 - \tau_1) \end{aligned} \tag{25}$$

where α is the input signal.

Lemma 2 *By choosing properly parameters r_1 and r_2, the following equalities are satisfied in finite-time:*

$$\begin{aligned} \xi_1 &= \alpha \\ \tau_1 &= \dot{\alpha} \end{aligned}$$

With the Levants HOSM differentiator, the derivative of the virtual control α_i can be easily given.

***Step i** ($2 \le i \le n-1$)*

By means of design of z_i, it is easy to obtain $\dot{z}_i$

$$\dot{z}_i = \hat{x}_i + 1 - \dot{\alpha}_i + l_i e_i + \theta_i^T \psi_i(\hat{x}) \tag{26}$$

Define another candidate Lyapunov function as

$$V_i = \frac{z_i^2}{2} + \frac{1}{2\sigma_i}\tilde{\theta}_i^T\tilde{\theta}_i + V_{i-1} \tag{27}$$

Similar to the preceding process, with Lemma 1, the following inequality is derived

$$\begin{aligned} \dot{V}_i \le -\lambda_i\|e\|^2 - \sum_{k=1}^{i} c_k z_k^2 + \sum_{k=1}^{i}\frac{1}{\sigma_k}\tilde{\theta}_k^T\theta_k + \sum_{k=1}^{i-1} z_k z_{k+1} \\ + \sum_{i=1}^{n}\frac{N^2}{2}\tilde{\theta}_i^T\theta_i + \sum_{j=2}^{i} N^2\tilde{\theta}_i^T\tilde{\theta}_i + N^2\sum_{j=2}^{i}\|\theta_j^*\|^2 + M_1 \end{aligned} \tag{28}$$

Step n

In this step, the following candidate Lyapunov function is chosen as

$$V_n = V_{n-1} + \frac{1}{2}z_n^2 + \frac{1}{2\sigma_n}\tilde{\theta}_n^T\tilde{\theta}_n \tag{29}$$

The same as the previous analysis, we can obtain $\dot{V}_n$ *as*

$$\dot{V}_n = \dot{V}_{n-1} + z_n \dot{z}_n - \frac{1}{2\sigma_n} \tilde{\theta}_n^T \dot{\theta}_n \tag{30}$$

With Young's inequality, one can obtain

$$z_k z_{k+1} \leq z_k^2 + \frac{1}{4} z_{k+1}^2 \tag{31}$$

$$\tilde{\theta}_k^T \theta_k \leq -\frac{1}{2} \tilde{\theta}_k^T \tilde{\theta}_k + \frac{1}{2} \tilde{\theta}_k^{*T} \tilde{\theta}_k^* \tag{32}$$

Substituting the inequalities into (30) *yields*

$$\dot{V}_n = -\lambda_n \|e\|^2 - \sum_{k=1}^{n} (c_k - 1) z_k^2 - \sum_{k=1}^{n} \left(\frac{1}{2\sigma_k} - \frac{3N^2}{2} \right) \tilde{\theta}_j^T \tilde{\theta}_j + M_n \tag{33}$$

where $M_n = \delta_1^2 + 0.5\|\delta\|^2 + \sum_{i=1}^{n} (N^2 + \frac{1}{2\sigma_i}) 0_i^{*T} \theta_i^*$
Then, $\dot{V}_n$ *can be rewritten as*

$$\dot{V}_n = -rV_n + M_n \tag{34}$$

where $r = min \frac{\lambda_n}{\lambda_{max}(P)}, 2(c_i - 1), \frac{1}{\sigma_i} - 3N^2$ *which means*

$$V_n \leq \frac{M_n}{r} + (V_n(0) - \frac{M_n}{r}) e^{-rt} \tag{35}$$

Therefore, the proof is completed.

4 Numerical Example

In the end, we design a numerical example to verify the developed strategy. Consider the following example:

$$\begin{cases} \dot{x}_1 = f_1(\bar{x}) + x_2 \\ \dot{x}_2 = f_2(\bar{x}) + x_3 \\ \dot{x}_3 = f_3(\bar{x}) + u \\ y \ = x_1 \end{cases}$$

the unknown functions are set as $f_1 = 0.03 \sin(x_1 x_3) - 0.03, f_2 = 0.05 \cos(x_1) - x_2 x_3 - 0.05, f_3 = 0.1x_2 - 0.03x_3$.

In the proposed method, initial values of the parameters are: $\sigma_i = 0.025$, $\delta = 0.05$, $r = 0.1$, $c = 1.5$, $c_2 = 2.0$, $c_3 = 1.0$, $\gamma_i = 0.1$, $\tau_i = 0.03$. The original values of the state observer are all 0. Initial values of states are set as $x = [0.5, 0, 0]^T$ and $\theta_i = 0$.There are 4 neurons in the first network. The centers are set to be uniformly distributed in the region from −6 to 6. And the widths of the network are 3. The second network has 5 nodes. The centers are set to be uniformly distributed in the

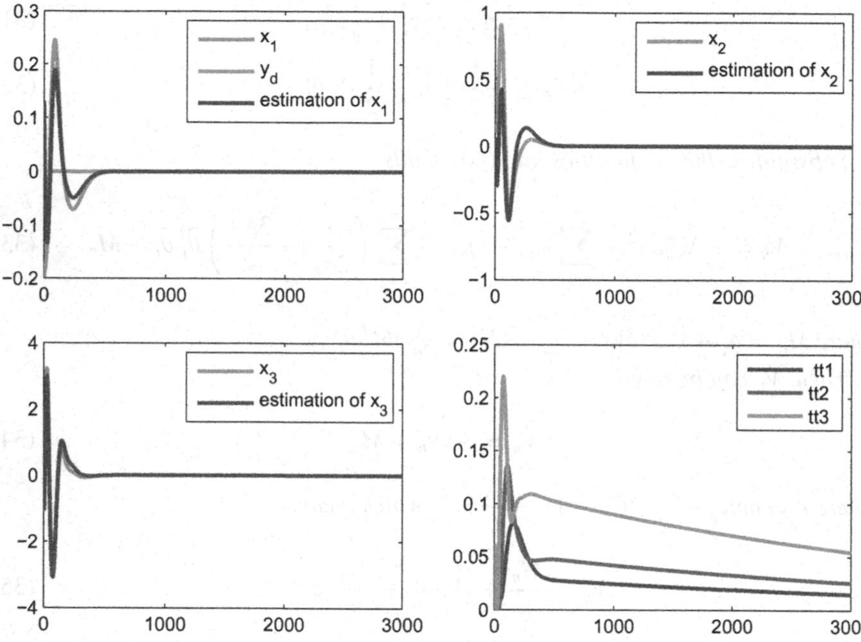

Fig. 1 Estimation of state variables and the 2-norm of the weight vectors

Fig. 2 Control input u

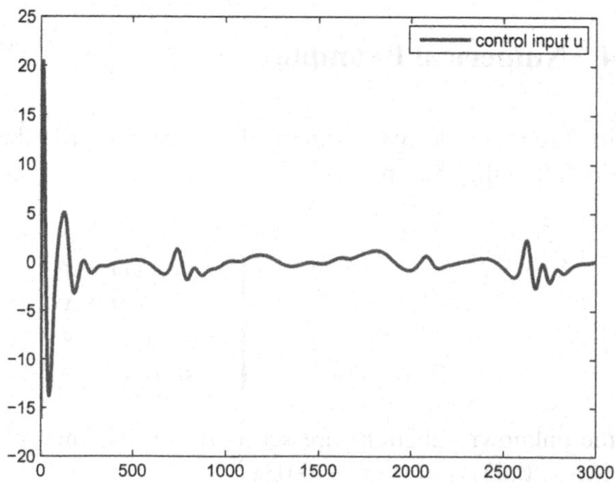

region from −7 to 7. And the widths of the network are 3. There are 6 neurons in the third network. the centers of the network are set to be uniformly distributed in the region of $[-4, 4]$. And the widths of the network are 2. The numerical results are collected in Figs. 1 and 2. As shown in these results, one can draw the conclusion that the proposed scheme is valid.

5 Conclusions

This paper aims to research on the topic of observer-based ANC method for a kind of specific nonlinear systems. In this work, the considered system consists of the non-strict-feedback form and unmeasurable states. By combining ANC technique with the Levants HOSM differentiator, a novel ANC algorithm is developed. It is worth to note that the proposed scheme settles the problem of explosion of complexity. In the stability analysis, it is proved that all signals in the considered system are SUUB and the system output can be remained in arbitrary small scale. Finally, numerical simulation is conducted. By means of the results, it is further verified the usefulness of the developed algorithm.

References

1. P.V. Kokotovic, I. Kanellakopoulos, A.S. Morse, Systematic design of adaptive controllers for feedback linearizable systems. IEEE Trans. Autom. Control **36**(11), 1241–1253 (1991)
2. T. Zhang, S.S. Ge, C.C. Hang, Stable adaptive control for a class of nonlinear systems using a modified Lyapunov function. IEEE Trans. Autom. Control **45**, 129–132 (2000)
3. M. Chen, S.S. Ge, Direct adaptive neural control for a class of uncertain nonaffine nonlinear systems based on disturbance observer. IEEE Trans. Cybern. **43**(4), 1213–1225 (2013)
4. X. Zhao, X. Zheng, B. Niu, L. Liu, Adaptive tracking control for a class of uncertain switched nonlinear systems. Automatica **52**(12), 185–191 (2015)
5. H.B. Ji, H.S. Xi, Adaptive output-feedback tracking of stochastic nonlinear systems. IEEE Trans. Autom Control **51**(8), 355–360 (2006)
6. S. Tong, Y. Li, H. Zhang, Adaptive neural network decentralized backstepping output-feedback control for nonlinear large-scale systems with time delays. IEEE Trans. Neural Netw. **22**(7), 1073–1086 (2011)
7. D. Swaroop, J.K. Hedrick, P.P. Yip, J.C. Gerdes, Dynamic surface control for a class of nonlinear systems. IEEE Trans. Autom. Control **45**(10), 1893–1899 (2000)
8. T. Zhang, S.S. Ge, Adaptive dynamic surface control of nonlinear systems with unknown dead zone in pure feedback form. Automatica **44**(7), 1895–1903 (2008)
9. Y. Li, S. Tong, Y. Li, Observer-based adaptive fuzzy backstepping dynamic surface control design and stability analysis for MIMO stochastic nonlinear systems. Nonlinear Dyn. **69**(3), 1333–1349 (2012)
10. Q. Shen, B. Jiang, V. Cocquempot, Adaptive fuzzy observerbased active fault-tolerant dynamic surface control for a class of nonlinear systems with actuator faults. IEEE Trans. Fuzzy Syst **22**(2), 338–349 (2014)

Printed in the United States
By Bookmasters